Stiffs, Skulls & Skeletons

Medical Photography and Symbolism

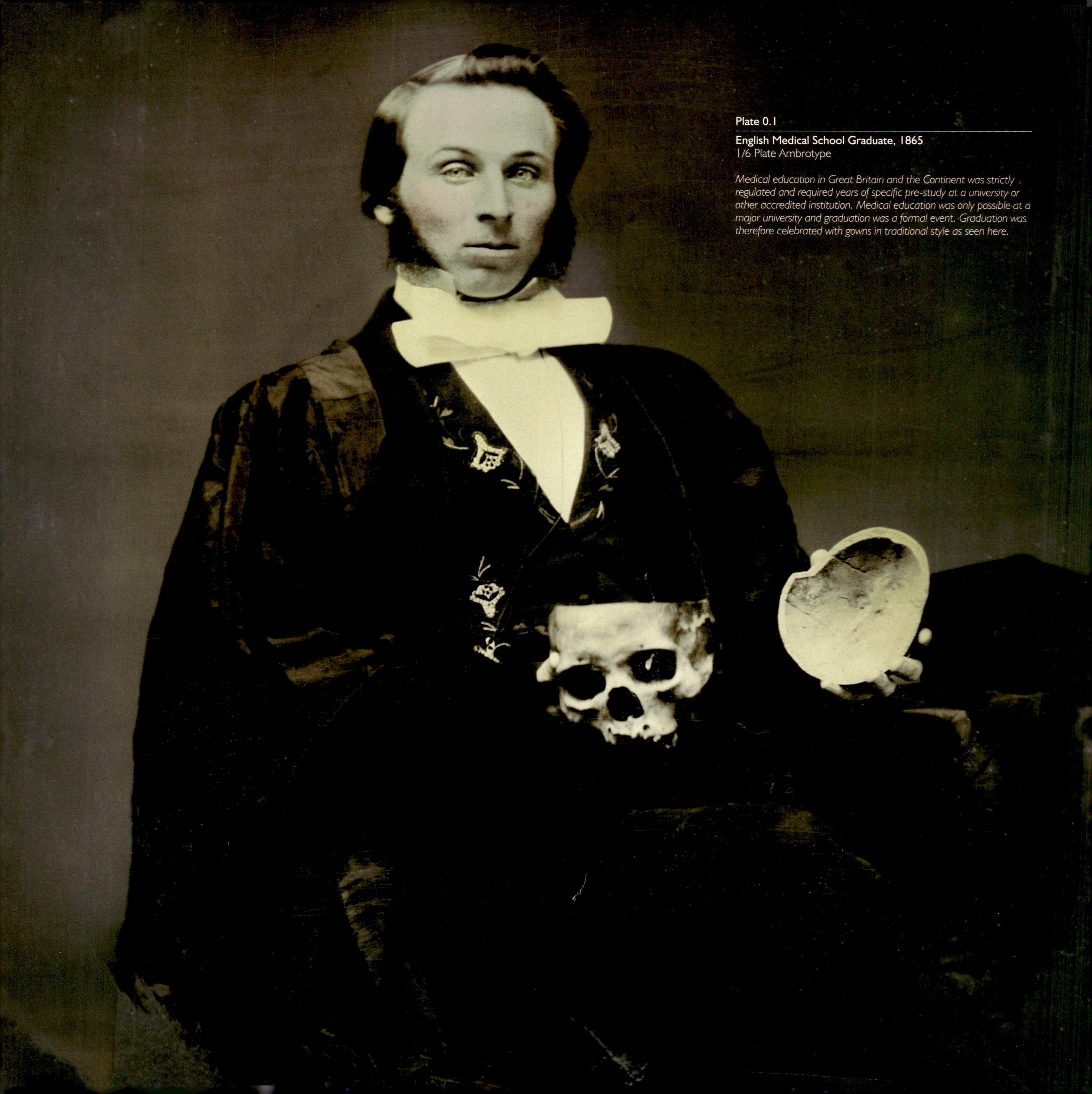

Plate 0.1

English Medical School Graduate, 1865
1/6 Plate Ambrotype

Medical education in Great Britain and the Continent was strictly regulated and required years of specific pre-study at a university or other accredited institution. Medical education was only possible at a major university and graduation was a formal event. Graduation was therefore celebrated with gowns in traditional style as seen here.

STIFFS, SKULLS & SKELETONS

Medical Photography and Symbolism

Stanley B. Burns, MD
Elizabeth A. Burns

Foreword by James M. Edmonson, PhD
Chief Curator, Dittrick Medical History Center,
Case Western Reserve University

Schiffer Publishing Ltd
4880 Lower Valley Road • Atglen, PA 19310

A BURNS ARCHIVE BOOK
Schiffer Special Edition

We dedicate this work to Alice Lease Dana, whose kindness, enthusiasm, and generosity of spirit facilitated the creation of this book.

Contents

READER'S NOTE:

icon denotes an enhanced caption which can be found beginning on page 290

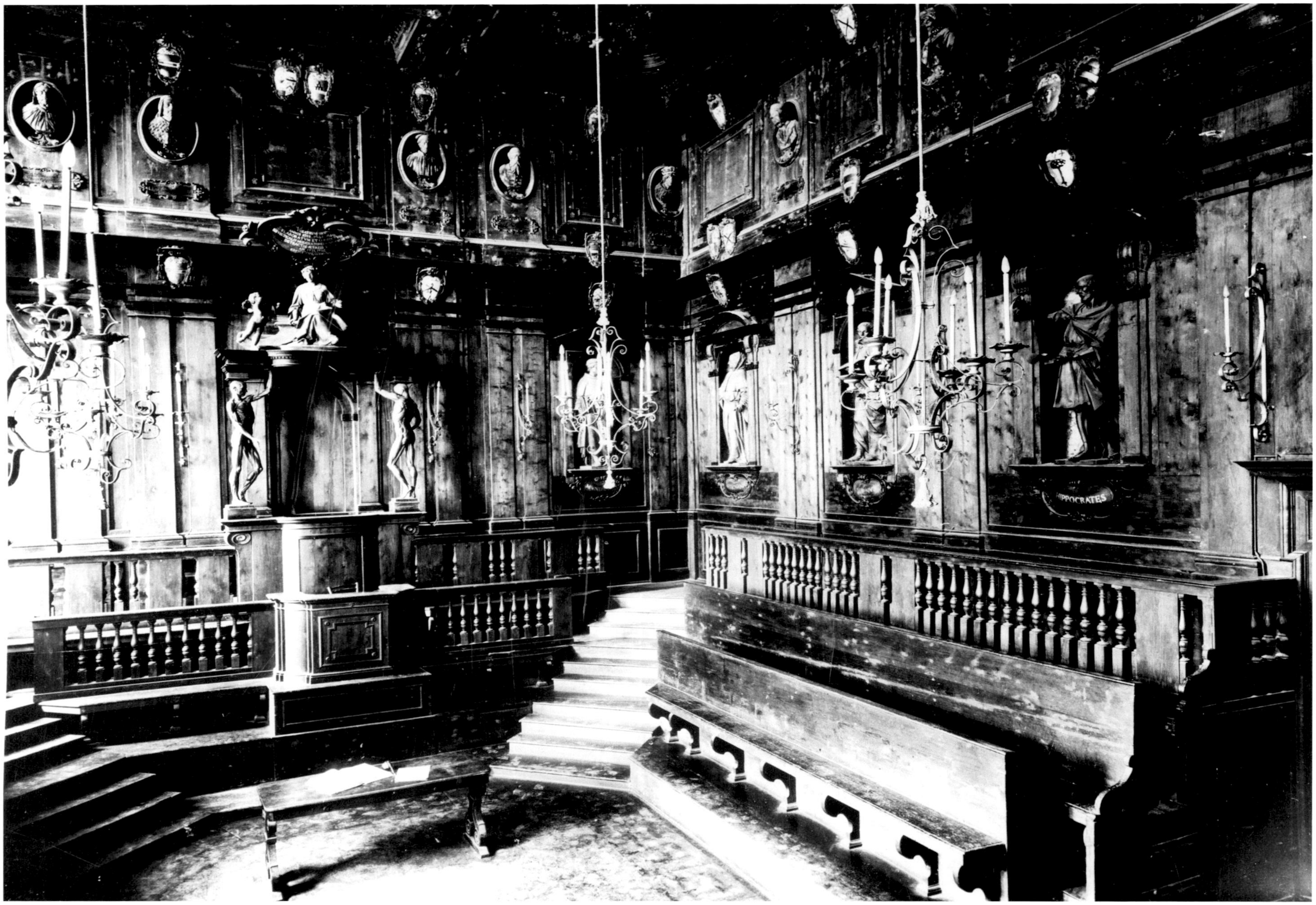

Plate 0.2

Anatomical Theatre of Archiginnasio, Bologna University
Silver Print, circa 1930

By the end of the sixteenth century, as a result of renewed interest in human anatomy and of notable publications, the teaching of anatomy became the basis of medical education. Medical universities began building special rooms for dissections. These rooms were magnificent palatial enterprises such as this anatomical theater in Bologna. The amphitheater style was usually an oval shape. Rows of seats surrounded a large white table on which the human and animal dissections were presented. Often a desk and an imposing chair were present from which the professor lectured. When the rooms were functional, they typically had skeletons of humans and various animals displayed. Originally, the theater was used for public dissection of criminals.

Foreword

This remarkable book is something both decidedly fresh and novel, and yet marks a culmination of decades of collecting and study of the medical image. I first encountered Stanley Burns's work almost thirty years ago when I consulted his magisterial history of medical photography, *Early Medical Photography in America (1839–1883)*. There was nothing like it in print, and it opened the arcane world of the medical past, as captured in photographic images, to the reading public.

On several occasions I visited Dr. Burns in New York City to view his amazing collection and to discuss topics of shared interest, ranging from the depiction of the doctor's office and the physician-patient encounter, to the dissection portrait. He has never ceased collecting, both in medical imagery of long-standing interest and new genres, including crime, death, and dying issues, Judaica, African American history, the Civil War, and early photography. The Burns Collection, now with over one million photographs, is recognized as the most important private collection of early vernacular photography (1840–1950).

Dr. Burns followed his path-breaking study of medical photography with a succession of arresting publications, most notably *Sleeping Beauty: Memorial Photography in America* (1990) and *A Morning's Work: Medical Photographs from The Burns Archive & Collection* (1998). These works established medical and memorial photography as subjects of scholarship and analysis, spawning and inspiring a number of complementary studies including Jay Ruby's *Secure the Shadow: Death and photography in America* (1995) and my own *Dissection: Photographs of a Rite of Passage in American Medicine* (2009). Dr. Burns has also lent important images from his collection to numerous exhibitions at leading art museums, notably *Le Dernier Portrait* (2002) at the Musée d'Orsay and most recently in *Photography and the American Civil War* (2013) at the Metropolitan Museum of Art.

In *Stiffs, Skulls & Skeletons*, Dr. Burns with his daughter, Elizabeth, showcases much clinical documentary, or notational medical imagery, that has remained unpublished until now. These images were created to be seen by the initiates of the medical profession and not by laypersons. While the thrill of the new accompanies seeing such unpublished photographs, one gazes upon them with an alternating sense of disquiet or poignancy. Such images are revelatory enough in their own right, but are perhaps upstaged here by the sheer range and variety of performative images, in a number of categories, conveying the formation and projection of professional group identity. Finally, Dr. Burns brings us some completely new variants of medical images, notably the exceedingly rare autopsy photograph and the more ubiquitous but not as yet studied X-ray photograph, both in its clinical incarnation as well as its pop culture variants. In sum, we have in hand a marvelous volume that brings to light images in familiar categories, providing fresh relief from the clichéd pictures that all too often populate many medical history works. But Dr. Burns steps out still further, introducing essentially new sub-genres of medical imagery. And to paraphrase the author, you may not take note of nor remember my comments here, but the images presented will most certainly leave an indelible, lasting impression on the viewer.

James M. Edmonson, PhD
Chief Curator, Dittrick Museum of Medical H story
Case Western Reserve University
Cleveland, Ohio

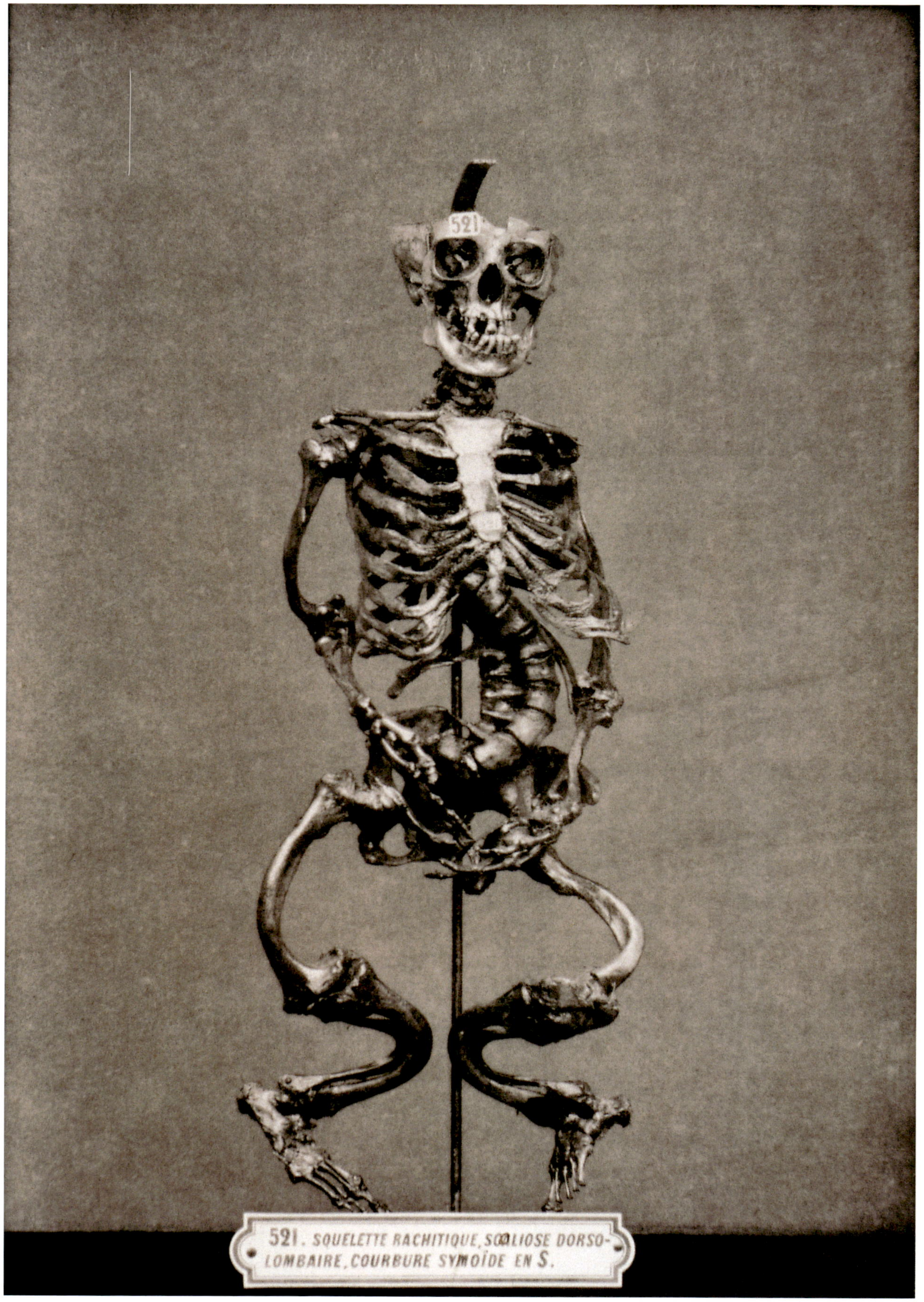

Plate 0.3

Deformed Skeleton Due to Rickets
Catalogue des pièces du Musée Dupuytren, 1879

This photograph is of a rickets skeleton specimen at Paris's Musée Dupuytren. The deformities of the skeleton were a phenomenon that fascinated physicians, and skeletal remains of patients with the extremes of disease were preserved in various anatomical collections. Rickets was first described by Daniel Whistler, MD, (1619–1684) in his doctoral thesis at Leiden, Netherlands, in 1645. Just four years later, Arnoldus Bootius, MD, (1600–1653), noted the widespread presence of rickets in Ireland. Francis Glisson, MD, (1597–1677) provided an accurate account of the condition and is frequently given credit for its discovery. In 1650, he published De rachitide sive morbo puerili, qui vulgo. *While Glisson was not the first to describe rickets, in this publication, he gave the first description of scurvy. A major step in the management of the disease was the 1824 report of D. Schutte, MD, who recognized the value of cod liver oil in treatment. About a century later, in 1919, Kurt Huldschinsky, MD, (1883–1940), developed the use of ultraviolet light to treat rickets. Around the same time, 1918–19, Sir Edward Mellanby proved rickets to be a dietary deficiency disease, curable with a proper diet. During the twentieth century, scientists revealed the metabolic, physiologic, and genetic causes of the disease.*

Preface

My lifelong interest in exploring unappreciated areas of photography or history is expanded upon in this publication. I have written forty-three photographic historical books and over one thousand articles adding significantly to the base of a second history of photography. The classic history of photography promoted by art historians is the history of the work of noted photographers and the innovations of photography. It emphasizes the history of photography as an art. The second history of photography, which I promote, is the history of how professions and people used photography. In this work I continue that view and present the first comprehensive overview in one volume of the use of bodies and body parts by physicians, scientists and the public.

The Burns Collection of about one million photographs is completely subject oriented. During the past four decades, my interest has been solely in subject matter. I am not as much interested in who took the photograph as I am in what was photographed. If the Collection contains a Mathew Brady or Lewis Hine photograph, it is because of the topic. Most of the photographs in the Collection were taken as documentation, however, due to the passage of time, they are now perceived as art. My intention in acquiring photography was not to collect photographs, but to collect the visual evidence of various subjects. I have always been interested in history and medicine, and my collecting endeavors mirrored these interests. I studied history approaching it as a scientist, to uncover truth, and as a key to understanding the present.

Stiffs, Skulls & Skeletons continues my series on the body in photography. To put into historic perspective the use of bodies and body parts in photographic history, I collected crime, war, and medical photographs, along with images of skulls used in medicine and popular culture. In prior texts, I presented the three major aspects of the subject: medical, crime scene, and postmortem photography. This book complements these publications by expanding the visual history of the body to include all the major significant photographic subjects, including images of scientists, artists, and the public, as well as photographs of physicians who used skulls or a cadaver for documentation of the profession.

The nineteenth-century study by physicians and scientists of skull features, skull capacity, and the brain as a basis for racial and intellectual theories was documented by photography. No prior work has presented these images, and I wanted to show them in context to the general use of body parts in photography. Another stimulus for presenting this work was to differentiate and illustrate the various ways physicians used skulls and bodies to indicate medical status.

This volume vividly illustrates that many of the photographs of skulls and dissections were artistically taken in the nineteenth century. Medical specimens have a universal appeal and significant attraction among artists and photographers. Many contemporary photographers frequent medical museums to photograph skulls and specimens, producing works of modern photographic art. Presented are vintage images of skulls and specimens that challenge modern photographers' efforts. These photographs are published for the first time for the general audience.

I have been a lifelong collector and historian. When I first started acquiring photographs in 1975, I chose images based both on their importance as historic documents and as evidence of medicine's rich past. What I soon

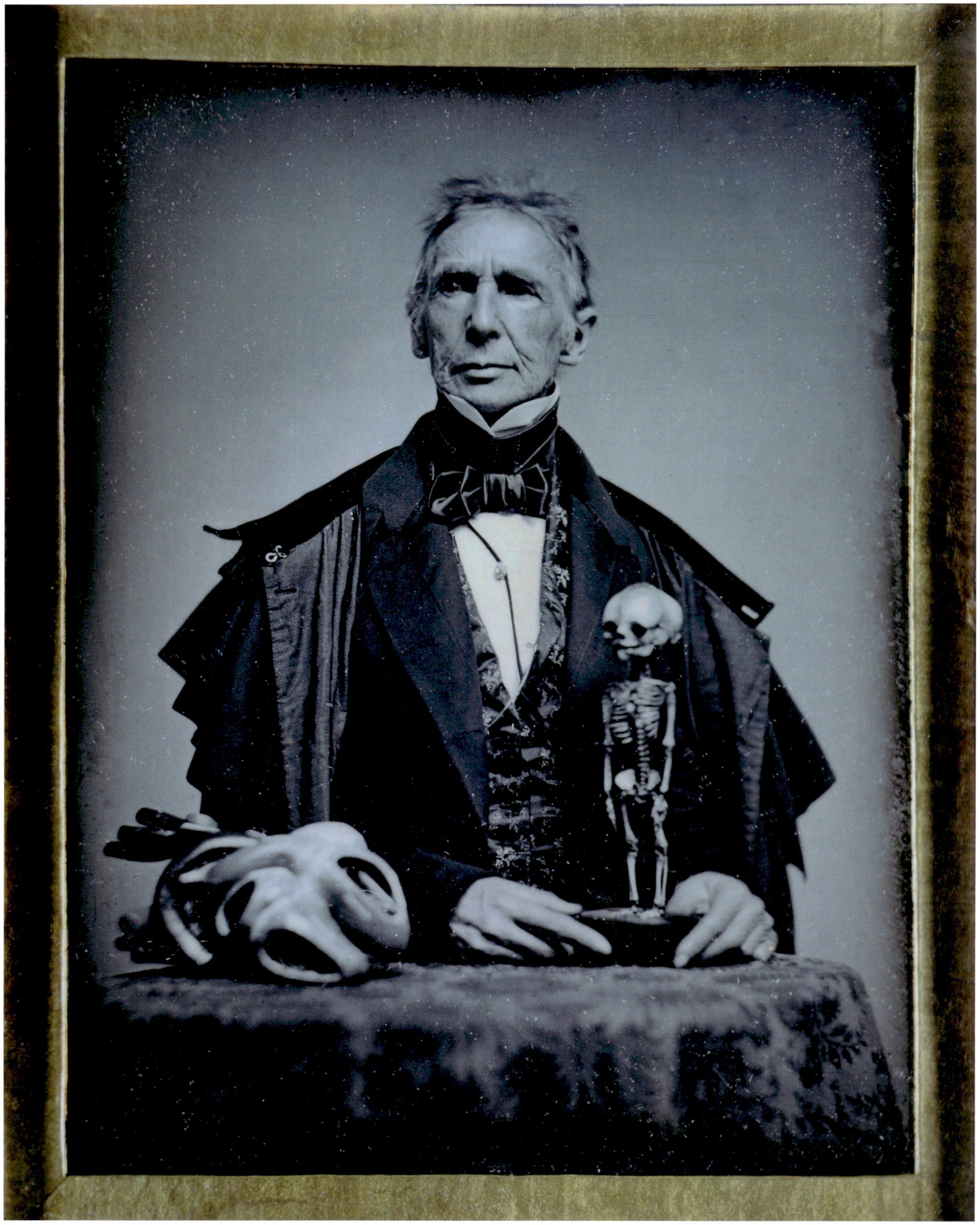

Plate 0.3x

John Collins Warren, MD (1778-1856) with a Fetal Human Skeleton and Model of a Human Heart in Wax
Full Plate Daguerreotype, American, circa 1848

Warren was the leading American surgeon, the first Dean of Harvard Medical and a founding member of the Massachusetts General Hospital. In 1846, Dentist William Morton, demonstrated the general anesthetic effects of sulfuric ether to Dr. Warren and his staff revolutionizing the practice of surgery.

recognized was their artistic strength. In 1977, I created The Burns Archive, which is dedicated to preserving medical photographs and producing publications on the history of medical photography. By the mid-1980s, noted curators and artists became interested in medical photography as art. In 1984, Marvin Heiferman curated *In the Picture of Health*, an exhibit of more than 140 photographs from The Burns Collection. This was the first exhibition of medical photographs displayed in a public art institution. In 1987, Joel-Peter Witkin edited *Masterpieces of Medical Photography: Selections from The Burns Archive*. The associated exhibitions have been showcased by museums and galleries in several countries and continue to be shown. In 1998, Jack Woody published *A Morning's Work: Medical Photographs from The Burns Archive & Collection, 1843–1939*, and a new series of medical photographic exhibitions was developed. In the first decade of the twenty-first century, several leading pharmaceutical companies supported production of twenty-five medical photographic historical volumes.

My most well known publications are photographic books documenting the death of loved ones, *The Sleeping Beauty* series. The three volumes on postmortem-memorial photography are landmark publications in the field, and have inspired exhibitions on memorial photography every year since 1990. The books vividly illustrate the use of photography in documenting personal death and raised the awareness of death, dying, and grief issues for the public and in the health care community. They have also stimulated symposiums and supported the development of professional organizations of photographers who take photographs of deceased children for grieving parents.

In the 1990s, recognizing the artistic value of medical images, numerous major museums began collecting and exhibiting medical photographs. Several institutions now display vintage medical photographs of patients, procedures, and practitioners to the general public.

In 2013, surgeon R. B. Bontecou's photographs of wounded soldiers from The Burns Collection were showcased at the Metropolitan Museum of Art in the exhibition *Photography and the American Civil War*. The exhibit travelled to the Gibbes Museum of Art and in 2014 to the New Orleans Museum of Art. Other medical photographs related to war are in the exhibition *War Photography*, travelled by the Museum of Fine Arts Houston.

My daughter Elizabeth grew up surrounded by historic medical photographs and became familiar with book and exhibition production. In 1999, she joined me in my publishing ventures. Since then, she has helped edit more than 200 of my medical photographic history articles written for a variety of medical journals. In 2002, we prepared and published *Sleeping Beauty II: Grief, Bereavement and the Family, European and American Traditions* for the exhibition *Le Dernier Portrait* at Paris's Musée d'Orsay. As co-author, Elizabeth was responsible for editing and preparing the photographs, content research and editing, as well as design and layout of the book. In 2003, we began to publish The Burns Archive Medical Specialty Series, in which she oversaw complete creation and production, including going on-press in China. By 2008, she helped publish twenty-two medical texts in the specialty series. In 2006, Elizabeth and I produced *Geisha: A Photographic History, 1872–1912*. We have since created several exhibitions on Japanese Meji era photography.

Elizabeth and I are continually researching and exploring topics for publication. We have long admired the work of Peter Schiffer of the Schiffer Publishing Company, appreciating and collecting their publications. We were compiling a definitive volume of the use of skulls and bodies in photography, as well as upcoming books on *Ambrotypes, Masked Fear: The Psychology of Gas Warfare, German Police Units of World War II*, and *The Battle of Nomanhan: Russia vs Japan 1939*. After a meeting with Peter and his enthusiastic staff, we chose *Stiffs, Skulls & Skeletons* as our first publication together. Almost all the photographs in the book are from The Burns Collection. Several are from fellow collectors and museum personnel who we requested supply their best photographs. The result is a comprehensive survey of the topic and a visual treat.

Our hope is that this survey will stimulate others to continue the exploration of this topic and add to the history of photography. We have enjoyed exploring this subject and plan exhibitions to expand our audience. Some of the photographs are beautiful examples of the art of photography and its ability to portray unpleasant subject matter artistically.

Stanley B. Burns, MD, FACS
September 2014

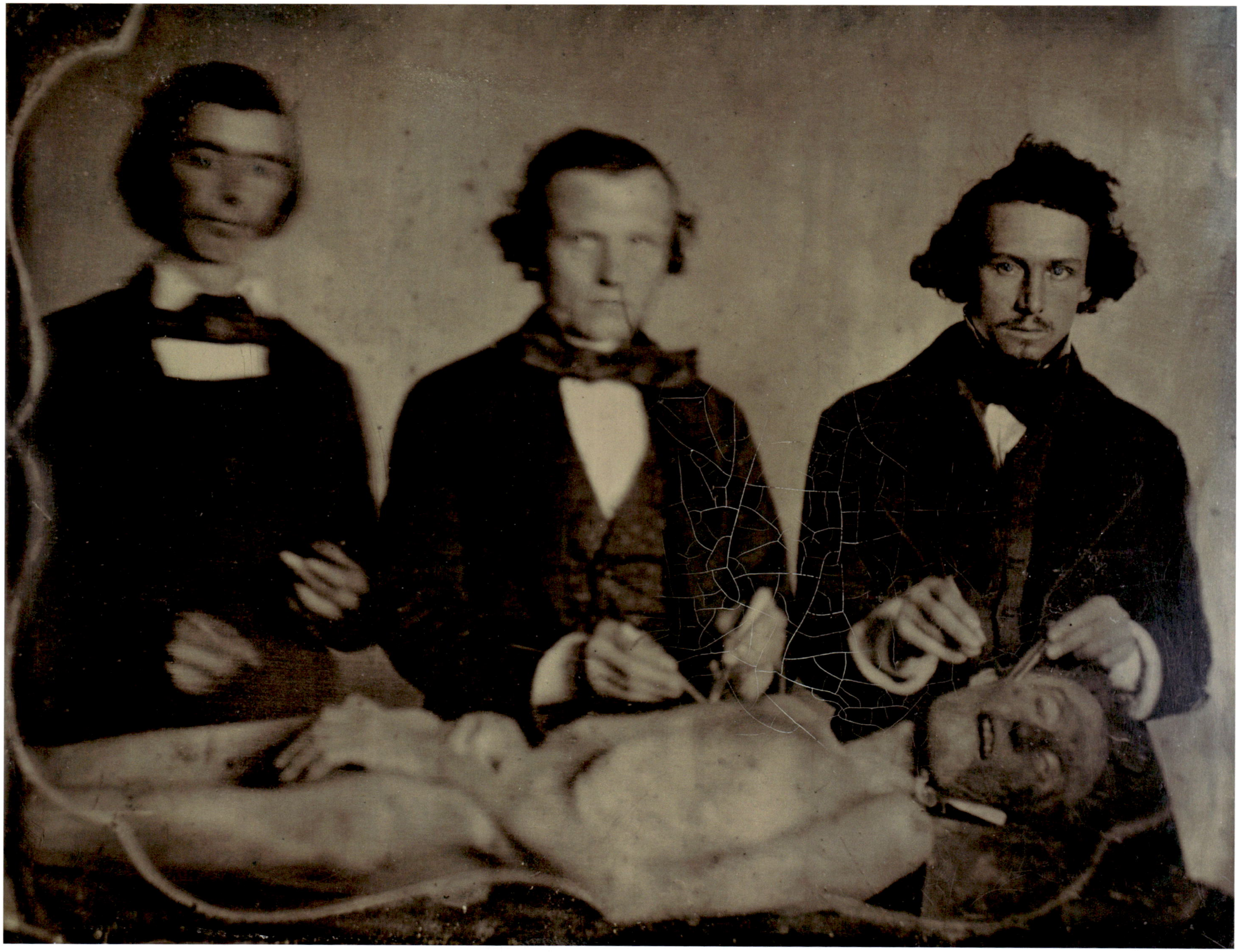

Plate 0.4

Earliest Dissection Scene
Daguerreotype, circa 1844

This is an 1850s full-plate tintype of an earlier daguerreotype. It is the only surviving representation of a dissection scene from photography's earliest era. The next earliest scene is from the late1860s. As is typical in this time period, the school and individuals are not identified. Taking a dissection photograph in this early era was not related to the standard 'rite of passage' images that began in the 1880s. It is most likely related to occupational photography popular in the 1840s. Photographing a dissection in this time period in the Midwest was quite dangerous due to the public concern with grave robbing.

Introduction

The skull is the most significant and time-honored prop for self-validation in a profession. The physician, the actor, the scientist, the explorer, the holy man, and the warrior all pose with skulls to convey accomplishment and status. Romantic dreams and glorious achievements are inherent in many of these poses. *Stiffs, Skulls & Skeletons* is the first publication to present a comprehensive photographic historical overview of the use of anatomical props during the first century of photography. In these images, the attitude of those posing with body parts changes over time, as each generation acquired new knowledge and approached the sanctity of the body with new cultural considerations. For example, up until the middle of the twentieth century, there was little concern for the identity of a medical subject; a patient's face, private parts, and disease were photographed without regard to the patient's wishes. Strict Victorian sexual mores were abandoned for medical subjects. In contrast, today under modern HIPPA rules, a patient cannot be identified out of concern for their privacy. Individuality and privacy are the standard of today and are a far cry from the time when bone yards were the final resting place of millions and hunting skulls was portrayed as a glorious and popular pastime.

Faced with the mysteries of the end of life, perhaps the most direct way of expressing one's self is through images. They move us deeply and have meaning that words cannot express. This series of iconic photographs document the nineteenth and early twentieth century's fascination with the dead body and body parts. It expands the classic visual iconography of dissection, bone, and postmortem photography to include early autopsy images and X-ray studies. The X-ray offered a view inside the living body and it revolutionized medical as well as popular thinking. Visualization of the skeleton opened up a window not only into the condition of the body, but also into how the body functioned. Several chapters document the manner in which physicians posed with body parts and span from serious portraits with skulls to whimsical tableaus involving skeletons and cadavers. The book ends with a wide range of images that document the public's use and fascination with these symbols of mortality.

Voices haunt the pages of every book and photograph. The dead speak to us from crime scenes, the diseased from hospital clinics, and loved ones from memorial images. In volumes on these subjects, the murdered, diseased, or deceased are often known. In this book, the voices are mostly unknown. Cadavers, skulls, and body parts were harvested in the name of medicine, science, religion, and art. These anatomical parts are anonymous props. Many were stolen from graves or removed postmortem from a diseased body by enterprising surgeons in the name of education and a quest for knowledge. One exception to lost identity is

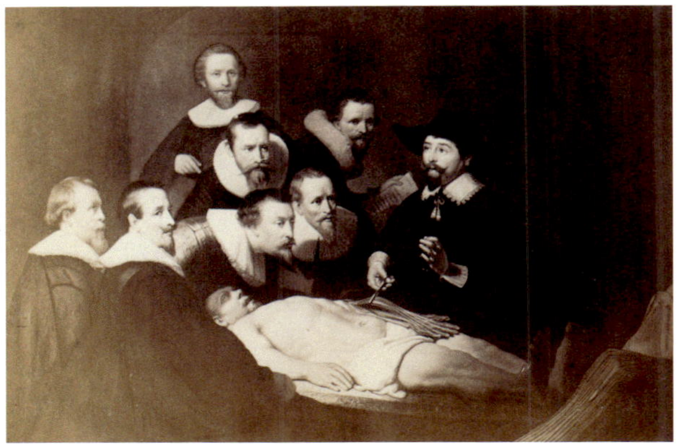

Plate 0.5

The Anatomy Lesson of Dr. Nicolaes Tulp
Rembrandt Harmenszoon van Rijn, 1632

that skulls of notables were often acquired as symbols of triumph and conquest.

Photography makes the inner sanctum of professions and cultural institutions available for scrutiny and exposure. That exposure often leads to changes in perception of people, places, and cultural practices. Cultural inferiority, mental incapacity, and criminality have been proven unrelated to general visible anatomical features. Many of the studies that posited these philosophies are illustrated in this work. The photographs provide evidence of the level of knowledge, intellectual thinking, and technology of the time. During the twentieth century, a remarkable transformation of appreciation of diverse cultures, as a result of increased social sensitivity to the blight of indigenous peoples and modern scientific studies of the value inherent in various cultures. There has been a dramatic shift in rhetoric and ethics, which resulted in a political change in attitude toward people and groups previously defined as 'the other.'

Photography provides knowledge independent of our own experience. It permits us entry into the personal world and affairs of others. Studying a series of similar photographs provides a platform to understanding specific cultural phenomena. Photographs provide clear evidence how our society has changed. Many cultural phenomena disappear from collective memory, as changing cultural taboos dictate the past be hidden. So it was that postmortem photographs of loved ones and the dissection photographs of medical students became closeted. During the early twentieth century, death was removed from everyday life and dead bodies were removed from personal photographs. Photographs are now focused more on life's celebrations.

THE POWER AND ART OF PHOTOGRAPHY

Photographs have two types of power, usually described as denotation and connotation. The denotation is simply what the image physically depicts. It is the reproduction in one moment from the camera's point of view. Connotation is the implied power. It suggests that photographs offer symbolic meaning, separate from the physical scene shown. For most, the magic of photography is the image seen. The symbolic nature of an image is intriguing; many are interested in not what is shown, but what is implied.

Many images indelibly etch in the mind. An intriguing subject remains in the psyche, and one wants to up the ante and see something fresh. Most photography enthusiasts want to see images and imagery they have not seen. Most of the photographs in this volume are previously unpublished; they offer a new palette of visual iconic images. Many of the images are symbolic representations of humankind's confrontation with mortality.

All great photographs, whether documentary or scientific, belong to the world of art. Many of the photographs in this volume clearly blur the distinction between art and documentary photography. Some photographers working day in and day out photographing similar objects—whether patients, body parts, or pottery—rise above the ordinary and create an artistic rendition of a seemingly mundane or horrific subject. One great example are the photographs of Civil War surgeon Reed Brockway Bontecou (1824–1907), whose renditions of Civil War wounded and of their diseased excised body parts have resulted in their exhibition at the Metropolitan Museum of Art on several occasions.

MEDICAL HUMANITIES

A current movement in medical education and practice is in the interdisciplinary field of medical humanities. It provides insight into the human condition and a historical perspective through social science, the arts, and humanities. Medical artifacts play a significant role in this discussion. Unquestionably, photography is one predominate source of education. From its inception in 1839, each decade saw changes in educational formats and each generation was educated by a new generation of photographic phenomena. From daguerreotypes, stereophotographs, motion pictures, videos, and now computer-generated visual formats, learning becomes easier and faster. The historic photographs presented in this work encourage a dialogue with our past and stimulate health care professionals to contemplate the effects of what they are doing today. Our predecessors were just as smart and innovative as we are, however they toiled to help and heal their patients with vastly inferior technology and knowledge. Health care practitioners believed that they were offering their patients the best care available. One hundred years from now future practitioners may look upon our labors with the same discerning eye.

RESURRECTING RARE ICONIC PHOTOGRAPHS FROM EARLY MEDICAL PUBLICATIONS

Included in this historical survey are images from perhaps the most significant nineteenth- and early twentieth-century photographic medical publications. These rare publications offered the period practitioner original photographs that provided a visual key to learning and understanding anatomy, surgery, and disease. They contain a variety of imagery, ranging from tipped-in albumen and silver-gelatin prints to woodburytype and heliotype mechanical photographs. Some of the photographs were iconic representations of medical conditions and were reproduced in other publications for decades. The list of publications include:

American Atlas of Stereoroentgenology, 1914–1920

Hôpitaux d'Alger: Iconographie Médicale Algérienne, 1905–07

*Catalogue des pièces du museé Dupuytren. Atlas du museé Dupuytren, (4 volumes)*1877–79

Cliniques Médicales Iconographiques (Nancy), 1891

La Photographie au laboratoire de Médicine Légale, 1906

Revue Photographique des Hôpitaux de Paris, 1869–72

Revue Médico-Photographique des Hôpitaux de Paris. 1873–1875

Photographs of Surgical Cases of the Army Medical Museum, 1870–1881

Use of the Roentgen Ray by the United States Medical Department of the United Sates Army in the War With Spain (1898), 1900

The Edinburgh Stereoscopic Atlas Of Anatomy, 1905-6

POSTMORTEM PHOTOGRAPHY: PRIOR TO THE TWENTIETH CENTURY THE AVERAGE CITIZEN WAS QUITE FAMILIAR WITH DEAD BODIES

It is important to present a discussion of death and dying concepts prior to the twentieth century in order to put into perspective the dissection and anatomical photographs. Today, few people ever see a dead body or are comfortable with dying rituals and behavior. However, prior to the twentieth century, the average person was very familiar with dying and dead bodies; it was not just

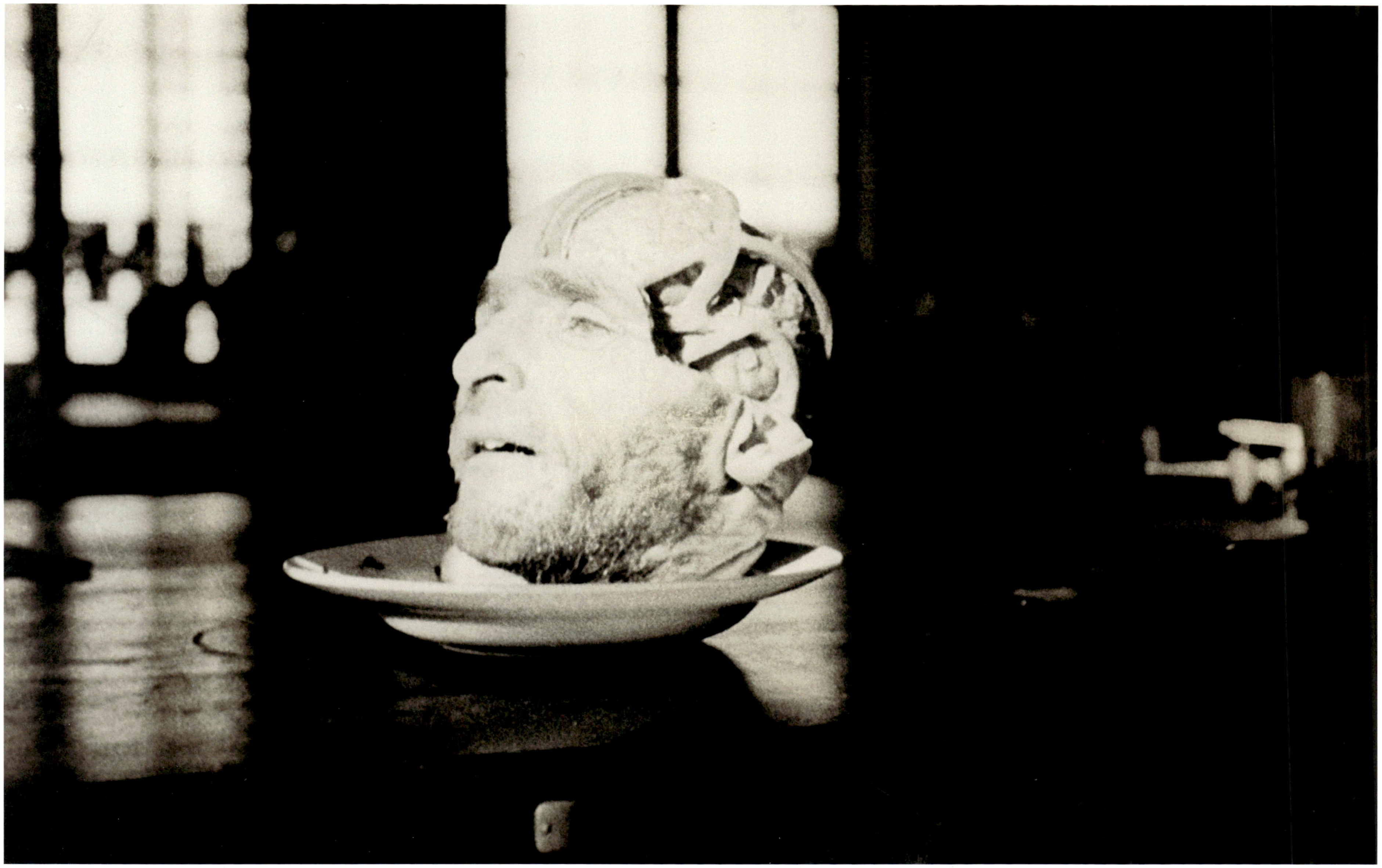

Plate 0.6

Dissected Head in a Soup Plate
Photographer & Dissector: Howard Brundage, Columbus Ohio, circa 1910

Brundage was a prolific photographer of his Columbus Ohio Medical University. He documented the wards, nurses, and personnel of this Columbus hospital and took many photographs of dissection scenes over the years during his early career.

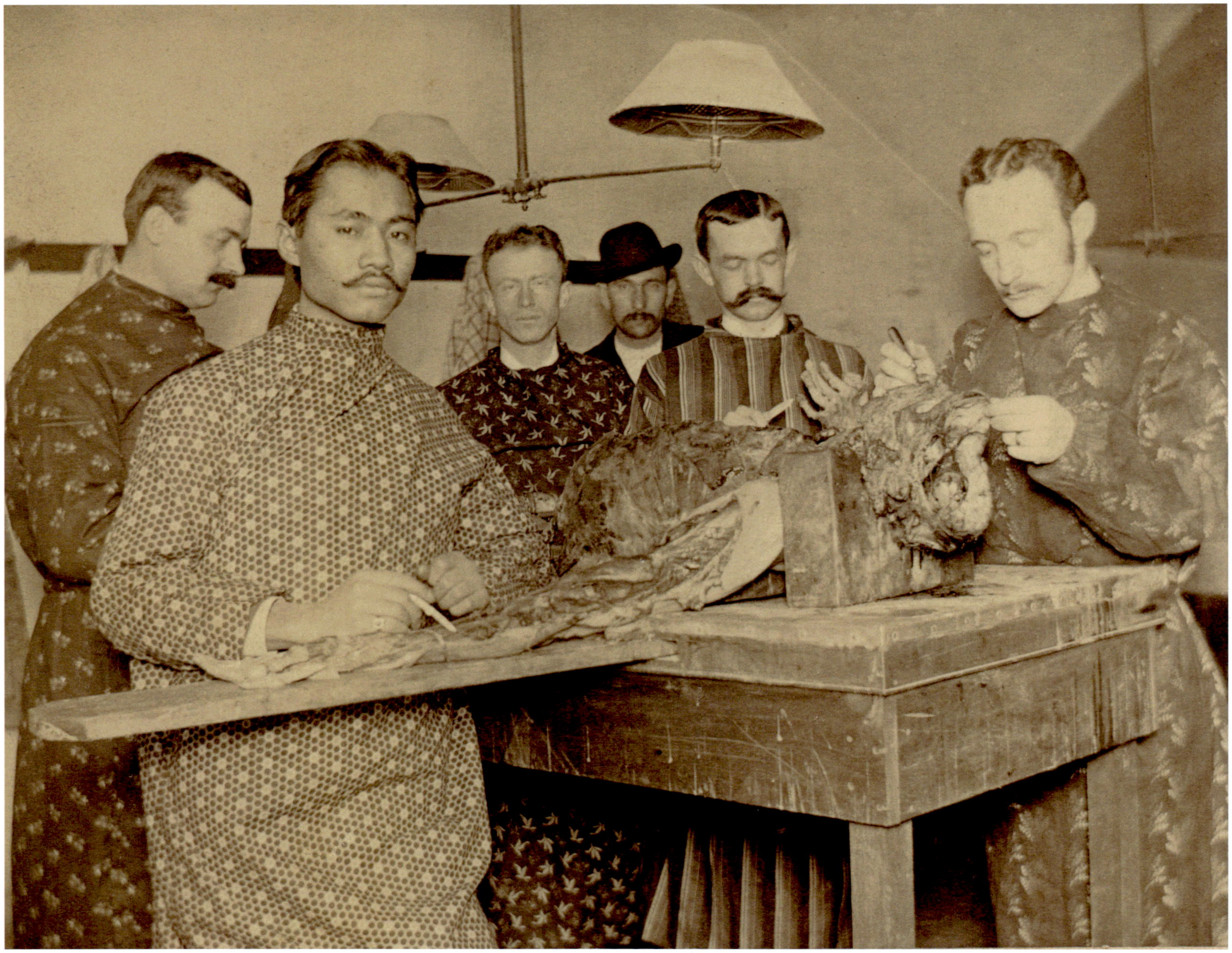

Plate 0.7

Adorning Colorful Robes for Dissection
Circa 1890

An Asian student poses with his classmates in this classic 'rite of passage' photograph. Minorities and women were rarely integrated into American medical schools during the nineteenth century. Special medical institutions for women and African Americans were established. As with all photographic genre posing and garb clearly changed each decade. In dissection, scenic changes also included statements written on the table and paraphernalia included. The chapters on teaching anatomy and dissection photography show these transitions.

the realm of the health care worker or mortician. Hospitals were not sanctuaries of healing, but dreaded institutions where one easily died from 'hospitalism,' infections endemic to the institution. People were cared for at home and died at home surrounded by their family. After death, prior to burial, family members of the deceased prepared the body at home and preserved the body on ice until the funeral. Until the end of the century, the average person was not embalmed. Therefore, the aromas and facts of death were known to even the youngest of family members. They often were posed for photographs with the bodies of parents and siblings in an effort to create a visual memento of the family bond. People were comfortable with the dead in the nineteenth century and readily shared photographs of their dead loved ones. The power and mutability of photographs, when used to mourn and commemorate the dead, provided a healthy form of the bereavement process, which has only recently been recognized. Photography mitigates the finality of death and offers survivors a memento.

In the twentieth and twenty-first centuries, sharing postmortem photographs or grief is frowned upon, as the emphasis is on living and living well. Death and mortality concerns have been relegated to a minimum of time. However, for centuries, in paintings and sculpture, icons of death permeated the arts. The changing role of death images and symbols from ancient times to the twentieth century and the use of death representations in various cultures have been well documented by Philippe Aries in his *Images of Man and Death*. During the nineteenth century, the emergence of death on the surface of daily life became evident. Aries notes, "Daily life was never free of death, for death was everywhere present and cultivated. The fashion was not to fear it but to cultivate it." With the deadly epidemics that ravaged Europe in the fourteenth century, artists began portraying death represented by a skeleton in various disguises taking young women and especially children.

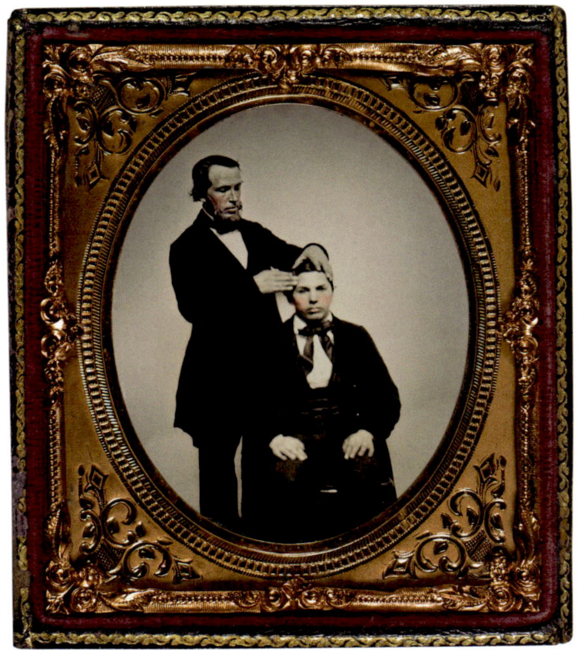

Plate 0.8

Phrenologist at Work
Ambrotype, 1856

The nineteenth century belonged to the middle class. The rise of cities and small business offered a comfortable living that, coupled with the development of photography, allowed people for the first time to have immortality previously available only to the rich. The common person used photography to express personal emotions, which expanded the use of the dead and death symbols. Industrialization, which concentrated people in cities, resulted in epidemics and pandemics that brought premature death into almost every household. Photographing the dead became an established norm in American and European culture. The camera allowed people to steal one of the great stings of death; it allowed the common man the ability to preserve forever a visual image of a loved one. It froze forever the faces of children, wives, and husbands, if not in life then in death.

As a result of general acceptance and usage, the memorial/postmortem photograph became the most common genre of photography in nineteenth-century America. The practice is documented and detailed in our *Sleeping Beauty* memorial book series. During the twentieth century, death and personal mourning was removed from everyday life. Only newsworthy death was subject for discussion. Today postmortem/memorial photography is the private act of mourners and serves some in their attempt to deal with the death of loved ones. It has been a quarter of a century since *Sleeping Beauty: Memorial Photography in America* was published. In that time, as a result of articles, exhibitions, and other books, postmortem/memorial photography has become increasingly accepted by health care professionals in dealing with bereavement and grief. Many institutions provide access to a photographer for a postmortem photograph, and some institutions have in-house photographers. These photographs are kept close to the heart and are personal mementoes. In the future, historians will analyze these images as a cultural expression of our current society, as we do the past.

DEATH IN THE PUBLIC EYE

Photographs of death were present in everyday life by the recording of various aspects of criminal punishment. Photographs of dead criminals and public executions were routine expressions of a society trying to rein in some of the terrors of the era. Photography changes consciousness more radically than any other medium, and seeing the swift and just punishment of offenders was hopefully thought to prevent criminal behavior. Some offenders were executed for their crimes in as little as a month.

Perhaps the best example is Guiseppe Zangara, who in 1933 attempted to assassinate President-Elect Franklin Delano Roosevelt and instead killed Chicago Mayor Anton Cermack. It took just thirty-three days from the time Zangara fired his fatal shots for him to die in "Old Sparky," Florida's electric chair. During those thirty-three days he had two trials, one for attempted murder when Cermack was wounded, and then another for murder when Cermack died. Death was felt to be a just punishment for capital crimes, and public executions were common occurrences. The last public execution was of Rainey Bethea in Owensboro, Kentucky, on August 15, 1936. About 10,000 attended.

THE PLATES, CHAPTERS & APPENDIX

The book is divided into ten chapters, each beginning with an informative introduction. The photographic plates in the chapters are presented with a title, and several have extended captions in the back of the book. The extended captions help flesh out the subject of each chapter.

A brief overview of each chapter: Chapter One, "Posing with Skulls & Bones," presents an exposé of physicians posing in time-honored manner with bones; Chapter Two, "Teaching Anatomy & Osteology," documents dissection and osteology classes, as well as laboratory scenes of teachers posed with their students; Chapter Three, "Dissection Photography," contains a series of images illustrating the mostly American phenomena of fledgling medical students photographed posing with their cadavers; Chapter Four, "Special Bone Studies," presents a series of images that document how nineteenth-century physicians studied normal and pathological bone anatomy; Chapter Five, "Skull Studies," gives a detailed history of the scientific use of skulls to define race; Chapter Six, "Special Dissection — Anatomical Studies," shows interesting painstakingly detailed dissections of various parts of the body as study aids; Chapter Seven, "Autopsy & Forensic Photography," explains the differences between dissection and autopsy photographs; Chapter Eight, "X-Rays — The Skeleton Becomes Visible in the Living," presents early X-rays and stereographic radiographs; Chapter Nine, "Gallows Humor," highlights some of the classical posing conventions of the doctors and nurses 'playing' with cadavers; Chapter Ten, "Popular Skeletal Views," includes a wide range of photographs of skulls and skeletons with which the public was familiar.

The Appendix supplements the text with original documents. Within the Appendix, original papers are republished for a greater understanding and appreciation of critical topics, as well as enjoyment of first-hand concepts of the time and the writer's style. *The Dying of Death* (1899) clearly explains the change in philosophy and

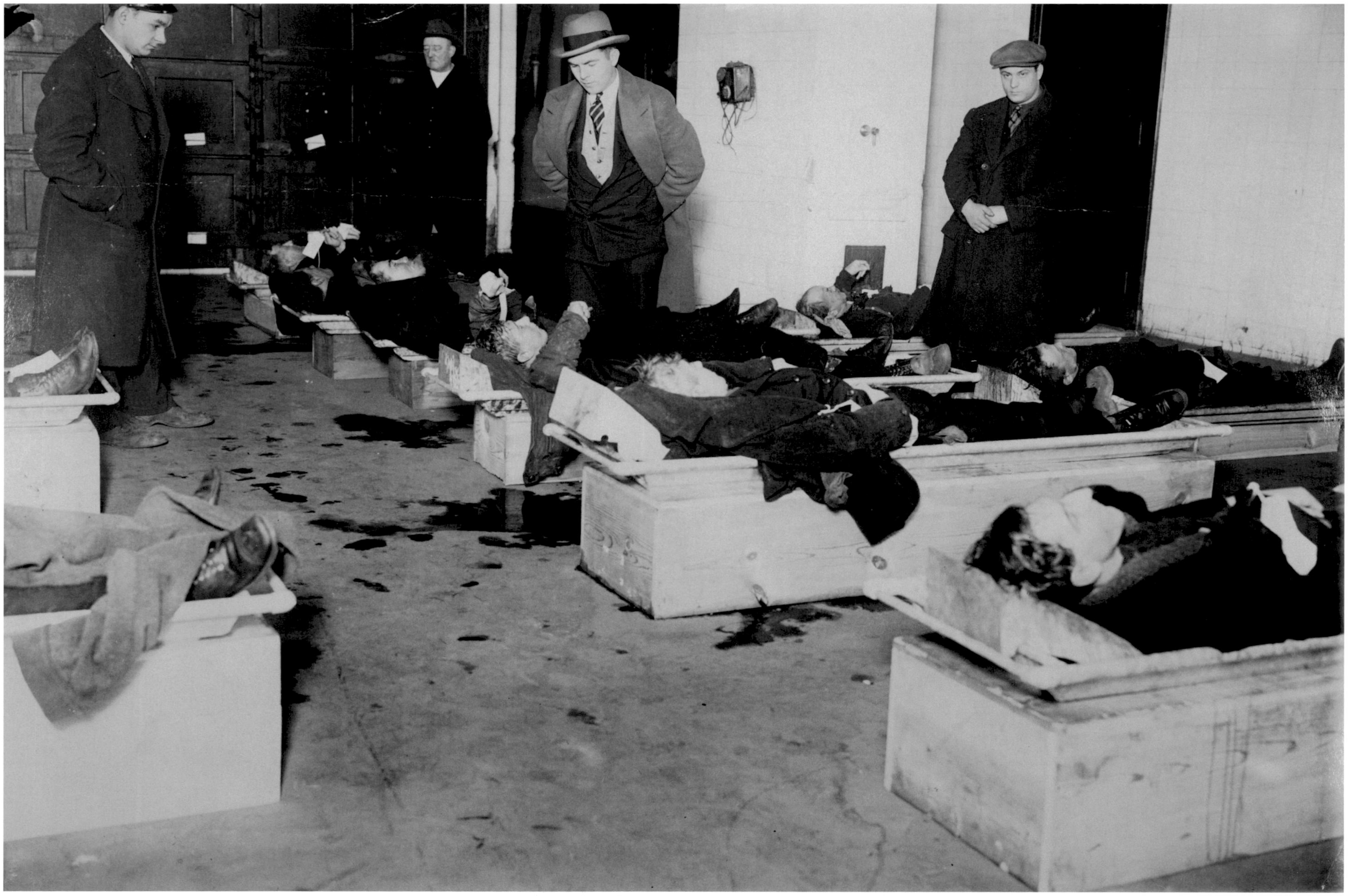

Plate 0.9

The Murdered Bugs Moran Gang, Chicago Morgue

Al Capone's Victims of the St. Valentine's Day Massacre, 1929

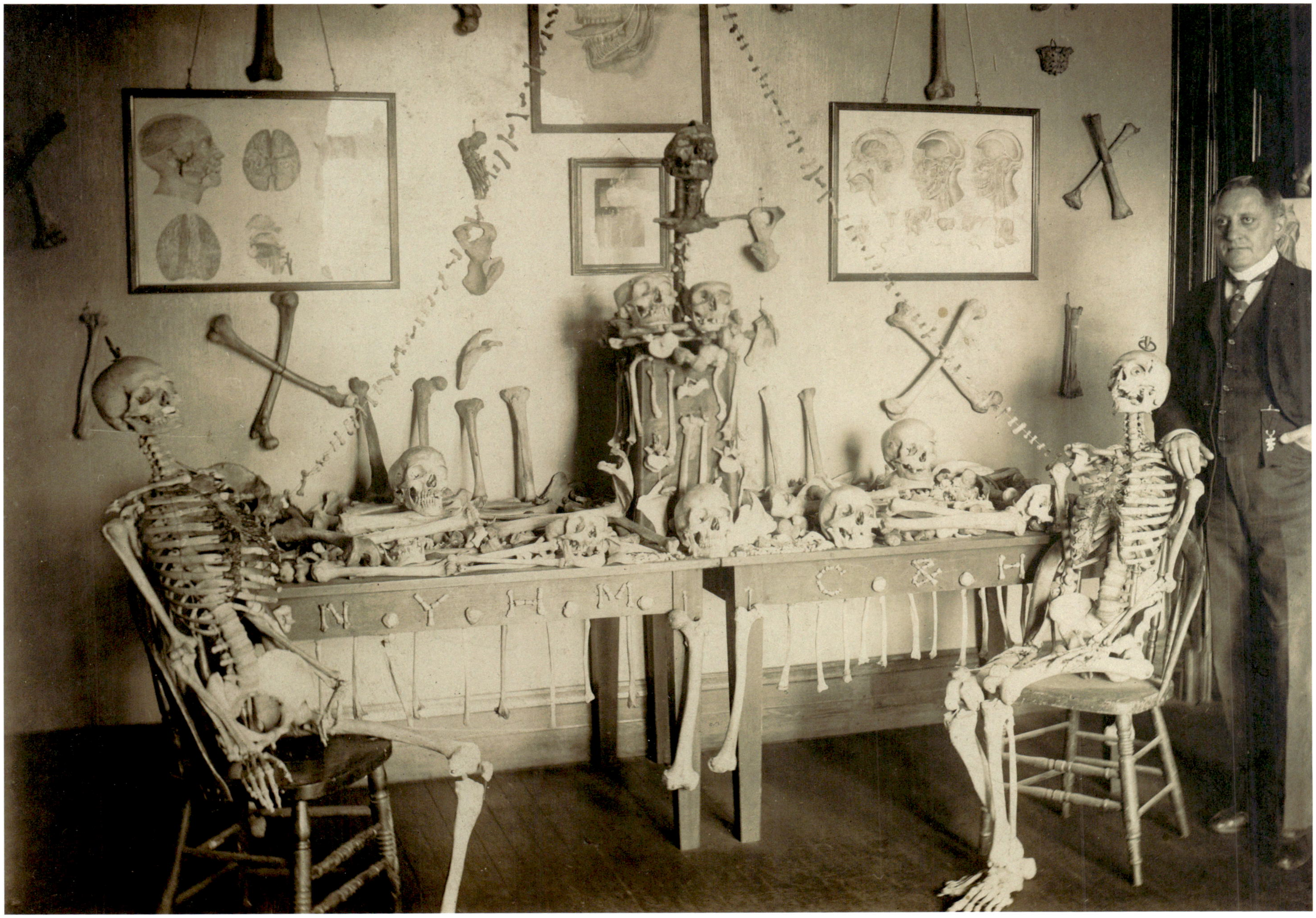

Plate 0.10

New York Homeopathic Medical College & Hospital
Osteology Display Advertising the Institution, 1908

attitude of death and dying principles at the turn of the century, which resulted in the removal of death from everyday life. *The Barred Road to Anatomy*, Dr. Howard Kelly's 1908 paper, puts in historical perspective the problems of those seeking to 'anatomise' in the nineteenth century through vivid descriptions of grave robbing and dissection. An essay from *Medicine-Men of the Apache*, 1892, by Captain John Bourke clarifies 'the magic' of the Cheyenne finger necklace and the use of body parts in cultures around the world. The 1916 introduction to *The American Atlas of Stereoroentgenology* describes the use of the stereophotographic X-ray. The Appendix also includes an outline explaining the various formats of photography and their years of popularity, 1840–1940.

The book is an encyclopedic historic overview of the use of cadavers, body parts, and skeletons in medical and popular photography. This compilation emphasizes iconic and artistic photographs that illustrate the transition of the use of the body during photography's formative century through over 400 photographs.

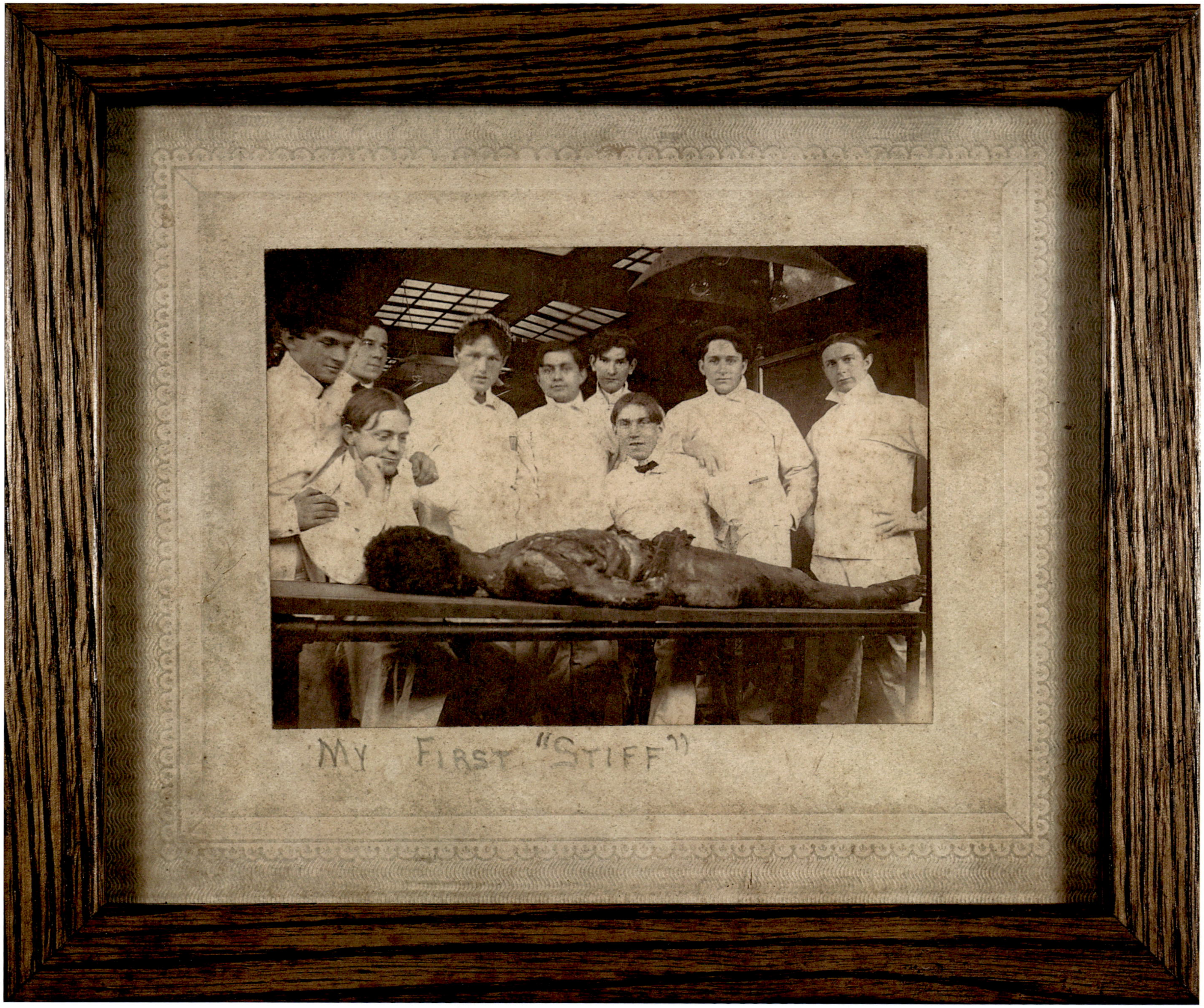

Plate 0.11

My First "Stiff"
Unidentified medical school, circa 1910

Nine medical students proudly pose around their cadaver. This photograph was framed, indicating it hung on a medical student's wall as a proud statement and visual record of his entering the medical profession.

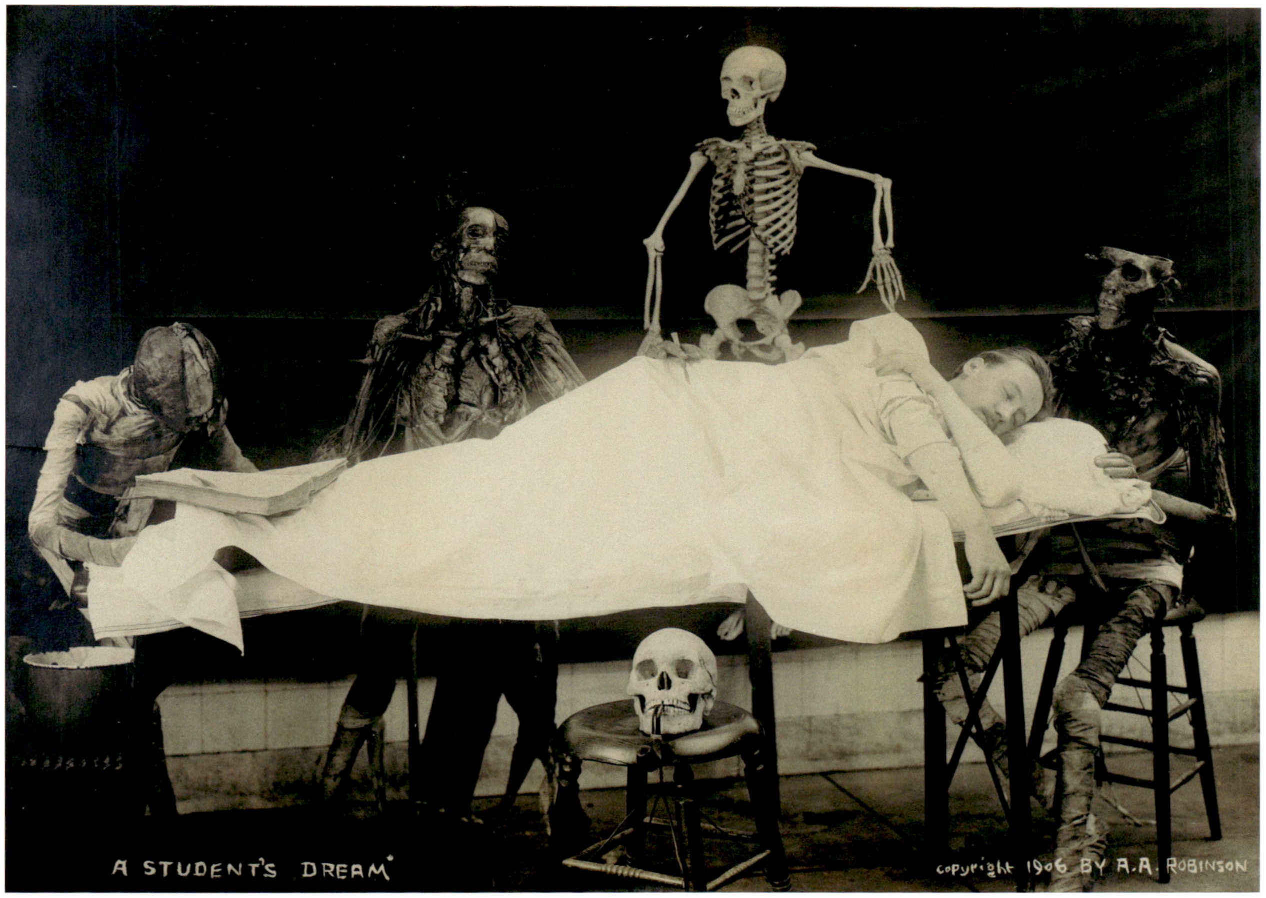

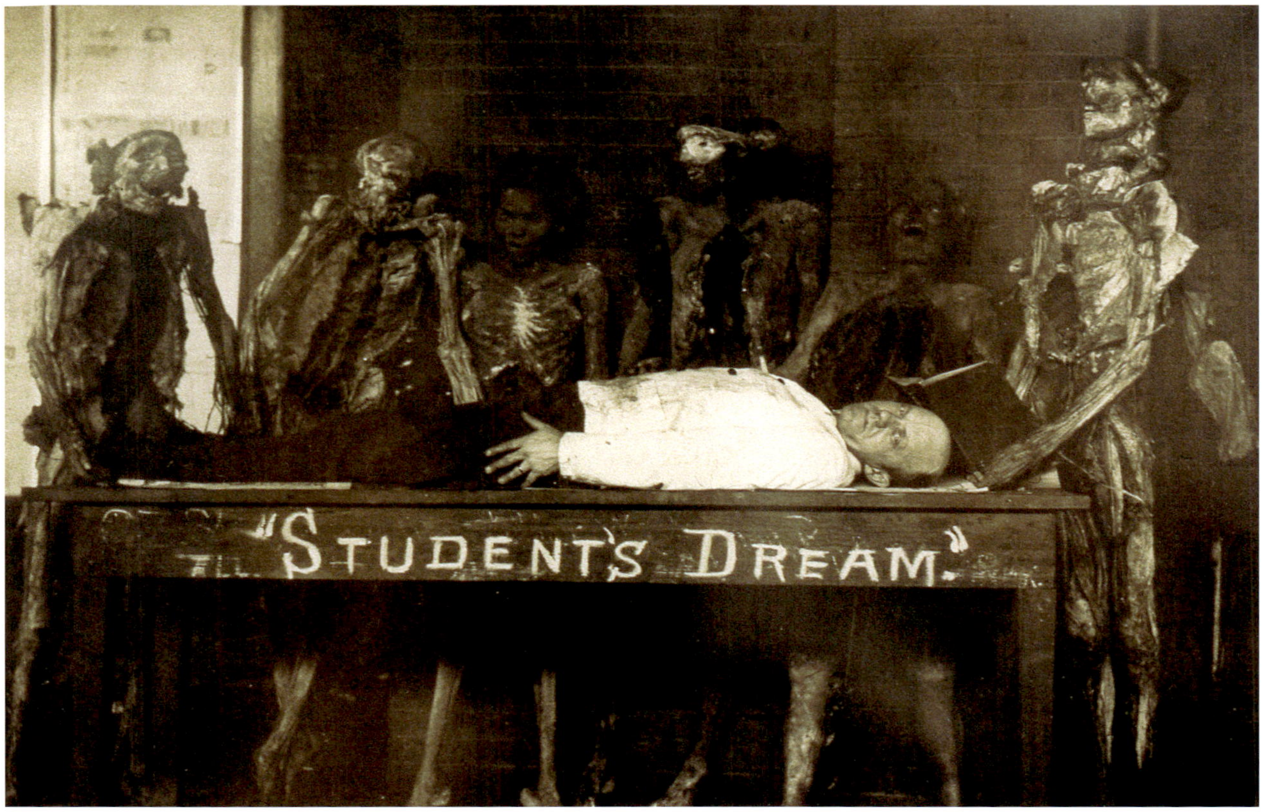

Plate 0.12 – 0.13

A Student's Dream
Photographer: A. A. Robinson, 1906

Robinson made a series of 8" x 10" photographs depicting students on the dissecting table surrounded by cadavers and/or skeletons. The photographs were popular images of their time and were sold to students throughout the United States. The images graphically represent one of photography's theoretical concepts—that of 'magical substitution.' Magical substitution is the phenomenon when the viewers place themselves in the depicted scene. It is one source of empathy, as one contemplates even fleetingly what it would mean to be in that situation.

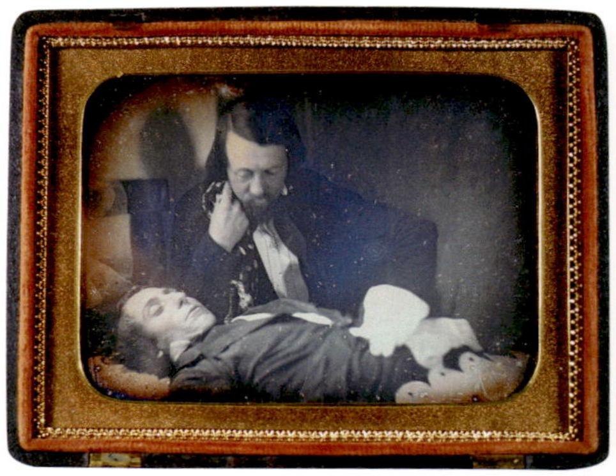
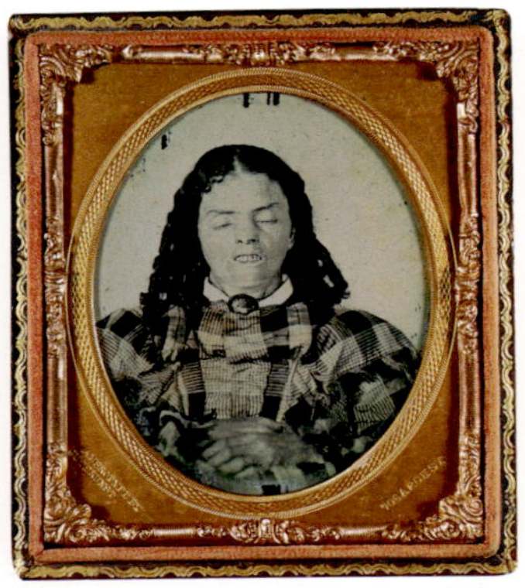
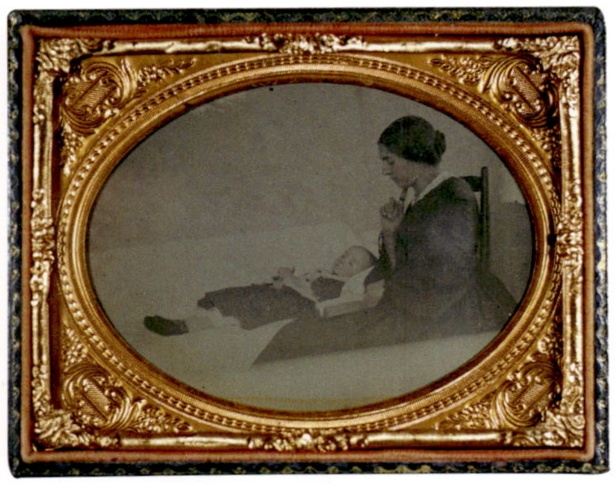
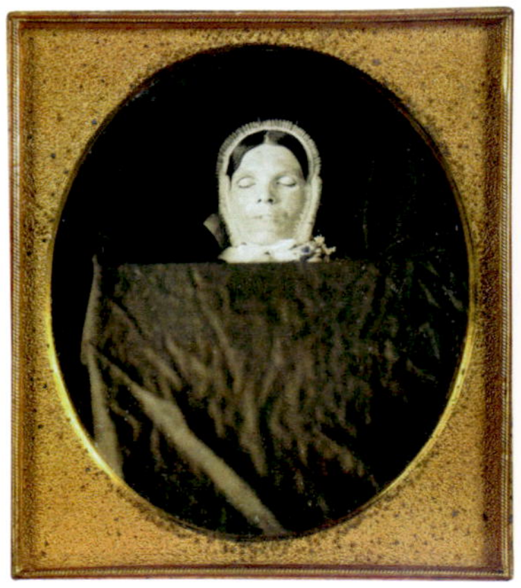
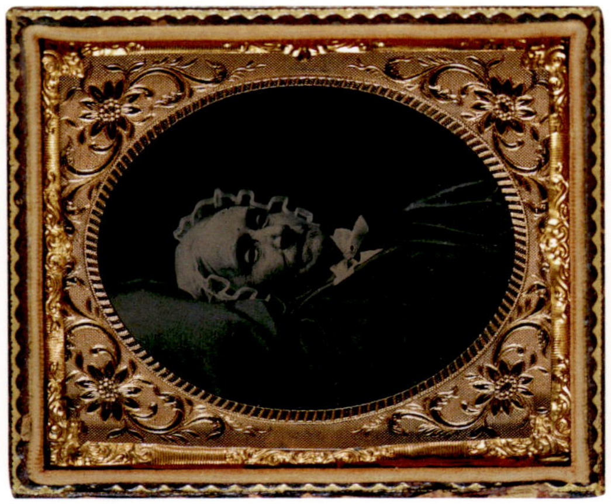
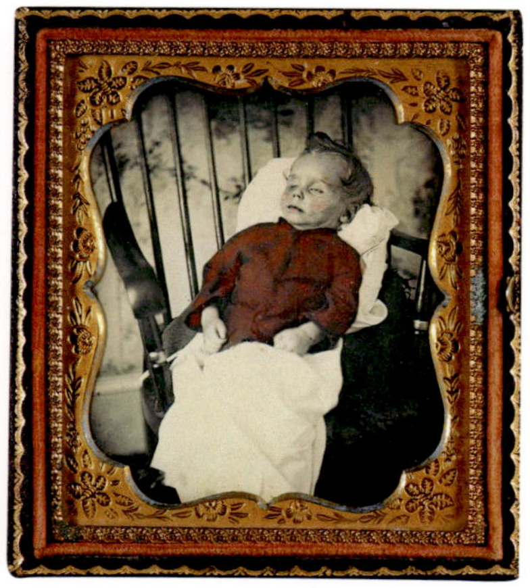
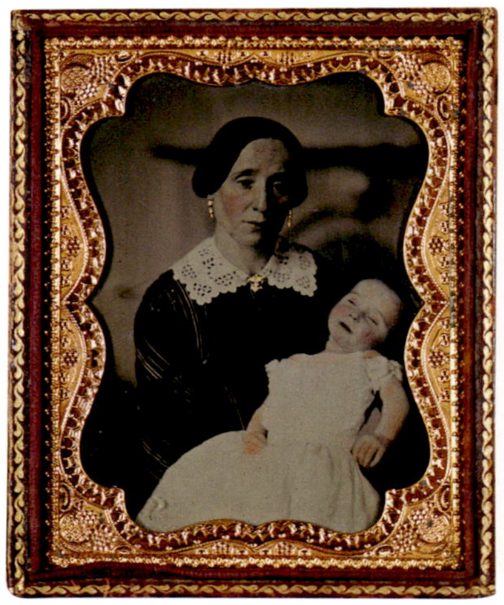
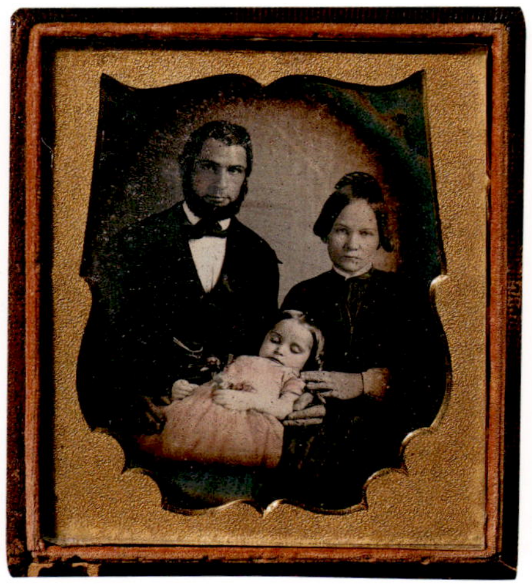
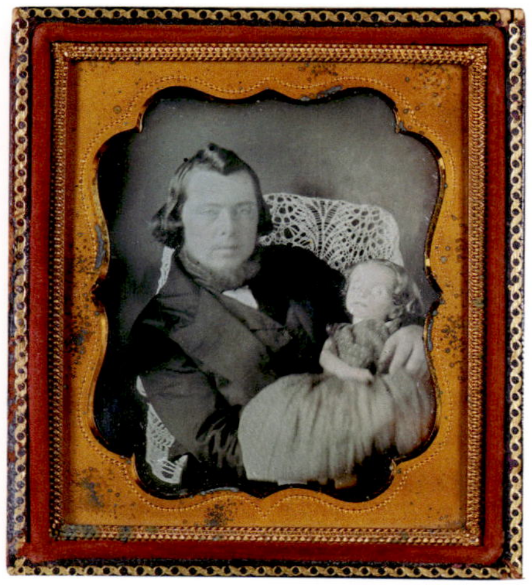

Plate PM.1

Jonathan Letterman, MD, at a Patient's Bedside
Philadelphia, Daguerreotype 1/4 plate, 1849

Plate PM.2

Girl Dead Several Days
Frederic Gutekunst, Philadelphia, PA
Ambrotype 1/6 plate, circa 1865

Plate PM.3

Mother Prays at the Side of Her Dead Daughter
Minnie Idell Hamlin, aged 1 year 11 months
Kendrick, Manlius, NY
Ambrotype 1/4 plate, January 7, 1862

Plate PM.4

Elizabeth Cooper in Coffin
Geneva, NY
Daguerroeotype, circa 1843

Plate PM.5

Old Woman with Discharge from Nose
Tintype 1/6 plate, circa 1857

Plate PM.6

Child in Red Shirt on Windsor Rocking Chair
Daguerreotype 1/6 plate, Tinted, circa 1856

Plate PM.7

Mother with Child, Dead from Measles
Ambrotype 1/4 plate, circa 1857

Plate PM.8

Couple Holding Daughter Wearing a Pink Dress
Daguerreotype 1/6 plate, Tinted, circa 1846

Plate PM.9

Father with Daughter on Chair Covered with Crocheted Lace
Daguerreotype 1/6 plate, circa 1847

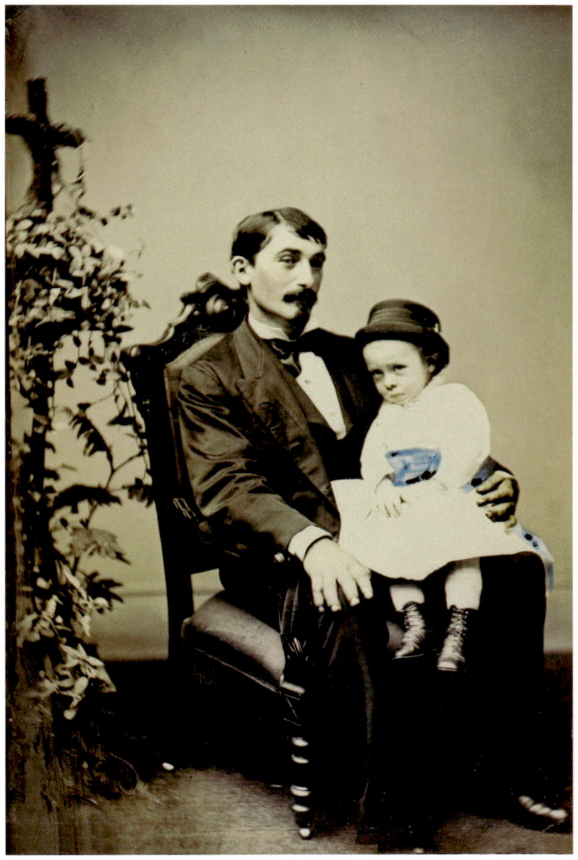

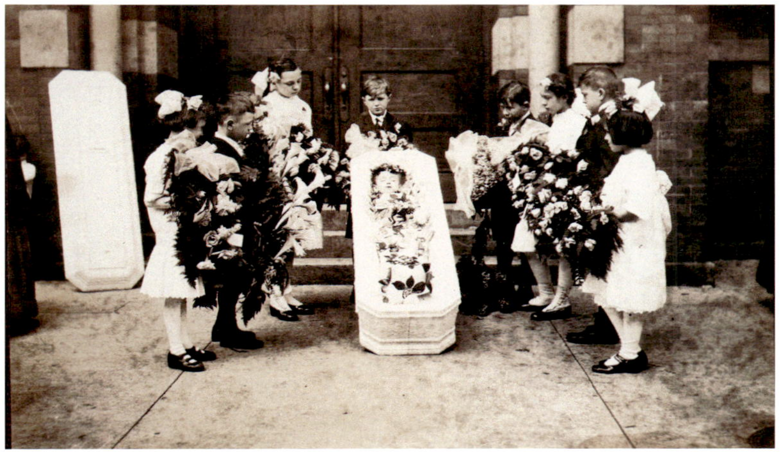

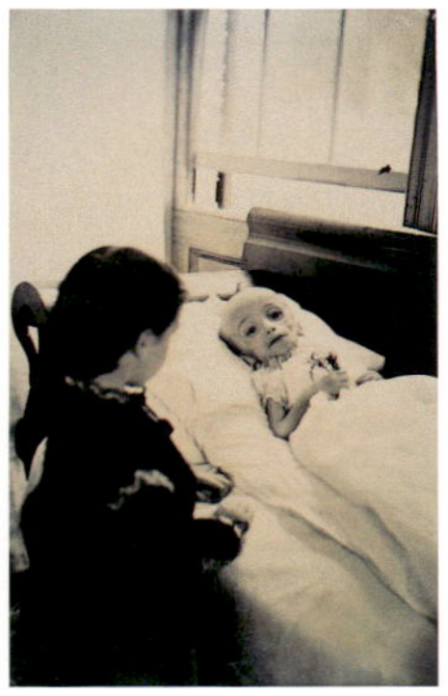

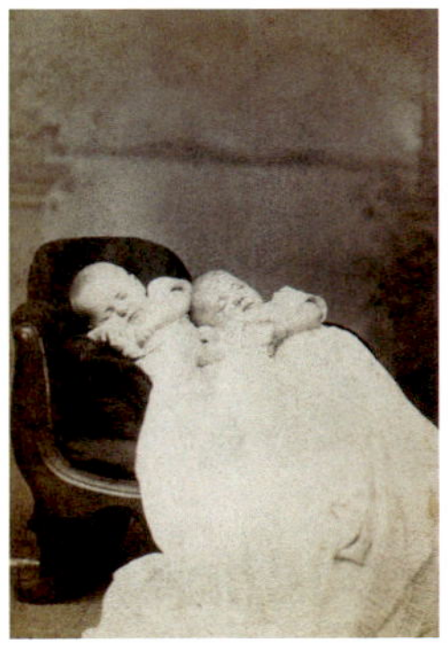

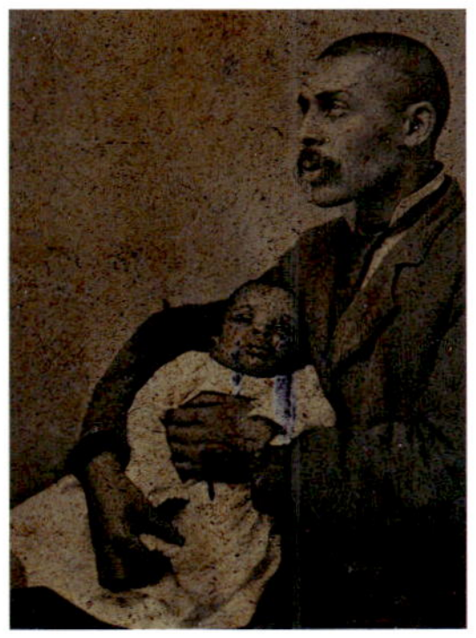

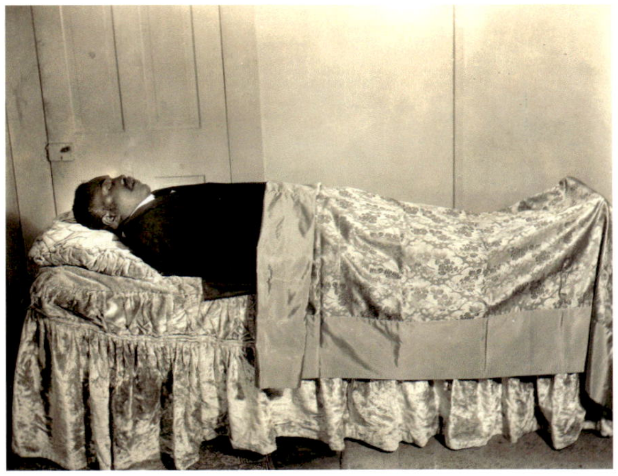

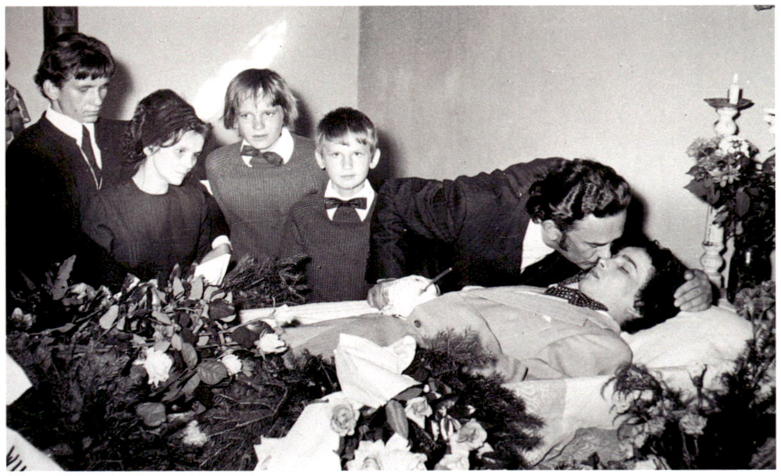

Plate PM.10

"The Last Bond"
Child on Dead Father's Lap
Tintype, circa 1875

Plate PM.15

Man Laid Out at Home
Circa 1920

Plate PM.11

School Children and Their Dead Classmate
Vcelik's New Studio, Northeastern United States
Real Photo Postcard, circa 1910

Plate PM.12

Postmortem of a Dehydrated Infant
Tintype 2 1/2" x 4", circa 1885

Plate PM.13

Twins Posed on Miniature Chaise
B.H.L. Dabbs, Pittsburgh, PA
Carte de Visite, circa 1872

Plate PM.14

African American Man Clutching His Infant Daughter
South Carolina origin
Tintype 1/6 plate, circa 1871

Plate PM.16

Father Kisses his Son Goodbye
Eastern Europe
Silver Print 3 1/2" x 5 1/2", circa 1960

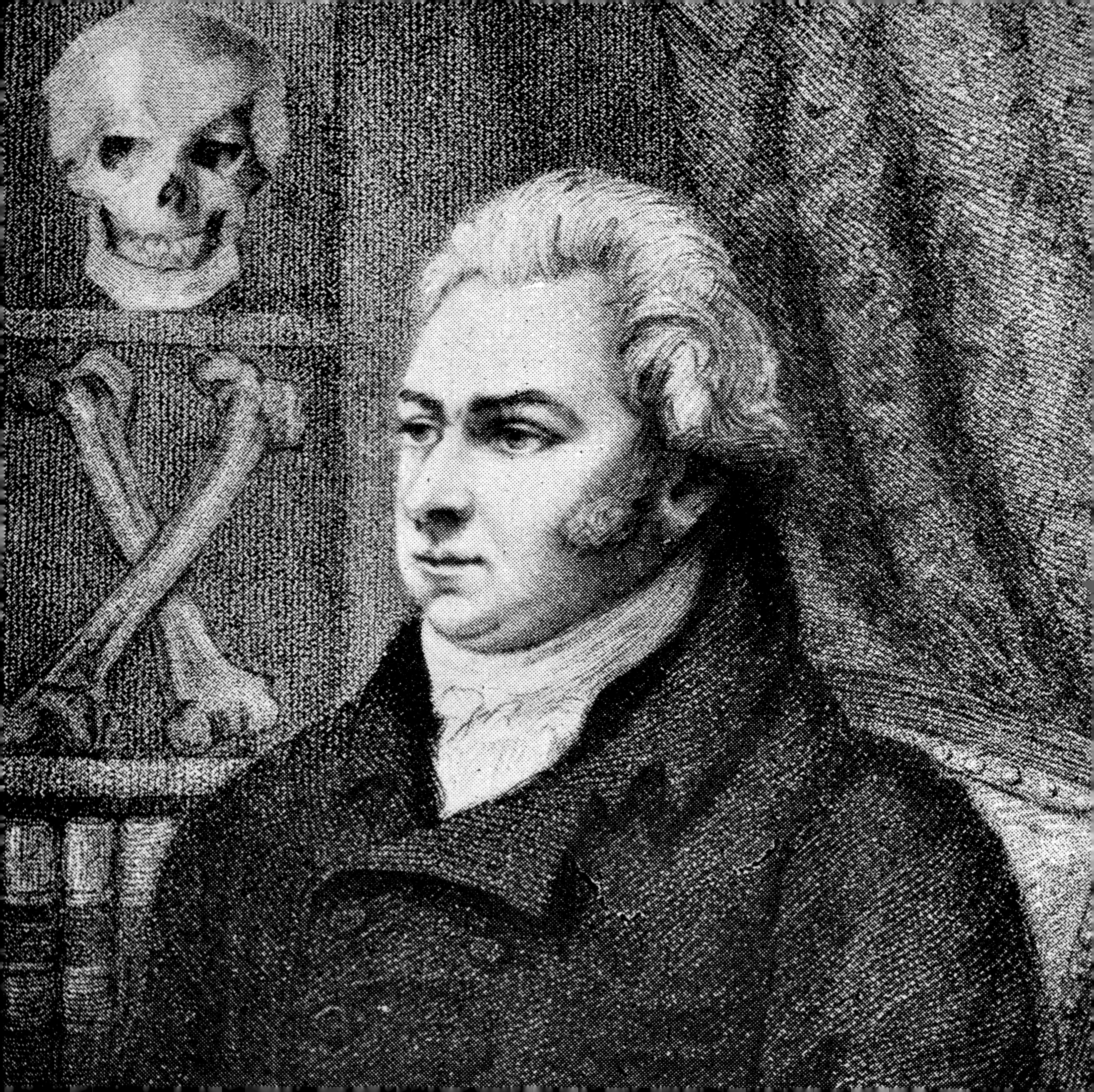

CHAPTER ONE

Posing with Skulls & Bones

Establishing Professional Identity

In 1839, with the presentation of photography to the world, for the first time, the common man had access to visual representation and immortality previously reserved for the rich and famous. People had to learn to pose for the camera. The obvious solution to photographic portraiture was to create posing conventions that followed painterly traditions. Poses with a skull or a skeleton, the classic icons of death, had been painted and drawn by artists, and commented upon by writers almost since the beginning of recorded history. The skull or skeleton symbolized the finality of worldly existence, and the individuals depicted with these props were those concerned with life and its problems. For centuries, noted physicians, scientists, and the intelligentsia were painted with skulls in serious portraits symbolizing their unique position and knowledge.

By the mid-1840s, occupational portraiture had become firmly established as an American and European posing genre. Subjects posed with the tools and, if appropriate, the special garb of their trade. Physicians, scientists, and medical students began posing with skulls and skeletal parts. These were serious portraits in emulation of time-honored painterly traditions and intended to convey the seriousness of their study or work. The skull photograph became one of the visual icons of the medical profession and representative of physicians at all levels of the occupation. The 1850 daguerreotype of America's leading physician, Harvard University Medical School and Massachusetts General Hospital founder, John Collins Warren, MD, (1778–1856) posing with a skull is the best known of this genre of photographs. From Dr. Warren to unidentified medical students, the skull and bone pose was the first photographic iconographic symbol of the profession.

In this chapter, a wide array of poses are presented. The daguerreotype on the back cover was taken in 1855 of Charles Linnaeus Allen, MD, (1820–1890) and his colleague Daniel Avery, MD. It is the most elaborate image of this photographic genre for it contains all of the intellectual elements of the medical profession: books, bones, skull, and illustrations. Also, the physician is studying a bone. Allen had the photograph taken when he was appointed professor of chemistry at Vermont's Castleton Medical College.

The range of images in this chapter of those posing with skulls or bones includes noted personalities such as Lorenzo Fowler, America's leading phrenologist, and Andrew Taylor Still, MD, DO, founder and father of osteopathy, the only nineteenth-century medical sect incorporated into mainstream (allopathic) medicine. Two other significant images include one taken by noted photographer William Notman of Richard Clarke Cabot, MD, (1868–1939) of Boston posing with associates studying a skull, and an image of Sir William Osler, MD, posing with staff, a skull casually on a table.

In Europe, physicians were rigorously selected and university trained. Therefore, there are photographs of physicians graduating from medical schools in the nineteenth century, and they are often in formal dress or gowns. However, there are no graduation photographs of American medical students until the last years of the nineteenth century, as the American method of physician education was not university based. Until the first decade of the twentieth century, American medical training consisted of a two-year apprenticeship with a physician and then some courses at a medical school, which until about 1910, was often a small local proprietary establishment. Posing with medical-related material was the only signifying photographic evidence of one's medical education.

In the 1840 to 1875 time period, posing with bones and skulls was common for medical practitioners. In the 1880 to 1920 era, the medical student photographic tradition changed. Photographs were taken around the dissection cadaver with one's classmates. These images were for private use among friends, colleagues, and family. These so-called 'rite of passage' dissection table photographs are the subject of chapters two and three. Photographs at dissection tables remained popular until the 1930s, however, by 1920, they were not taken by professional photographers and had become, for the most part, snapshot images for private consumption. It was not until the last years of the nineteenth century that traditional medical graduation photographs were taken. As in Europe, these photographs were associated with graduation from a university medical school program. Johns Hopkins Hospital (founded 1889) and other medical universities created strict standards of medical education that involved attending an established formal program at a university. The 1910 Flexner report on medical education noted the poor quality of proprietary medical schools. The report was part of the movement started by universities to end substandard medical education. In 1905, there were 163 medical schools in the United States. By 1920 there were only eighty-eight, in 1925, eighty, and for almost the next forty years the number of medical schools varied only between seventy-six and eighty-six.

Toward the end of the nineteenth century, it was popular in medical environments to have a skull in a photograph, even if it was casually placed. Physicians' offices, medical students' dorm rooms, and even hospital staff used the skull as a prop. Several of these photographs are shown in a variety of interiors. By the end of World War I (1918), the symbols of death disappeared from physician culture and portraiture. Medicine was focused on saving lives, and the accomplishments of medical science had removed death from everyday life (see Appendix, *The Dying of Death*). The hospital was seen not as a place to die, but as an innovative life-saving mecca. By the 1930s, medical achievements dictated that the physician pose with the latest technological tools or technology, a practice that continues to this day.

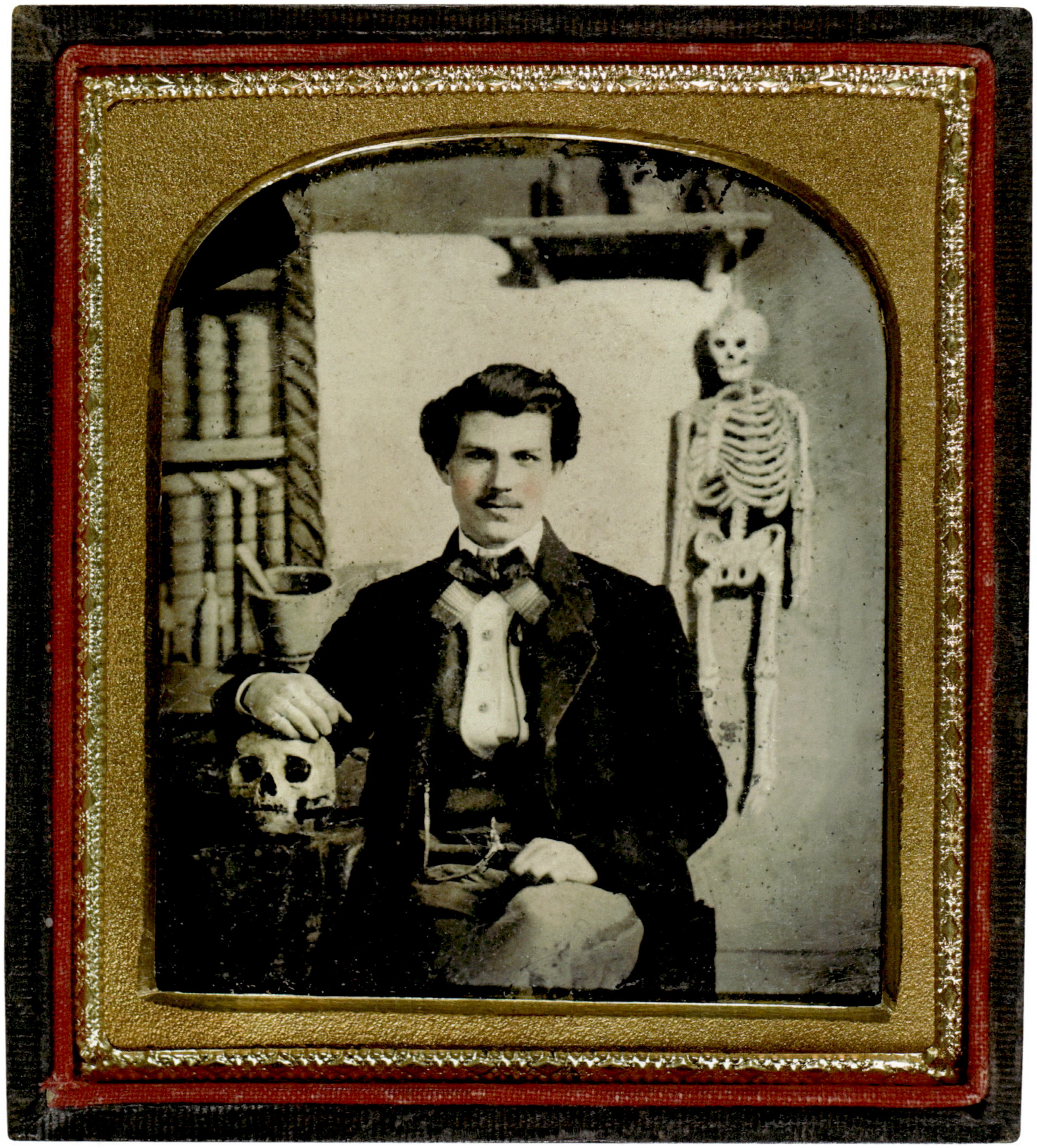

Plate 1.1

Philadelphia Medical Student
1/6 Plate Ambrotype, circa 1858

Classic photographic background used in Philadelphia.

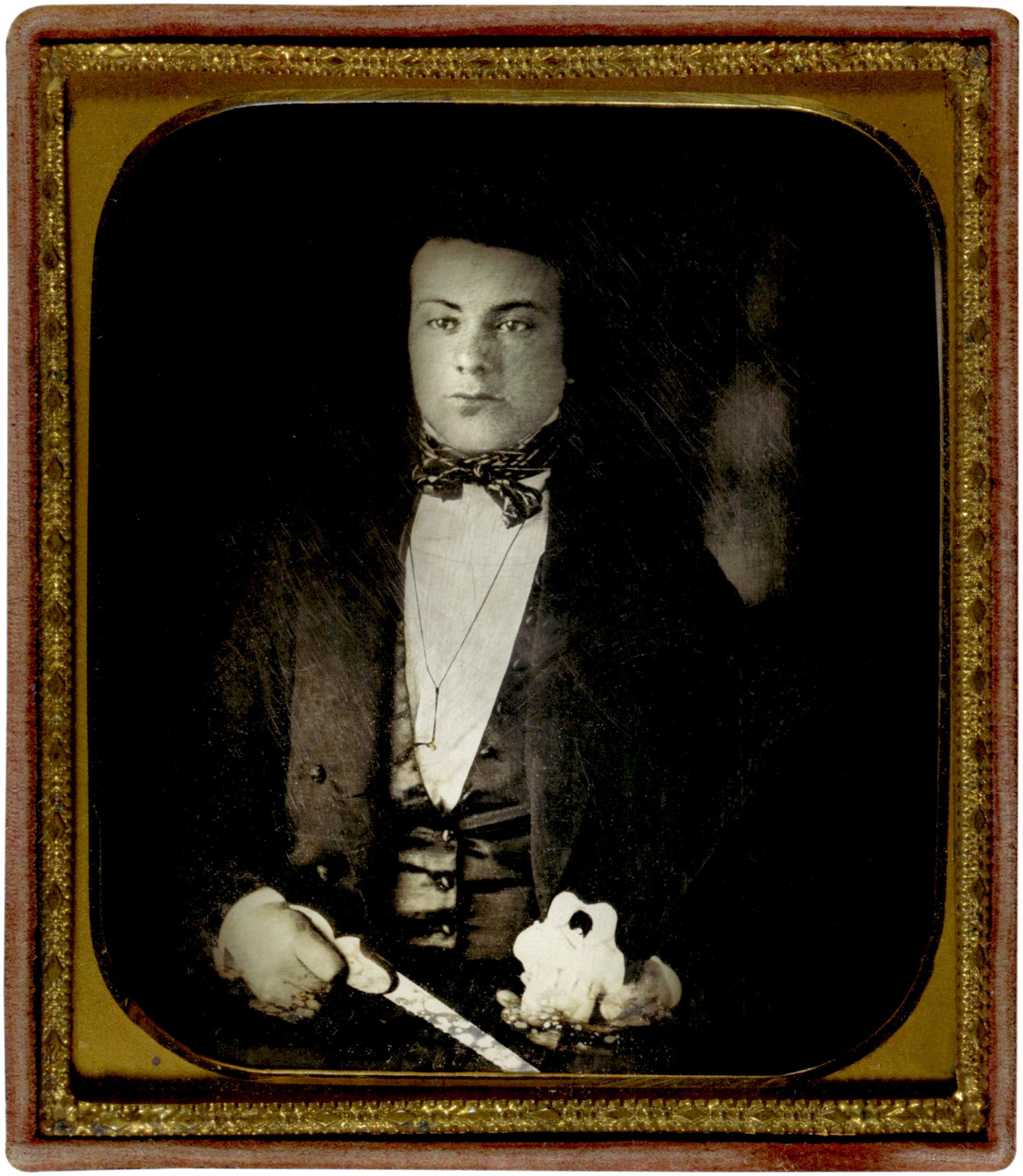

Plate 1.2

Saw Bones
1/6 Plate Daguerreotype, circa 1848

Classic symbolism, as a physician poses with saw and skull. In the nineteenth century, 'saw bones' was a popular slang term for a physician, especially a surgeon.

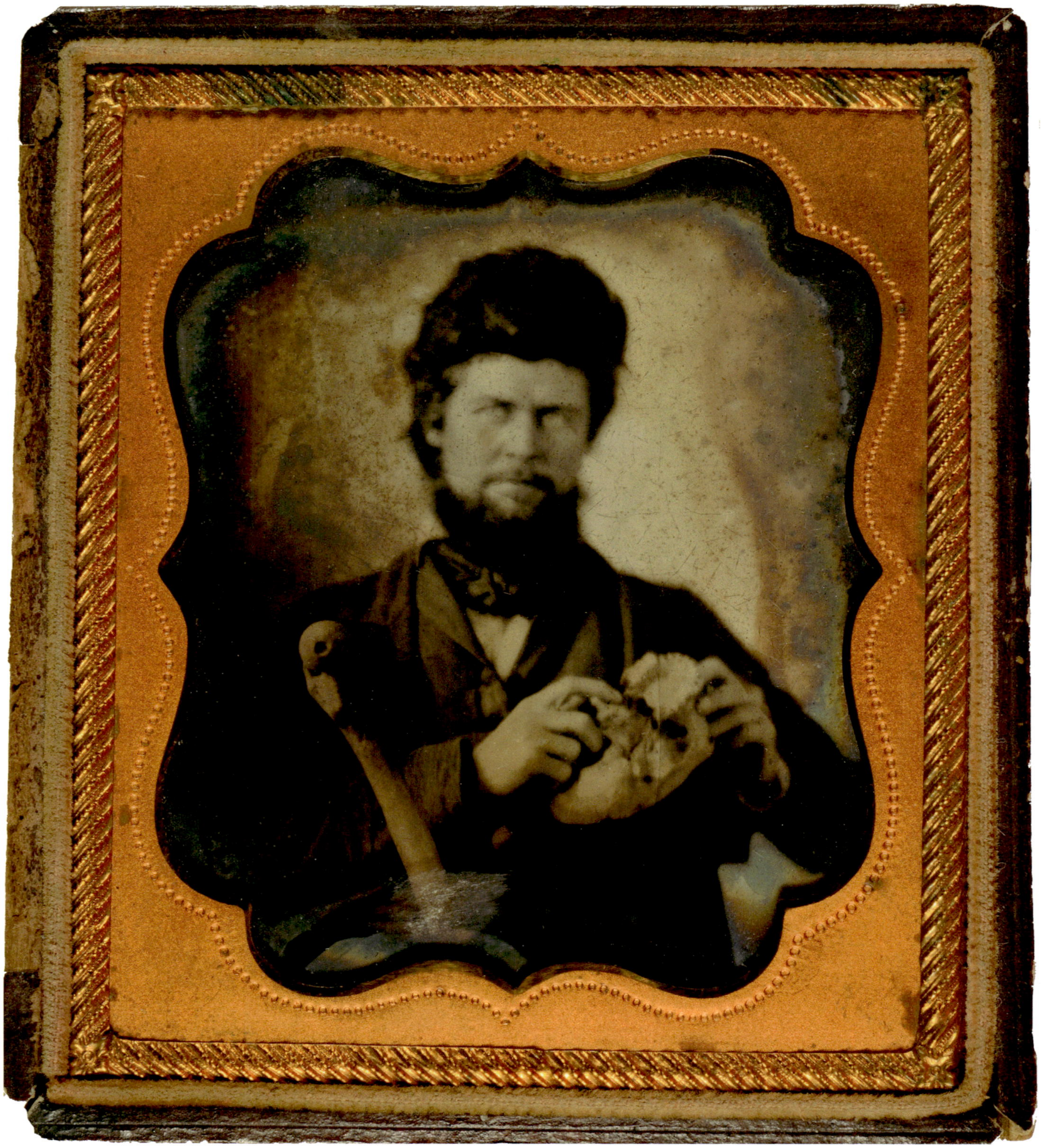

Plate 1.3

Frontier Physician
1/6 Plate Ambrotype, circa 1856

Pointing out an interior structure of skull.

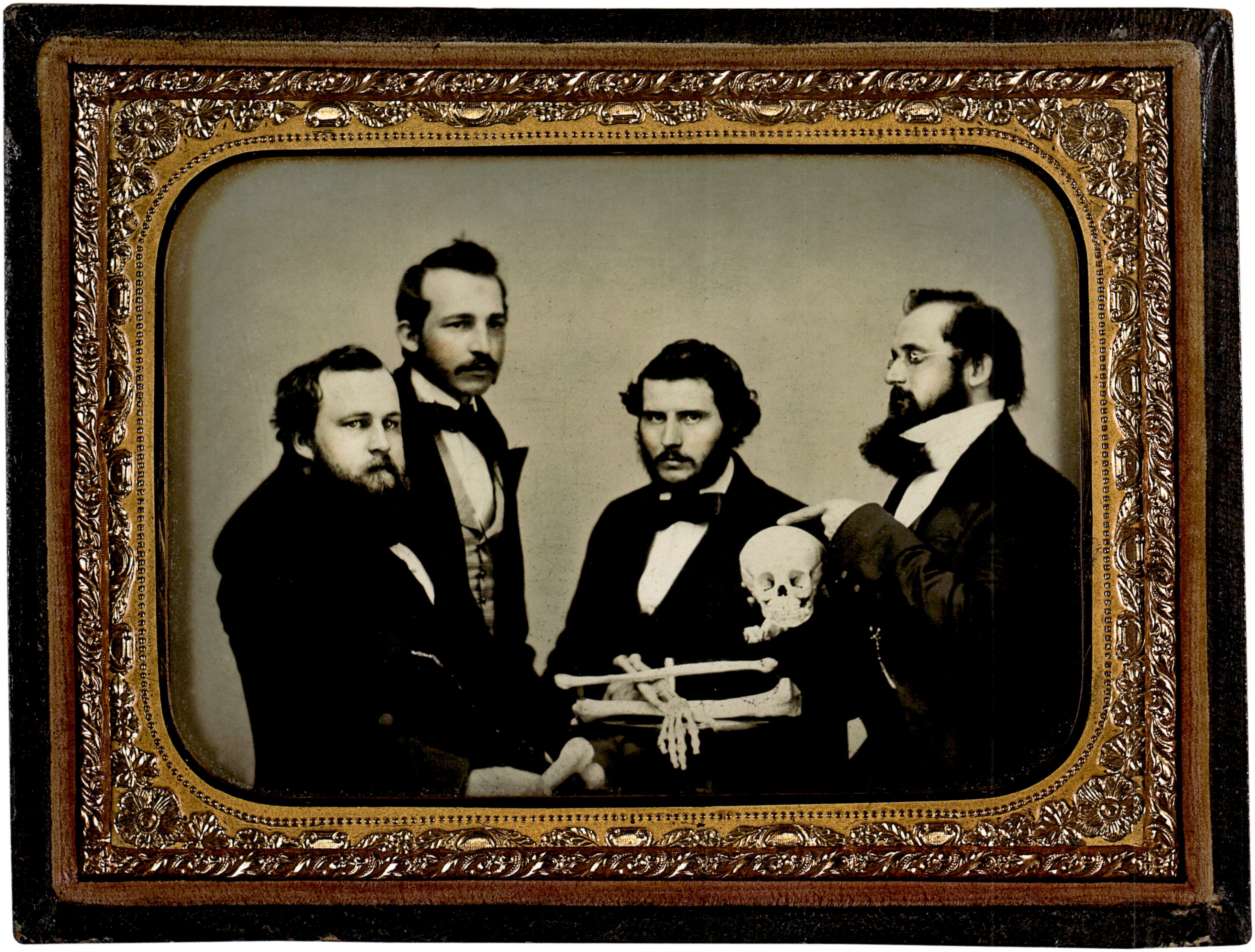

Plate 1.4

Graduate Physicians, Albany Medical School
1/4 Plate Ambrotype, May 23, 1857
Photographer: H. McBride, 51 State St., Albany, NY

Study group in a rare identified image. From left to right: Dr. G.H, Chadwick, Portland, ME; Dr. J.J. Myres, Utica, NY; Dr. S. Traves, Luzerne, NY; Dr. C.S. Spencer, New Bedford, Mass

Plate 1.5

Studying Anatomy
1/4 Plate Tintype, circa 1868

Two men pose with book and bone.

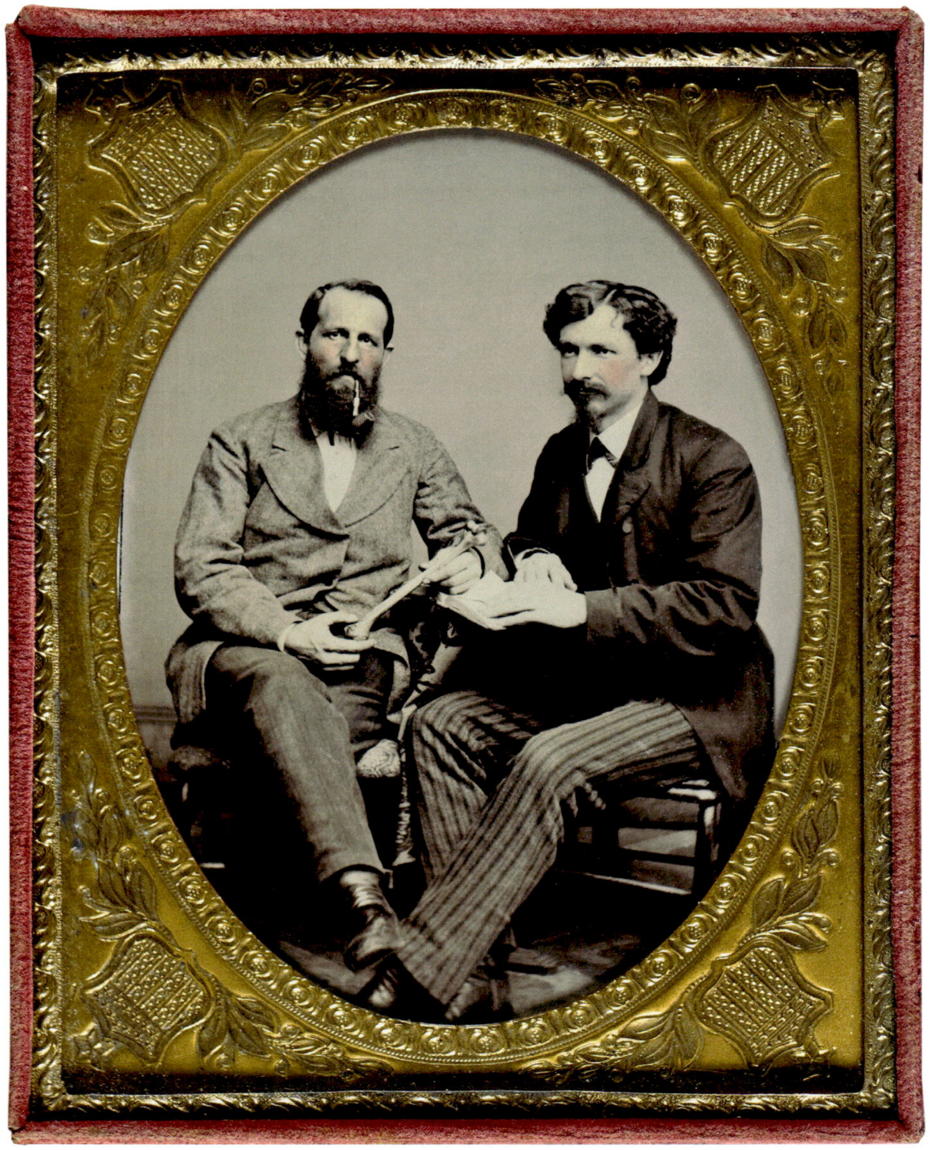

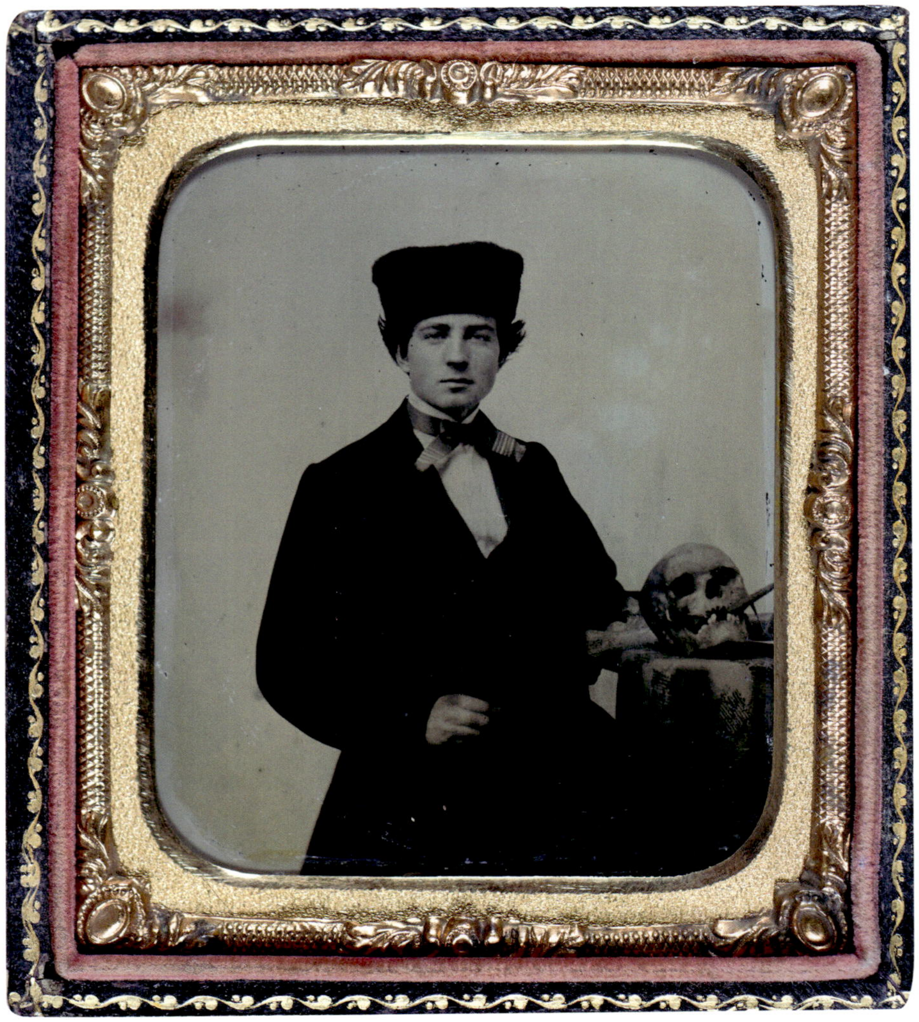

Plate 1.6

Dr. Belden, New York City
1/6 Plate Ambrotype, circa 1858

A cigar is placed in the jaw of the skull.

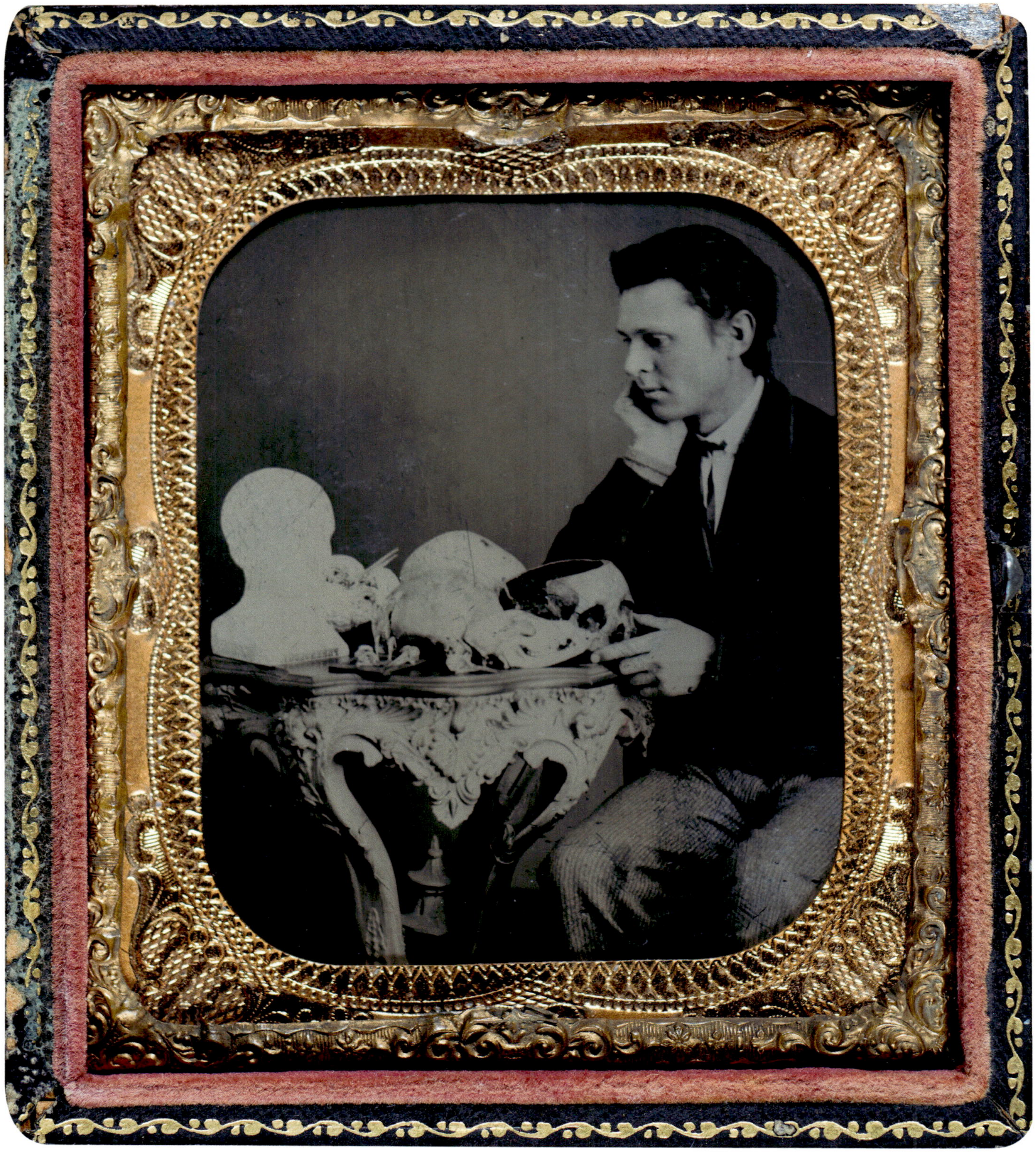

Plate 1.7

Phrenologist Studying Skull Anatomy
1/6 Plate Tintype, circa 1860

Dr. Charles Van Nette, born 1835.

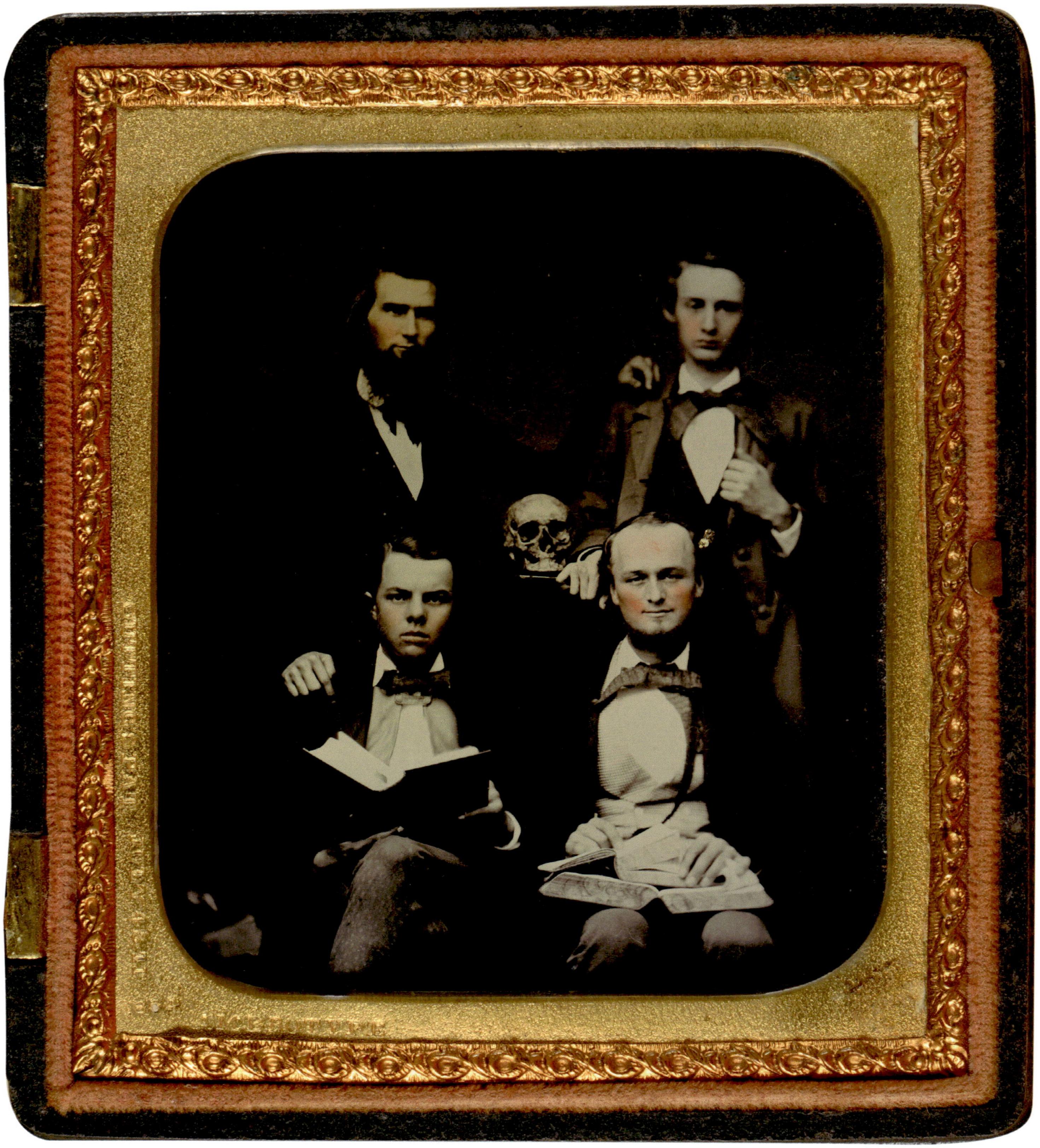

Plate 1.8

Four Medical Students
1/6 Plate Tintype, circa 1875

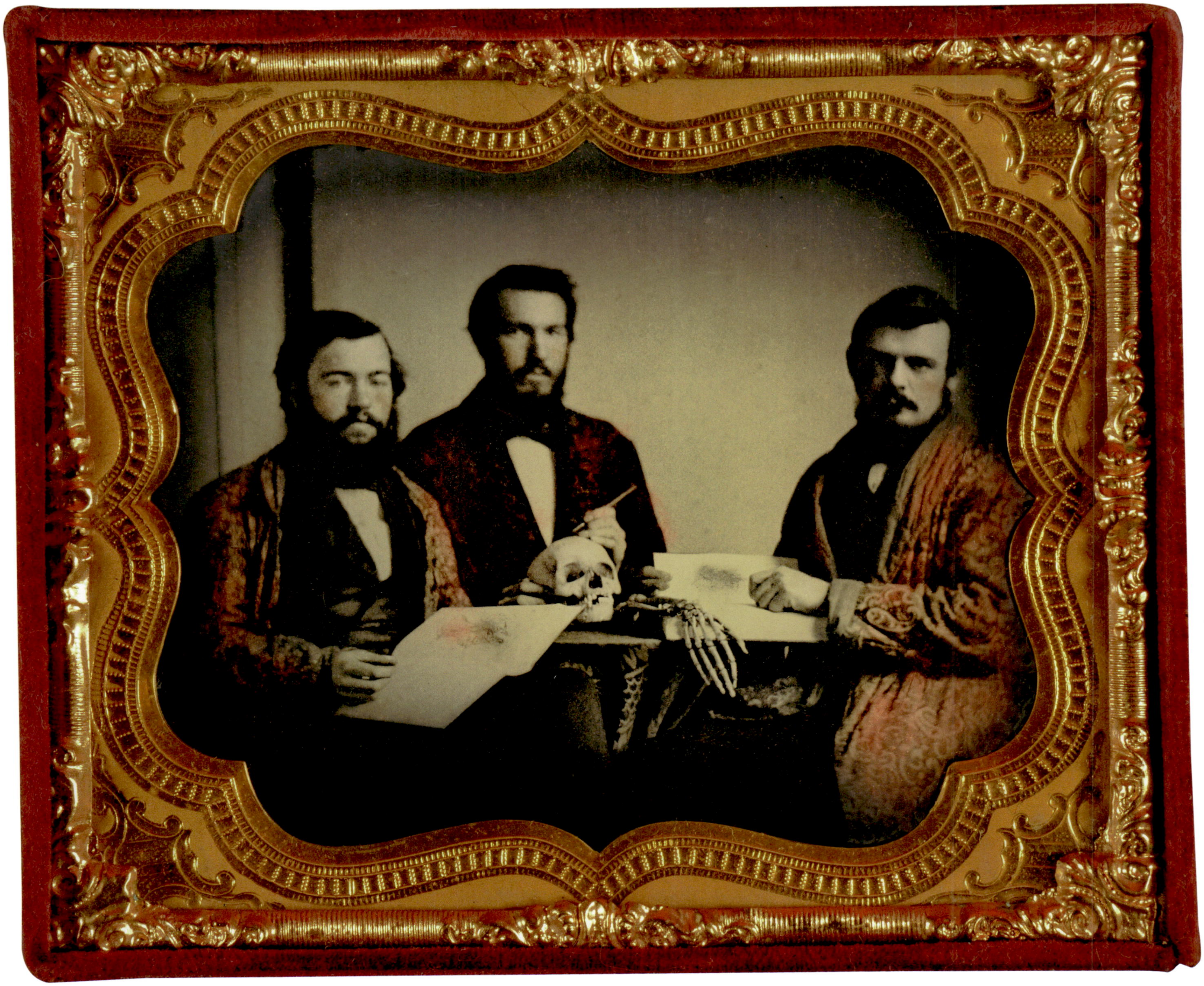

Plate 1.9

New York City Students in Fancy Robes
Hand-Colored 1/6 Plate Ambrotype, circa 1860
Photographer: Meade Brothers, 235 Broadway, New York City

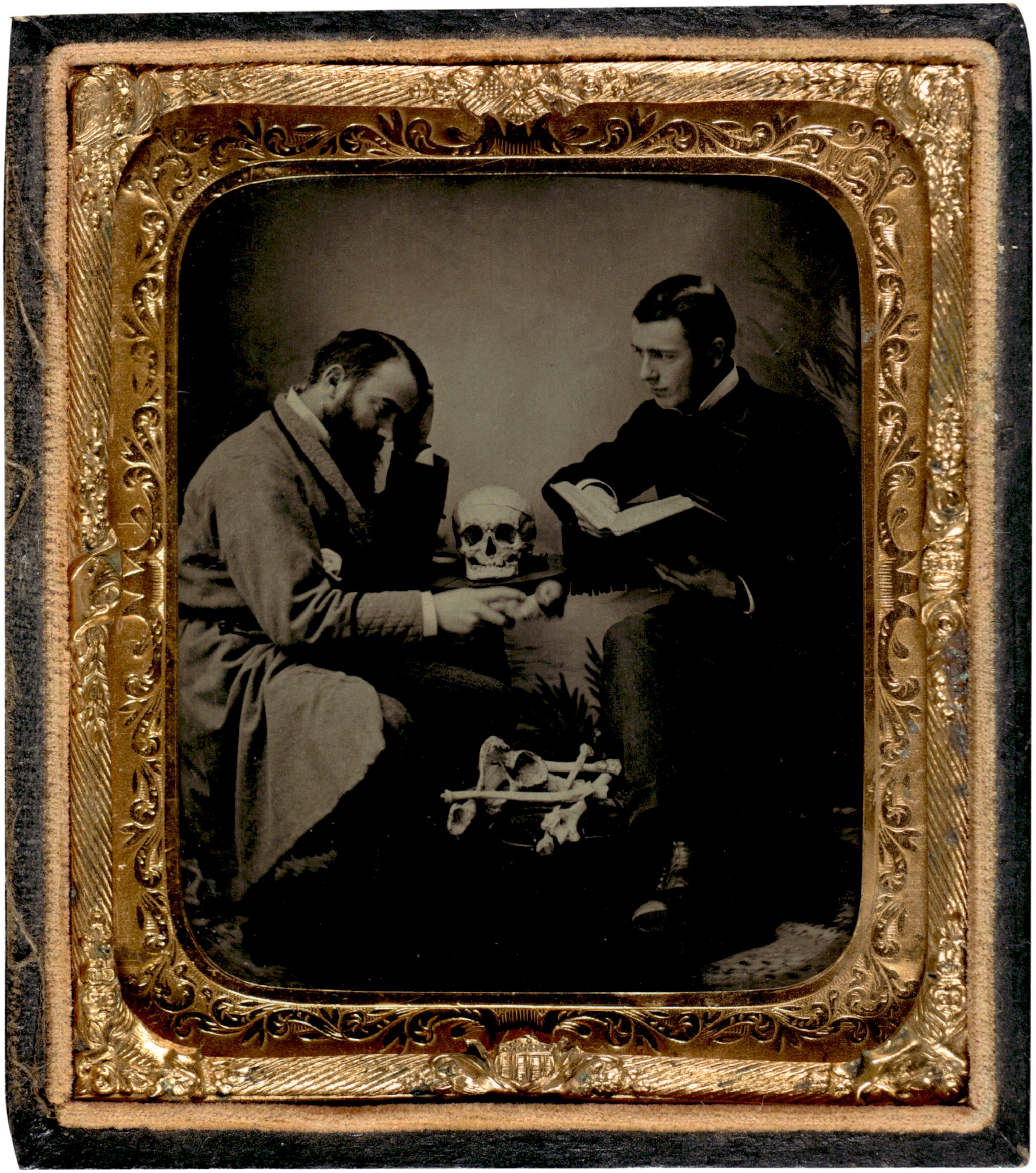

Plate 1.10

The Meade Brothers
1/6 Plate Tintype, circa 1878

Physicians in Victor, NY

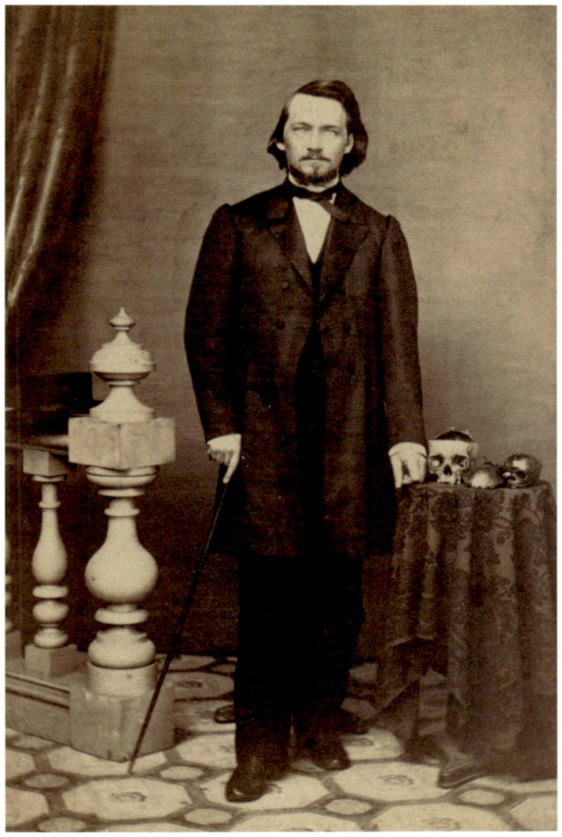

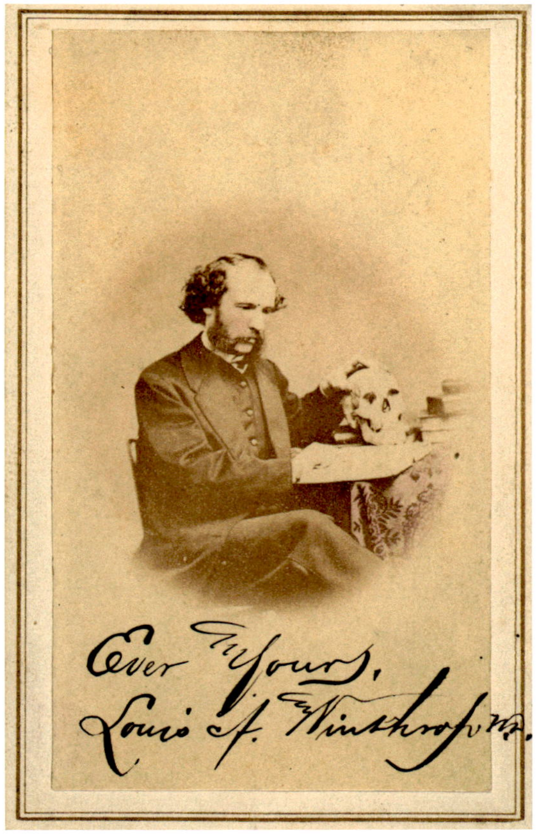

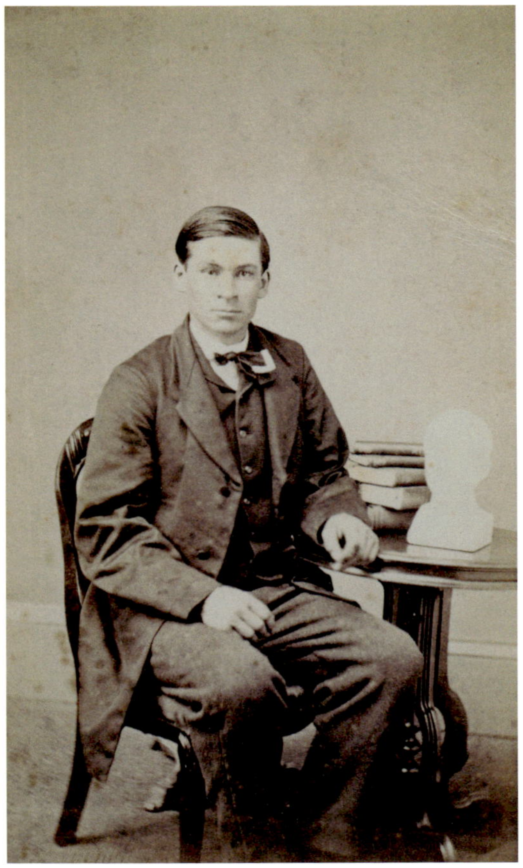

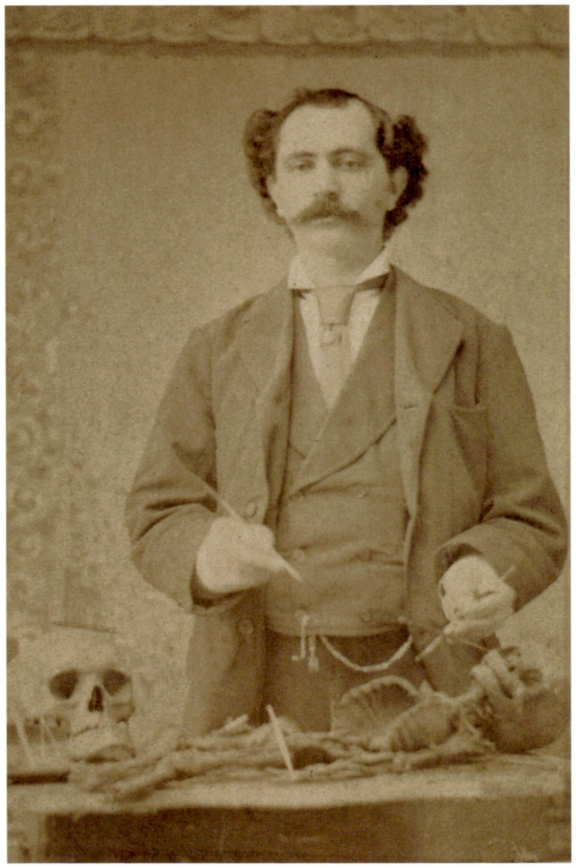

Plate 1.11

Dr. Henry Mathew, New York City
Carte de Visite, circa 1865
Photographer: Maurice Stadtfield, 711 Broadway, NY

Plate 1.12

Louis A. Winthrop, MD, Souvenir Photograph
Carte de Visite, June 1869
Photographer: Edgar Moran, Main St., Ackley, IA

Plate 1.13

John Crisfell With A Phrenology Head
Carte de Visite, circa 1875
Photographer: Lewis' National Photographic Galleries, Kingston, NY

Plate 1.14

Anatomist J. Connover With Child's Skeleton
Carte de Visite, August 13, 1883
Photographer: Andrews & Frommyer, York Street, Hanover, PA

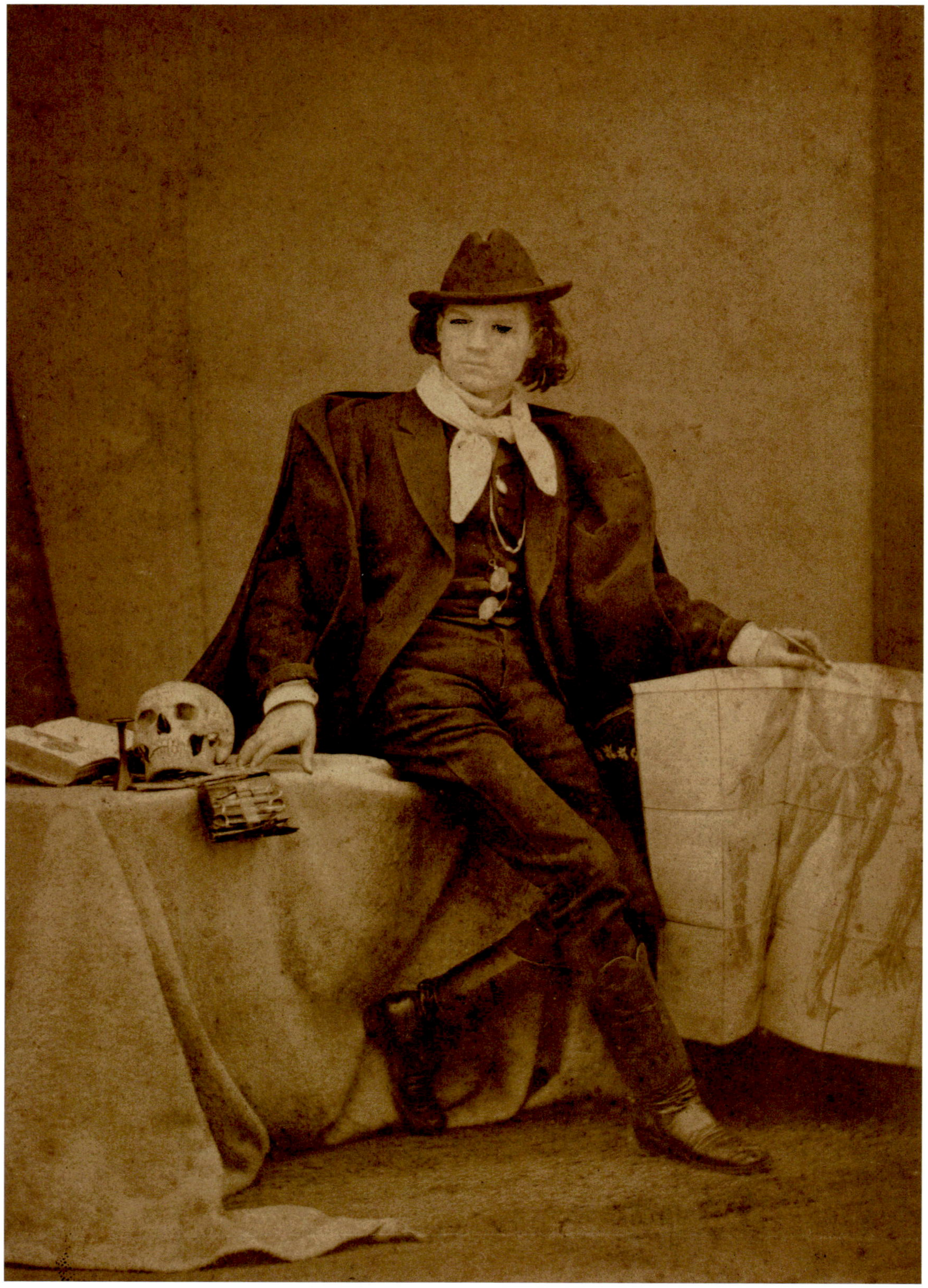

Plate 1.15

Parisian Medical Student
Cabinet Card, Paris, January 6, 1870

Rare pose with an operative kit, book, skull, and anatomy poster.

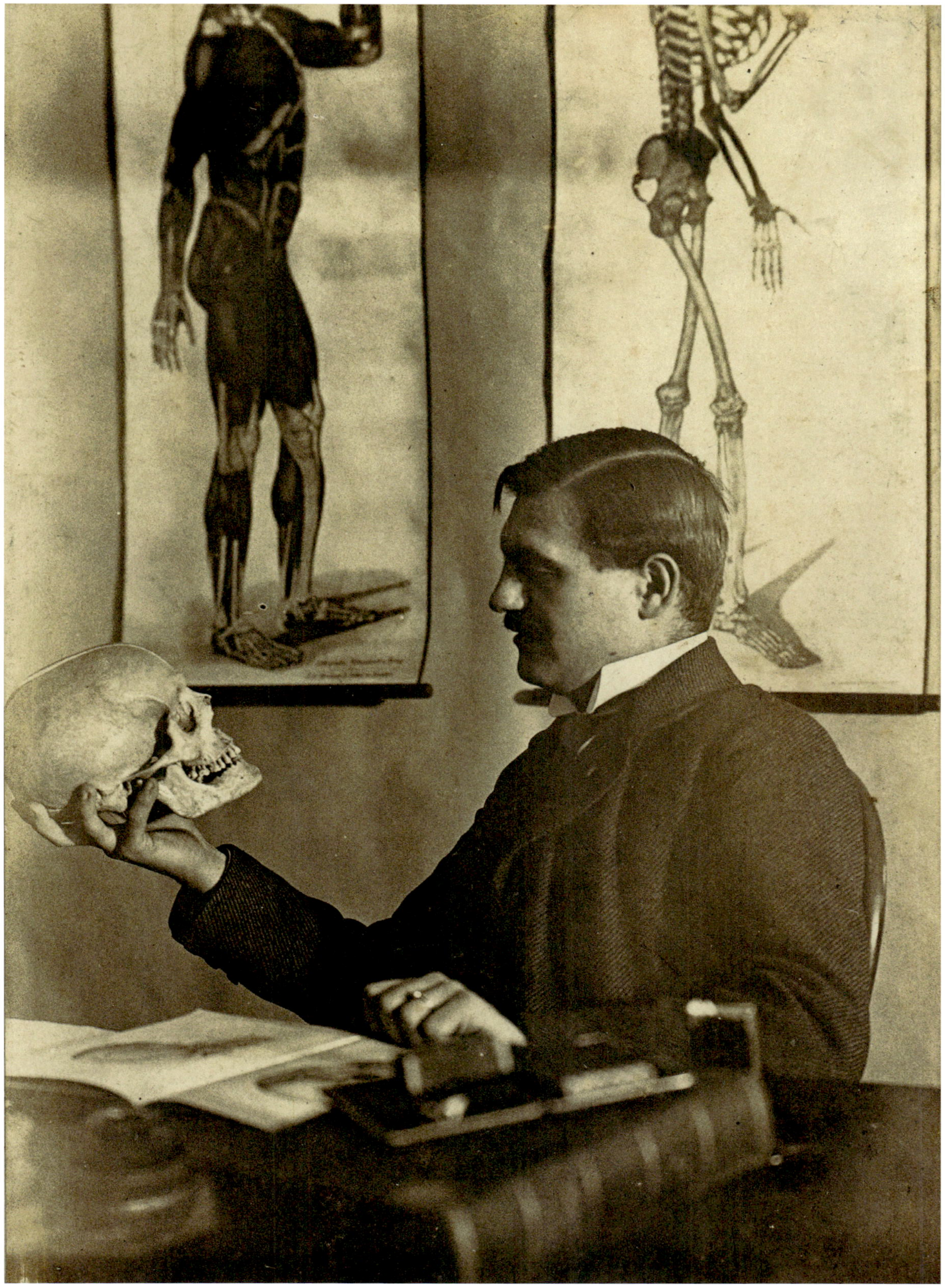

Plate 1.16

Viennese Medical Student with Charts and Skull
Cabinet Card, circa 1888

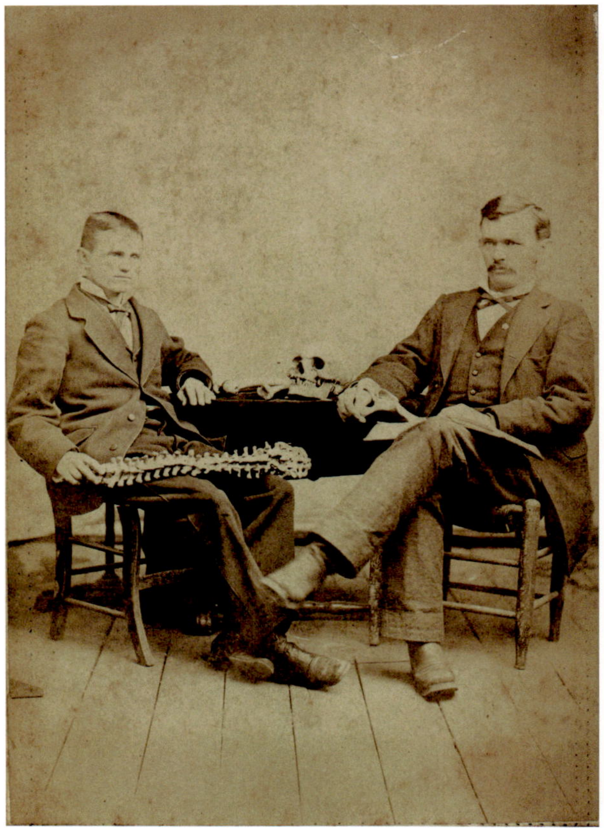

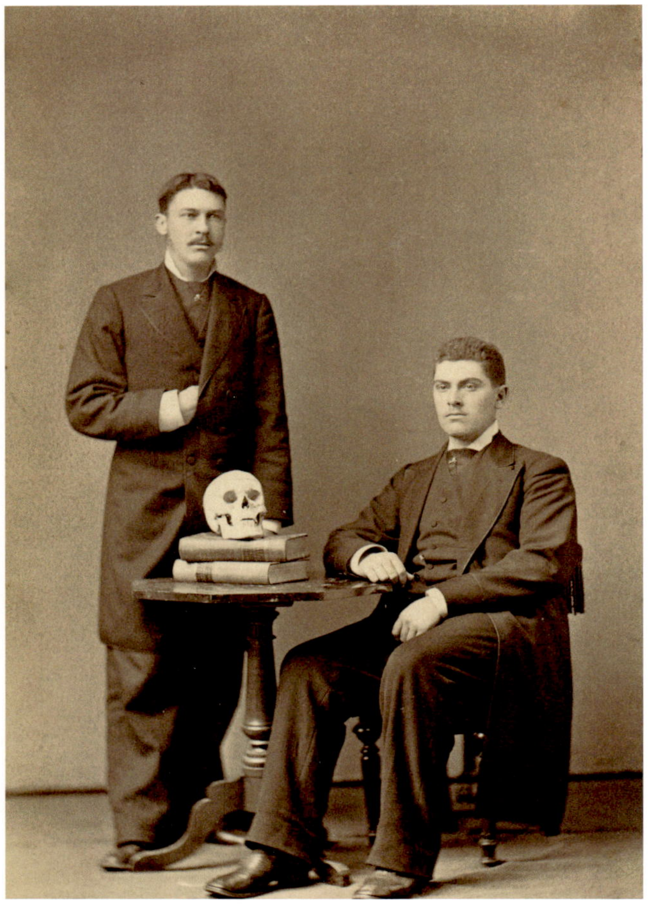

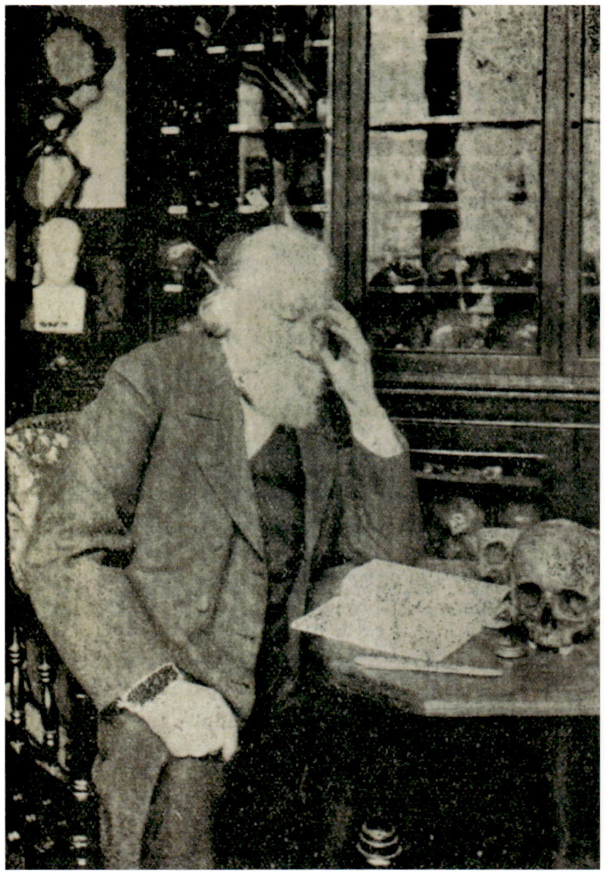

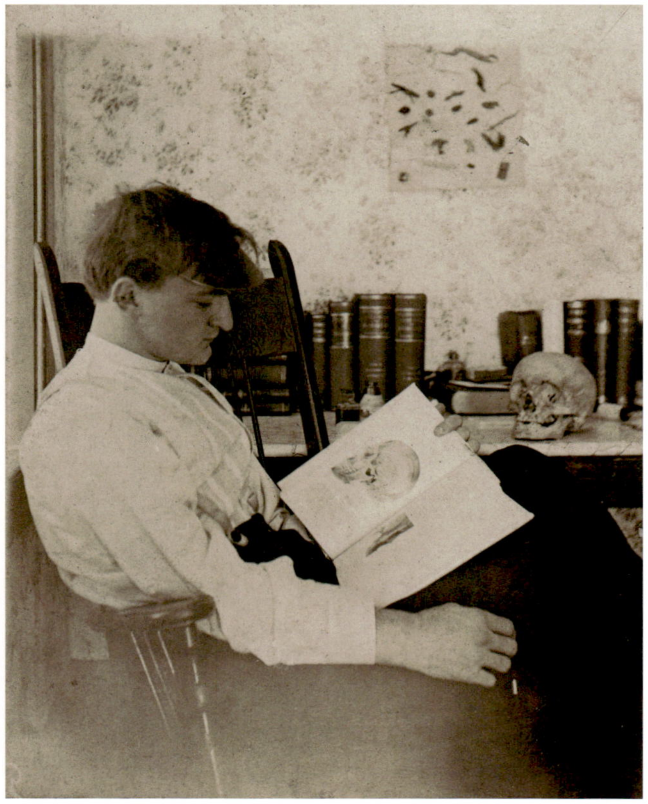

Plate 1.17

Studying the Spinal Column
Cabinet Card, circa 1880

Plate 1.18

Yale Students William B. Graves and W. S. P.
Cabinet Card, circa 1885
Photographer: Warren's, New Haven, CT

Plate 1.19

Lorenzo Fowler, Leading Phrenologist of His Time
Carte de Visite, circa 1860

Fowler developed the famed porcelain phrenology head.

Plate 1.20

Studying A Textbook with Skull Pictures
Silver Gelatin Print, circa 1910

Identified as 'Brian.'

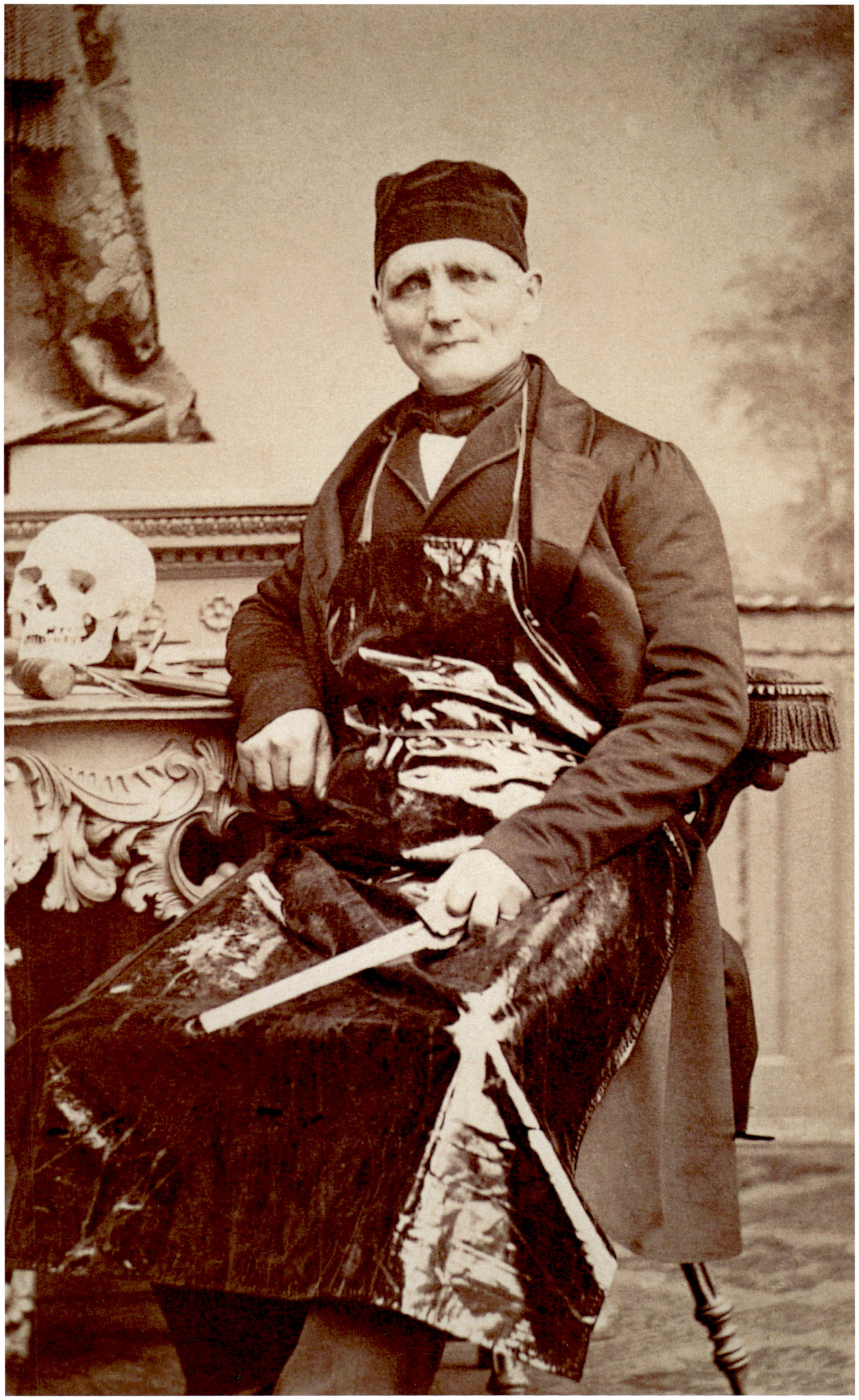

Plate 1.21

German Anatomist
Cabinet Card, circa 1880

Oilcloth apron and amputation knife.

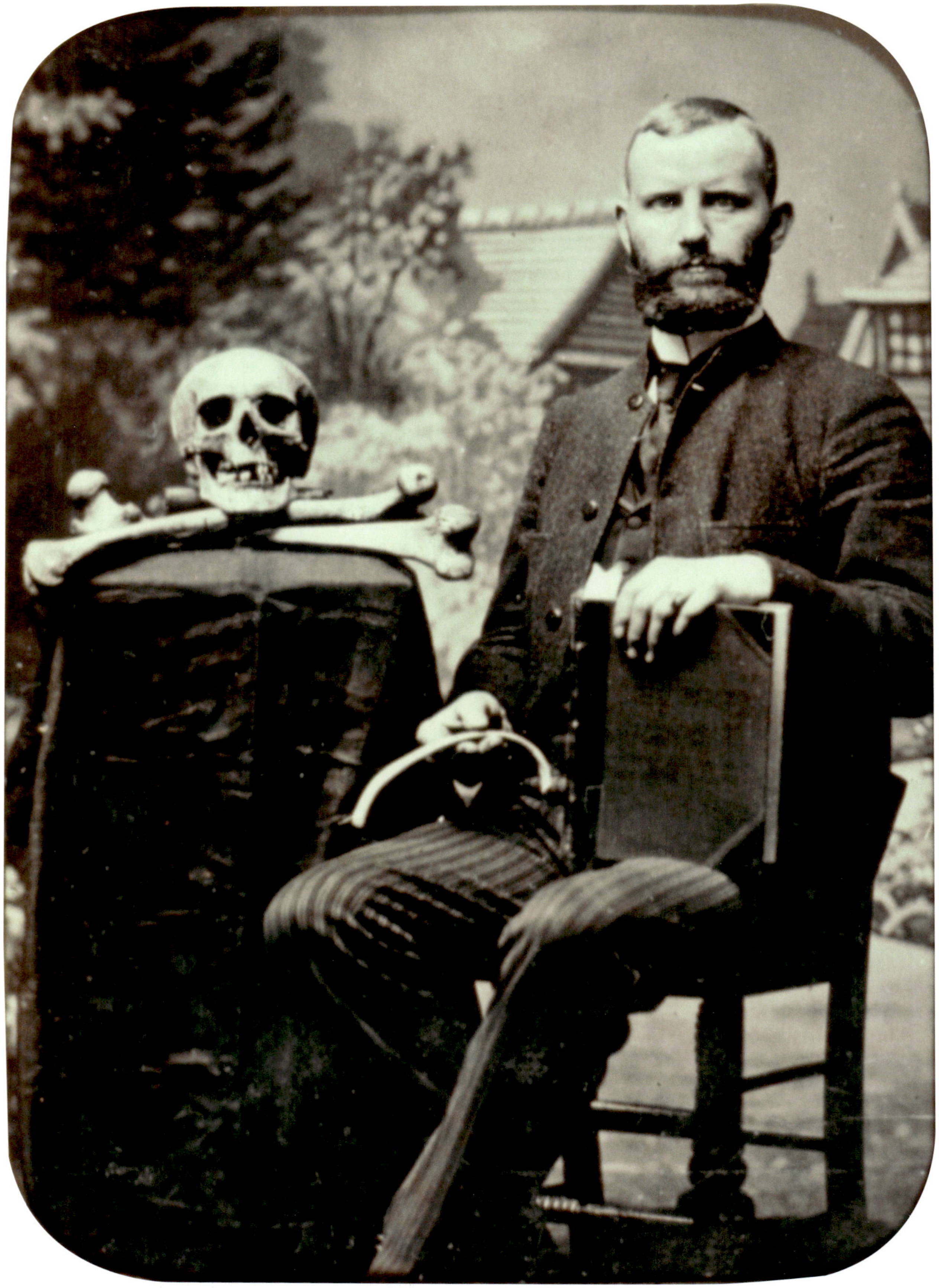

Plate 1.22

Medical Student with Skull and Crossed Bones
Tintype, circa 1875

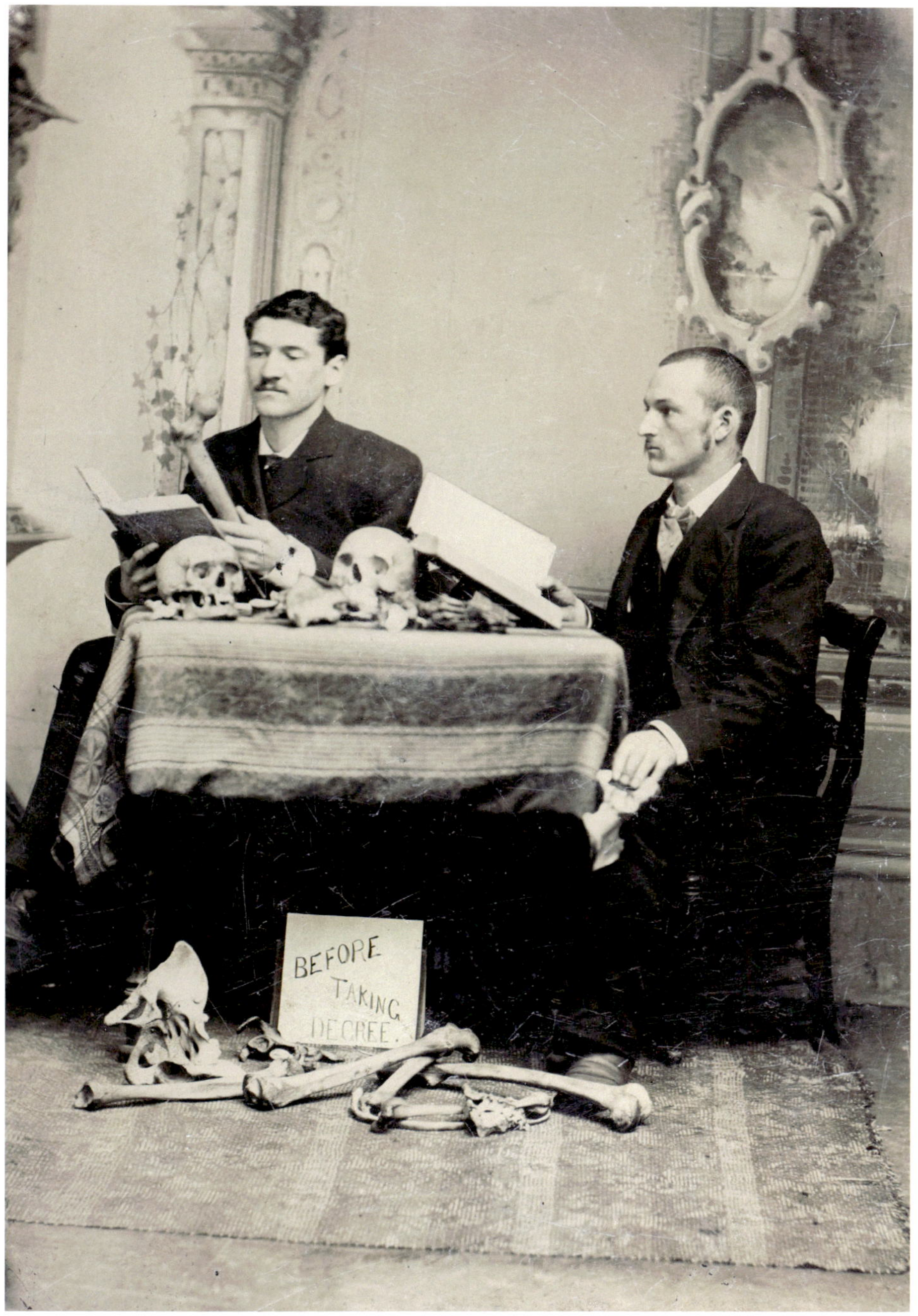

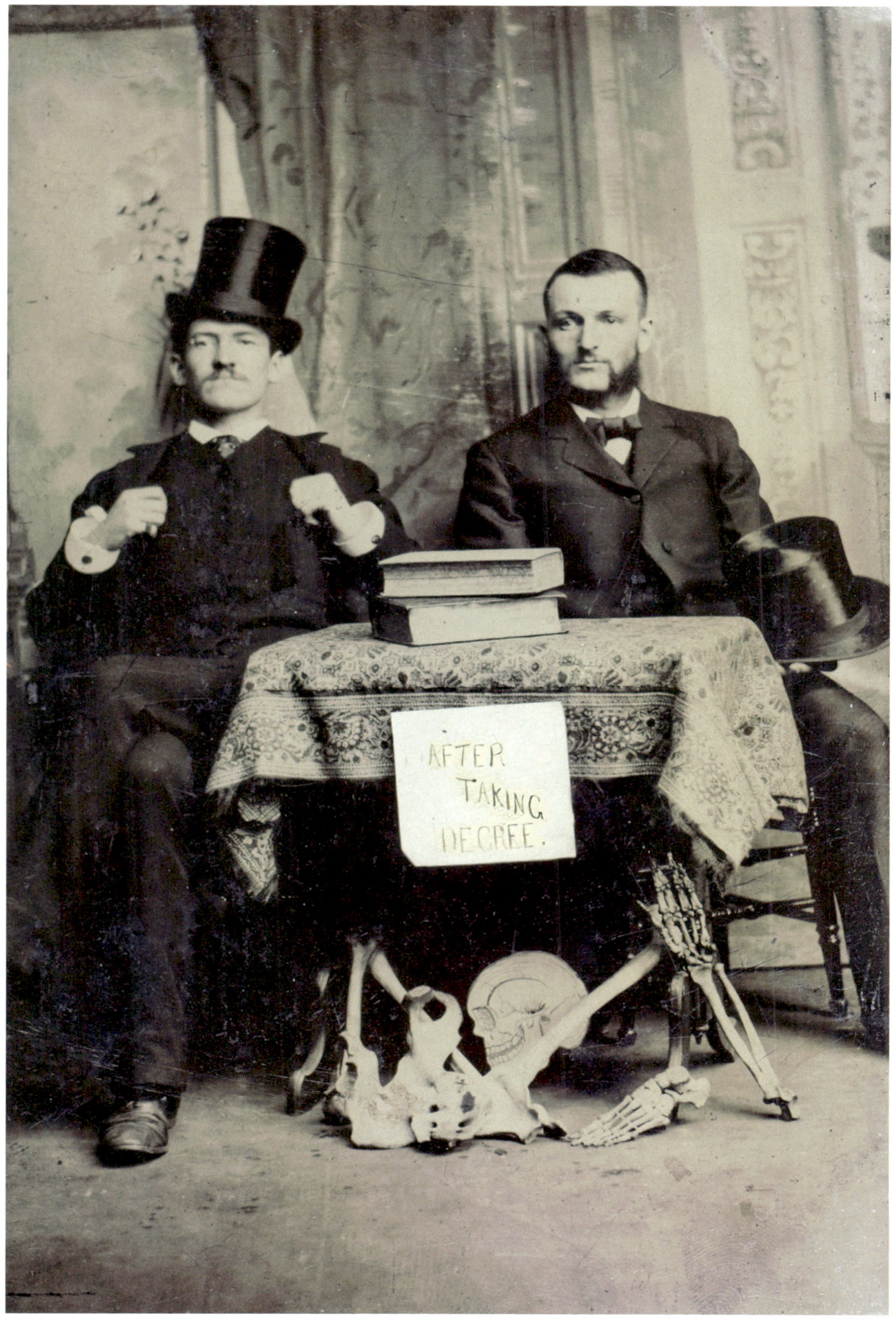

Plate 1.23

F.R. Reynolds Before Taking Degree
1/2 Plate Tintype, 1883

Rush Medical College

Plate 1.24

Dr. Reynolds After Taking Degree
1/2 Plate Tintype, 1884

Rush Medical College

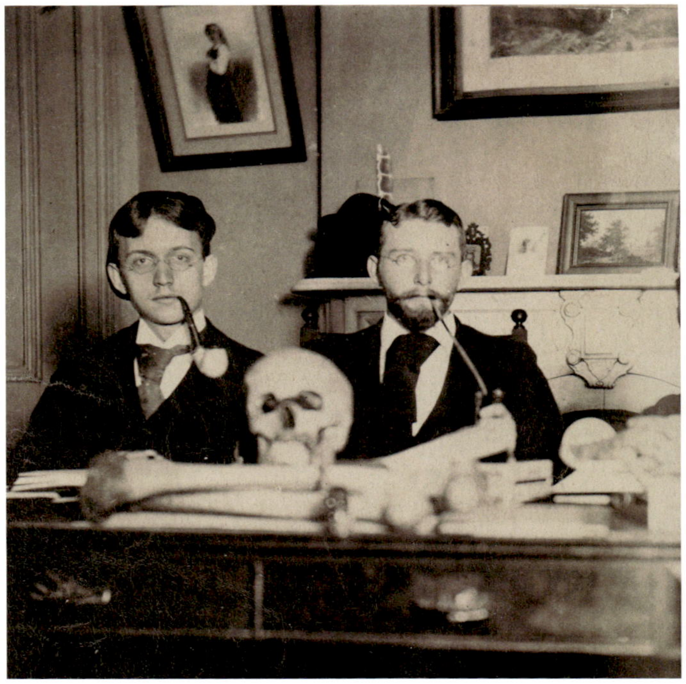

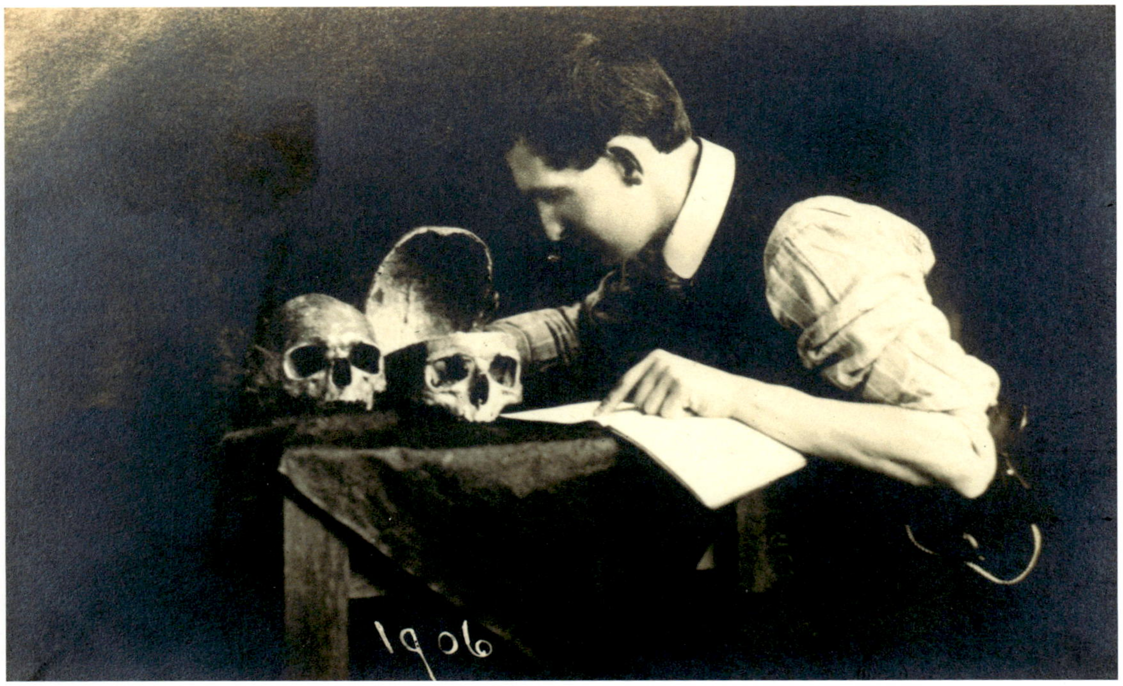
1906

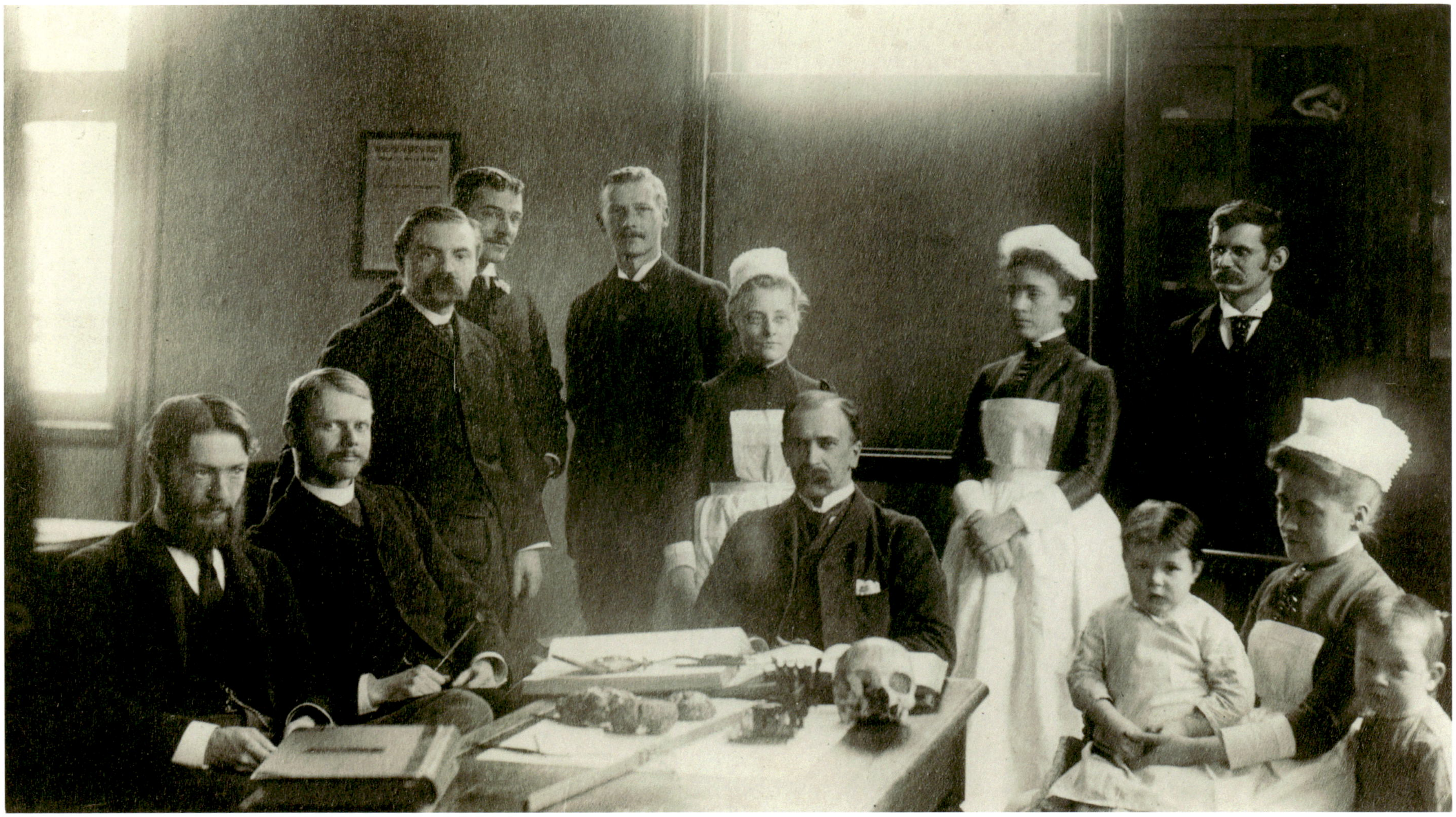

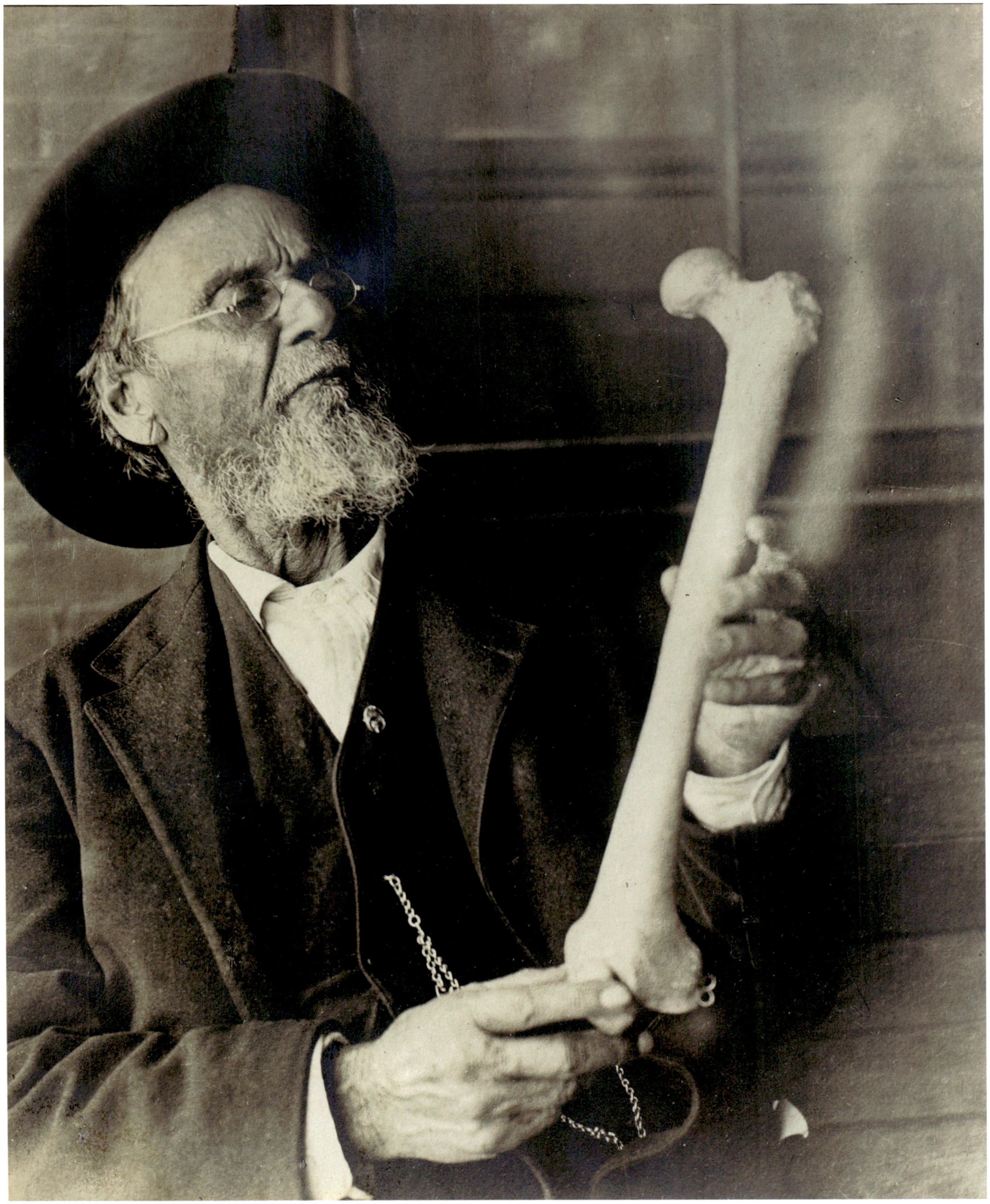

Opposite page:
Plate 1.25

"The Dorm Room"
Silver Gelatin Print, circa 1900

Plate 1.26

Student Studying a Skull
Silver Gelatin Print, 1906

Plate 1.27

William Osler, MD, at University of Pennsylvania
Orthopedic Hospital and Infirmary for Nervous Diseases, Summer 1887

Plate 1.28

Andrew Taylor Still, MD, Founder of Osteopathy
Silver Print, circa 1890

Dr. Still opened the first school of osteopathy in Kirksville, MO, in 1892.

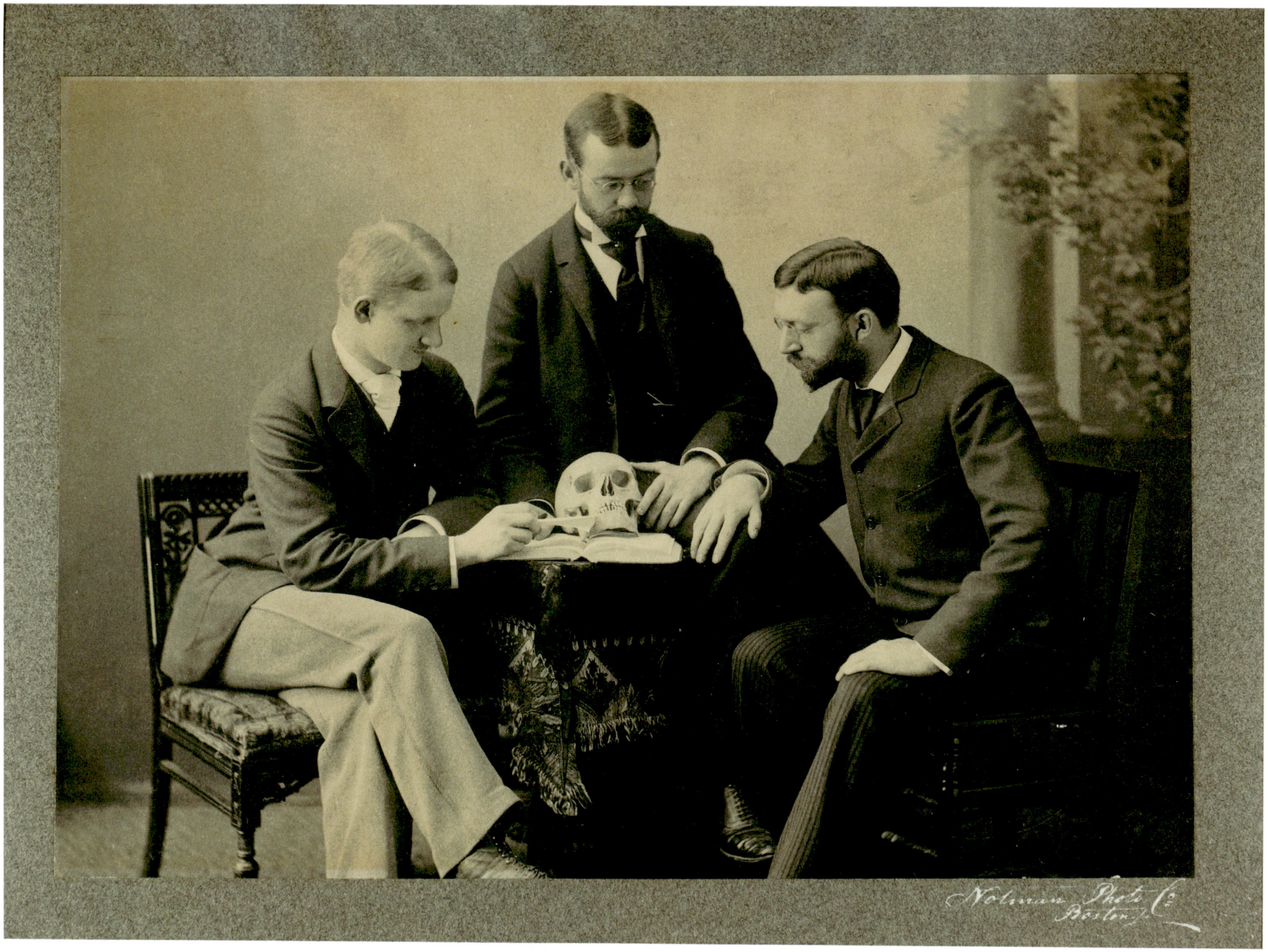

Plate 1.29

The Doctors Cabot of Harvard
Silver Gelatin Print, circa 1903
Photographer: Wm. Notman, Boston

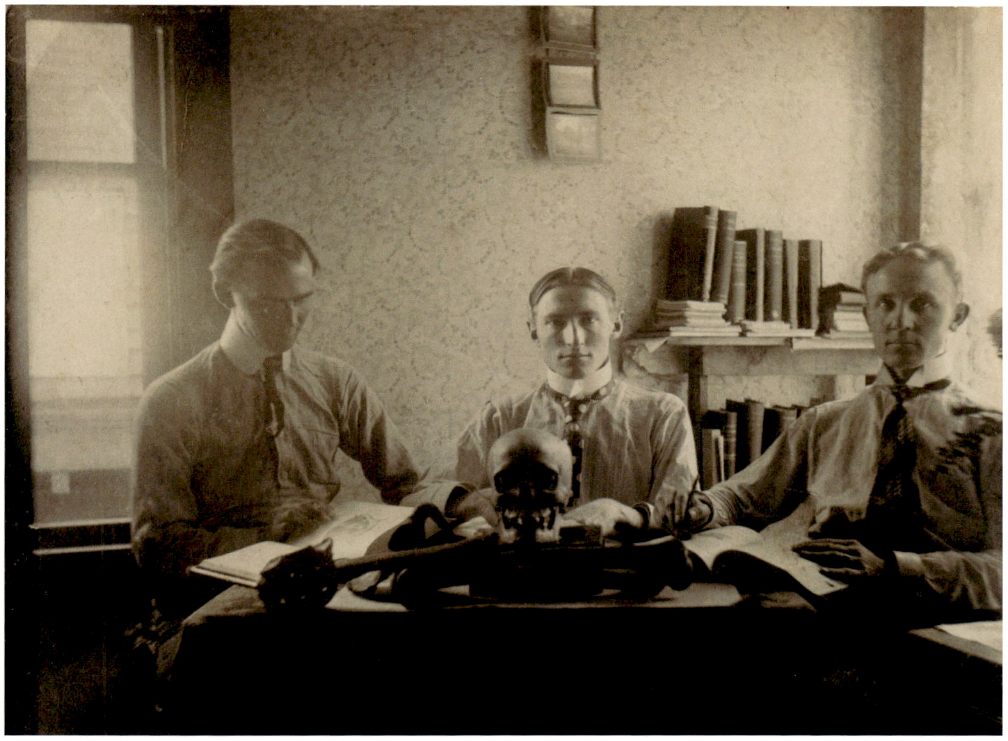

Plate 1.30

"In My Room"
Silver Gelatin Print, circa 1915

University of Washington medical students.

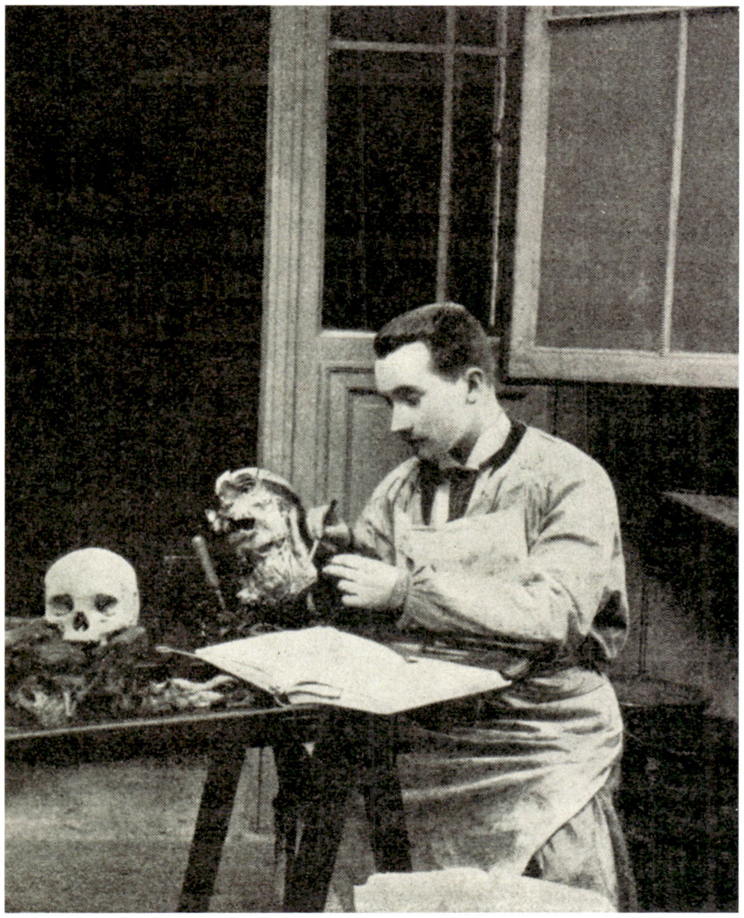

Plate 1.31

Studying Osteology in Paris
Heliotype, 1898

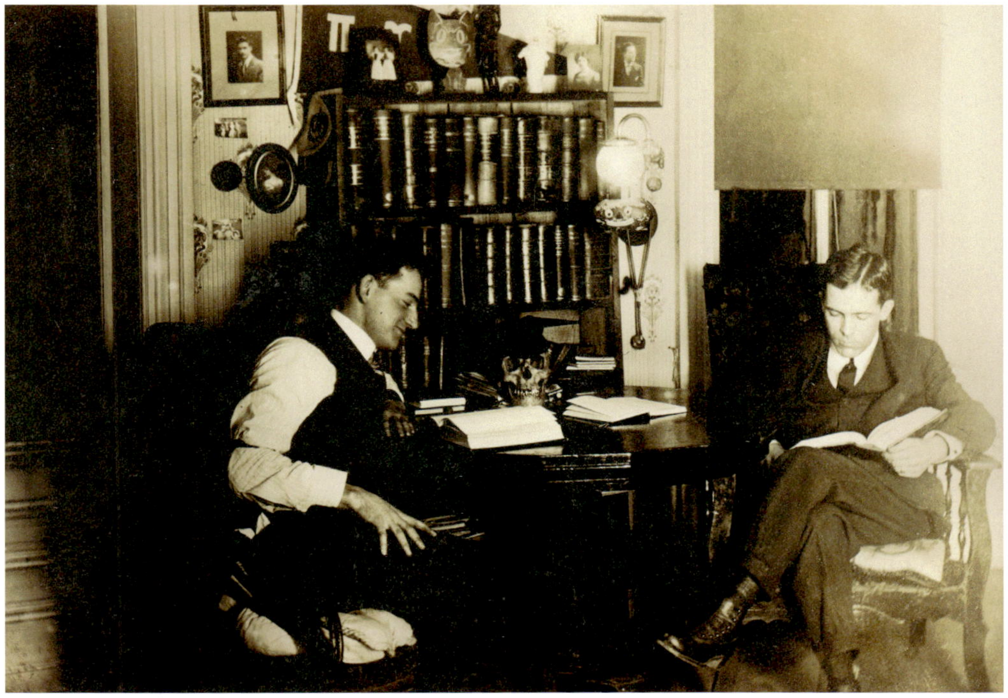

Plate 1.32

Dorm Room Study
Postcard, circa 1910

Blackened skull on desk.

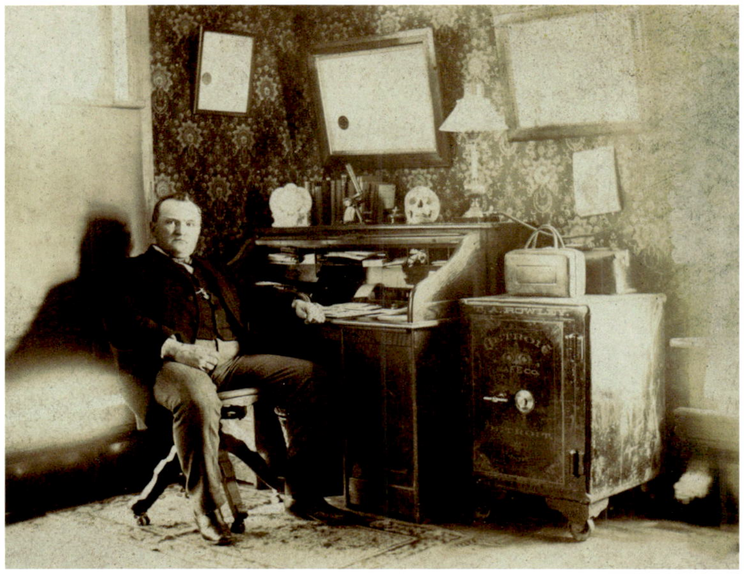

Plate 1.33

Doctor in His Office With Large Safe
Silver Gelatin Print, circa 1895

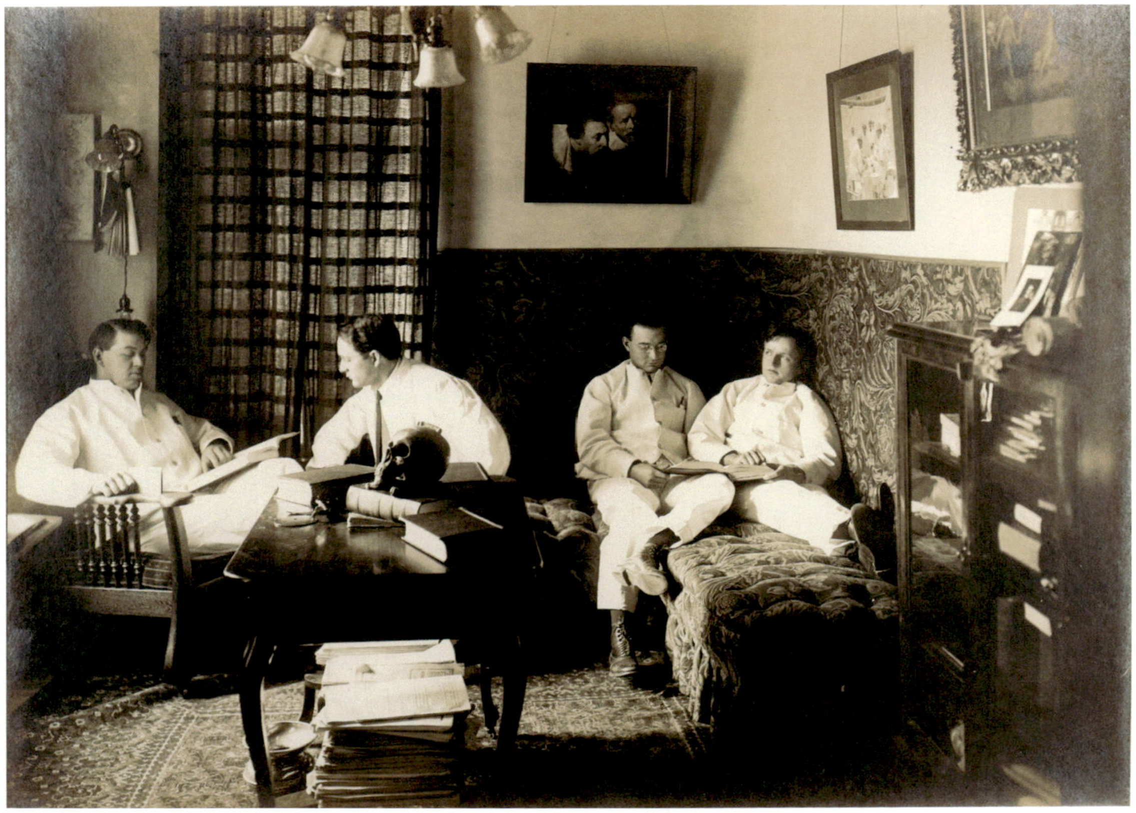

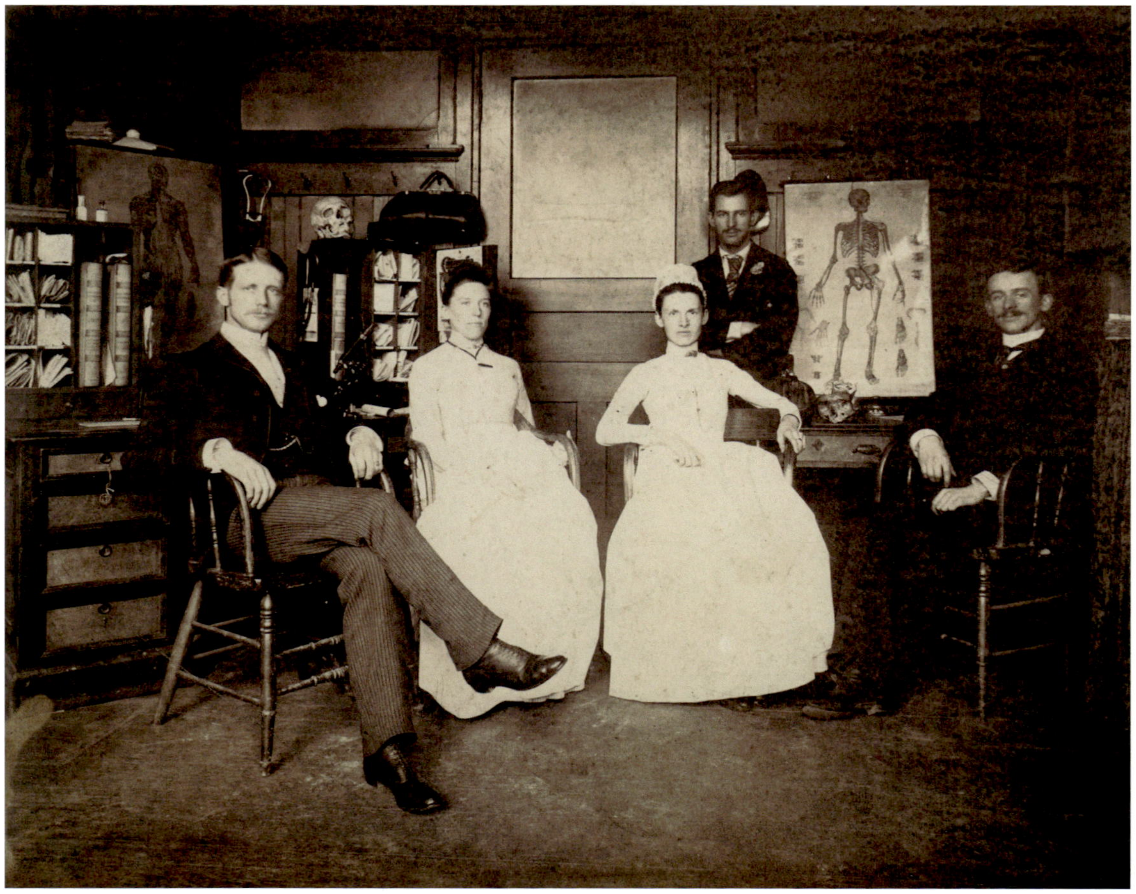

Plate 1.34

House Staff in Lounge, Good Shepherd Hospital Syracuse, NY
Photograph of section of Rembrandt anatomy painting on wall
Silver Gelatin Print, circa 1908

Plate 1.35

Physicians and Nurses in Lounge
Room 7, Fitch Hospital, MA
Silver Gelatin Print, May 1889

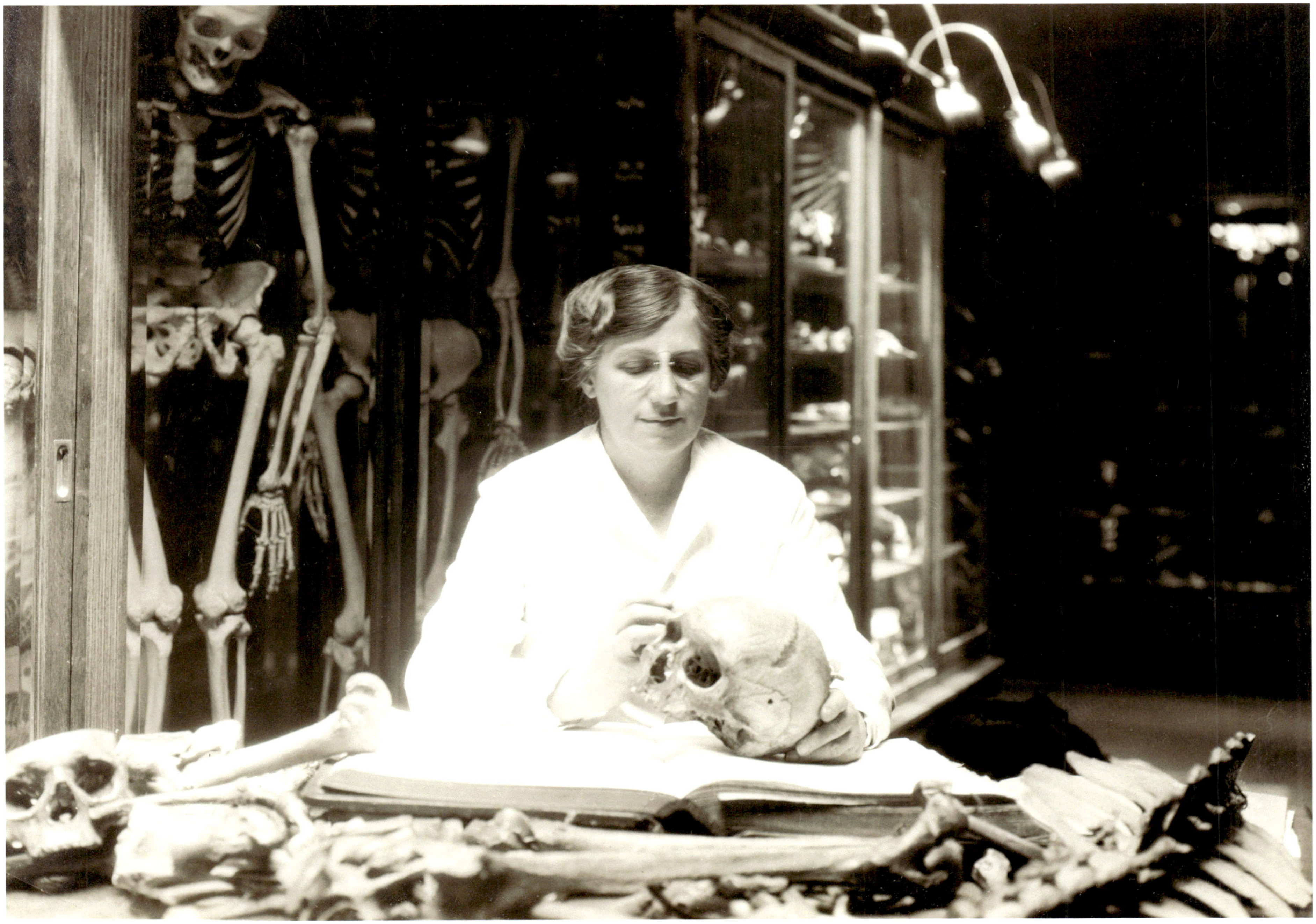

Plate 1.36

Female Chiropractor In Osteology Laboratory
Palmer College of Chiropractic, Davenport, IA
Silver Gelatin Print, 1921

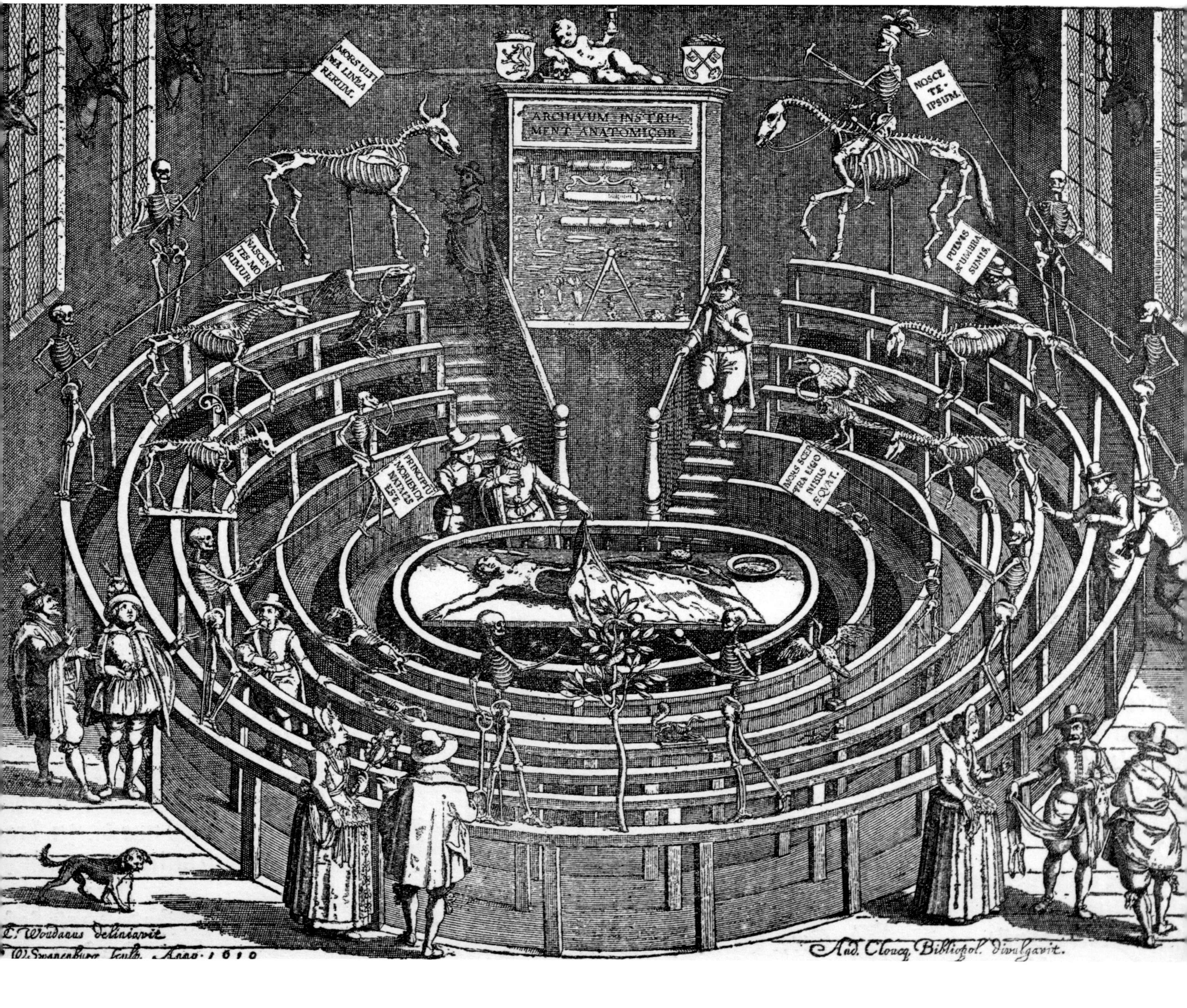

ARCHIVUM INSTRUS MENT ANATOMIÇOR
MORS ULTI MA LINEA RERUM.
NOSCE TE IPSUM.
NASCEN TES MO RIMUR
PULVIS & UMBRA SUMUS.
PRINCIPIU MORIENDI NATURA EST.
MORS SCEP TRA LIGO NIBUS ÆQUAT.
C. Woudanus deliniavit
Anno 1610
And. Cloucq Bibliopol. divulgavit.

CHAPTER TWO

Teaching Anatomy & Osteology

Master Classes and Personal Lessons

Teaching anatomy, the cornerstone of medical education for centuries, was elevated to an art form with the construction of anatomical amphitheaters. The emergence of scientific teaching during the sixteenth century brought new heights to medical education with these imposing structures. In 1594, the first anatomical theater was built at the University of Padua, Italy and was soon followed by one in Leiden, Holland, in 1596. The Anatomical Theatre of the Archignnasio at Bologna University opened in 1637 (see Foreword photograph, p. 6) and was constructed in a building built in 1563. These theaters created a dramatic setting in which to learn medicine.

Ultimately, operative theaters, not as elaborately decorated, were built in hospitals mimicking the anatomical theater design and made surgery seem like a dramatic performance. This was especially true in the pre-anesthetic era when speed in surgery was essential and patients were often restrained by trained attendants. Surgery became a fine art with the demonstration of general anesthesia in Boston in 1846, as operative speed and patient anguish was no longer a part of the procedure. This enabled surgeons to construct and develop new techniques and approaches. Unfortunately, the ignorance of the role germs played in disease and surgery led to deadly infections that killed and that spread in hospitals by dirty hands and instruments. But anesthesia allowed even the timid to operate, and in order to operate accurately, knowledge of anatomy was paramount. Advanced surgical anatomy dissection courses became necessary to learn specific techniques and remain necessary to this day.

Dissection photographs are divided into two categories, those with a professor present and those without. In this chapter, dissection photographs with a teacher are presented, and the classic 'rite of passage' photographs without teachers are in the following chapter. These dissection or anatomical images document the education of the physician in a more serious atmosphere and are devoid of the shenanigans of students establishing their individual and team identity. Particularly significant are the images at Philadelphian medical colleges, which are dramatic examples of the highest caliber of American medical education of the time. European medical training in the nineteenth century required an undergraduate university degree, and the imposing lecture halls of some of these universities are shown. Among the photographs in this chapter are several from a St. Petersburg military medical school album issued in 1904, as well as overviews of dissection rooms. These photographs of dissection laboratories were taken to show the wealth of cadavers available for students. Throughout much of medical history, acquiring cadavers involved stealing the bodies or waiting for a criminal execution.

This chapter illustrates advanced anatomy classes for graduate students, which were usually given at reputable medical institutions, such as the New York Post-Graduate Medical School. A tradition for these photographs was to record the names of classmates on a board (see Plates 2.6 and 2.17). Images in this chapter also document some of the American and European teaching amphitheaters, as well as graduate physician classes and medical students with their cadavers and teachers.

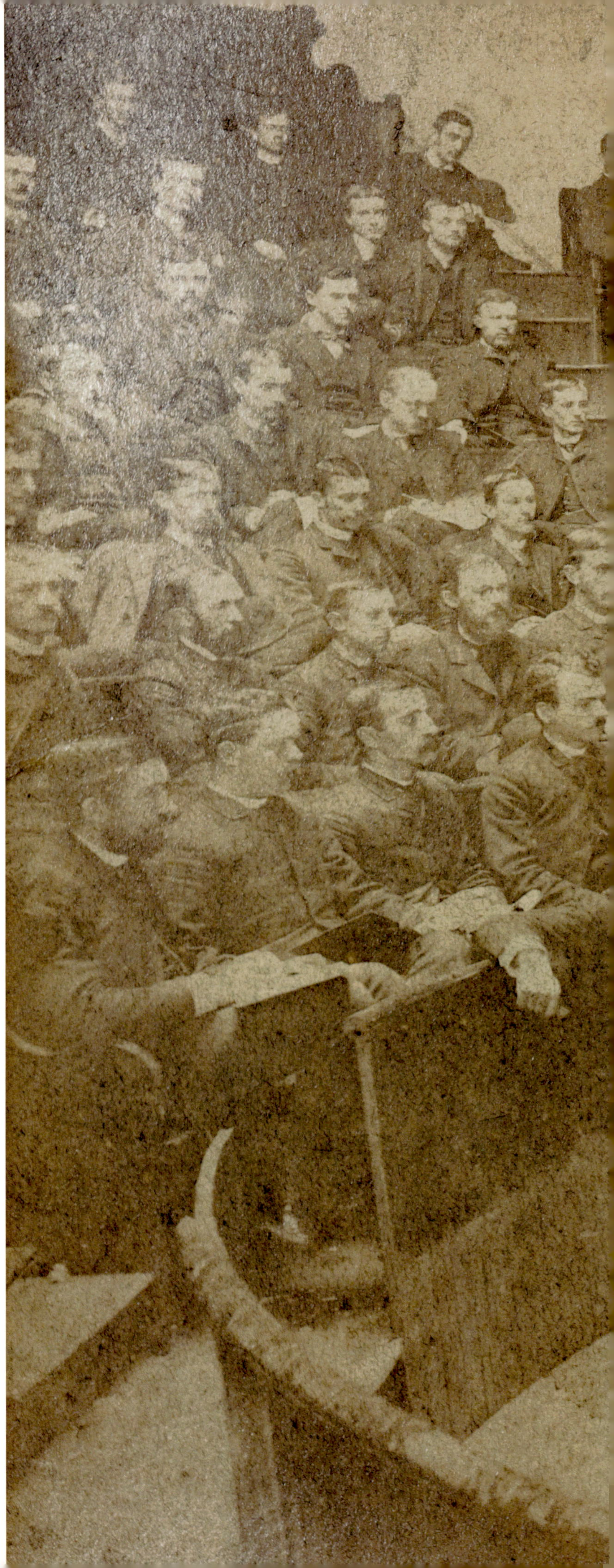

Plate 2.1

Joseph Leidy, MD, Anatomy Lecture
University of Pennsylvania, Philadelphia, April 5, 1886
Medical Class of 1887 & 1888, Dental Class of 1887 & 1888

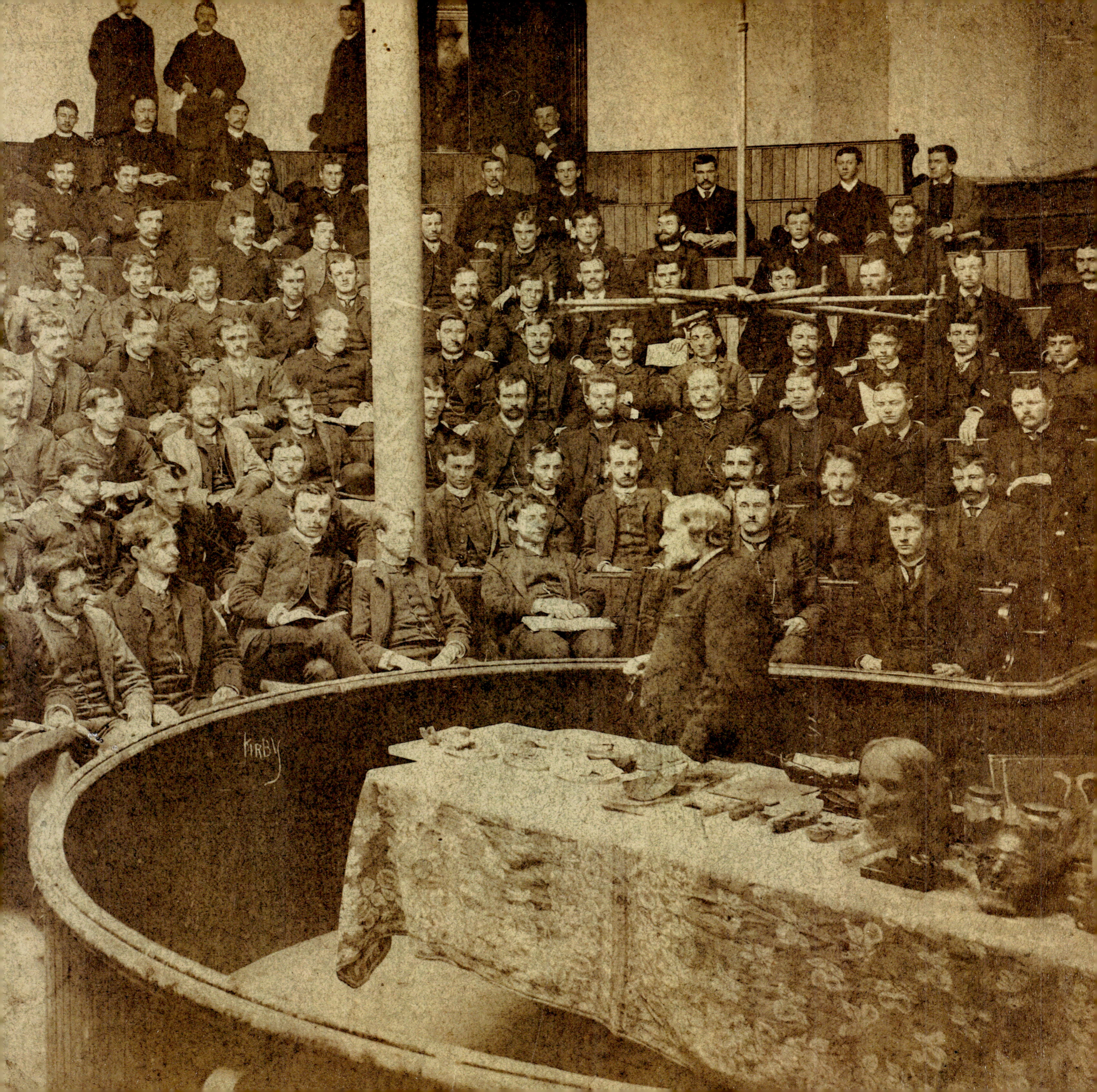
KIRBY

Plate 2.2

Anatomy Class Philadelphia Medical College
Albumen Print, 10 1/2" x 25 1/2", circa 1898

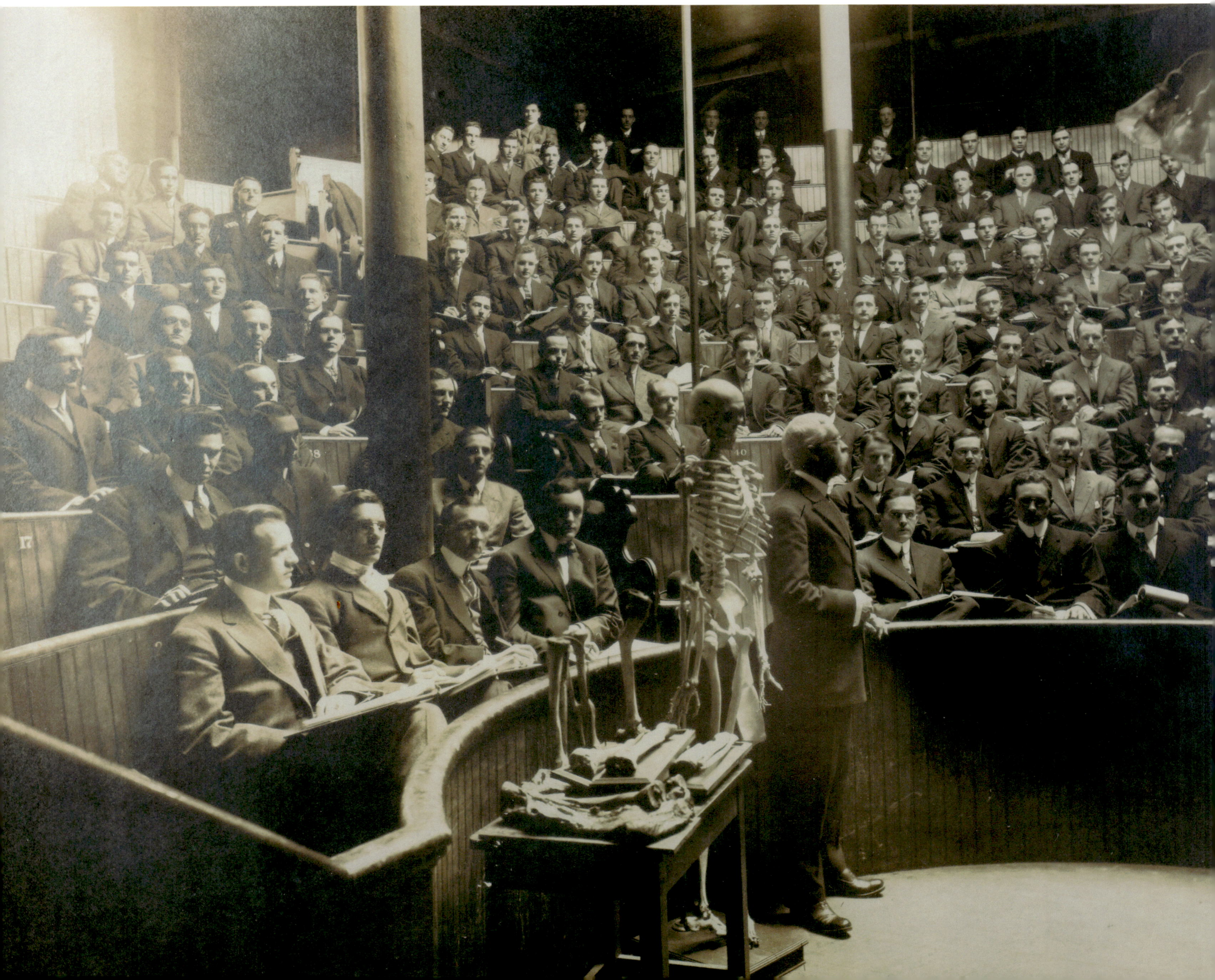

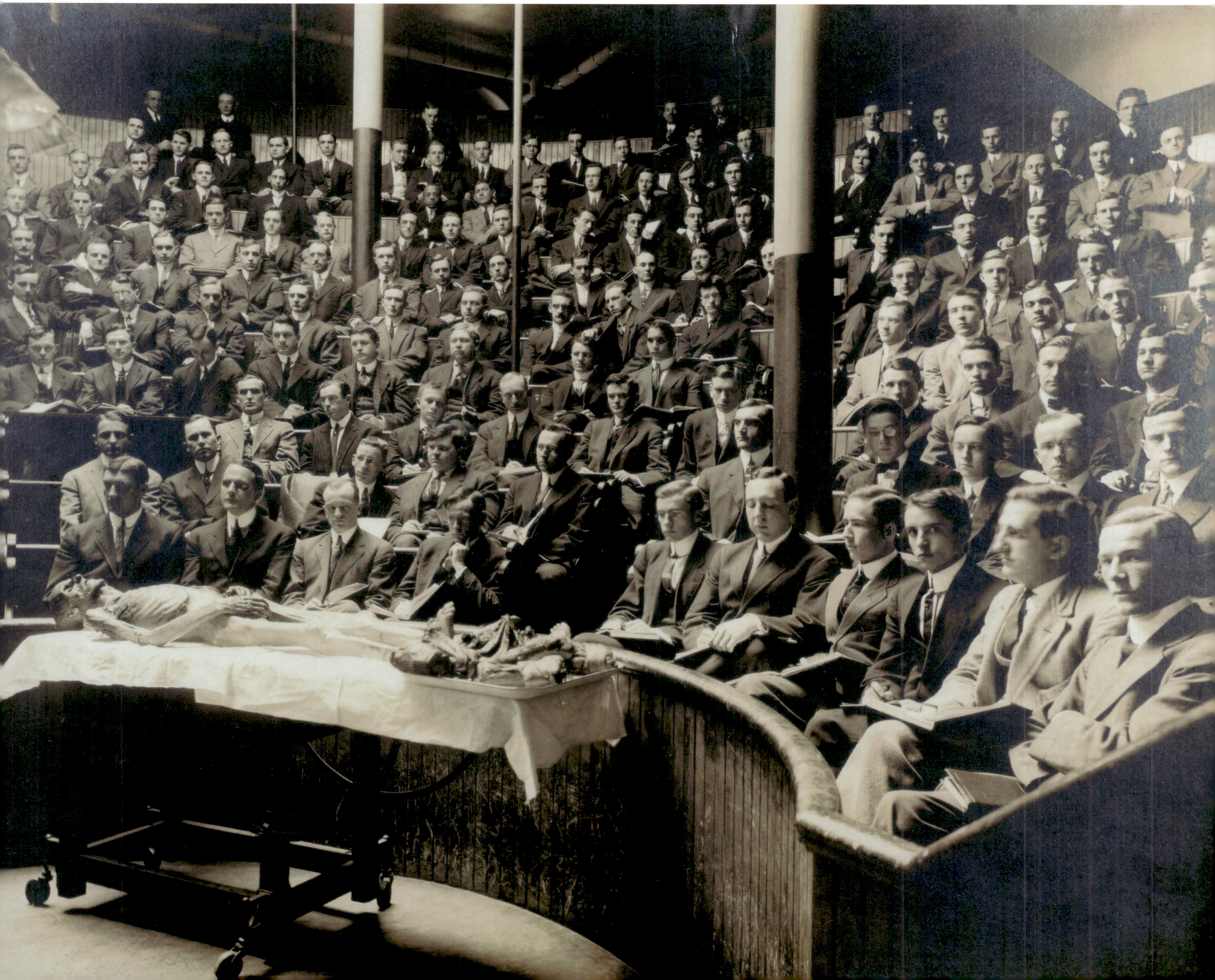

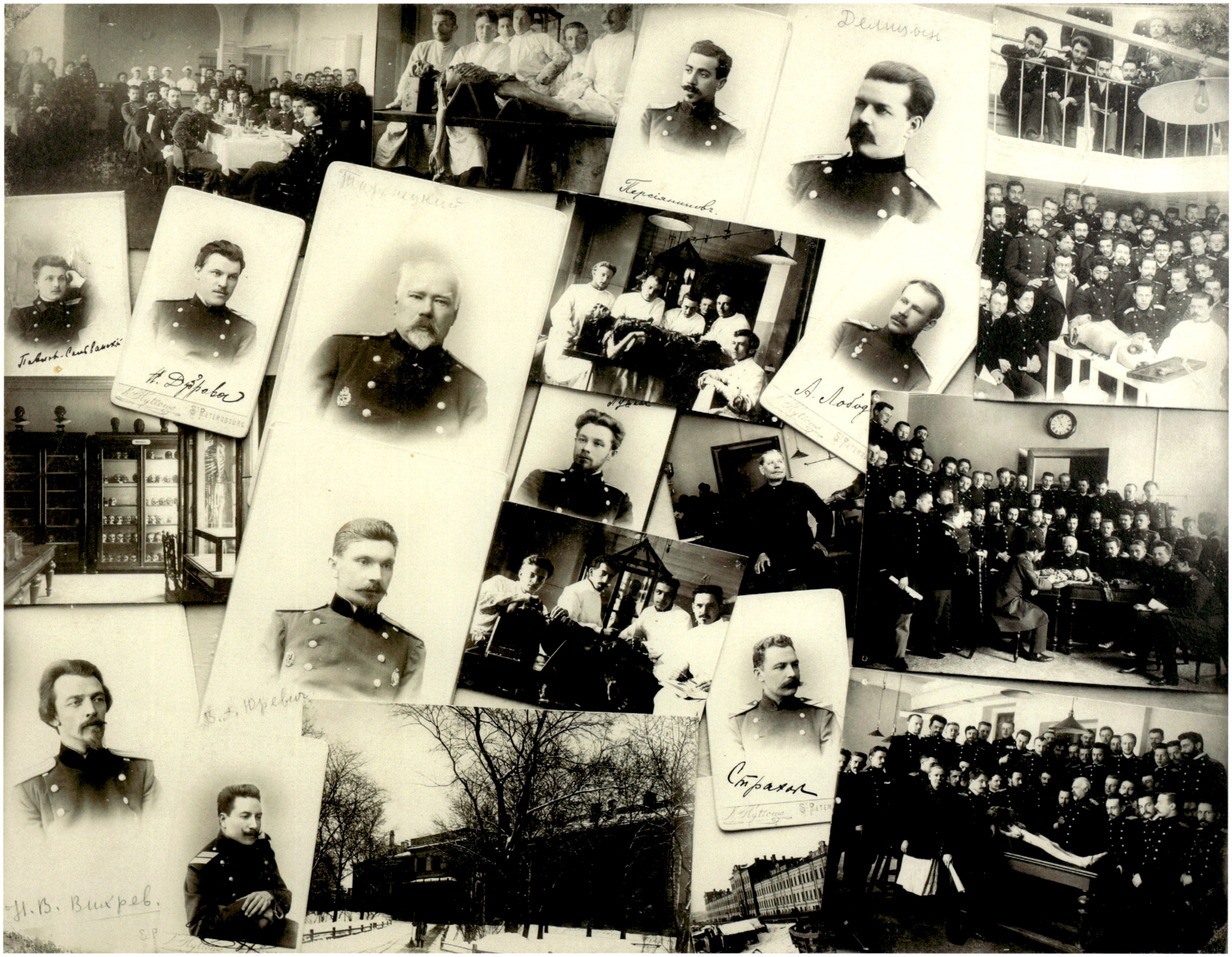

Plate 2.3

Military Medical-Surgical Academy of St. Petersburg
The Anatomy Department, 1904

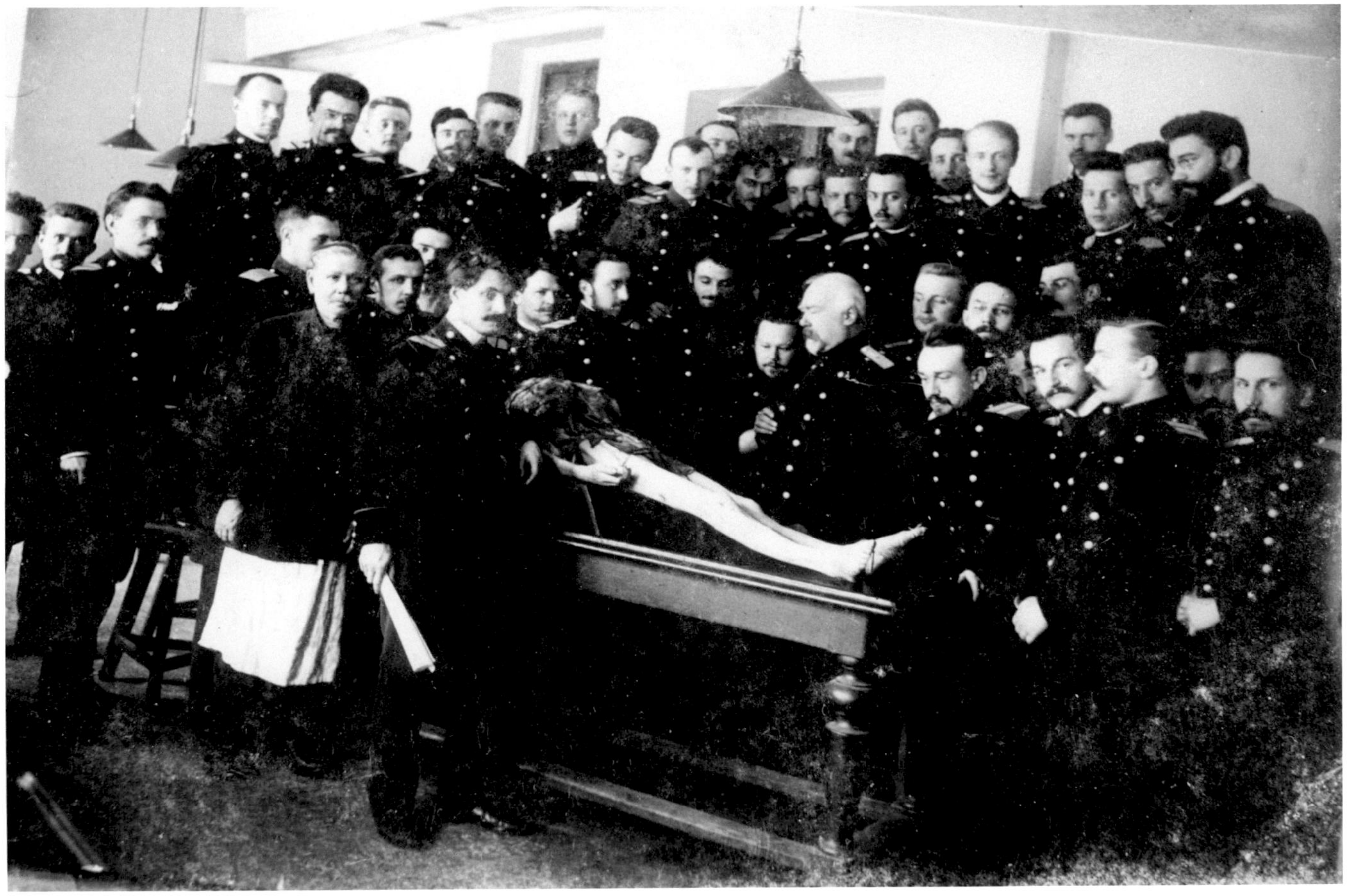

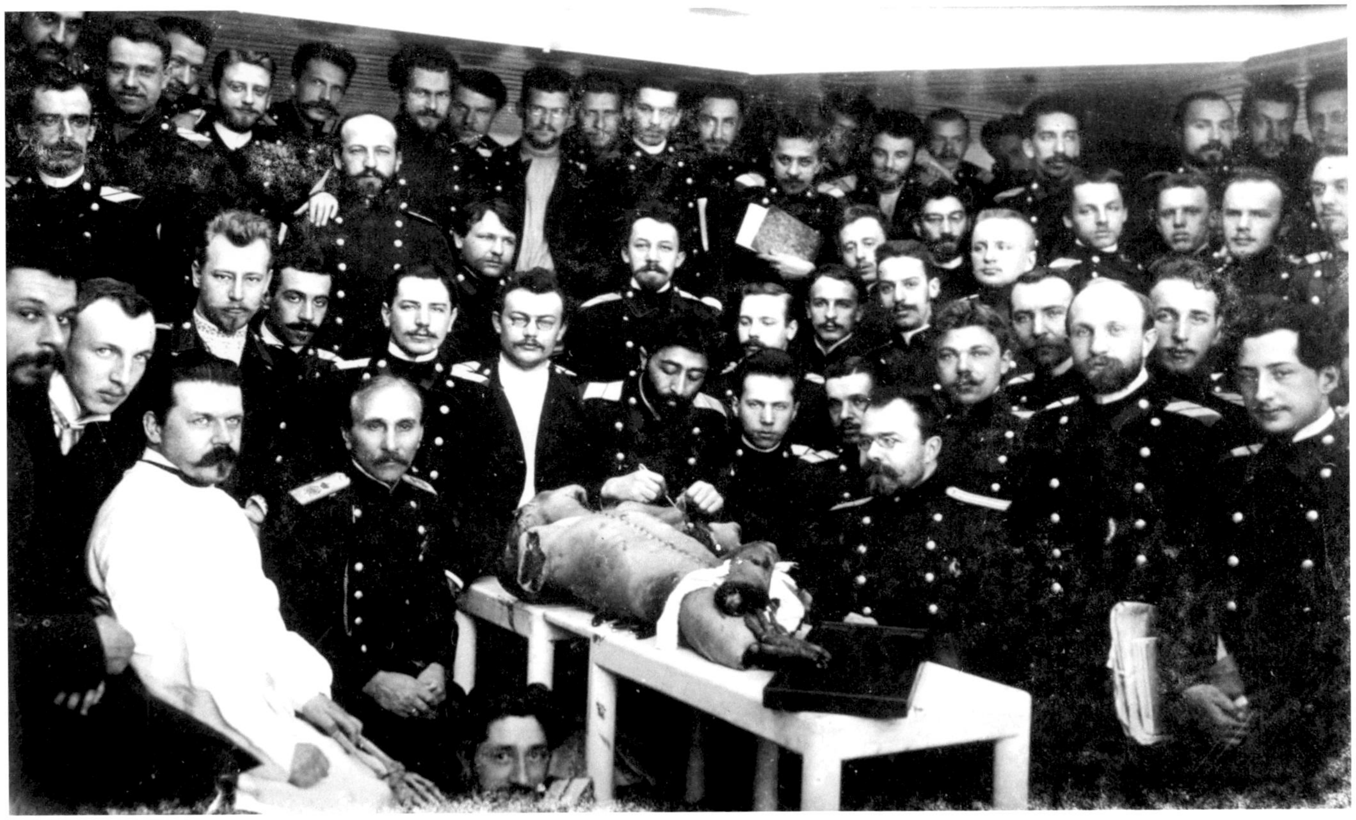

Plate 2.4

Dissection Class
Military Medical-Surgical Academy of St. Petersburg, 1904

Plate 2.5

Dissection Class
Military Medical-Surgical Academy of St. Petersburg, 1904

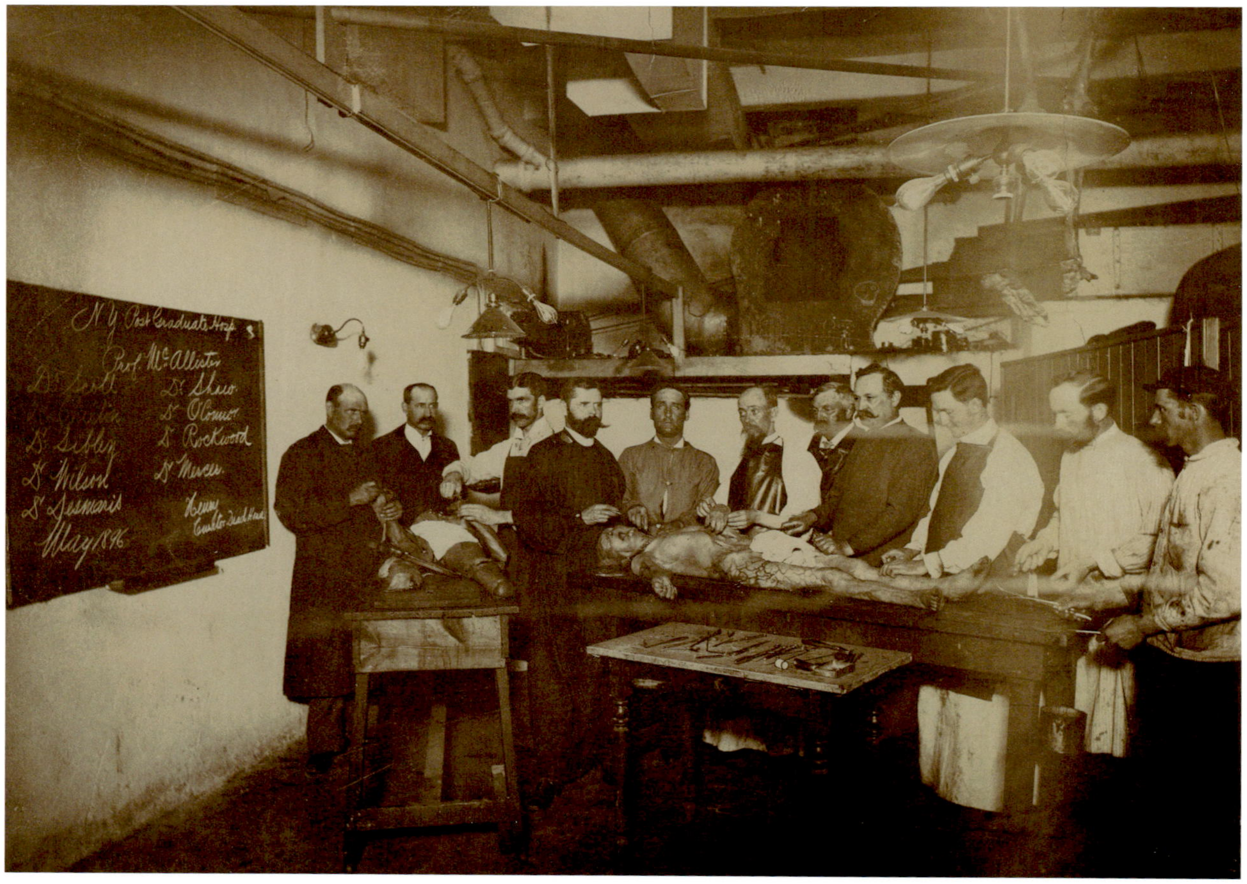

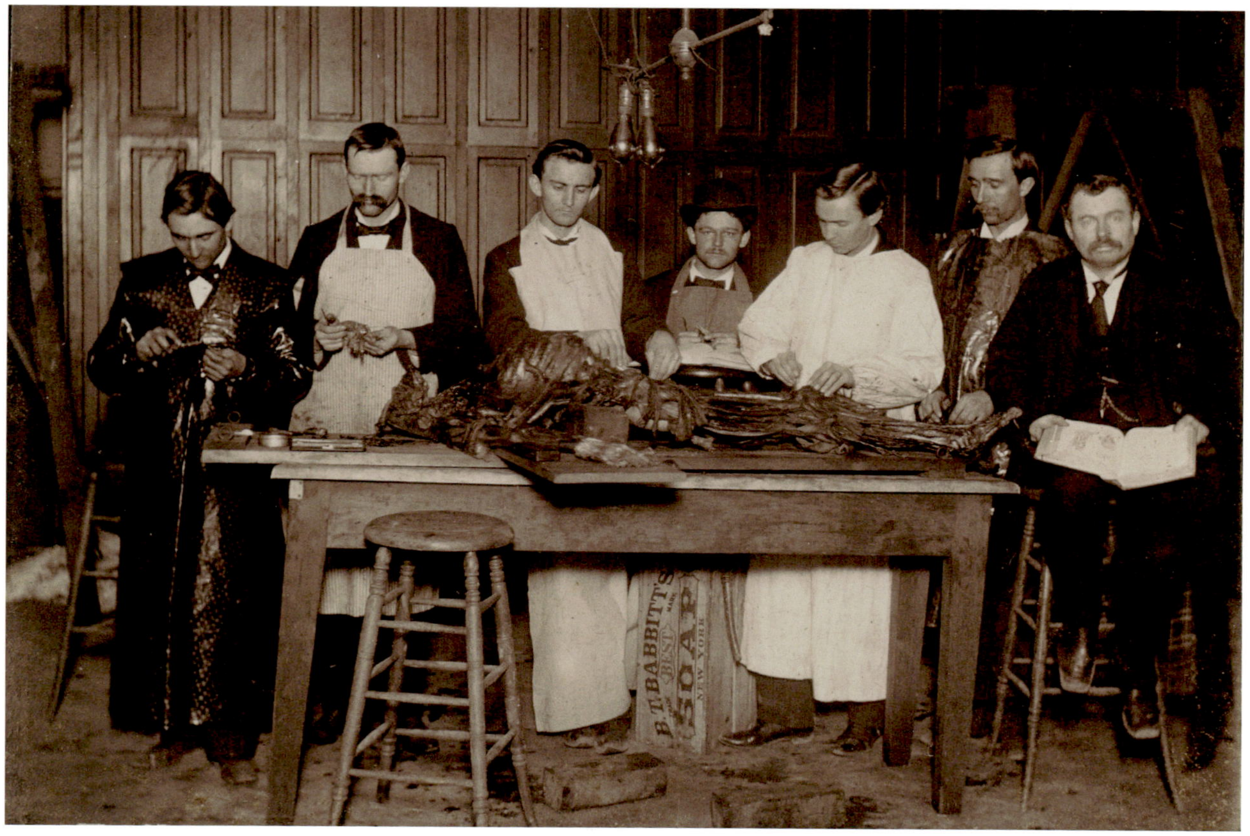

Plate 2.6

Professor McAllister Teaching Anatomy
New York Post Graduate Hospital, May 1896

Plate 2.7

Dissection with Participants in Varied Garb
Unidentified School, circa 1900

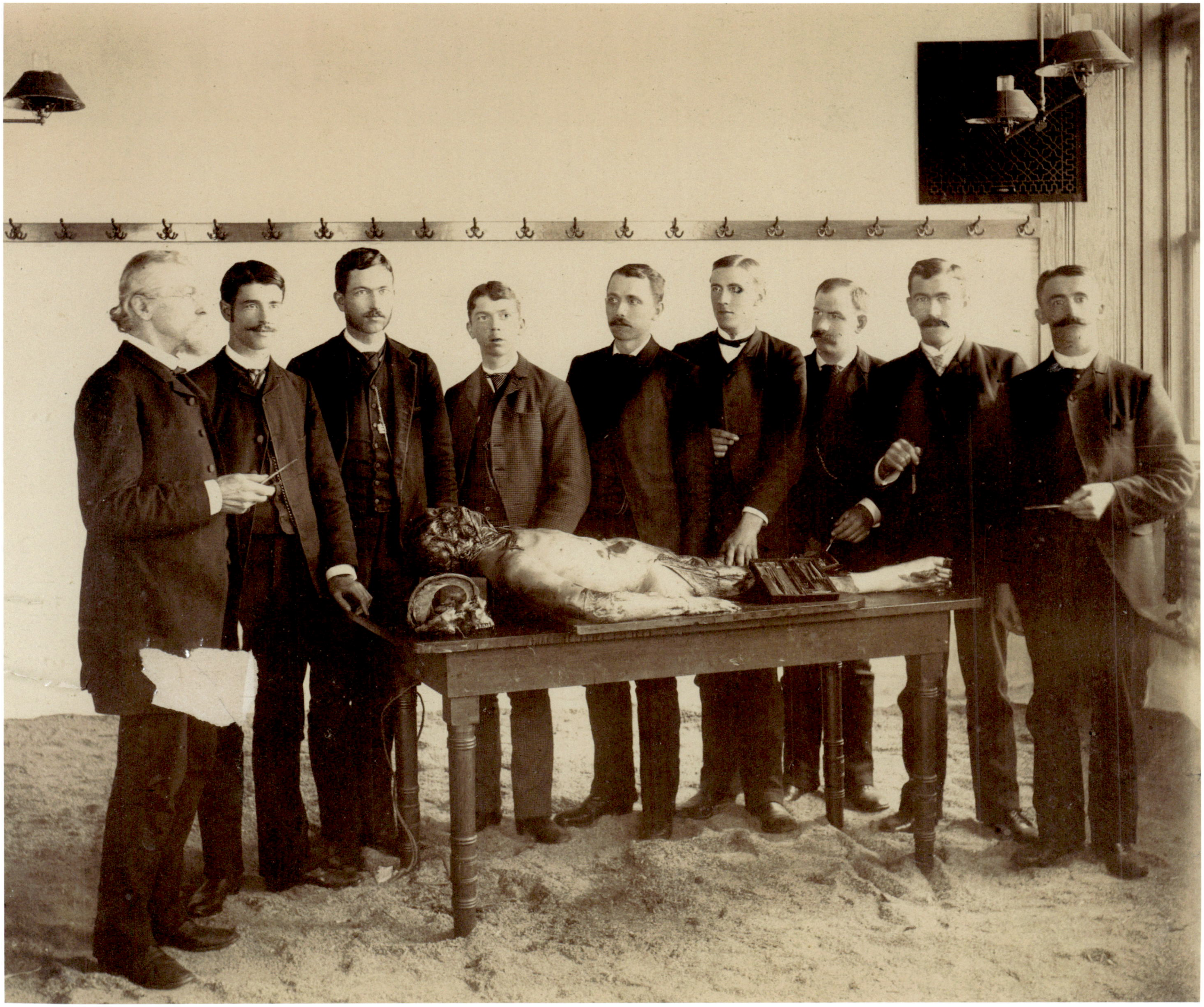

Plate 2.8

Dr. Jacob Laisy Teaching Anatomy
Western Reserve School of Medicine, 1890
Photographer: H. M. Fowler, 93 Public Square, Cleveland, OH
On back: "The photograph was taken in the northwest corner of the dissecting room after the building at East 9th and St. Clair was erected in 1887 and before Dr. Laisy retired from teaching in 1893."

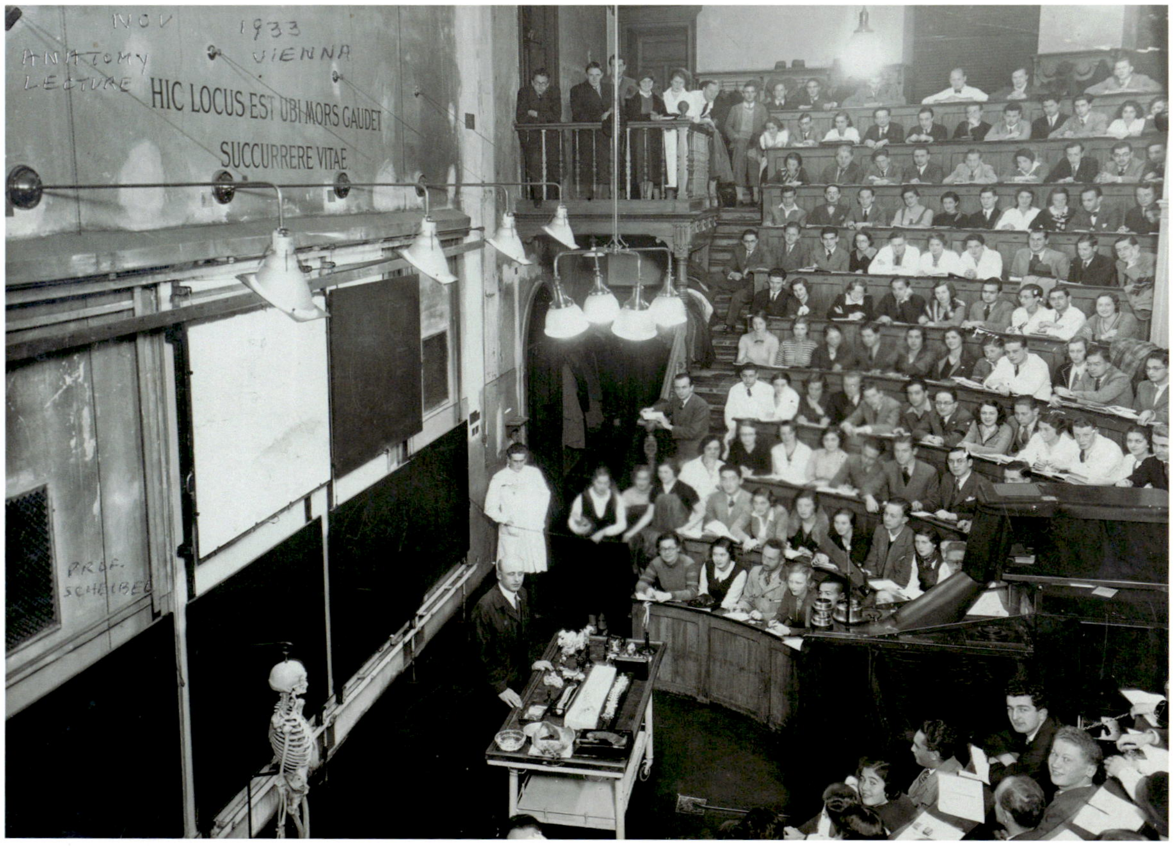

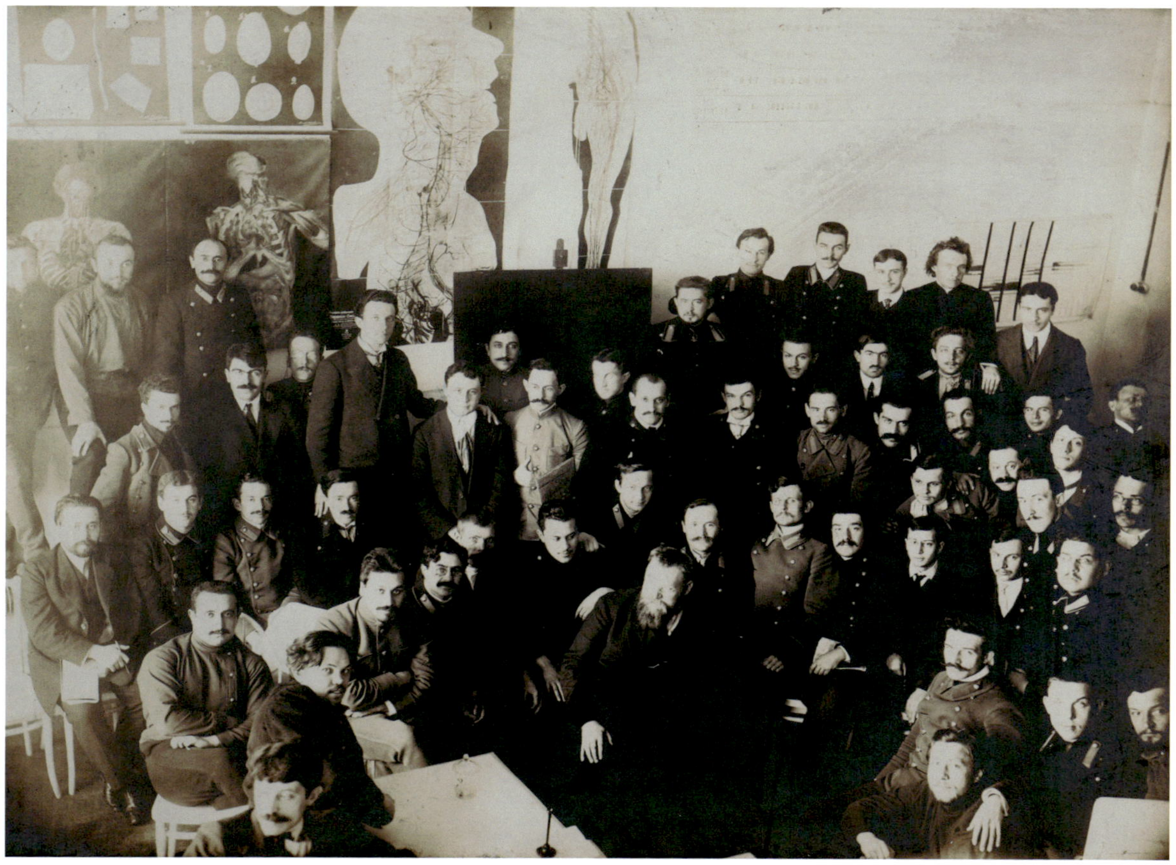

Plate 2.9

Professor Scheibel Teaching Anatomy Class
University of Vienna, November 1933
Photographer: Simonis

Plate 2.10

Anatomy Class with Large Posters
European Medical School, circa 1925

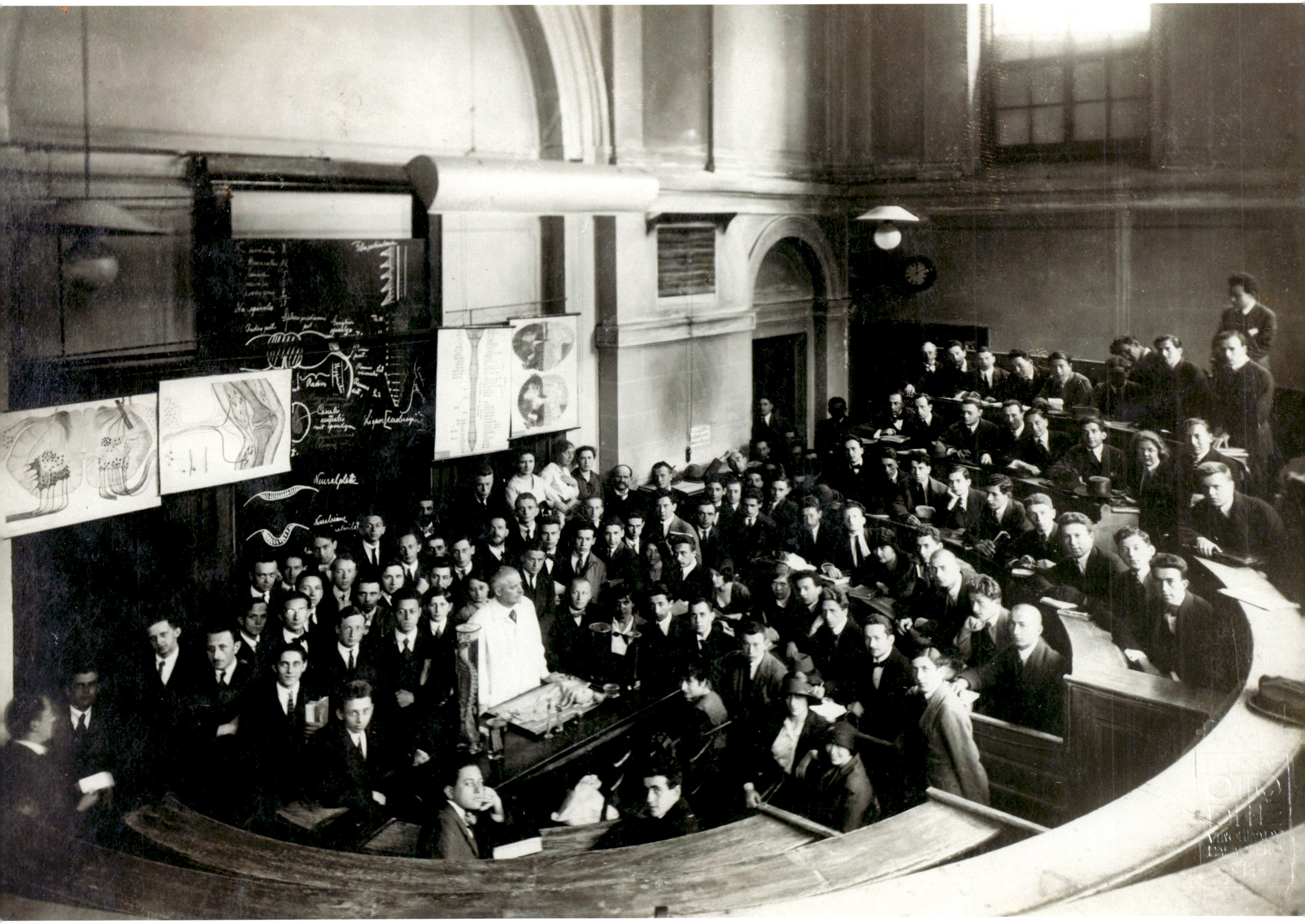

Plate 2.11

Anatomy Class Studying Neural Plate and Neuroanatomy
Unidentified European School, circa 1910

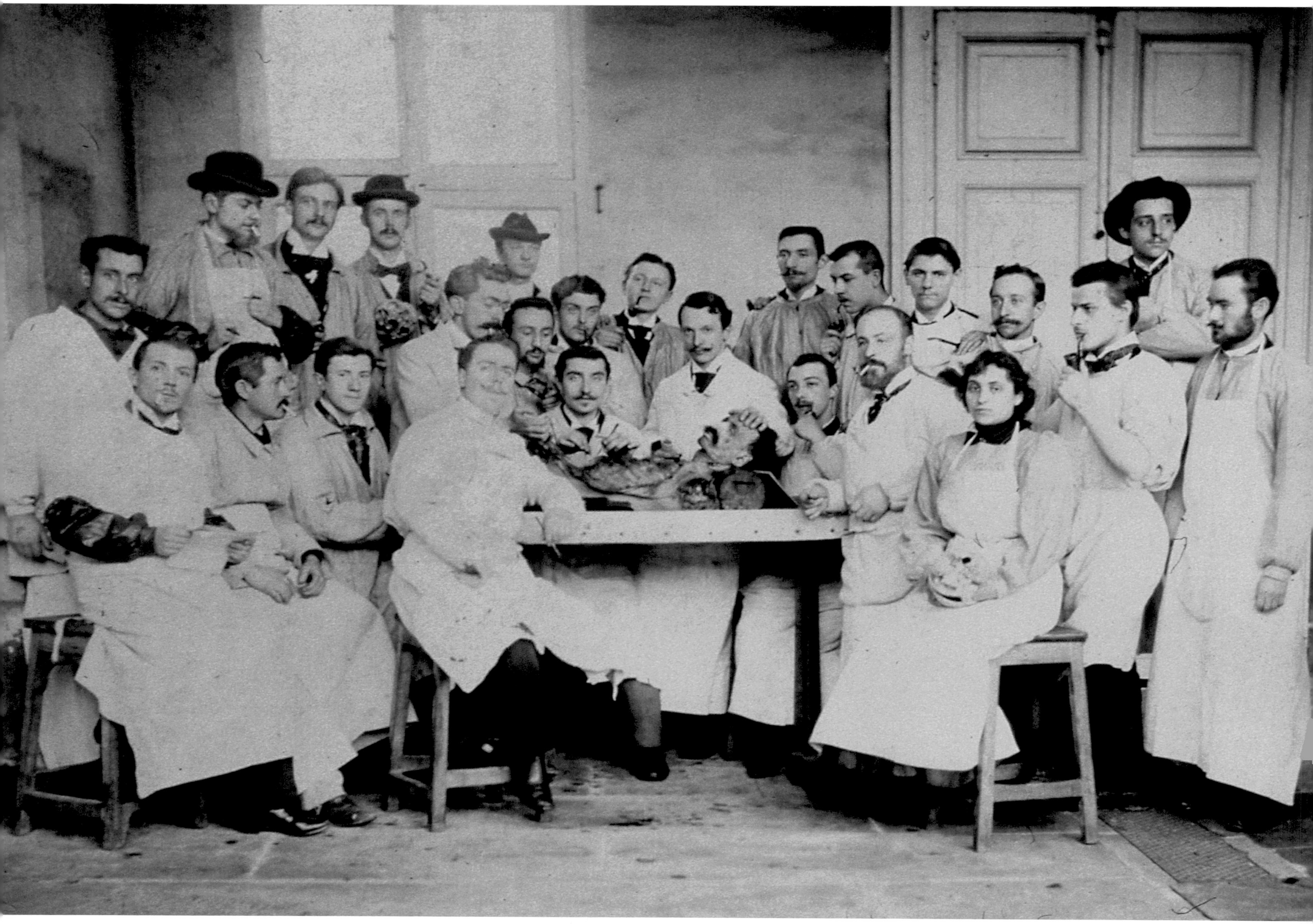

Plate 2.12

Anatomy Class, Many Smoke to Disguise Cadaver Odor
Unidentified School, circa 1905

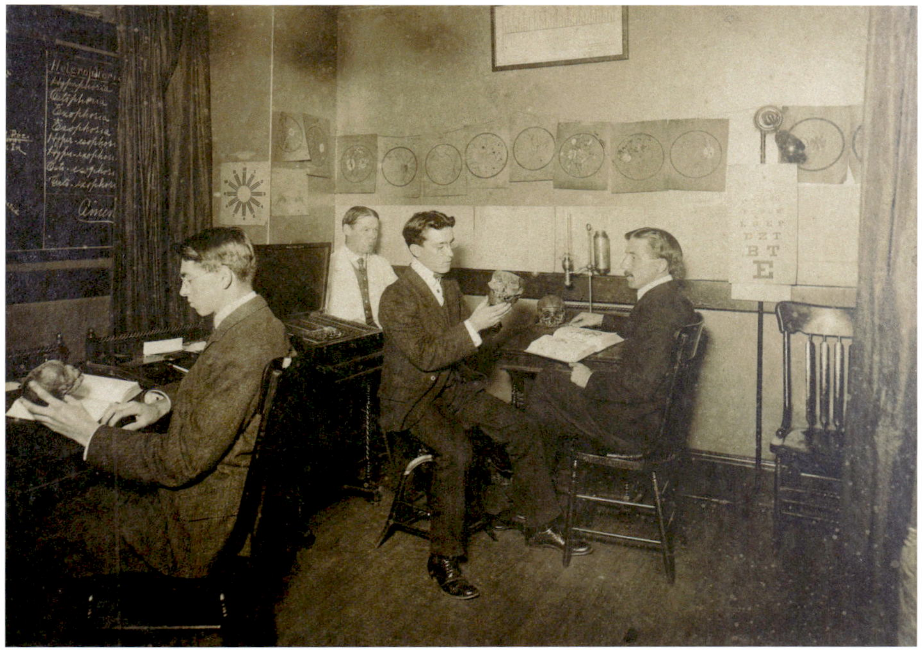

Plate 2.13

Studying The Orbital Anatomy
Unidentified New York Institution, 1936

Advanced class in eye muscle disorders, examining the orbital attachments of the extra ocular muscles. On the blackboard is a list of eye positions.

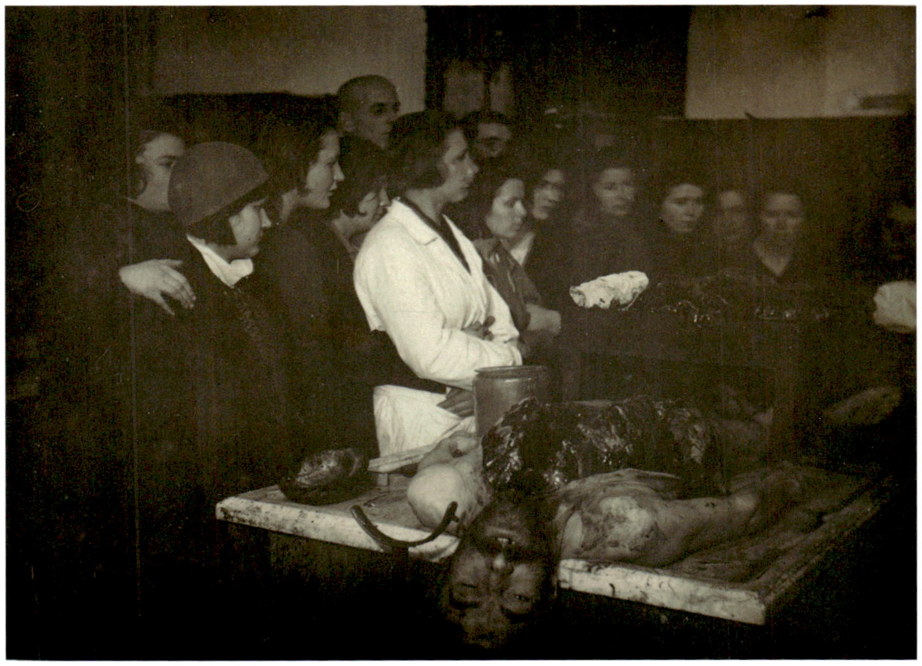

Plate 2.14

Women Listen to Anatomy Lecture
European Medical School, circa 1930

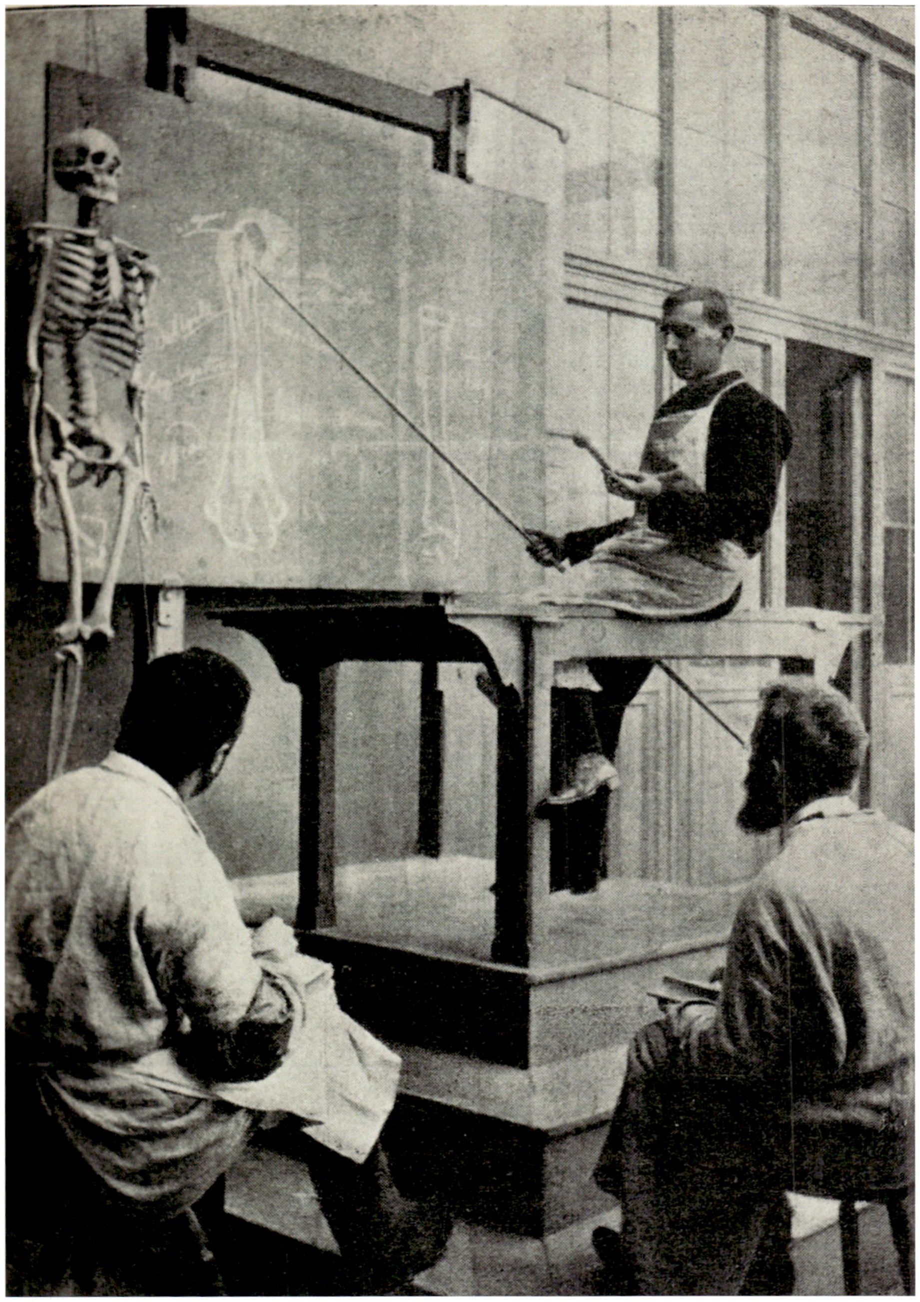

Plate 2.15

Teaching Osteology,
Le leçon d'ostéolgie, Paris, 1898

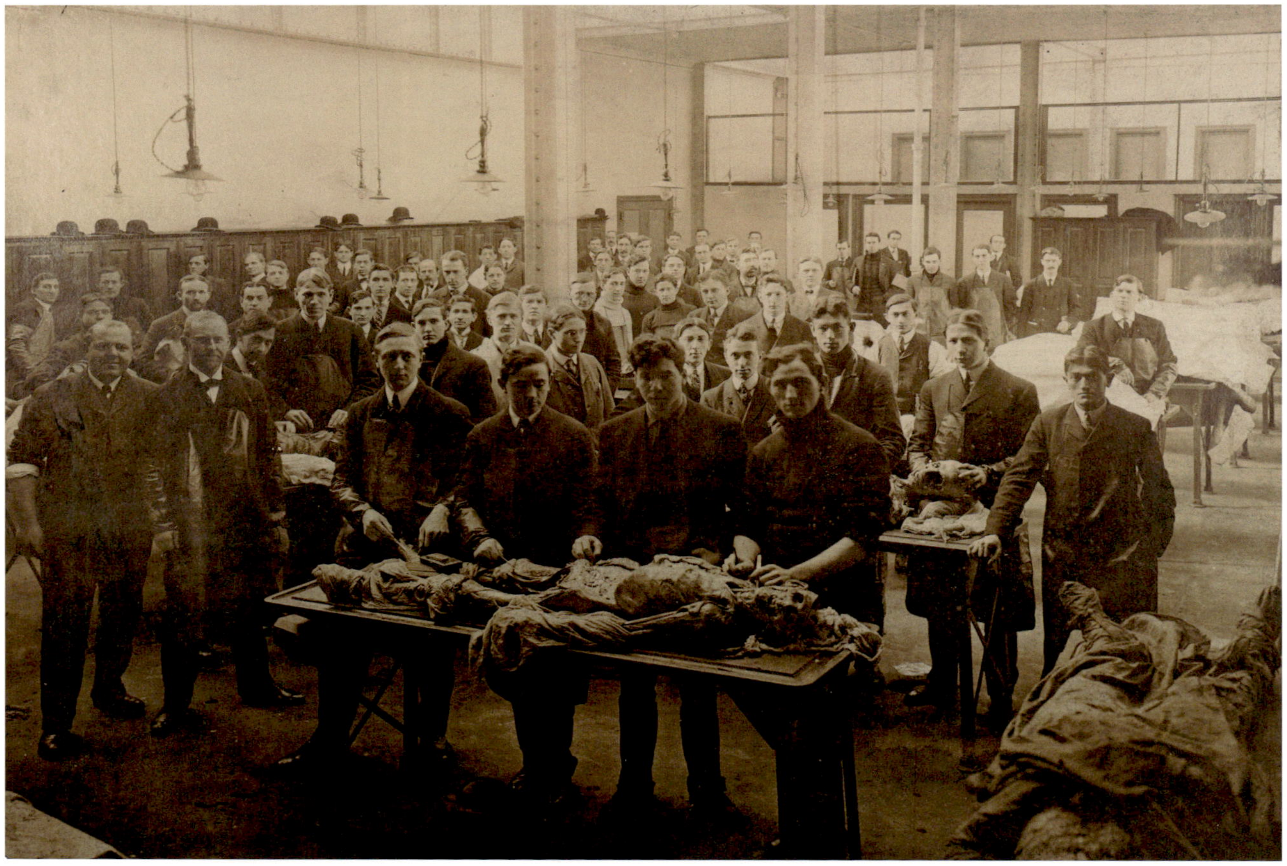

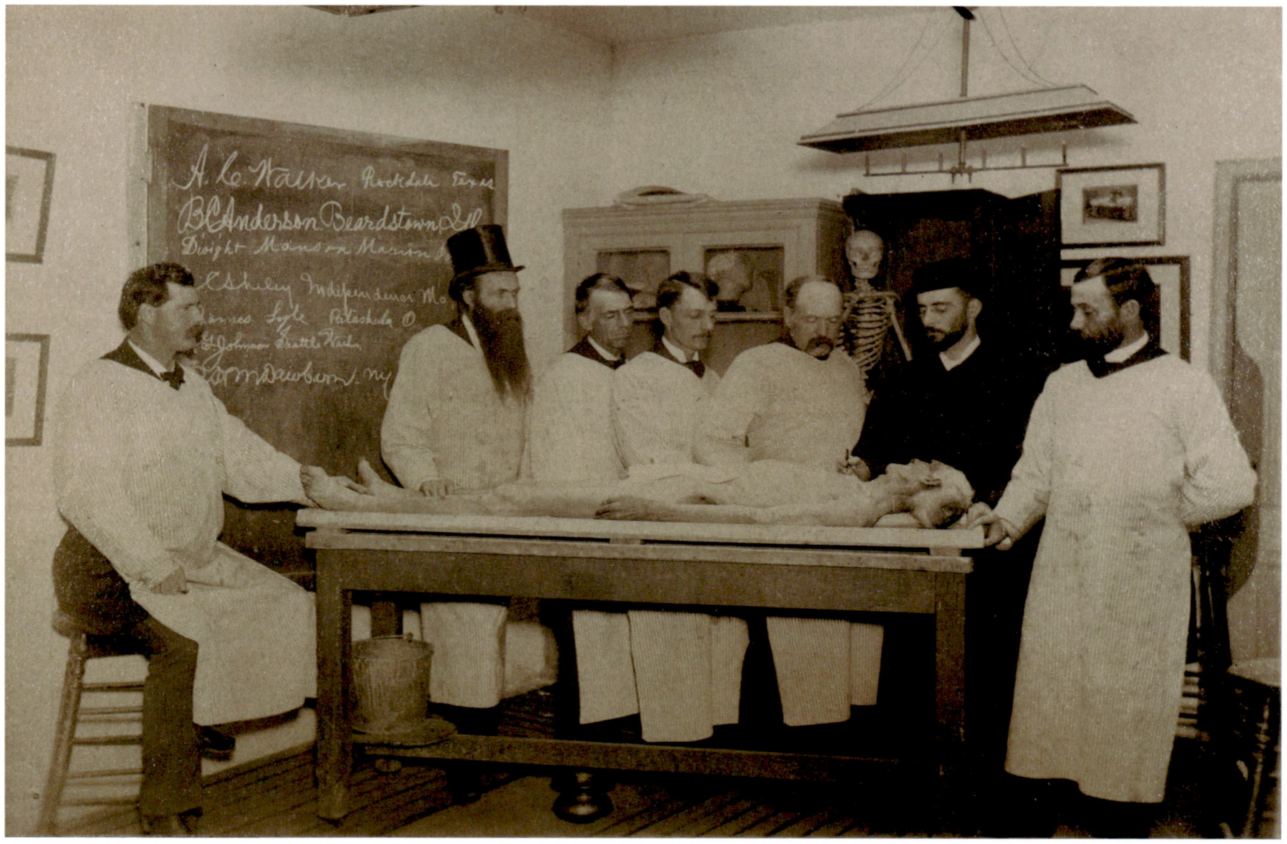

Plate 2.16

Two Professors Pose with Large Class
Unidentified Chicago Medical School, circa 1910

Plate 2.17

Graduate Anatomy Class
New York City, circa 1885
Photographer: Long & Heppner, 225 East 34th St., NY

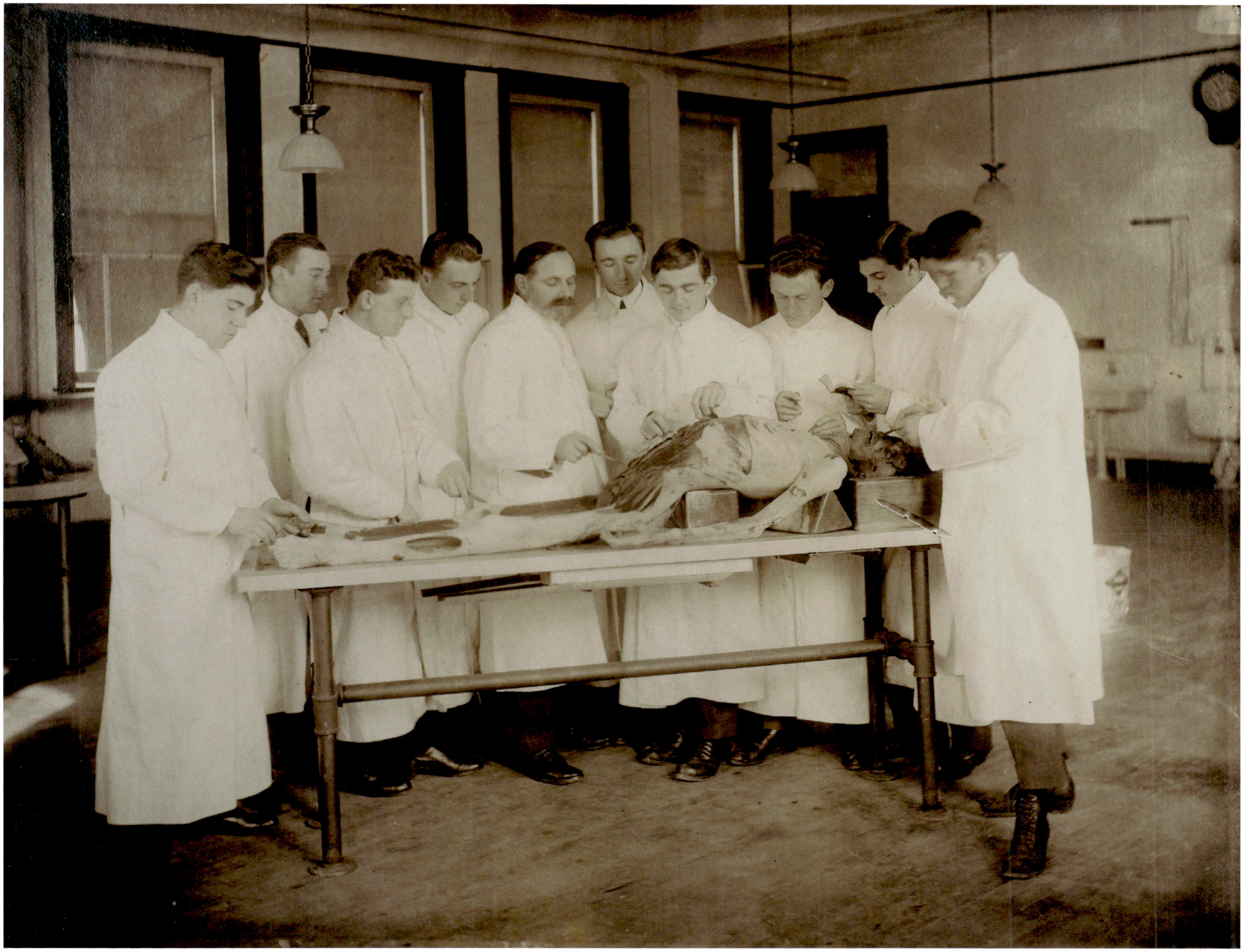

Plate 2.18

Anatomy Class in White Garb, Professor in Center
Unidentified School, circa 1917

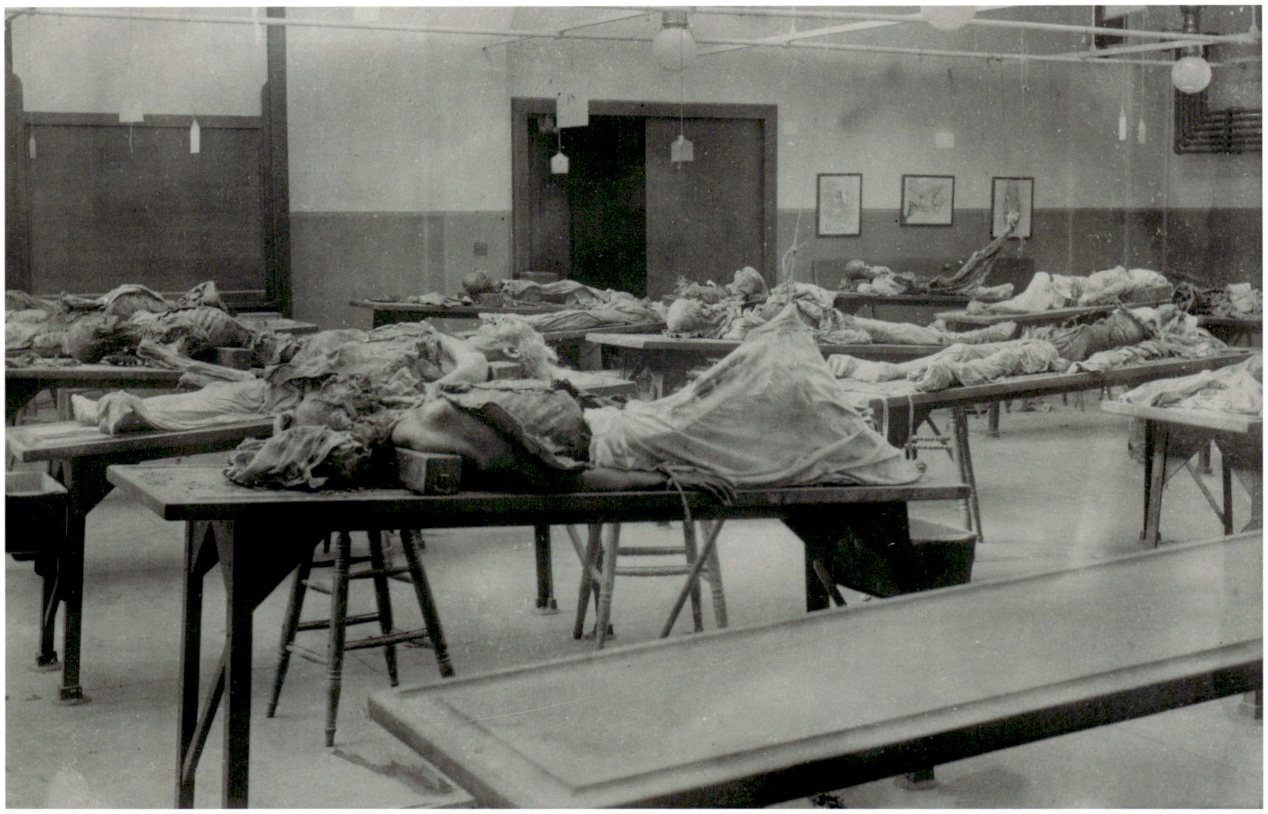

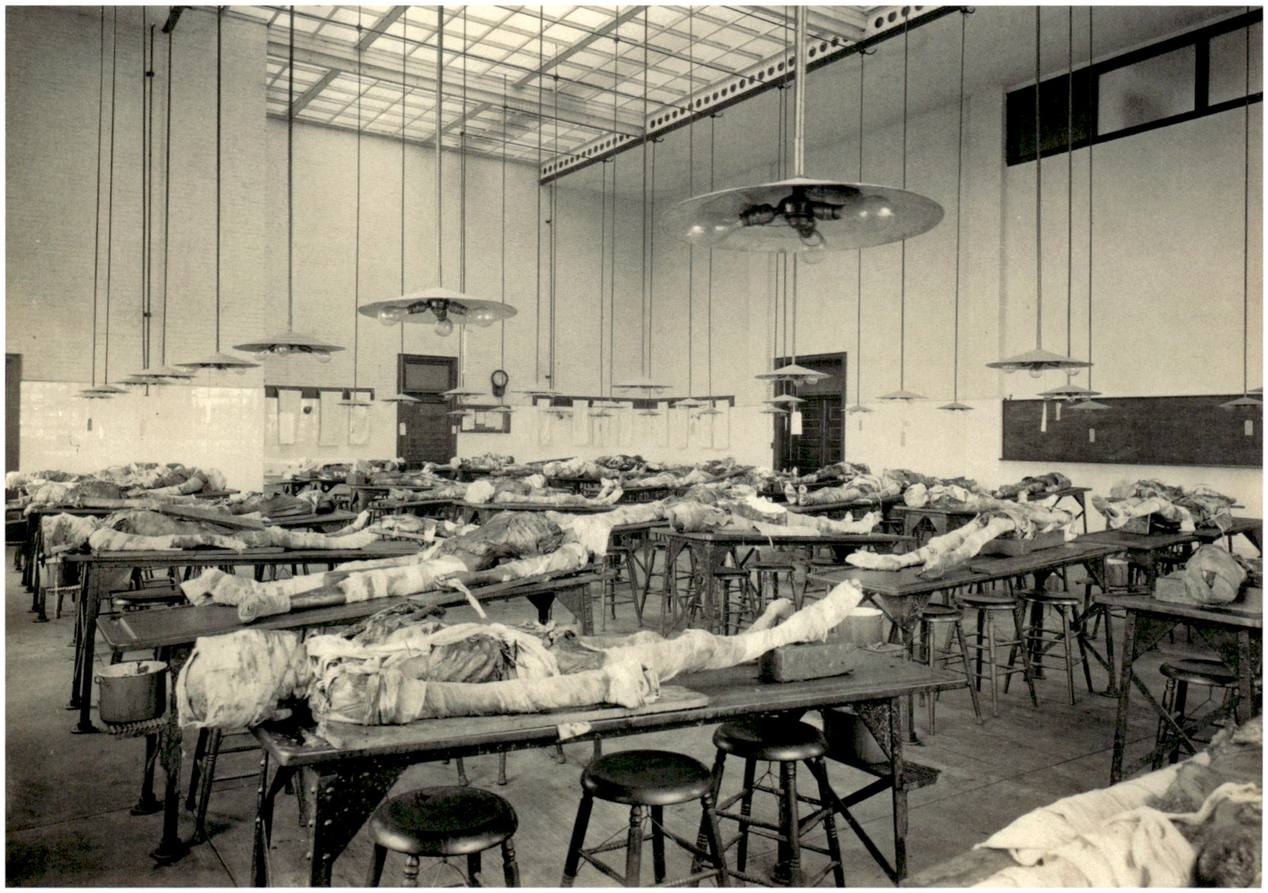

Plate 2.19

Overview of Anatomy Laboratory
Circa 1920

Plate 2.20

Overview of Anatomy Laboratory
Circa 1930

Parts of the cadavers are wrapped in bandages indicating dissections in progress.

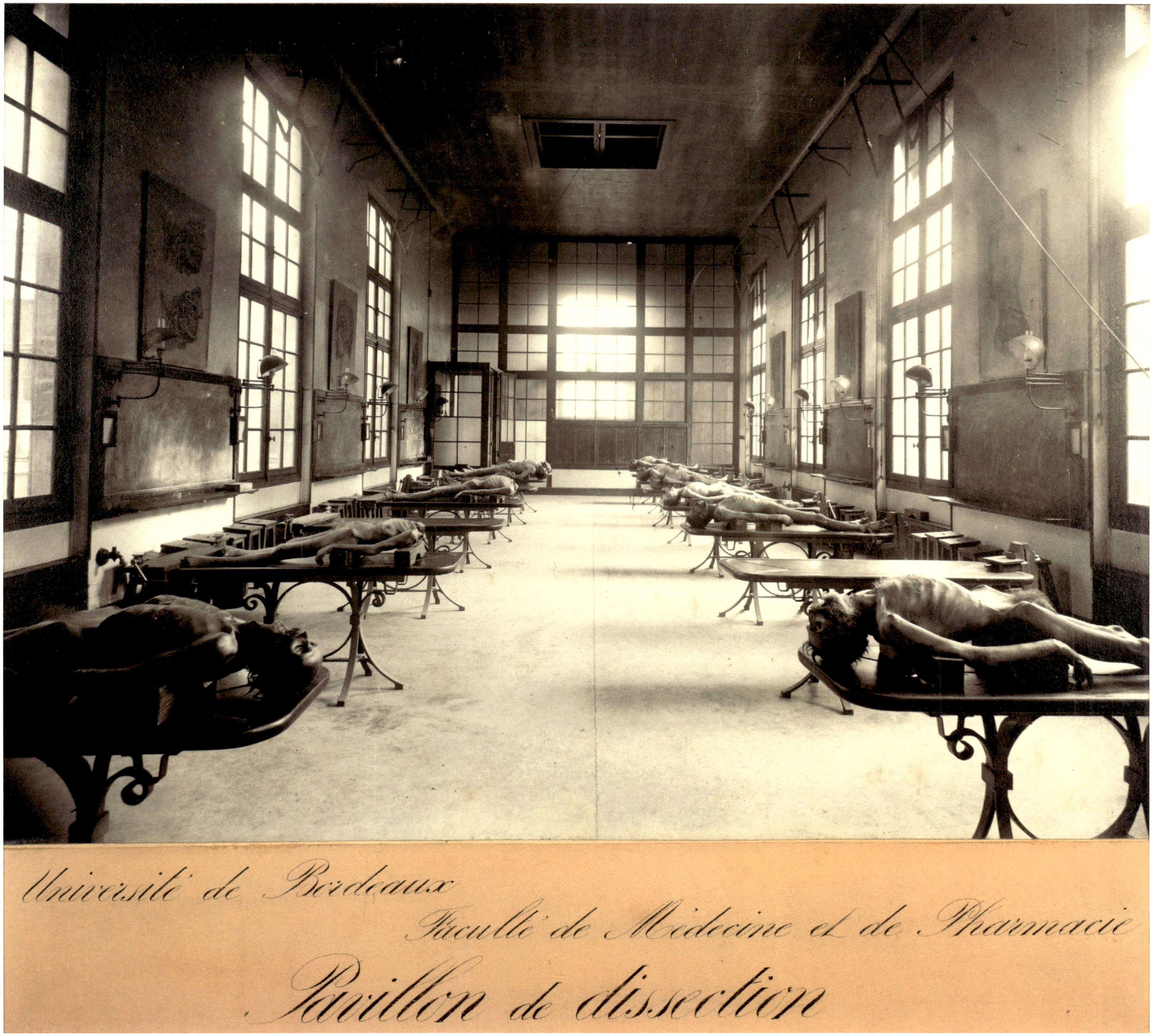

Plate 2.21

Pavillion de Dissection
Université de Bordeaux, Faculté de médicine et de pharmacie, 1895

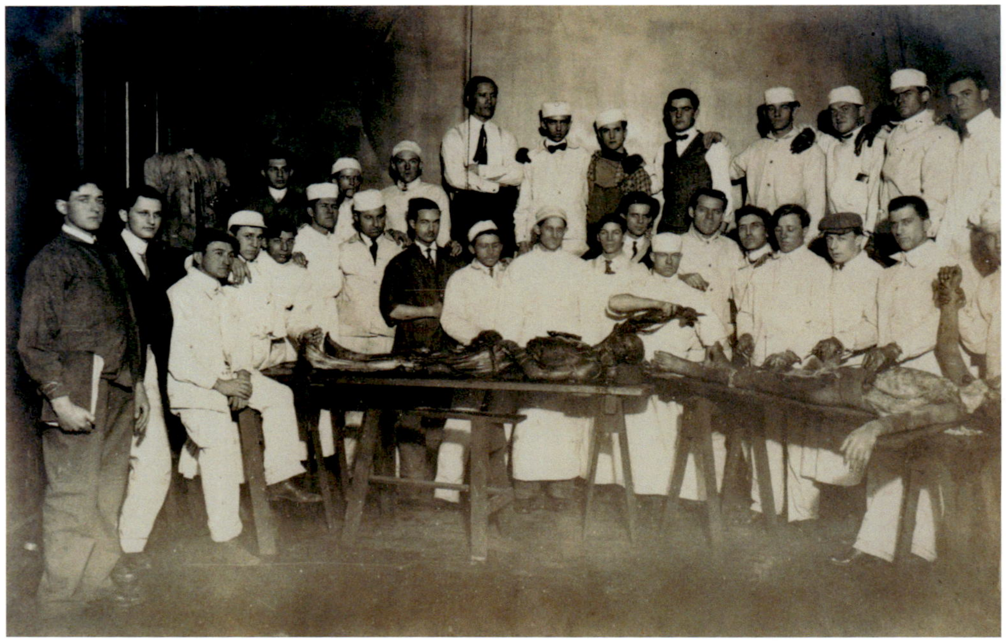

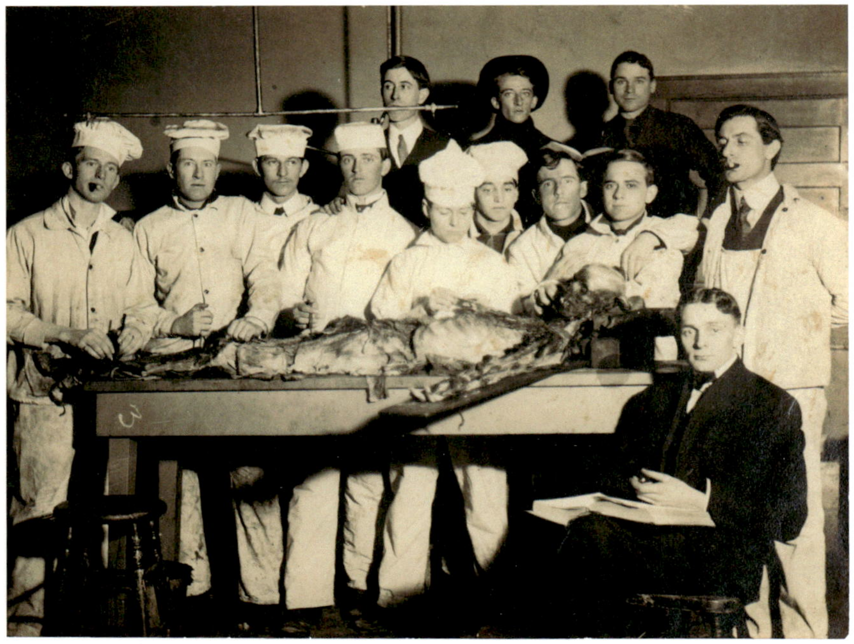

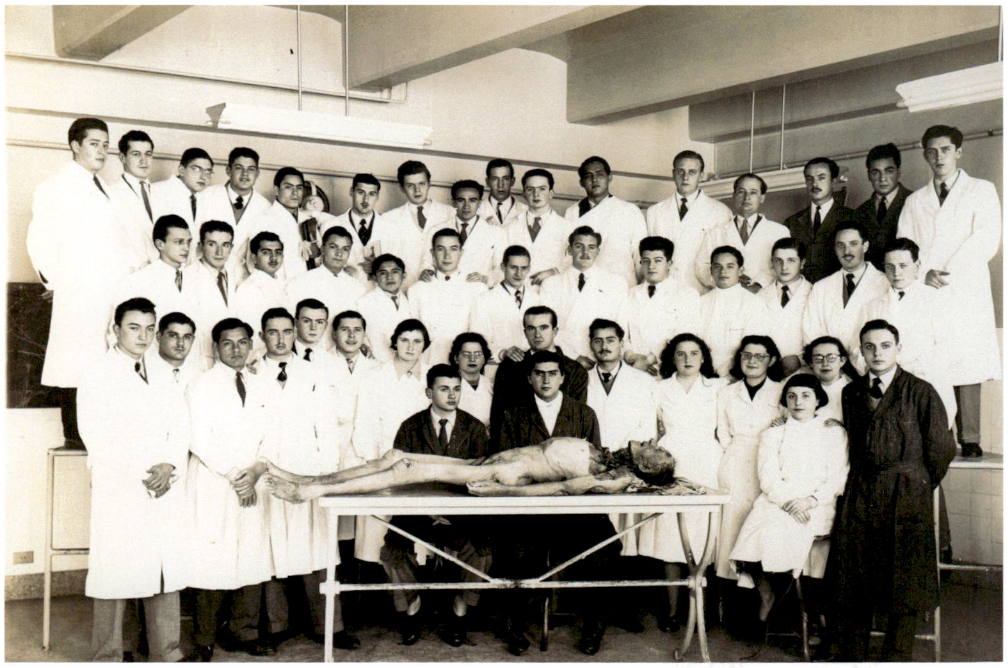

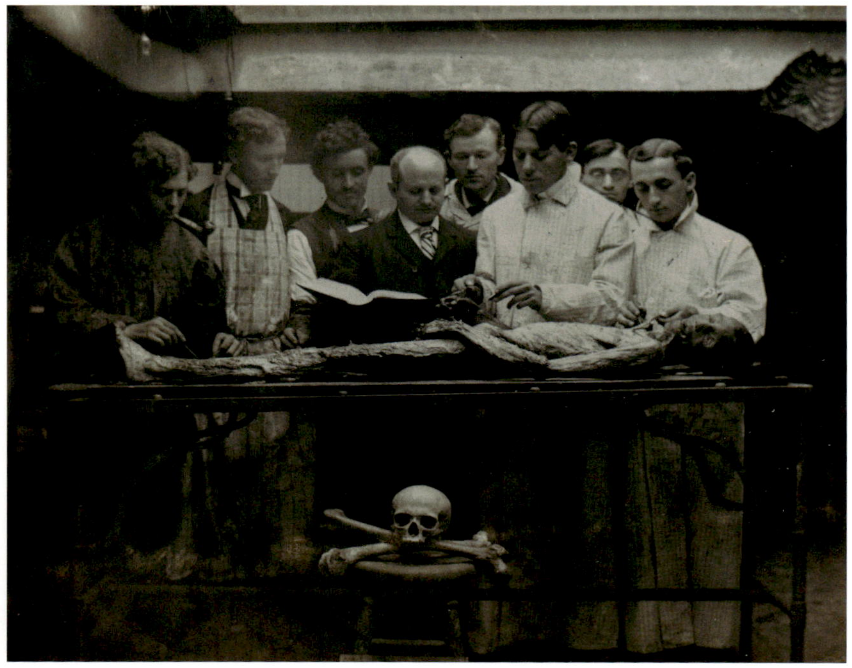

Plate 2.22

Class with Two Bodies
Unidentified School, circa 1915

African American attendant stands with light bulb in the top row.

Plate 2.23

Professor with Students in Baker-style Hats
Unidentified School, circa 1920

Plate 2.24

Anatomy Class
Unidentified School, 1938

Plate 2.25

Anatomy Class with Skull and Crossed Bones as Decoration
Unidentified School, circa 1912

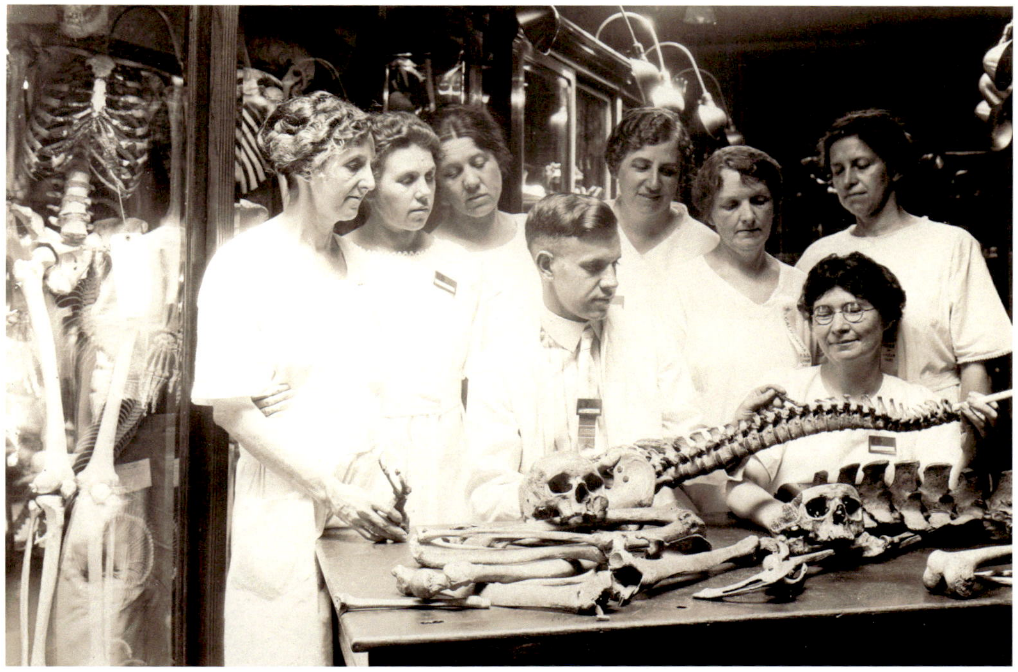

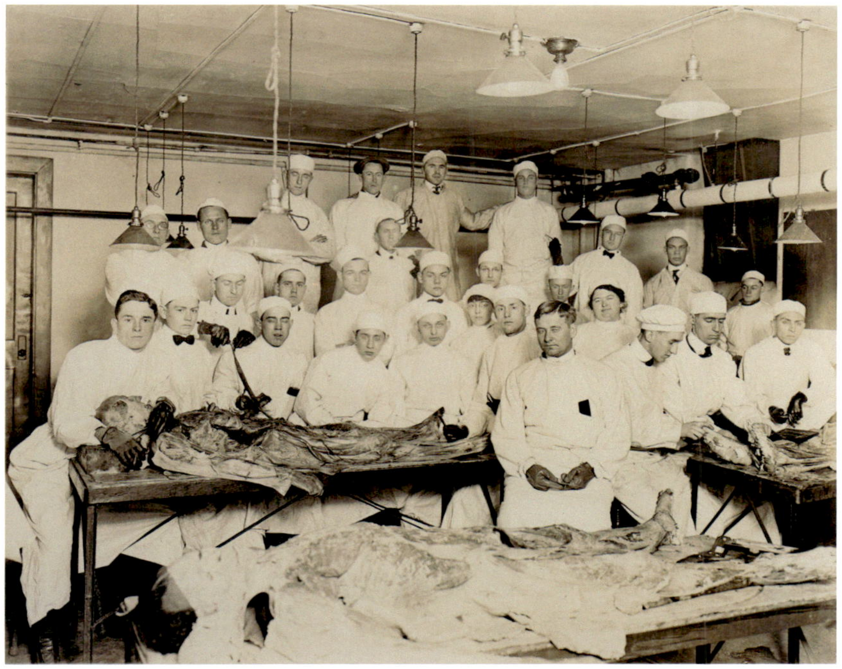

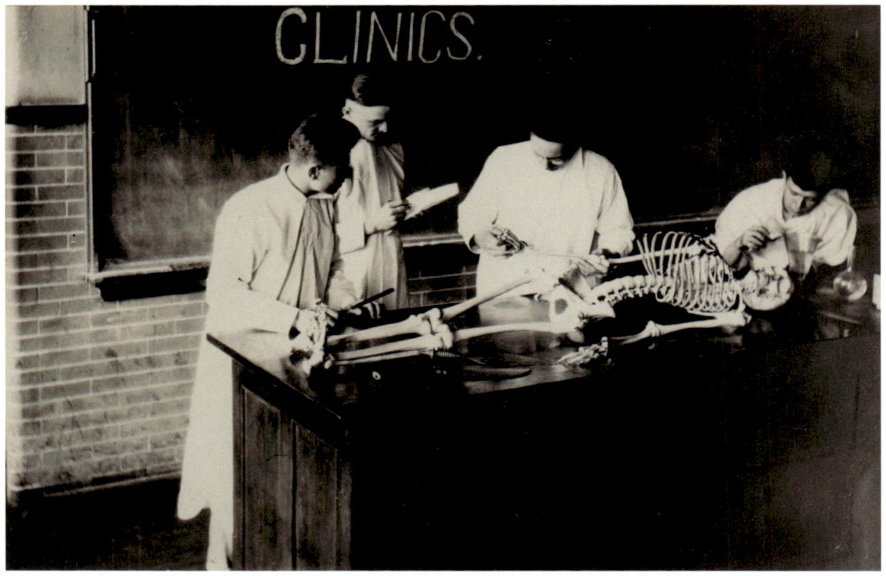

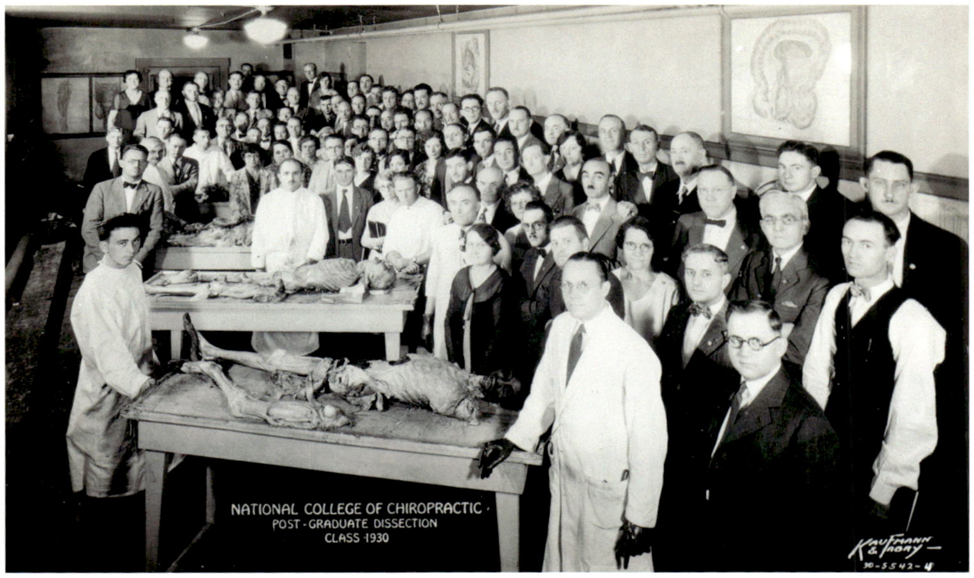

Plate 2.26

Explaining Spinal Anatomy to Graduate Practitioners
Palmer College of Chiropractic, Davenport, IA, 1921

Plate 2.28

Osteology Class
Unidentified School, circa 1930

Plate 2.27

Class Portrait Dissectors in White Gowns, Hats and Rubber Gloves
Unidentified School, circa 1925

The Professor sits prominently in front. Two women and one African American man are in the class.

Plate 2.29

Post-Graduate Dissection Class, Chicago
National School of Chiropractic, 1930
Postcard by Photographer: Kaufman and Faory

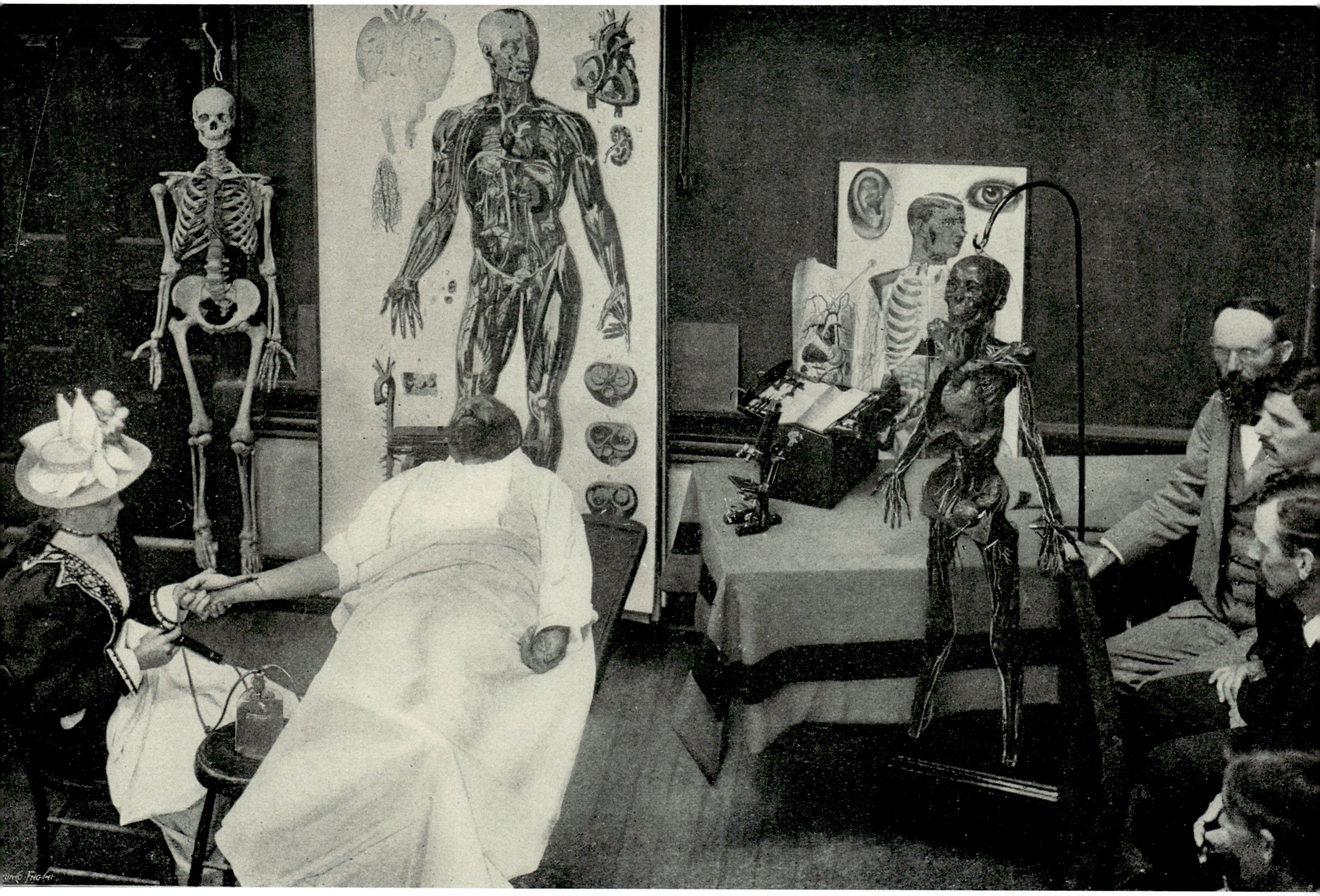

Plate 2.30

Woman Mortician Learning Embalming
Injecting the Arterial System Through the Radial Artery
Champion College of Embalming, Springfield, OH, circa 1899

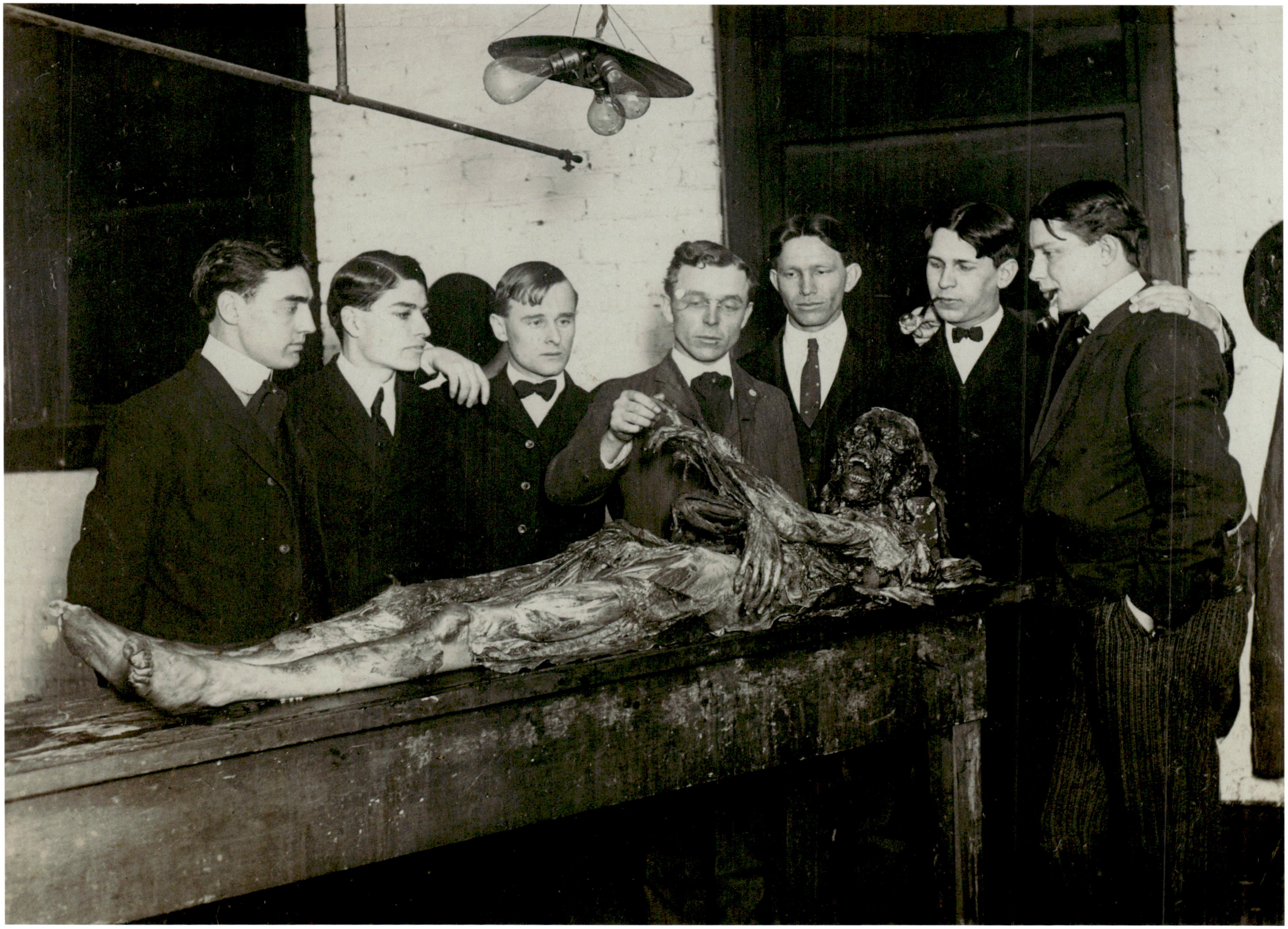

Plate 2.31

Dr. Howard Brundage Teaching Anatomy
Columbus, OH, circa 1910

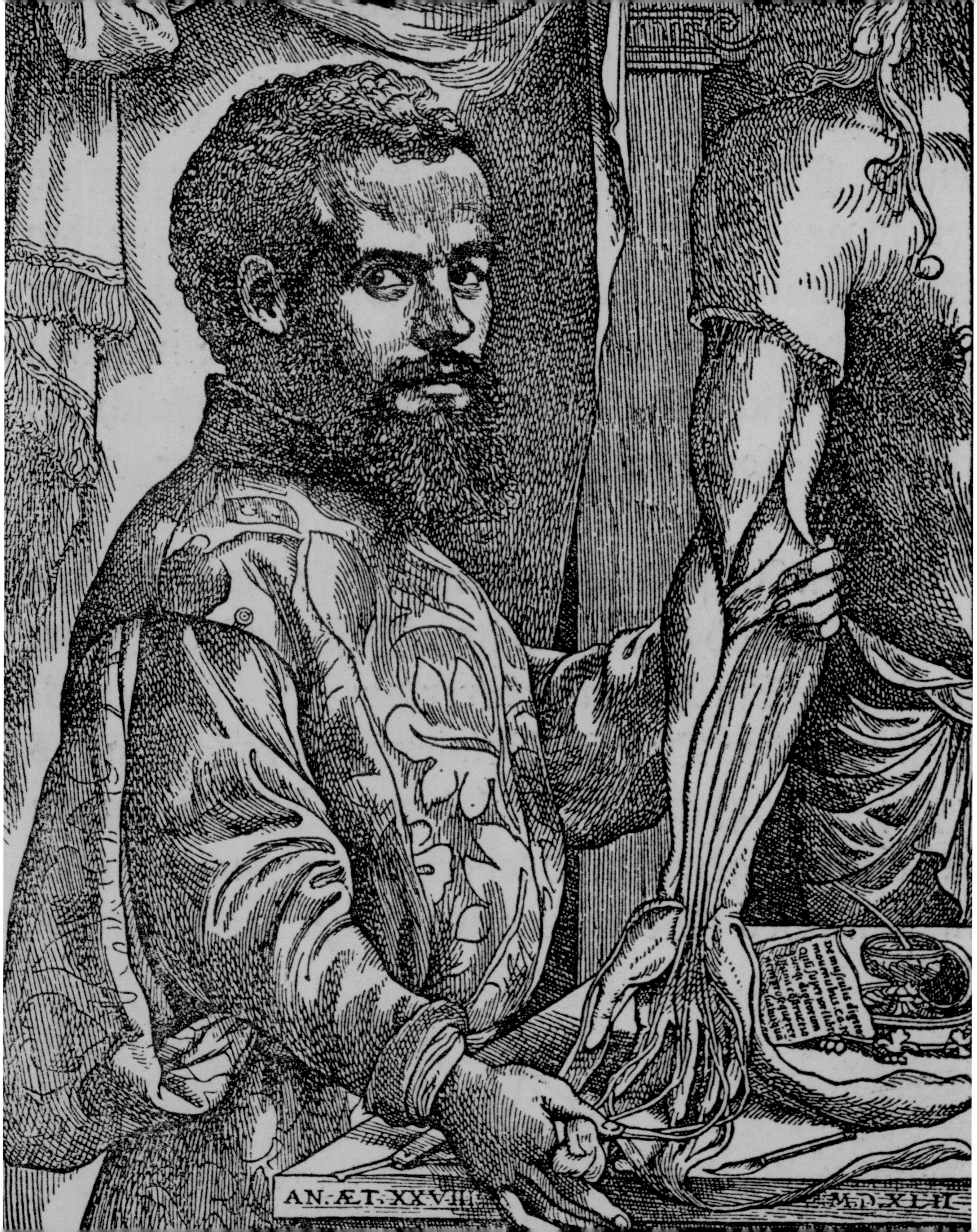
AN. ÆT. XXVIII
M.D.XLII

CHAPTER THREE

Dissection Photography

Rite of Passage & Medical Identity

Posing with cadavers was mainly an American practice of young medical students. These are classic 'rite of passage' photographs, which students took for their initiation into the practice of medicine. The genesis of dissection photography derives from two important sources—the engravings of dissection scenes presented in anatomical texts and the paintings of sixteenth-century masters. Stylized scenes of dissections appeared prominently in several anatomical texts beginning in the fifteenth century. One primary work is the dissection class engraving in Andreas Vesalius's 1543 *De Humani Corporis Fabrica*. The frontispiece of this landmark text has been reproduced for centuries in various formats and displayed in medical schools. Rembrandt Harmensz van Rijn's (1606–1669) *The Anatomy Lesson of Dr. Nicolas Tulp,* painted in 1632, has been widely reproduced as an engraving. With the discovery of photography, it was also reproduced as a photograph, not only of the entire scene, but of sections, which were enlarged, then framed, and can be seen in photographs of physicians' offices.

Dissection photographs were popular and ubiquitous from 1880 to 1930. Knowledge of anatomy and dissection was the primary distinctive aspect of the medical profession since the Renaissance, and the camera offered late nineteenth-century American medical students a visual memento of their entrance into the profession. The study of anatomy was usually the first course in medical schools, and, in many ways, the most significant. Surviving the dissection experience was a major hurdle to entering the profession. The camaraderie and teamwork inherent in dissection scenes is evident in many of these images. As these young men and women embarked on a lifelong professional journey, the bond they acquired around the dissection table was also lifelong.

Over time, photographic posing conventions changed with the improvement of photographic technology. All the early dissection scenes until the mid-1890s were mainly formal compositions taken by professional photographers. From the daguerreian era (1840-1860), only one dissection scene has survived. There are no known ambrotype era (1854-1865) dissection scenes. Only one 1860s tintype scene remains and no cartes de visite paper prints survive until the mid-1860s. From the 1870s, there are a few albumen prints. Dissection photography began to change in 1878 with the commercial availability of the dry plate process and development of silver gelatin paper for prints. The dry plate freed the photographer from plate making and developing. 1880 became the year that dissection scenes started becoming a tradition in American medical schools. For the first time, images could be easily and inexpensively taken. Safe flash powder techniques also increased the availability of indoor photography. In 1888, the development of Kodak celluloid roll film and a special camera allowed amateurs to take photographs easily. In 1900, the Kodak Brownie camera made everyone a photographer. All one had to do was "press the button." The camera took one hundred photographs and was then sent back to the Eastman Kodak Company, who returned the developed film and the camera with a roll of fresh film. All these photographic innovations are seen in the images in this chapter.

The early photographs are formal; the later ones are more relaxed, as people became more comfortable in front of the camera. Time relaxes the subjects, and they become more creative, producing classic yet unique imagery. In each decade, medical students became more inventive with props, cadaver positioning, and inscriptions, as well as gowns and hats to distinguish the photographs. In the first decade of the twentieth century, the concept of the clean sterility-conscious physician became established, and white coats and garb became traditional at major medical schools. It was only during the early twentieth century that medical students worldwide began taking the 'rite of passage' photographs, and they can be found from many European, Asian, and South American medical schools.

The nature of the practice of photography and its professionalization made a significant difference in medical portraiture in the United States and in Europe. The democratization of photographic portraiture in the United States was in sharp contrast to European practices. In Europe, early photography was an elitist practice and its early practitioners were artists, scientists, and wealthy entrepreneurs. In England, licenses were necessary to practice daguerreotypy. Photography in Europe in the 1840s and early 1850s was by, for, and of the wealthy. There are few photographs of European workmen in photography's earliest era, and they were not commissioned by the subject but were taken as social or anthropological studies. In America, anyone could become a photographer, and by the 1850s, photography was ubiquitous—everyone could afford to have a photograph taken. If not while alive, when dead. There are no known dissection photographs from Europe until the last years of the nineteenth century.

In Europe, strict guild rules and travel restrictions hindered the advancement of those wanting to enter almost any profession. In the United States, just as anyone could become a photographer, anyone could become a physician or lawyer or practice almost any trade. University graduation was not a prerequisite for advanced legal or medical training. No formal education was required. President Andrew Jackson's administration in 1828 liberalized American licensing laws, resulting in the democratization of professions. In Europe, professions were very strictly regulated as centuries-old guilds and academic organizations made the road to professionalism a lifelong educational and professional struggle. It was in opposition to the European circumstance of the lack of opportunity that our fledgling country opened the doors to all professions. American government regulation of occupations and professional practice standards were nil. People could say anything and do anything; it was a century of a Wild West mentality. It was up to the consumer to be aware of, and wary of, every aspect of consumption and trade.

Medical historian Henry Siegrist, in his foreword to William Frederick Norwood's 1944 classic study, *Medical Education in the United States before the Civil War*, notes that America "was a society that worshipped the principles of free enterprise and unlimited laissez faire." Norwood exposed the complete disregard of educational standards as a feature of prewar medicine. As a corollary, it has been noted by medical historian Richard Shyrock and others that American medical education up to the Civil War had little to offer the world of medical science compared to European accomplishments. Nonetheless, medicine was an important factor in American life and the development of the nation.

Apprenticeship with a tradesman or professional was the American road to many careers, including medicine. A few lectures at a medical or a law school was an added bonus. Any professional in his trade could open a school and give lectures. Hundreds of American medical schools in the nineteenth and early twentieth centuries were small proprietary establishments. Many were simply diploma-mills. Chicago's infamous part-time medical night schools of the early twentieth century are an example of this practice carried to an extreme.

However, in Europe, a university degree was required to enter medical school. Training in Europe was arduous with strict regulations, examinations and supervision. It was especially difficult to gain advancement in a university-associated hospital. In Germanic countries, a strict ladder of accomplishment for hierarchical advancement meant lifelong yearning to achieve a professorship. Most often, a physician had to wait until a professor died and a position opened. Until the 1930s, American physicians had to go to Europe for postgraduate study in order to be considered a specialist or expert in a field due to the superiority of European medical education. The lack of formal education in America may be reflected in the casual attitude in some of the American dissection scenes.

THE CADAVER AND GRAVE ROBBING

The history of dissection is fascinating. There are a multitude of books on grave robbing, dissection, and the history of anatomy laws. Most of the subjects of dissections were executed criminals or the unclaimed bodies of paupers. This was the standard practice for obtaining bodies in both the United States and Europe. In the seventeenth century, colonists recognized the need for anatomy; Native Americans and those convicted of capital crimes were dissected. Recent works have noted that in the United States, African Americans were perhaps the most frequent 'pauper' anatomized in the nineteenth century. Dissection photographs substantiate that premise.

Grave robbing was a big business in the United States in the mid-nineteenth century, as bodies were stolen in one area and shipped in vats to another by train. In many small localities, medical students were 'elected' by their professors to be grave robbers. It was not an easy job as Dr. Howard Kelly, John Hopkins's brilliant surgical innovator, explains in his 1908 essay, *The Barred Road to Anatomy* (see Appendix). Dr. Kelly provides a vivid description of what anatomizing was like in the United States and also includes several detailed episodes of resurrectionist activities in the United States and Great Britain that resulted in the establishment of anatomy/dissection laws.

To combat grave robbing, some cemeteries placed iron cages, called mort-safes, around fresh graves, while others stored corpses in vaults until the bodies decomposed to the point that they were safe from robbers. More commonly, a guard was hired or family members took turns watching over the graves for the first few nights after burial. Social position was no bar to whether or not your body was stolen, and one noted outrageous grave robbing episode dramatically changed laws nationwide—the stealing of the body of former United States Senator John Scott Harrison (1804–1878).

Harrison was the son of U.S. President William Henry Harrison (1773–1841) and the father of future U.S. President Benjamin Harrison (1889–1893). He died on May 25, 1878, and was buried in Congress Green Cemetery in North Bend, Ohio. While attending Senator Harrison's funeral, family members noted that the ground around a nearby grave of a recently deceased friend was disheveled. The grave was examined, and the body of the friend, August Devlin, had indeed been stolen. The Harrison family was not too concerned about the Senator's grave, as the future president and his younger brother took serious preventive measures. The grave was eight feet deep, eight feet long and wide. The Senator's body was in a metal casket and placed on a stone

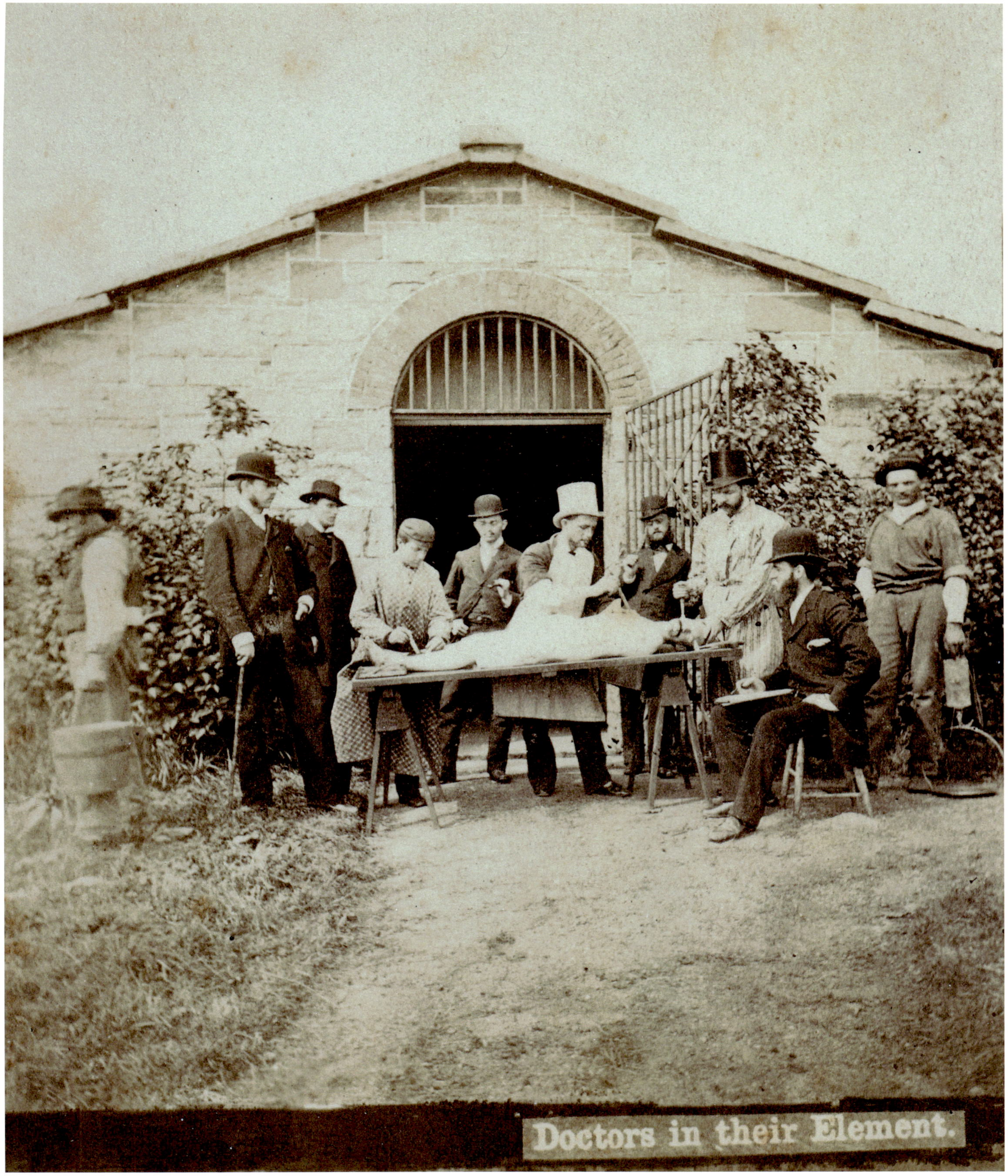

Plate 3.1

"Doctors in their Element"
A very popular stereoview issued for the public, 1870

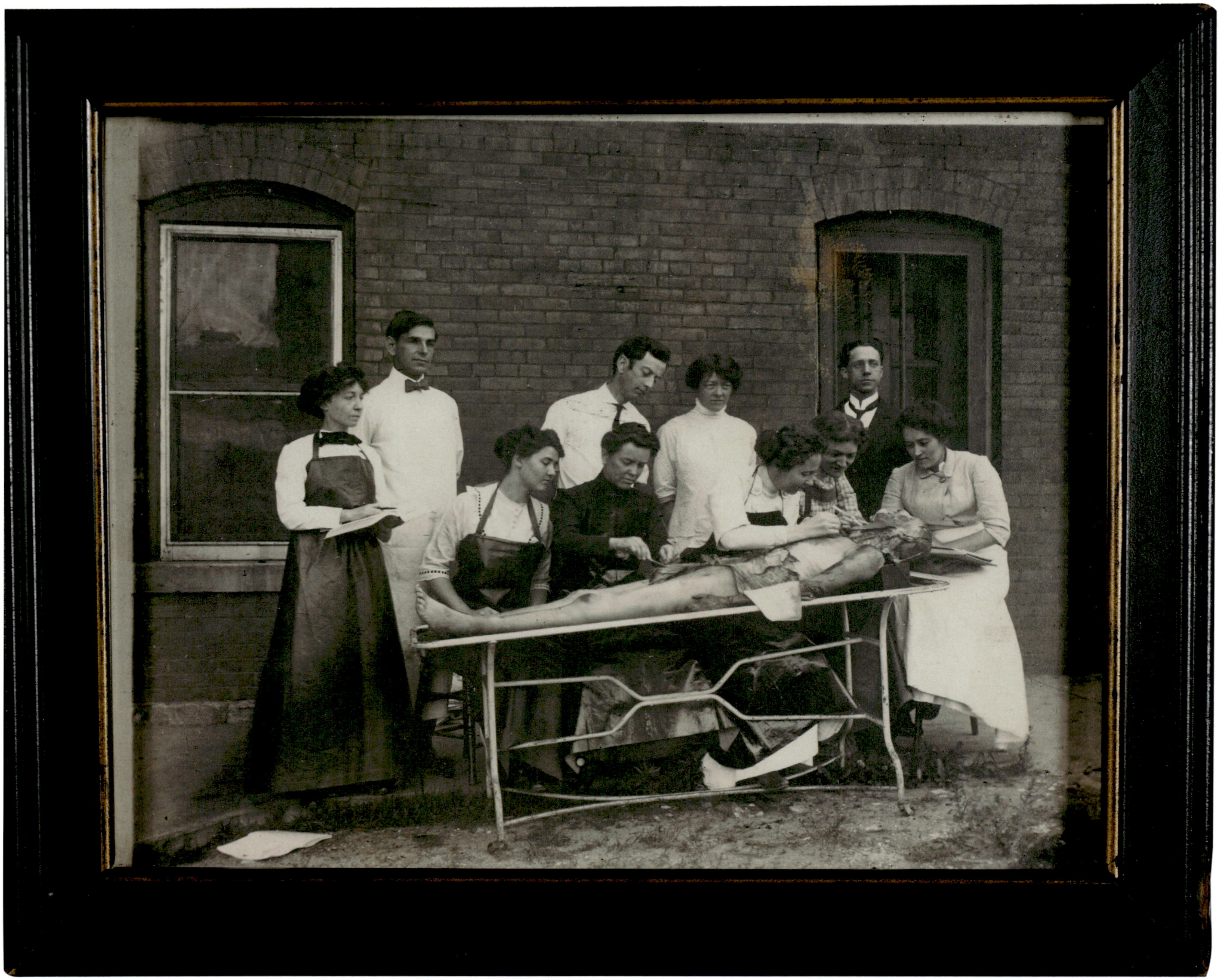

Plate 3.2

Outdoor Dissection

American School of Osteopathy, Kirksville, MO, circa 1900

base with bricks placed around it. Three eight-inch-thick flat stones were placed on top of the bricks enclosing the casket. The largest of the stones was placed at the head end, as bodies were usually hoisted up headfirst when grave robbing occurred. Cement was poured on top of the three stones, left to set for several hours, and then the grave was filled in. A guard was hired to watch the grave for thirty nights.

Friends and family travelled to Cincinnati to search the medical schools to look for Devlin's body. They obtained a warrant and had detectives accompany them. Their first stop was the Ohio Medical School. They searched the school and janitor A.Q. Marshall told them that there were no bodies there. As they were about to leave, a constable noticed a taut rope was attached to a windlass in an elevator chute. They pulled the object up and found a dead body, that of an old man. It was not young Devlin. They removed the cloth sack over the man's head. Harrison caught sight of the face, he stepped back in horror and exclaimed, "My God, that's my father!" Harrison came looking for his young cousin and instead found the corpse of his father, who had been buried only twenty-four hours earlier. The grave robbers had been present at the burial, moved the small stones at the foot of the casket, and then removed the body feet first, contrary to the usual practice. The hired watchman had run away when it became dark. The uproar in Congress and state legislatures resulted in the passage of the Anatomy Acts in most states. The Ohio legislature passed a law providing serious penalties for grave robbing.

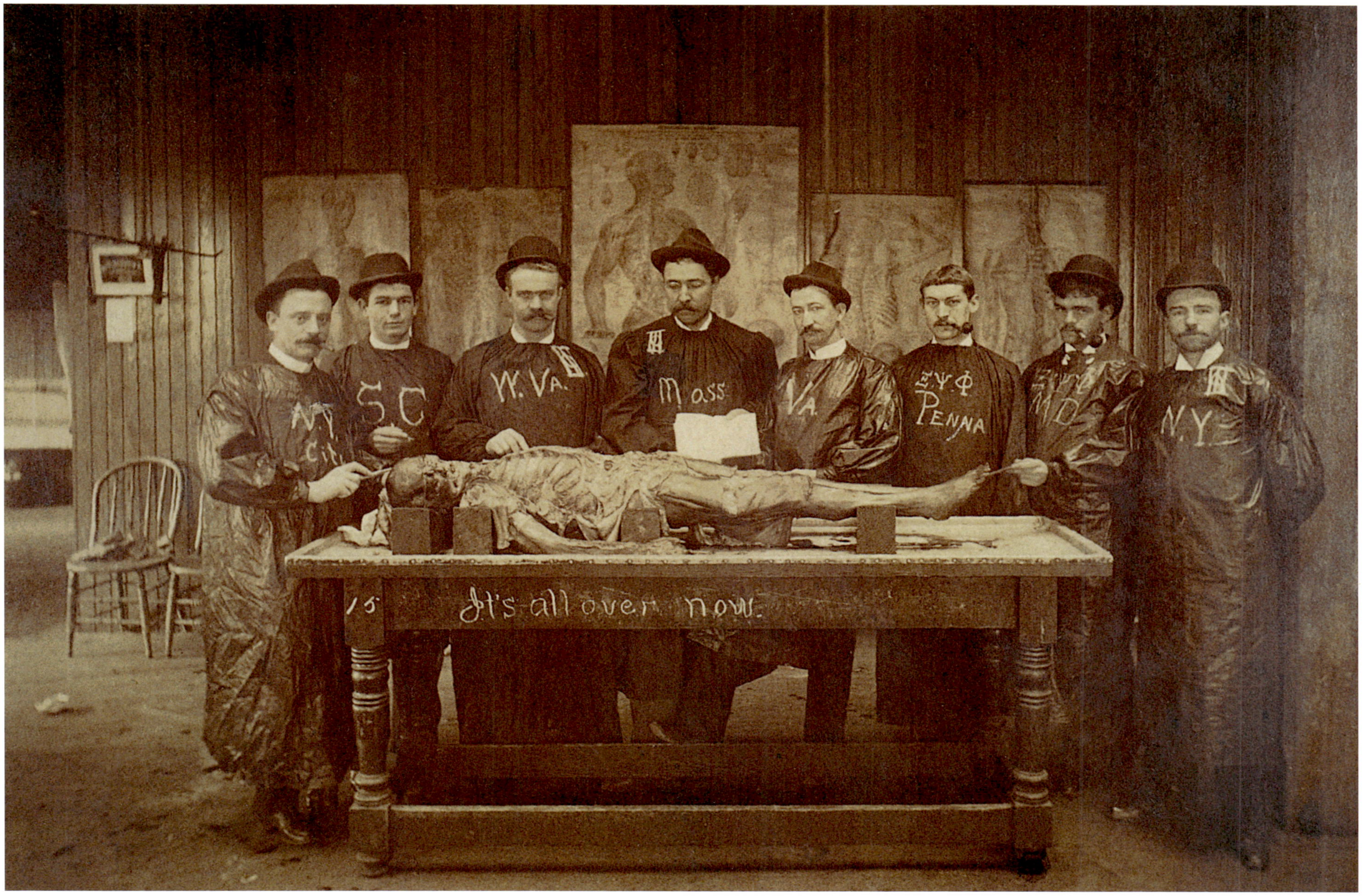

Plate 3.3

Eastern States Students in Baltimore
University of Maryland School of Medicine, circa 1915
Epigraph on table: "It's All Over Now"

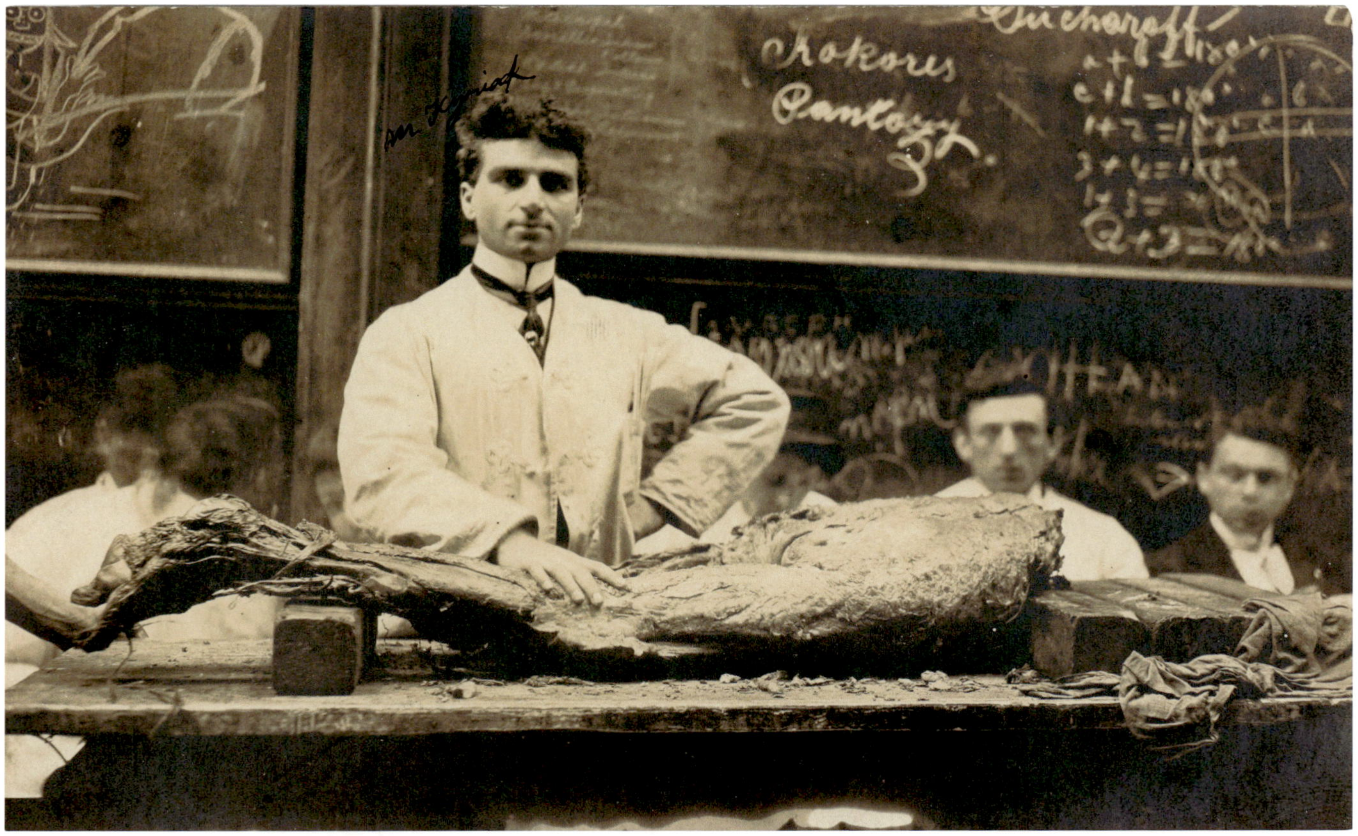

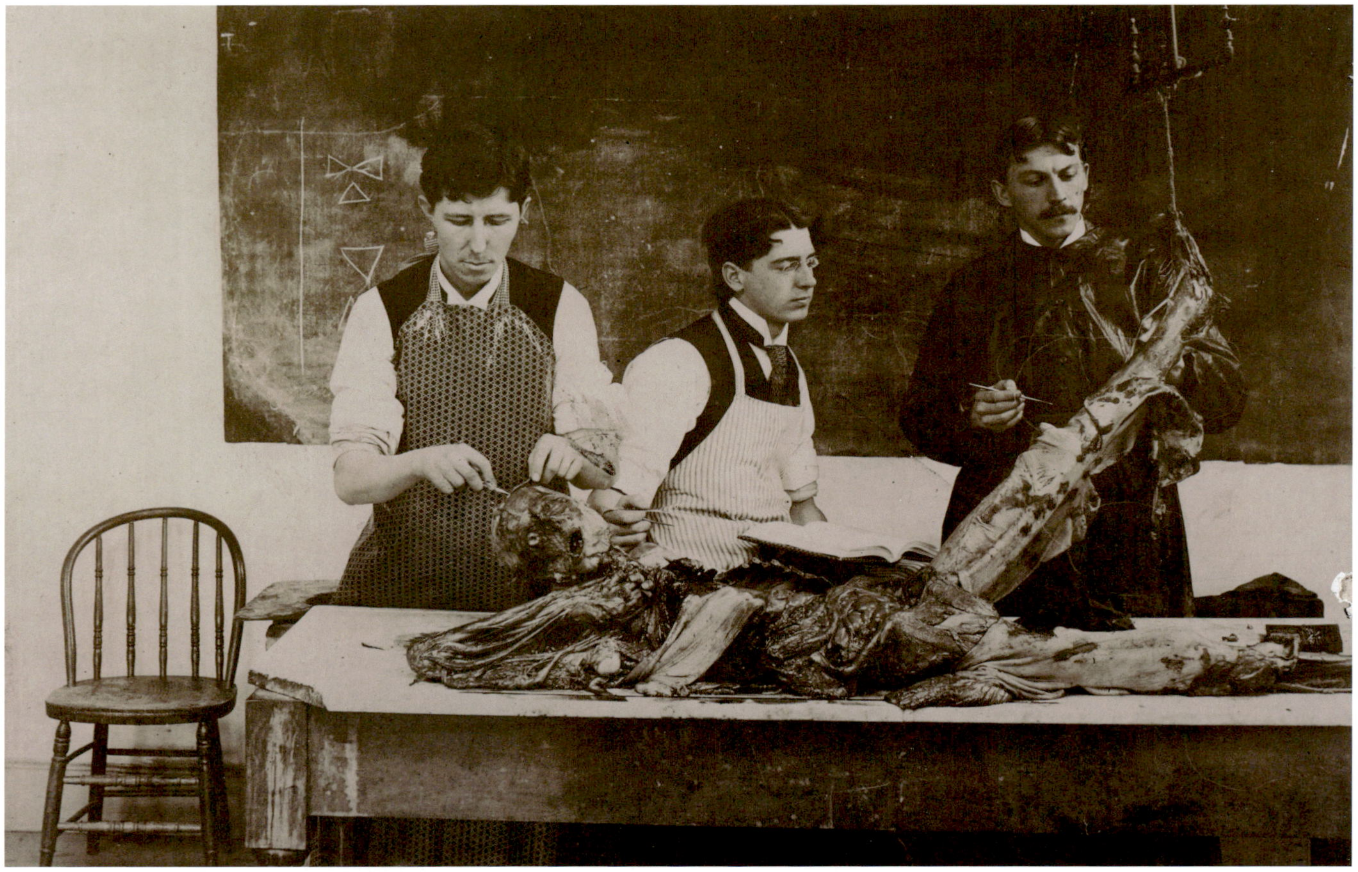

Plate 3.4

Posing Alone with a Cadaver
Europe, circa 1910

Plate 3.5

Dissecting Head and Leg
American School, circa 1905

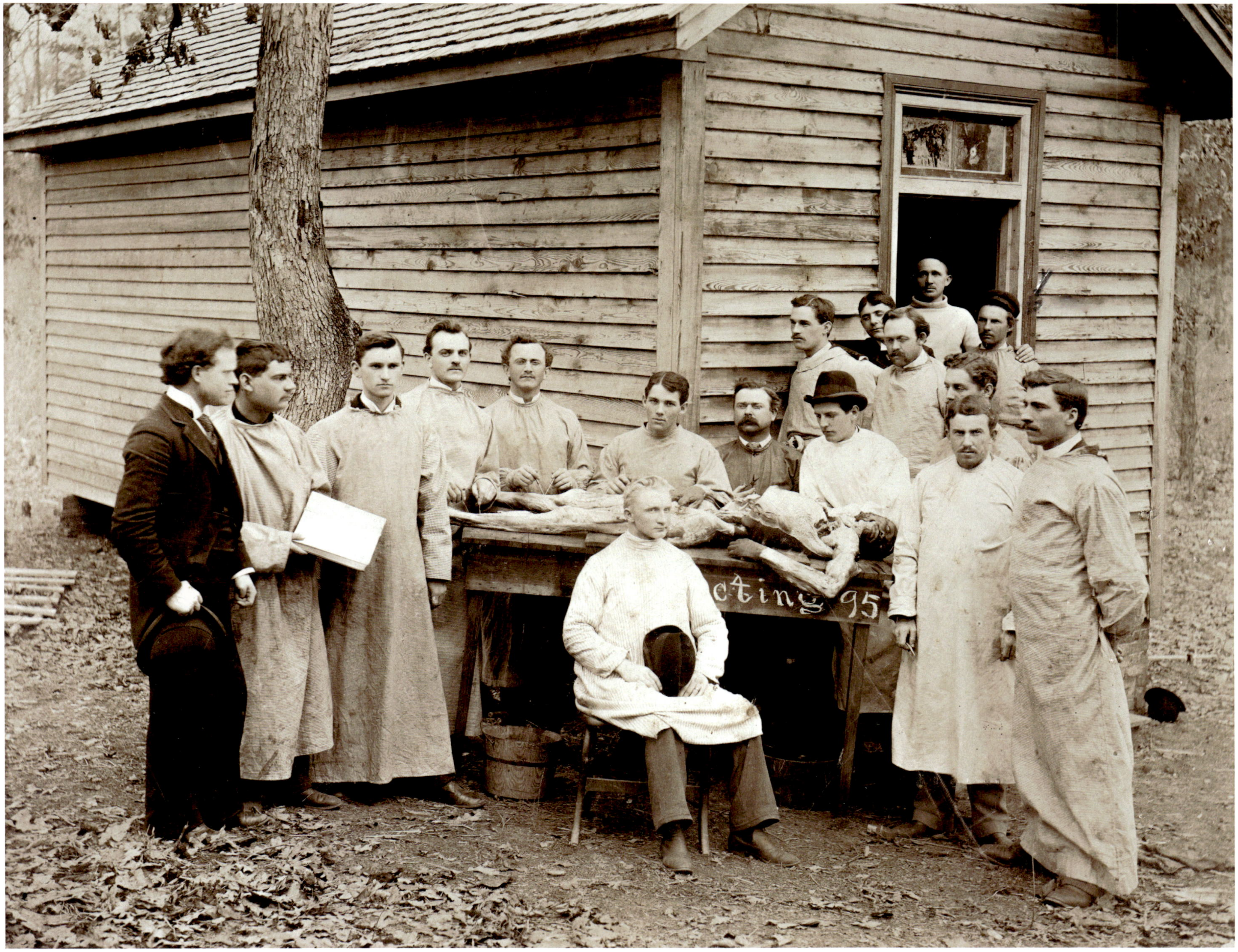

Plate 3.6

Dissecting Outdoors, University of North Carolina
Photographer: J.H. Van...Charlotte, NC, 1895

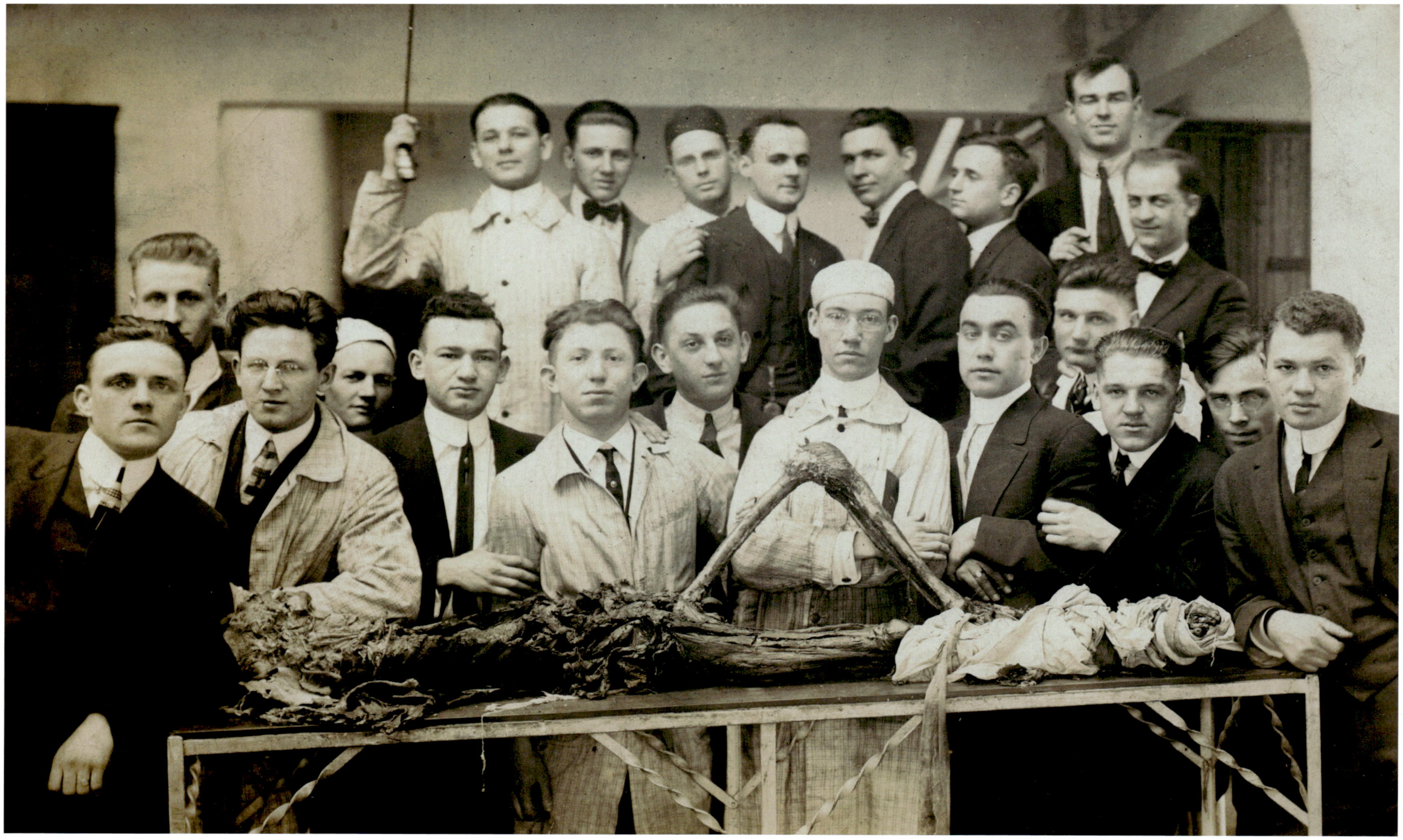

Plate 3.7

Seriously Posed Anatomy Class
Unidentified School, circa 1917

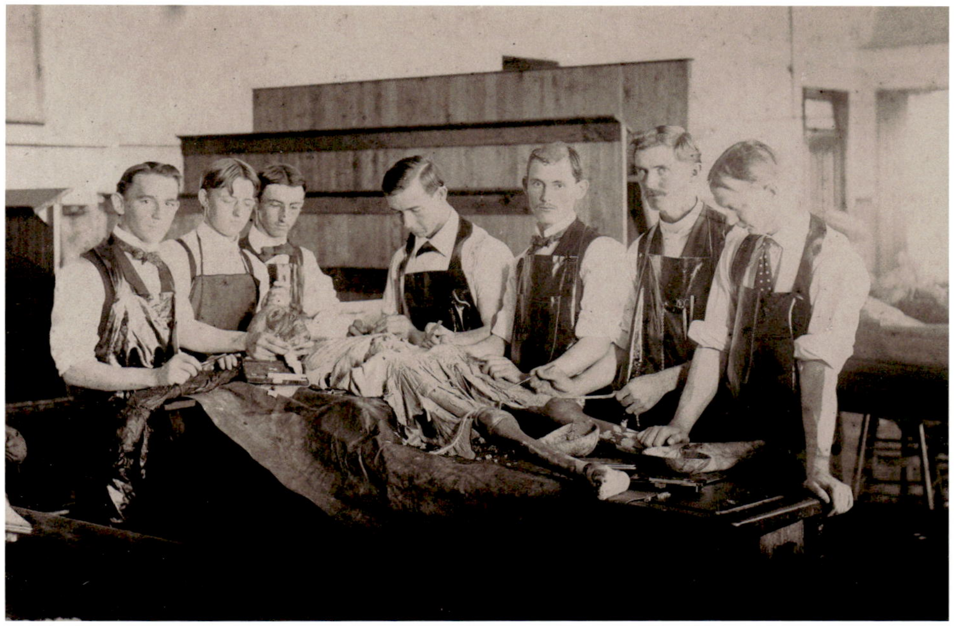

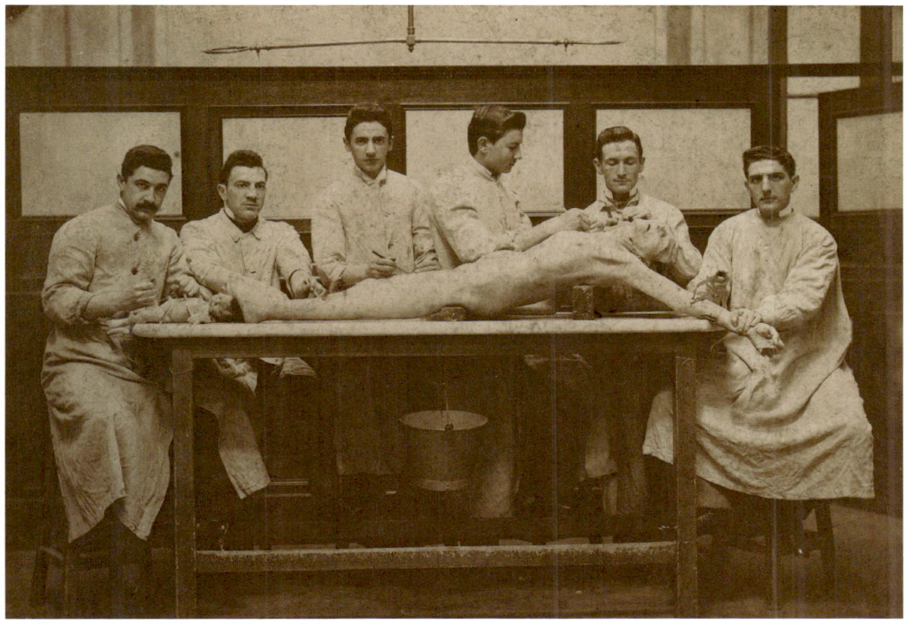

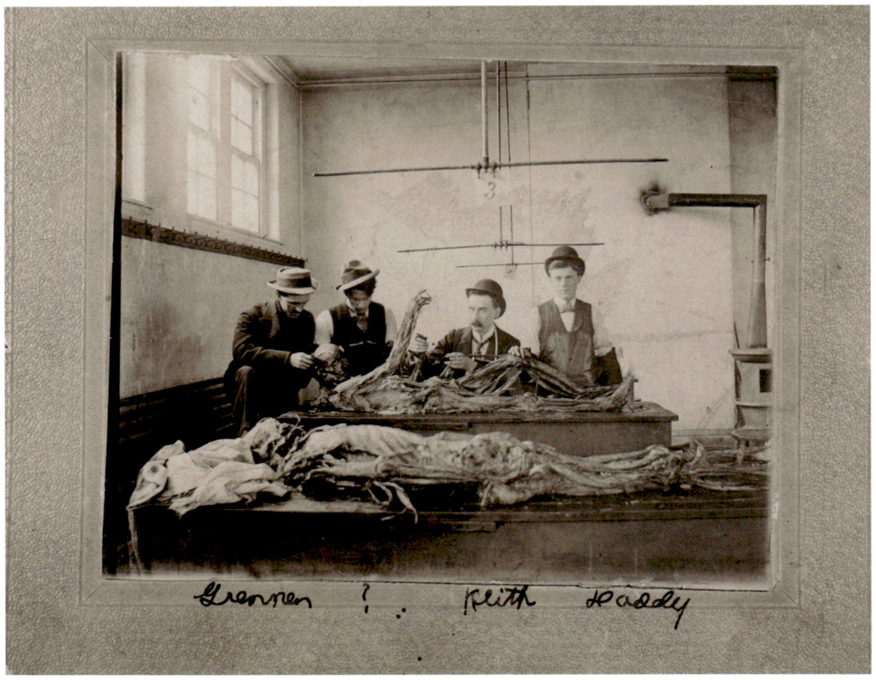

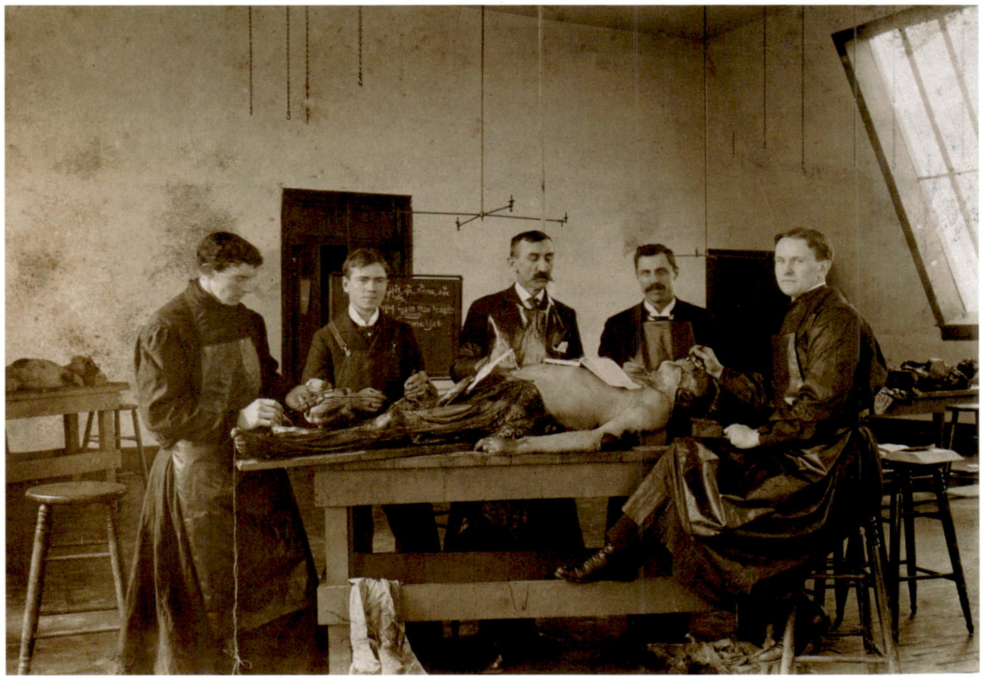

Plate 3.8

Unidentified School
Circa 1885

Oilcloth aprons were the most popular protective wear for dissection and surgery until the 1890s.

Plate 3.10

Casual Dissecting Team, Only One Man Wears a Gown
American School, circa 1905

Plate 3.9

French Medical Anatomy Laboratory, circa 1920
Circa 1920

Plate 3.11

Dissection Class Using Natural Light for their Work
American School, 1890

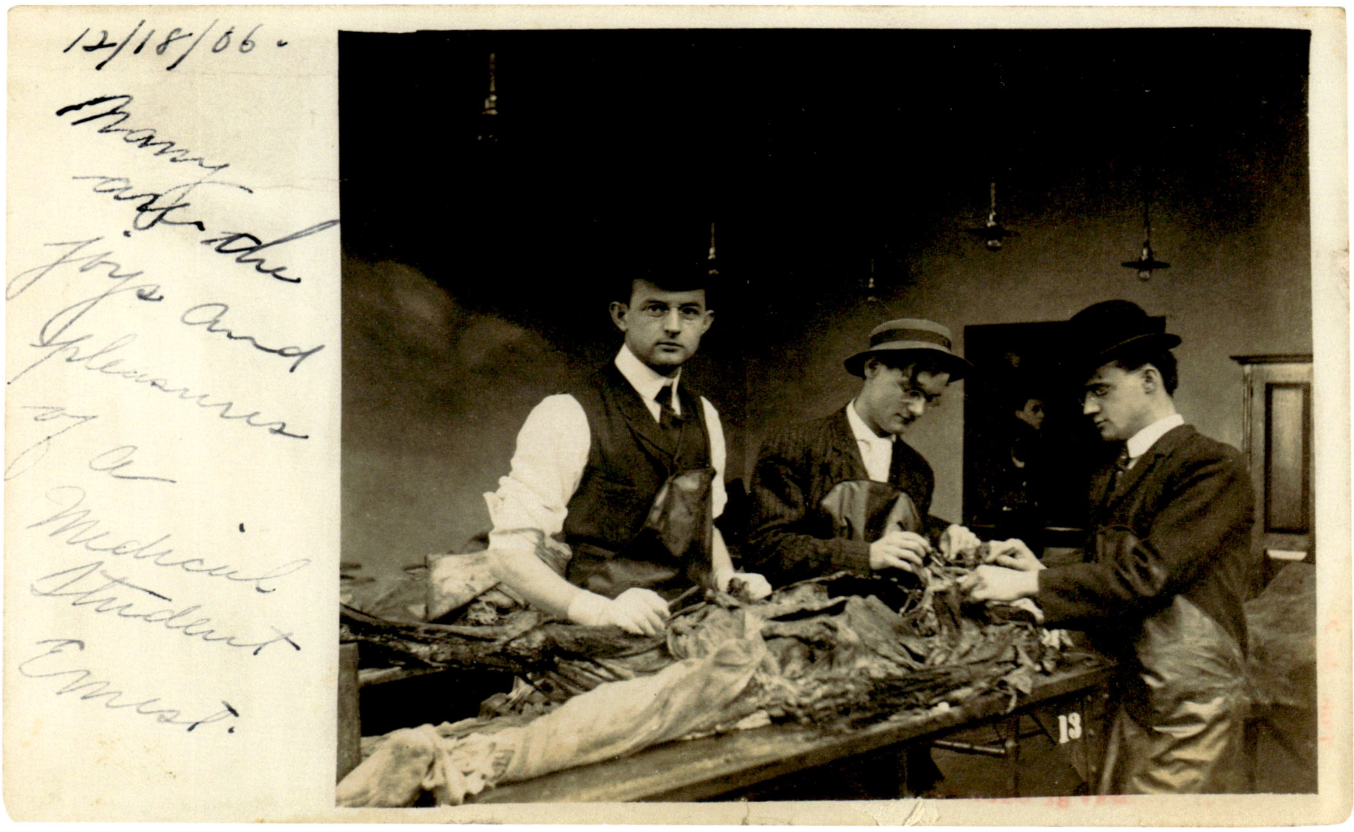

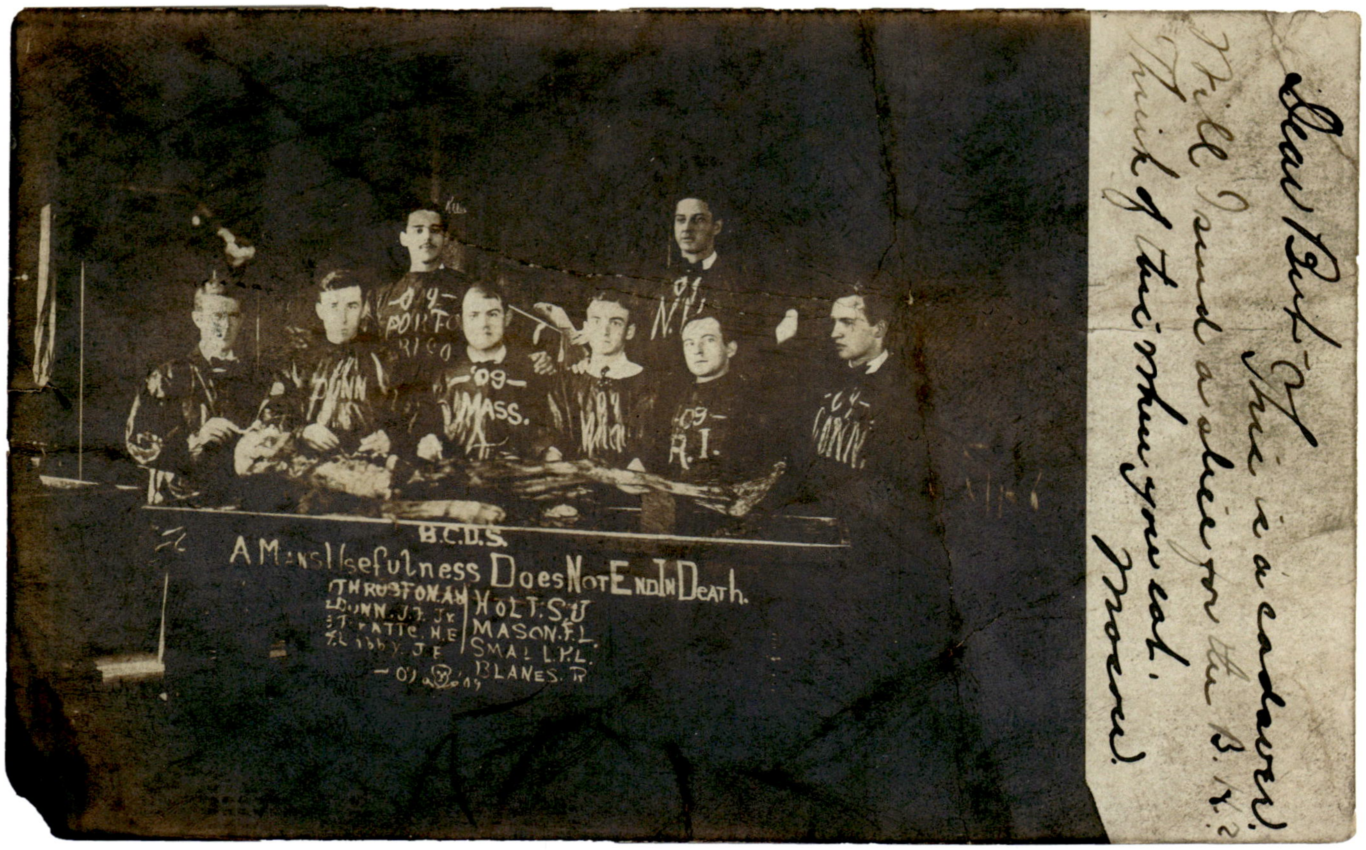

Plate 3.12

"Many Are the Joys and Pleasures of a Medical Student"
Postcard, Pennsylvania Medical School, December 1906

Plate 3.13

"A Mans Usefulness Does Not End In Death"
Postcard, Baltimore College of Dental Surgery, November 1909

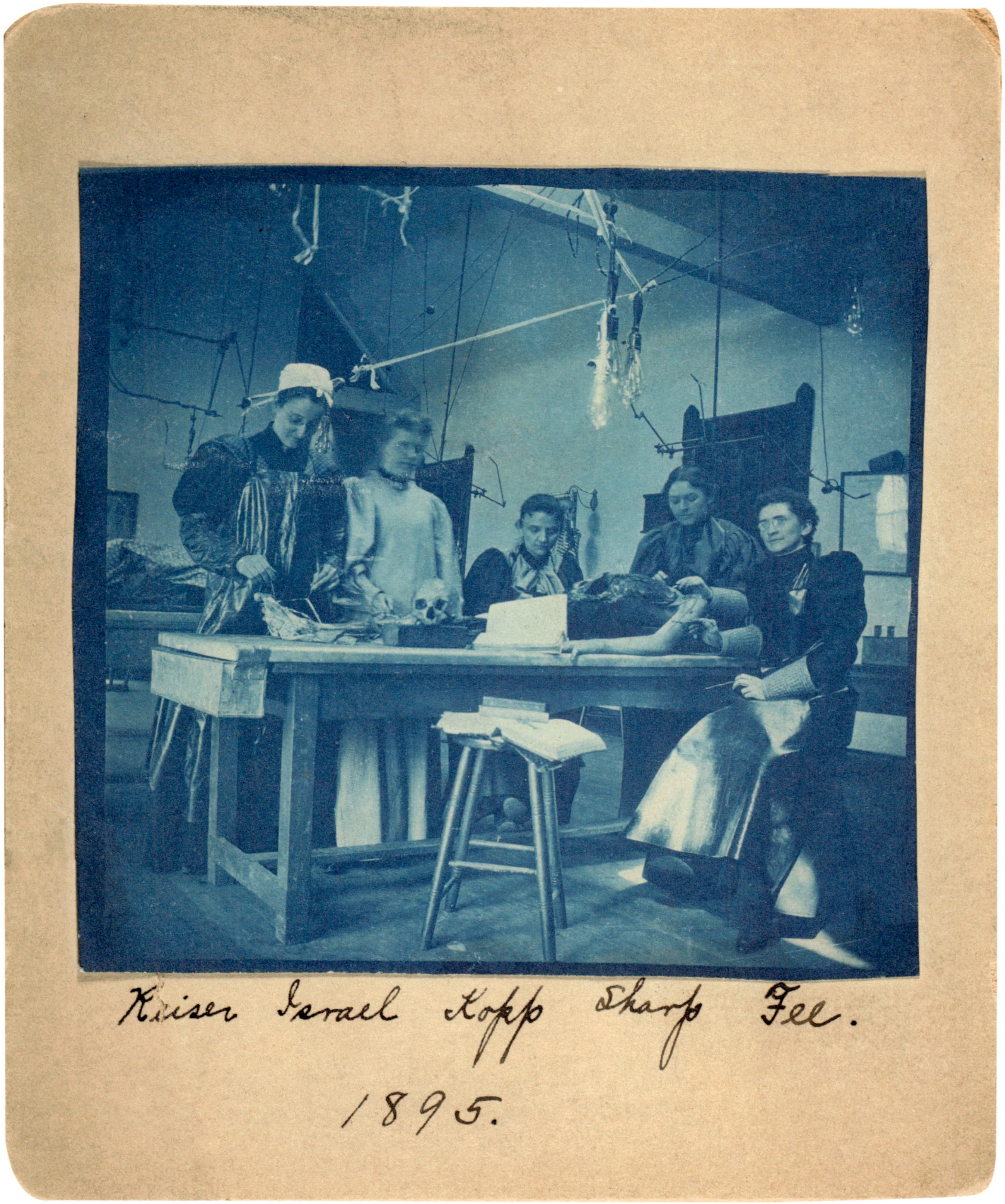

Plate 3.14

Women at Work
Cyanotype, 1895

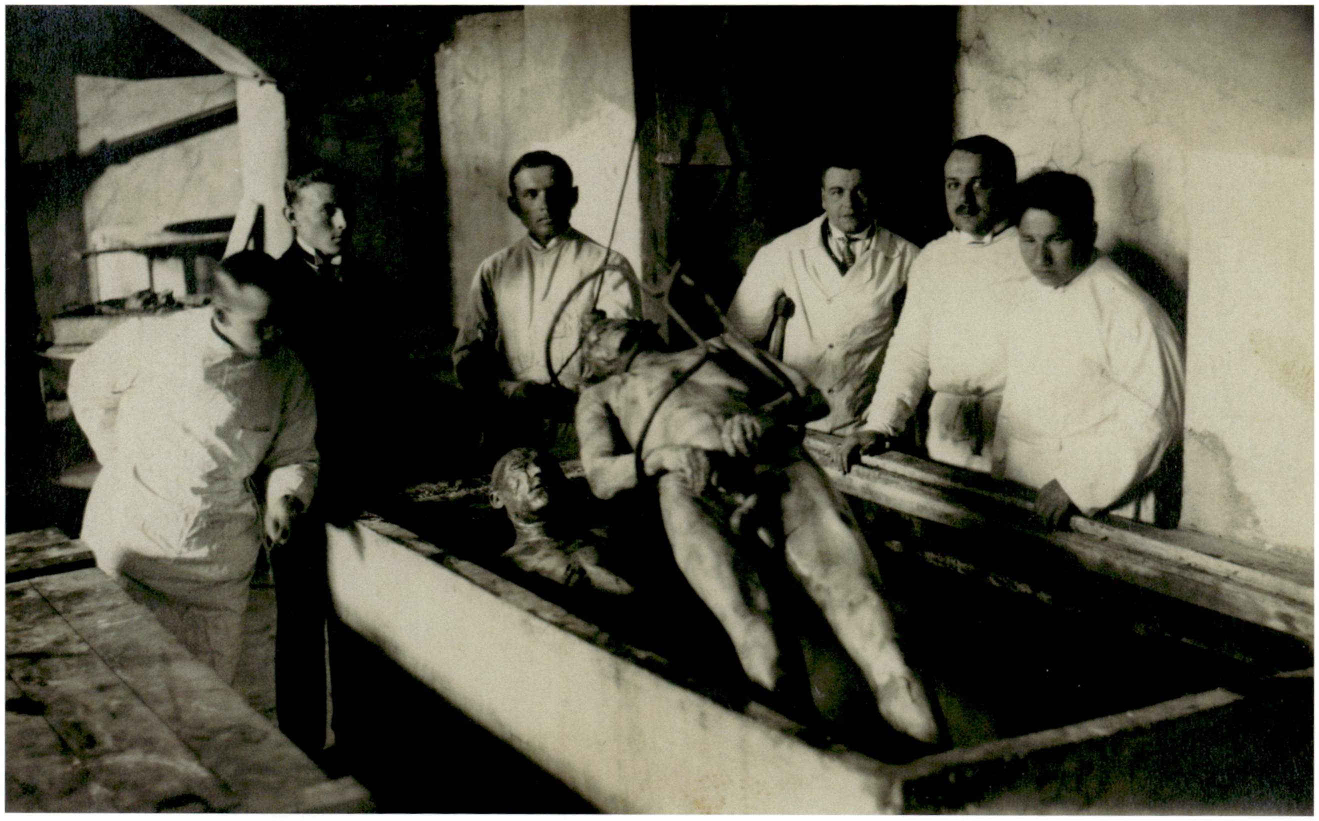

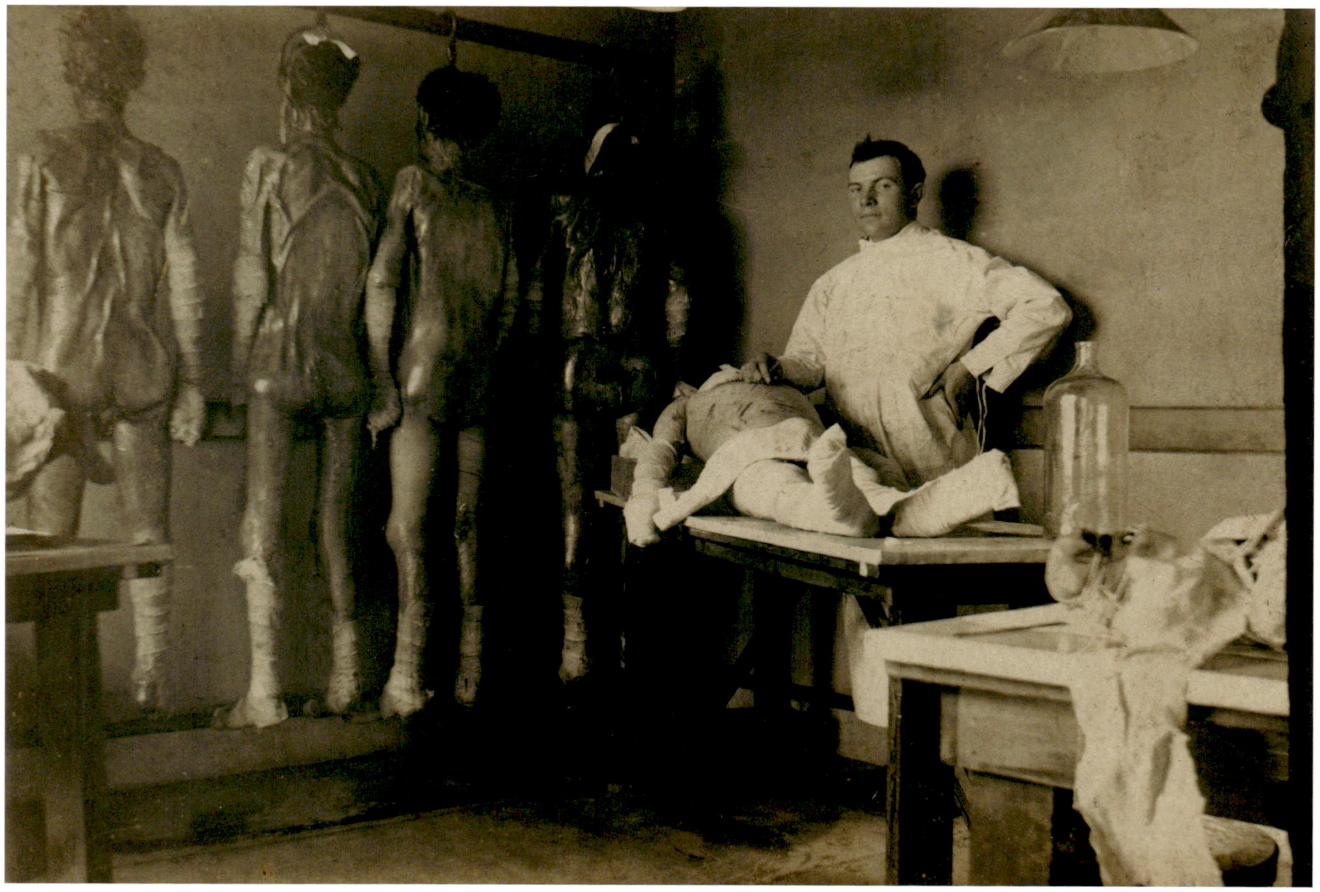

Plate 3.15

Cadavers Removed from Chemical Vat
Postcard, French Medical School, circa 1920

Plate 3.16

Cadavers Hooked and Ready To Be Delivered
Postcard, American School, circa 1920

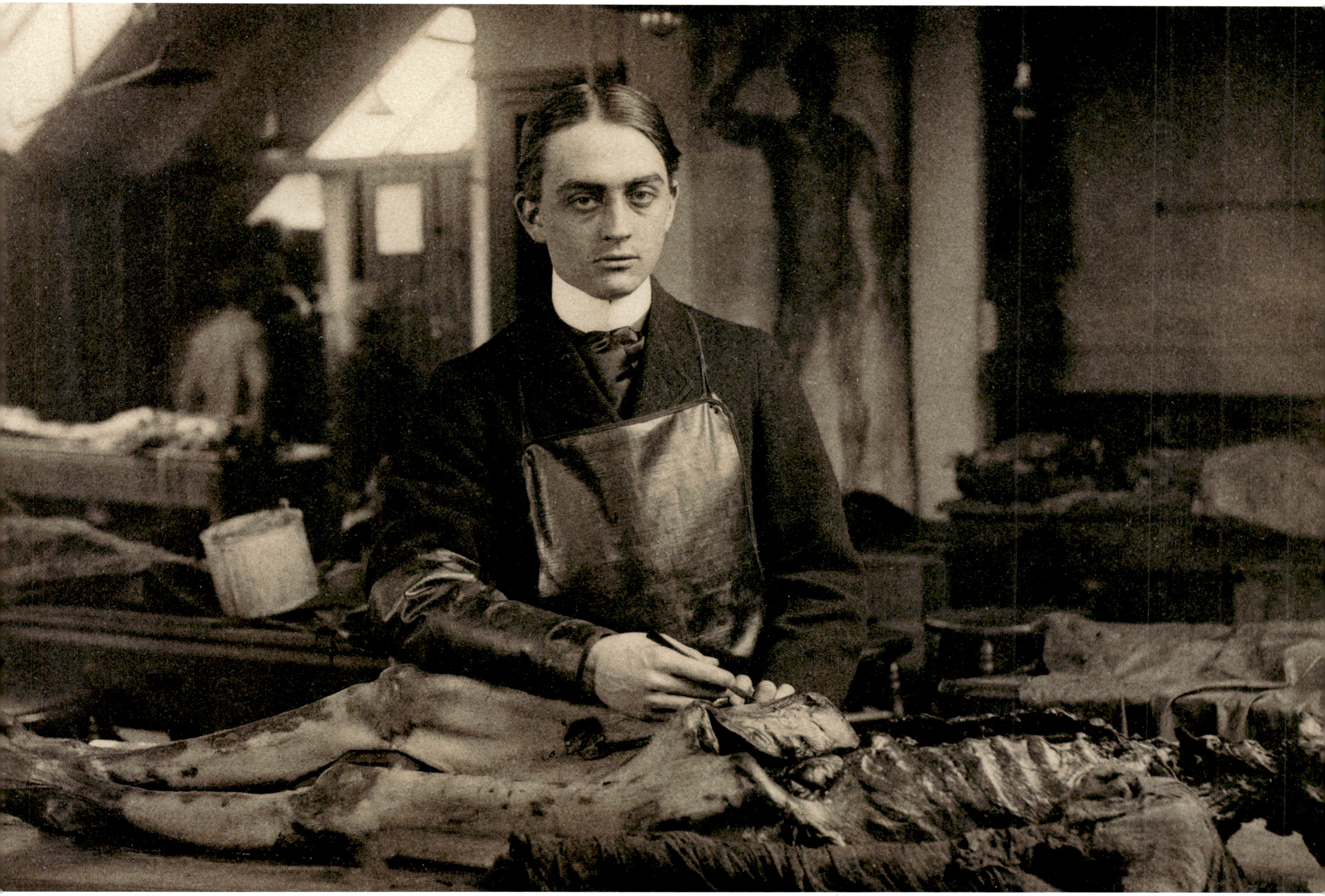

Plate 3.17

Formal Individual Portrait, Philadelphia Medical School
Photographer: Hecker, circa 1905

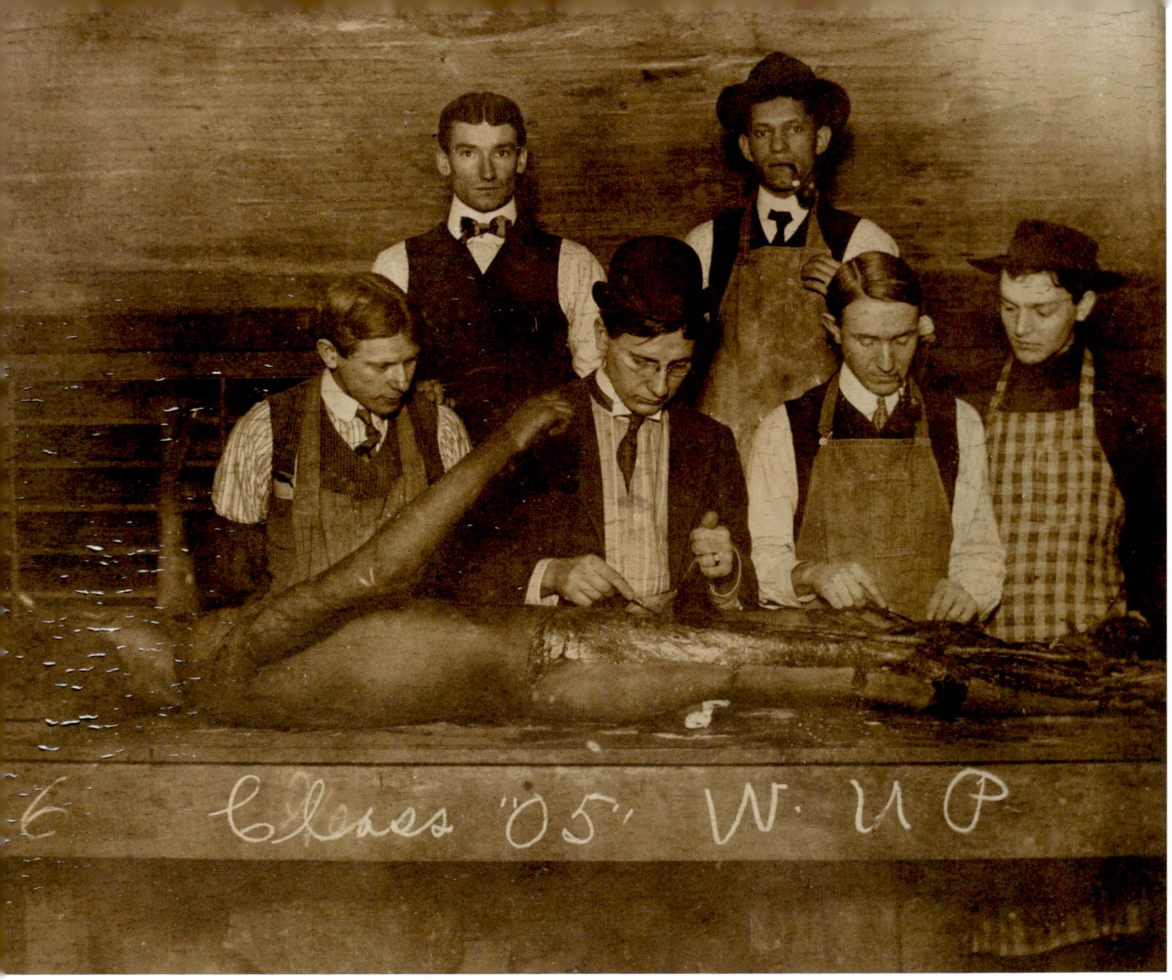
Class "05" W. U P

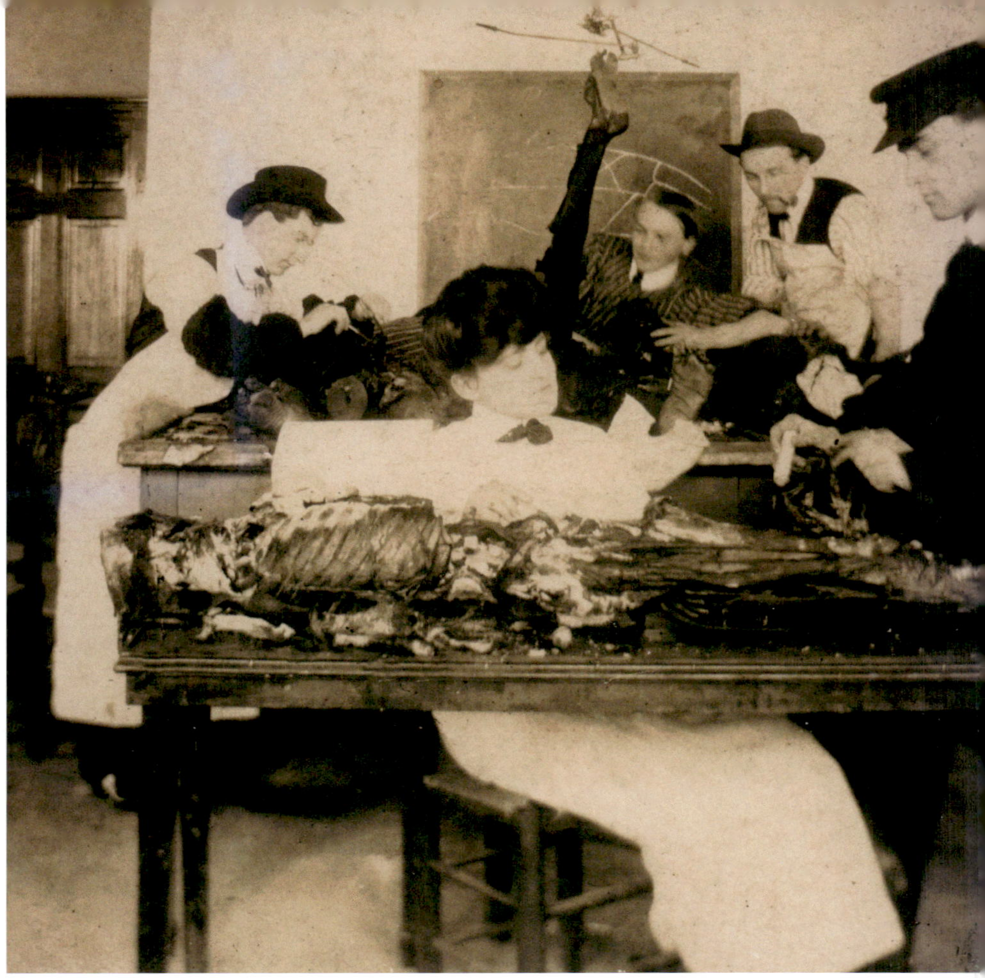

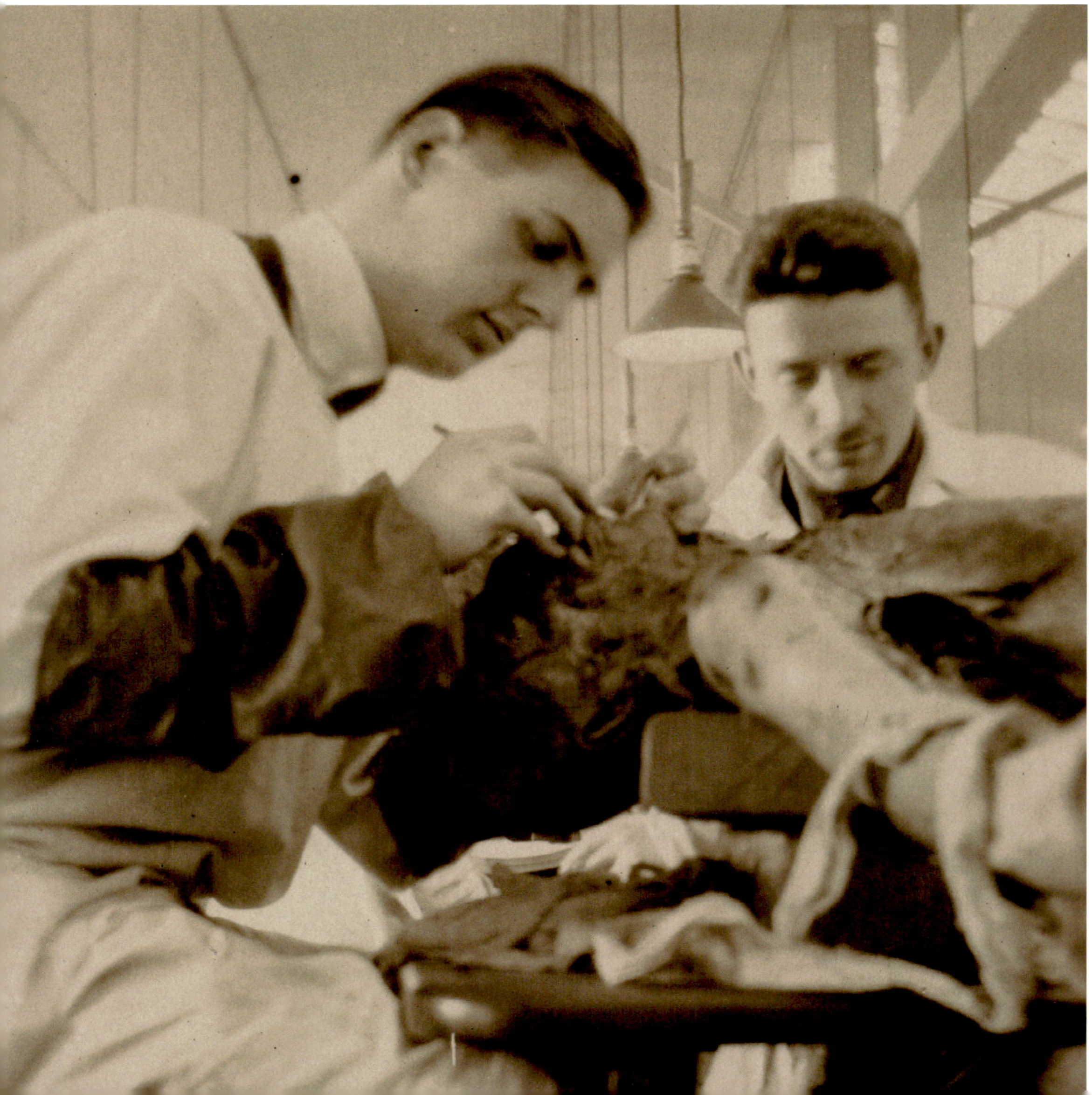

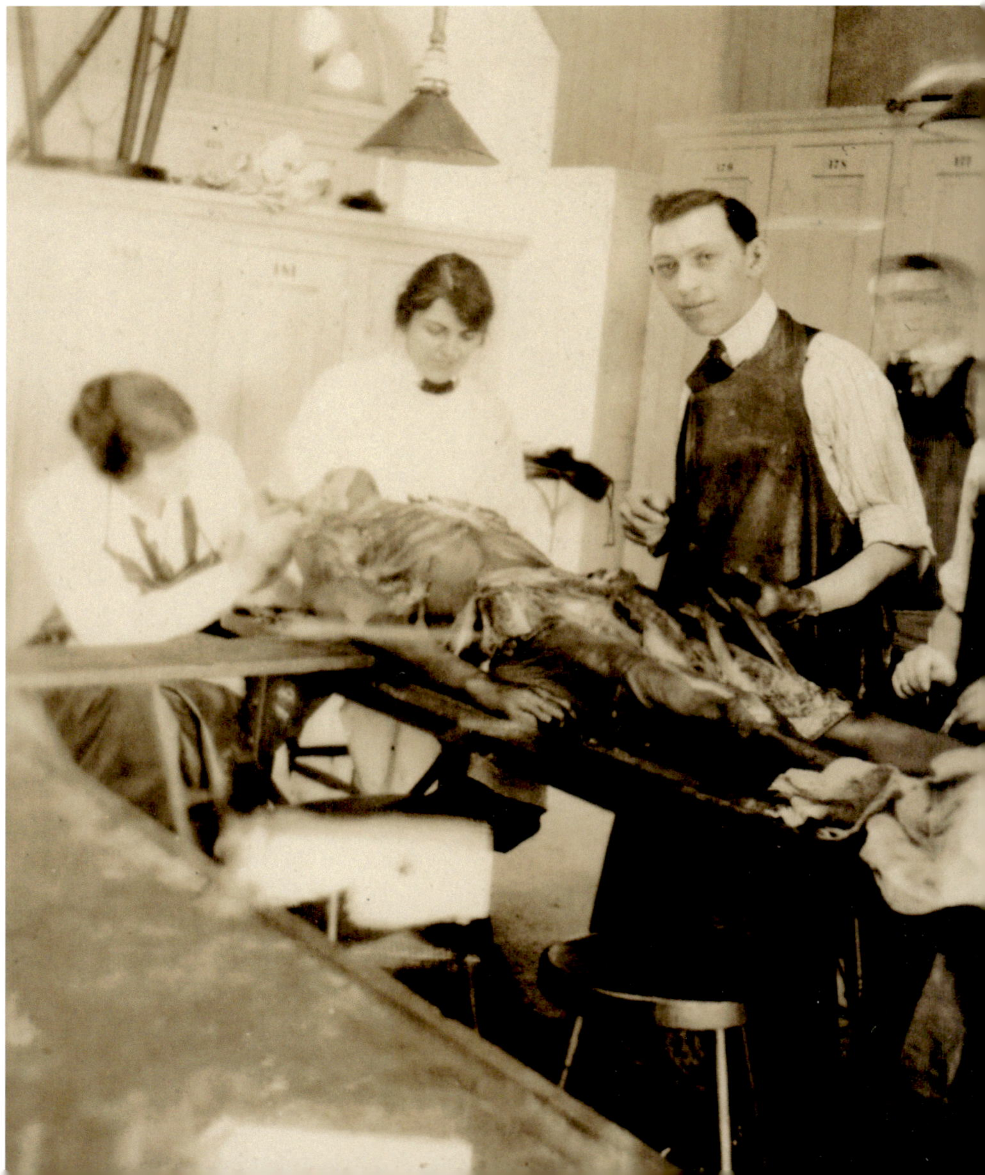

OPPOSITE PAGE:
Plate 3.18

Western University of Pennsylvania, Pittsburgh, Class of 1905

In 1908, the school was renamed the University of Pittsburgh Medical School.

Plate 3.19

Susan California Fitt, MD
Circa 1880

First woman physician to practice in Tennessee.

Plate 3.20

Two Serious Students
Snapshot, circa 1925

Plate 3.21

Women Dissectors as Part of the Team
Snapshot, circa 1905

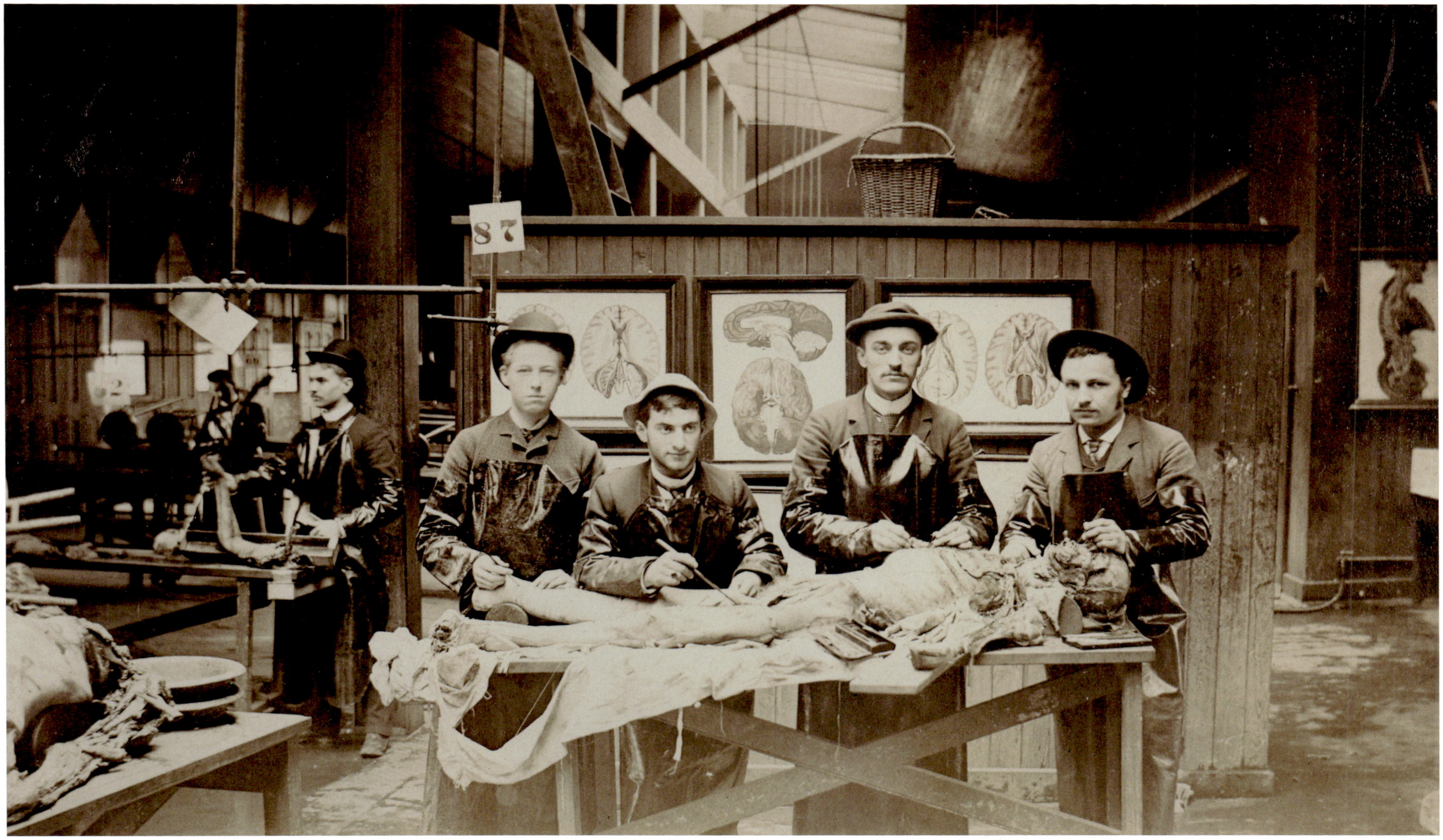

Plate 3.22

Anatomy Laboratory, University of Pennsylvania Medical School
Dr. William Huber (on right), 1884

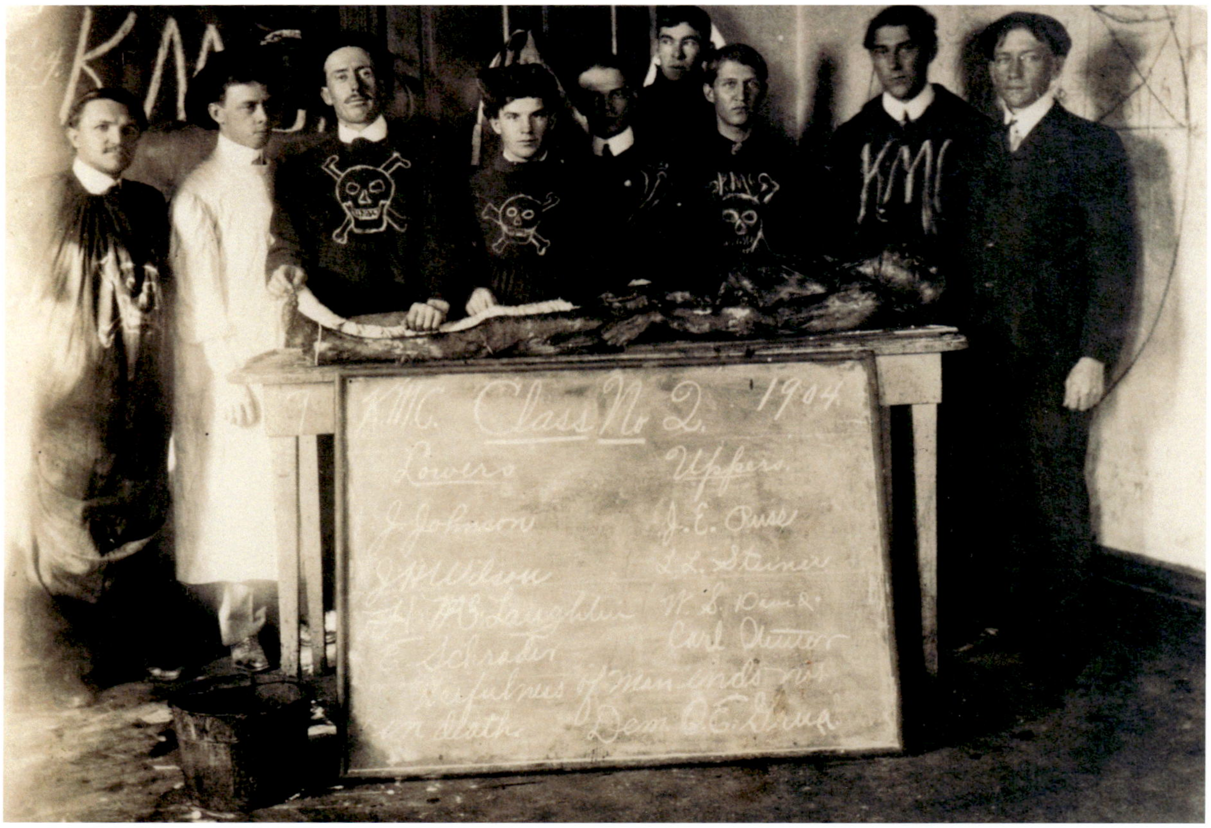

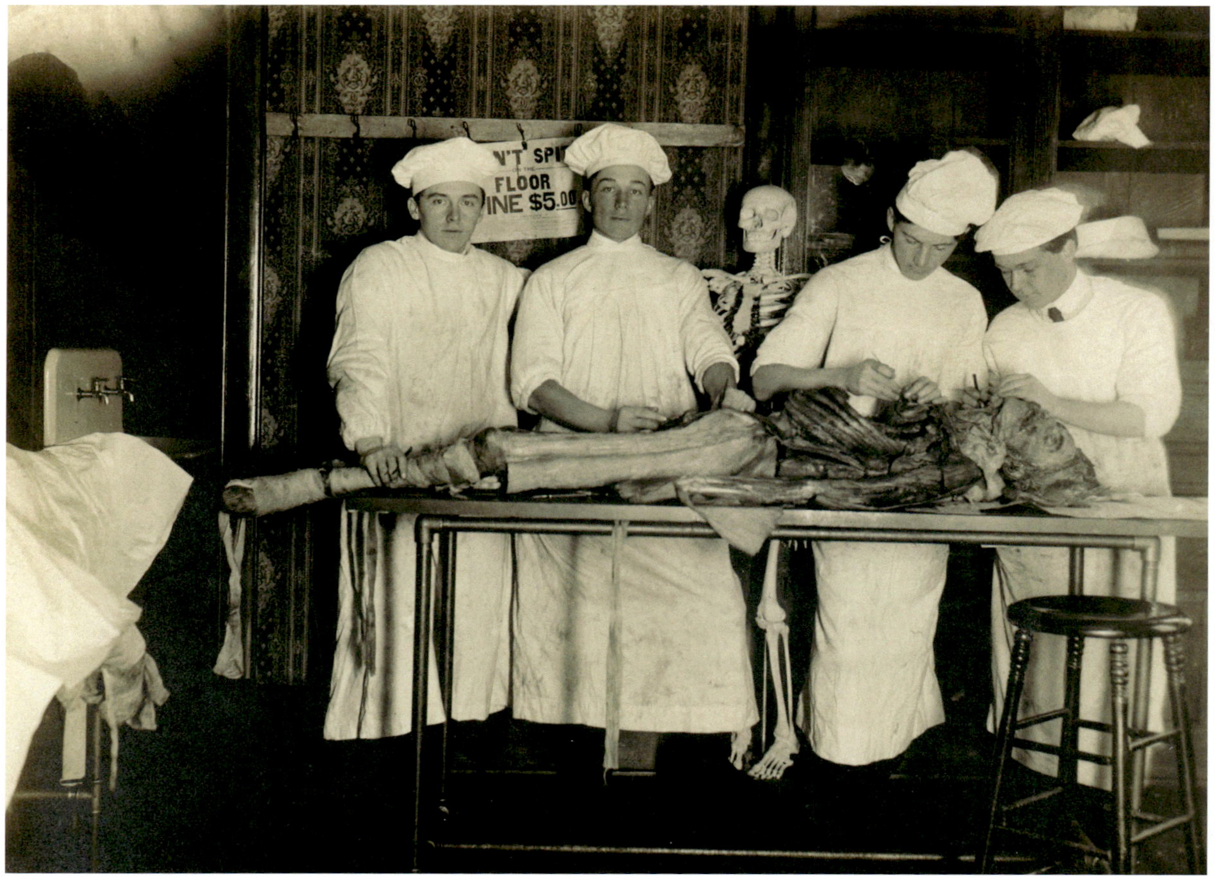

Plate 3.23

"Usefulness of Man Ends Not in Death"
KMC (Probably Kirksville Medical College), 1904

Plate 3.24

"Don't Spit on the Floor, Fine $5.00"
Postcard, 1915

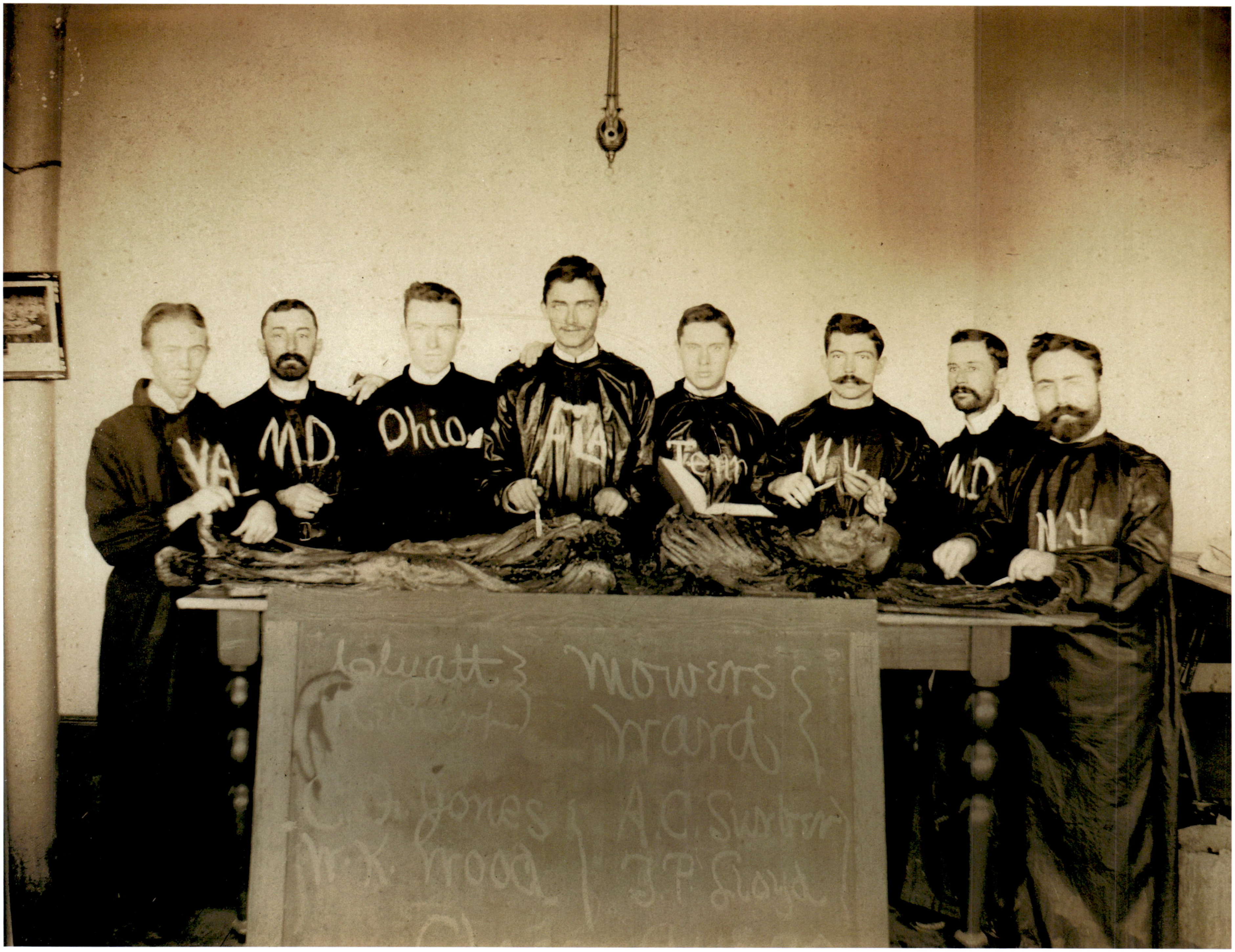

Plate 3.25

Framed Dissection Scene, Hangs on Wall
School with wide range of Eastern States students: VA, MD, OH, FL, TN, NY, MD, NY, circa 1895

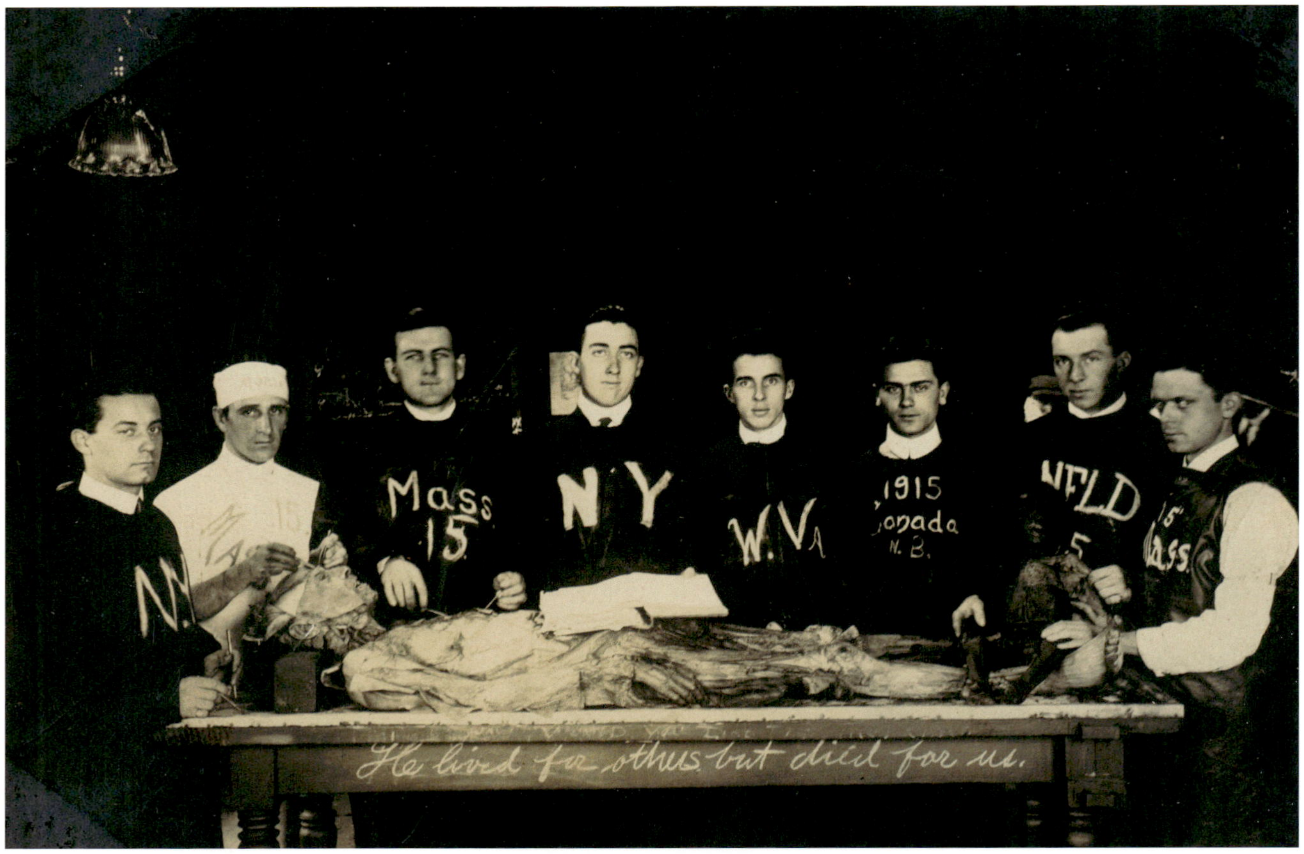

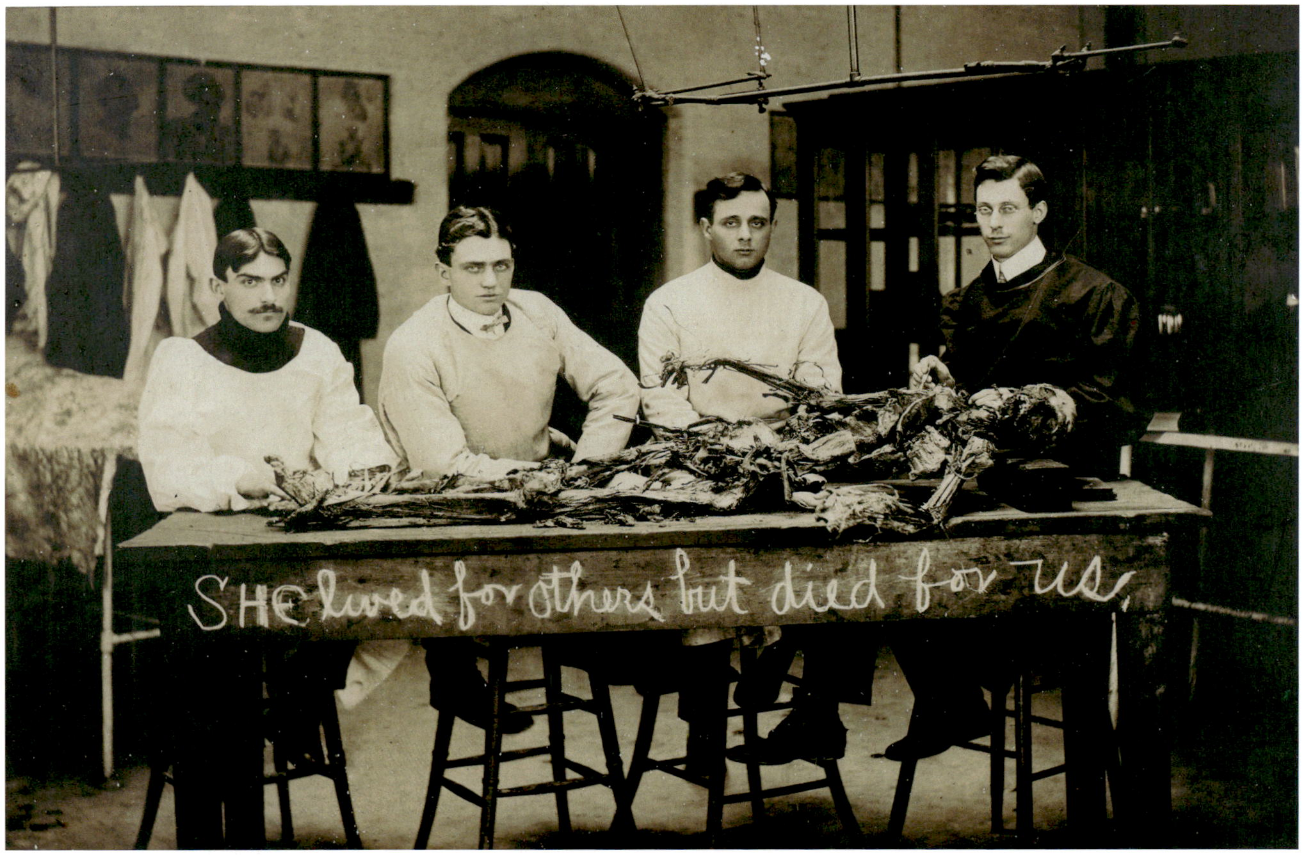

Plate 3.26

"He Lived For Others But Died For Us."
Postcard, Baltimore College of Dental Surgery, 1915

Plate 3.27

"She Lived For Others But Died For Us."
Photographer: H. O. Fuller, Nashville, TN, circa 1917

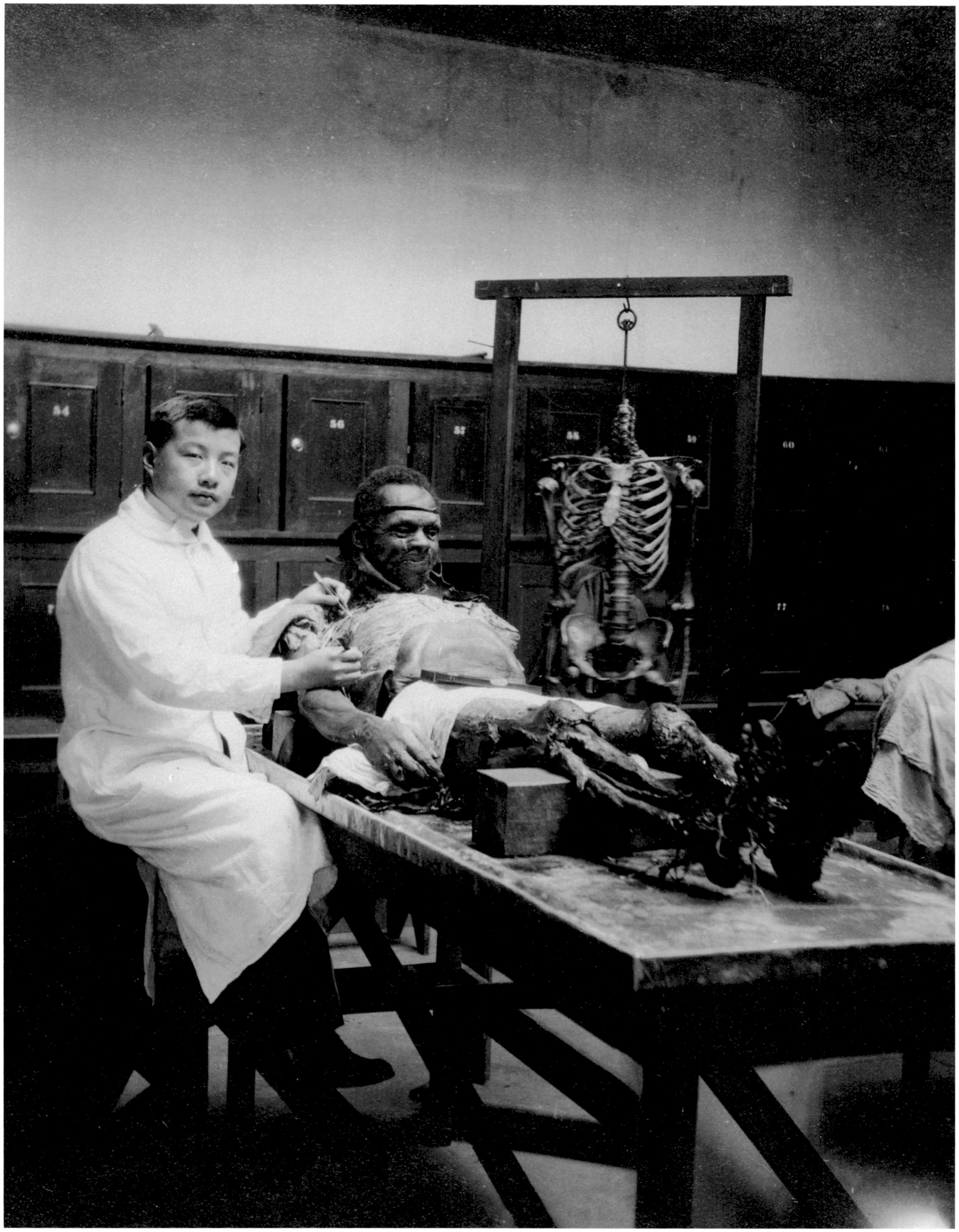

Plate 3.28

Asian Dissector with African American Cadaver
American School, circa 1920

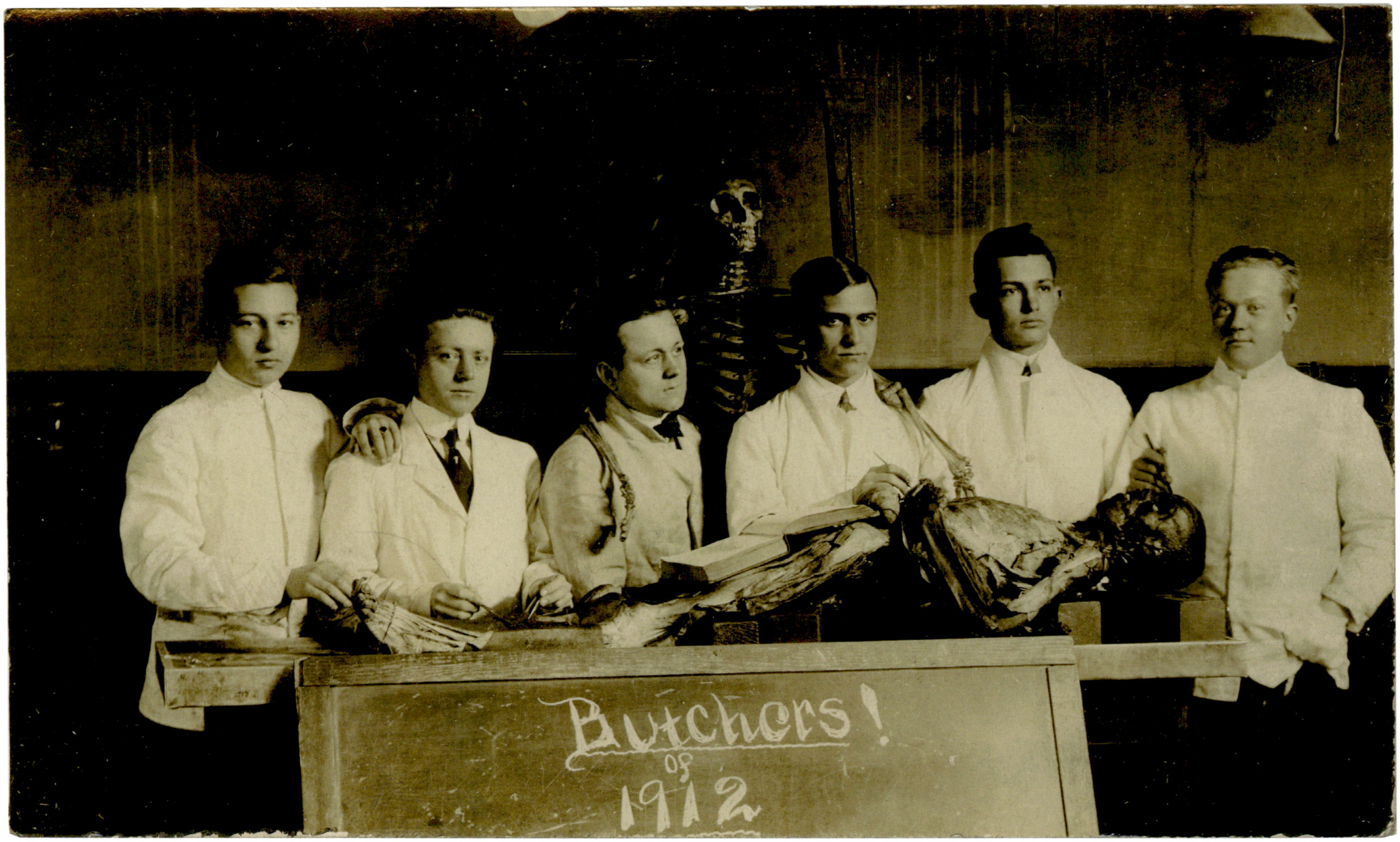

Plate 3.29

The "Butchers! of 1912"

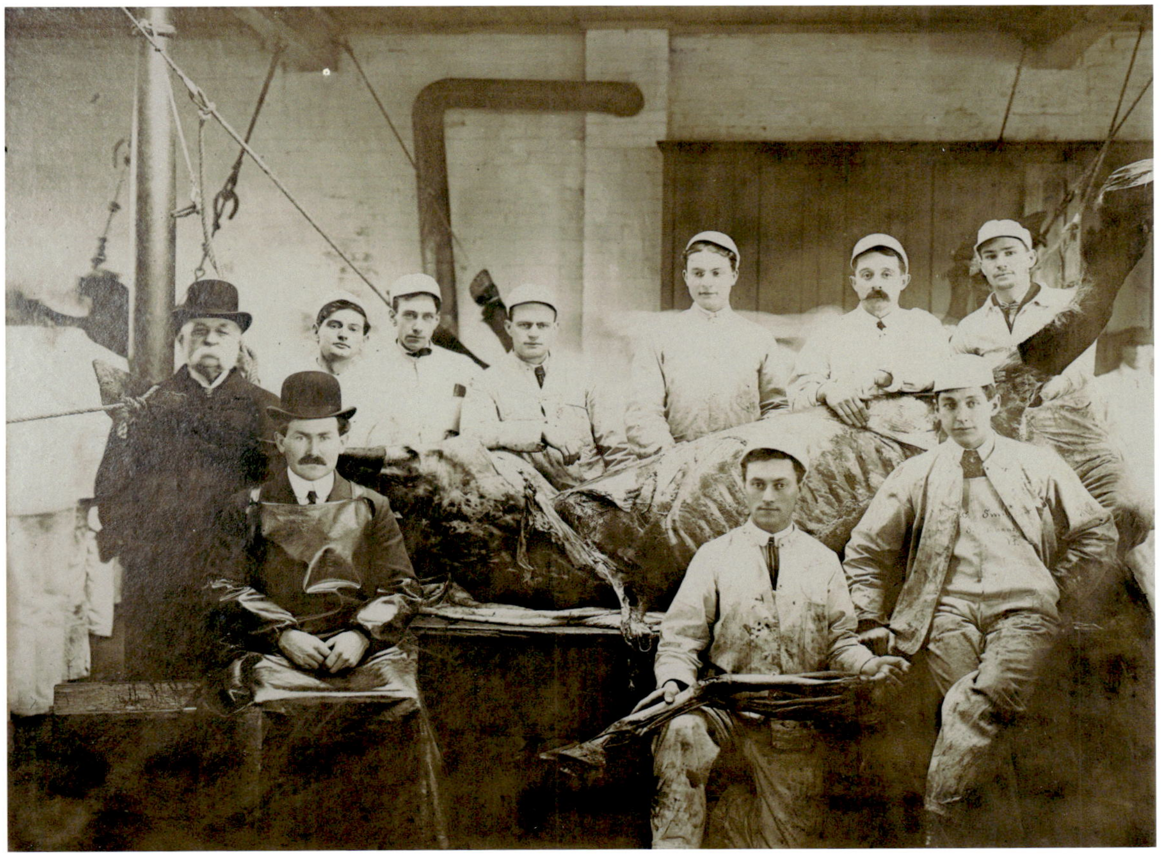

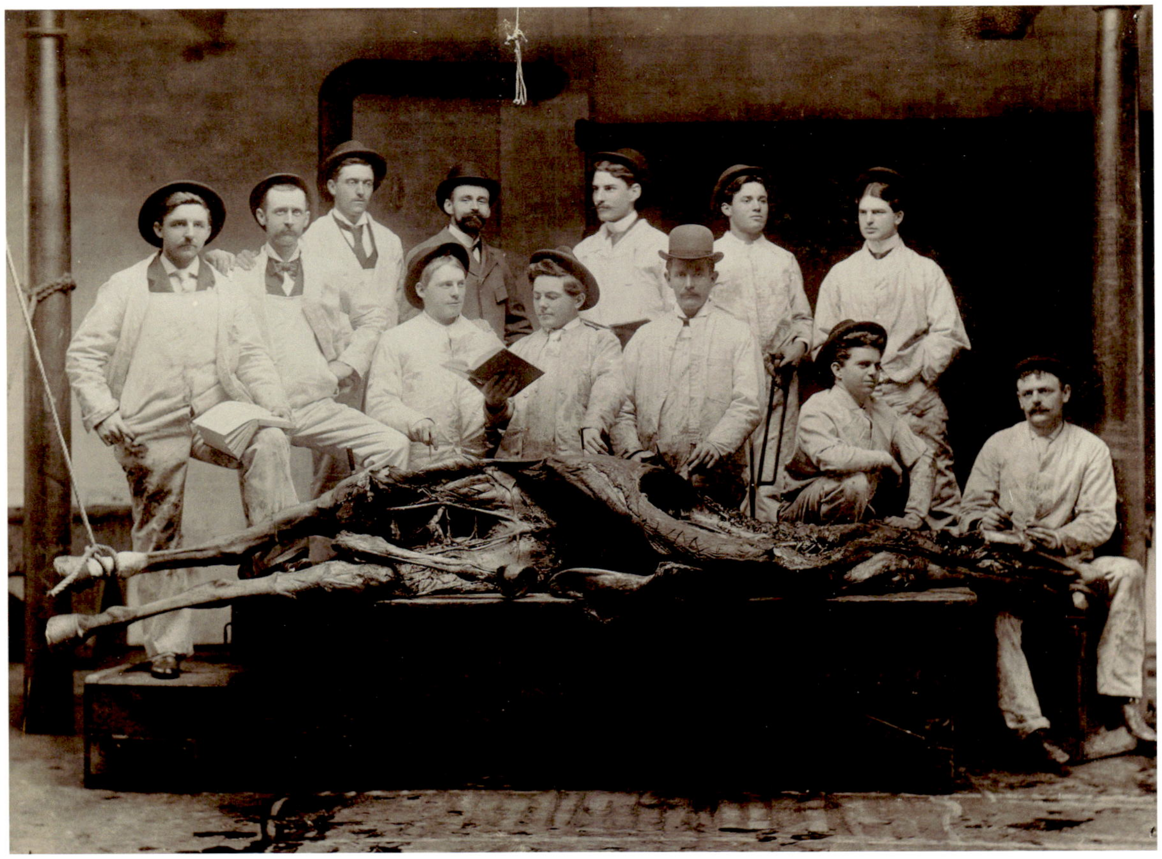

Plate 3.30

Unidentified Veterinary School with Horse
Circa 1905

Plate 3.31

Equine Study Group
Photographer: E. P. Coulmar, 79 Bay St., Toronto, Canada
Circa 1900

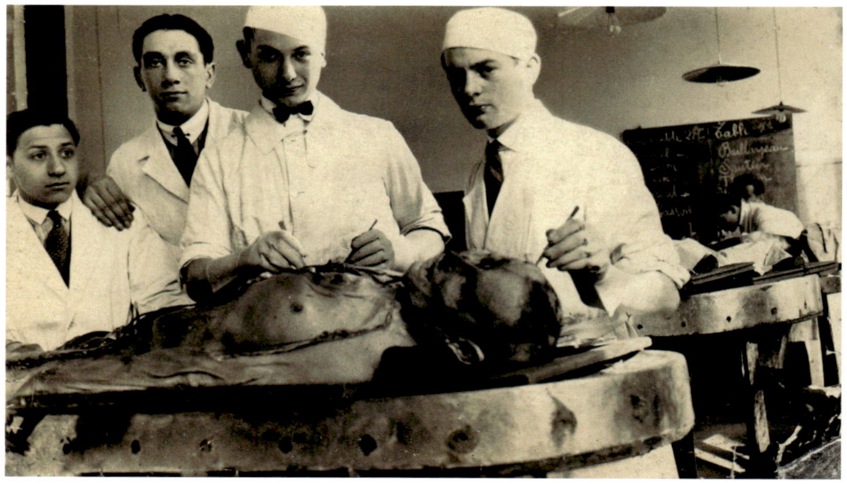

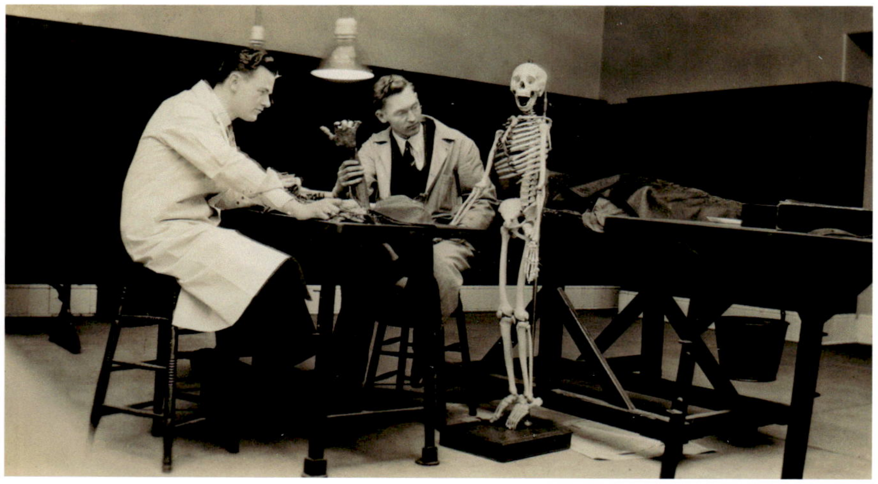

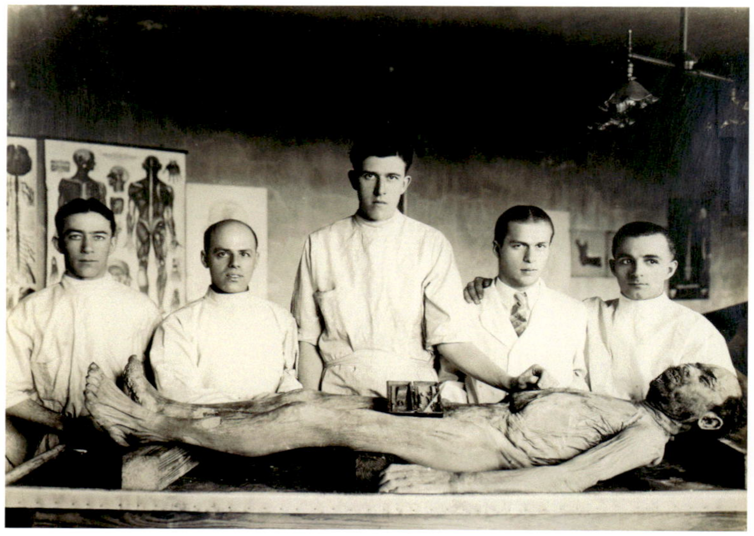

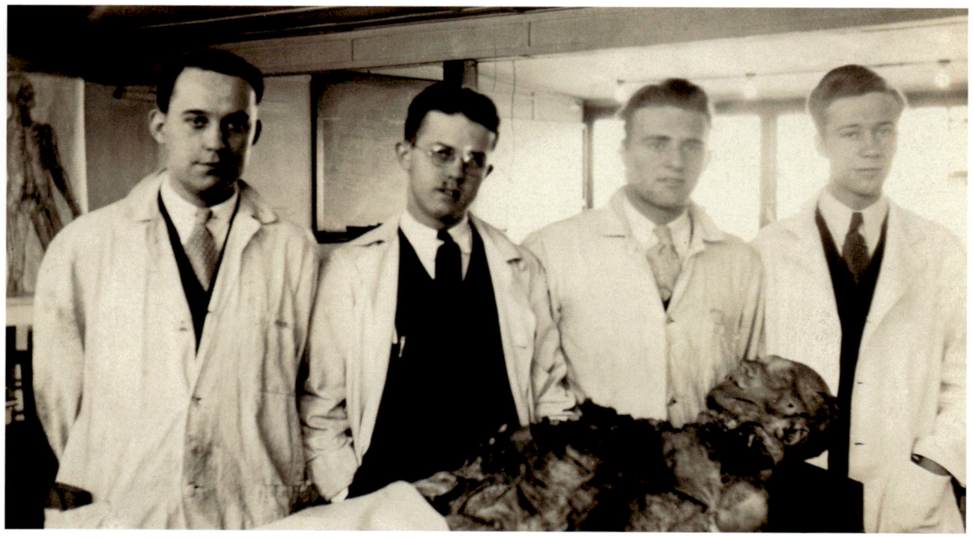

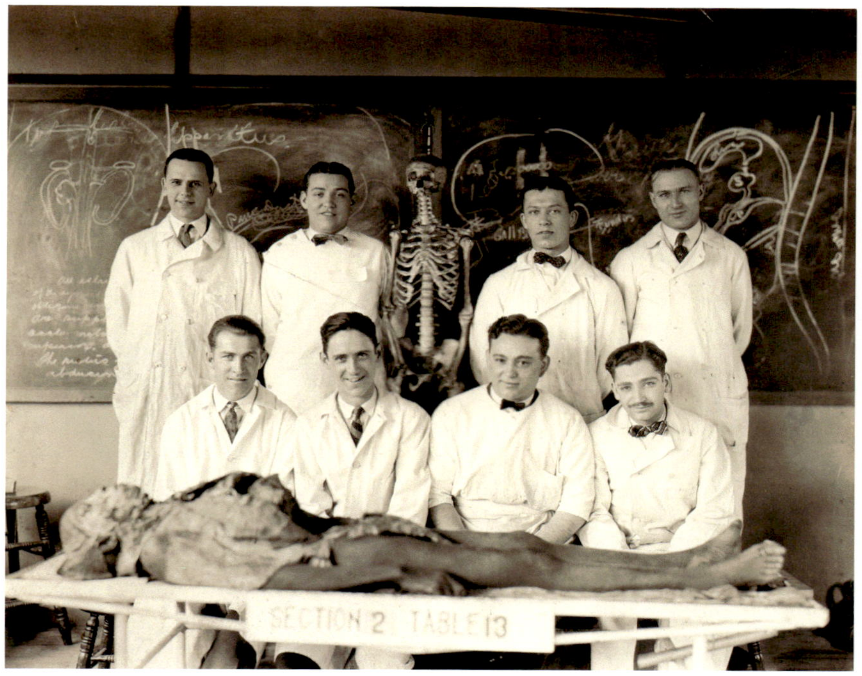
SECTION 2 TABLE 13

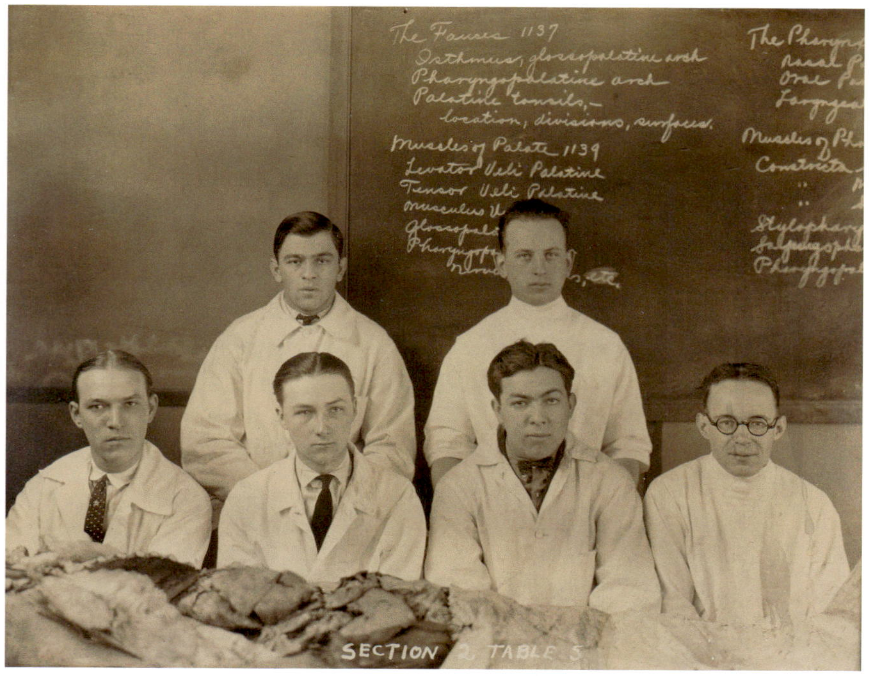
The Fauces 1137
Isthmus, glossopalatine arch
Pharyngopalatine arch
Palatine tonsils,–
location, divisions, surfaces.
Muscles of Palate 1139
Levator Veli Palatine
Tensor Veli Palatine
SECTION 2 TABLE 5

Opposite page:

Plate 3.32

Parisian Dissection Team
Circa 1930

In Europe, white caps and gowns were standard by the first decade of the twentieth century.

Plate 3.33

Dissection Duo with Skeleton
Snapshot, circa 1936

Plate 3.34

Dissection Team with Instruments on Cadaver
Unidentified School, circa 1935

Plate 3.35

Dissection Partners
1932
Left to Right: Robert Succop, Jack McParland, Kenneth Leonard, and Charles Oakley

Plate 3.36

Class with Skeleton and Cadaver
Section 2, Table 13, Unidentified School, circa 1940

Plate 3.37

Pittsburgh School of Dentistry
Section 2, Table 5, 1923

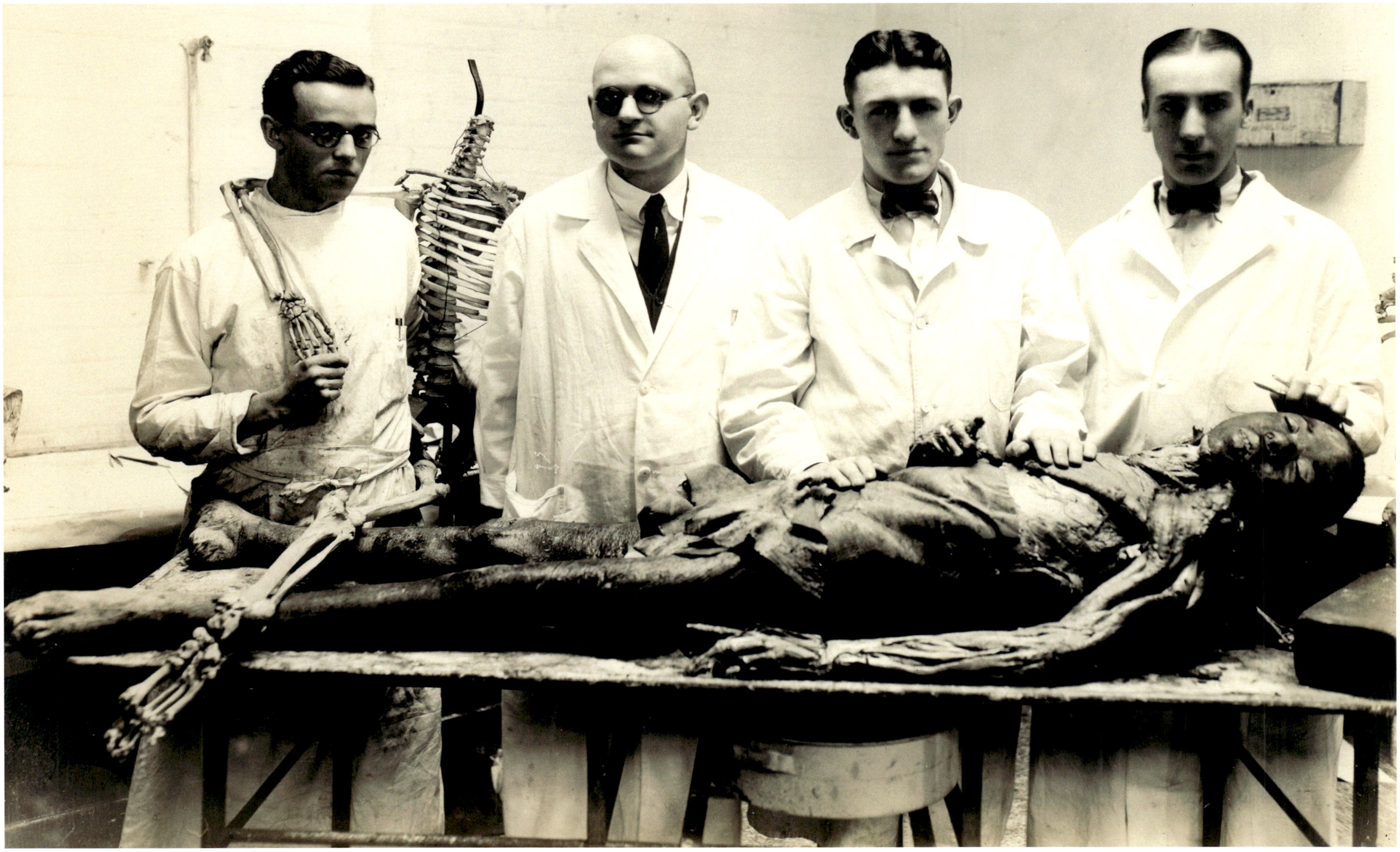

Plate 3.38

Dissection Team with Headless Skeleton Draped Across Cadaver
Gallows humor in a formal dissection scene, circa 1940

Plate 3.39

Proudly Signed Dissection Scene
Circa 1930

Plate 3.40

Class 25, circa 1908
Circa 1908

Casual dress is indicative of a small proprietary medical school.

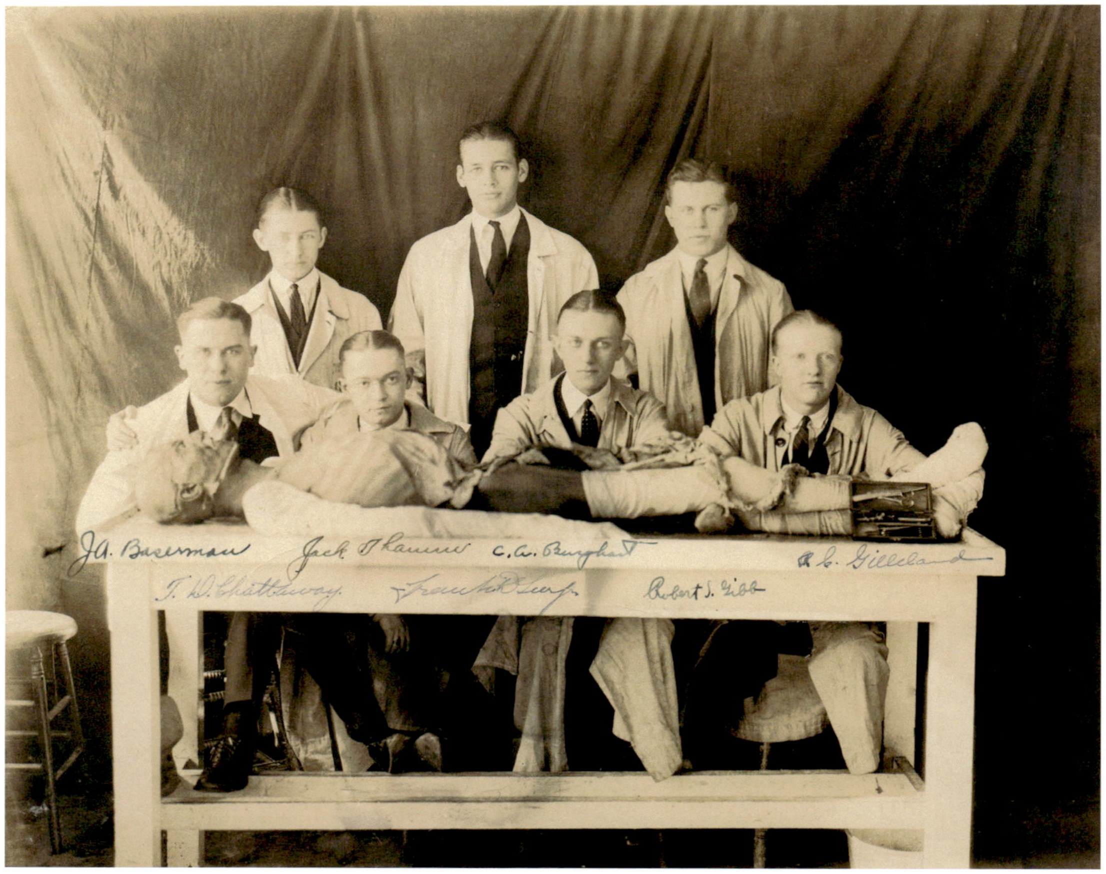

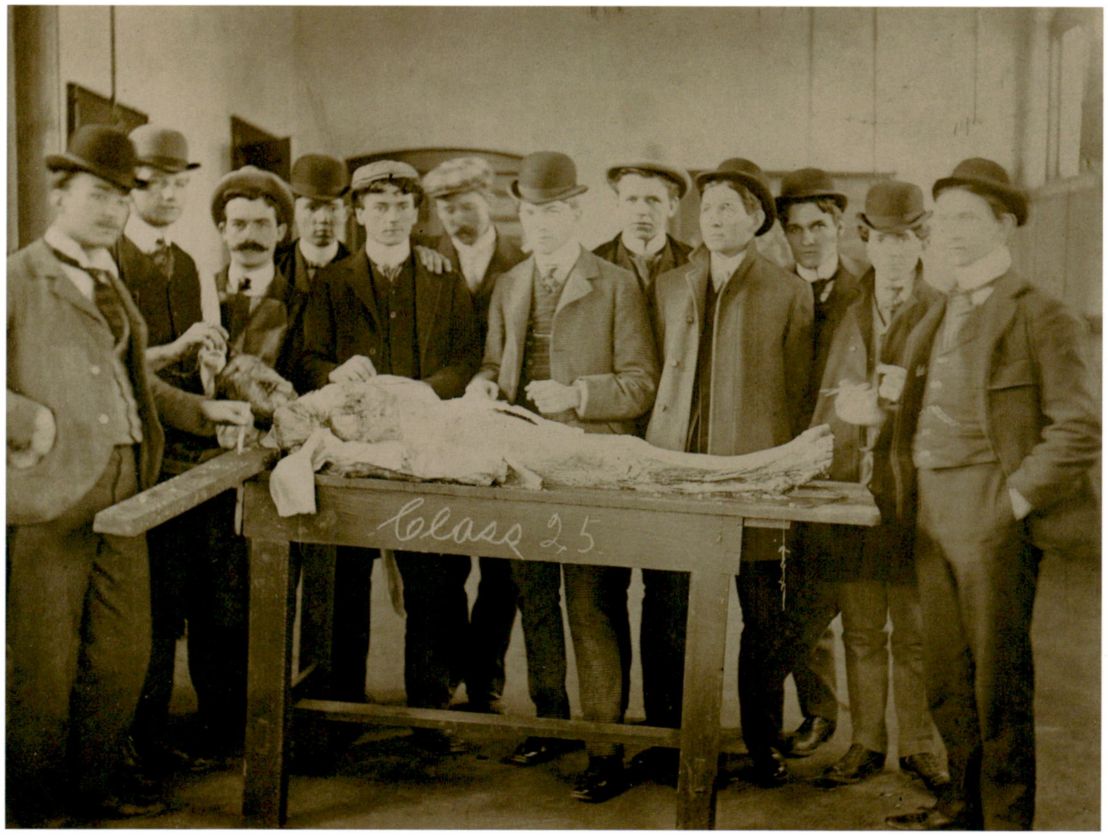

OPPOSITE PAGE:
Plate 3.41

Unusual Positioning of Cadaver Legs for Dissection
College of Physicians and Surgeons, Columbia University, 1905

Plate 3.42

Parisian Medical Students Experiment on a Dog
La Sorbonne, Physiology Dissection Room, 1910

Plate 3.43

Parisian Medical Student
Signed, circa 1940

Plate 3.44

Dissection Duo
Circa 1920

The comradeship of working on a cadaver was a special bonding experience.

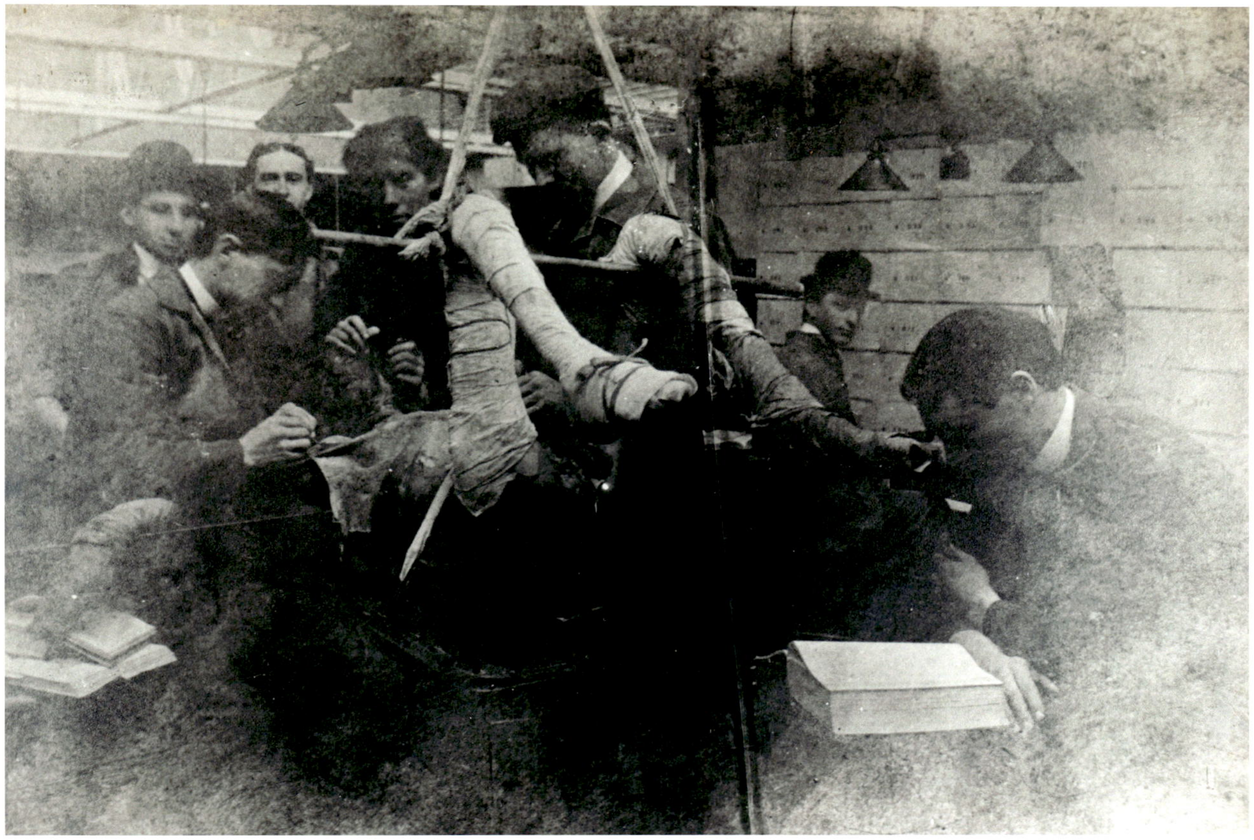

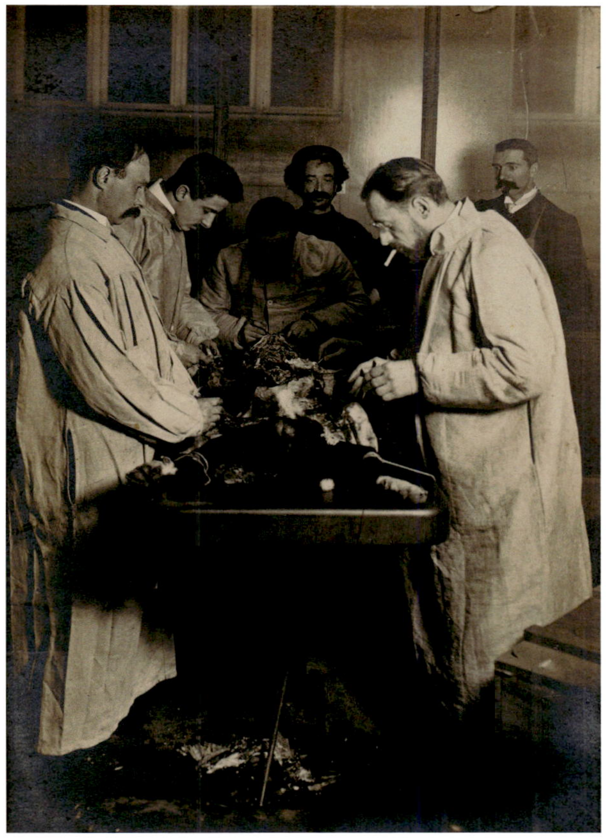

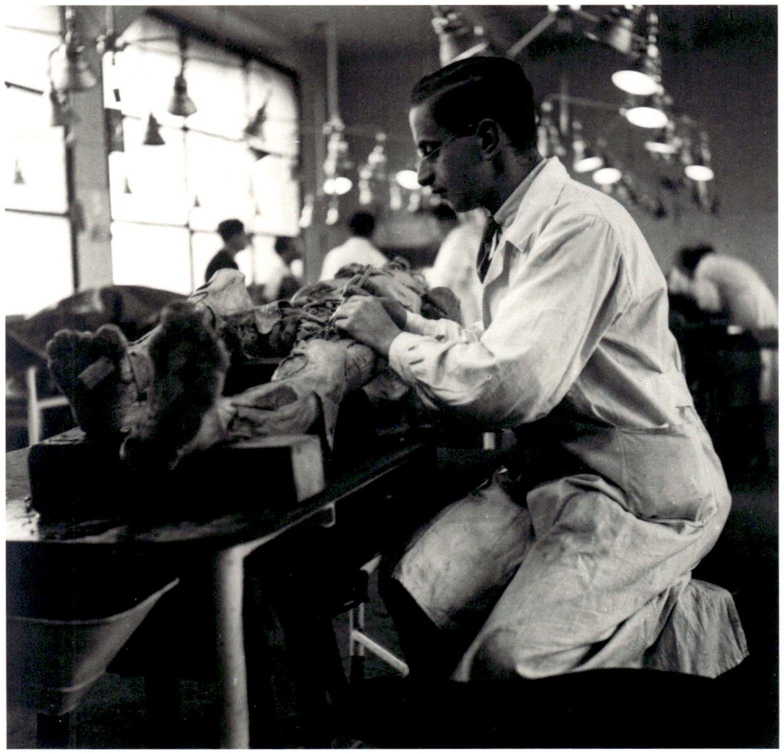

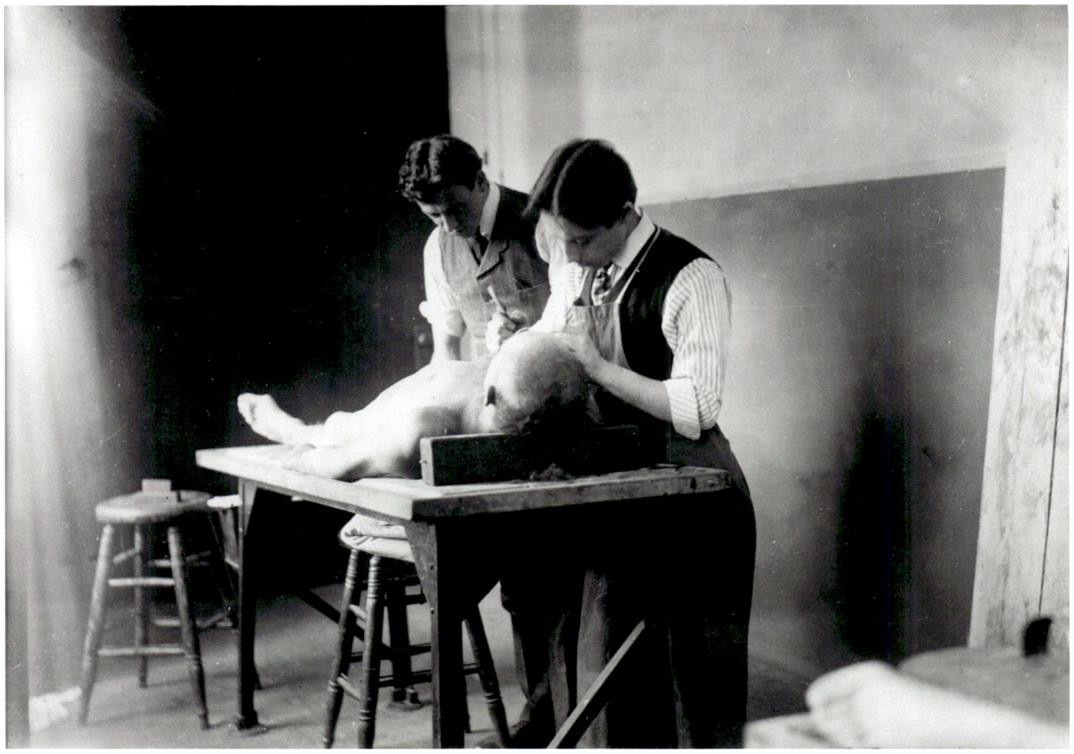

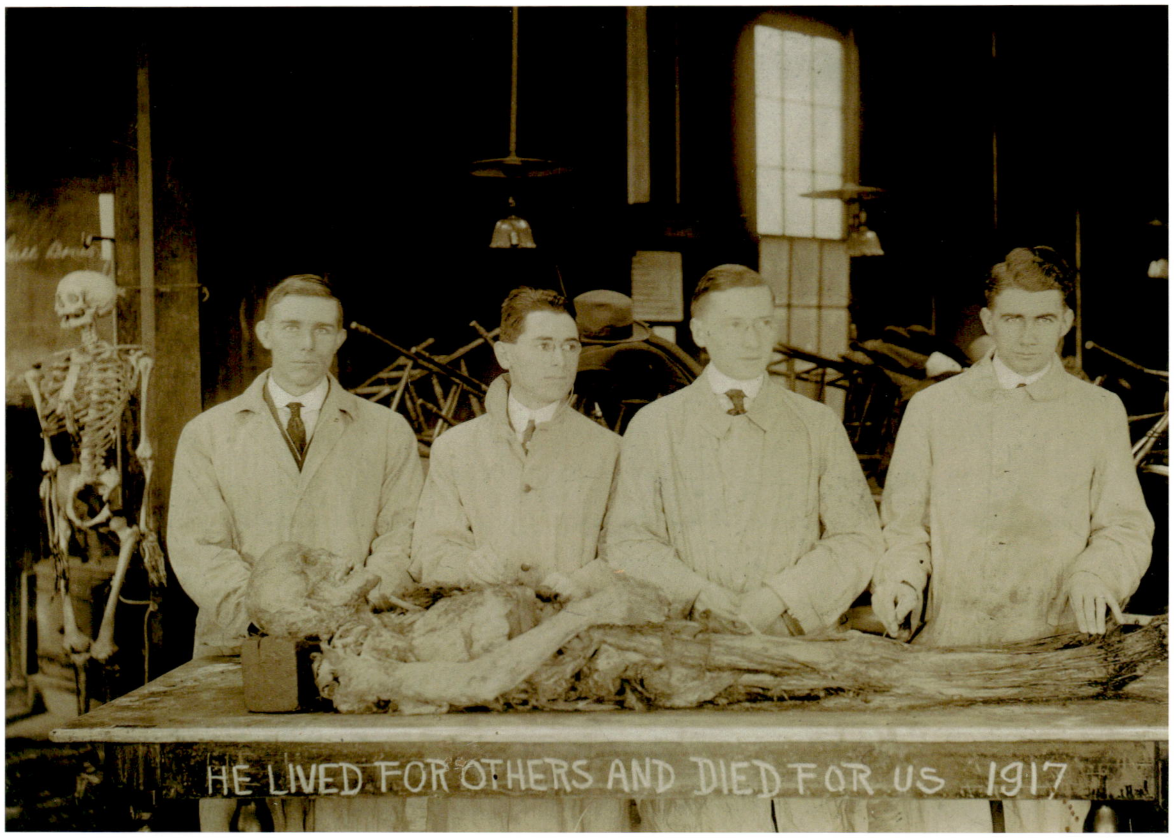

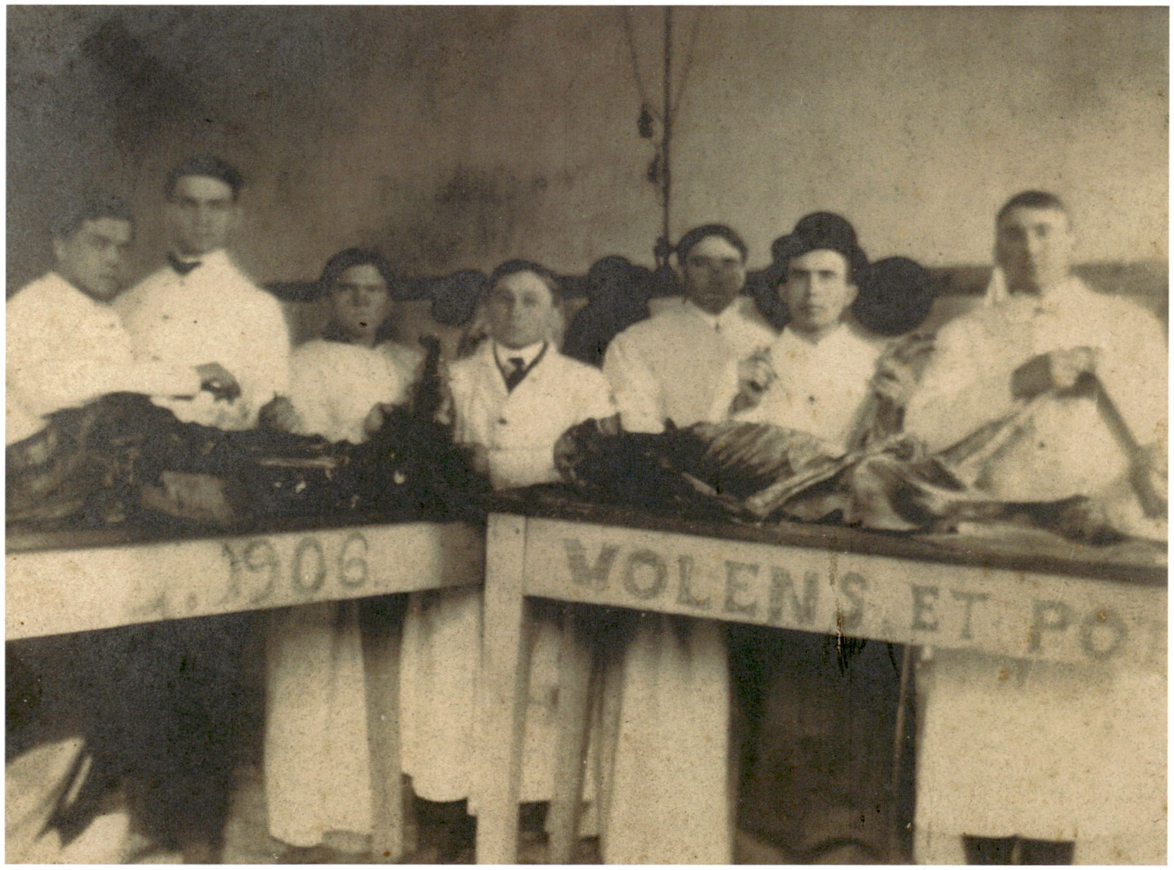

Plate 3.45

"He Lived For Others And Died For Us"
1917

Plate 3.46

"Volens et Potens" (Willing and Able)
1906

A popular slogan that represented the student's attitude instead of the cadaver's contribution.

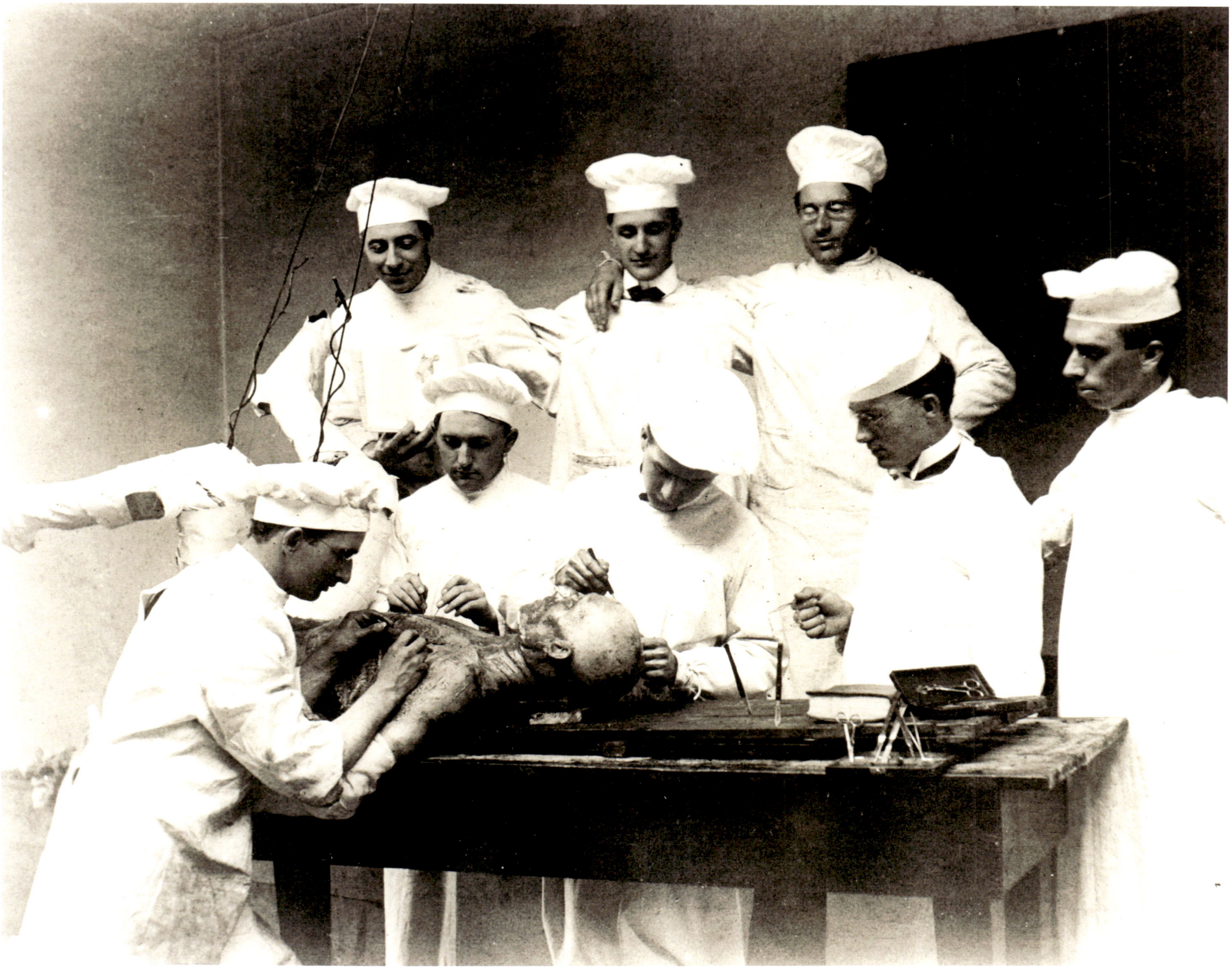

Plate 3.47

Hamline University School of Medicine, St. Paul, MN
1905

Theodore Paulson on the far right. A leg suspension cadaver photo.

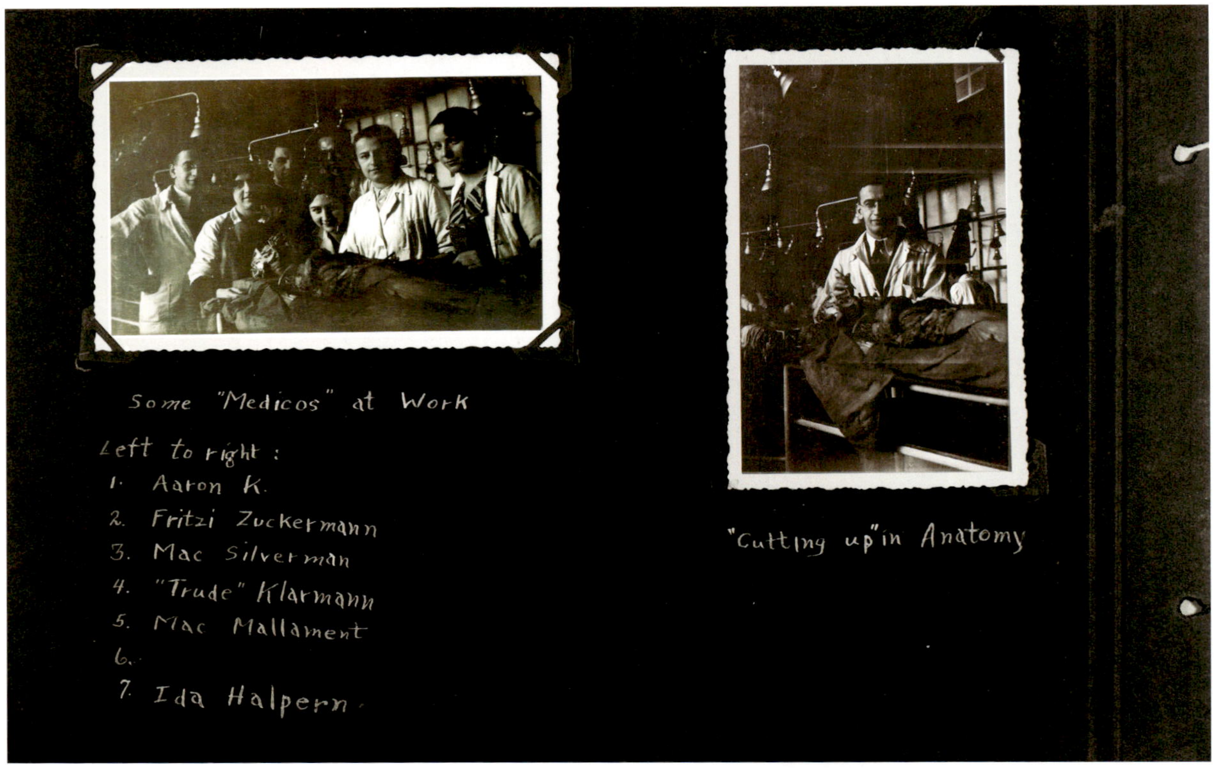

Plate 3.48

"Some 'Medicos' at Work", Vienna Medical School
Album Page from American Post-Graduate Medical Student, 1934

1934 was the last year American students attended anatomy classes in Vienna. Jewish students were no longer welcome in Germany or Austria. The students witnessed the civil uprising, and photographs of the conflict are in the album.

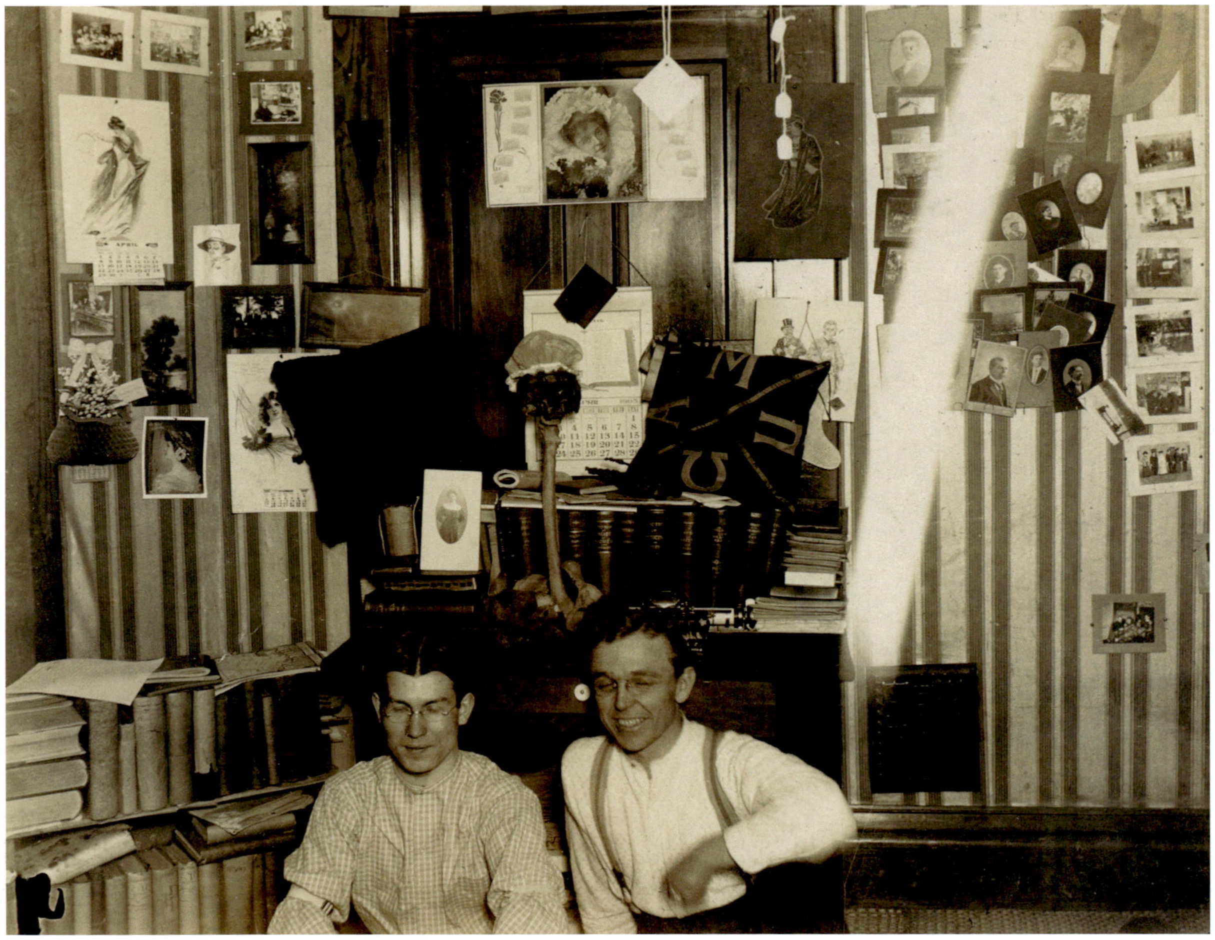

Plates 3.49 – 3.55

A Group of Dissection Scene Photographs of and by Dr. Howard Brundage
Columbus, OH, 1905–1915

Howard Brundage was an avid photographer documenting his life in medicine beginning from 1905. His collection is unusual as it illustrates the life and career of a city physician. Here he is in his room at Ohio Medical University with a dissection class photograph on the wall behind him. Brundage is seated on the right. Brundage took photographs of his classmates and also appears in several scenes as a instructor. His photograph of rearranged cadaver parts is an unusual 'artistic' presentation. Brundage is also seen in the chapter on autopsy, as he is the pathologist busy at work. He continued his photographic endeavors when he worked at the Protestant and St. Francis Hospital taking photographs on the ward and of hospital staff.

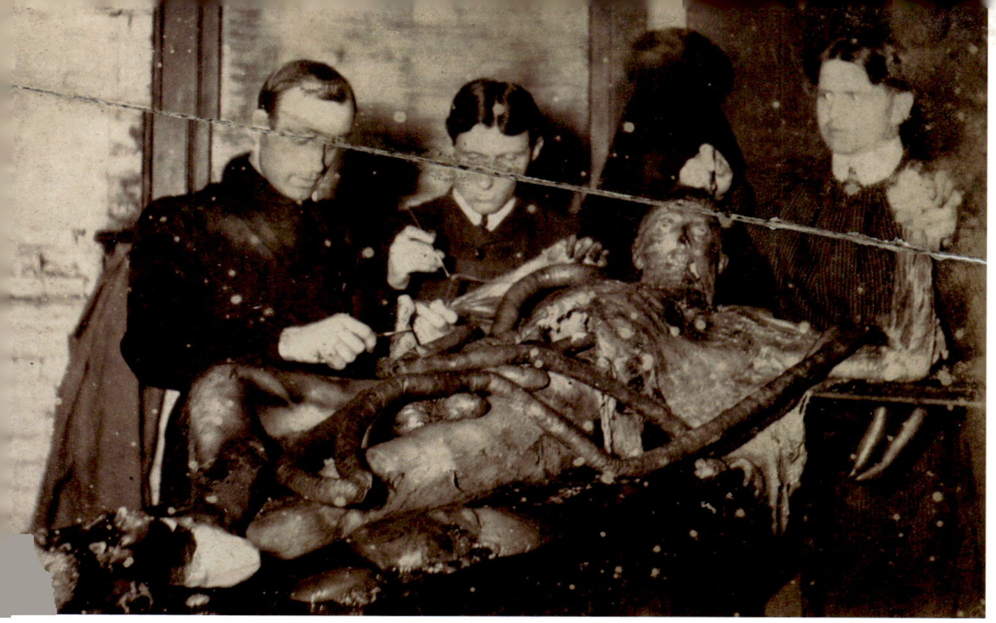
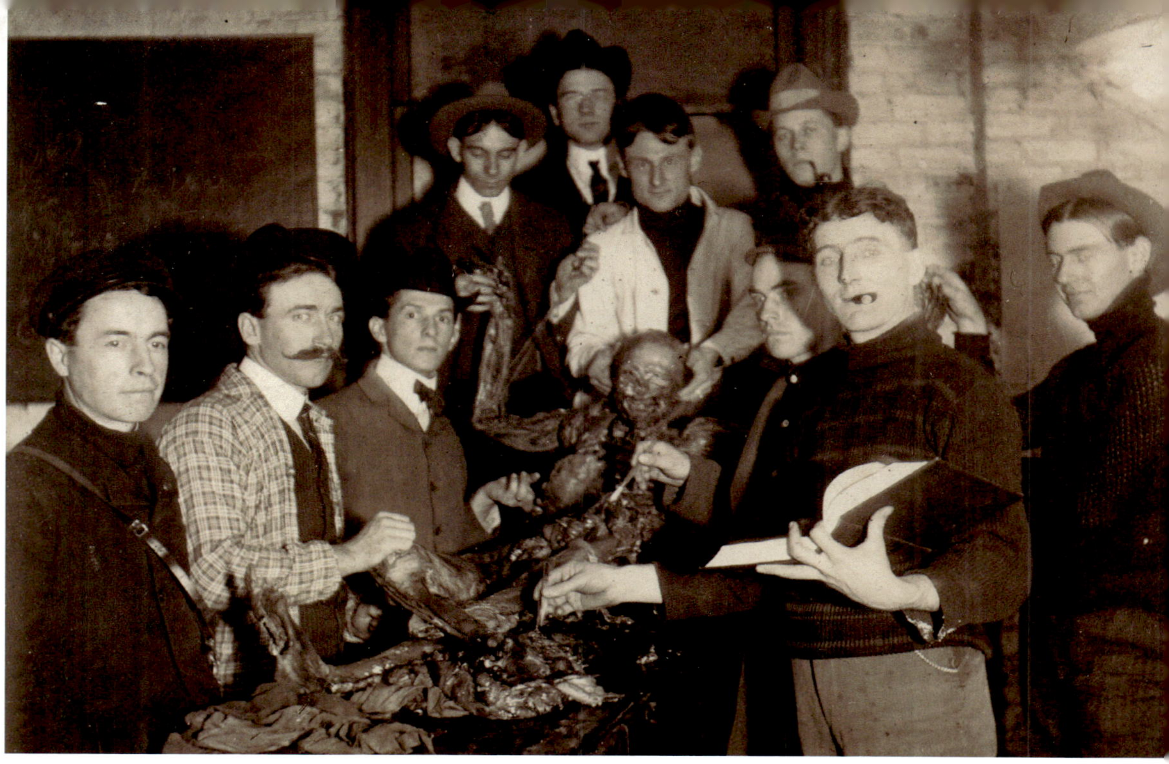
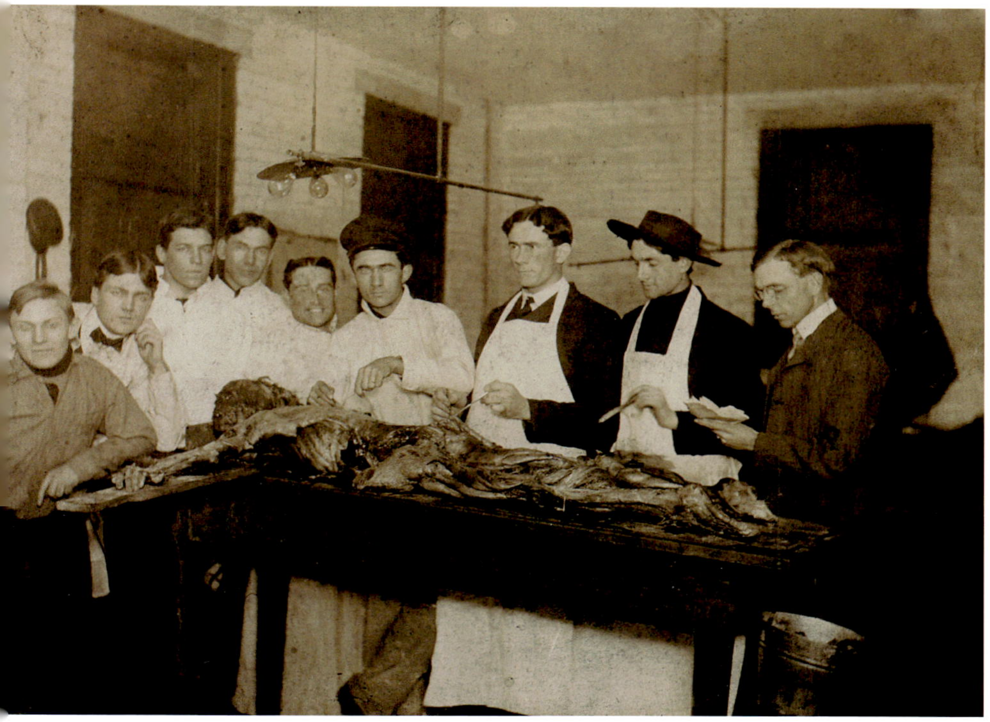
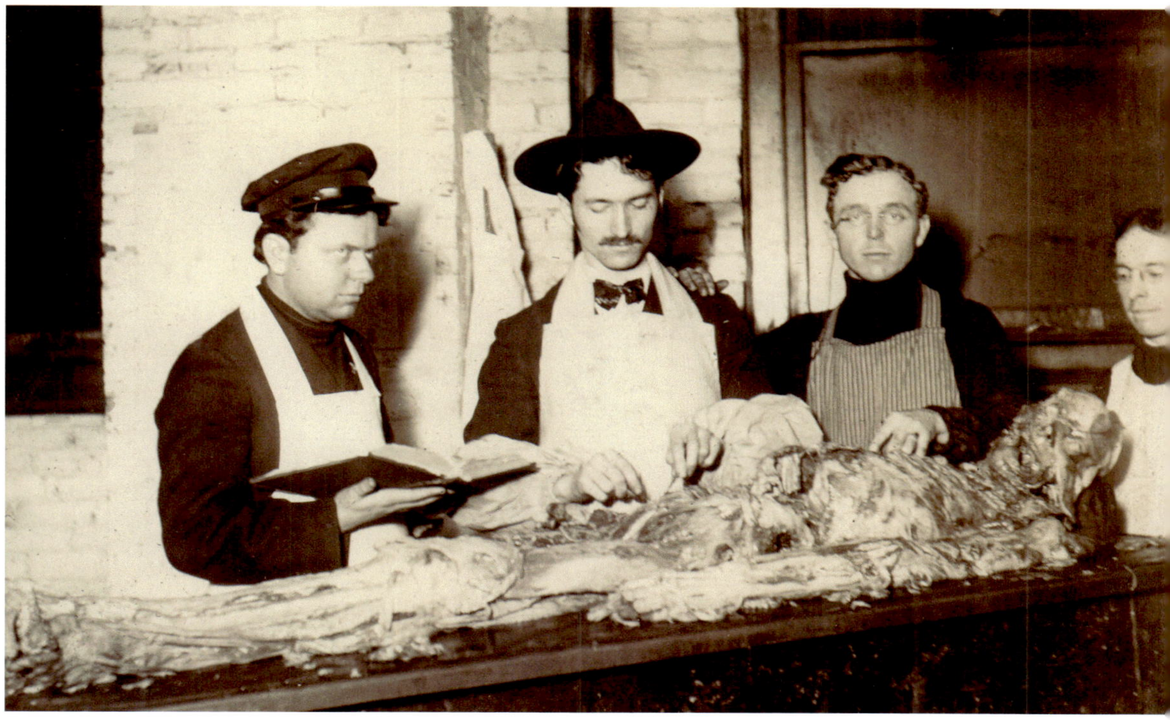
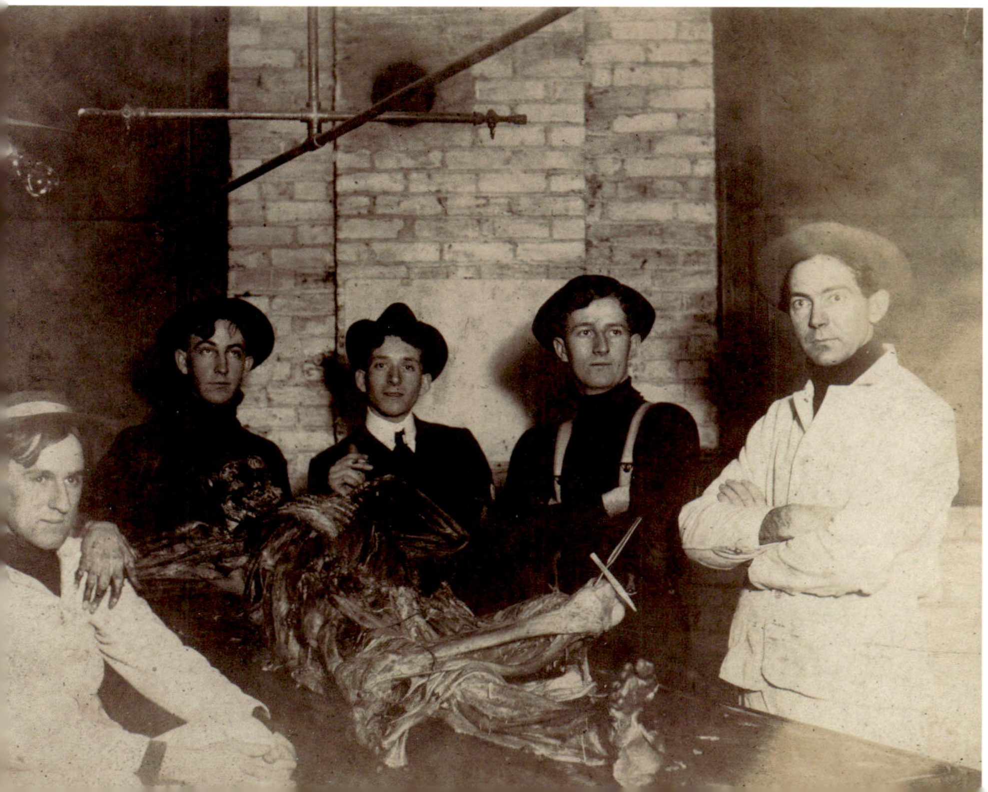
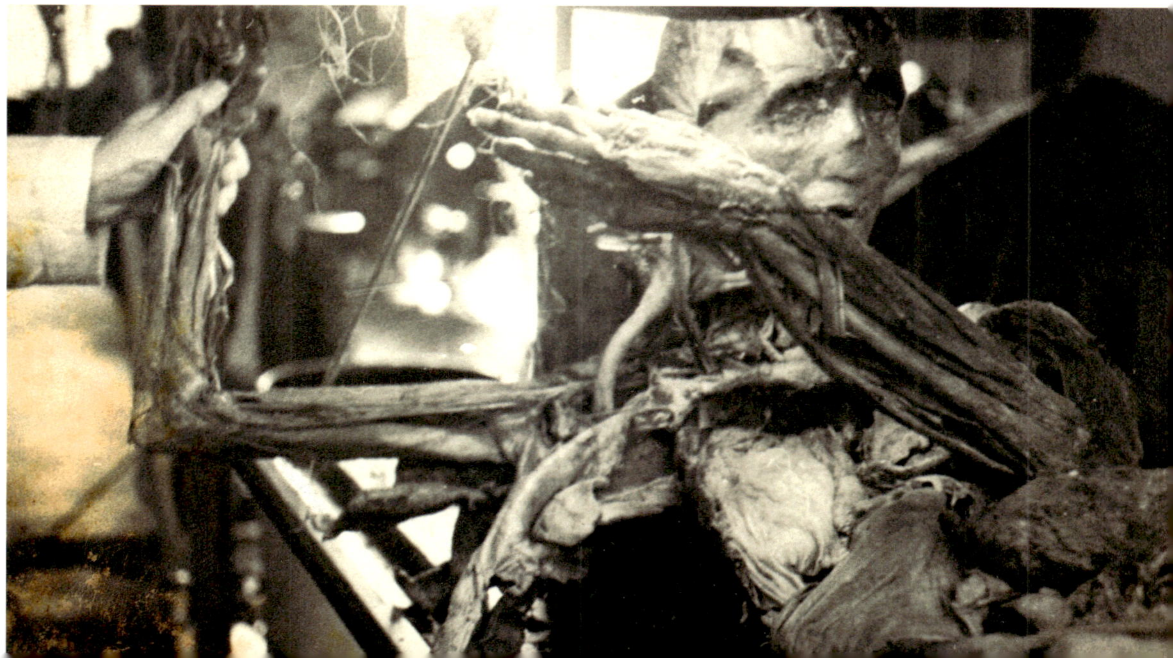

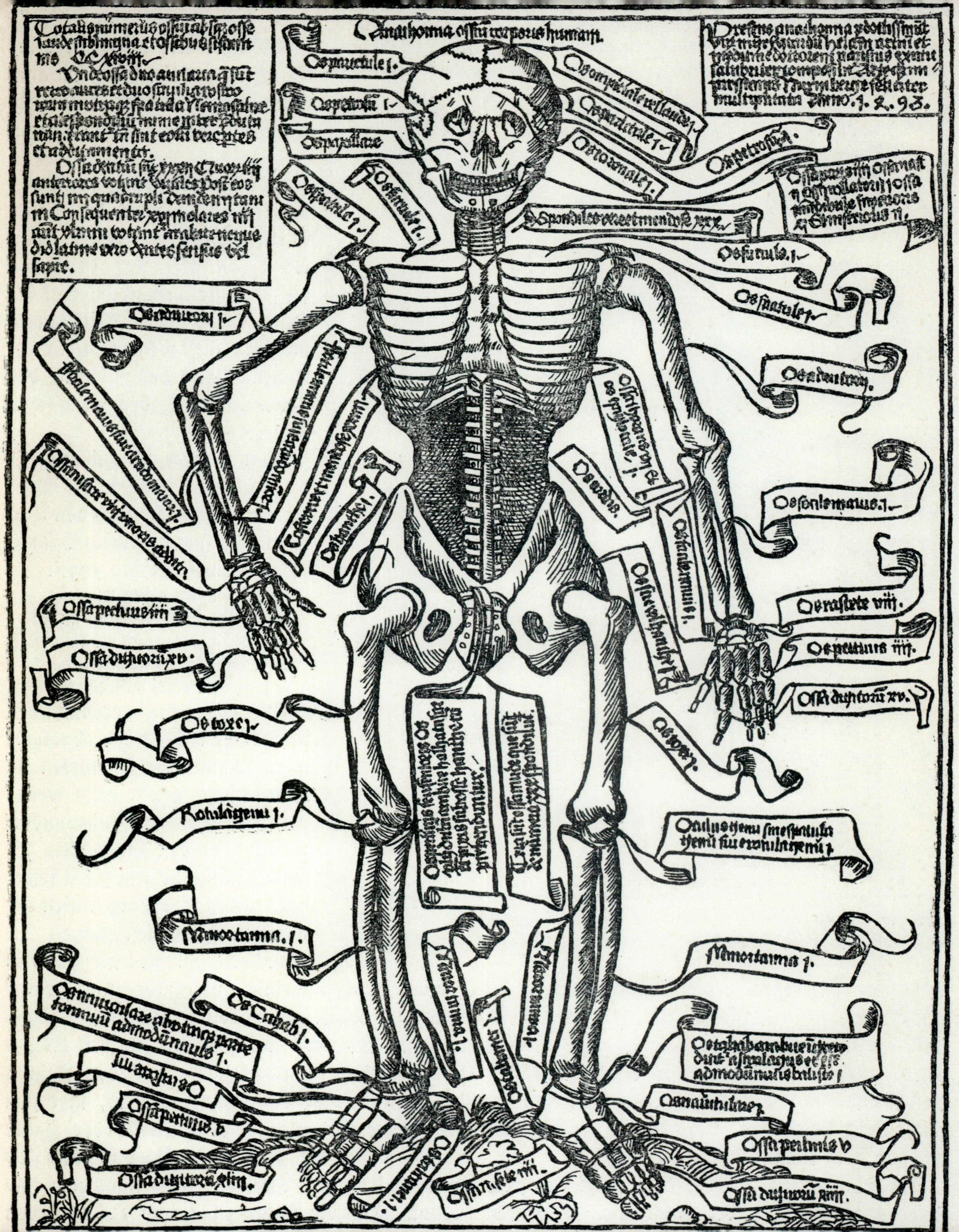
Anathomia ossiu corporis humani.

CHAPTER FOUR

Special Bone Studies

Anatomical & Pathological

The photographs of pathological bone anatomy document the extremes of disease and trauma. In the nineteenth and early twentieth centuries, patients endured many common and unusual conditions. Many of the conditions have since been conquered, and the photographs are a legacy that vividly portrays the advances of medicine. Bony deformities were commonplace due to nutritional, inflammatory, infectious, and congenital diseases. Fractures were one of the few injuries physicians were confident in treating. However, compound fractures, those with an open wound, often resulted in infections, which frequently destroyed their best efforts as osteomyelitis often set in. If the bone infection didn't kill the patient by a complicating blood borne sepsis, the infection created abnormal bone overgrowth and deformities. The study of bony abnormalities was a very popular aspect of medical research, as all physicians had patients with some type of bony deformity. Vitamin deficiencies or congenital skeletal diseases affected the young, arthritis, the elderly, and perhaps more importantly, tuberculosis and syphilis affected all. These were among the most common diseases of the time that resulted in bone destruction.

For centuries, physicians collected normal and pathological skeletons, skulls, and bones. The objects formed the basis of noted doctors' personal collections, as well as the core of medical school and medical museum collections. With the invention of photography, physicians and institutions had the objects in their collection photographed. It also dramatically allowed expansion of bone collections by simply photographing the object, and not keeping the specimen, unless it was extremely unusual. As was often the case with medical photography, the job was routine work and as time went on some photographers came to create artistic photographs. Such was the case with Reed Brockway Bontecou, MD, who photographed the extracted bones of his Civil War patients.

This chapter concentrates on showcasing some of the most significant illustrated publications of the nineteenth century that published skeletal photographs. One of the most interesting is the *Catalogue des pièces du museé Dupuytren, Atlas du du museé Dupuytren,* by M. Houel, conservator des collections de la Faculté de la Médicine de Paris. Published in 1879, the atlas presented some of the most fascinating skeletal preparations ever photographed. The catalog's emphasis is on skeletons with a few examples of special dissections and pathological specimens. The section on skulls shows the common diseases that destroyed the skull, especially infections such as syphilis. The array of images of deformities illustrates the role of trauma and infections common in the early nineteenth century. The most significant nineteenth-century American publication with skull and bony studies is *Photographs of Surgical Cases and Specimens Prepared by Direction of the Surgeon General.* This publication consists of eight volumes, each with fifty photographs. Issued between 1869 and 1881, the emphasis is on the most unusual Civil War bone injuries. Bone specimens make up the majority of the 400 photographs in the eight volumes. Skeletal deformities are documented, as are skull wounds by Indian arrowheads. The photographs chosen represent some of the most interesting cases as well as artistically rendered images. The non-fatal spinal-area wound suffered by President James Garfield in 1881 is shown, and the President's death by physician maltreatment is explained.

Among the photographs presented are rare images of patients who had bones removed surgically and were photographed post-operatively with their bones. These images represent a significant chapter in the history of medical photography from the 1864–1878 time period. The fact that a diseased bone could be removed and the patient survive the operation and heal without complication was the result of antiseptic surgical principles established after 1867. The chapter also includes images of skeletal positioning studies popular in the mid-twentieth century that were taken to create guidelines for people working at desks or playing sports.

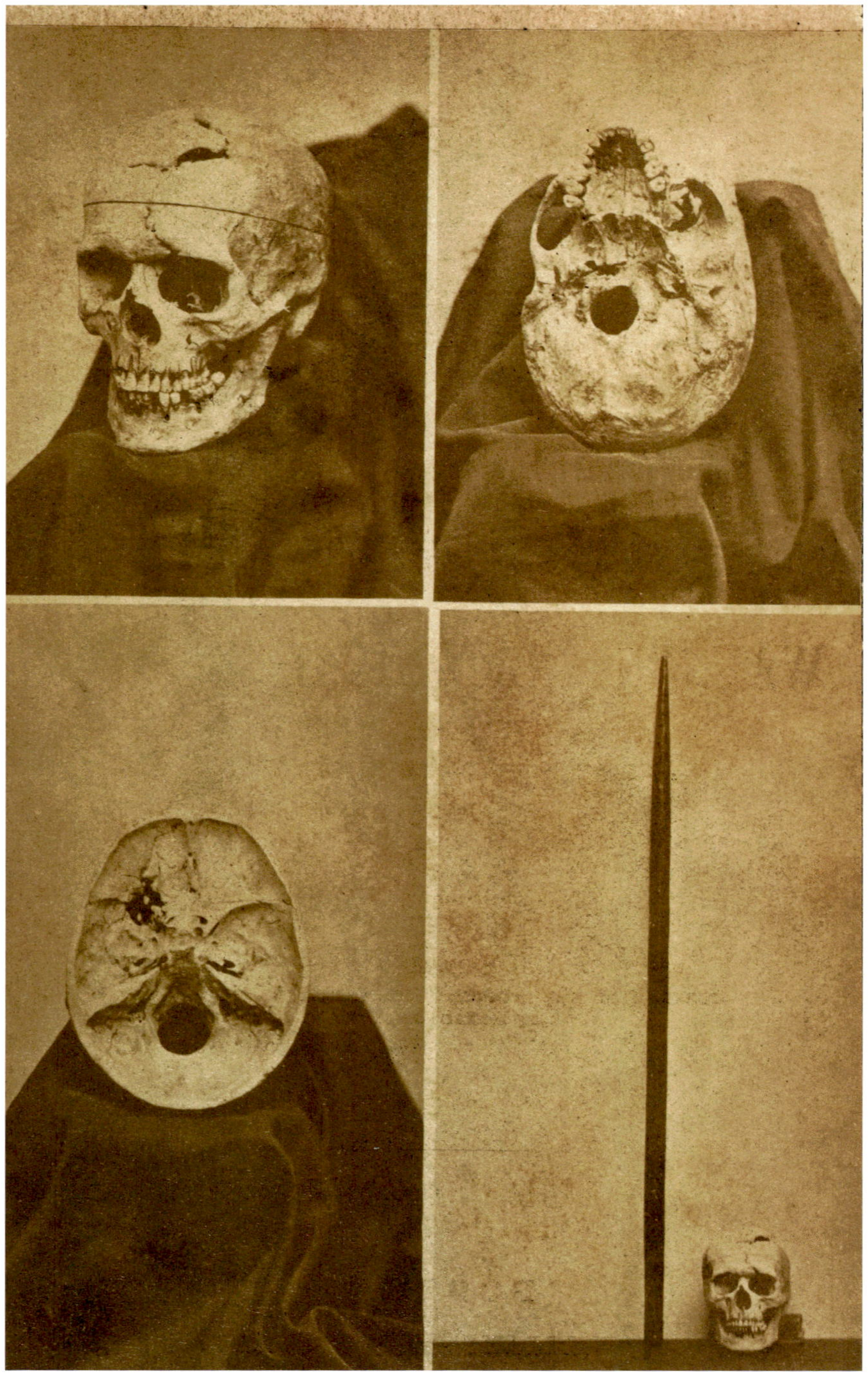

Plate 4.1

Phineas Gage's Skull and Tamping Bar that Flew Through His Head

Descriptive Catalogue of the Warren Anatomical Museum, Frontispiece, 1870

The most studied skull of the nineteenth century.

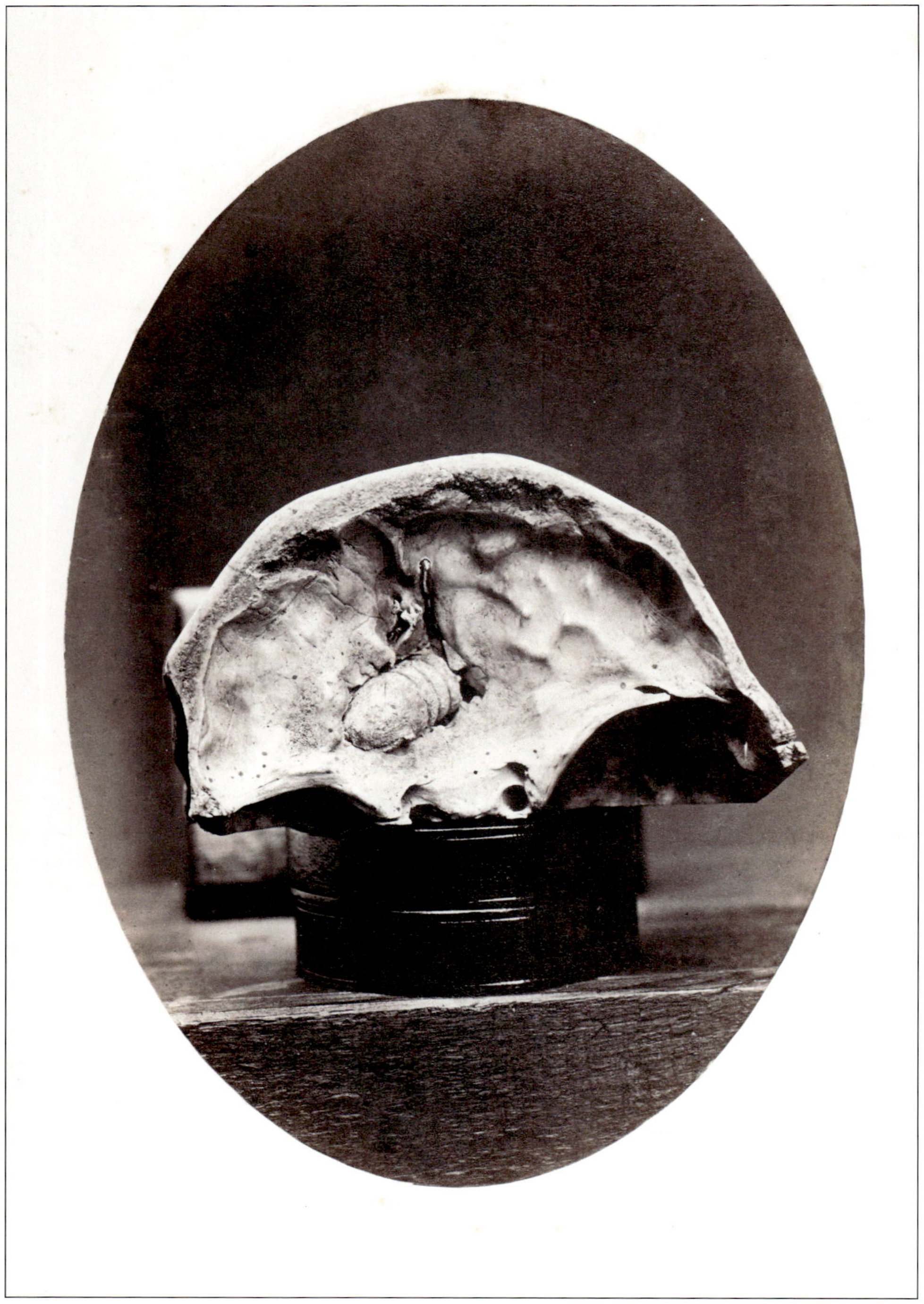

Plate 4.2

Conoidal Bullet in Ethmoid and Frontal Bones

Specimen 1108, Civil War Collection, Army Medical Museum, 1863

Wounded Dec 19, 1862, no cerebral signs except for headache until February 10, died February 11.

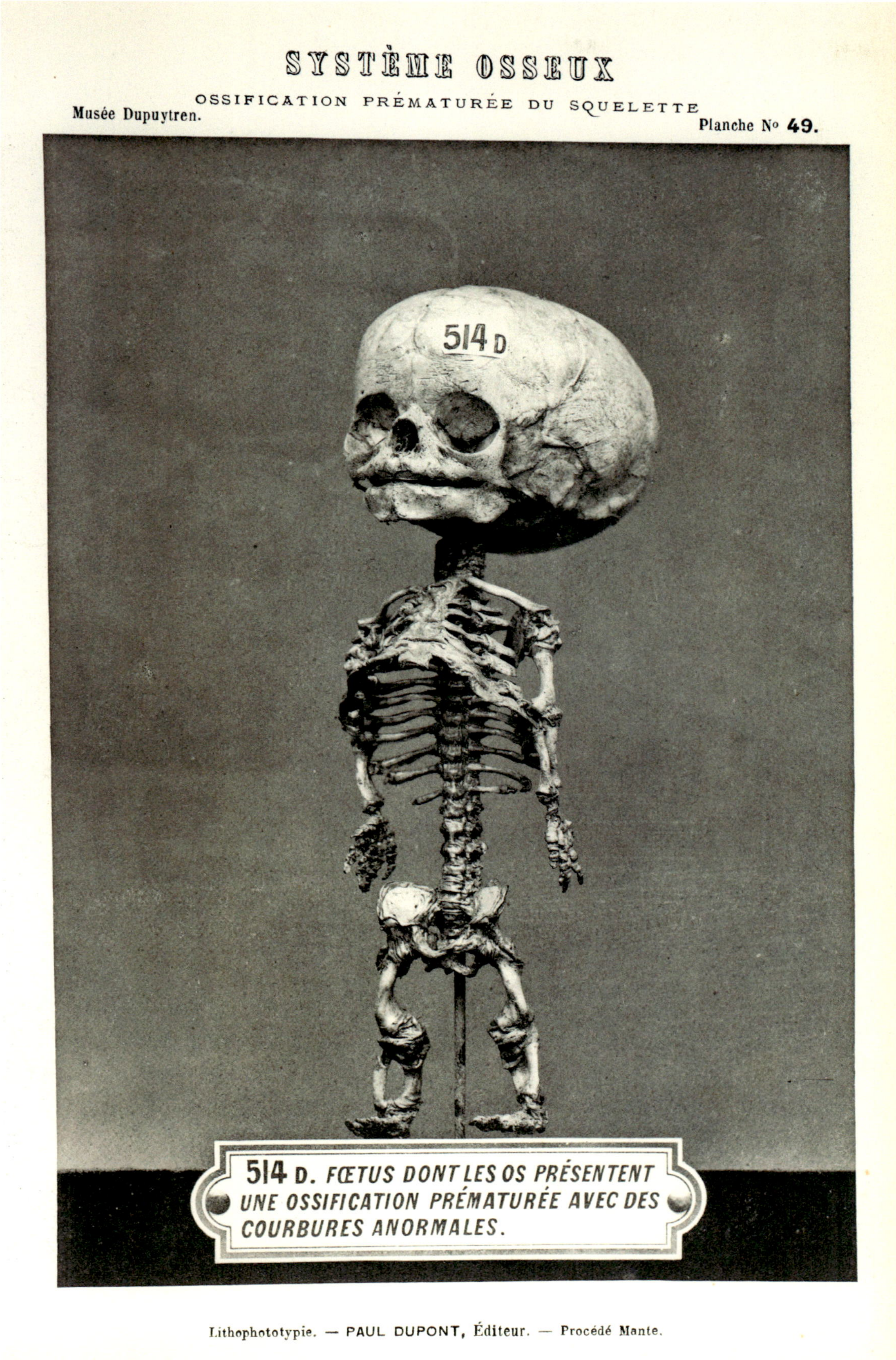

Plate 4.3

Infant With Premature Closure of the Fontanels
Catalogue des pièces du museé Dupuytren, Plate 49, 1879

A well-prepared infant skeleton with unusual head pathology.

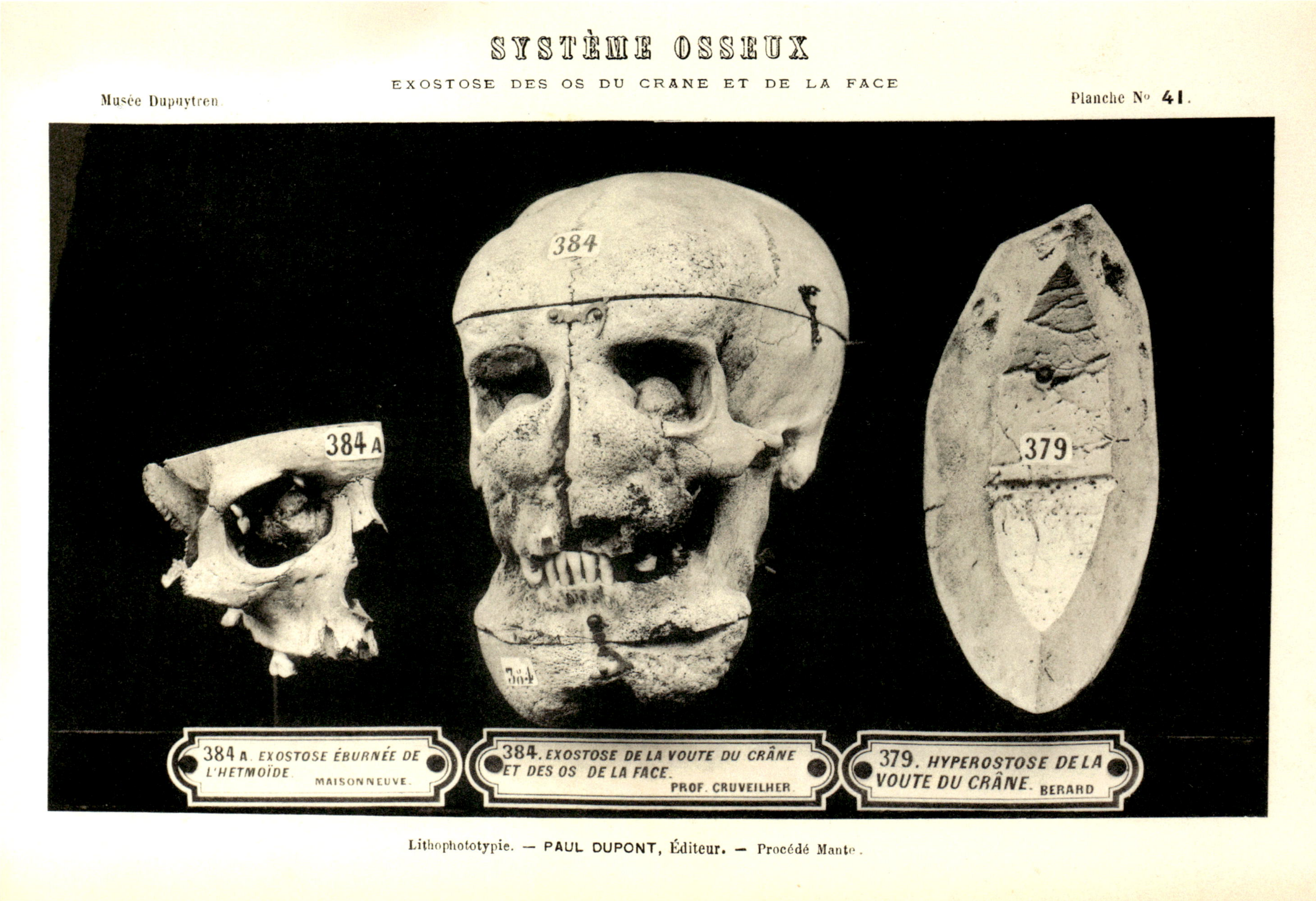

Plate 4.4

Multiple Exostosis of the Skull with a Case of Leontiasis (Lion Face)
Catalogue des pièces du museé Dupuytren, Plate 41, 1879

The abnormal bone formation often destroys facial features.

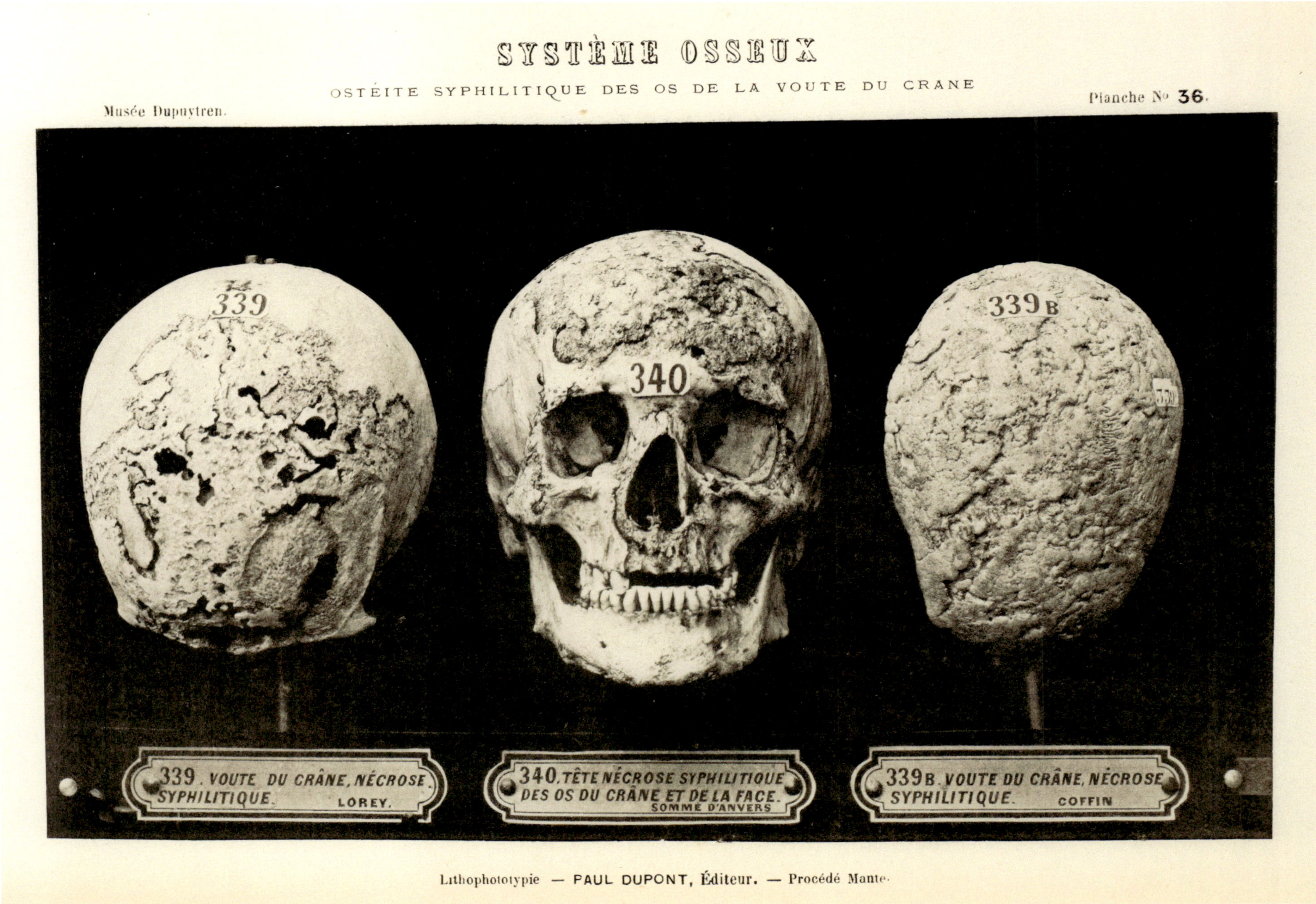

Plate 4.5

Syphilitic Caries of the Skull

Bony destruction of various parts of the skeleton is common in syphilis.

Catalogue des pièces du museé Dupuytren, Plate 36, 1879

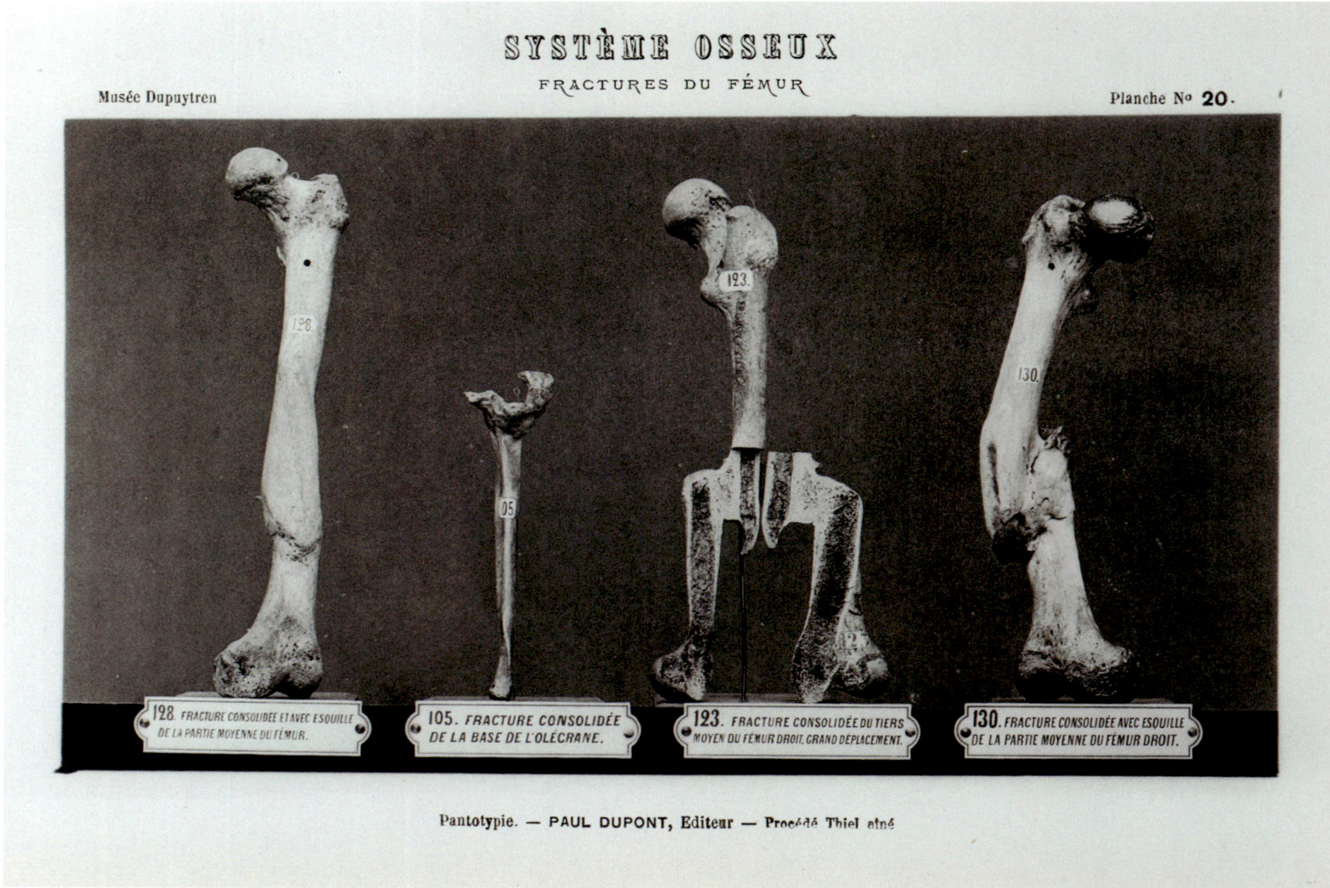

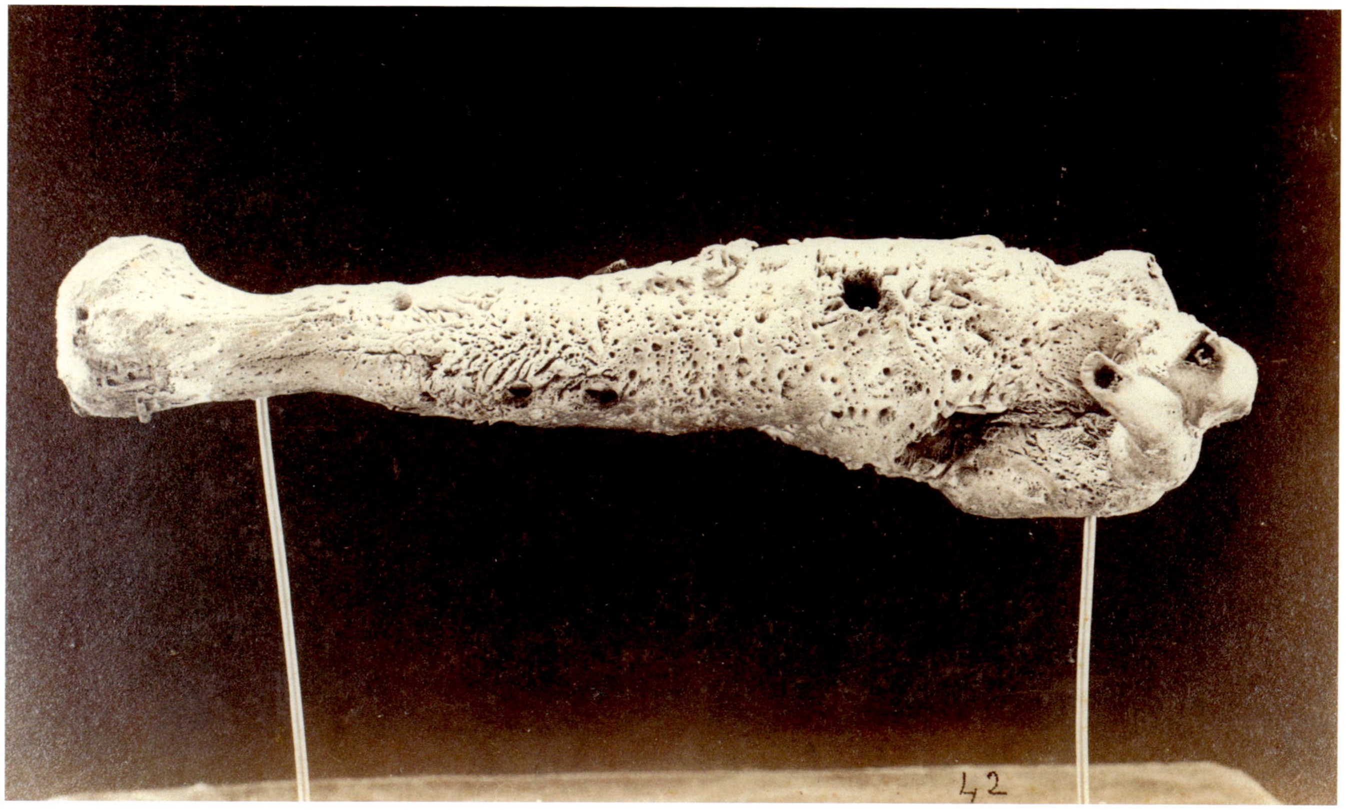

Plate 4.6

Examples of Four Healed Fractures: Three Femur and One Elbow
Catalogue des pièces du museé Dupuytren, Plate 20, 1879

Healed bones showing the degree of deformities with which patients lived.

Plate 4.7

Necrosis of the Entire Humerus Following Gunshot Fracture of the Epiphysis
Photographs of Surgical Cases and Specimens… Number 42

Private Barney White, 37th New York Volunteers, Wounded May 5, 1862.

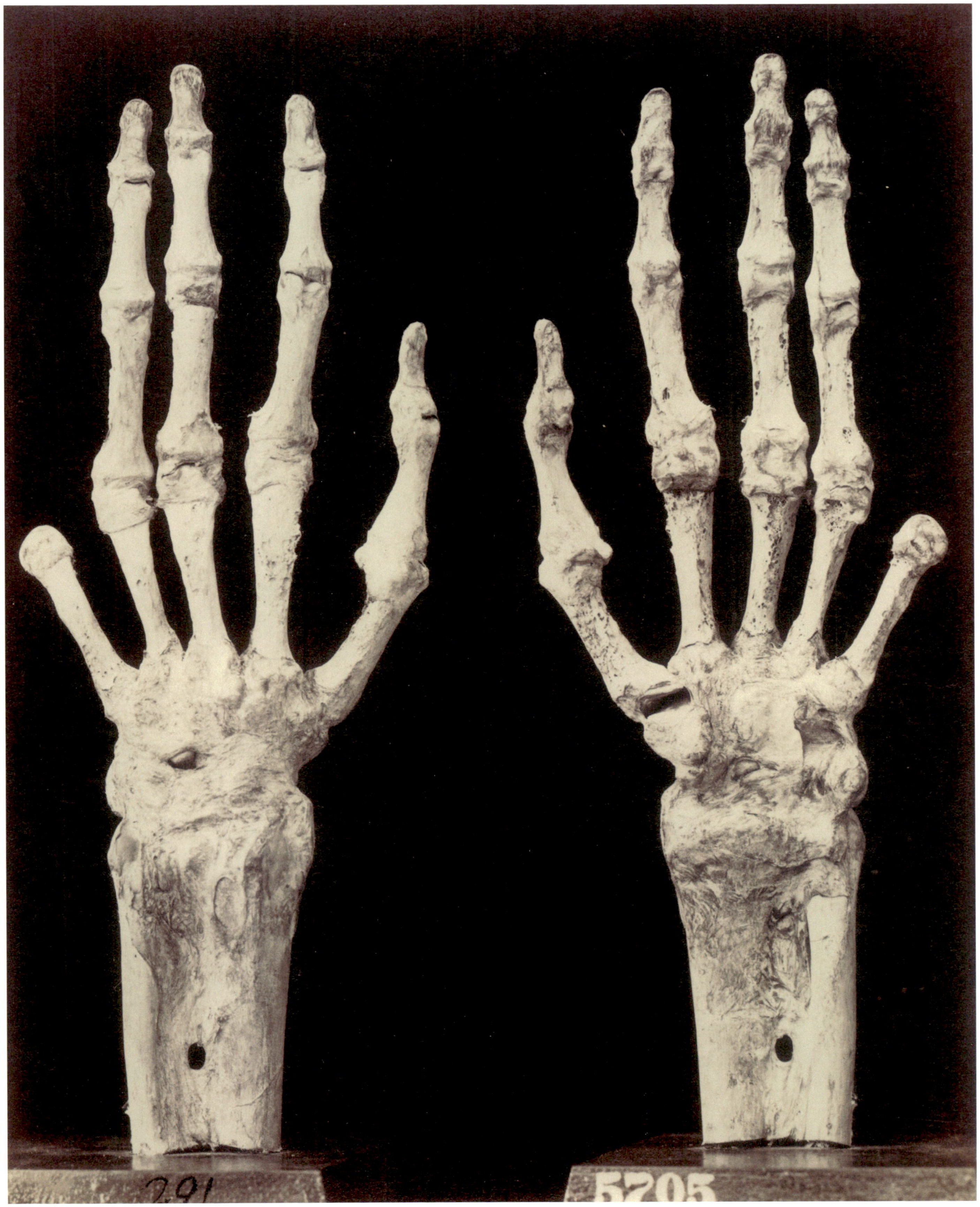

Plate 4.8

A Dry Preparation of the Left Hand and Lower Portion of the Fore-arm
Photographs of Surgical Cases and Specimens…
Number 291

Unusual cases were often sent to the Army Medical Museum for help in diagnosis.

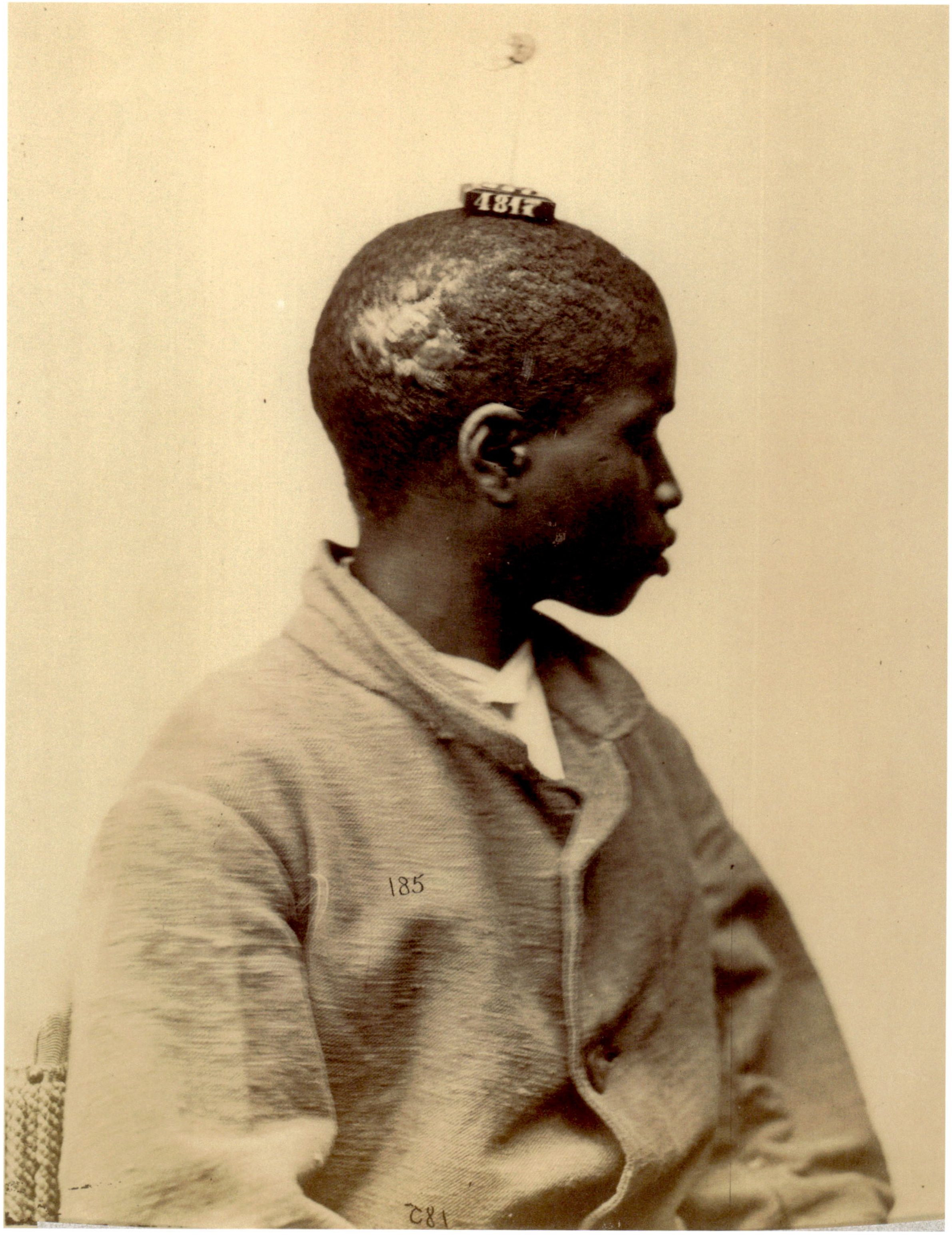

Plate 4.9

Posing With His Skull Bone Fragment on Top of the Head
Successful Trephining in a Case of Depressed Fracture Produced by a Fall
Photographs of Surgical Cases and Specimens… Number 185

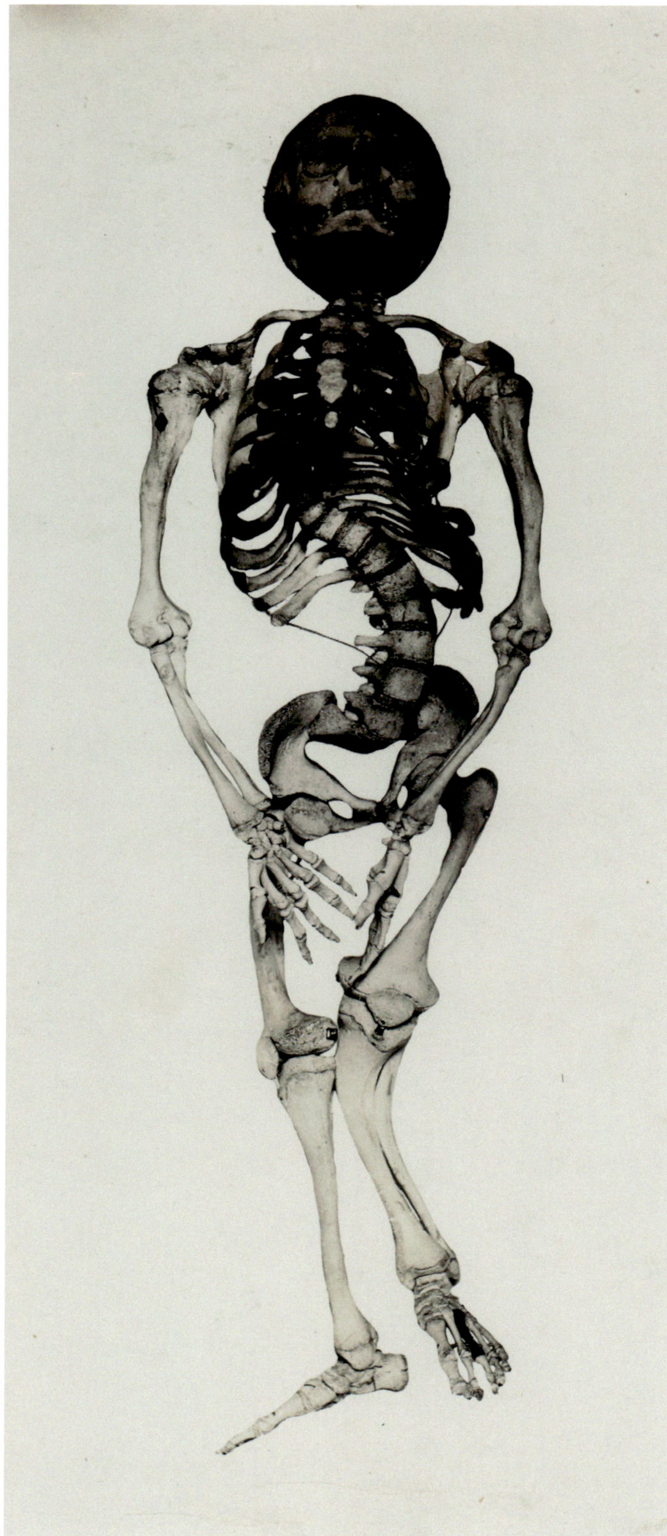

Plate 4.10

Case of Rickets in a 23-Year-Old Woman
Johns Hopkins Hospital, 1928

Photographs of the bony deformities of rickets remained a popular subject into the twentieth century.

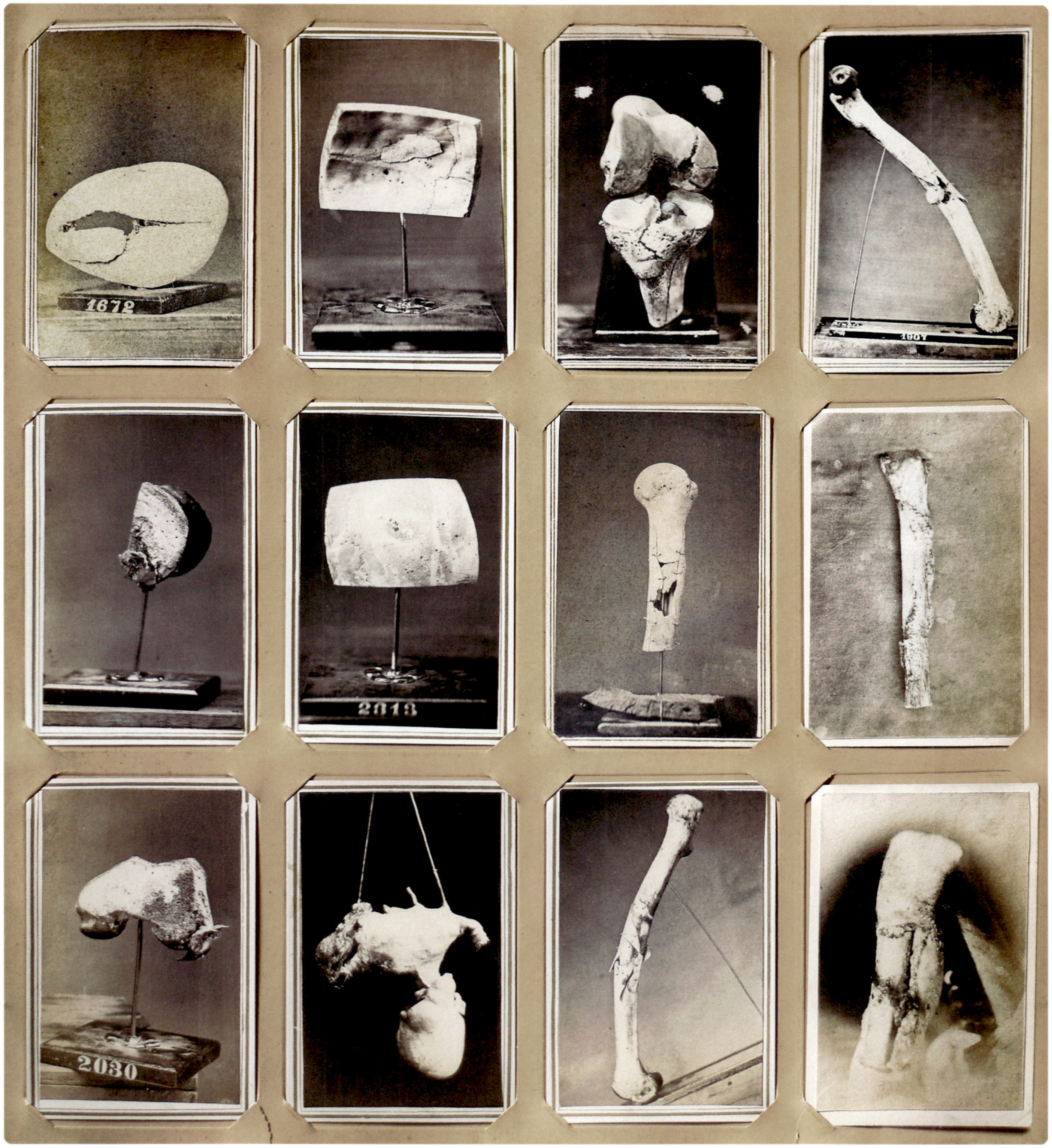

Plate 4.11

Page from US Army Surgeon Reed Brockway Bontecou's Harewood Hospital Album
1865

Bontecou prepared an album of 590 cartes de visite of Civil War wounded soldiers. Dr. Bontecou has been recognized as the war's leading medical photographer.

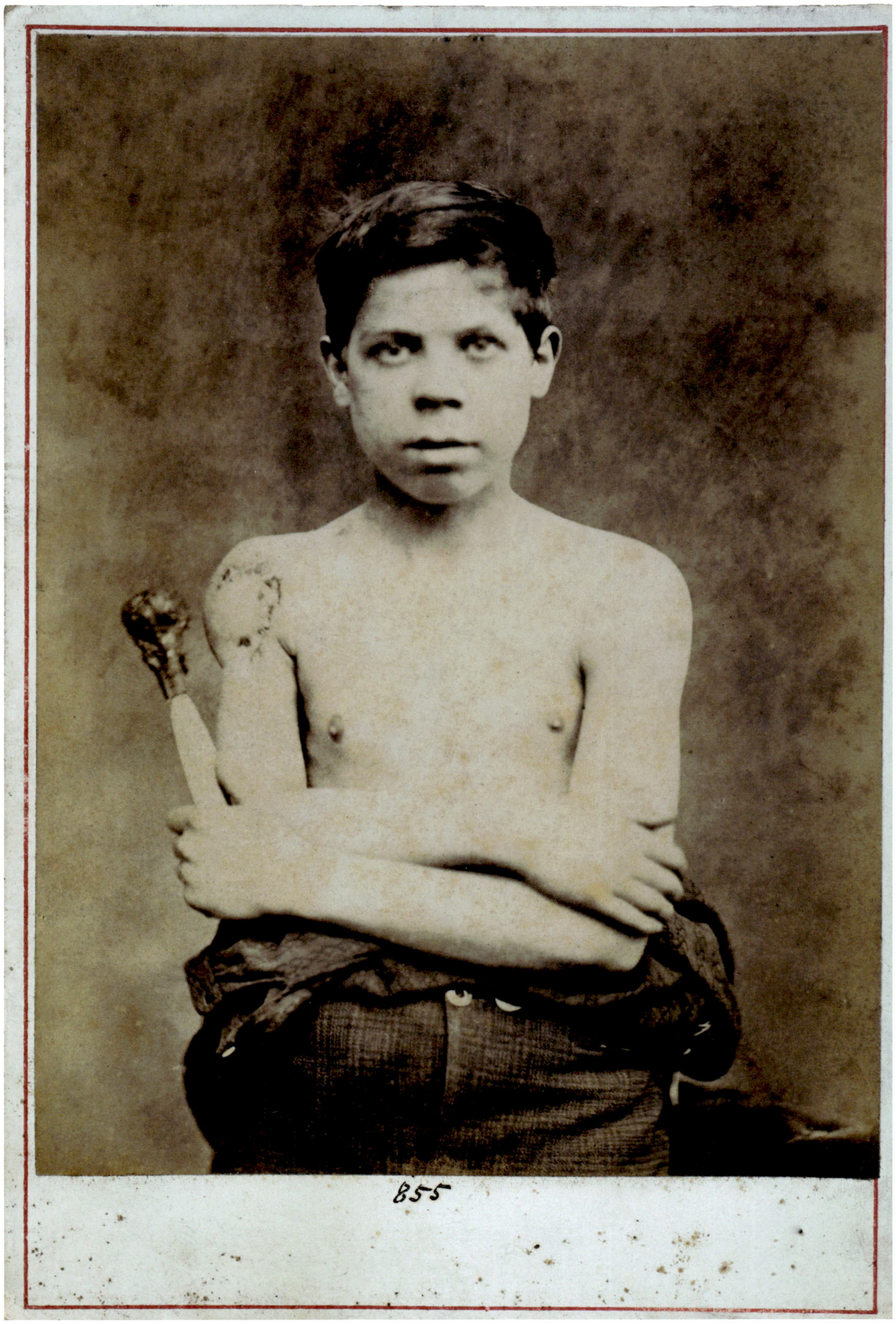
855

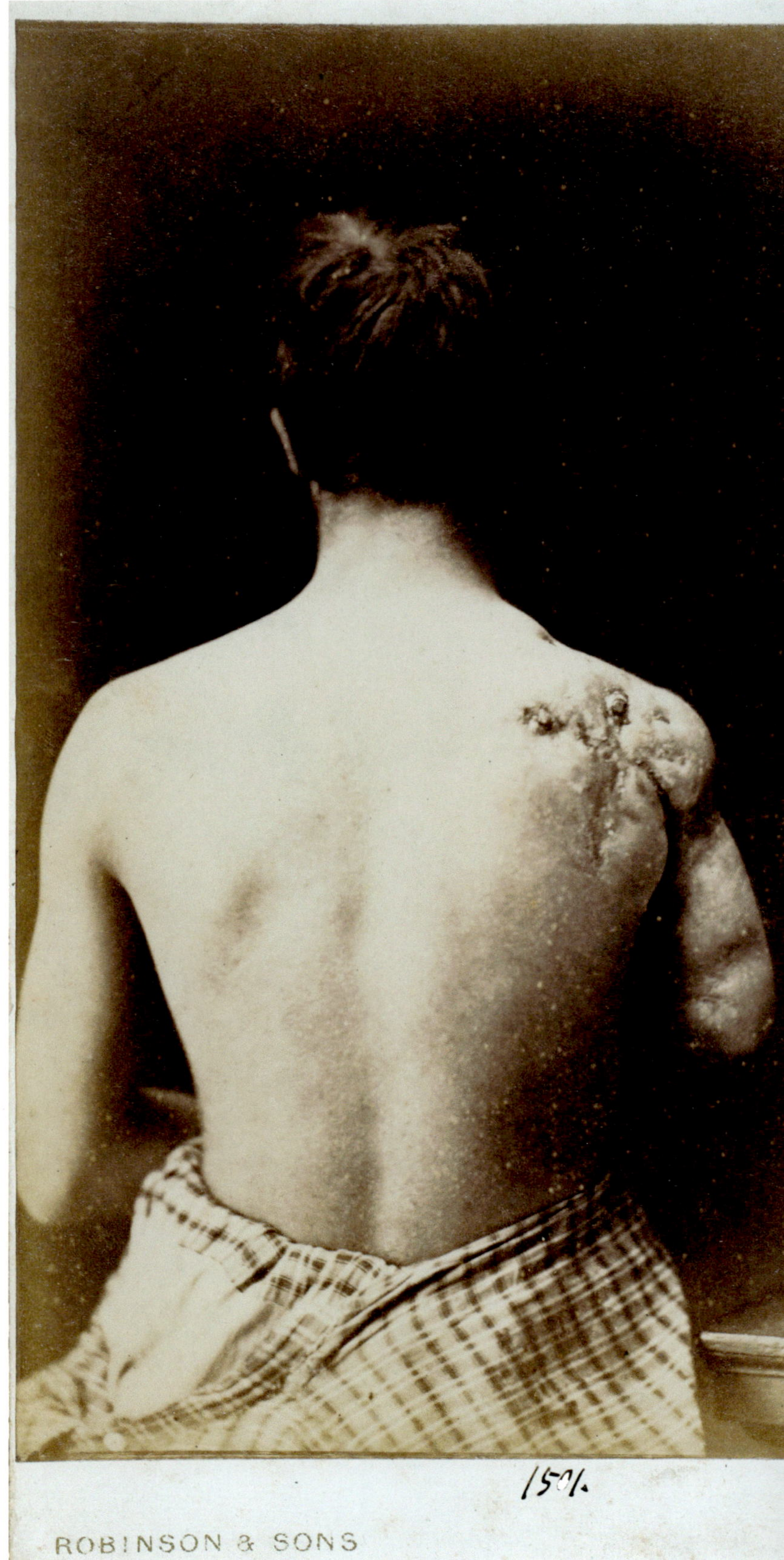
1501.
ROBINSON & SONS

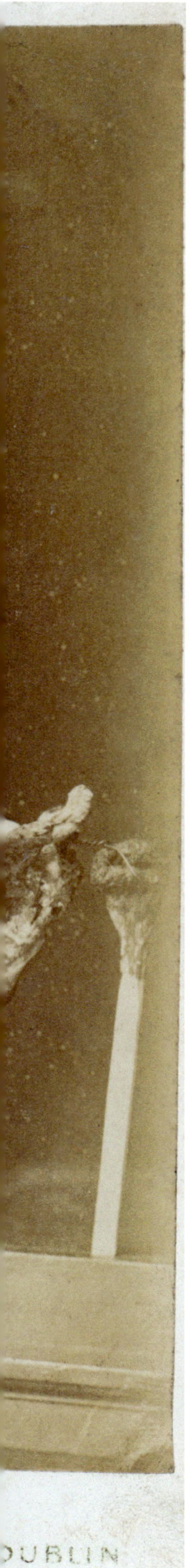

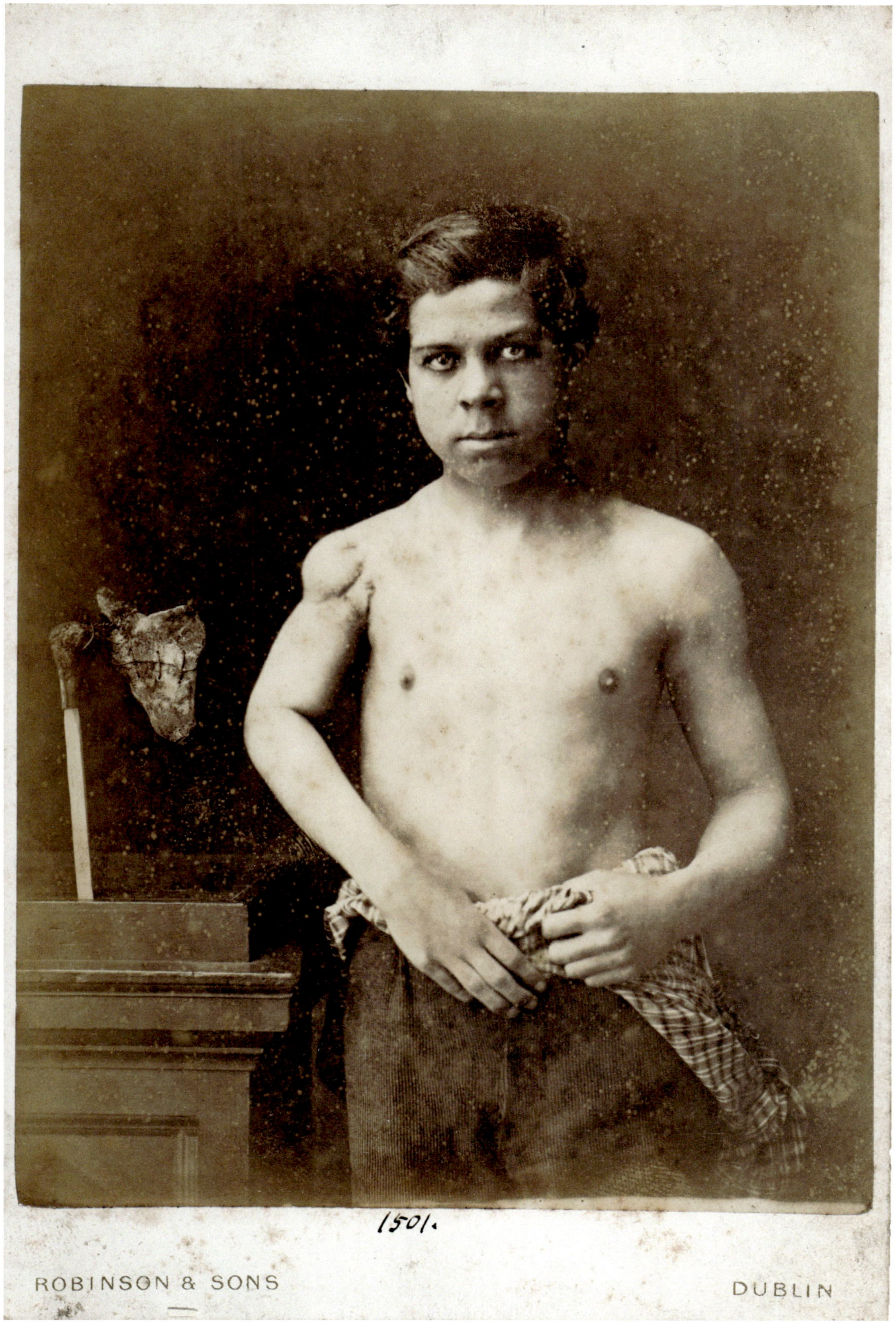

Plate 4.12

Patient Posing With His Excised Humerus
Circa 1870
Photographer: Thomas North, Dublin, Ireland

Posing with your removed bony body part was popular in the 1870 era.

Plates 4.13 – 4.14

Patient Posing with his Humerus and Shoulder
1874
Photographer: Robinson and Sons. Dublin, Ireland

Rare photo of a successful second shoulder procedure, Surgeon Edwin O'Grady.

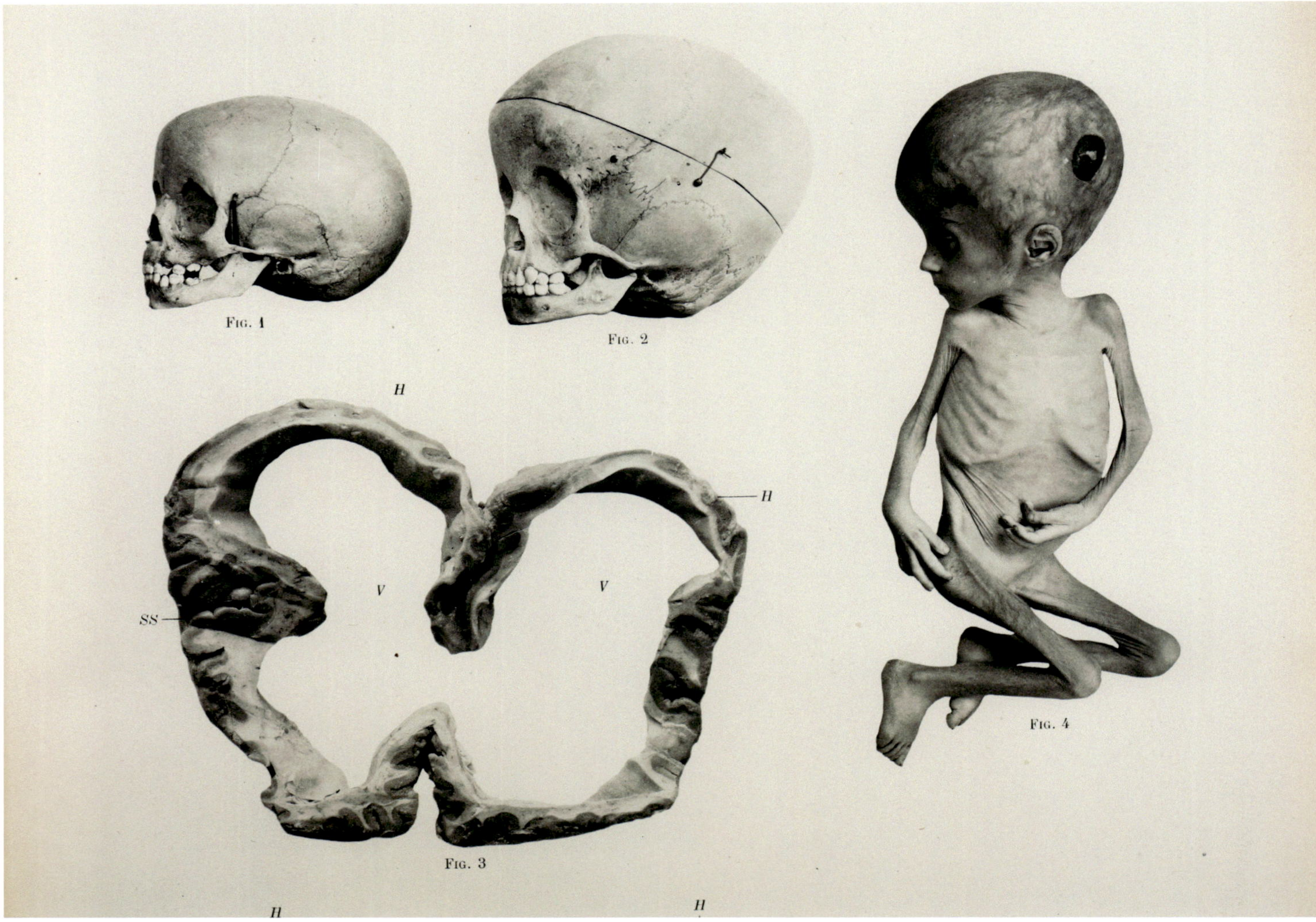

Plate 4.15

Hydrocephalic Skull Compared to Normal Adult Skull
Cliniques Médical Iconographiques by Haushalter, P. Spillman, L, et al., Plate 20, 1901
From the Faculty de Médicine de Nancy

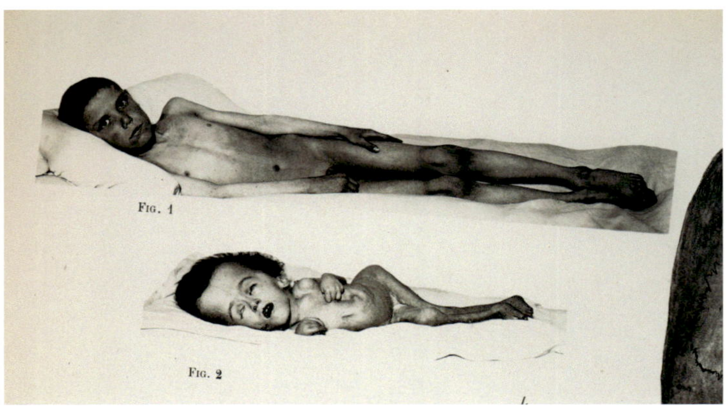

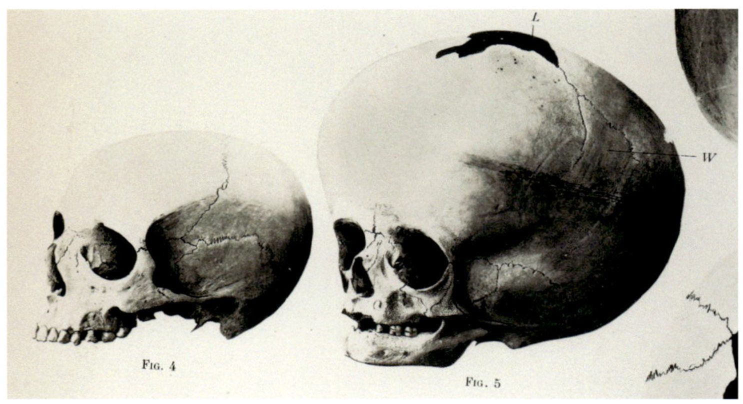

Plates 4.16 – 4.17

Ten-Year-Old Hydrocephalic Boy's Skull Compared to Normal Adult Skull
Cliniques Médical Iconographiques by Haushalter, P. Spillman, L, et al., Plate 19, 1901
From the Faculty de Médicine de Nancy

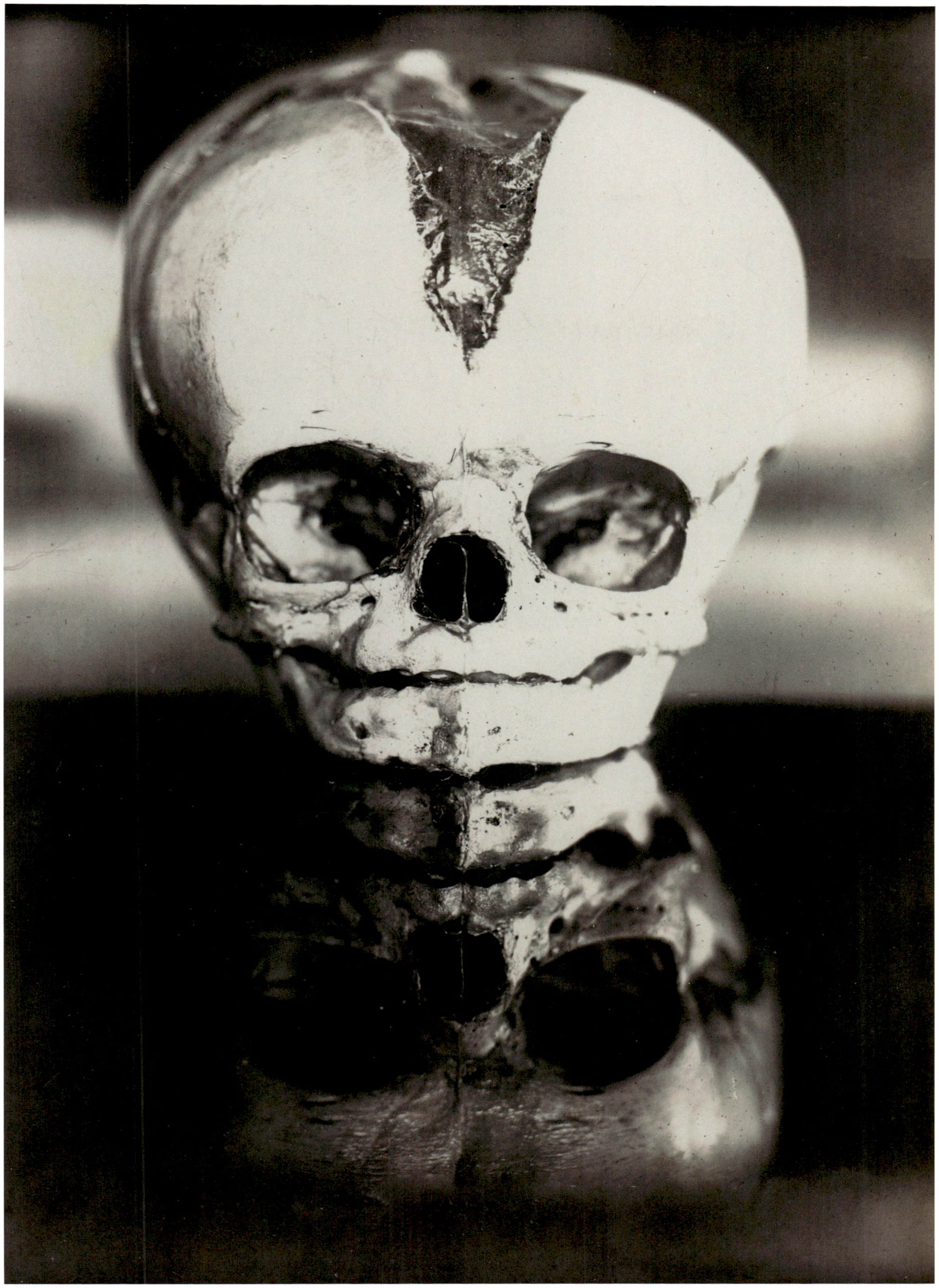

Plate 4.18

Infant's Skull Illustrating Open Fontanel
Circa 1910

The large frontal fissure makes a baby's skull flexible enough to pass through the birth canal. Six primary fontanels are present in humans and close as the skull grows.

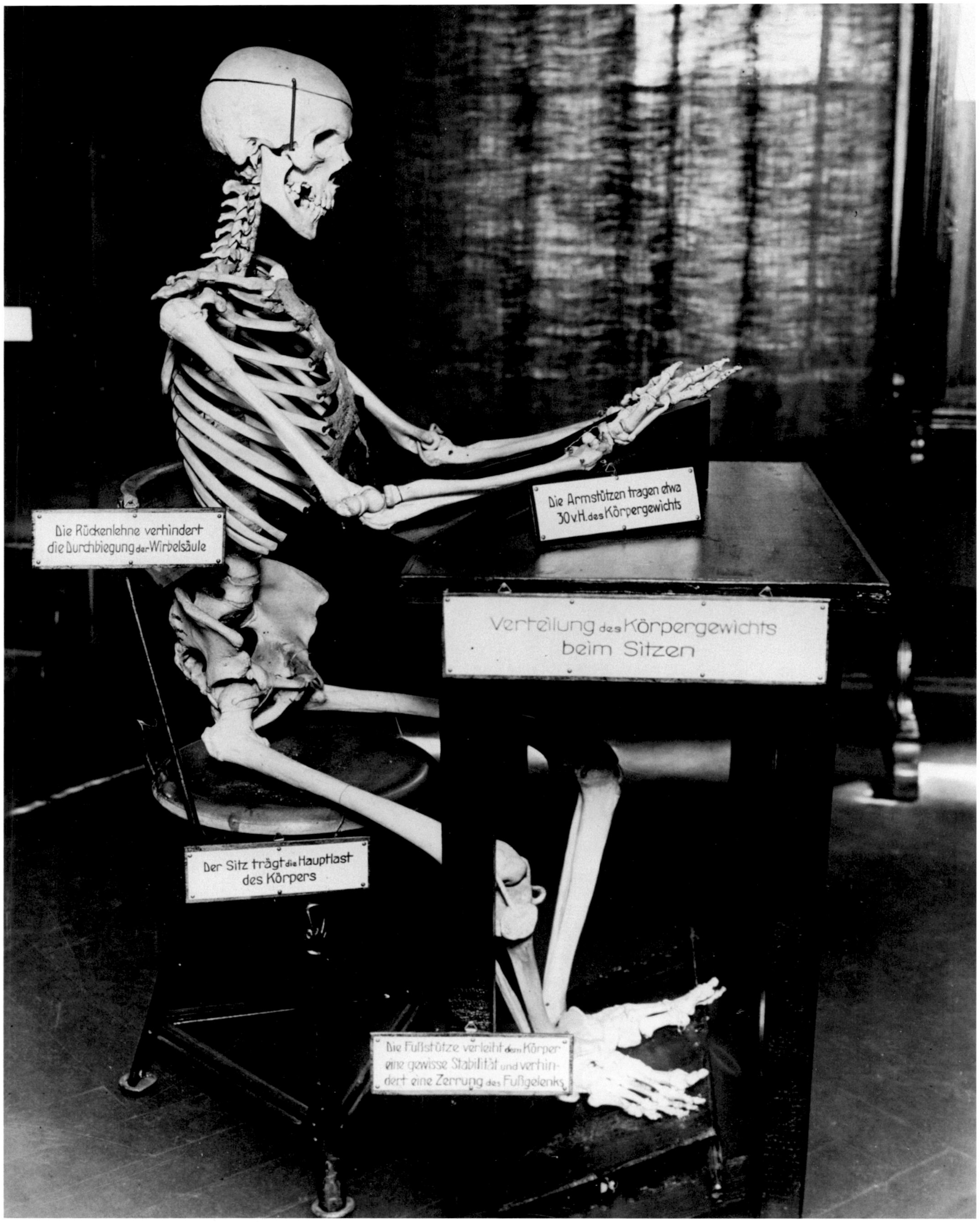

Plate 4.19

New German Chair Illustrates Ergonomic Posture for Desk Work to Prevent Fatigue

Display at the Society of Trade Hygiene Exposition, June 1929

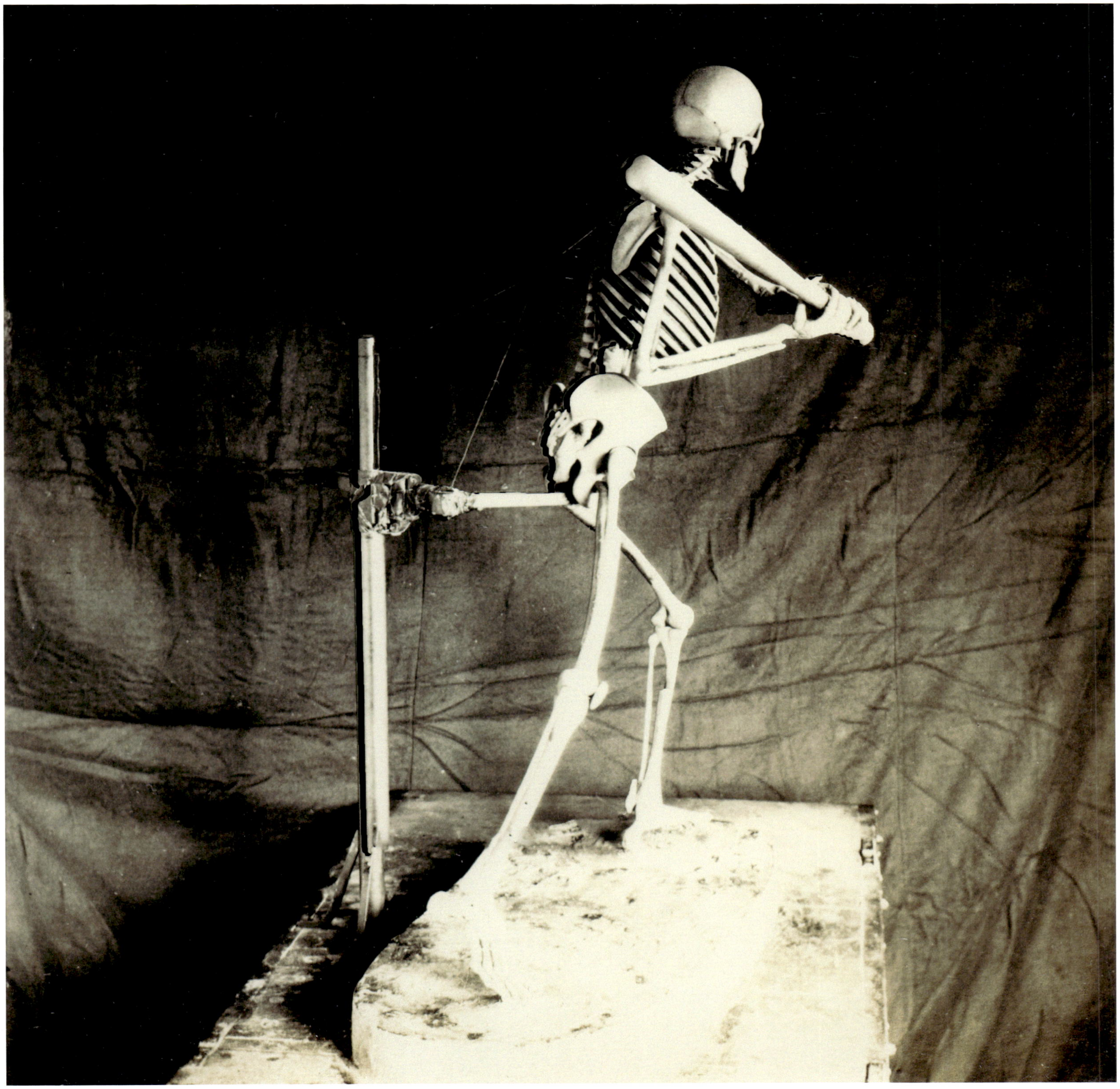

Plate 4.20

Skeletal Position at Bat
Stereoview, circa 1920

Part of a series of photographs of a skeleton in various baseball batting positions.

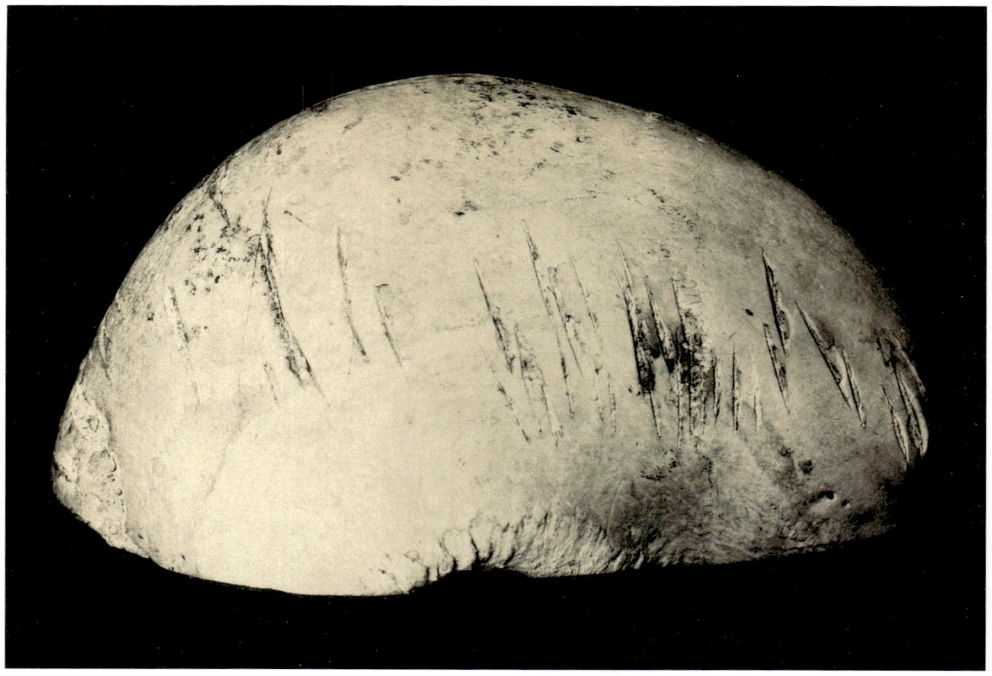

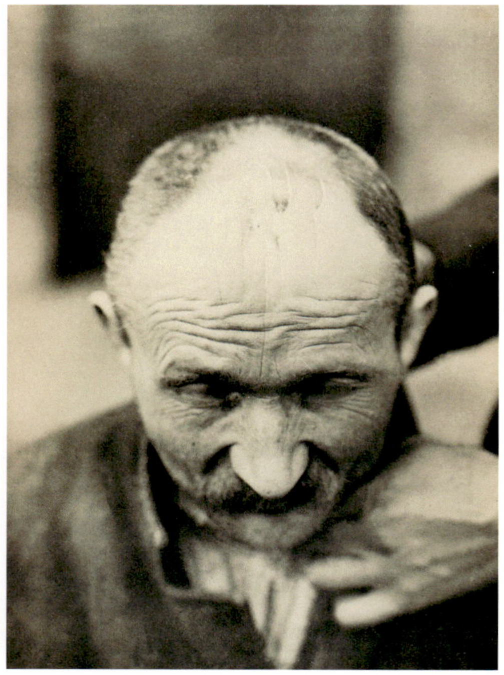

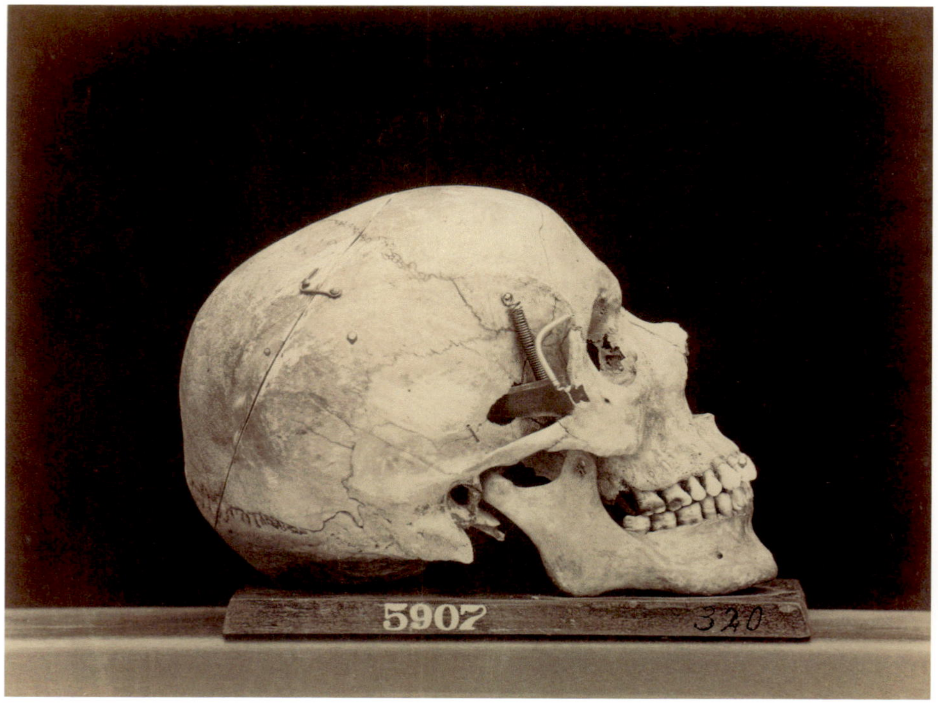

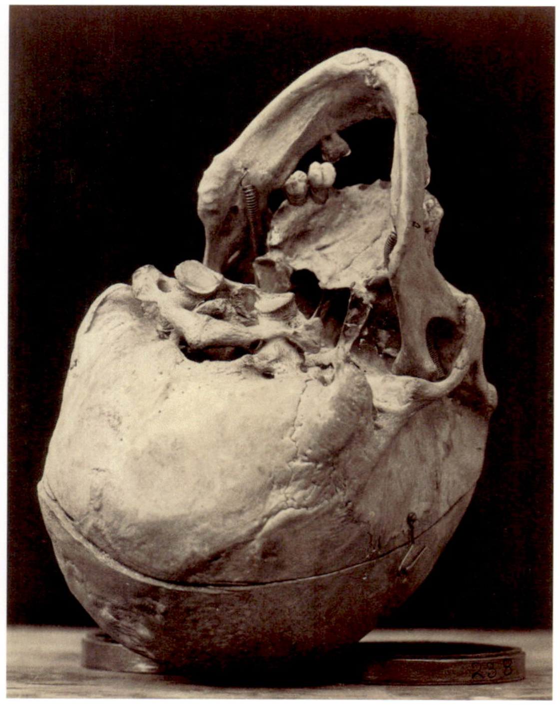

Plate 4.21

Skull with Evidence of 50 Suicide Attempts With An Axe
Alphonse Bedin, MD, *La Photographie au laboratoire de médecine légale*, Plate 32, 1908

This psychotic patient died from an unassociated infection. The skull findings were incidental at autopsy.

Plate 4.23

Fatal Arrow Wound of the Skull
Wounded Sept 1, 1870 died Sept 19
Photographs of Surgical Cases and Specimens… Number 320

Plate 4.22

'Melancholic' Man Who Attempted Suicide by Repeatedly Hitting the Front of His Head with an Axe
Alphonse Bedin, MD, *La Photographie au laboratoire de médecine légale*, 1908

Plate 4.24

Cranium, with the Atlas Luxated to the Right and Firmly Ossified to the Occipital Condyles
Photographs of Surgical Cases and Specimens… Number 238

Unusual positioning of skull but necessary to show the fused condyles.

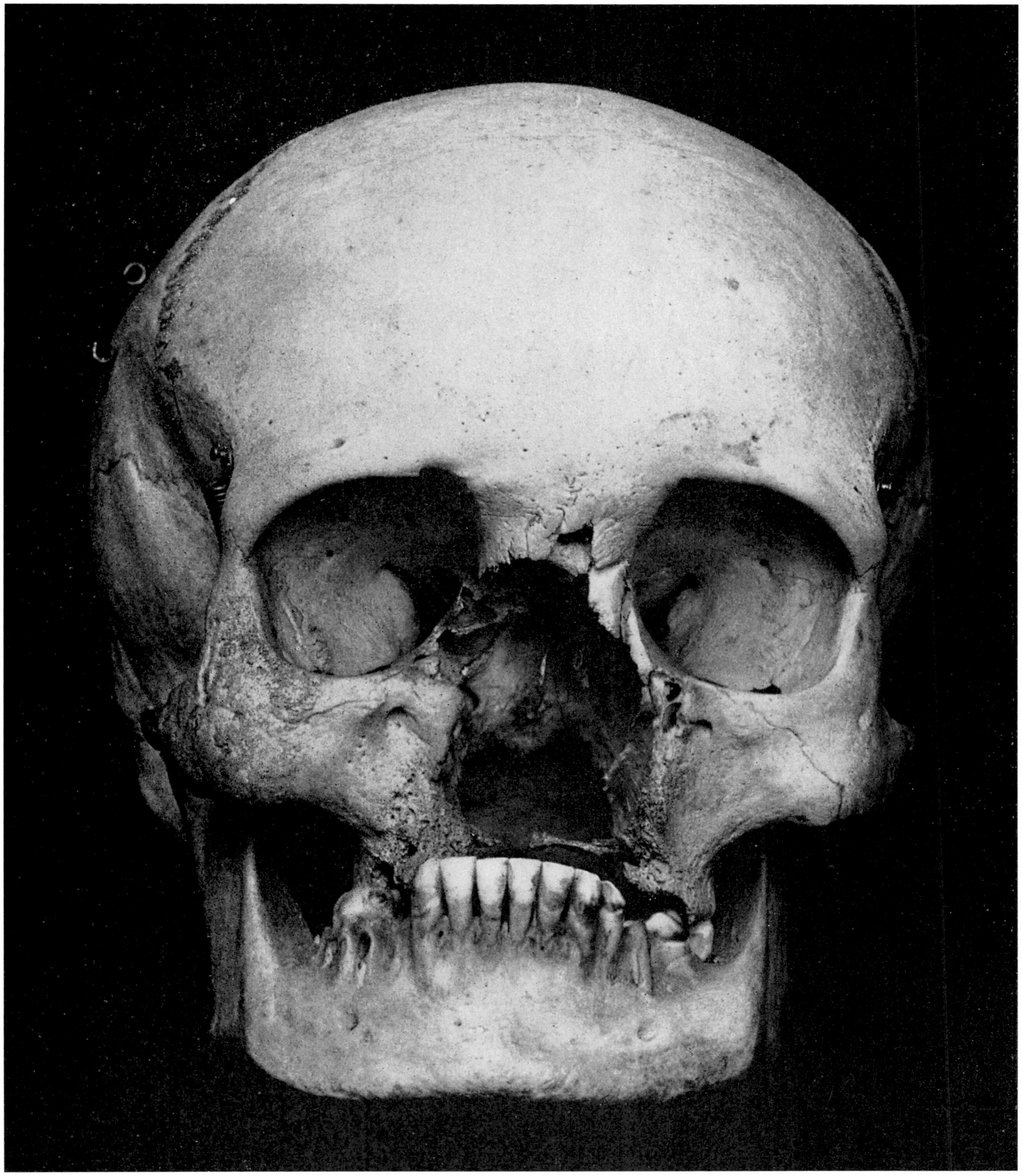

Plate 4.25

Skull Showing the Typical Facial Bony Destruction of Syphilis
1900

Syphilis destroyed the ethmoid, facial and nasal bones.

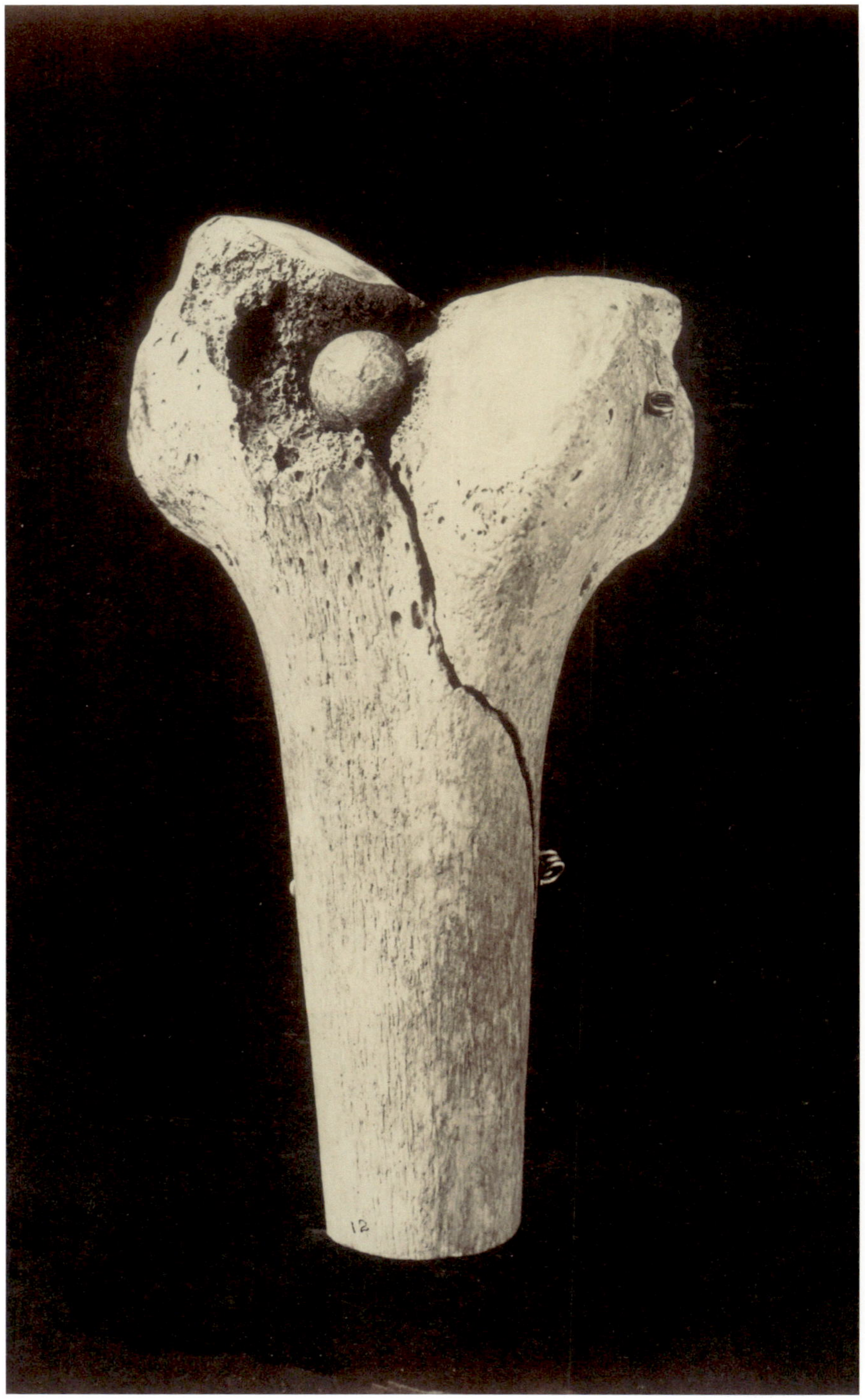

Plate 4.26

Bullet in Lower Third of Femur
Private Richard Williams, 28th Pennsylvania, Wounded Sept 16, 1862, Died October 2, 1862
Photographs of Surgical Cases and Specimens... Number 12

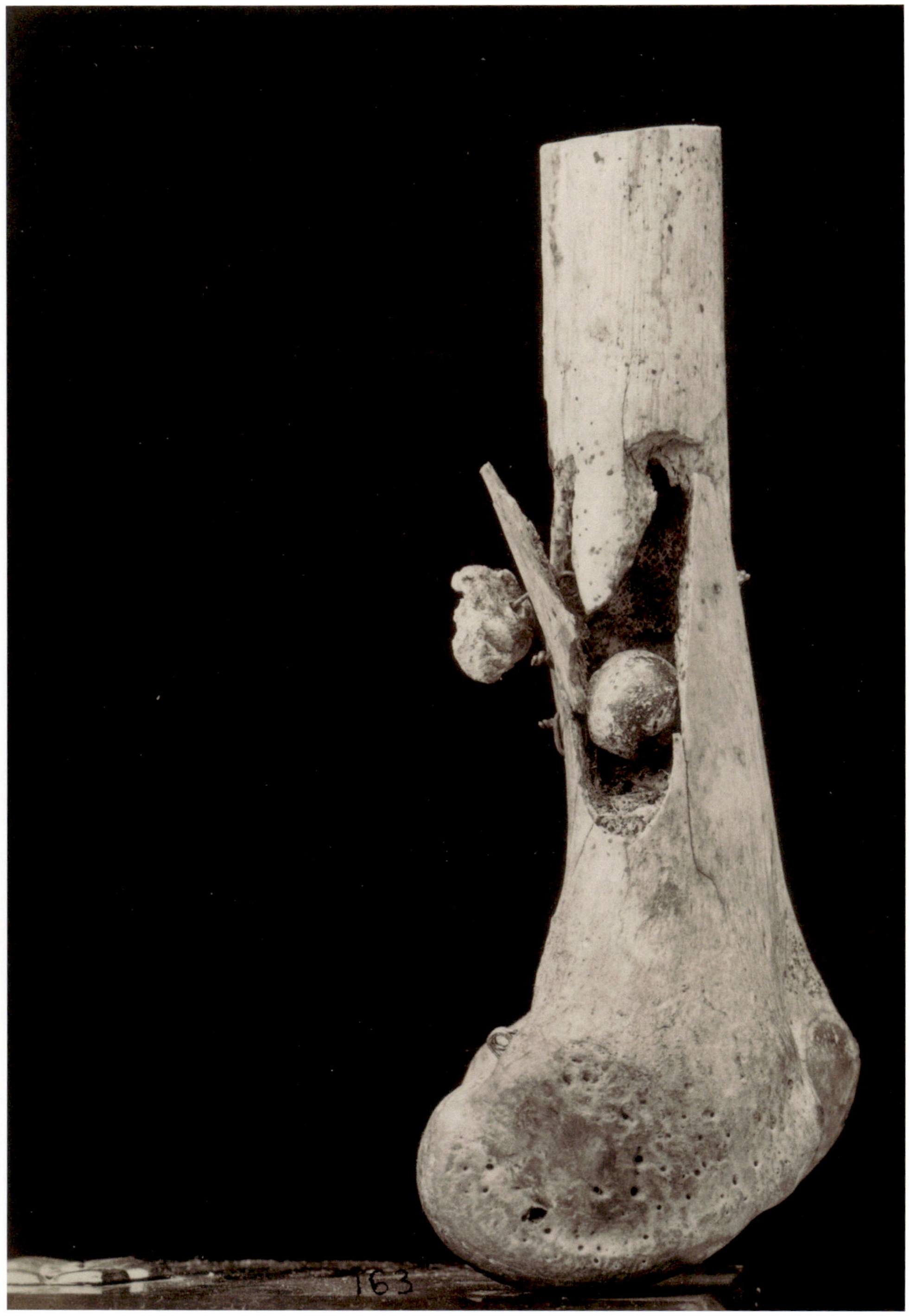

Plate 4.27

Amputated Portion of Left Femur with Leaden Ball in the Medullary Canal
Brigadier General Edmund K, 1st US Artillery, Wounded at Chancellorsville, May 3, 1863
Photographs of Surgical Cases and Specimens... Number 163

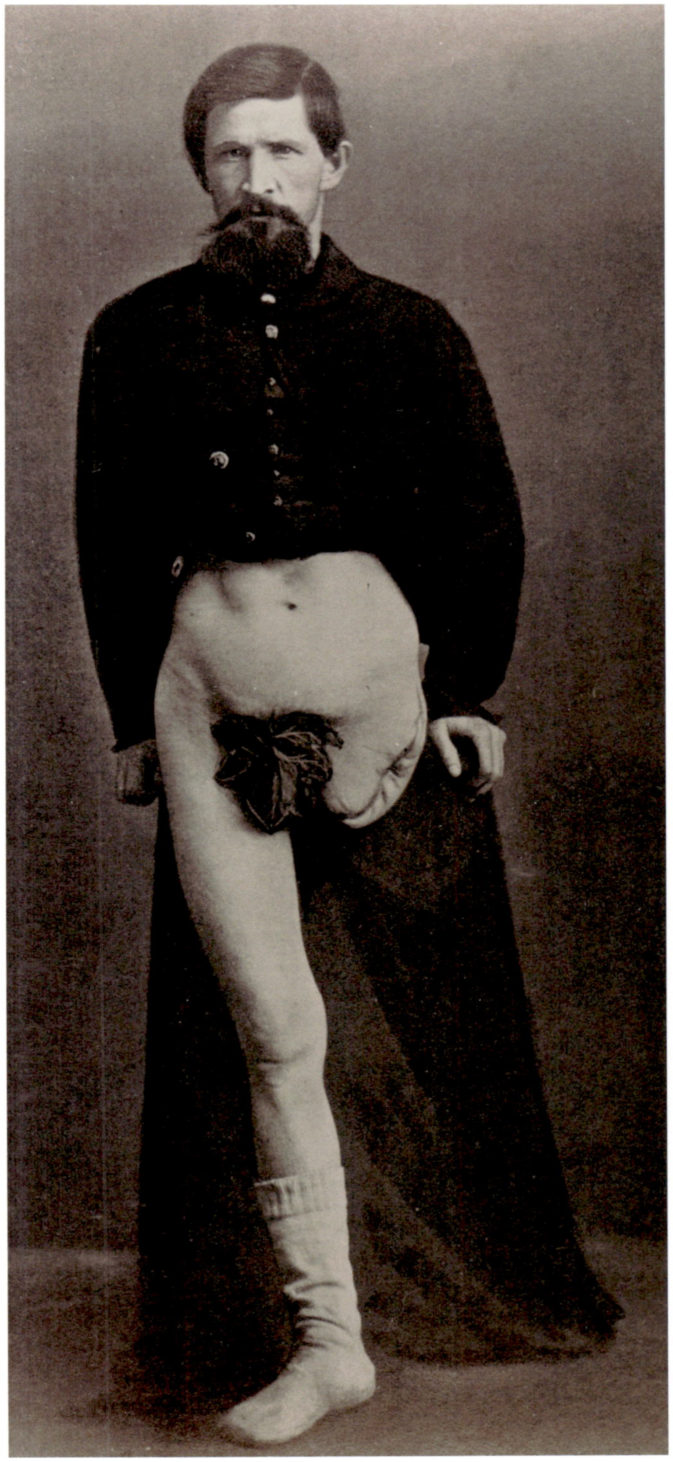

Plate 4.28

Successful Amputation of the Hip Joint, 17 Months after Wounding
Private George Lemon, 6th Maryland Volunteers, Wounded May 5, 1864 Operated October 12, 1865
Photographs of Surgical Cases and Specimens... Number 136

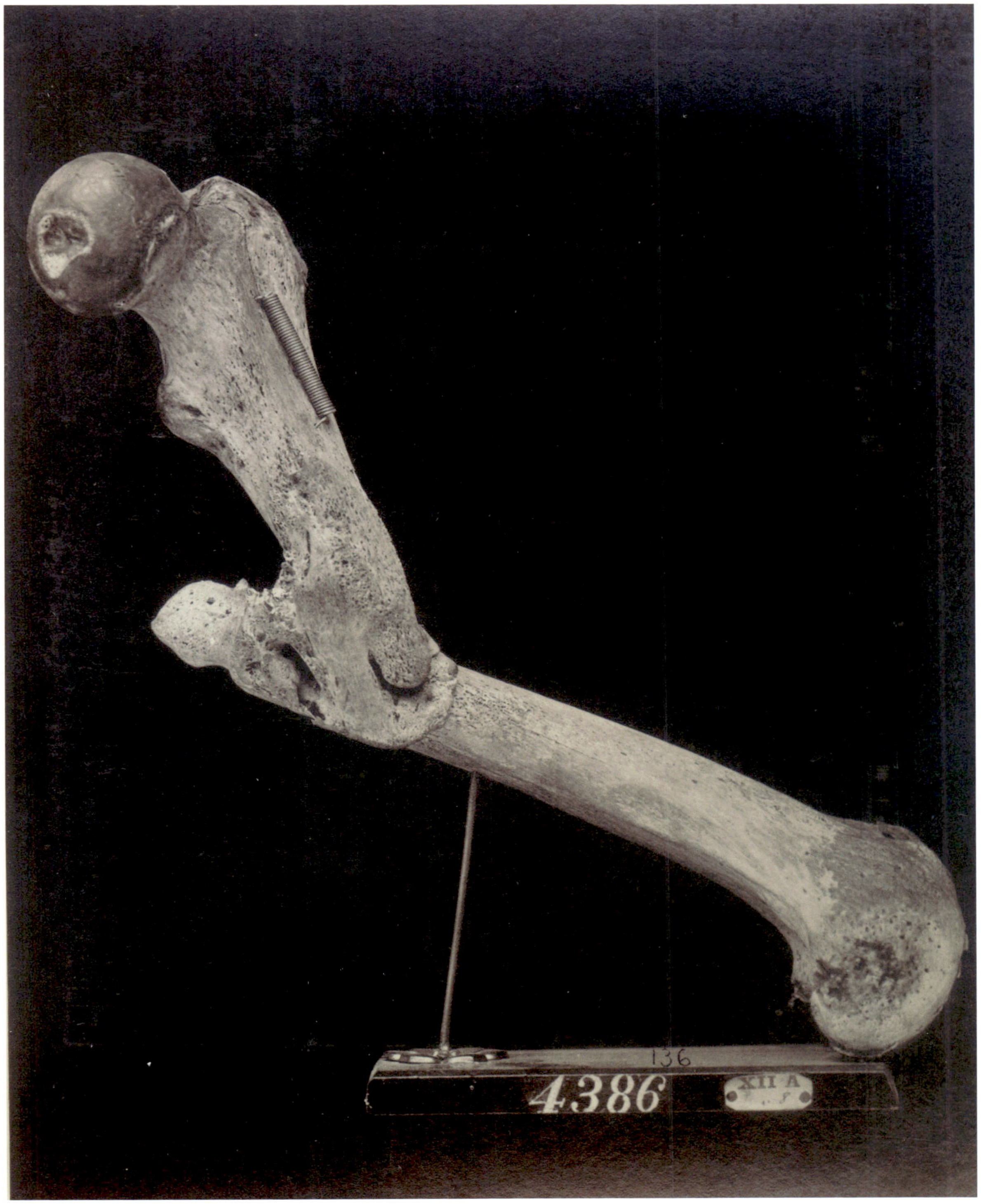

Plate 4.29

Specimen of Deformed Femur of Pvt. George Lemon, Survivor of Hip Amputation
Surgeon Edwin Bentley, Alexandria Hospital, VA, 1865
Photographs of Surgical Cases and Specimens... Number 137

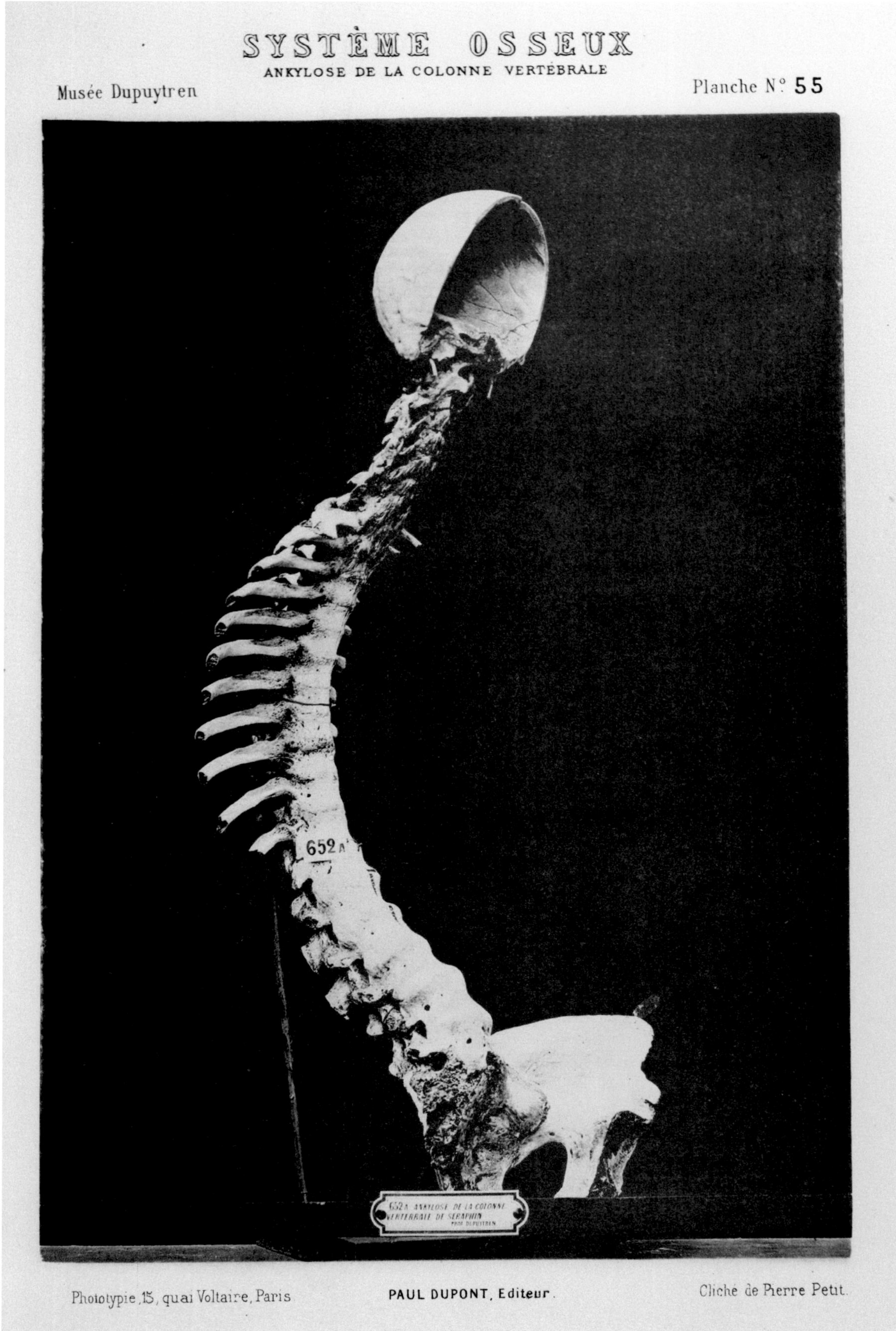

Plate 4.30

Ankylosis of the vertebral column
Catalogue des pièces du museé Dupuytren, Plate 55, 1879

One of Dr. Dupuytren's preparations.

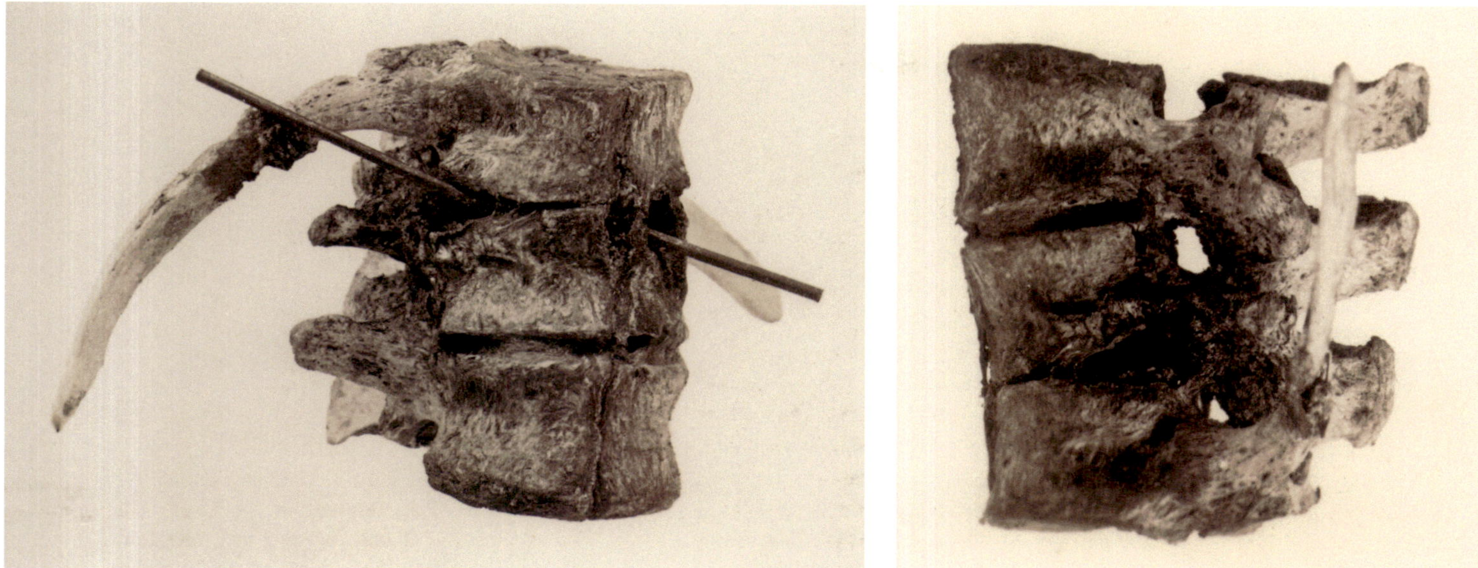

Plates 4.31

Vertebra of President James A. Garfield Shot by an Assassin, But Killed by Physicians
Pistol Shot Perforation of the First Lumbar Vertebra
Photographs of Surgical Cases and Specimens… Number 385

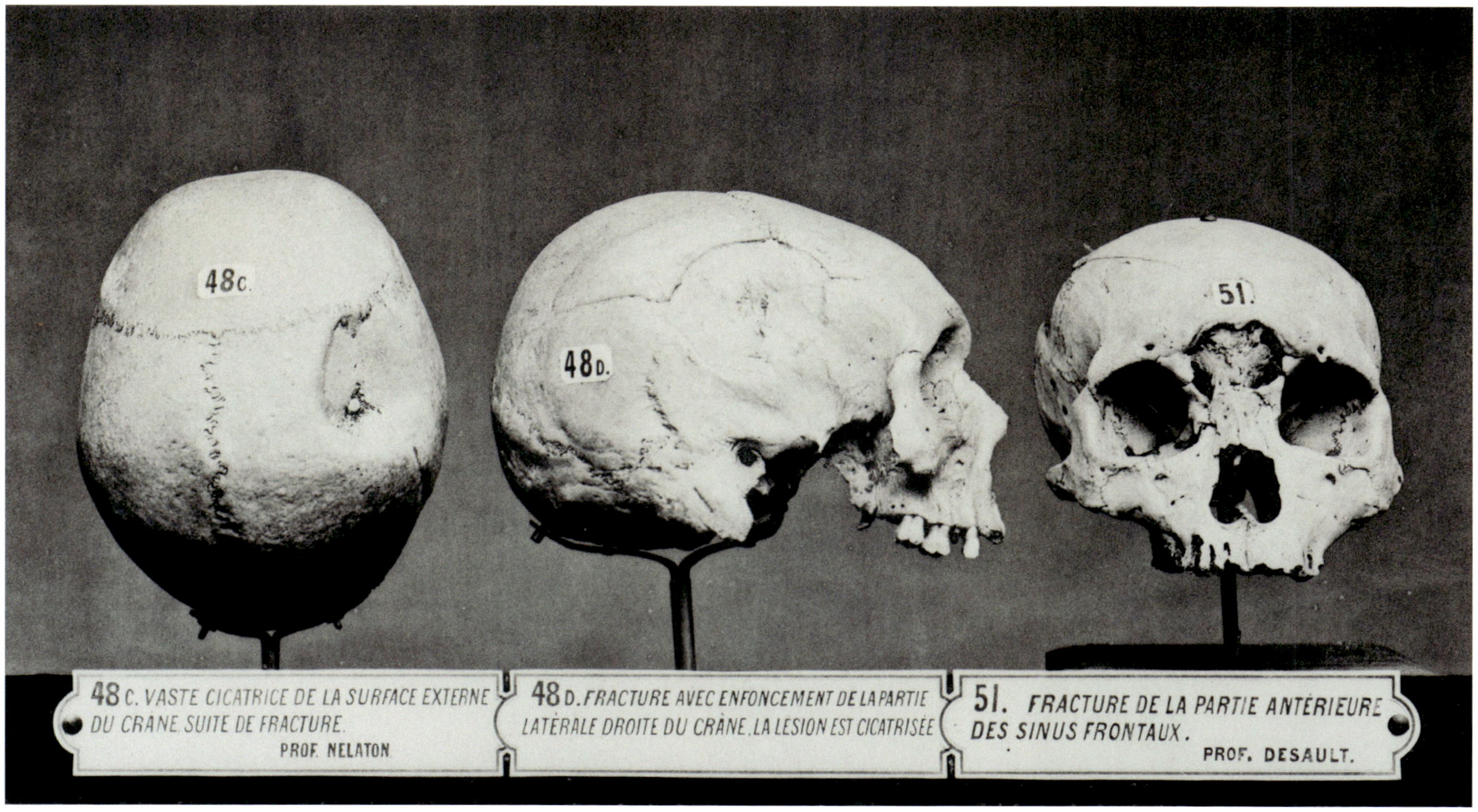

Plate 4.32

Fractures of the Skull
Catalogue des pièces du museé Dupuytren, Plate 14, 1879

A depressed fracture in the parietal bone and a fracture into the frontal sinus.

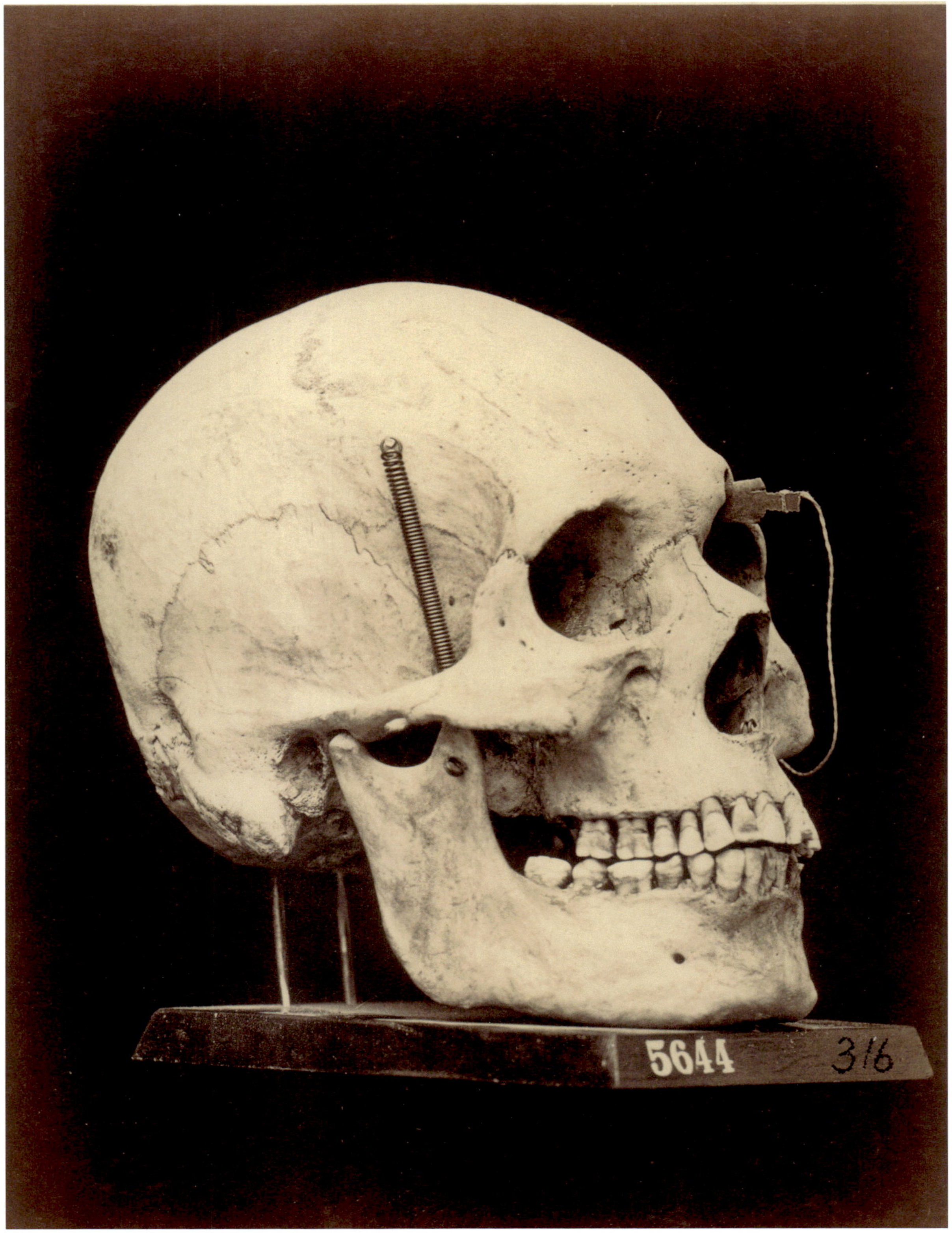

Plate 4.33

Cranium penetrated by an Arrow through the Left Superciliary Ridge
Photographs of Surgical Cases and Specimens... Number 316

Arrow wounds were just one category of head wounds collected by the Army Medical Museum.

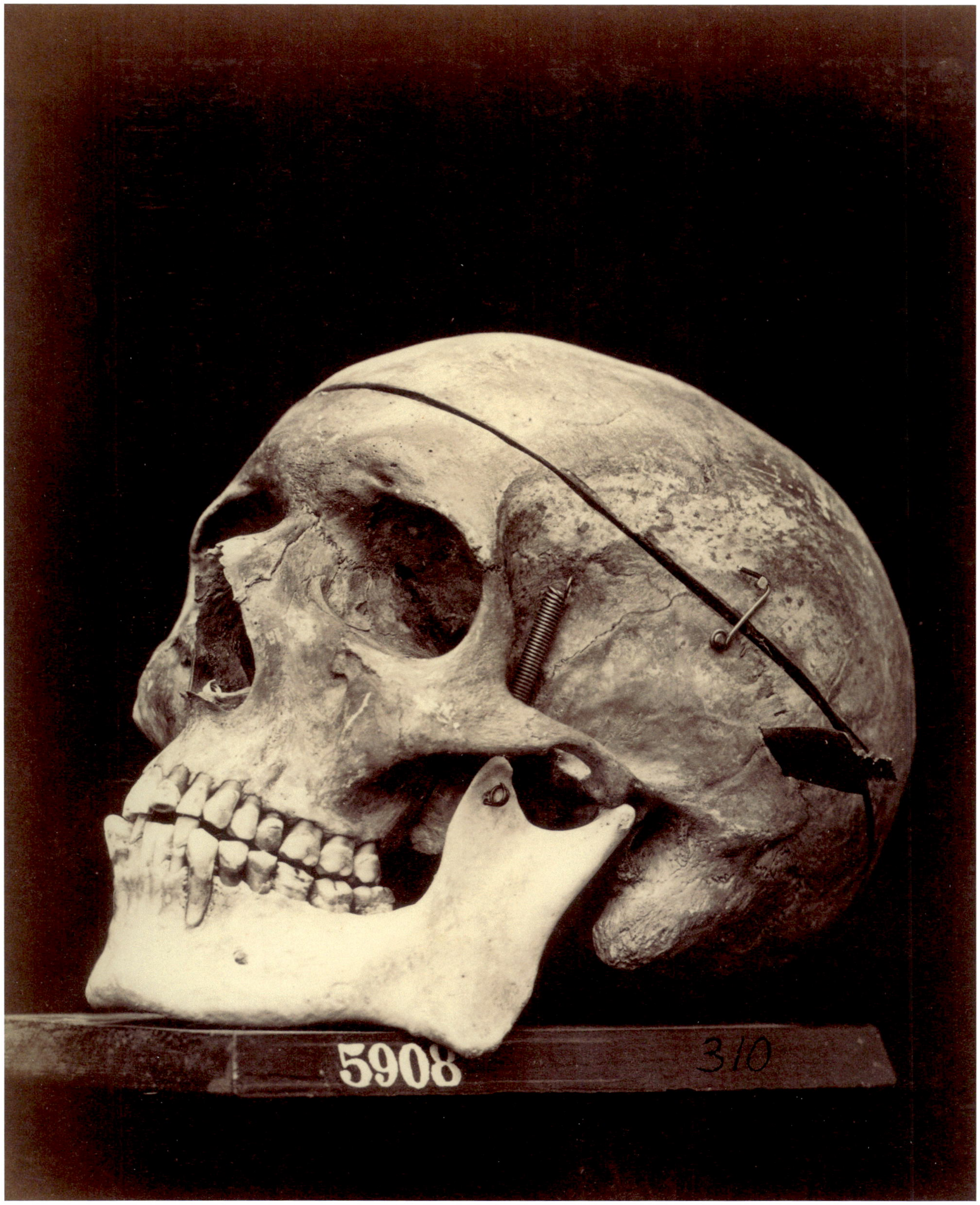

Plate 4.34

Cranium with an Iron Arrow-Head impacted in the Left Temporal Bone
Pvt Martin W. 4th Cavalry, Killed by Indians
Photographs of Surgical Cases and Specimens… Number 310

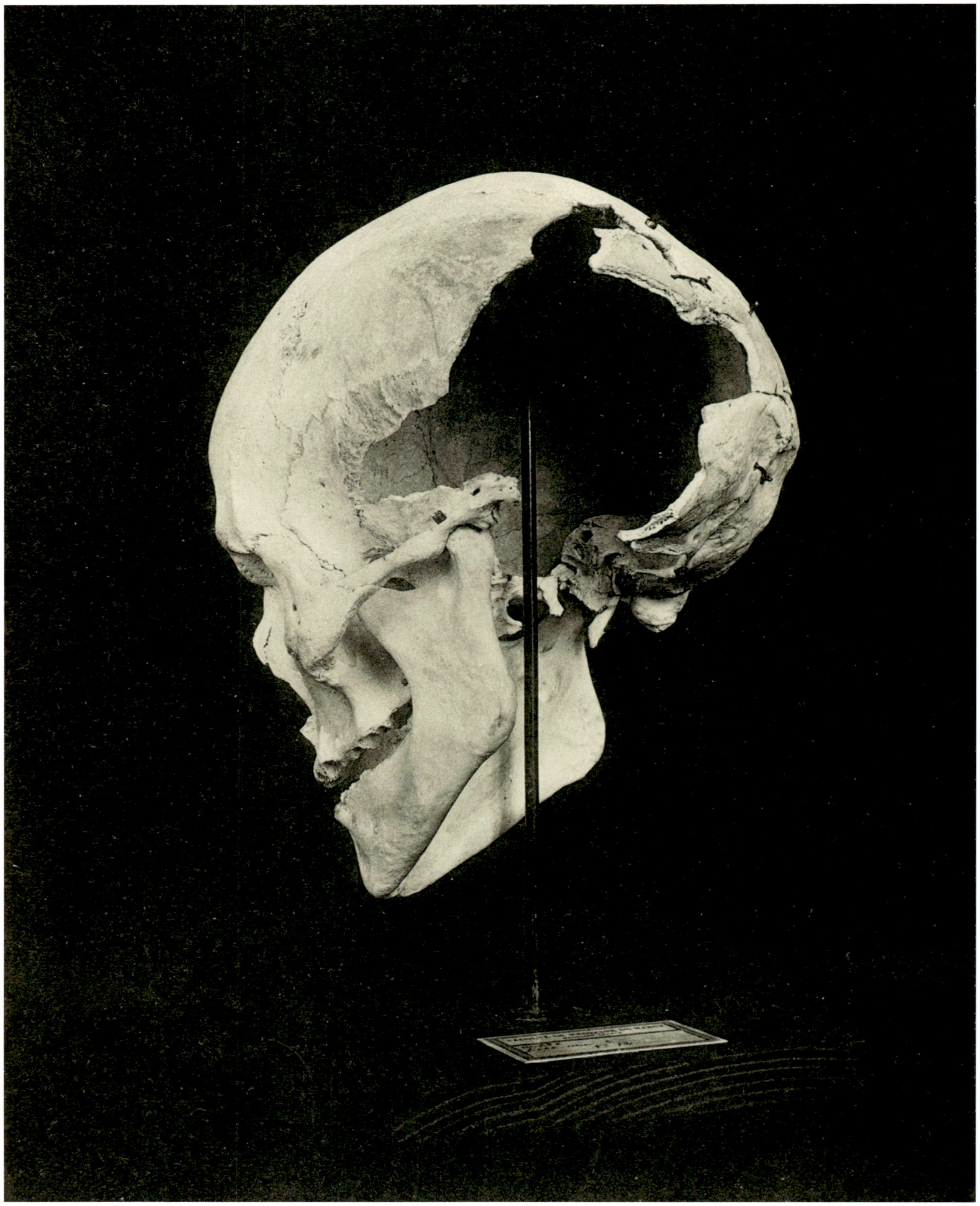

Plate 4.35

Fracture of the Cranium Due to An Assault
Alphonse Bedin, MD, *La Photographie au laboratoire de médecine légale*, Plate 34, 1908

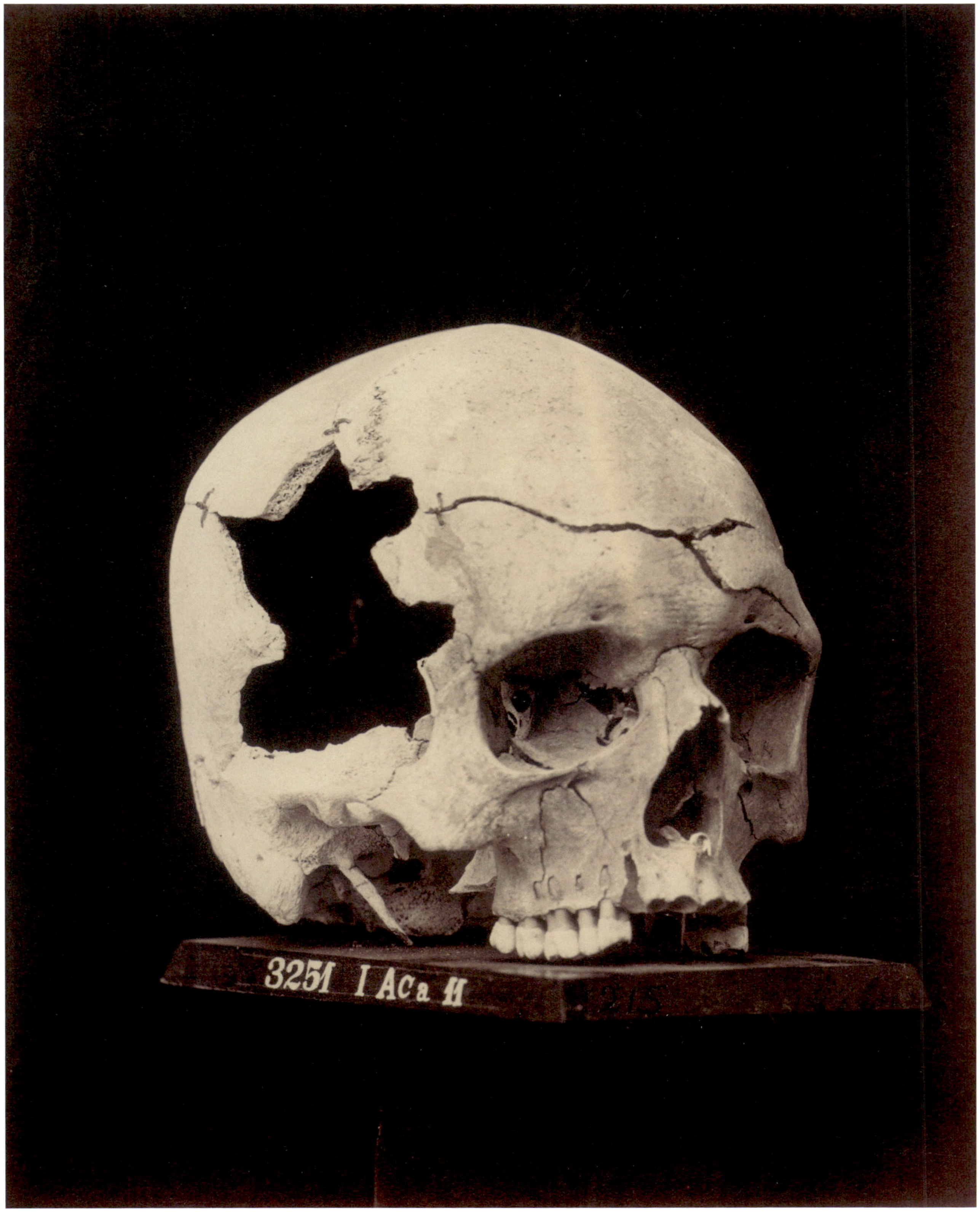

Plate 4.36

Bull Run Battle Field Find, Perforation of the Cranium by a Musket Ball
Skulls with interesting injuries were often picked up on battlefields.
Photographs of Surgical Cases and Specimens… Photograph No. 215

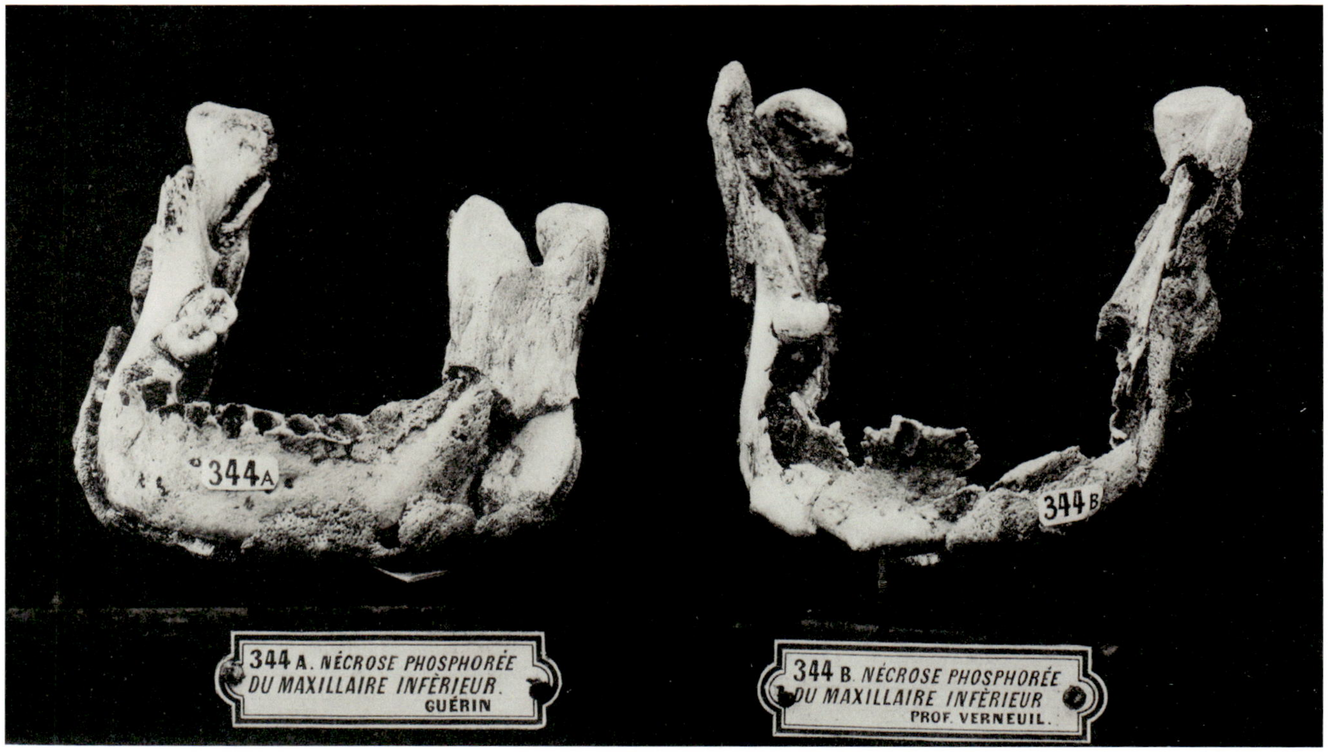

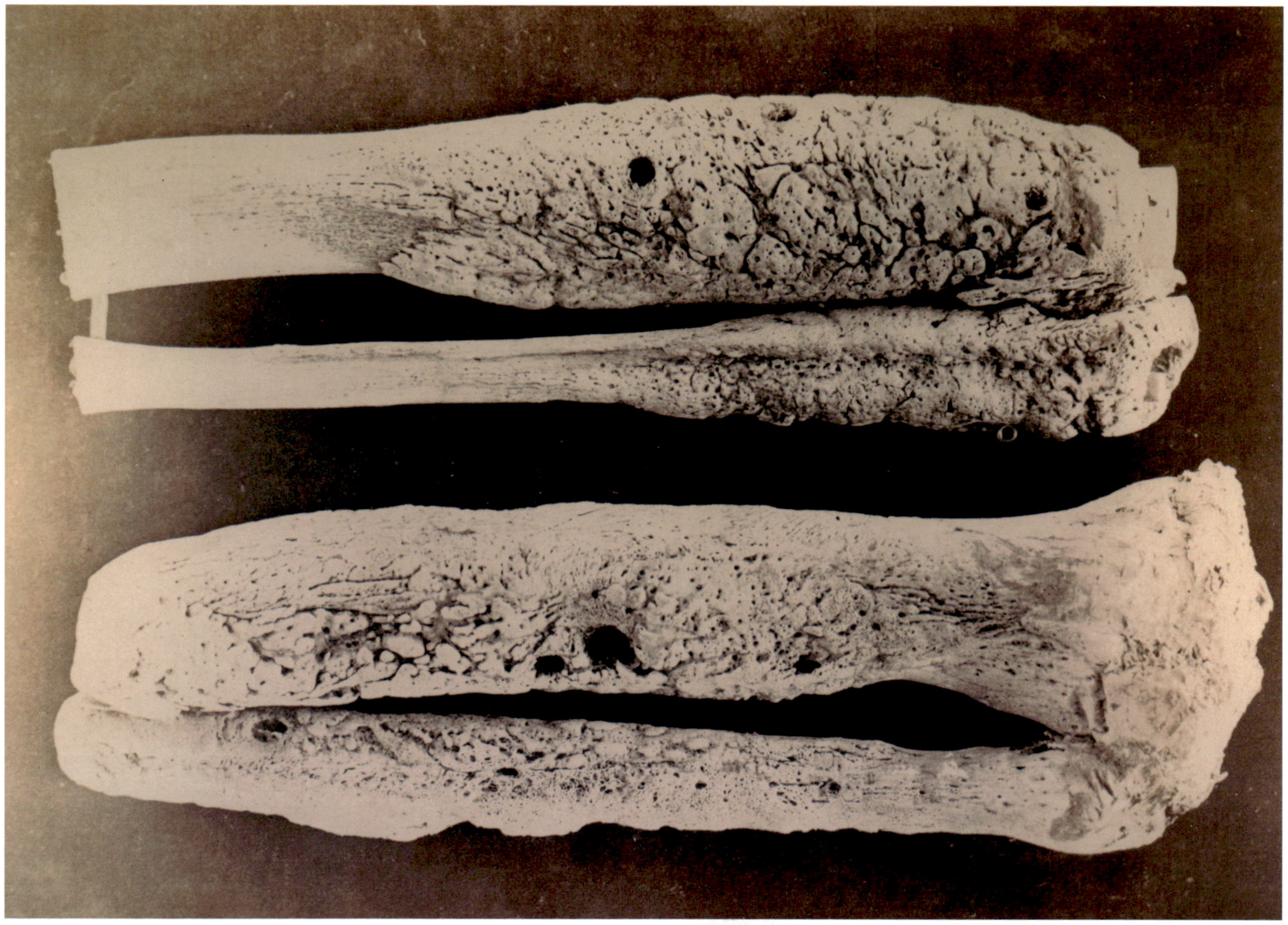

Plate 4.37

Necrosis of the Lower Jaw Bone Due To Phosphorus, Phossy Jaw
Catalogue des pièces du muse Dupuytren, Plate 396, 1879

A dreaded occupational disease of young girls in the match-making industry.

Plate 4.38

Leg bones from Two Civil War Soldiers, Second Amputation Due to Necrosed Bones
Photographs of Surgical Cases and Specimens… Number 396

Persistent bone infection after amputation often required another procedure or resulted in death.

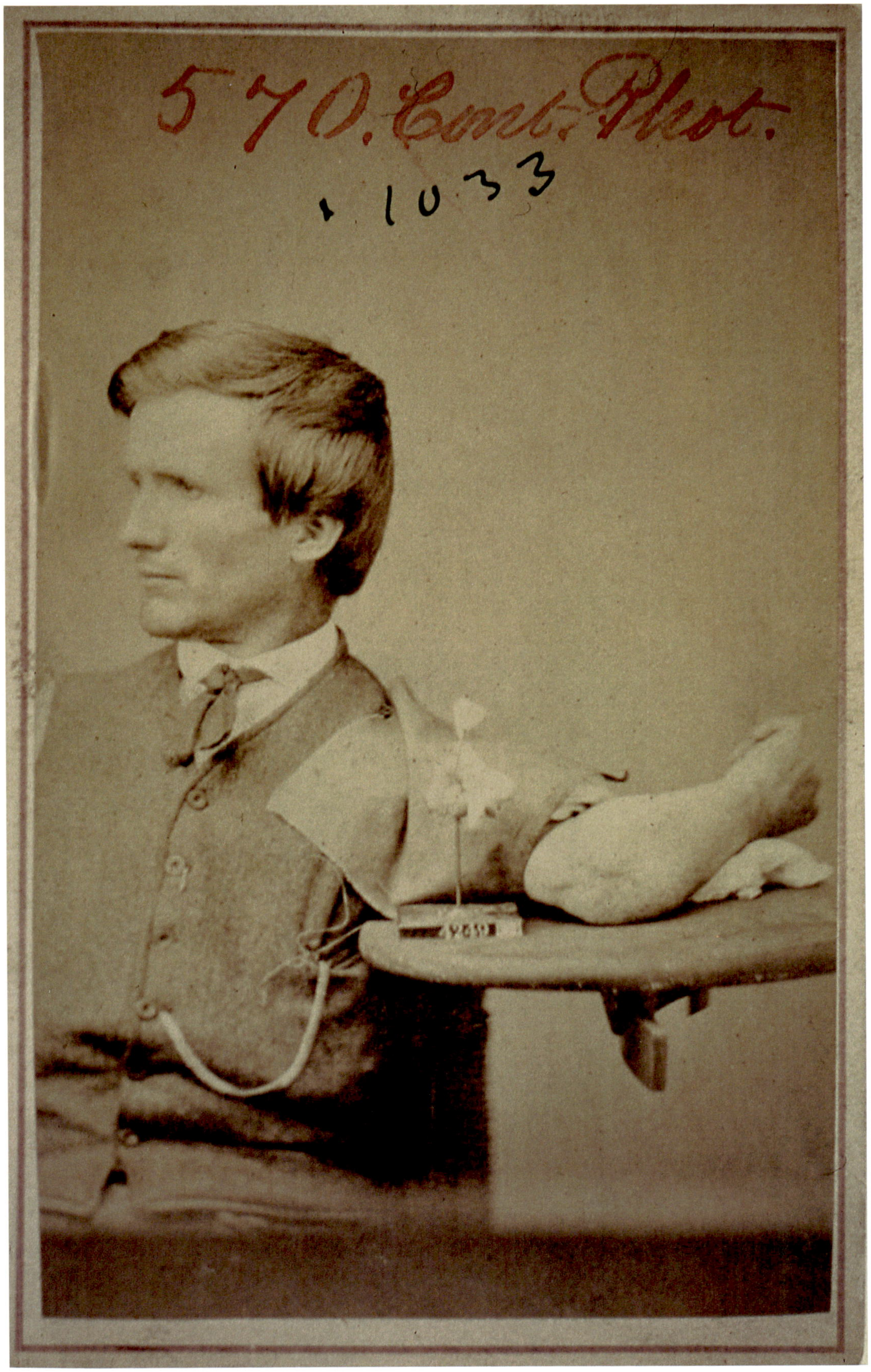

Plate 4.39

Corporal Nichol Posing with His Excised Elbow Joint
Army Medical Museum Contributed Photograph No. 570

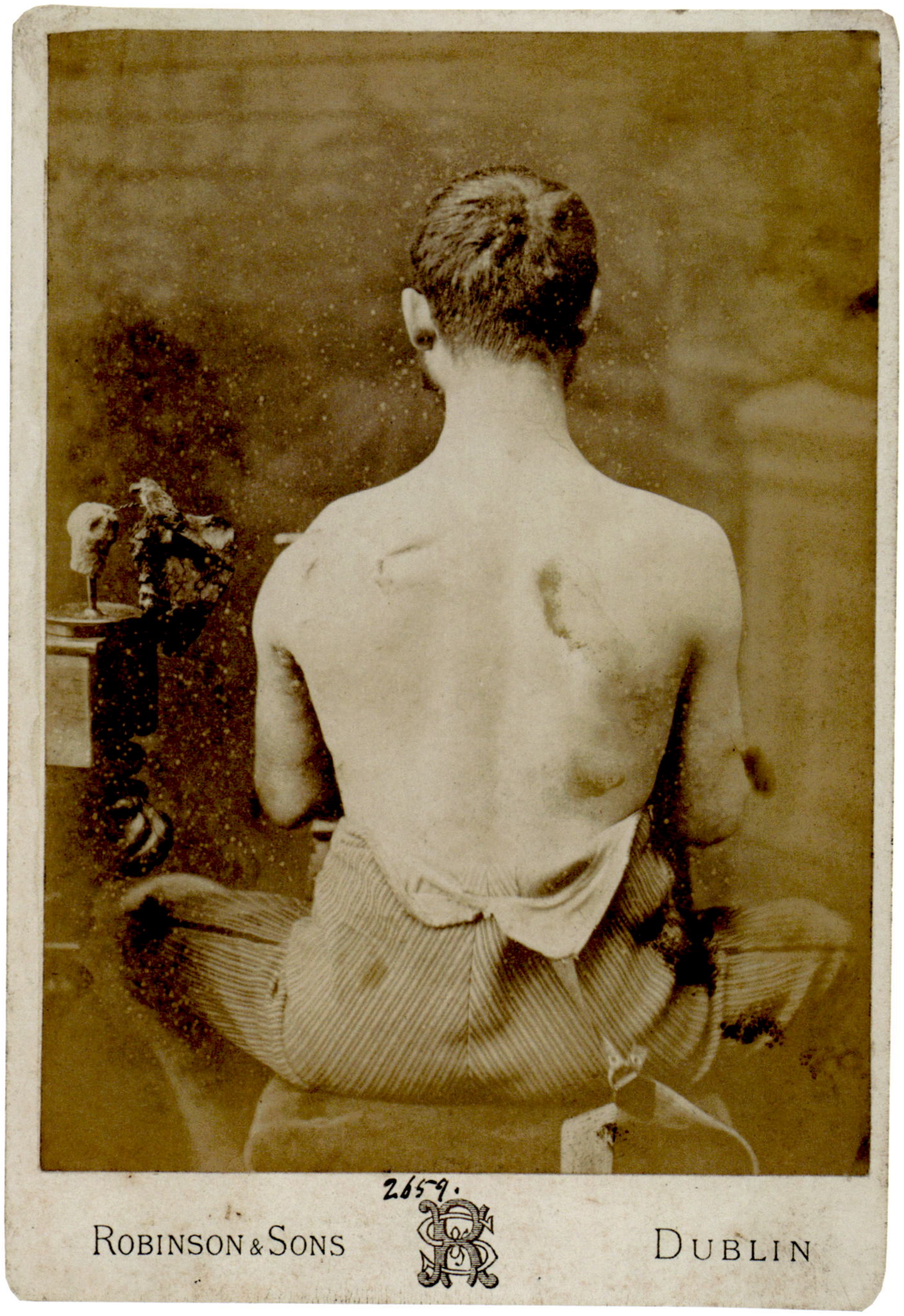

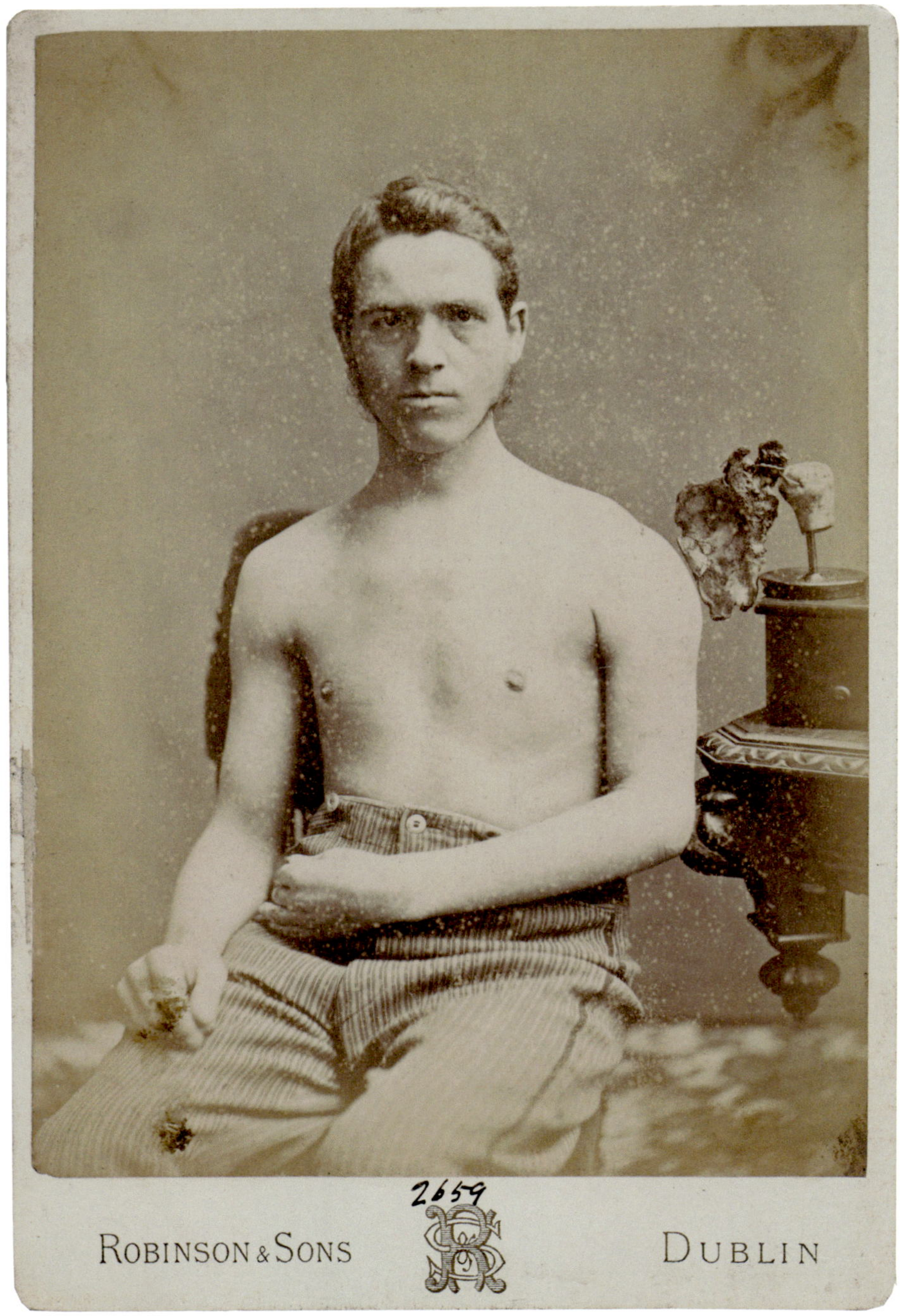

Plates 4.40 – 4.41

Patient posing with his Humerus and Shoulder
Circa 1878
Photographer: Robinson and Sons, Dublin, Ireland
Patient of Edward Stamer O'Grady, MD

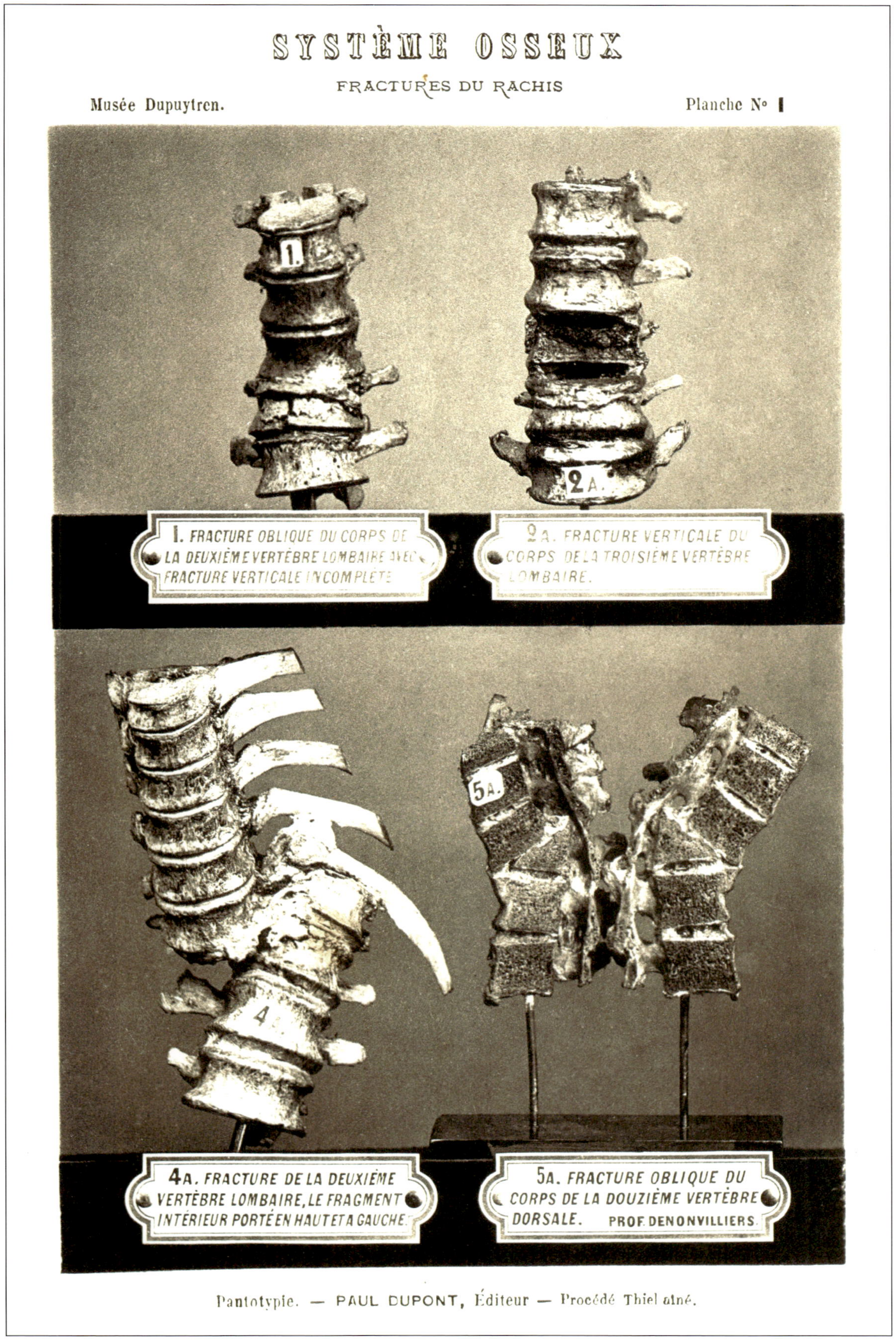

Plate 4.42

Various Fractures of the 12th and 13th Vertebra
Catalogue des pièces au museé Dupuytren, Plate 1, 1879

Duypuytren's museum has a wide variety of skeletal fractures and deformities. The museum is still open.

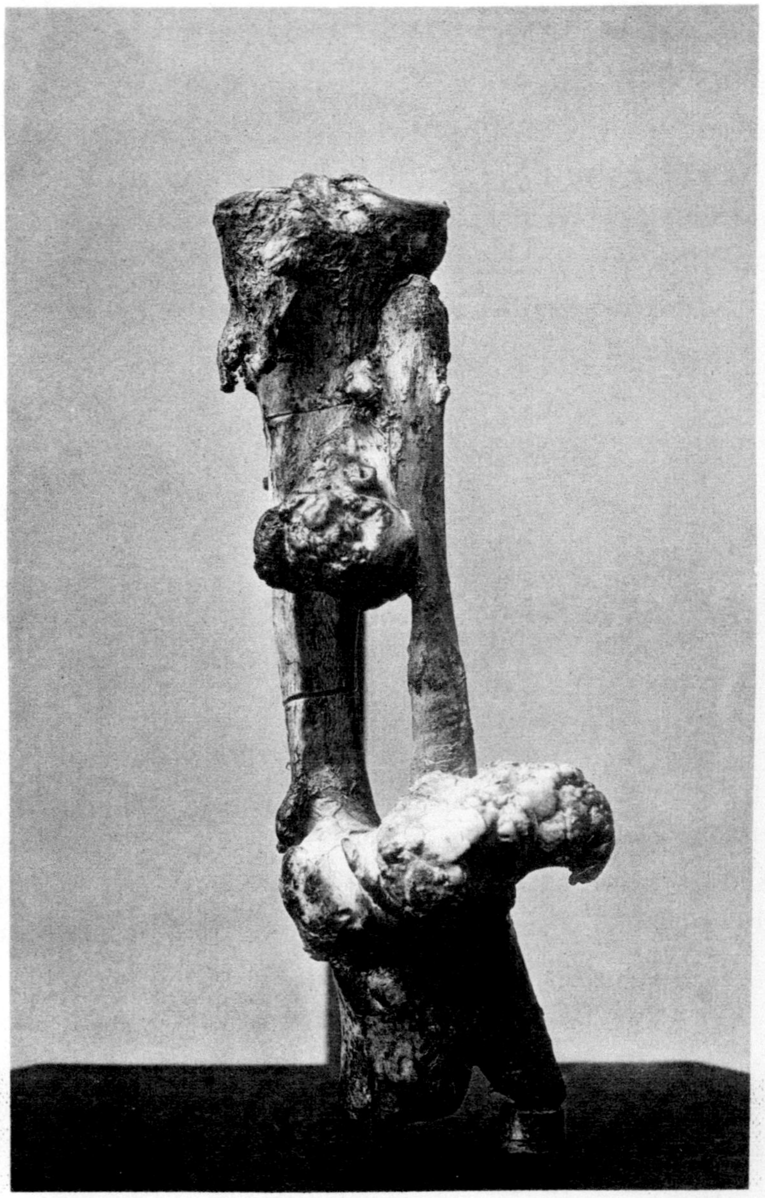

PLATE 12. MULTIPLE SPONGY EXOSTOSIS. XXIII.

Plate 4.43

Multiple Spongy Exostosis
Charles Brigham, MD, *Surgical Cases with Illustrations*, Plate 12, 1876

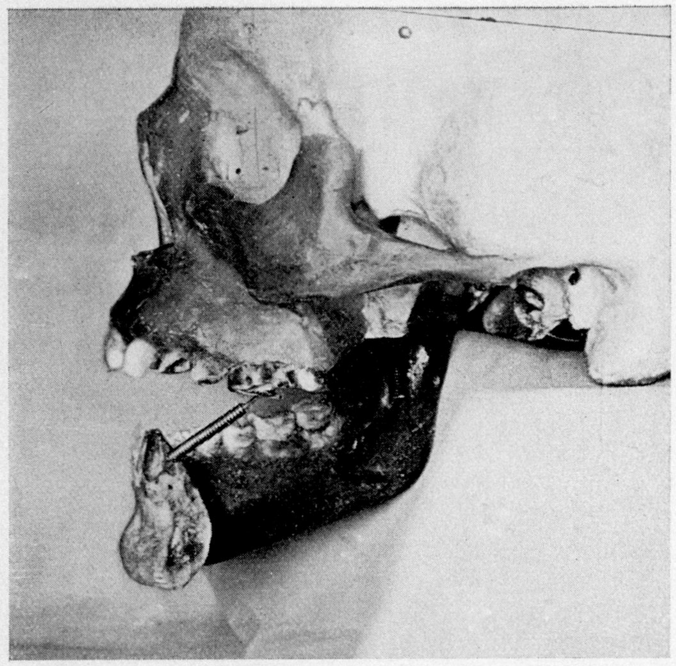

FIG. 2.—Spiral spring in position, side view.

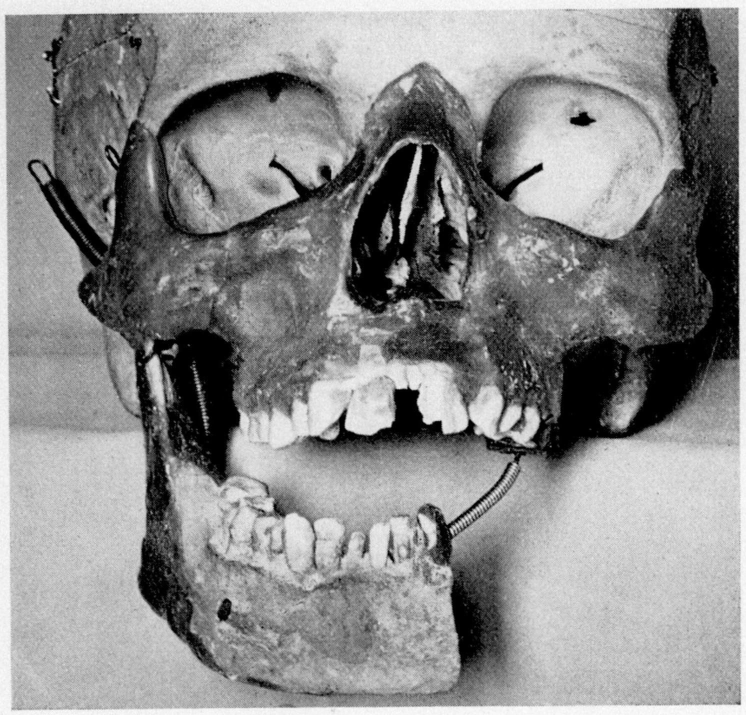

FIG. 3.—Spiral spring in position, front view.

Plate 4.44

Mechanical Appliance to Correct the Deformity after Removal of Half of the Lower Jaw
Patient of noted New York City surgeon, Charles McBurney, MD
Published in the *Annals of Surgery,* Vol XX, Jan.-June 1894

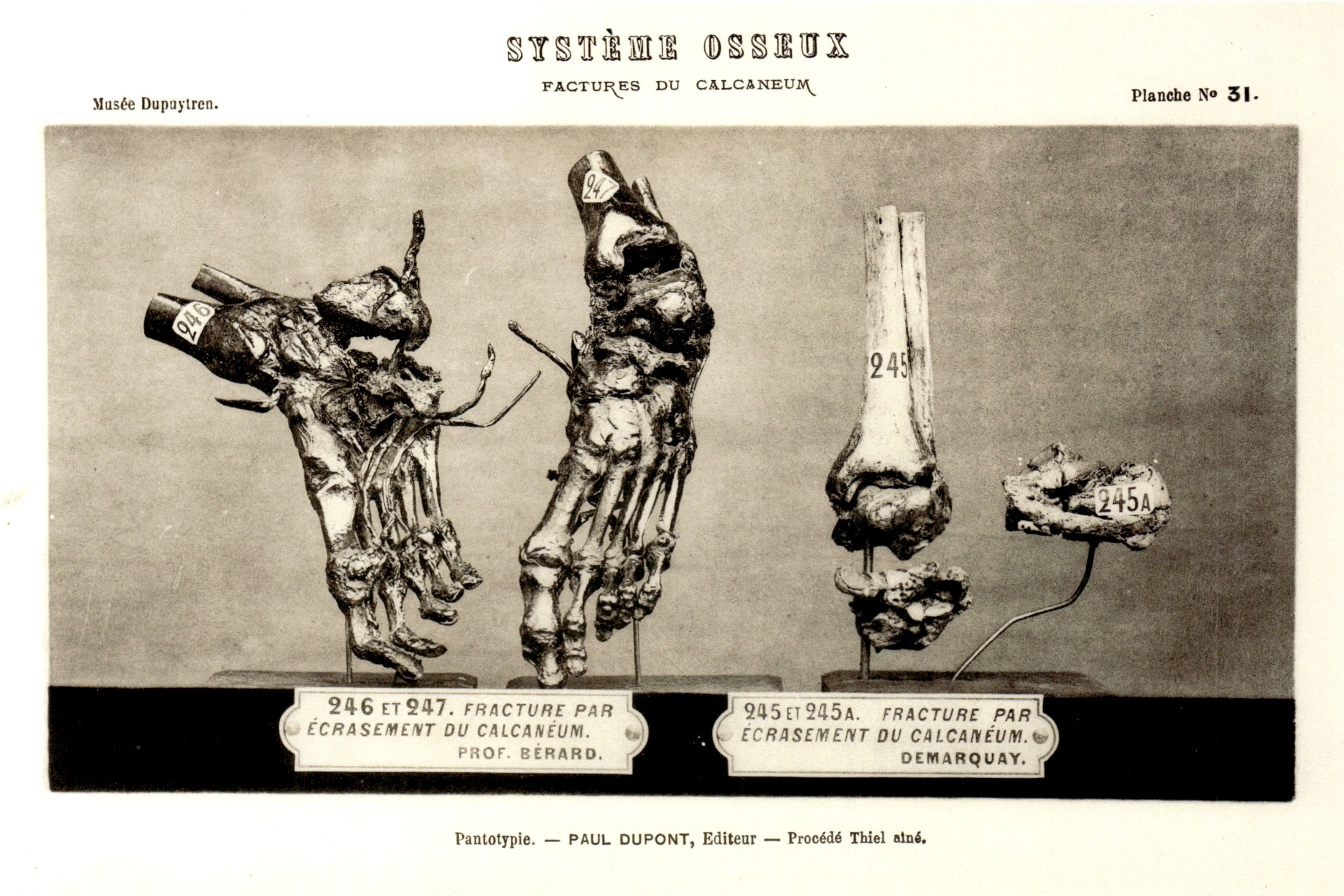

Plate 4.45

Various Crushing Fractures of the Calcaneum (Heelbone)
Catalogue des pièces du museé Dupuytren, Plate 31, 1879

Foot injuries and deformities often made a person unemployable in the nineteenth century.

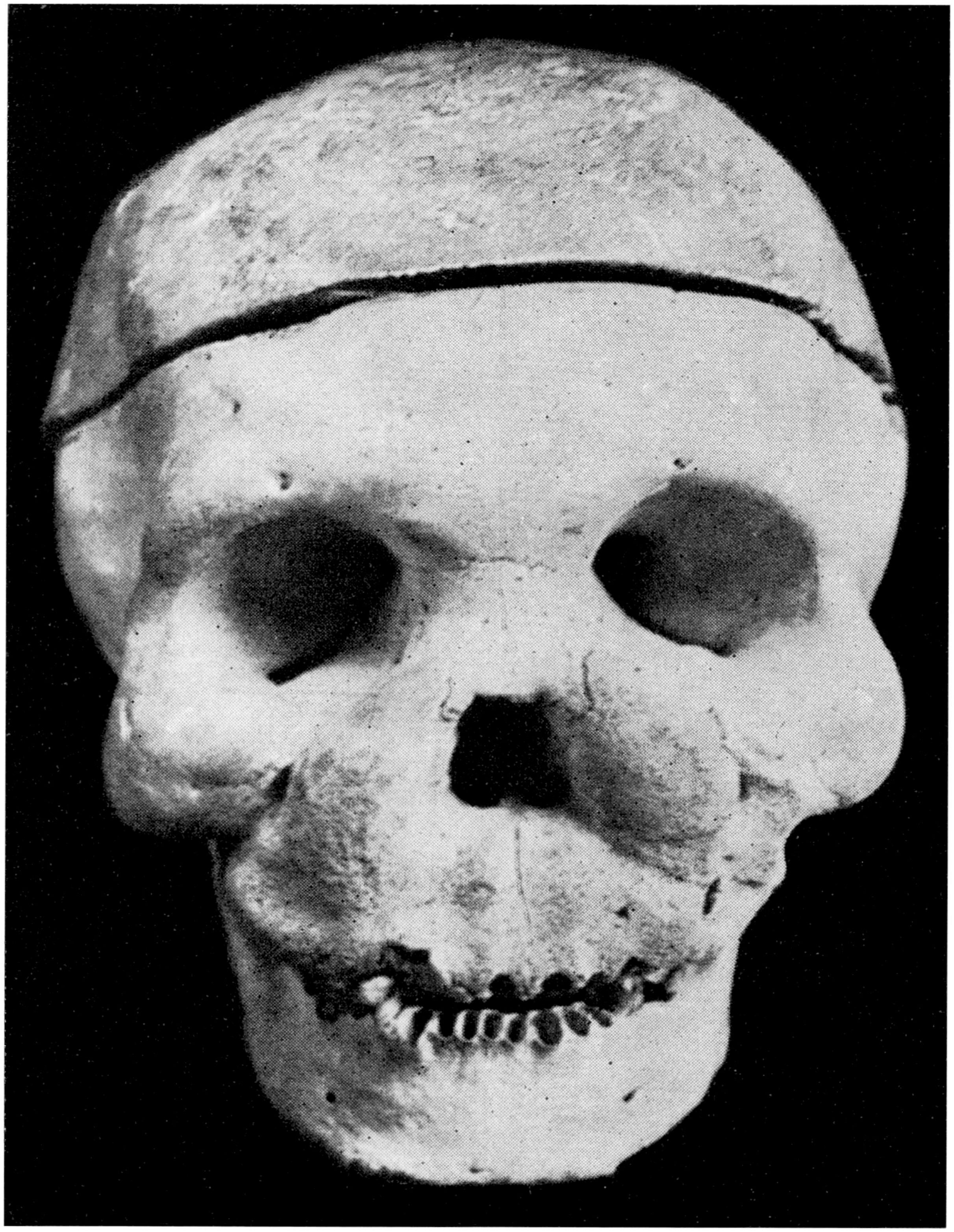

Plate 4.46

Case of Hyperostosis Cranii or Leontiasis Ossea
Published in the *American Journal of Medical Science*, July 1896

Overgrowth of facial and cranial bone destroys facial features. Death occurs from cerebral pressure.

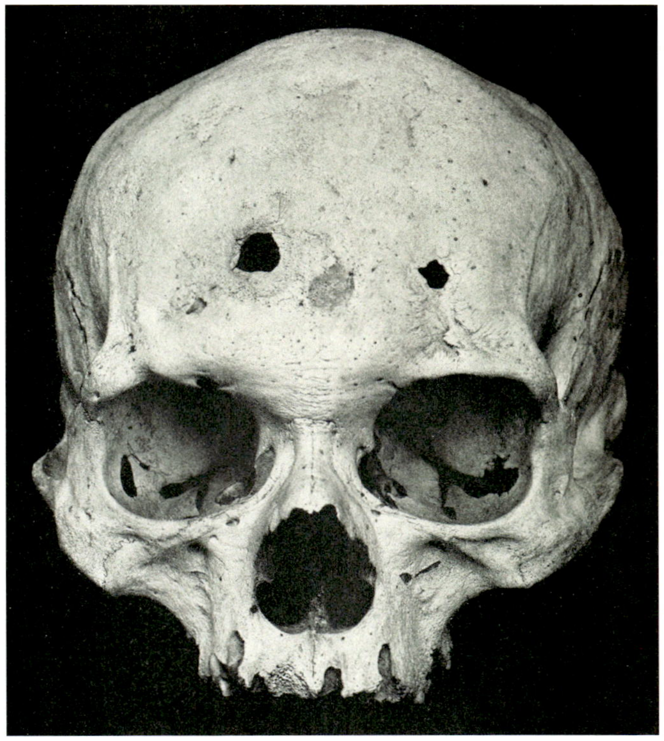

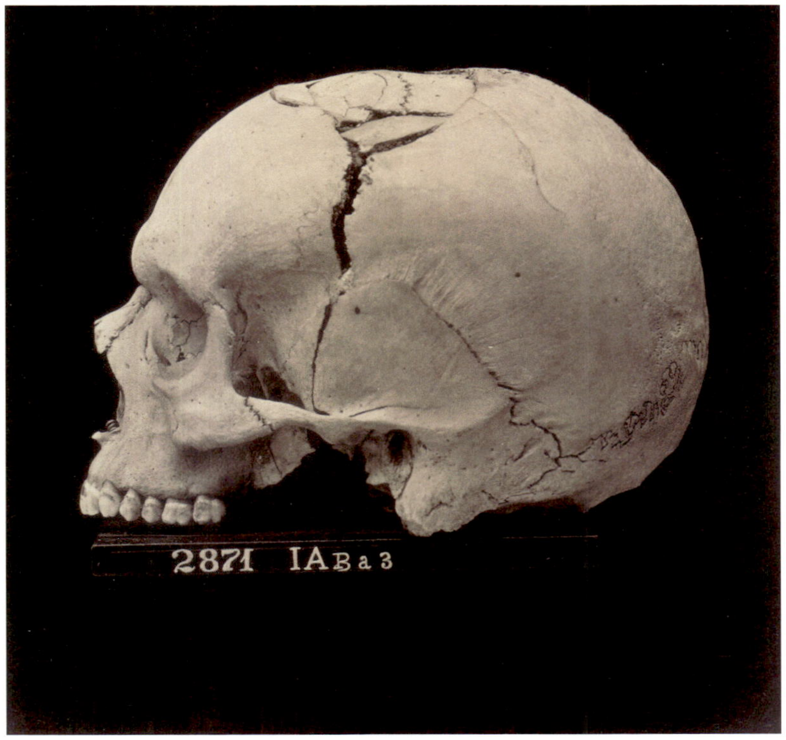

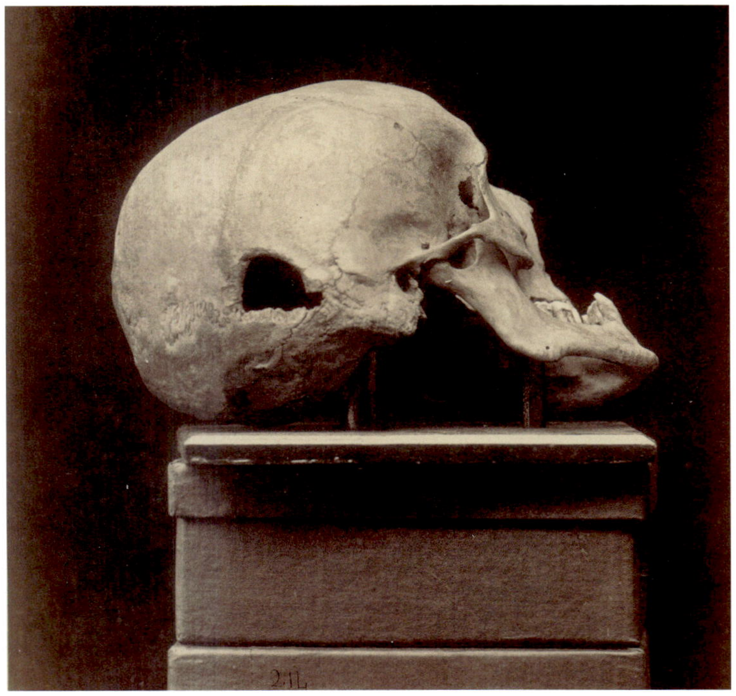

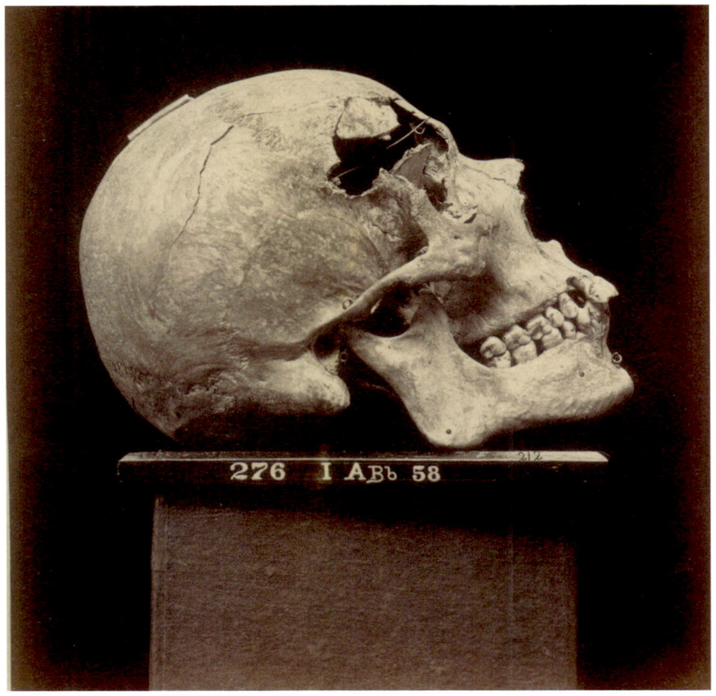

Plate 4.47

Cranium of Indian with Syphilitic Ulceration of the Skull
Circa 1890

Plate 4.49

Cranium Perforated by a Musket Ball
Unidentified Civ l War Soldier
Photographs of Surgical Cases and Specimens… Number 214

Plate 4.48

Cranium Extensively Fractured by a Shell
Confederate soldier wounded during demonstration in Washington, DC
July 12, 1864
Photographs of Surgical Cases and Specimens… Number 214

Plate 4.50

Gunshot Fracture of the Cranium Caused by a Musket Ball Striking with Great Obliquity
Private Edward Volk, D, 55th Ohio, Left on Field Six Days
Photographs of Surgical Cases and Specimens... Number 212

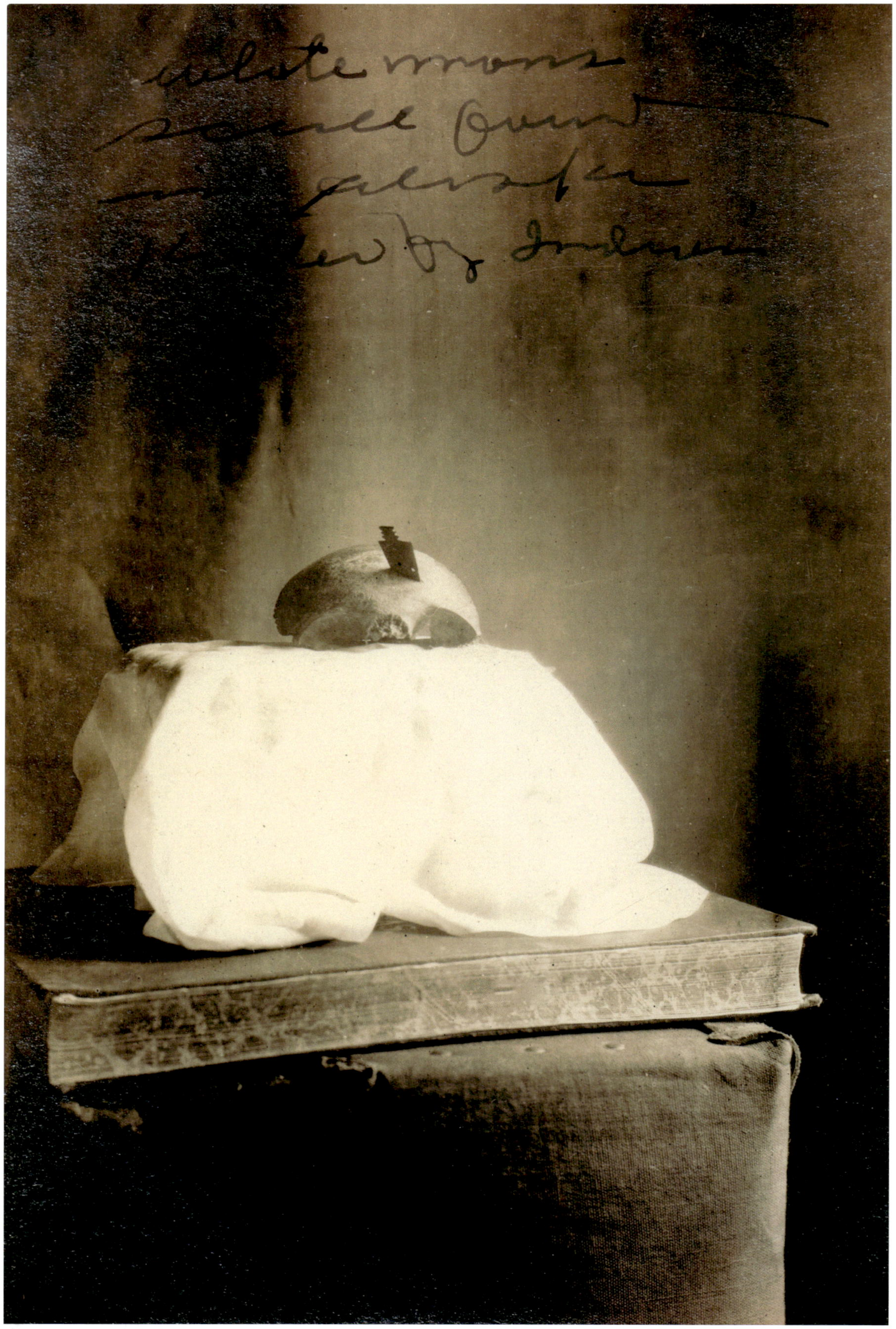

Plate 4.51

White Man's Skull Found, Killed by Indians
Postcard, circa 1905

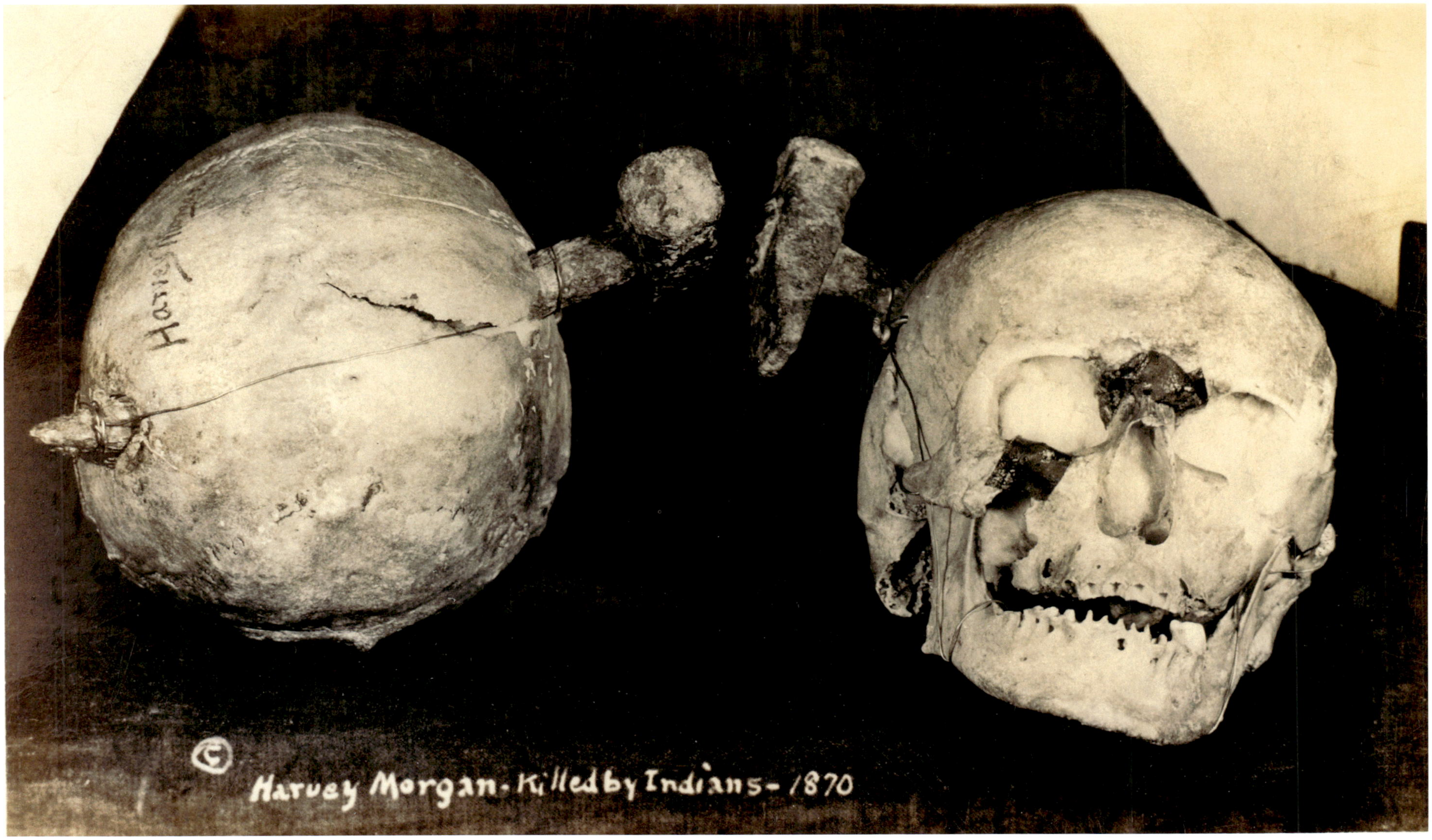

Plate 4.52

Harvey Morgan, 25 Years Old, Killed by Indians, June 25, 1870

Killed, then a wagon hammer was driven through his skull, six miles from Lander, WY. Morgan was scalped, then his back was split open and "the sinews of his back" removed.

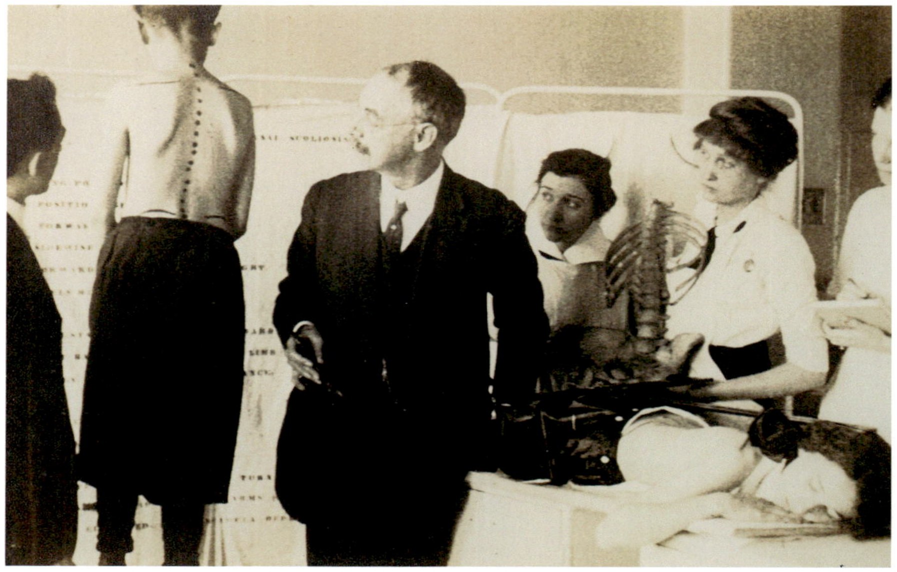

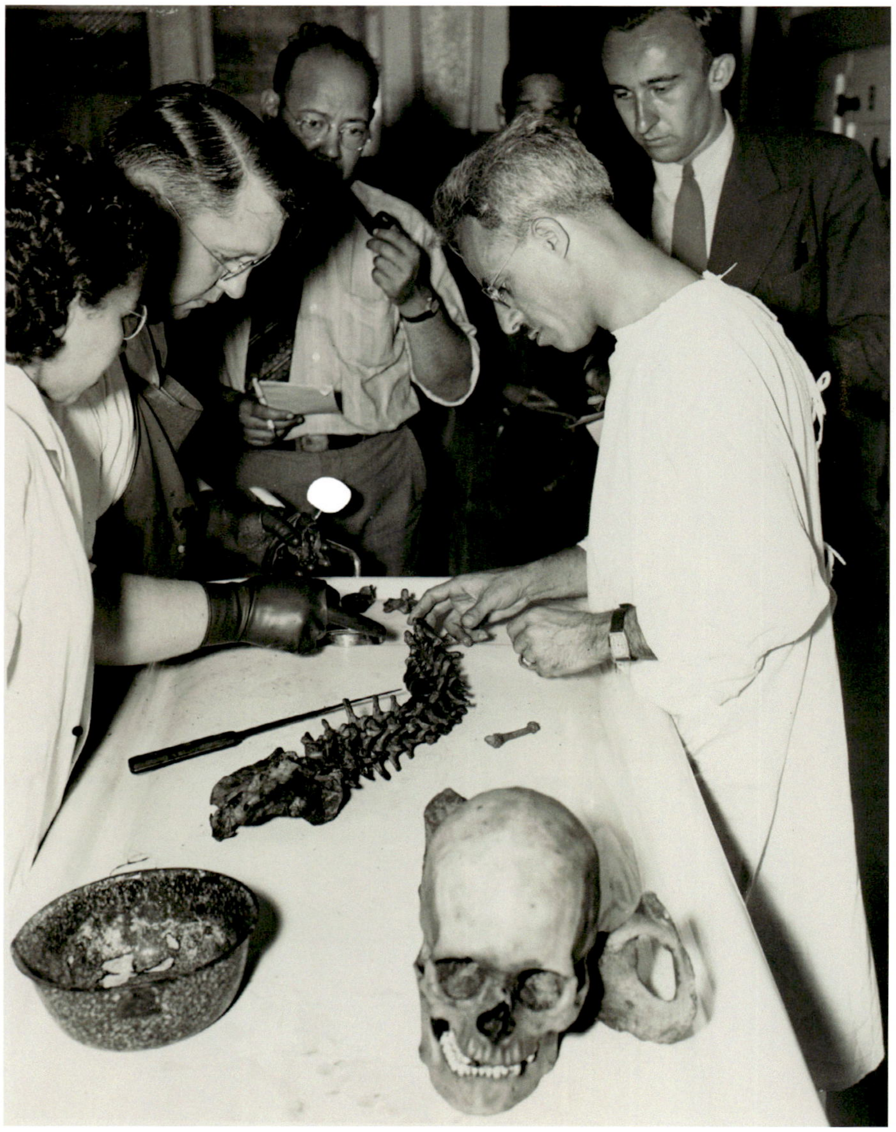

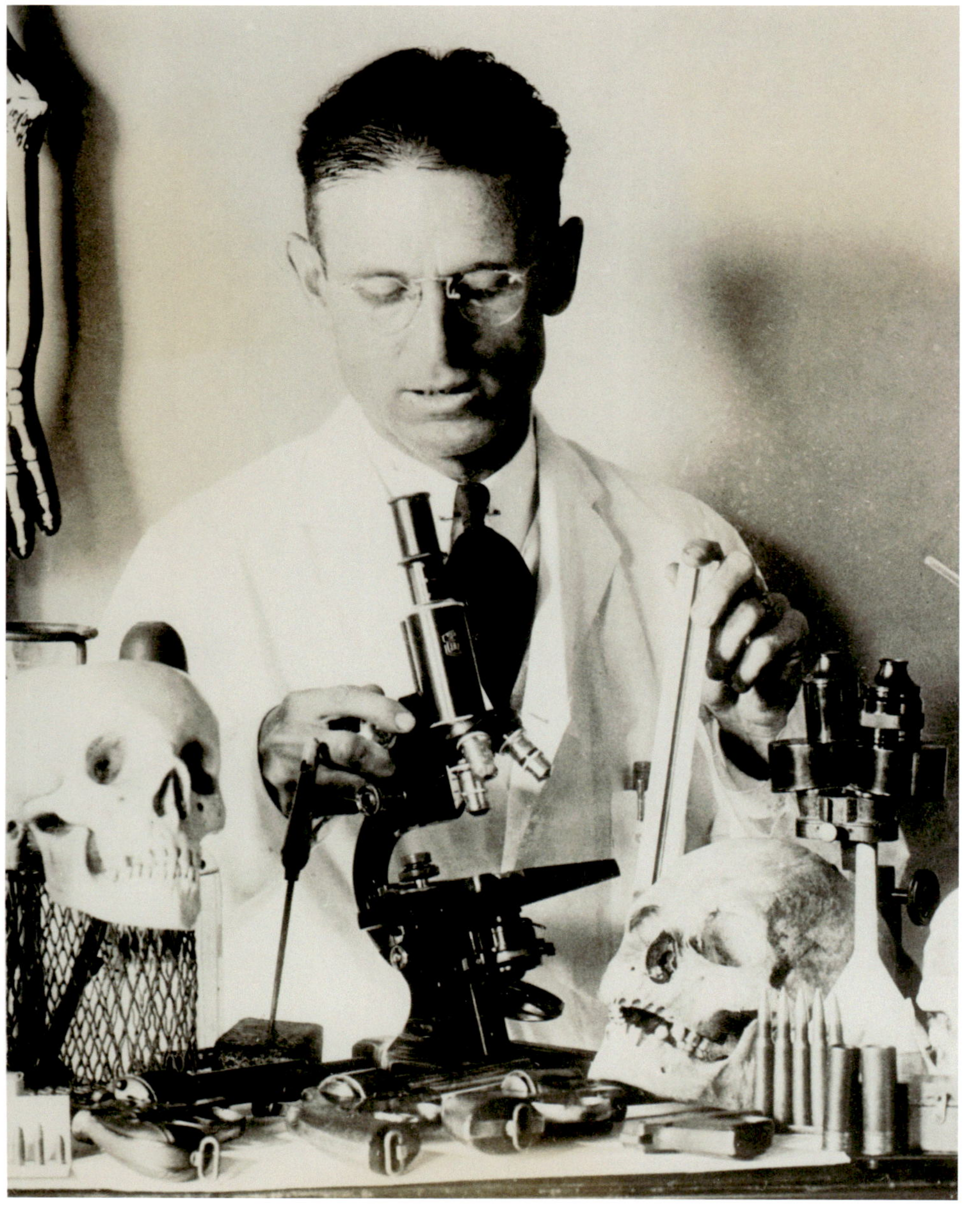

Plate 4.53

School Physician Checking for Spinal Curvature
1915

The physician has a model of the spinal cord for instruction for school nurses. At the turn of the century, scoliosis and other spinal curvatures were evaluated by school physicians.

Plate 4.54

Cleveland Coroner Dr. Gerber and Staff Examining Remains of Body
1938

Evaluation of skeleton parts is an important part of forensic pathology.

Plate 4.55

Criminologist Lt. Robert Hicks Working on Lindbergh Baby Murder Case
1936

Special assistant to Governor H. Hoffman to help determine Bruno Hauptmann's guilt.

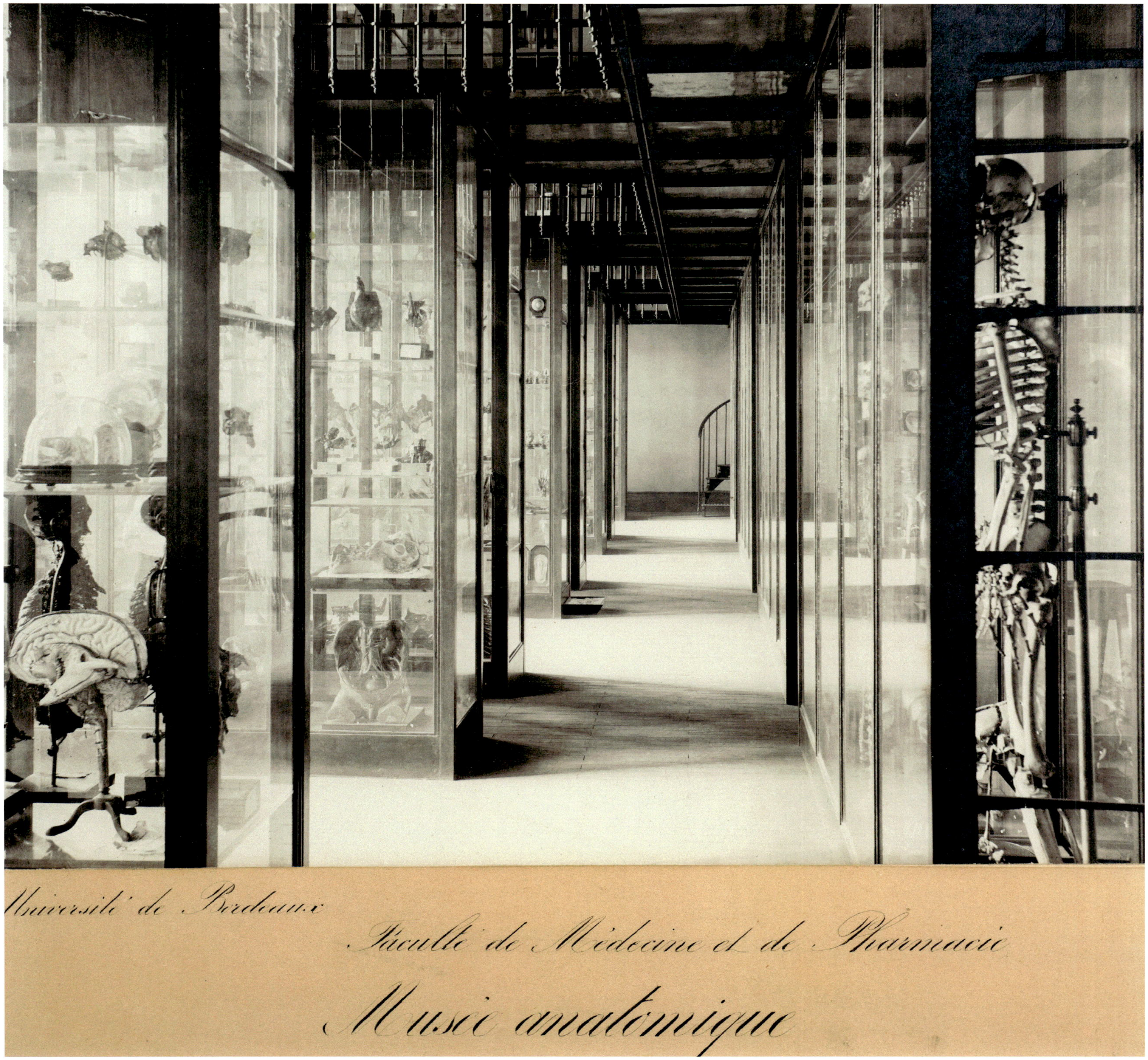

Plate 4.56

Skeletons in the Anatomy Museum at Bordeaux Medical School
From an album documenting the medical school, circa 1900

Skeletons and skeletal parts were always an important aspect of anatomical museums.

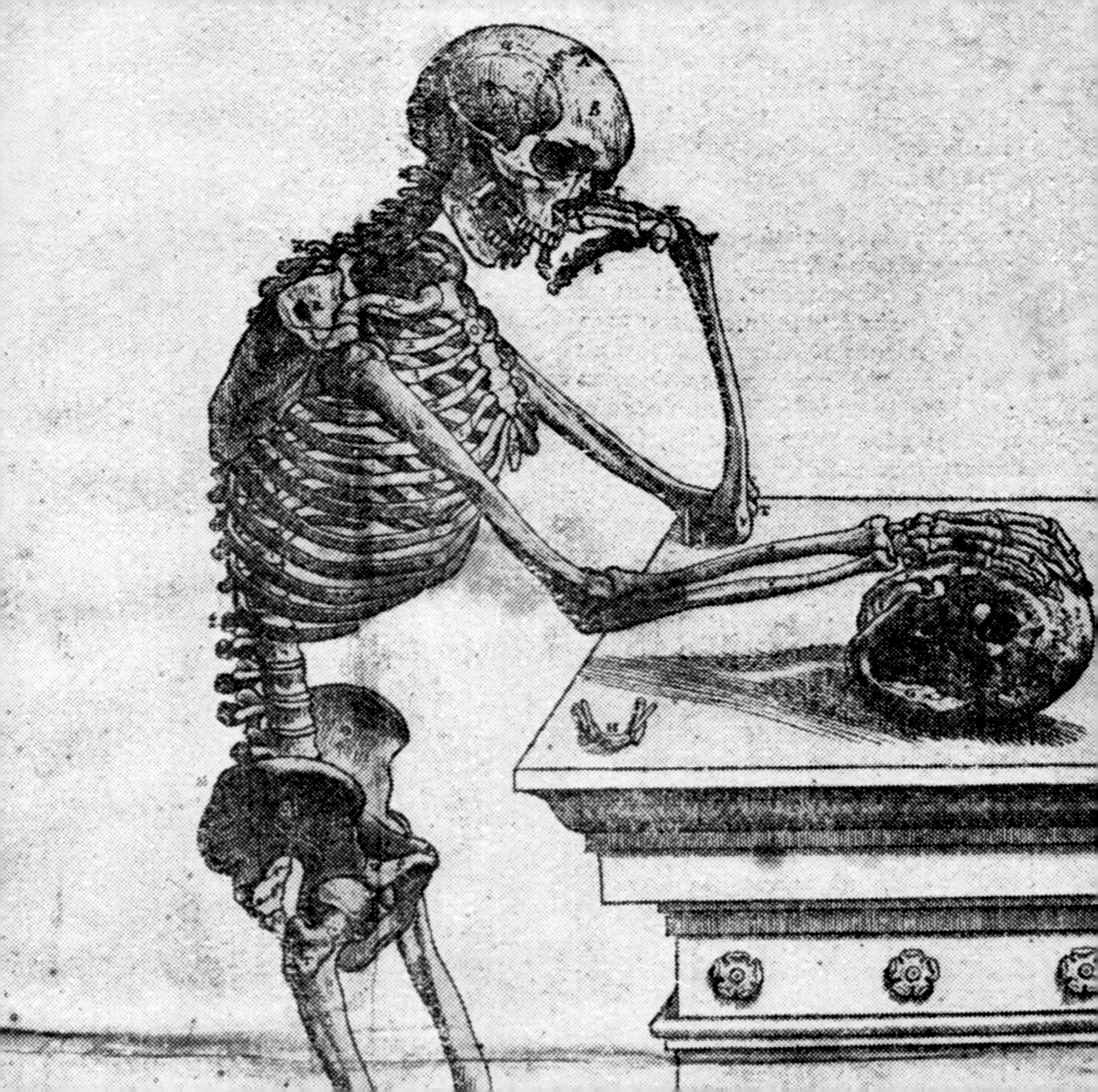

CHAPTER FIVE

Skull Studies

Anthropological & Scientific

We have a very large collection of skulls of murderers, who have been executed and of soldiers killed on battlefields, also of Indians, Africans, Egyptians, Chinese and Cannibals, but we have only a few from the higher class of minds, such as Reformers, Statesman, Scholars Etc. Of these we have hundreds of casts and busts from living heads, but not their skulls. What a treasure it would be if some plan could be devised, by which the leading "types" could be preserved, as specimens for scientific purposes.

—*The Phrenological Journal, May, 1855*

The most important aspect of skull collecting and analysis was, for centuries, focused on the scientific study of skulls as part of cultural identification, anthropology, psychiatry, and criminology. Today, studying skulls is focused on identifying early humans and humanoids in the quest to fill in the blanks of the evolution of mankind. Prior to this lofty goal, skulls were studied by a wide variety of scientists who were bent to establish racism on a scientific basis.

Skull measurement is a significant facet of physical anthropology, phrenology, ethnology, eugenics, and, especially the science of polygenism, the belief that human races were distinct, created separately, and thus were inescapably and irrevocably different and unequal. In this theory, the 'Caucasian' race is seen as having the highest intelligence and social development and the 'Black' races as inferior. Opposing the polygenist view was that of the monogenists who argued that all races were descended from Adam and Eve, and that the differences in race were due to 'degeneration.' The idea that there were several independent races of man dominated scientific study for much of the nineteenth century.

For almost three centuries, skull studies dominated anatomists and medical scientists in their quest for understanding the human race whether, polygenist or monogenist. Many of the past studies of skull size, brain size, and intelligence have long been proven to be pseudo-science and formulated to support racism. By the first third of the twentieth century, it was shown to be false. The acceptance of Darwin's theory of evolution stimulated the modern interest in skull analysis.

POLYGENISM

According to eighteenth-century science, there appeared to be an anatomical basis for racism: skull shape was one determinant. Scientific racism became the cornerstone of racial belief, as scientists seemed to be proving that there were scientific reasons for prejudice. Also in the late eighteenth century, physiognomy and phrenology became firmly established among a wide range of scientists as being valid determinants of personality, behavior, and intelligence. Both of these disciplines were based on skull shape, features, and configuration.

The idea of organizing humankind into specific racial groups was initially supported by skull study. German anatomist Johann Friedrich Blumbach, MD, (1752–1840) developed the polygenism basic theory of mankind's five races in 1779. During the next decade, he studied and collected skulls and clarified his racial divisions in 1793–1795 as: 1) the Caucasian or white race, 2) the Mongolian or yellow race, 3) the Malayan or brown race, 4) the Ethiopian or black race, 5) the American or red race. His classification of Mongolian race included all East Asians and some Central Asians.

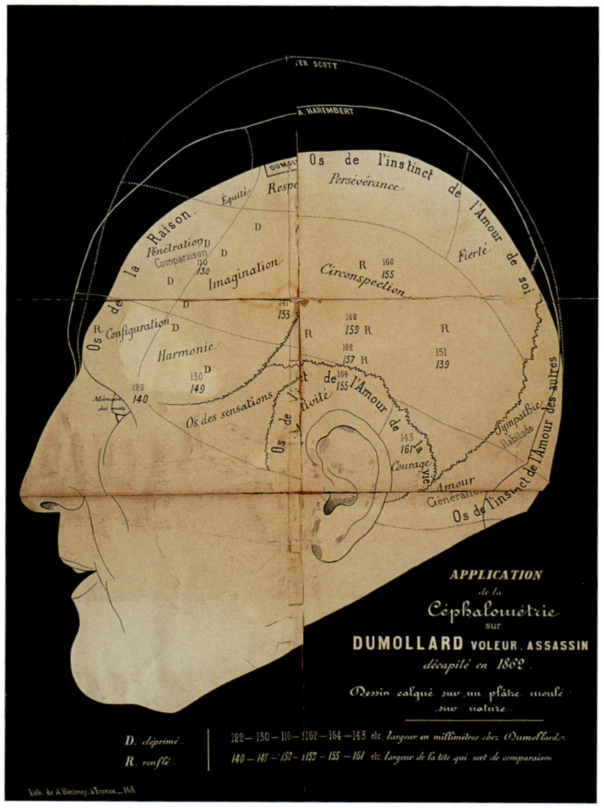

Plate 5.1

Application of Cephalometry to the Head of the Assassin, Martin Dumollard
1863

Study of the head of the "Monster of Montieul." He supposedly drank the blood of his victims.

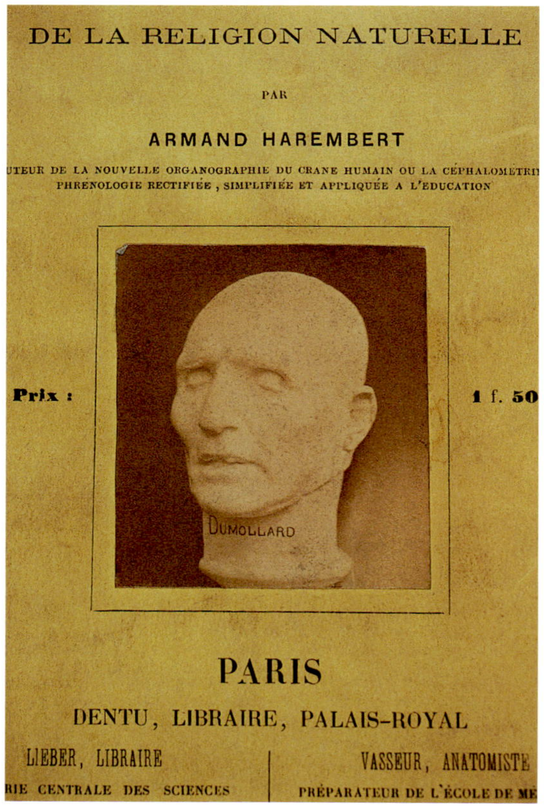

Plate 5.2

Phrenological Analysis Of The First French Serial Killer, Martin Dumollard
1863

Photograph of model of Dumollard's guillotined head.

During the nineteenth century, a wide group of scientists, especially anatomists, began to study skulls and collect examples of the world's crania. Philadelphia physician and professor of anatomy at the University of Pennsylvania Medical School Samuel George Morton, MD, (1799–1851) was the preeminent collector of skulls as he attempted to substantiate the existence of five races of mankind. In 1830, Morton presented a lecture explaining Blumbach's racial theory, "The Different Forms of the Skull as Exhibited in the Five Races of Men." While Morton had a small collection of skulls to illustrate his lecture, he realized he didn't have enough examples of the world's peoples to satisfy his own curiosity and scientific concepts. He began his hunt to acquire skull specimens from all around the world. As a respected and distinguished university anatomist and a member of the National Academy of Natural Sciences, Morton began a campaign to collect skulls from across the globe by writing to scientific colleagues, as well as missionaries and military personnel stationed at remote facilities. Over the next two decades, until his death in 1851, at age 52, his 138 contacts supplied him with 867 skulls. It was the largest collection of crania in the world.

After Morton's death, fellow Philadelphian anatomist James Aitken Meigs, MD, (1829–1879) continued adding to the collection. By 1872, it numbered 1,225 skulls. Meigs published several significant articles on ethnological and craniological subjects during his career. In 1857 he wrote "The Cranial Characteristics of the Races of Men" in Nott and Glydden's *Indigenous Races of the Earth*. The Morton collection had become known as "the American Golgotha," (Biblical meaning the place where Jesus was buried, however it is interpreted by the evangelists as meaning "the place of a skull" Matthew 27:33, Mark 15:22, John 19:17). Morton's crania were indeed representative of the world's people. His skulls were taken from fallen combatants in wars with indigenous peoples around the globe, and if no wars were in progress, Morton's aggressive scientific suppliers would simply rob the graves and tombs of noted identified local populace. Epidemics and local diseases also served as a source for specimens. For practical storage purposes, and because it was essentially the brain cavity and cranial shape that was of most interest, the majority of Morton's skulls are devoid of the mandible (lower jaw bone).

In 1839, Morton began his cranial publication project with, *Crania Americana: or A Comparative View of the Skull of Various Aboriginal Nations of North and South America: To which is Prefixed an Essay on the Varieties of the Human Species*. The book with seventy-one illustrative plates concludes with an essay by noted Scottish phrenologist George Combe (1788–1858). In 1844, Morton published, *Crania Aegyptiaca, or, Observations on Egyptian Ethnography, Derived from Anatomy, History, and the Monuments*. In this volume, Morton examined the ancient crania of Egypt and determined the ruling classes to be 'Caucasian' and the servant class to be 'Negro.' The book was used by racists to support the ideology of the American slavery system. Morton's work was not an act of simply collecting skulls. He marked each cranium with a number, the date, geographic place where found, and at times additional information such as who supplied the skull and circumstances of death. Morton preserved each skull with varnish and the collection seemed to prove there were separate human races.

The Morton Collection is currently housed at the University of Pennsylvania Museum of Archeology and Anthropology in Philadelphia. In 1990, the Native American Graves Protection and Repatriation Act (NAGPRA) was passed by Congress. It requires that all organizations receiving federal funding notify and work in conjunction with Native American, Hawaiian, and Alaskan groups in order to return skeletal materials and artifacts that derive from their ancestors. By 2009, the Morton collection returned about one hundred skulls, mainly of Hawaiian origin. Computed tomography has made the skulls available for future research.

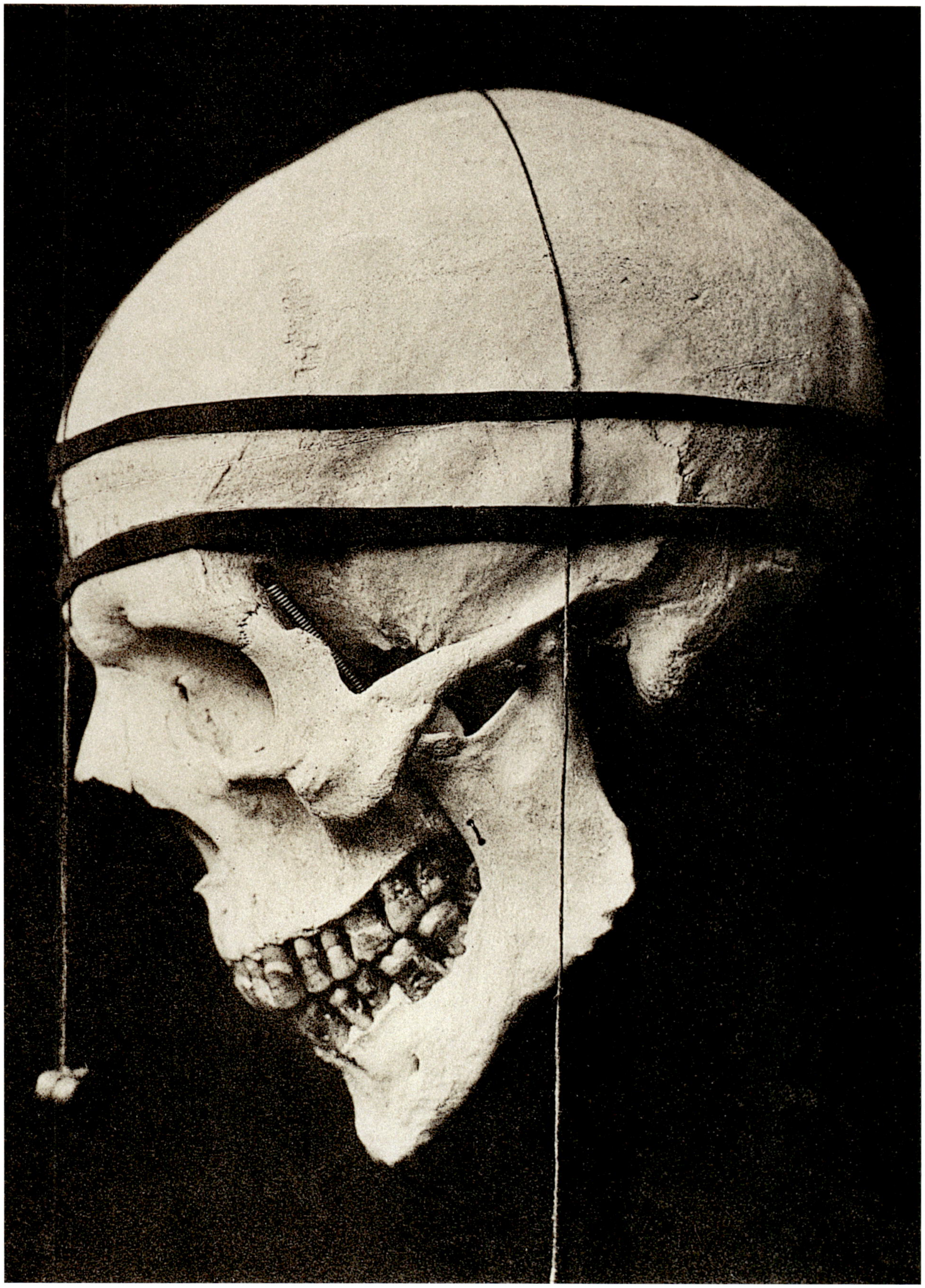

Harvard University professor Stephen J. Gould's (1941–2002) much acclaimed 1981 book, *The Mismeasure of Man*, criticized Morton, accusing him of bias and using his personal beliefs in dictating his analysis of crania. Gould stated that Morton manipulated samplings of skulls, as well as calculations, to prove that whites had the largest cranial capacities of all races. Professor Gould achieved much fame in the 1980s and '90s, an era in which racial profiling and social rights were in the forefront of political and social change.

In 1998, John S. Michael published in *Current Anthropology*, "A New Look at Morton's Craniological Research." Michael actually remeasured a selection of Morton's skulls and found that Morton's measurements were accurate and there was no evidence that Morton's personal beliefs racially biased his data. C. Loring Brace, in his 2005, *"Race" is a Four-Letter Word: The Genesis of the Concept*, and other authors have pointed out that it was Gould, not Morton, who represents a scientist who selectively has chosen his data to support an assertion. It was Gould who erred; he allowed his own personal beliefs and his perception that Morton was a racist to determine that Morton was a suspect scientist. Gould was the suspect scientist, as he never checked Morton's calculations. The latest data also supported Morton. In "The Mismeasure of Science: Stephen Jay Gould versus Samuel George Morton on Skulls and Bias," published in 2011 (*PLOS Biology* journal), six scholars agree that there was bias, but the bias came from Gould, who failed to examine, let alone remeasure, the crania to determine Morton's level of accuracy.

Plate 5.3

Measuring the Skull to Measure the Mind
George Konrad Rieger, MD, Plate 10, *Eine Exacte Method der Craniographie*, 1885

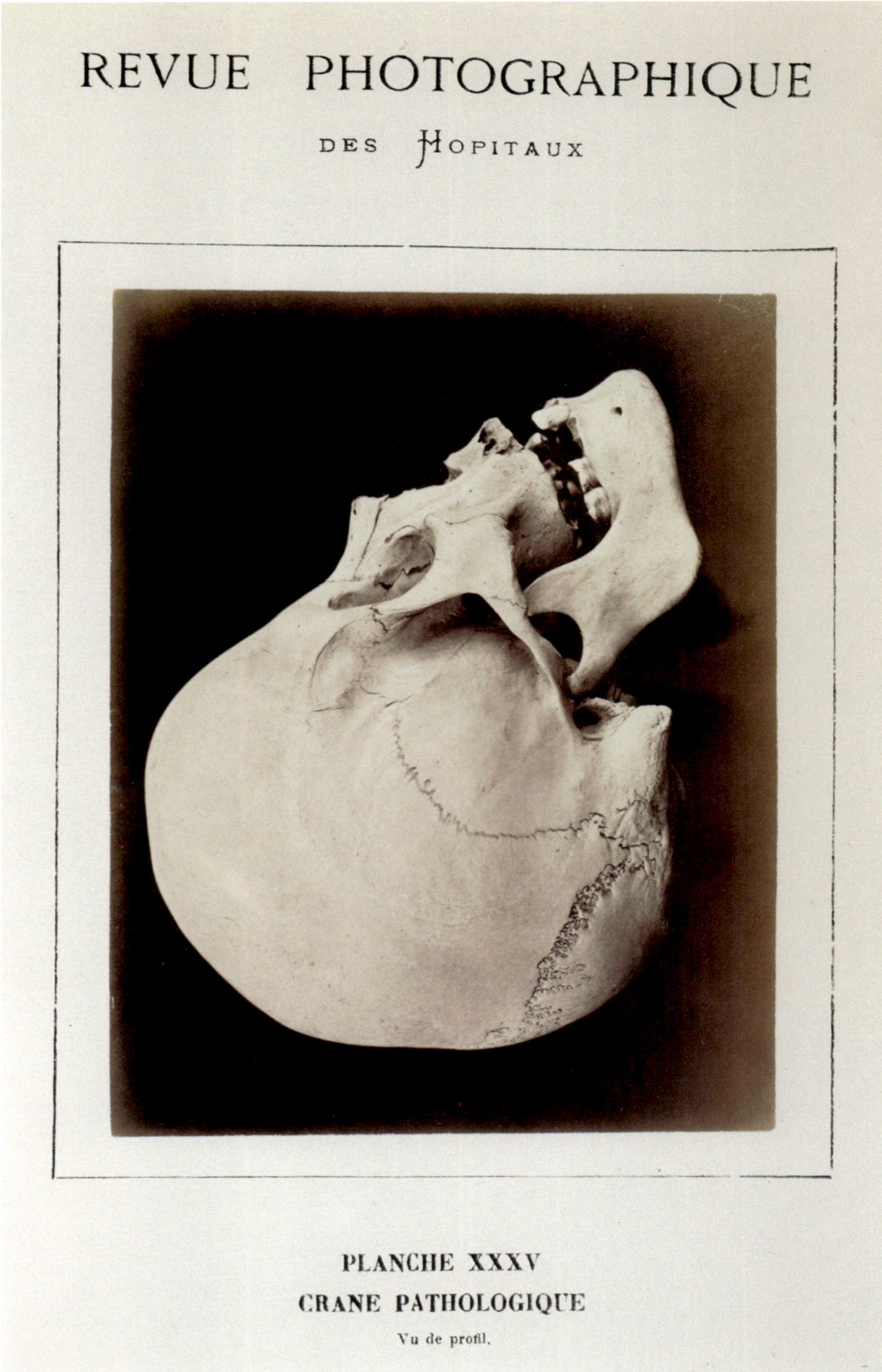

Plate 5.4

Lateral Study of the Head of an 'Idiot'
Revue photographique des hôpitaux de Paris, Vol. I Plate 35, 1869

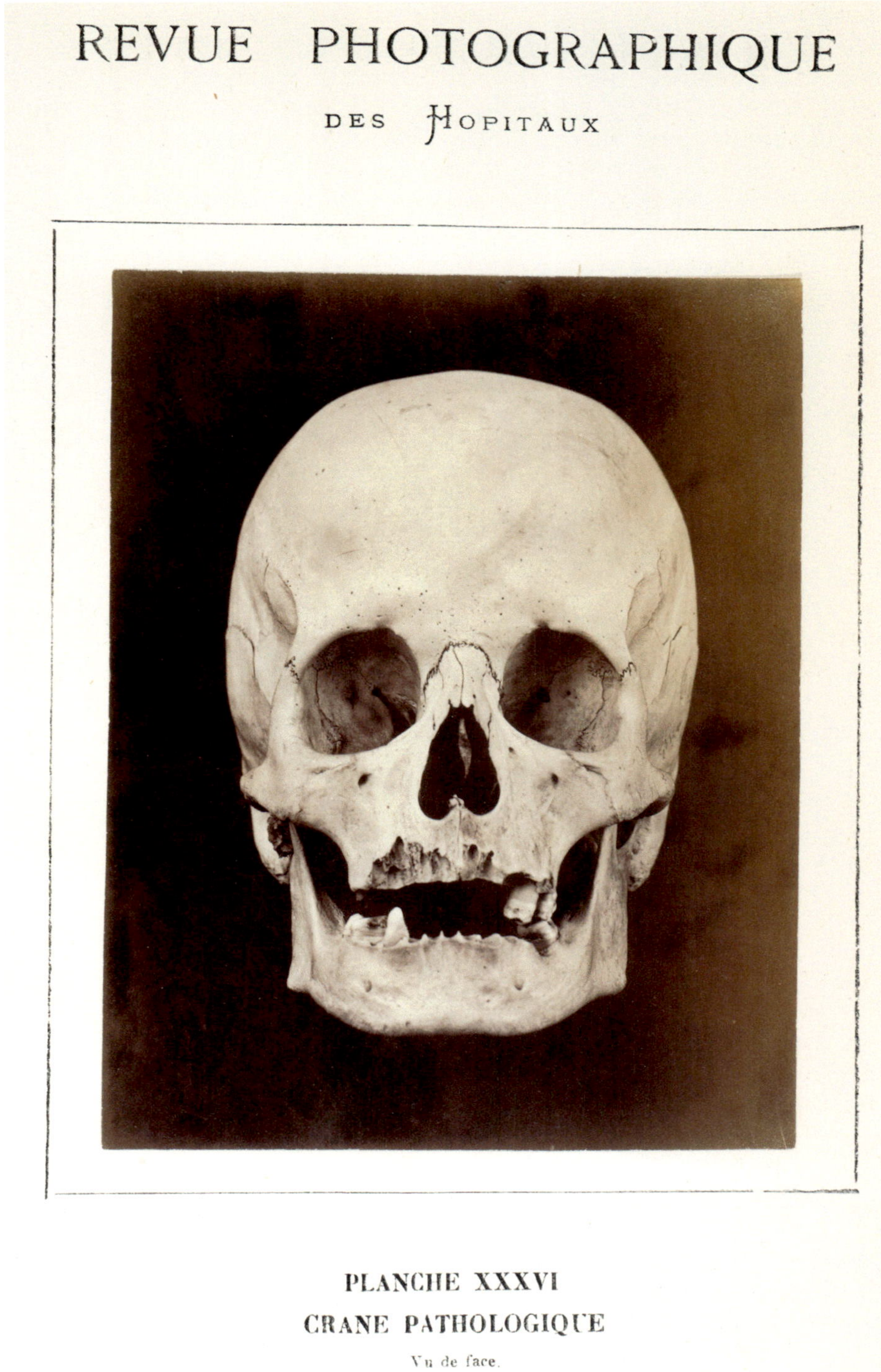

Plate 5.5

Frontal Study of the Head of an 'Idiot'
Revue photographique des hôpitaux de Paris, Vol. I,Plate 36, 1869

According to Morton and his followers, a large skull meant a large brain and high intellectual capacity, while a small skull indicated a small brain and decreased intellectual capacity. The skulls of extremely mentally defective individuals such as microcephalics are one source of the concept that brain size mattered. Modern scientists confirmed that there is a correlation between cranium size and intelligence as measured by IQ tests, although it is a weak correlation at about 0.2. Brain volume as measured with MRI scanners also finds a correlation between brain size and intelligence at about 0.4, as published in 2009 in "Whole Brain Size and General Mental Ability: A Review" (*International Journal of Neuroscience*) by Philippe Rushton and Davison Ankney.

Today, anthropologists, while noting several subgroups of cranial shapes, identify some easily observable craniofacial differences including, head shape (mesocephalic, brachycephalic, doliocephalic), breadth of nasal aperture, nasal root height, sagittal crest appearance, jaw thickness, brow ridge size and forehead slope. Using this skull-based categorization, anthropologists identify at least three large racial groups; Caucasoid, characterized by a tall dolichocephalic skull, receded zygomas, large brow ridge and projecting-narrow nasal apertures; Negroid, characterized by a short dolichocephalic skull, receded zygomas and wide nasal apertures; Mongoloid, characterized by a medium brachycephalic skull, projecting zygomas, small brow ridge and small nasal apertures. These categories are most helpful in identifying skeletal remains.

CRANIOMETRY

Since the eighteenth century, scientists have worked to establish a set of standardized skull measurements. Today MRI studies provide the most accurate tool for measurement. Pieter Camper, MD, (1722–1789) chair of anatomy and surgery at Amsterdam's Athenaeum Illustre, laid the foundation of craniometric measurements as a result of his interest in art. Camper, like most scholars of his time, was accomplished in several fields, among them comparative anatomy, zoology, and anthropology. He was also a gifted artist and, in August 1770, presented two lectures to the Amsterdam Drawing Academy on the facial angle. In these lectures, using various skulls, he created a new drawing method based on the 'facial angle.' According to Camper's technique, a 'facial angle' was formed by drawing two lines: one horizontally from the nostril to the ear and the other perpendicularly from the advancing part of the upper jawbone to the most prominent part of the forehead. Camper's measurements of facial angles were first made to compare the skulls of men with those of other animals. He found that in apes this angle is from 45 to 60 degrees, in an orangutan it's 58 degrees, in the skull of Central Africans about 70 degrees, and in Europeans from 75 degrees to 85 degrees. This finding reinforced the prevalent European racist hypothesis that Caucasians were superior and more advanced than Negros. In 1792, Camper published his work on physiognomy, which included his description of his craniometrical measurement system.

The next major influence in craniometric studies was Swedish professor of anatomy at Stockholm's Karolinska Institute Anders Retzius, MD, (1796–1860). In 1842, Rezius published *Om formen af nordboernes cranier* describing a method for classifying races according to a cephalic index. He defined the cephalic index, which is the ratio of width to length of one's head. He classified skulls in three main categories, which are still used today: "dolichocephalic" (long and thin), "brachycephalic" (short and broad) and "mesocephalic" (intermediate length and width), or simply long headed, short headed or moderate headed. Physical anthropologists first used the cephalic index to classify ancient human remains found in Europe. Retzius studied many different skull types from different races. Because the skulls were so different from each race, he believed that the races had a separate origin and he became an ardent polygenist.

Other scientists who advanced craniometric measurements were the French naturalist and professor of zoology at the Faculty of Sciences Paris, Étienne Geoffrey St. Hilaire (1722–1844), who described in 1795 a horizontal plane for studying skulls and French surgeon and anatomist Jules Germain Cloquet, MD, (1790–1883) who also described a horizontal plane in 1821. A final horizontal plane determination was established in 1882 by an agreement reached among anthropologists at Frankfurt-am-Main. In 1884, The World Congress of Anthropology decreed it as the anatomical position of the human head as it is carried in the living subject. The term is still in vogue and used as a reference point in dentistry, radiology, and other fields. The Frankfort plane (also called the auriculo-orbital plane or eye-ear plane) is determined as the highest point on the upper margin of the opening of each external auditory canal and the low point on the lower margin of the left orbit. It is used to orient a human skull or head so that the plane is horizontal. The skulls of hominids and primates may be studied in the Frankfurt plane, but it is not considered to be the anatomical position for other species.

The intriguing findings of Morton, Camper, Retzius, and others stimulated great interest in measuring exact cranial capacity. As a result of these pioneer scientists, a new generation of researchers, using new technology, opened the skull to study cerebral function and closed the orifices of the cranial cavity to measure brain volume.

Paul Broca, MD, (1824–1880), French physician, surgeon, and anthropologist worked not only on the skull, but also on the brain, helping to identify cerebral areas that involved specific functions. The region of the brain related to speech has come to be known as Broca's area. Broca was a pioneer neurologist and also contributed to the development of physical anthropology and especially anthropometry. He advanced the science of cranial anthropometry by developing many new types of measuring instruments (craniometers) and numerical indices. In 1859, Broca founded the Societé d'anthropologie and the journal *Revue d'Anthropologie*. He believed, as did many in his time, that man's intellectual qualities could be measured by the size of his brain. The brains of many of Broca's aphasic patients are still preserved in the Musée Dupuytren, and his collection of skull casts are in the Musée d'Anatomie Delmas-Orfila-Rouvière. French physical anthropologists, led by Broca, focused on craniometry to understand human behavior and culture. Among Broca's several devices that he developed was a three-dimensional tracing machine that could accurately trace the outline of a skull so that the traced outlines could be compared. Skull shape was a critical anatomical determinant of race.

France's arch 'intellectual' rival, the Germans, led by leading pathologist and anthropologist Rudolf Carl Virchow, MD, (1821–1902) emphasized the influence of environment and disease upon the human body as a determinant of human behavior. American scientific thought by the end of the century involved a 'four-field approach' to understanding human behavior based upon studying the remains of North American skeletons, artifacts, language, and culture. In the third quarter of the nineteenth century (1850–1875), it was the anthropologists who were the leaders in studying skulls. By the end of the nineteenth century, criminal anthropologists had developed an entire field based on physiognomy and skull shape. By the early twentieth century, the racial and behavioral ideology of the anthropologists had fallen into disrepute as laboratory and scientific testing offered better perspectives on race and behavior.

Curiously, Virchow, in direct opposition to polygenists' views and also to the policies that were to become Nazi Germany's racist philosophy, in 1885 launched a study of craniometry, which gave surprising results contradictory to contemporary scientific racist theories on the "Aryan race." It lead him to denounce the 'Nordic mysticism' in 1885 at the Anthropology Congress in Karlsruhe. A colleague, Josef Kollmann, stated in the same congress that the people of Europe, be they German, Italian, English, or French, belonged to a "mixture of various races," furthermore declaring the "results of craniology" erroneous in any theory "concerning the superiority of this or that European race" on others.

SKULL COLLECTORS

Attention to the value of the skull, as a key to understanding life, engendered a second group of skull collectors. These individuals were not scientists working to determine the nature of man, but collectors who desired skulls as status symbols or as indicators of their power and position. One attribute of human nature is the desire to collect and own a piece of society's heroes and celebrities. People want to own a piece of what defines the culture. Today, it is autographs, sports memorabilia, objects or clothing worn by the rich and famous, etc. Any object has increased value if owned or used by a star personality. Colin Dickey in his *Cranioklepty: Grave Robbing and the Search for Genius* describes the late eighteenth and nineteenth centuries' obsession to own the skulls of the famous. The desire was described as instinctual and irresistible. While a lock of hair would do as a memento for the average man, it appeared that the rich wanted to own a skull.

Some of the most recognized geniuses of the era were composers, and the skulls of two of the most famous, Franz Joseph Haydn (1732–1809) and Ludwig van Beethoven (1770–1827), were stolen. Some of these notables remained in personal collections into the twenty-first century. Beethoven was autopsied at his request, so that physicians could learn about his disease. His skullcap was damaged into several fragments by the soon-to-be notable Viennese pathologist Karl von Rokitansky (1804–1878). It was his first autopsy, and he would go on to perform 59,786 autopsies in his career. Beethoven died of liver failure, thickened bones of his skull, which indicated he probably had Paget's disease, perhaps the cause on his deafness. His skull was ultimately located in a private doctor's collection. California businessman Paul Kaufmann found the skull in his mother's belongings after she died in 1990. She was a descendant of one of the physicians involved with Beethoven's and Schubert's 1863 exhumation for scientific purposes. The skull was loaned in 2005 to the Ira F. Brilliant Center for Beethoven Studies at San Jose State University. DNA testing has proven it was indeed his skull.

Skulls could be obtained through legal channels or simply stolen from graves, especially after notables were exhumed. Noted nineteenth-century skull collectors, such as Joseph Carl Rosenheim, went to extraordinary lengths to acquire skulls by bribing officials and hiring grave robbers. Rosenheim's most famous acquisition was the skull of Joseph Haydn, who died in 1809. Wealthy collectors with no scientific interest simply wanted to have the best souvenir of a notable. It was only in 1954 that Haydn's skull was united with his remains in his tomb in the Bergkirche, Eisenstadt. Another example of a legal purchase was that of Duke Carl August who bought the skull of German poet and philosopher Friedrich Schiller who died in 1825. Twenty-one years later, when Schiller's body was exhumed, the duke was able to legally buy Schiller's skull. He mounted it in a special case in his library, the proper place for an intellectual. Selling human remains is very complicated. Contemporary auction houses are asked to offer skulls in their sales and some will, if no local, state or federal laws are violated.

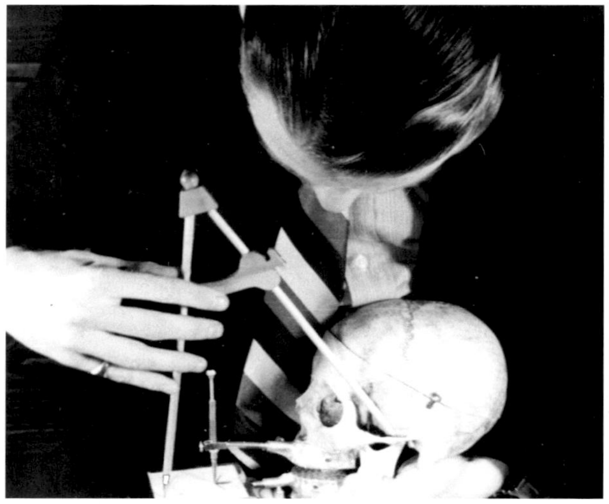

Plate 5.6

Anterior Measurements of the Skull with a Special Caliper
Cranial Studies, circa 1940

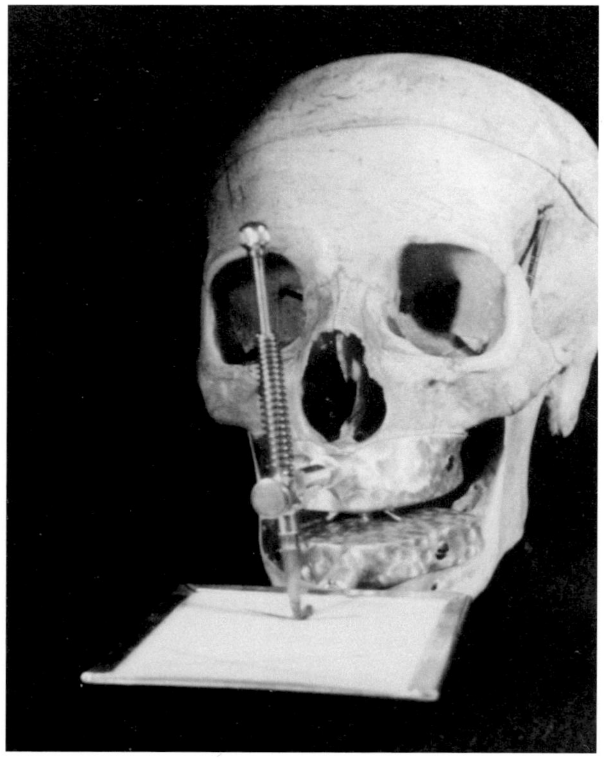

Plate 5.7

Anterior Measurements of the Skull, Frontal View
Cranial Studies, circa 1940

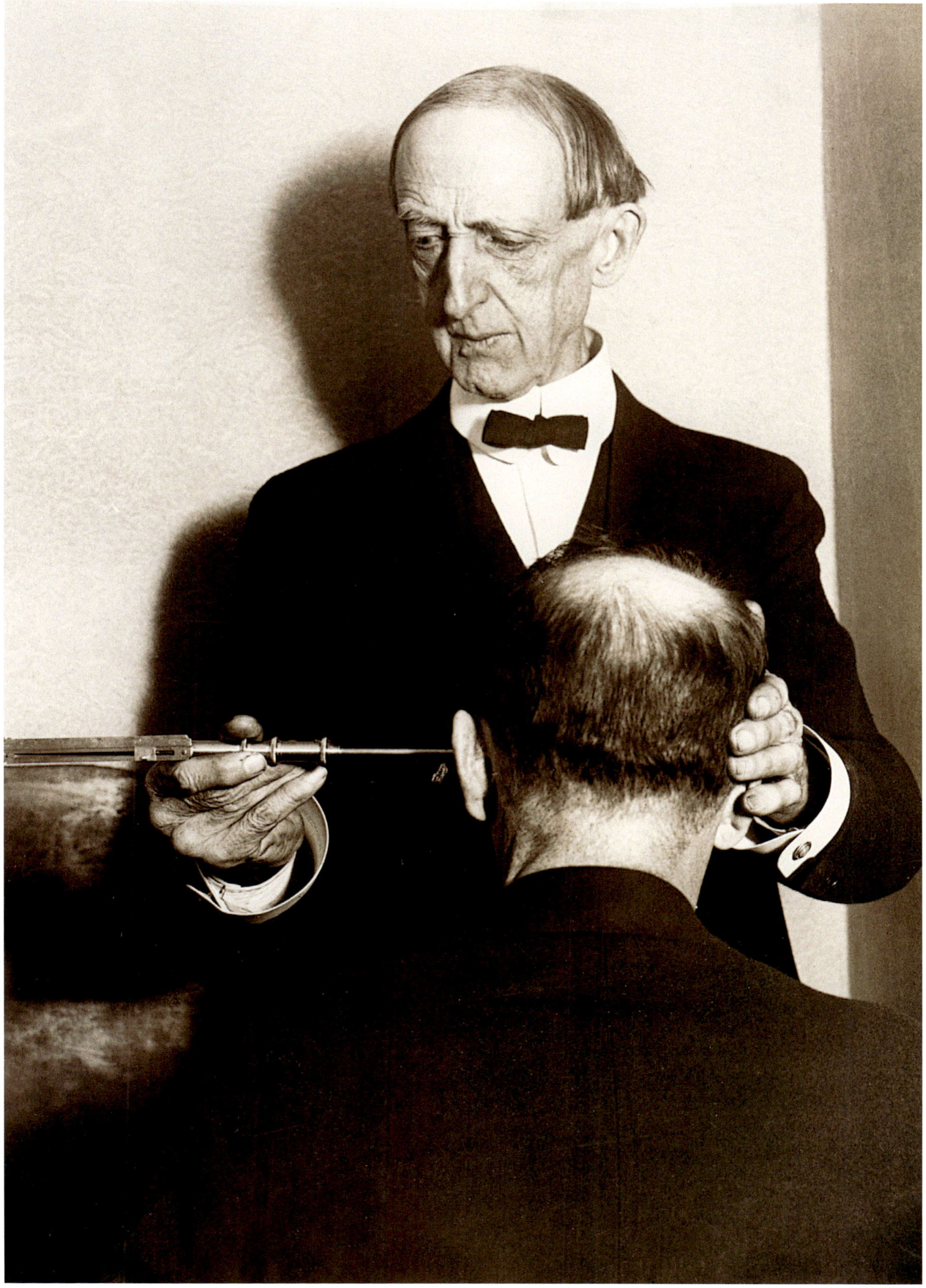

PHOTOGRAPHY

Aside from collecting skulls, to further the study of anthropology and ethnology, the *Manual of Ethnological Inquiry* published by the British Association for the Advancement of Science recommended, as early as 1852, the use of photography to help create an "accurate record of individual likenesses." The technical difficulties of the early photographic processes hindered application, and among the best known early ethnographic studies is that of photographer J.T. Zealy (1812–1893) of Columbia, South Carolina, who daguerreotyped slaves to support the theory of racial superiority, polygenesis, as promoted by Harvard's Jean Louis Rodolphe Agassiz (1807–1873). Agassiz hired Columbia physician and noted paleontologist Robert Wilson Gibbes to have Zealy take the photographs. The daguerreotypes are preserved at Harvard's Peabody Museum.

As paper print processes were developed and improved in the 1850s, it was recognized that photography could offer a standardized and convenient means of measuring and analyzing faces, bodies, and skulls. By the mid-1860s, photography was generally recognized as perhaps the best and most accurate means to preserve human races that were being annihilated across the globe. The photograph was a permanent visual record. It could be analyzed and reevaluated in the context of ethnology as a reliable source of comparative morphometric data. It was recognized that specific positions, lighting, and an external marked measuring scale, against which to photograph the person or object, must be developed. Joseph Bernard Davis photographed skulls in the 1850s without such a scale, and although his images are interesting and representative examples of indigenous peoples, they are useless for accurate comparisons. The point of the ethnological photographs was not only for documentation but also for comparison

Plate 5.8

Dr. Arthur MacDonald using his Brain Measuring Device on a U.S. Senator

Government Research on Intelligence: Evaluating the Brains of U.S. Congressmen & Senators, 1926

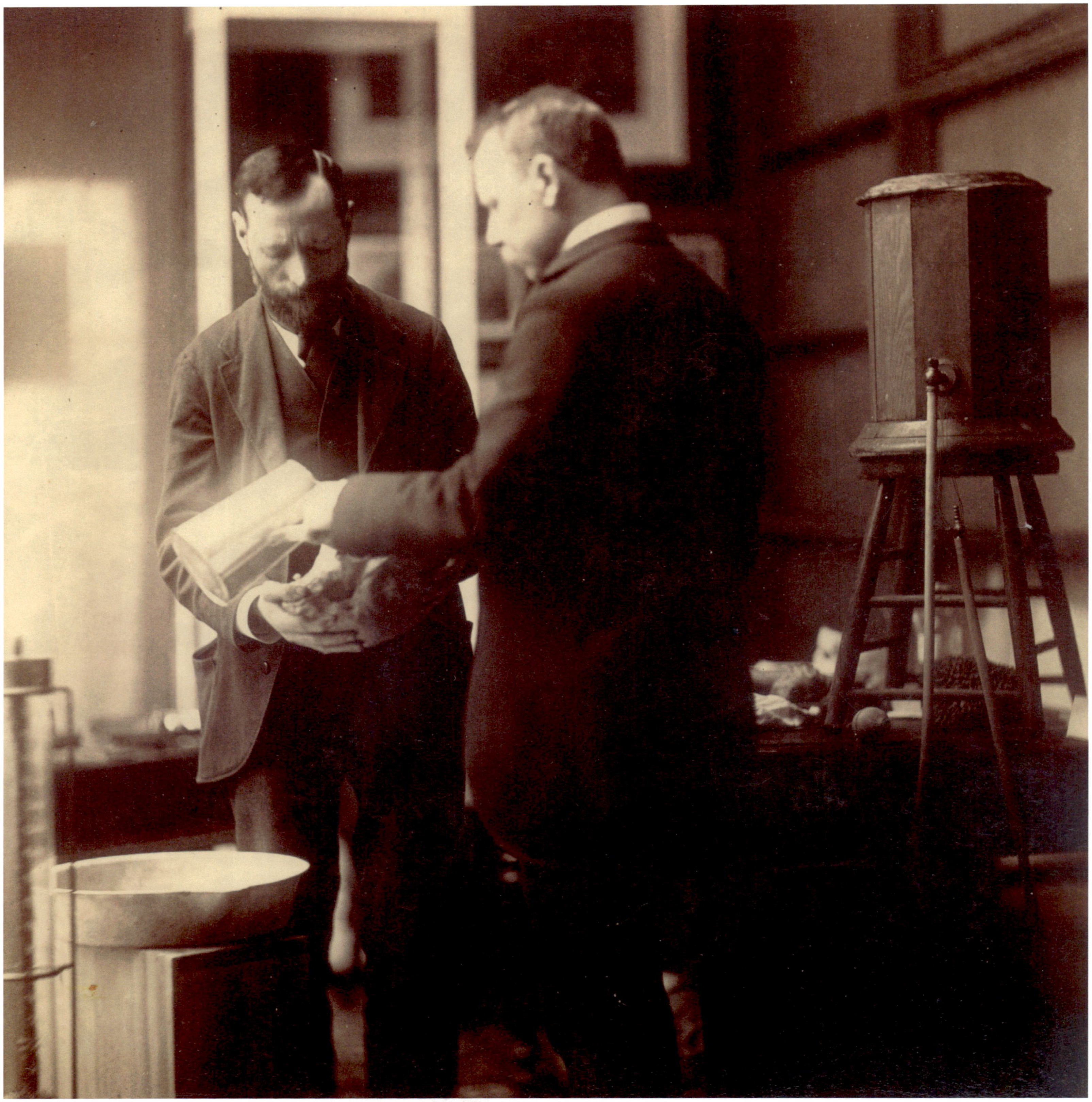

Plate 5.9

Ascertaining Capacity of Cranial Cavity by Means of Water, First Part

Washington Matthews, MD, holds skull as John Shaw Billings, MD, fills it with water from a pitcher, 1885

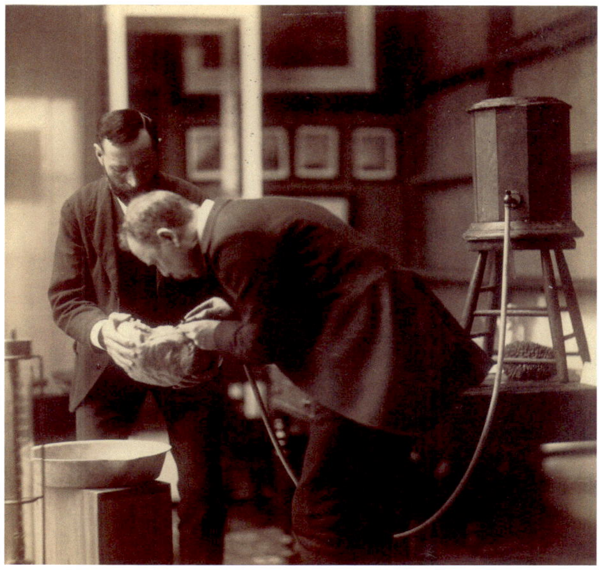

Plate 5.10

Ascertaining Capacity of Cranial Cavity by Means of Water, Second Part
1885

John Shaw Billings, MD, fills skull held horizontally with water from reservoir.

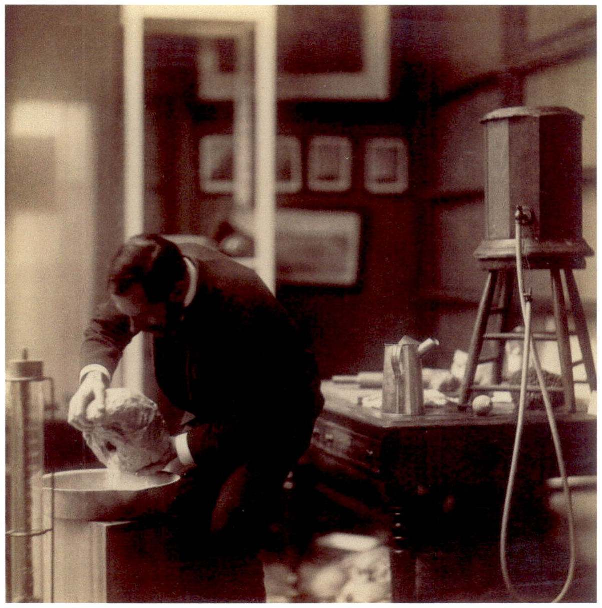

Plate 5.11

Washington Matthews, MD, Emptying the Skull, Second Position
1885

Simply pouring the contents in a measuring basin.

of races in the ongoing scientific evaluation of mankind and behavior. Great Britain's Thomas Huxley (1825–1895), best known as 'Darwin's Bulldog' for his advocacy and promotion of Charles Darwin's theory of evolution (*Origin of Species*, 1859), noted:

> Great numbers of ethnological photographs already exist... but they lose much of their value from not being taken upon a uniform and well considered plan. The result they are rarely either measurable or comparable with one another and that they fail to give the precise information respecting the proportions and the confirmation of the body, which... is of paramount worth to the ethnologist.

Huxley and others suggested a standardized photometric system should be employed to document all the peoples in the British Empire. Huxley produced a photometric grid in which subjects were to be photographed naked so reliable information could be recovered and analyzed. In 1869, John Lamprey introduced his fine grid system for photography. It was a grid with two-inch squares, against which a person or skull was placed in a specific position to be photographed. With the development of the dry plate process in the late 1870s ,the fieldwork of photographing races rapidly expanded. Those studying body types, somatomorphology, used the Lamprey system with various improvements well into the twentieth century.

Sir Francis Galton (1822–1911), the brilliant polymath and founder of the science of statistics and eugenics, was an advocate of composite photography in analyzing normal and disease states, as well as racial differences. Galton was the cousin of Charles Darwin, and was intrigued by the implications of Darwin's work on populations, heredity, and the human condition. Galton by 1878 had devised a technique called "composite portraiture," a photograph produced by superimposing multiple photographic portraits of faces centered on the eyes to create an 'average' face. In 1878, Galton published his technique in *Nature* (vol. 18), "Composite portraits made by combining those of many different persons into a single figure." Sir Francis Galton was a psychometrician and attempted to define physiognomic characteristics of health, disease, beauty, and criminality, via statistics and his method of composite photography. Galton invented the term eugenics in 1883, and set down many of his observations and conclusions in a book, *Inquiries into Human Faculty and Its Development*. He published several composite photographs as the frontispiece. Galton also photographed skulls, and in 1886 the *Journal of the Anthropological Institute* reported on an "Exhibition of composite photographs of skulls." He devoted much of the rest of his life to exploring variation in human populations and its implications.

UNITED STATES GOVERNMENT CRANIAL RESEARCH

Nineteenth-century science was dominated by the concept of measurements in the hope that there would be some correlation of a particular measurement with a disease or other condition. Accurate atmospheric temperature measurements were the duty of government physicians in their service stations. John Shaw Billings, MD, (1838–1913) one of America's leading physicians, presented on April 22, 1885, at the National Academy of Sciences, "On Composite Photography as Applied to Craniology: and on Measuring the Cubic Capacity of Skulls." Billings worked with Washington Matthews, who shared credit and is seen with him in a series of photographs demonstrating his methodology. Using Galton's composite photography, Billings studied and hoped to improve on the methodology for cranial photography.

The first American experimental research in a government laboratory were these cranial capacity studies. The research was under the direction of John Shaw Billings, MD, (1838–1913). Billings and Army Chief of Staff Leonard Wood, MD, were two of America's greatest physicians at the turn of the century and are relatively forgotten today. By the 1870s, Billings was America's leading government scientist and physician. During the Civil War, he organized hospitals. After the war, he organized and developed the (SGO) Surgeon General's Library, later to become the current National Library of Medicine. As director of the SGO Library, he figured out how to mechanically analyze medical and mechanical data by turning it into numbers and punching it into cardboard cards, which is the origin of the computer punch card system used until the 1970s. He created the Index Medicus and the SGO Index-Catalogue system allowing easy retrieval of biomedical data. He oversaw the 1890 census and used his punch card system for calculating and analyzing the American populations. Recognized as an expert in planning hospitals, Billings was asked to plan the hospital for Johns Hopkins University Medical School and Hospital. He oversaw its construction and helped recruit Hopkins's noted early physicians. Later, he planned six other medical institutions and was a professor of hygiene at the University of Pennsylvania. After leaving government service, he was on the Board of the Carnegie Foundation and convinced Carnegie to fund and build sixty-five branch libraries in New York and 2,059 libraries throughout the country. He helped to unite the various great library collections of New York notables (Tilden, Astor, Lenox) into the New York Public Library and was its first director. The library opened on May 23, 1911. Billings died in 1913.

Billings's credentials are presented in detail to emphasize the importance of cranial capacity research and cranial photography. He headed the United States government's research team, lending his expertise and prestige to the project. Stimulated by the work of Samuel Morton, the government had also aggressively collected indigenous peoples' skulls (see Plates 5.14 and 5.15), and he used some of these Native American skulls. He did his experiments at the Surgeon General's Office Army Medical Museum in Washington, D.C., starting in 1884. He presented a preliminary report on composite photography to the National Academy of Sciences at the annual meeting that year.

Billings prepared a series of experiments on composite photography of skulls. For one composite series he used six adult male (Hawaiian) Sandwich Island skulls, making a series of six different views of the six skulls. For his second series, he used six male Arapahoe Indian skulls for a series of six different composite views. Using standard focal distances and scale, the pictures were accurately superimposed. Billings presents his work, but does not seem to come to any positive conclusion as to the value of the procedure. He notes:

> It is well known to ethnologist that the distinctions of race are much more marked in the physiognomy of the living subject than in the differences of dried crania: and that the bones of the face with the relations which they bear to those of the calvarium, give more valuable race indications than do the calvaria alone. While something has been done in the study of the internal configuration of the cranial cavity... with reference to their difference in various races... it is necessary that sections should be made of the skulls, and before making such sections it is desirable that all measurements and especially the measurements of cranial capacity of these crania, should be made...From the results of some preliminary experiments upon the methods most in use for measuring the cubic capacity of crania, I became much dissatisfied with their accuracy, and accordingly requested Dr. Washington Matthews, my assistant at the Museum, to undertake a series of experiments for the purpose of obtaining, if possible some more accurate and reliable method of ascertaining the cubic capacity.

Washington Matthews, MD, read the report on the cranial capacity studies at the Academy meeting. He noted that anthropologists working in Europe use solid granular substances, such as millet seeds, as a filler for measuring cranial capacity, but the seeds are not an accurate measure due to the conditions and manipulations of the particles. He proposed filling the skulls with water to obtain an accurate measurement. Washington compared the system they devised with that of Paul Topinard, MD, (1830–1911), Broca's successor and director of Paris' École d'Anthropologie, who led France's study of cranial capacity studies.

Washington's system consisted of closing all the skull orifices with wax except for the occipital foramen, using distilled water at 14 degrees centigrade (57.2 degrees Fahrenheit). The water must be put into the skull in 45 seconds, maintaining all at 14 degrees C. Putty, also at 14 degrees C, applied and the orifices closed within fifteen minutes. The water must be drained out in fifteen seconds and measured immediately. To define the surface of the water, a fine impermeable powder is placed on the top of the water to determine the demarcation line of the filled skull. Matthew notes that "A mixture of two parts of lycopodium and one of charcoal floats well and gives a more distinct line...."

The objections to the process are that the sagittal, coronal, and squamous sutures do not empty quickly enough into the cranial cavity. The skulls are in different states of preservation. Water seeps into the walls of the skull and can greatly expand the crania capacity. Varnished and unvarnished skulls were used and the amount of water they retained were measured. Twenty-one skulls were immediately refilled with water to see the effect of the process and comparable retention of water. Expansion of the capacity was noted due to the swelling of the cranial cavity by the water. Washington presents detailed tables of their work at the end of his discussion. Washington states that if the museum's method is accurately followed by researchers around the world, then a more accurate assessment of cranial capacity will be achieved and results compared with more accuracy.

Billings and Washington co-authored a paper, "On a New Craniophore for the Use in Making Composite Photographs of Skulls." In it they wrote:

> The apparatus consists of an object-stand, with four hinged frames and a craniophore with two different attachments for holding the skull... The craniophore is of brass... One attachment is for supporting the skull, base downwards, when the facial, lateral and occipital views are taken. It has a cone which enters the foramen magnum, and a jointed arm elongated telescopically, which supports the palate. The other attachment is for holding the skull when the basal and horizontal views are taken...Mr. Ed Kubel built the craniophore and museum carpenters the object stand.

Four photographs of the device that were shown to the National Academy of Sciences meeting in 1885 are presented. Craniometric measurement standards were finalized in G. Papillault's report in 1906 in *L'Anthropologie* on "The International Agreement for the Unification of Craniometric and Cephalometric Measurements."

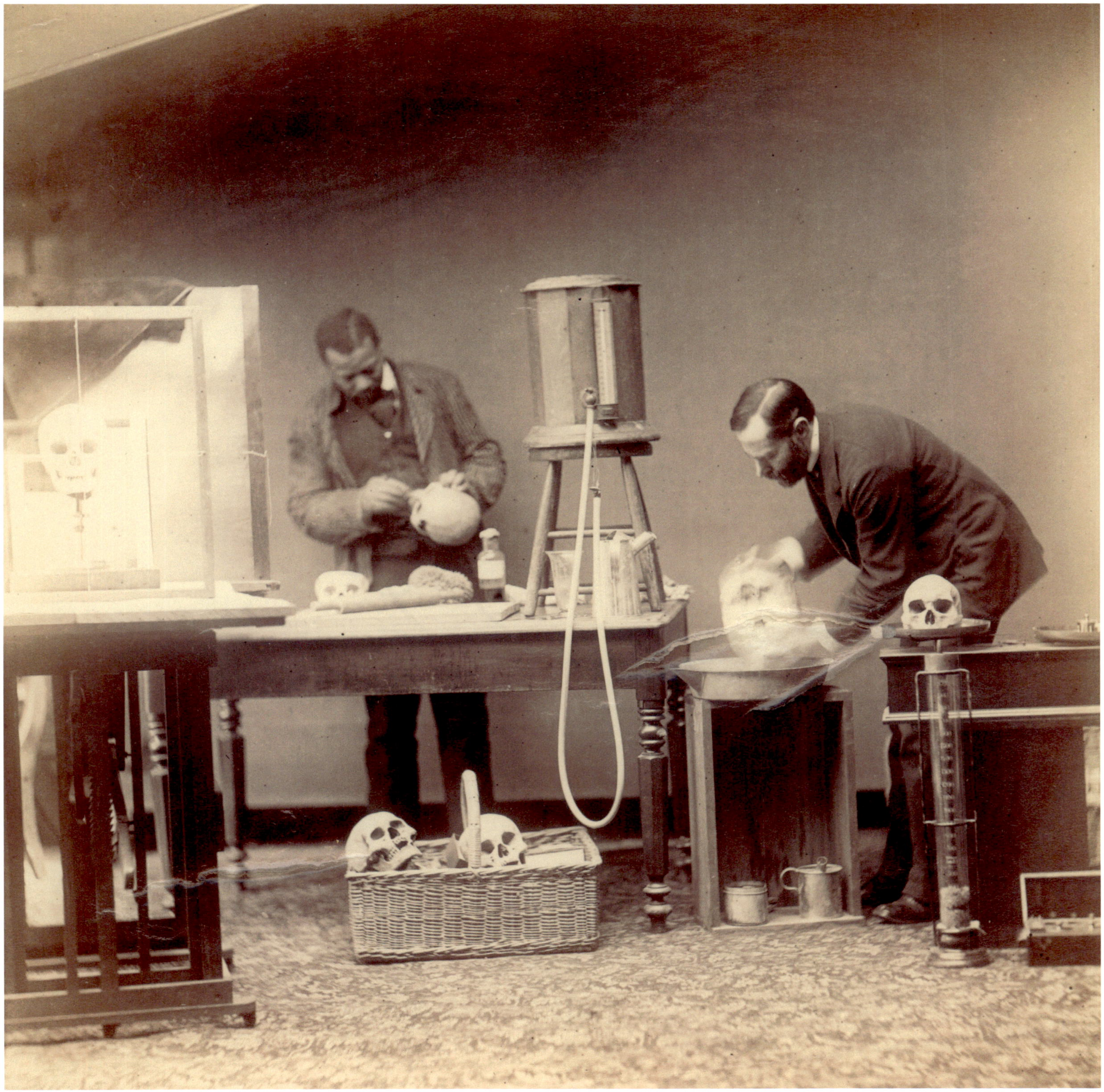

Plate 5.12

Washington Matthews, MD, Emptying the Skull, First Position
1885

An attendant is removing putty from skull.

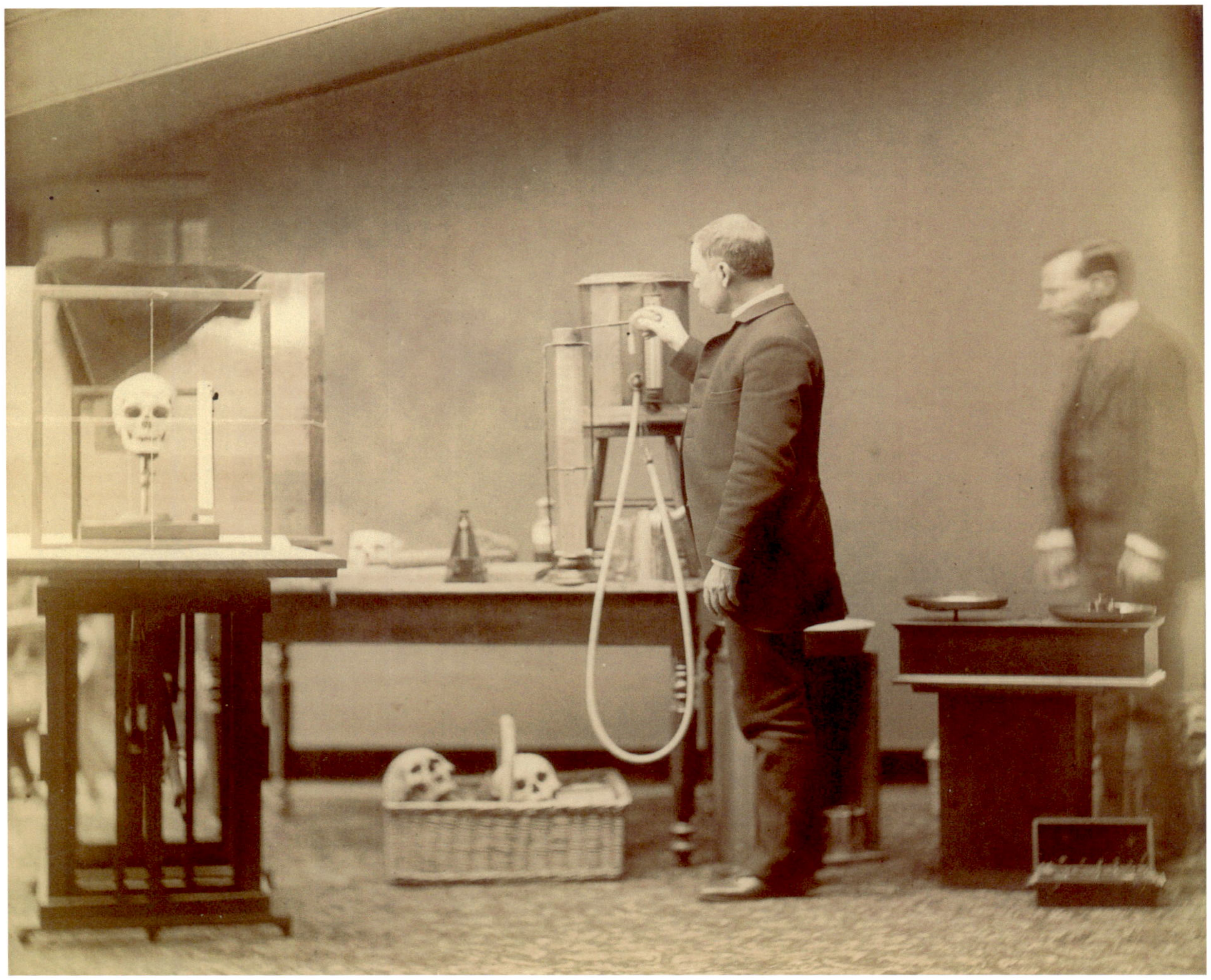

Plate 5.13

Using an Insufflator with Dry Powder to Define the Surface of the Water in a Skull
1885

The composite photograph machine is on the left.

While the scientific world was making landmark discoveries in many fields, the United States government plodded along supporting questionable governmental scientific projects and pseudo-science as late as the fourth decade of the twentieth century. Perhaps one of the most far-fetched and unbelievable examples is the work of Dr. Arthur McDonald (1856-1936), who attempted to measure the brains of sitting members of the U.S. legislature by inserting his supposed brain-measuring device into their ears. It was perhaps part of the aggressive government policy of the time to identifying mental defectives. In 1926, physician and U.S. Senator Royal Copeland introduced the bill under which McDonald worked. Dr. Copeland had been dean of the New York Homeopathic Medical College and director of Flower Hospital and president of the New York Board of Health. He believed it was necessary to have "anthropological, psycho-physical, and statistical examination of members of both houses." McDonald measured seventy-five members of the U.S. House of Representatives and twenty-five U.S. senators. The bill required the analysis to be made public but without identifying the subjects directly.

THE FUTURE

By the end of World War II, nineteenth-century pseudo-sciences had fallen into official disrepute and government support turned to emerging sciences and technologies. Today, skulls are still searched for by paleoarchologists seeking to identify the human and humanoid ancestral chain.

Scientists have not given up their quest to find an anatomical key to genius. Today, famous scientists, such as Albert Einstein, have had their brains preserved for research purposes. The brain has replaced the skull as the source of identifying humanity and intelligence. It is not unusual for scientists to seek out people with unusual traits and have their brain preserved. Advances of science have demonstrated amazing leaps in technology and comprehension in each generation, and advances are accelerating. Imaging techniques and modern neuroscientists are deciphering the neural and chemical mechanisms of behavior and thought processes. The study of cerebral neural mechanisms is being exploited as a new frontier. Pharmaceutical companies and other research laboratories are seeking cerebral-altering drugs to reverse Alzheimer's disease, multiple sclerosis and other degenerative diseases, as well as mental diseases, such as schizophrenia.

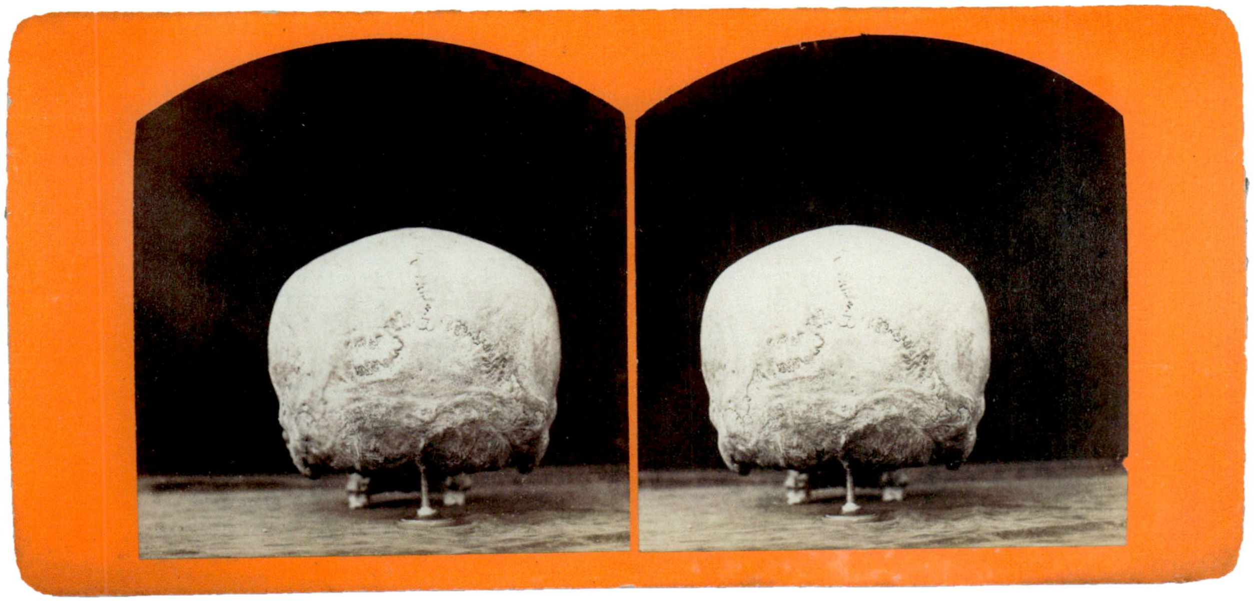

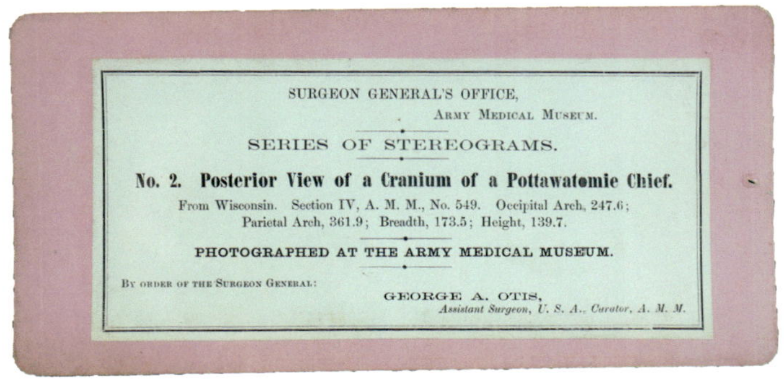

SURGEON GENERAL'S OFFICE,
ARMY MEDICAL MUSEUM.

SERIES OF STEREOGRAMS.

No. 2. Posterior View of a Cranium of a Pottawatomie Chief.

From Wisconsin. Section IV, A. M. M., No. 549. Occipital Arch, 247.6; Parietal Arch, 361.9; Breadth, 173.5; Height, 139.7.

PHOTOGRAPHED AT THE ARMY MEDICAL MUSEUM.

BY ORDER OF THE SURGEON GENERAL:

GEORGE A. OTIS,
Assistant Surgeon, U. S. A., Curator, A. M. M.

Plate 5.14

Posterior View of Cranium of a Pottawatomic Chief from Wisconsin
Circa 1880
Photographic Study of Native American skulls at the SGO Army Medical Museum

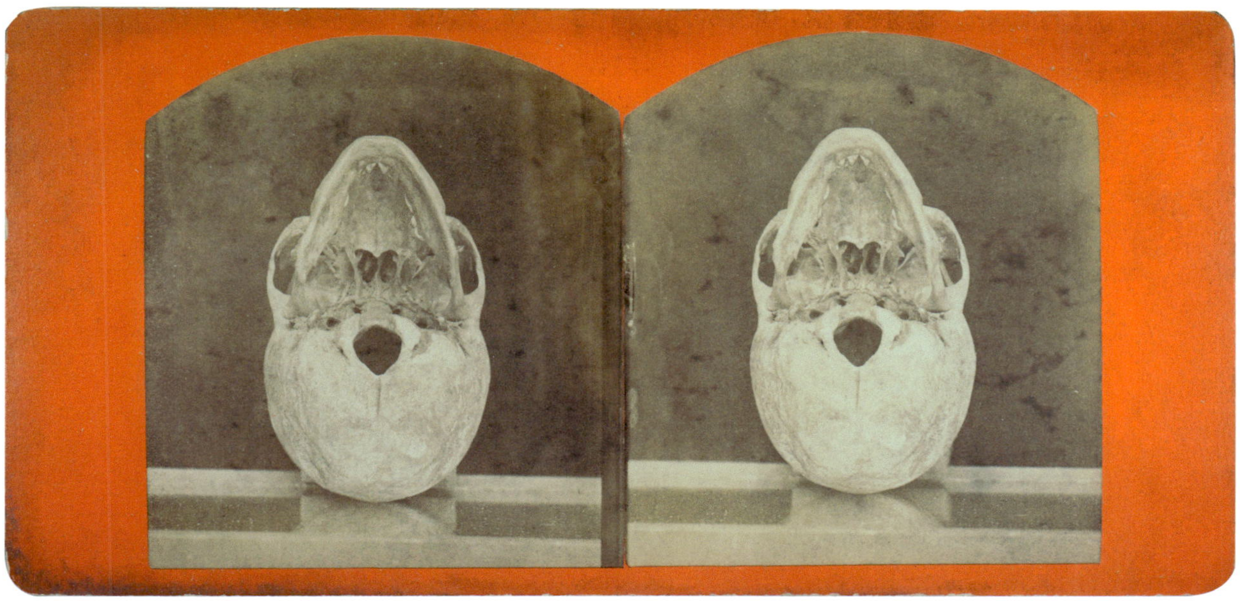

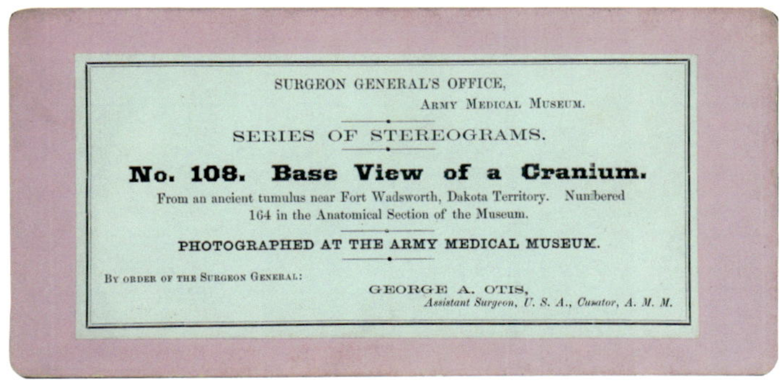

SURGEON GENERAL'S OFFICE,
ARMY MEDICAL MUSEUM.

SERIES OF STEREOGRAMS.

No. 108. Base View of a Cranium.

From an ancient tumulus near Fort Wadsworth, Dakota Territory. Numbered 164 in the Anatomical Section of the Museum.

PHOTOGRAPHED AT THE ARMY MEDICAL MUSEUM.

BY ORDER OF THE SURGEON GENERAL:

GEORGE A. OTIS,
Assistant Surgeon, U. S. A., Curator, A. M. M.

Plate 5.15

Base View of a Cranium from an Ancient Tumulus, Ft. Wadsworth, Dakota Territory
Circa 1880
No. 164 in Anatomical Section of Army Medical Museum, Collection of Native Americans

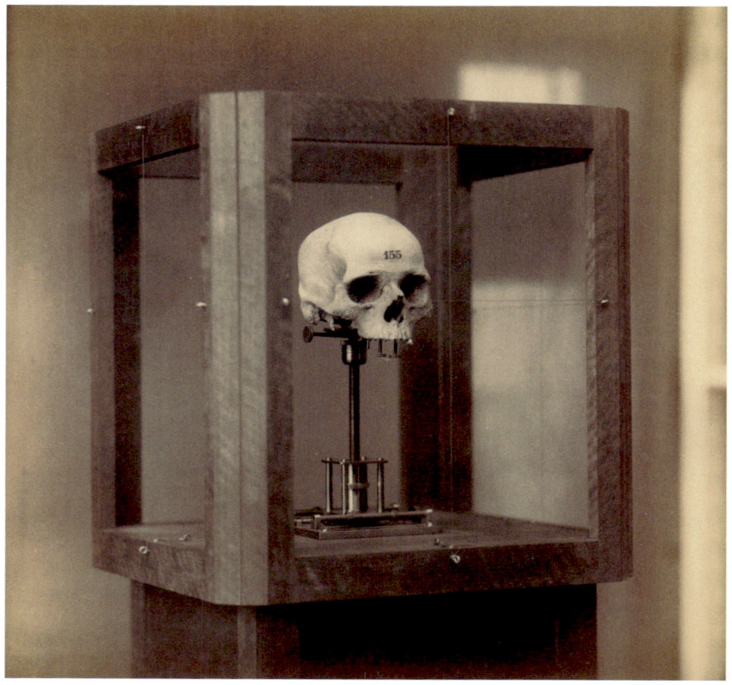

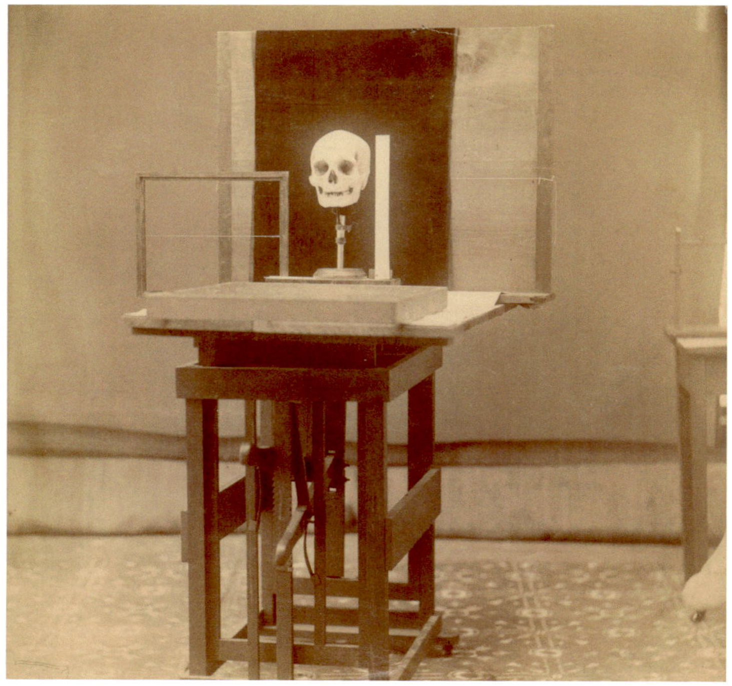

Plate 5.16

A New Craniophore Apparatus for Taking Composite Photographs of Skulls
John Shaw Billings, MD, and Washington Matthews, MD, 1885

Shows everything ready to take the photograph.

Plate 5.18

Apparatus for Taking Composite Photographs of Skulls, No 7. The Exposure
1885

Two sides of the wood frame removed, black backing added and a white card to illuminate skull.

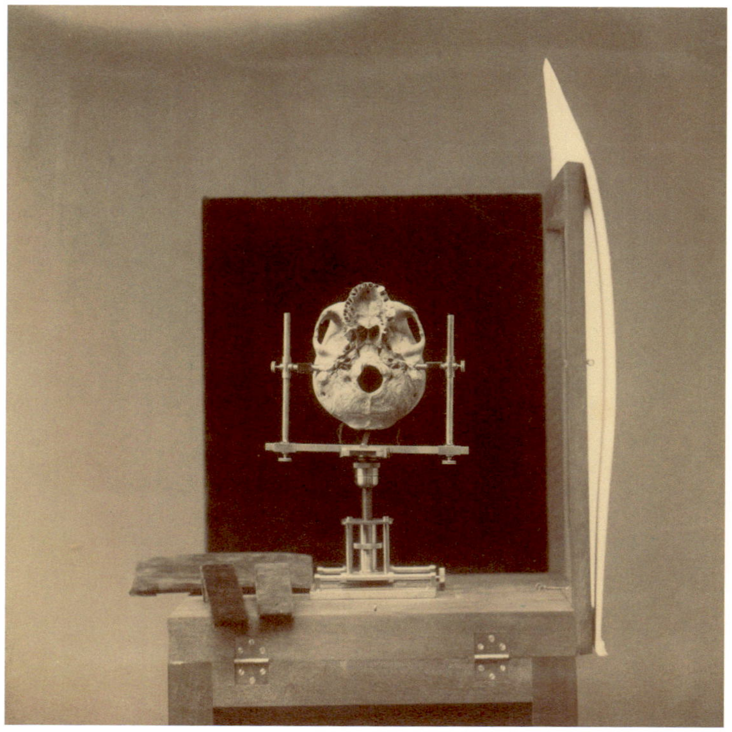

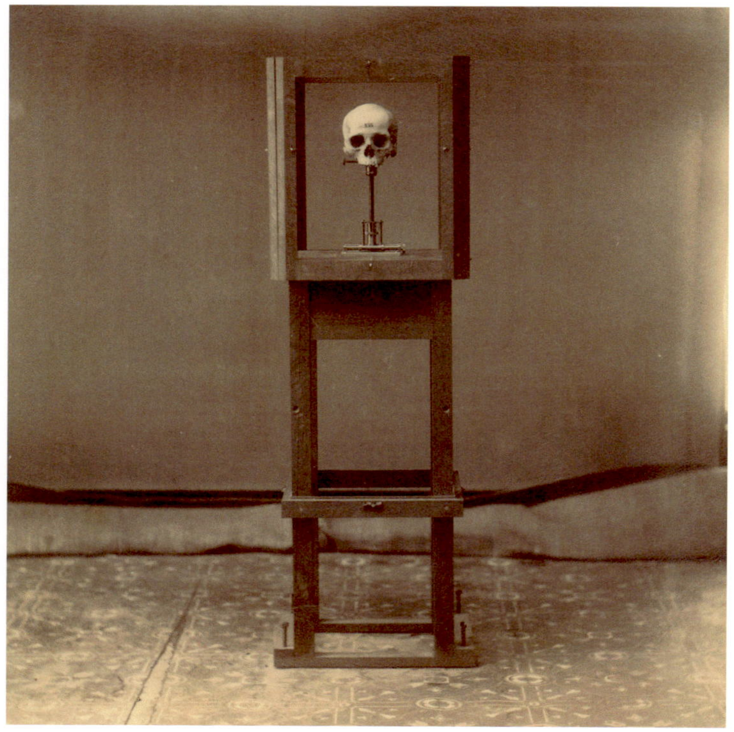

Plate 5.17

Apparatus for Taking Composite Photographs of Skulls
1885

Diagonal view with the skull on the newly designed brass craniophore, stand, and frames.

Plate 5.19

Apparatus for Taking Composite Photographs of Skulls, No 5. The Adjustment
1885

The skull in position for a photograph of the 'normal facial view.'

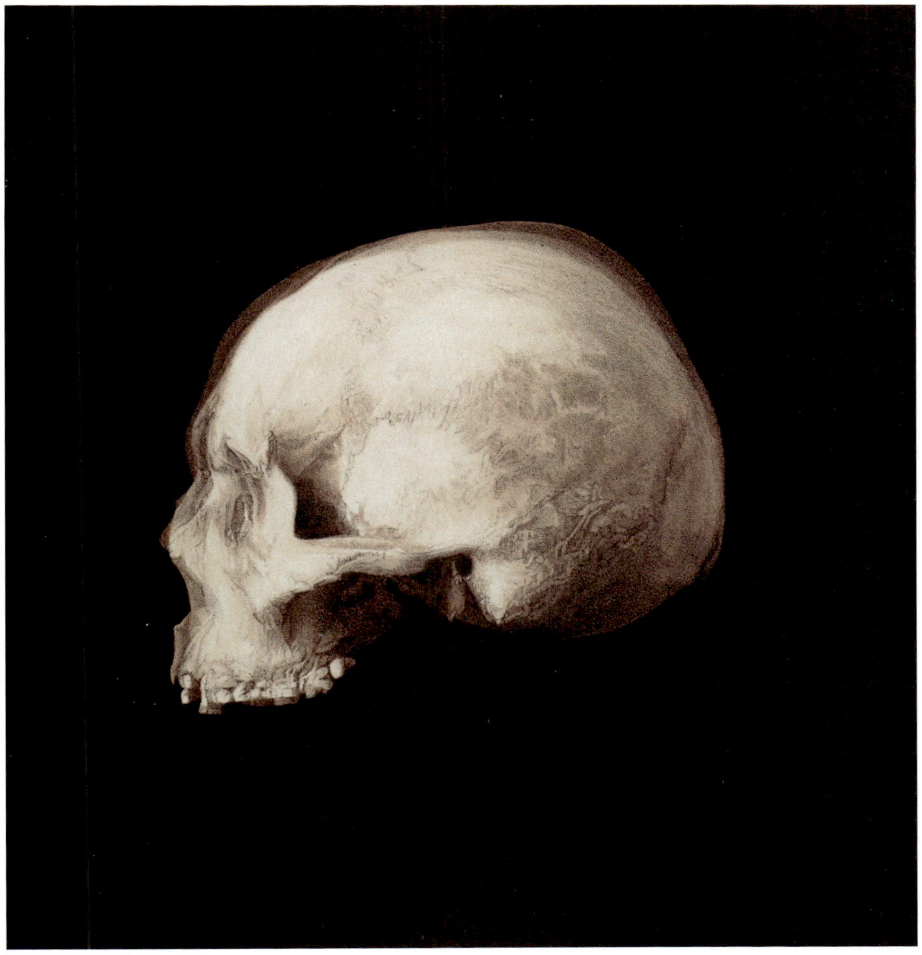

Plate 5.20

Composite Photograph of Six Male Arapahoe Indian Skulls, Basal View
John Shaw Billings, MD, and Washington Matthews, MD, 1885

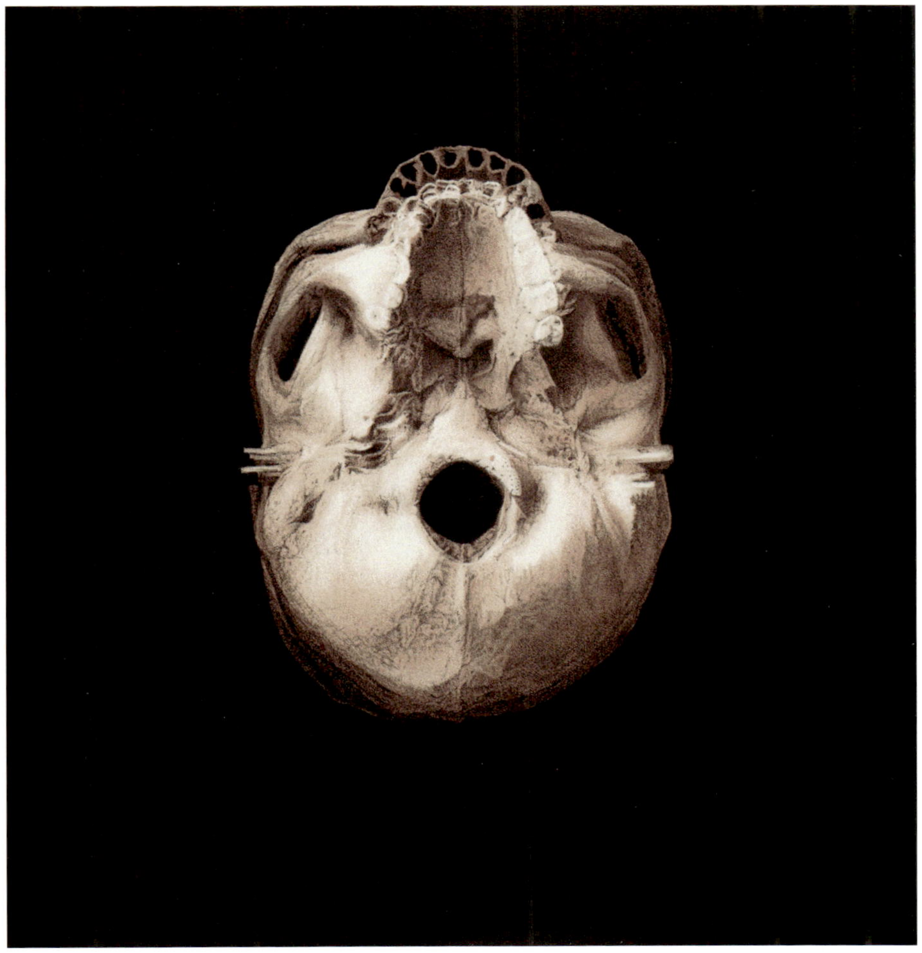

Plate 5.21

Composite Photograph of Six Male Arapahoe Indian Skulls, Lateral View
John Shaw Billings, MD, and Washington Matthews, MD, 1885

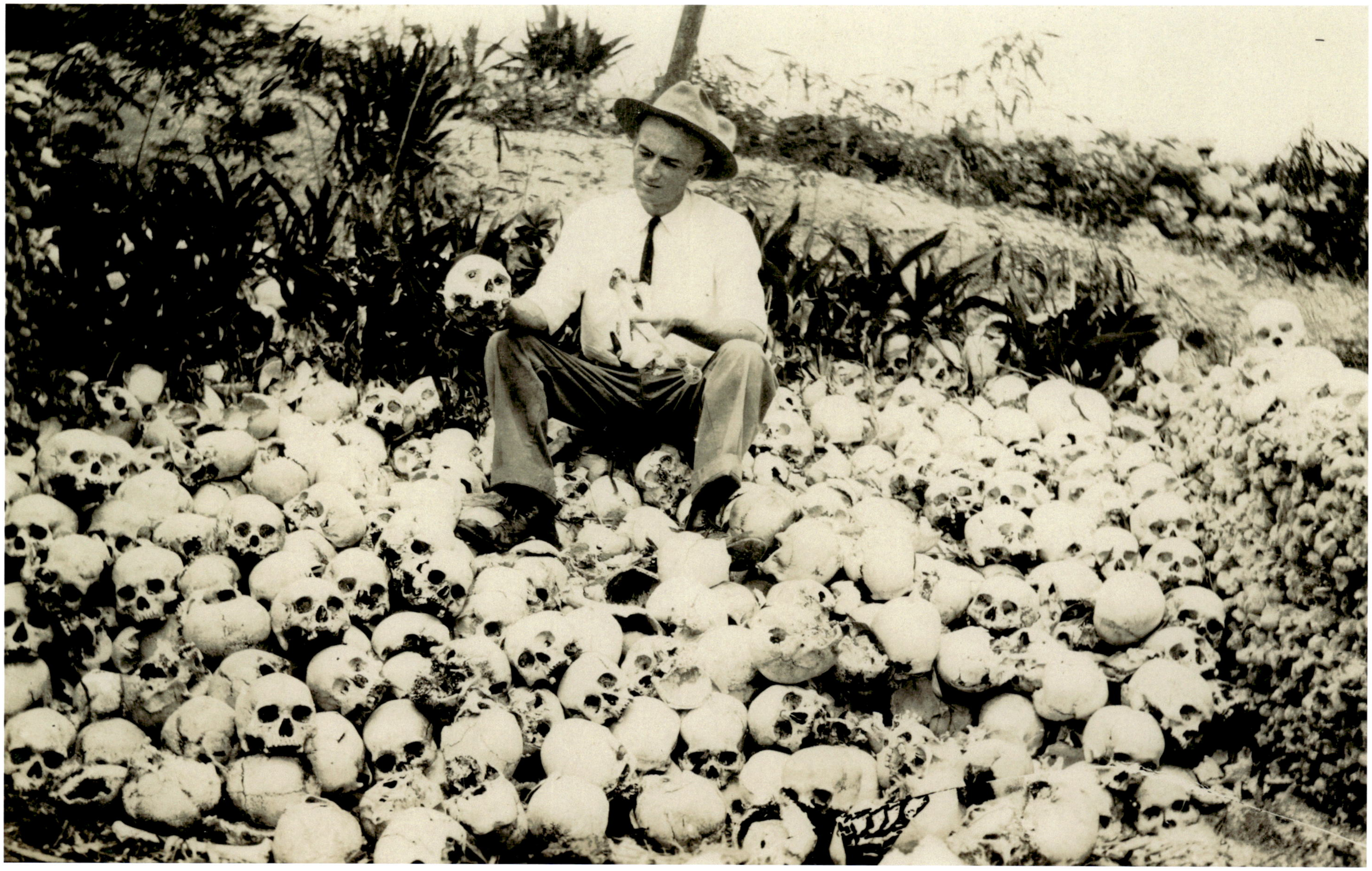

Plate 5.22

Exploring a Pile of Skulls
Circa 1915

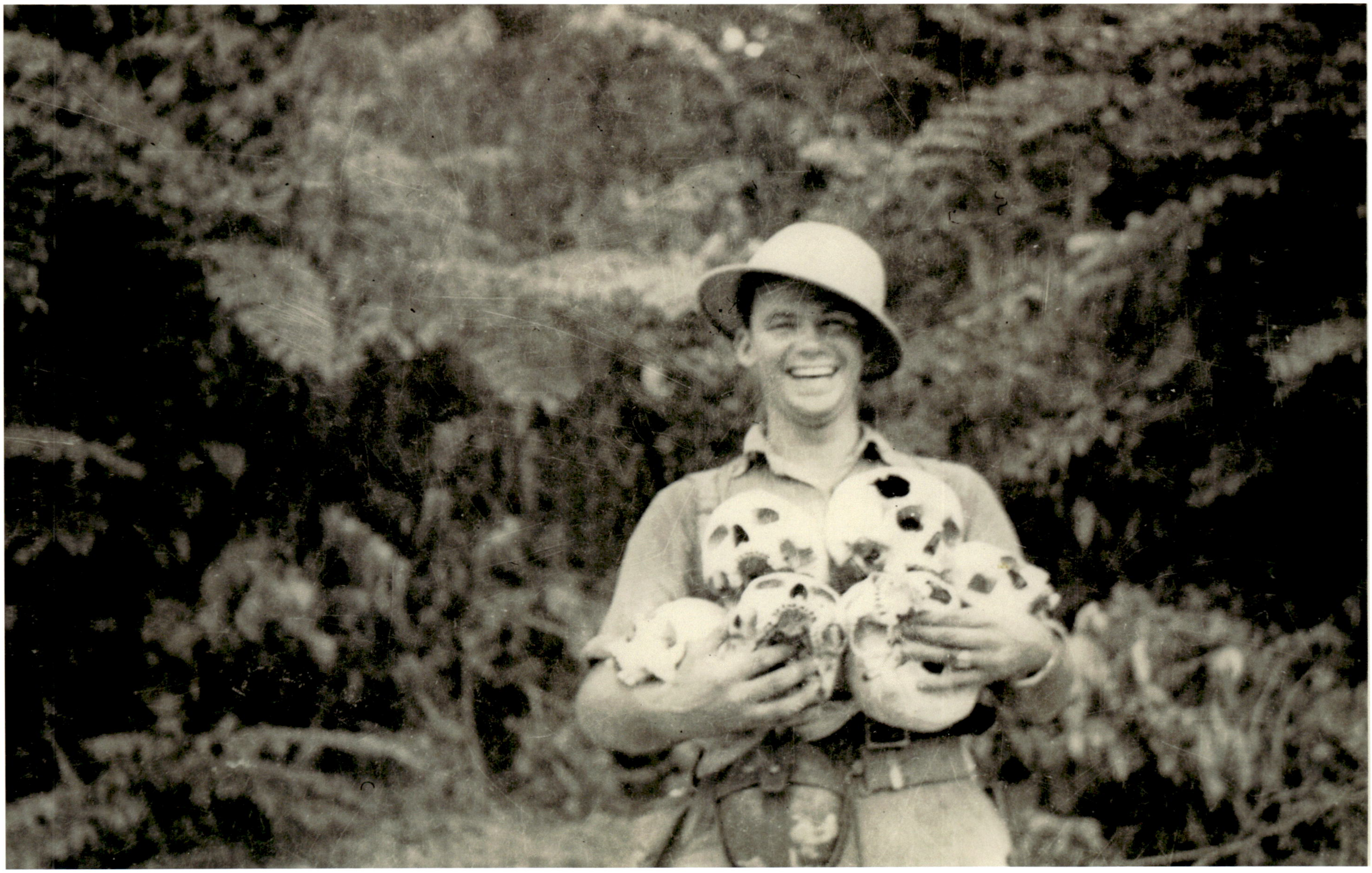

Plate 5.23

Happy Hunter, A Day's Treasure Trove of Skulls
Circa 1905

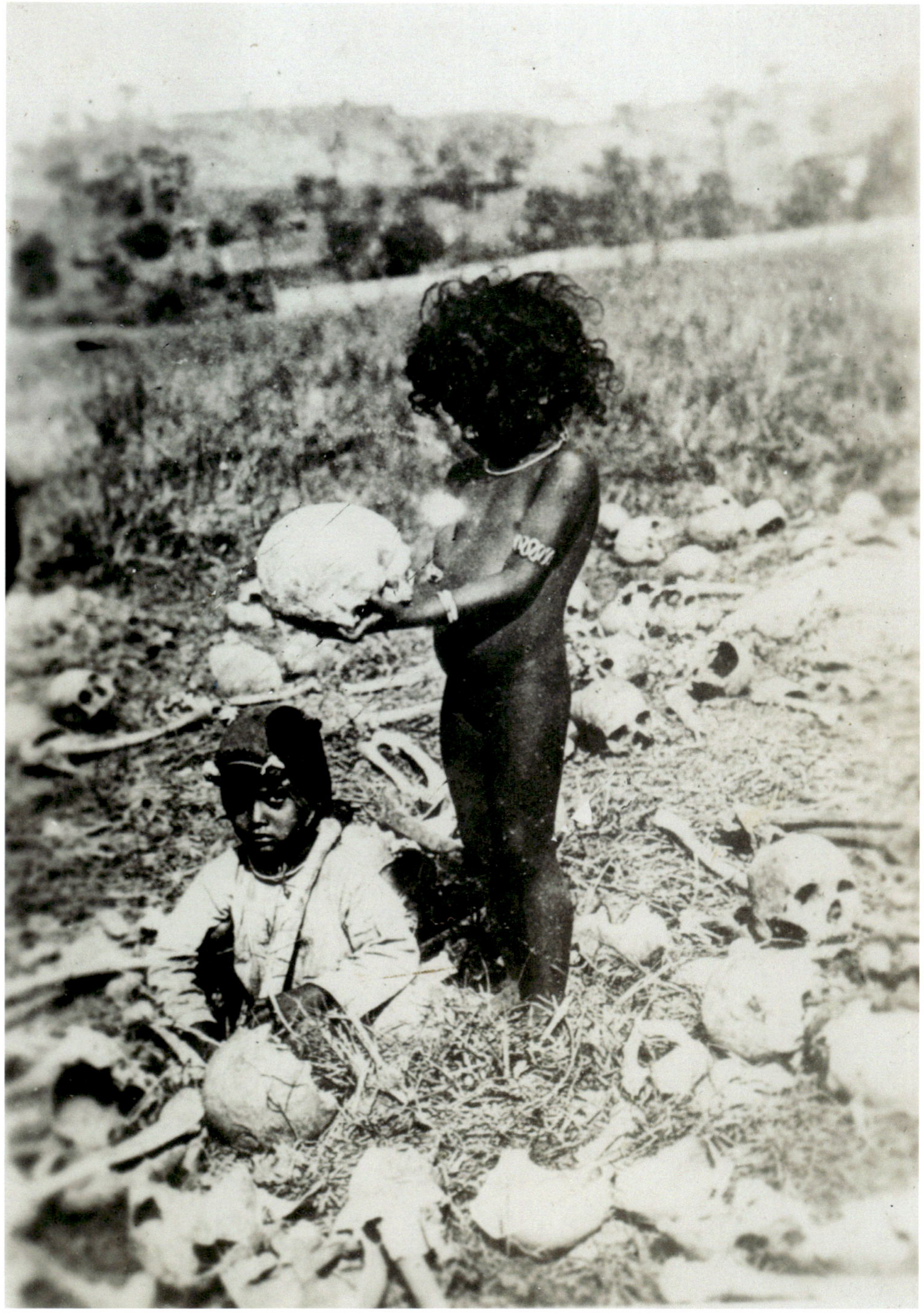

Plate 5.24

South Asian Indigenous Children Examining Skulls in a Cemetery
Circa 1920

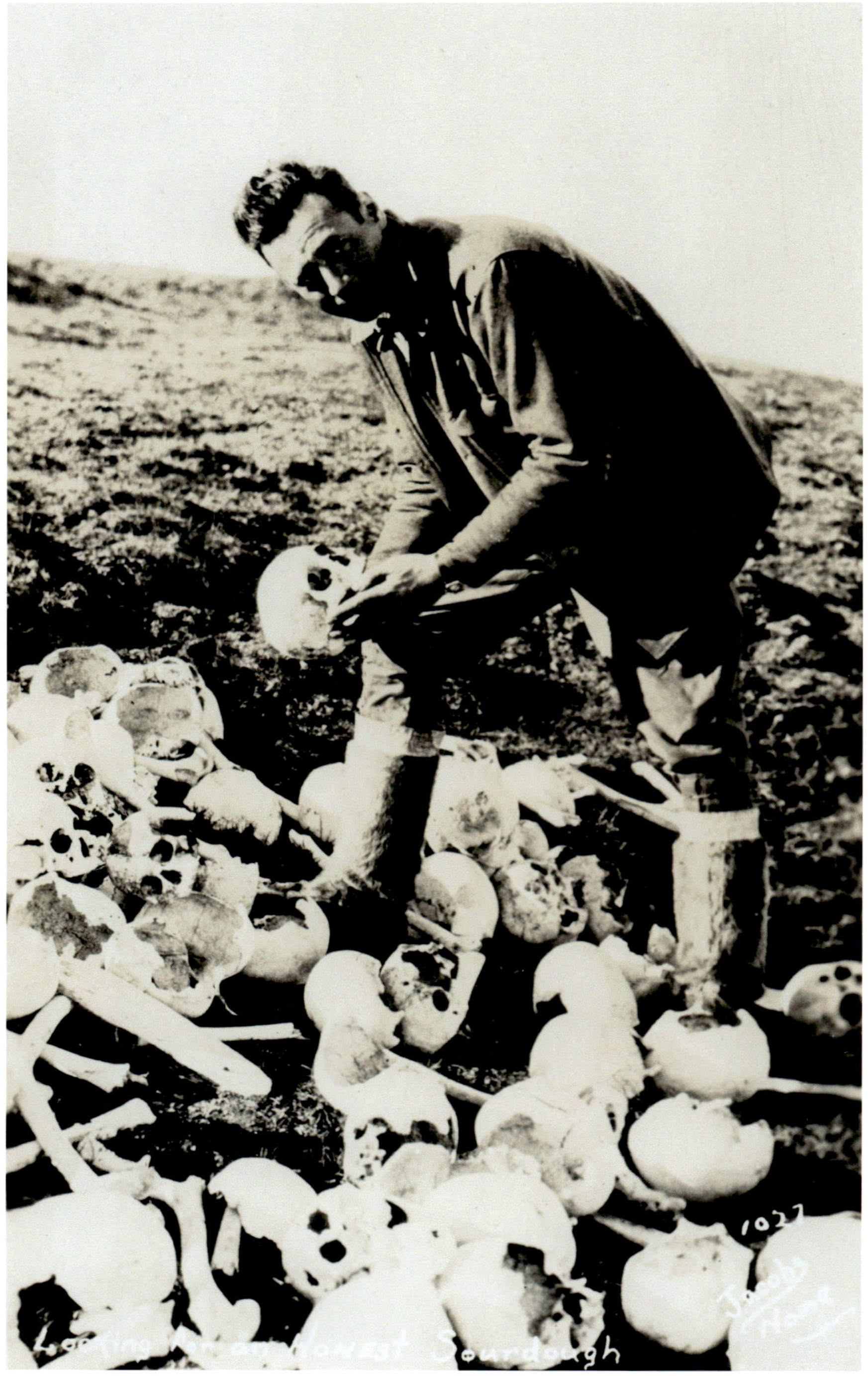

Plate 5.25

Looking for a Good Specimen
Circa 1915

Plate 5.26

Souvenir Hunters with Cliff Dweller Indian Skulls and Bones
1890
Cabinet Card with Handwritten Note, one of fifteen rooms explored

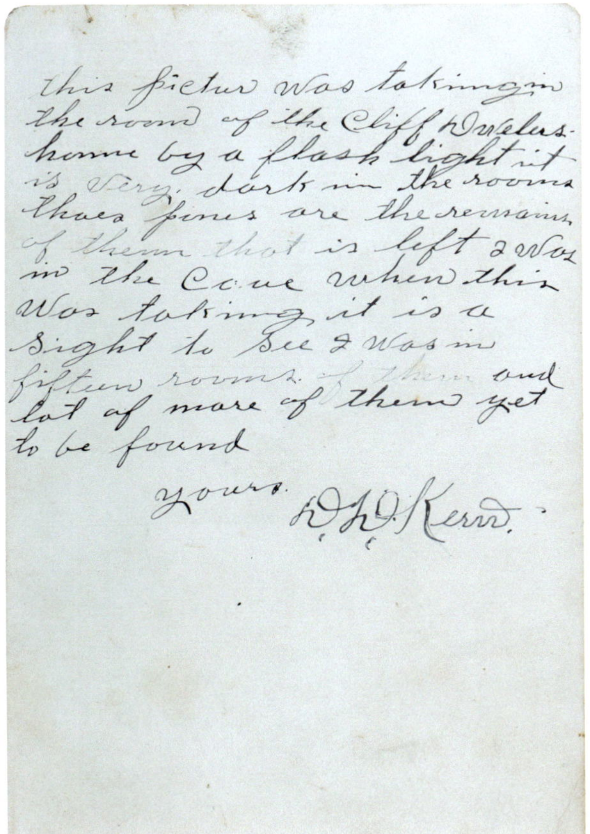

this picture Was taking in
the room of the Cliff Dwelers-
home by a flash light it
is very dark in the rooms
thoes pines are the remains
of them that is left I Was
in the Cave when this
Was taking it is a
Sight to See I Was in
fifteen rooms of them and
lot of more of them yet
to be found

yours. D. L. D. Kerr.

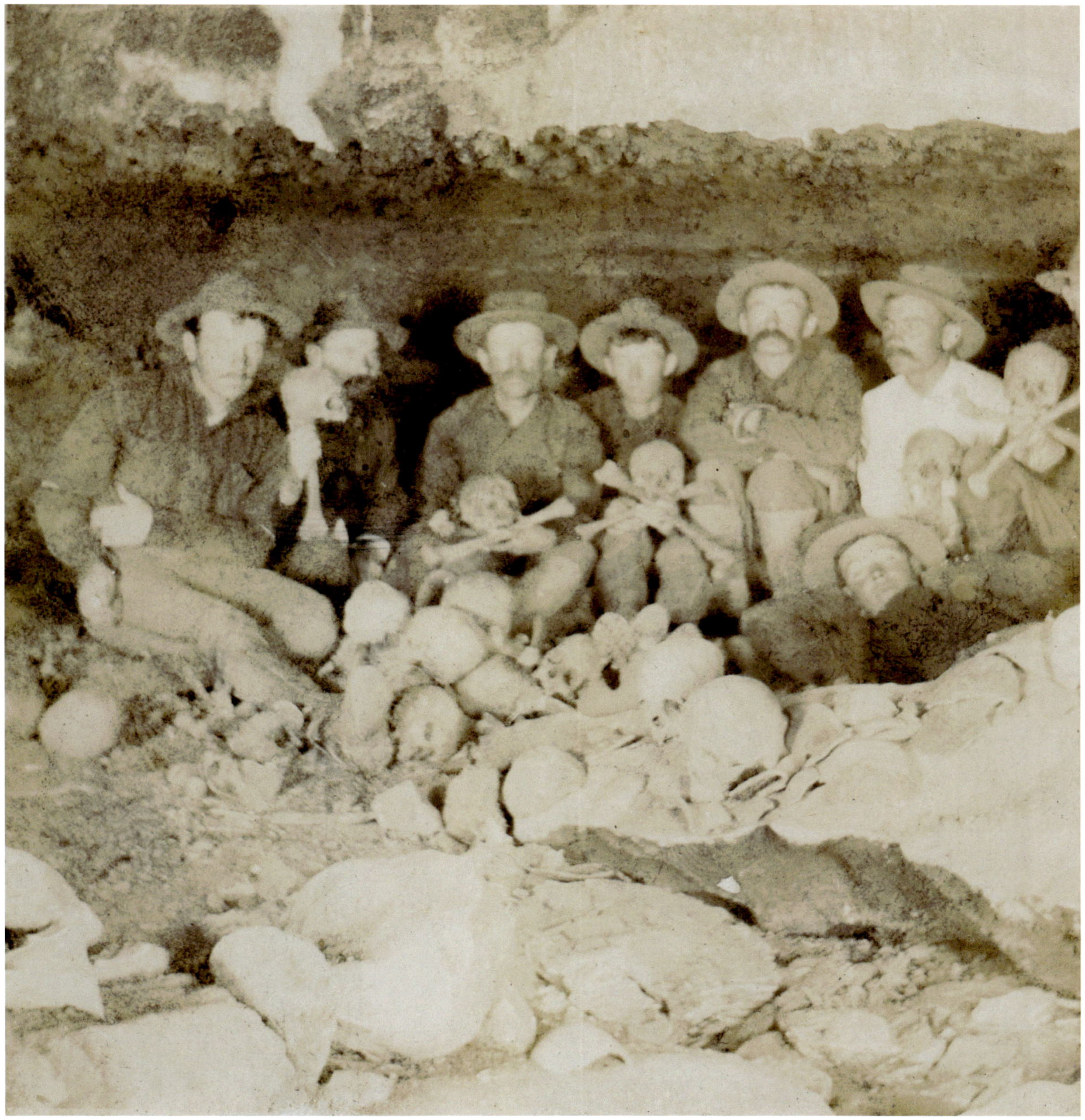

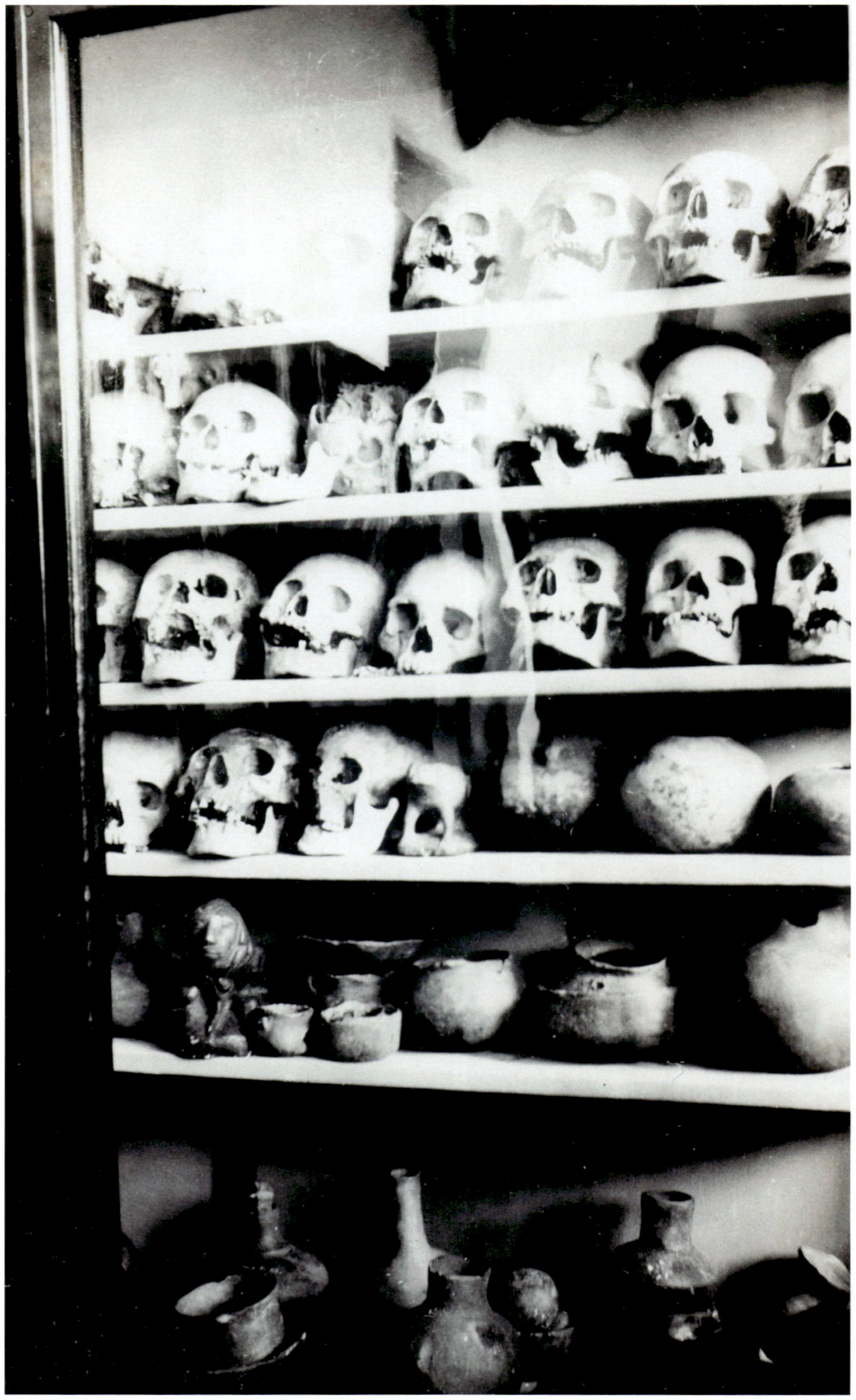

Plate 5.27

Shelf of Native American Skulls, Stanford University Museum
Snapshot, circa 1930

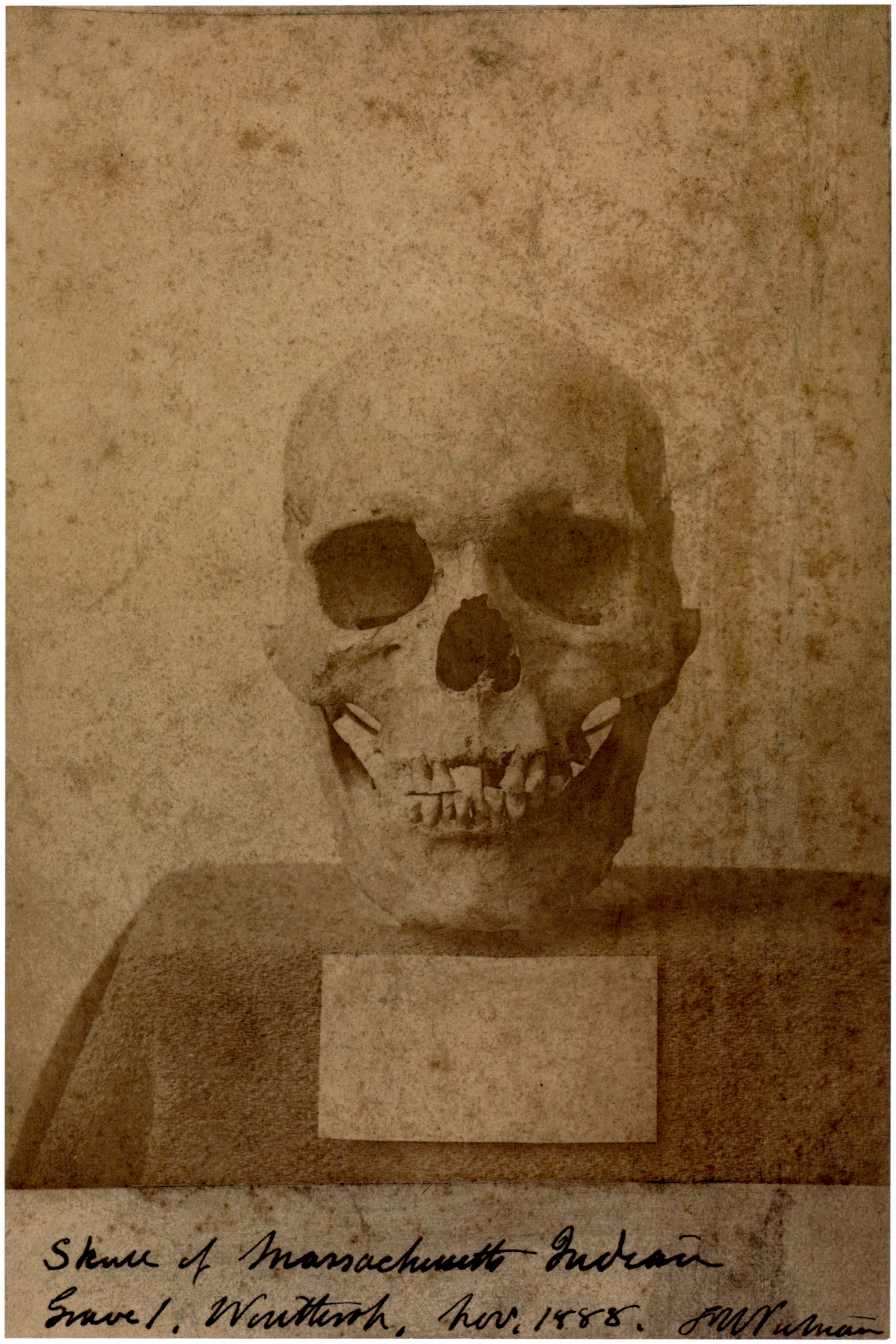

Plate 5.28

Skull of a Massachusetts Indian, Grave 1, Winotauk
Cabinet Card, November 1885

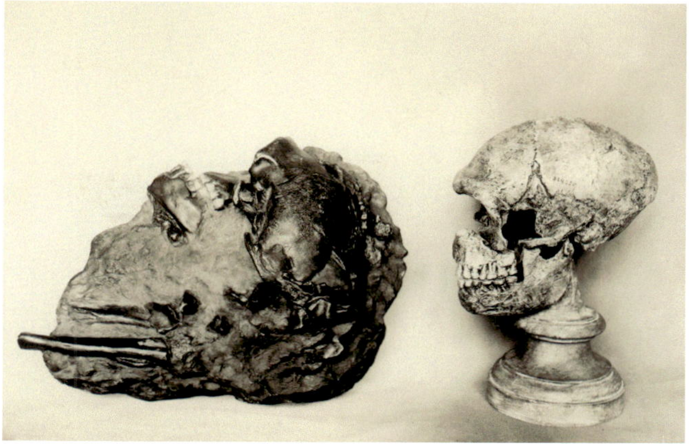

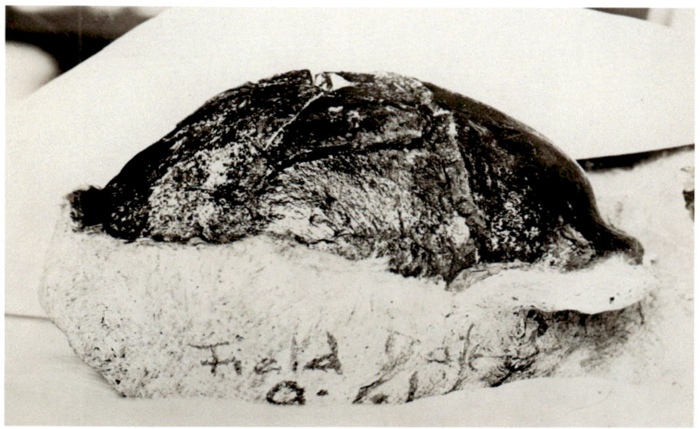

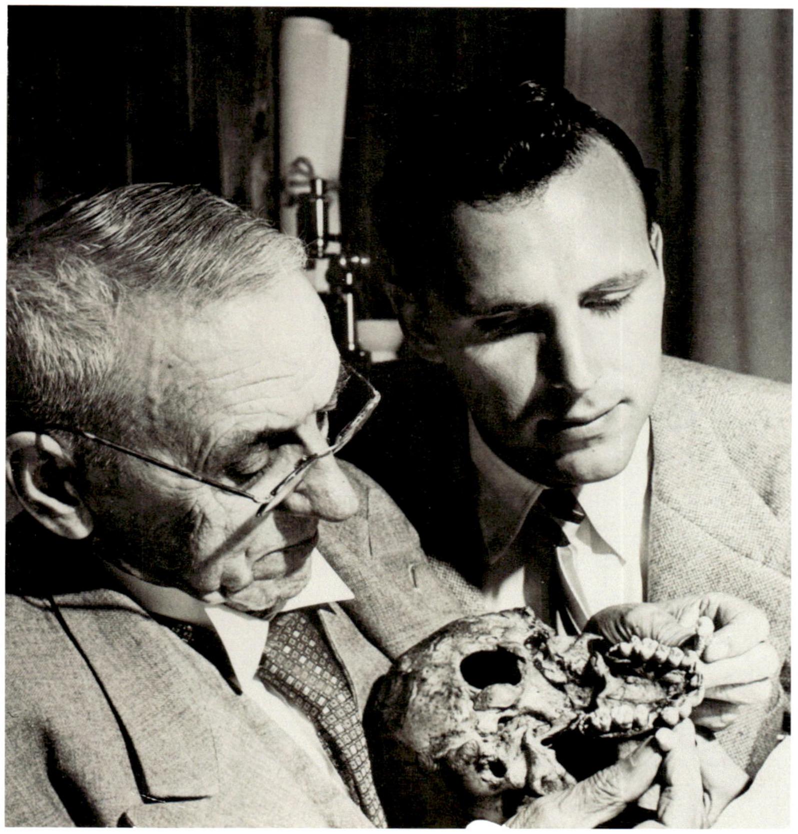

Plate 5.29

Skull of the Earliest Known Prehistoric Man
Specimen unearthed in La Quina, Southern France, 1923

On the left is the skull before restoration.

Plate 5.30

Discovery of the "Peking Man," Found near Peking in Choukouton, China
1930

The dome of the skull partially freed from the rock.

Plate 5.31

Robert Broom, MD, (holding skull) and Explorer William Terry Examine a Prehistoric Skull
1948

Dr. Broom discovered the million-year-old early hominid 'Sterkfontein Man.'

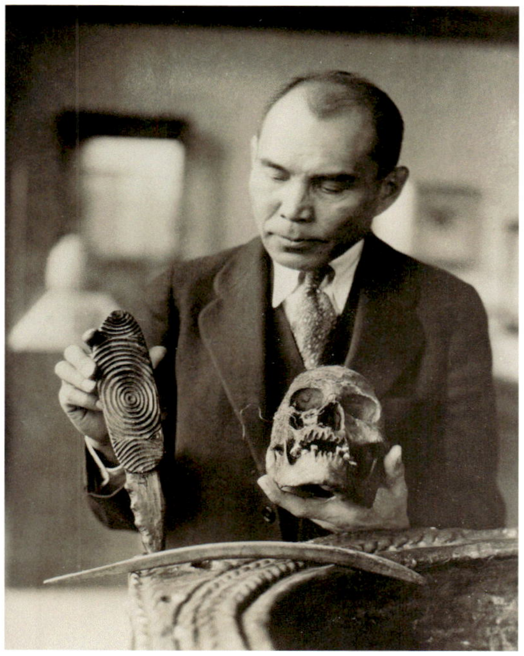

Plate 5.32

Louis S. Shotridge Examining Thule Skull and Artifacts Found Near Point Barrow, Alaska
1923

Shotridge is a Tlingit Indian and was an assistant at University of Pennsylvania Museum.

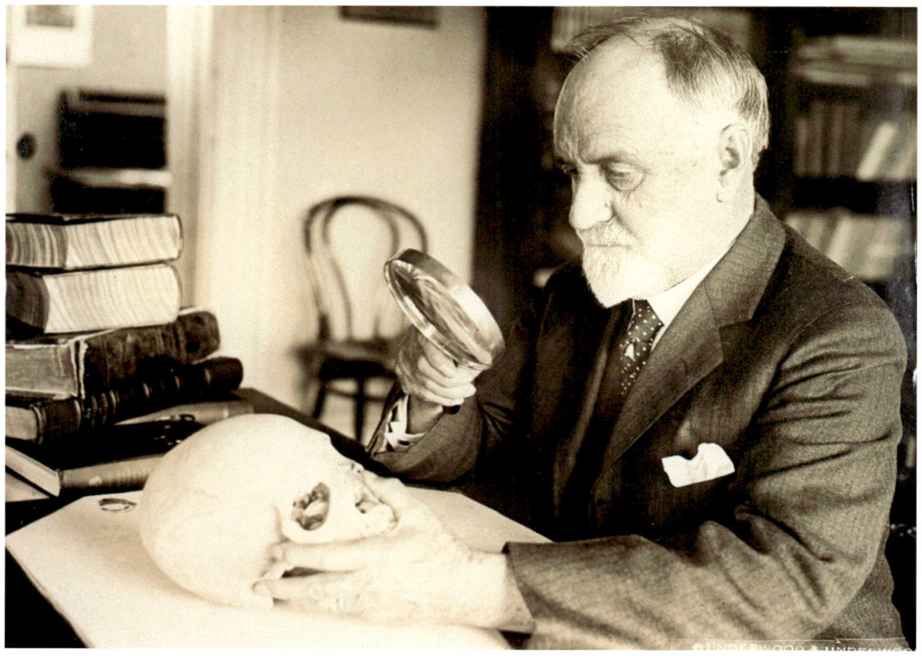

Plate 5.33

"Searching for America's Oldest Man," Examining Indian Skull Found Near Washington, DC
1924
Dr. Jesse Walter Fewkes, Chief of the Bureau of Ethnology, Smithsonian Institution

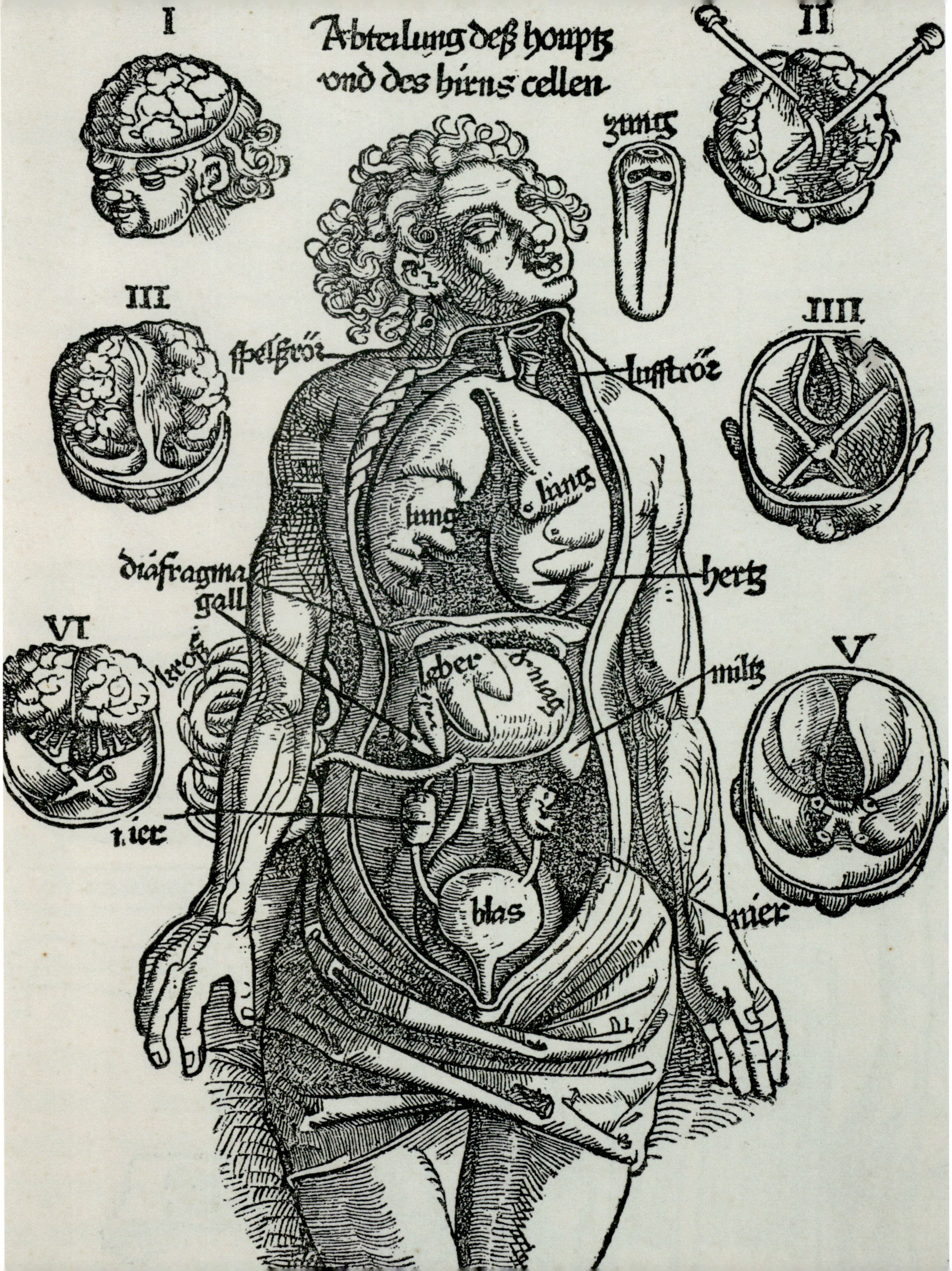

Abteilung deß houptz
vnd des hirns cellen
I
II
zung
III
IIII
spelßrör
lufftrör
lung
lüng
diafragma
gall
hertz
VI
V
leber
mag
miltz
blas
nier

CHAPTER SIX

Special Dissections – Anatomical Studies

Organ Identification & Procedure Anatomy

It is in the area of special dissections that photography particularly enhanced the study of anatomy. In preparation for dissections, ancillary materials are beneficial as study aids. However, studying anatomy from books with engravings or lithographs, which are only two-dimensional secondary renditions, does not provide an accurate visual representation of a body. Photography allowed permanent preservation of fresh dissections in all its detail and provided a more physical experience of an actual body. Publication of photographed dissections enabled health care professionals worldwide access to the images. As new visual technologies are developed they are applied to anatomical study. The modern 3-D virtual interactive photographic dissection allows realistic anatomy studies that some say rival actual cadaver dissection because anatomical variations can be presented easily and realistically. Virtual dissections and other modern applications will become standards in the future.

To learn and appreciate specific body structures, such as that of the eye, ear, brain, hand, nervous, or lymphatic system, requires specialized dissection. For centuries, prosectors created painstakingly detailed dissections of various parts of the body as study aids. They preserved these body parts, which were housed in prestigious medical institutions for instruction. Engravings made from these dissections were presented in texts. Only a relative handful of medical personnel had access to the original dissections. Unfortunately, most specimens deteriorated over time, losing the fine details. Photography allowed permanent visual preservation of the fresh dissections.

Photographic formats and technology are always in flux, providing each generation with a new visual tool for learning. At the end of the nineteenth century, the manufacture of stereoscopic images was inexpensive and became the leading educational medical device. Three-dimensional stereo photographs were a way to show the spatial relationships of gross anatomy that aided comprehension. Physicians and surgeons in the late 1890s began producing photographic stereographic texts and card atlases in many specialties, especially anatomy. Among the first early texts was that of Leipzig dermatologist Albert Neisser (1855-1916). On December 3, 1898, the *British Medical Journal* (p. 1697), published an article extoling the virtues of stereo photographs in facilitating understanding in clinical and anatomical study.

In 1905, Edinburgh's David Waterston, MD, published *The Edinburgh Stereoscopic Atlas of Anatomy*, a set of detailed stereo cards. The set was quite popular and became a standard tool to learning anatomy; it was the 'Bible' for anatomy students and was published into the 1930s. The cards had dissections with labels and titles, as well as general descriptions. They came packaged with the viewer. The original 1905 set contains five sections of fifty stereo cards, each 9" x 7 1/8" with descriptive text. The volumes include: 1) abdomen; 2) pelvis and thorax; 3) thorax, upper limb, cranio-cerebral topography; 4) central nervous system, brain, head, and neck; 5) lower limb. Several examples are shown. By 1910, D.J. Cunningham and other editors expanded the set to ten volumes. Over the next two decades, the set continued to expand as organ systems were focused on and

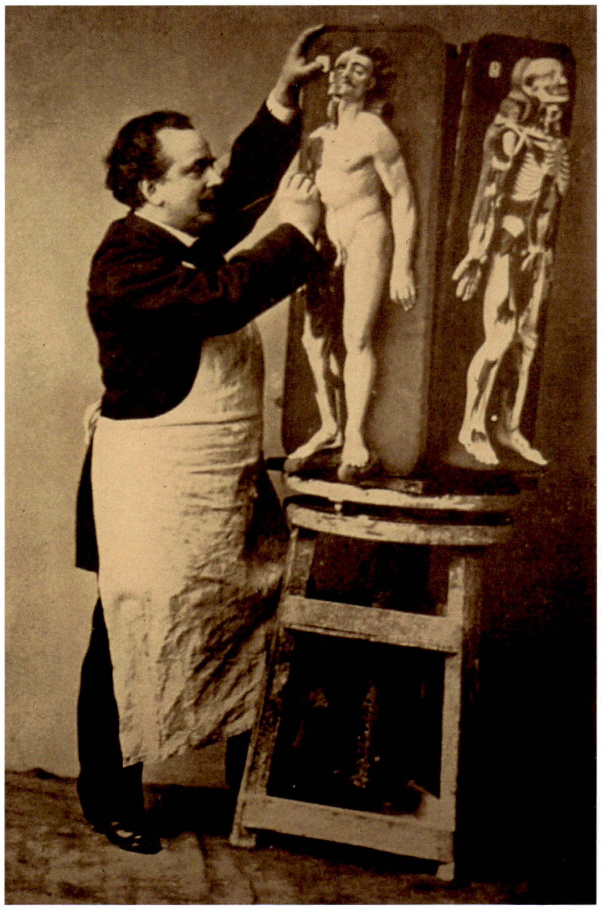

Plate 6.1

Mr. Jules Talrich, Parisian Anatomical Model Maker, 1866
Models made for L'Exposition Universal de Paris, 1867

studied in detail. Several of these later Edinburgh stereographs are also shown including a series on head anatomy.

It was extremely difficult to dissect the eye and internal ear because of their minute structure, and by mid-century several texts were produced using stereo imagery. In 1912, Dr. Arthur Thompson published a set of sixty-seven stereoviews of the eye that set a standard in ophthalmology. One of the stereo images from his text is reproduced. In 1939, at New York's World Fair, photographer William B. Gruber presented his invention, the View-Master, which uses a rotating disc of seven stereo images. This stimulated the production of easy-to-make stereographic texts, as the discs were easily put in a book in a slipcase. A photograph of David Lee Bassett, MD, (1913–1966) taken in 1948 is presented as he is performing a dissection for his acclaimed *Stereoscopic Atlas of Human Anatomy*, which consisted of 221 View-Master reels with 1,554 color stereo views of dissections of every body region.

Bodies decayed over time with the preservation methods in use prior to the late twentieth century. To augment anatomical and pathological study, artisans created detailed papier-mâché models and wax models called moulage. Many are considered 'beautiful' works of art. Ecorché statuettes, flayed anatomic models, became popular during the sixteenth and seventeenth centuries. Bolognese artist Gaetano Guilo Zumbo was the first to make colored wax models. Clemente Susini, aided by the anatomist Paolo Mascagni, carried on this specialized craft in the eighteenth century. The wax figures were individually carved, and it took up to two hundred cadavers to make one model, due to the lack of adequate preservation. Many eighteenth-century wax models can still be seen on display at the Specola Museum in Florence, Italy. In Paris, during the nineteenth century, Jules Baretta (1834–1923) made more than 2,000 wax models of pathological and anatomical conditions, concentrating on dermatological conditions. These and about 2,000 other wax models were collected at Paris's Hôpital Saint-Louis where they can be seen to this day. Photographs and engravings of moulage were published in texts and were particularly popular study aids.

Paris in the nineteenth century became noted for its manikin and model makers, and three photographs of these artists with their models are presented. The best known is Louis Thomas Jerôme Auzoux, MD, (1797–1880). In the 1820s, he saw the need for highly accurate anatomical models as an alternative to cadaver study and started to manufacture hand-painted papier-mâché models. His company is still in business. Also depicted is Mr. Jules Talrich with a model he is preparing for an exhibition in 1867. The effort to create the 'best'—that is, the most accurate, realistic, and permanent anatomical model—was perhaps solved by German anatomist Gunther von Hagens. In 1977, he invented a technique for preserving biological tissue specimens called plastination. Until the 1990s, the process was used for small specimens. By 1993, equipment was developed that allowed whole body plastination. Von Hagens' first exhibit was in Japan in 1995. His Body Worlds exhibition of whole bodies plastinated in life-like poses is currently on tour around the world. Each body is dissected to show a particular organ system or function and has been acclaimed by medical personnel as a creative tool for learning anatomy.

"Harriet" is an unusual dissection presented in this chapter. Named in honor of the donor, Harriet, a cleaning woman at Hahnemann Medical School. It is a dissection of the complete cerebro-spinal nervous system by Rufus Weaver, MD, (1841–1936) in 1888. Also presented are some of the interesting and artistic mid-nineteenth century dissections images published in noted texts such as *Revue Photographique des Hôpitaux de Paris,* 1869-72, *Photographie Médicale des Hôpitaux de Paris,* 1873-75, and *Catalogue des pièces du musée Dupuytren, Atlas du tome,* 1877-79. The chapter ends with photograms of a serial body dissection printed as cyanotypes.

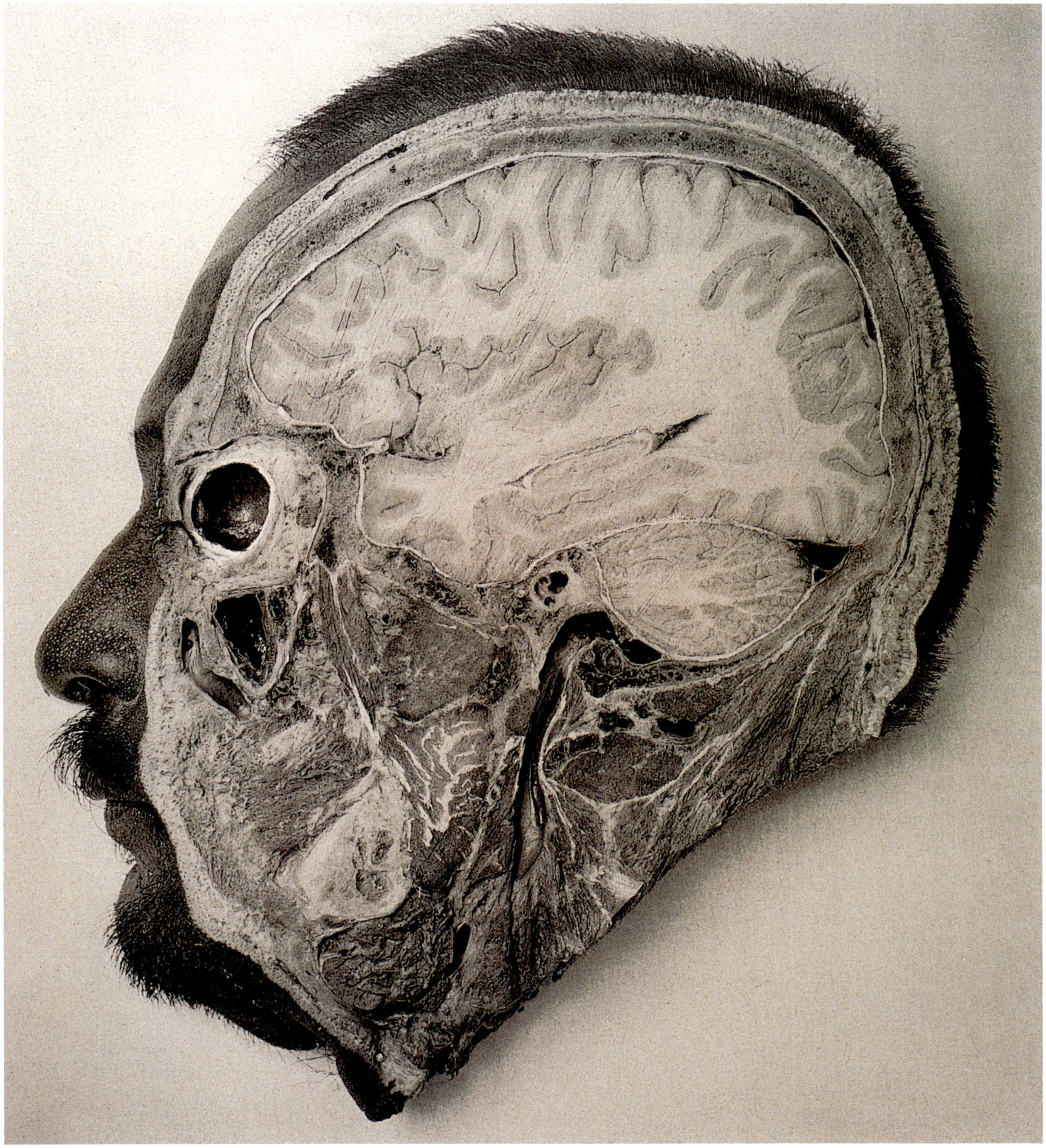

Plate 6.2

Serial Section of Head, Executed Argentinian Criminal
1904

Searching for an anatomical cause of criminality.
Published in Buenos Aires by Pathologist F. Perez, MD

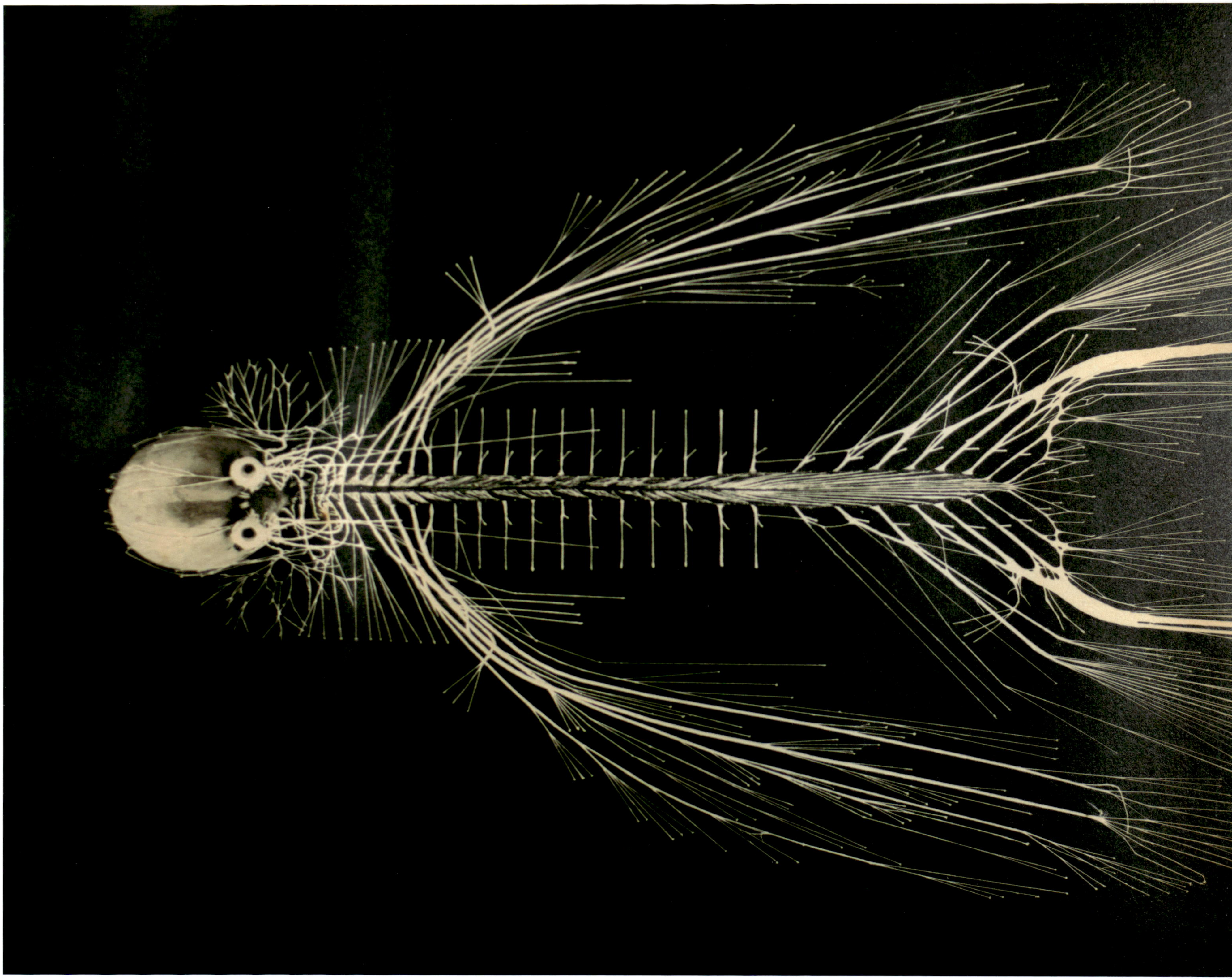

Plate 6.3

"Harriet" Full Body Cerebral-Spinal Nervous System Dissection
1888

Rufus B. Weaver, MD, dissected Harriet Cole, a dedicated anatomy laboratory employee. Dissection completed in five months at Hahnemann Medical College.

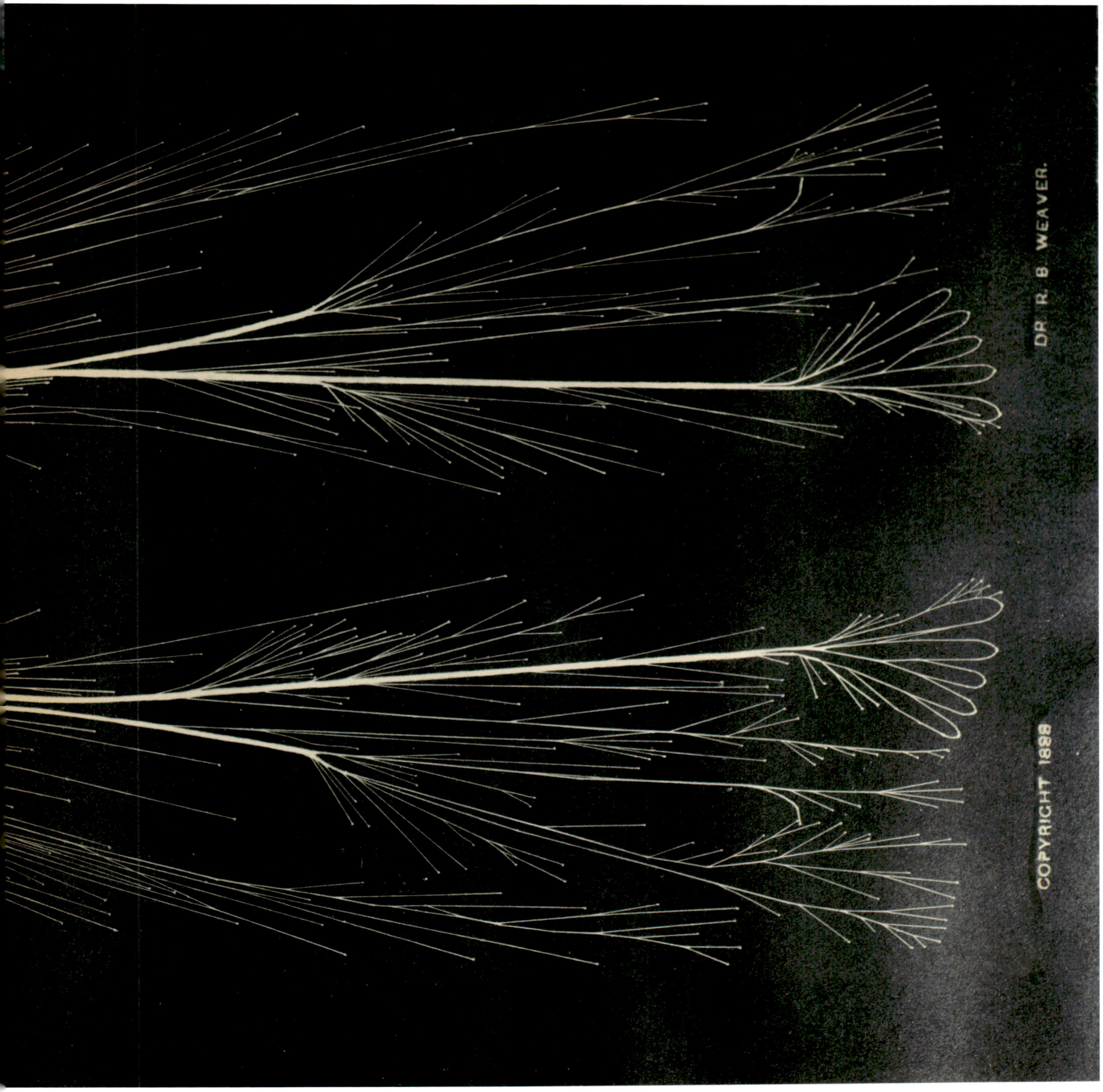
DR. R. B. WEAVER.
COPYRIGHT 1888

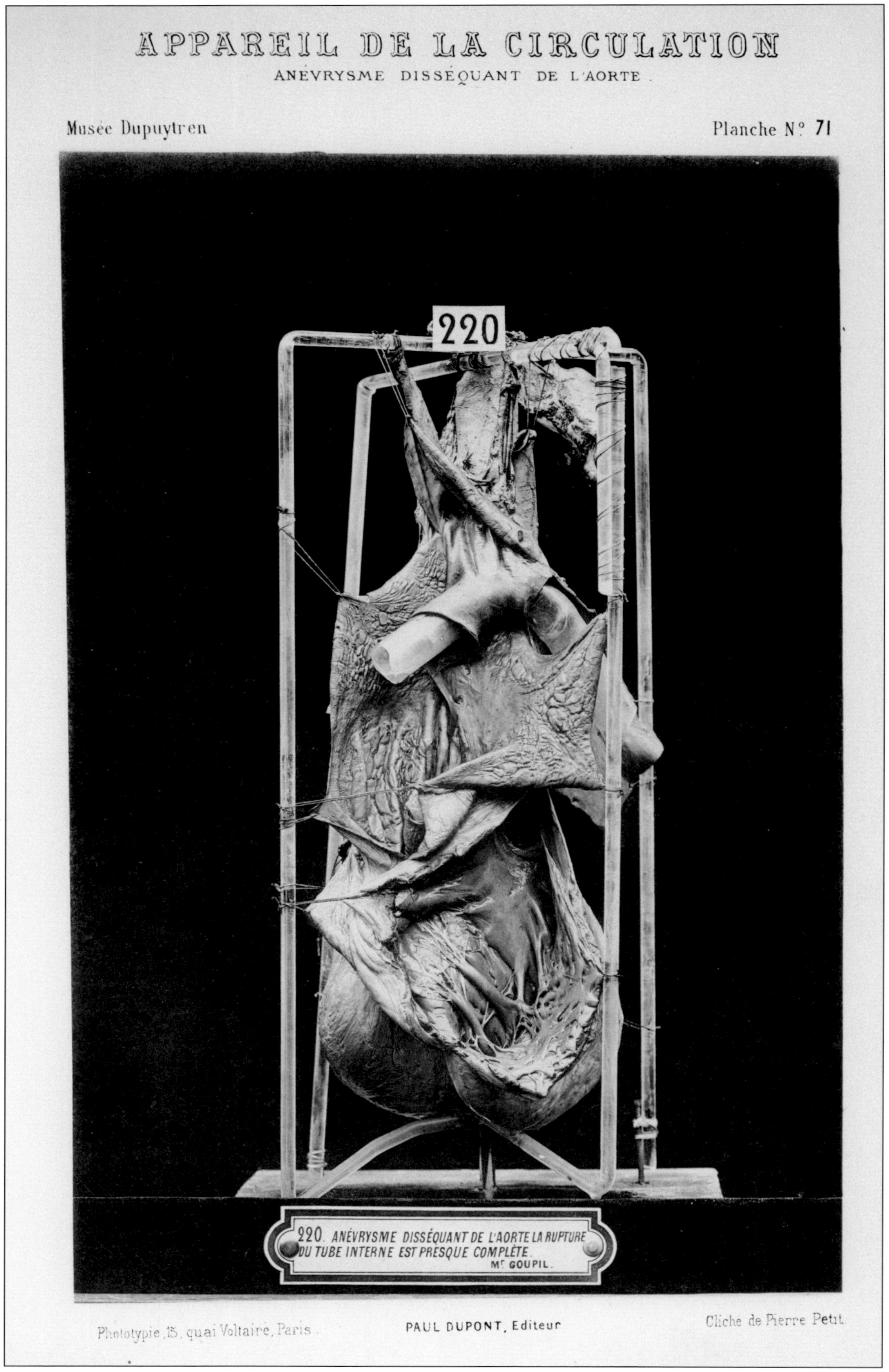

Plate 6.4

Ruptured Dissecting Aortic Aneurism
Catalogue des pièces du museé Dupuytren, Atlas du tome quatriéme, Plate 71, 1879

A tube placed in the lumen of the aorta.

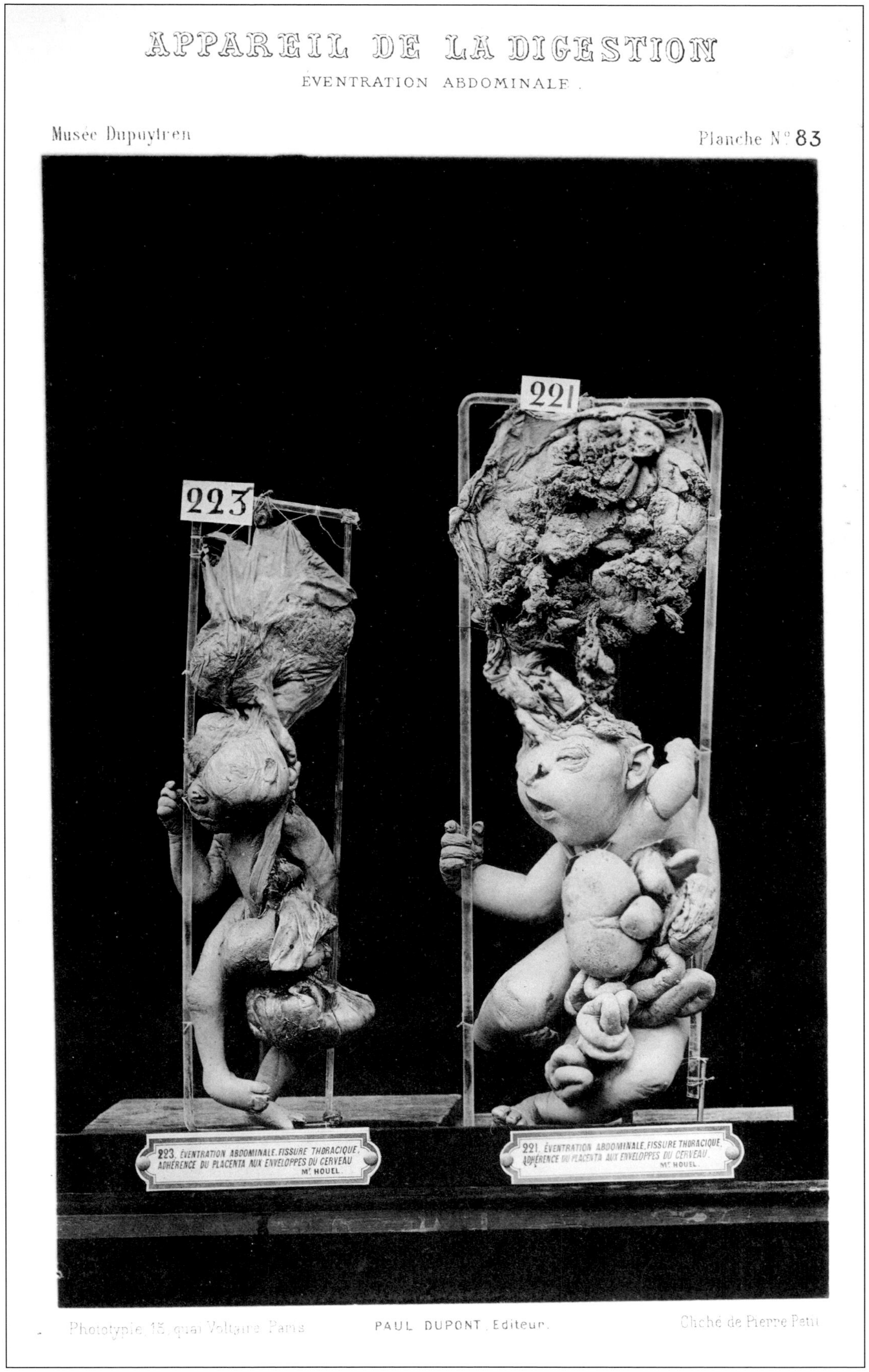

Plate 6.5

"Abdominal Eventration" Herniation of Intestines from Abdomen
Catalogue des pièces du museé Dupuytren, Atlas du tome quatriéme,
Plate 83 from 1879

A congenital anomaly, the diaphragm and muscles fail to develop and abdominal contents are released.

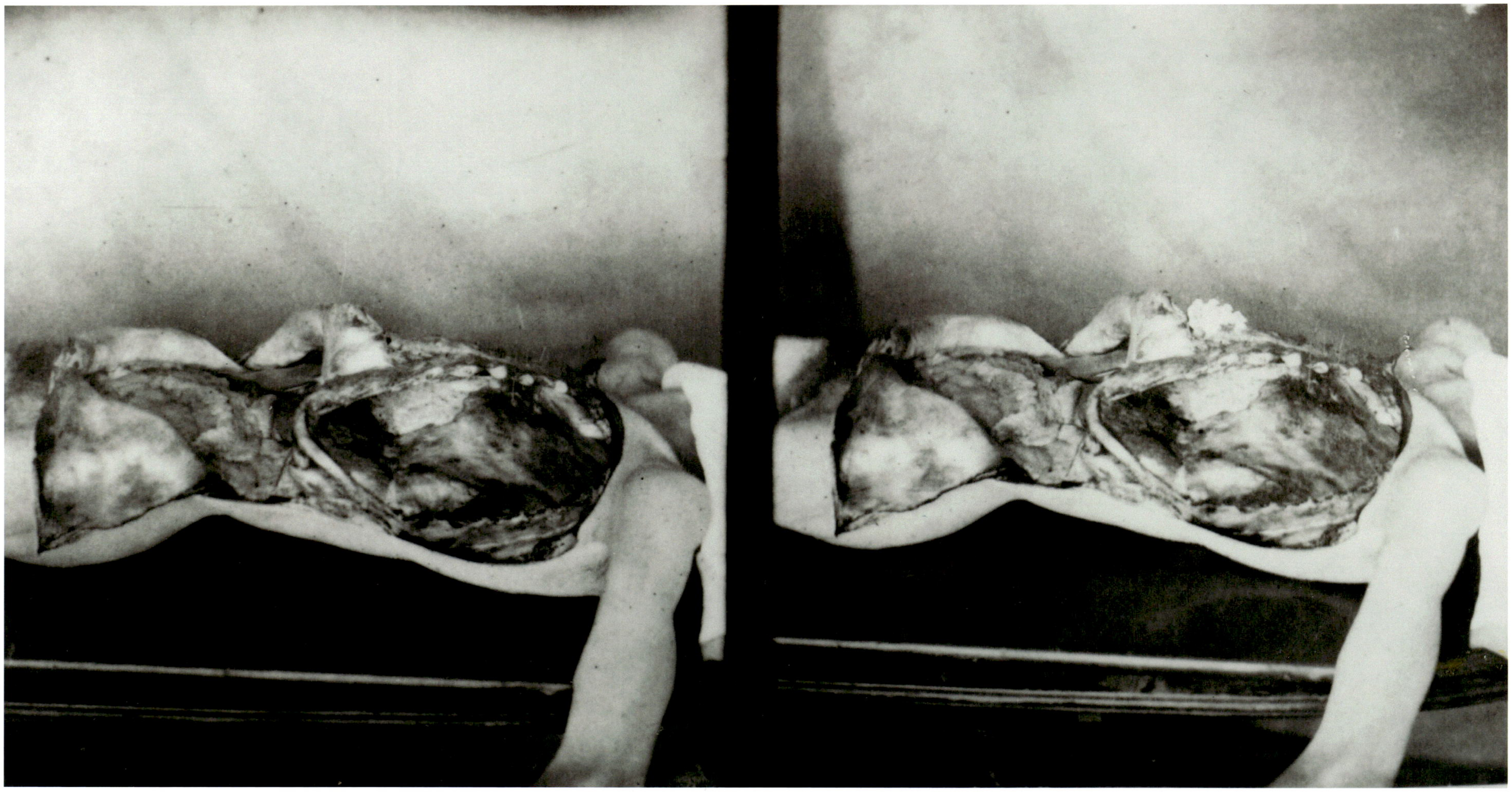

Plate 6.6

Dissection by Oliver Wendell Holmes, MD
Stereoview, circa 1865

Holmes was a proponent of stereoscopic libraries.

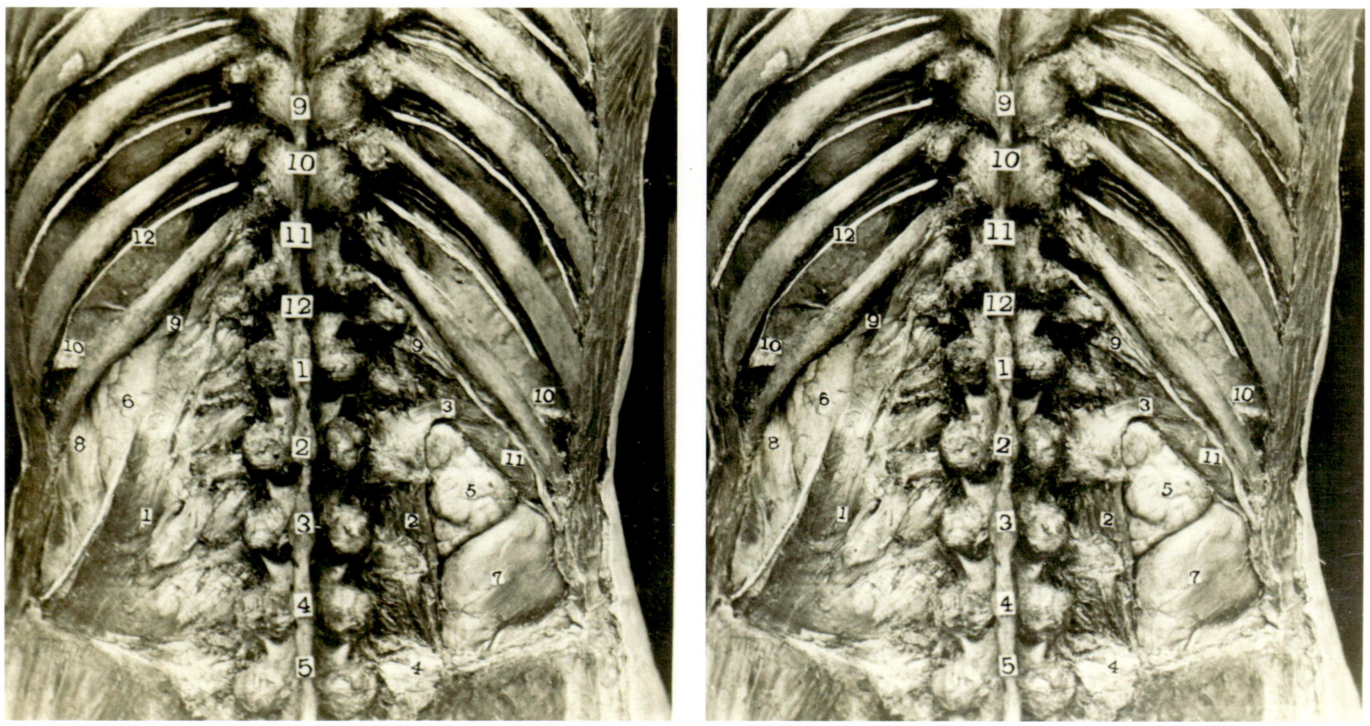

Plate 6.7

Dissection of the Lumbar Region to Show the Lumbar Vertebrae
Edinburgh Stereoscopic Atlas of Anatomy, Section V, No. 27, Lumbar Region 3, 1915

The muscles of the back have been removed down to the Quadratus Lumborum on the left.

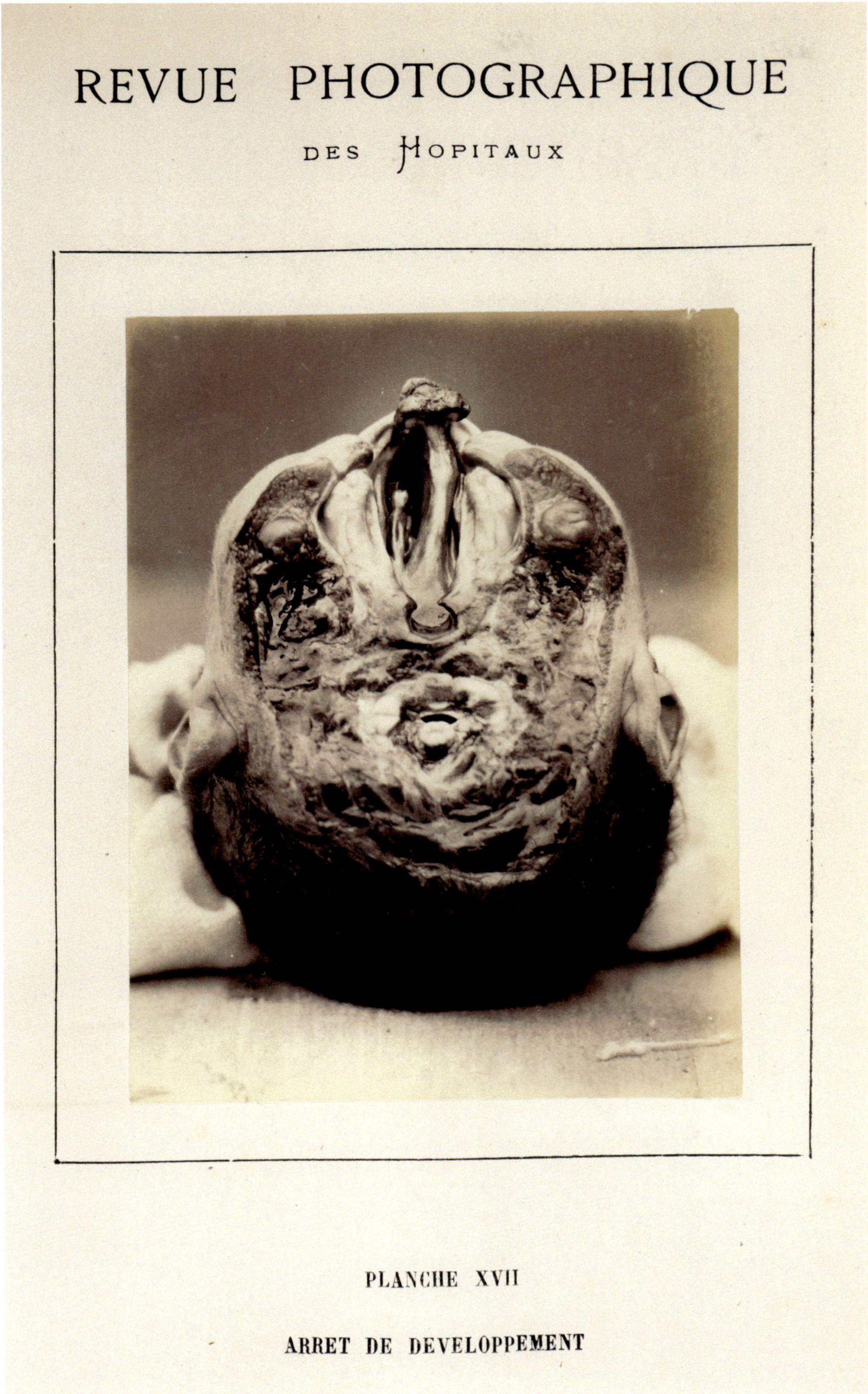

Plate 6.8

Arrested Development, Head Dissection
Revue photographique des hôpitaux de *Paris*, Plate 17, Volume 1, 1869

Dissection through the head to show the abnormalities.

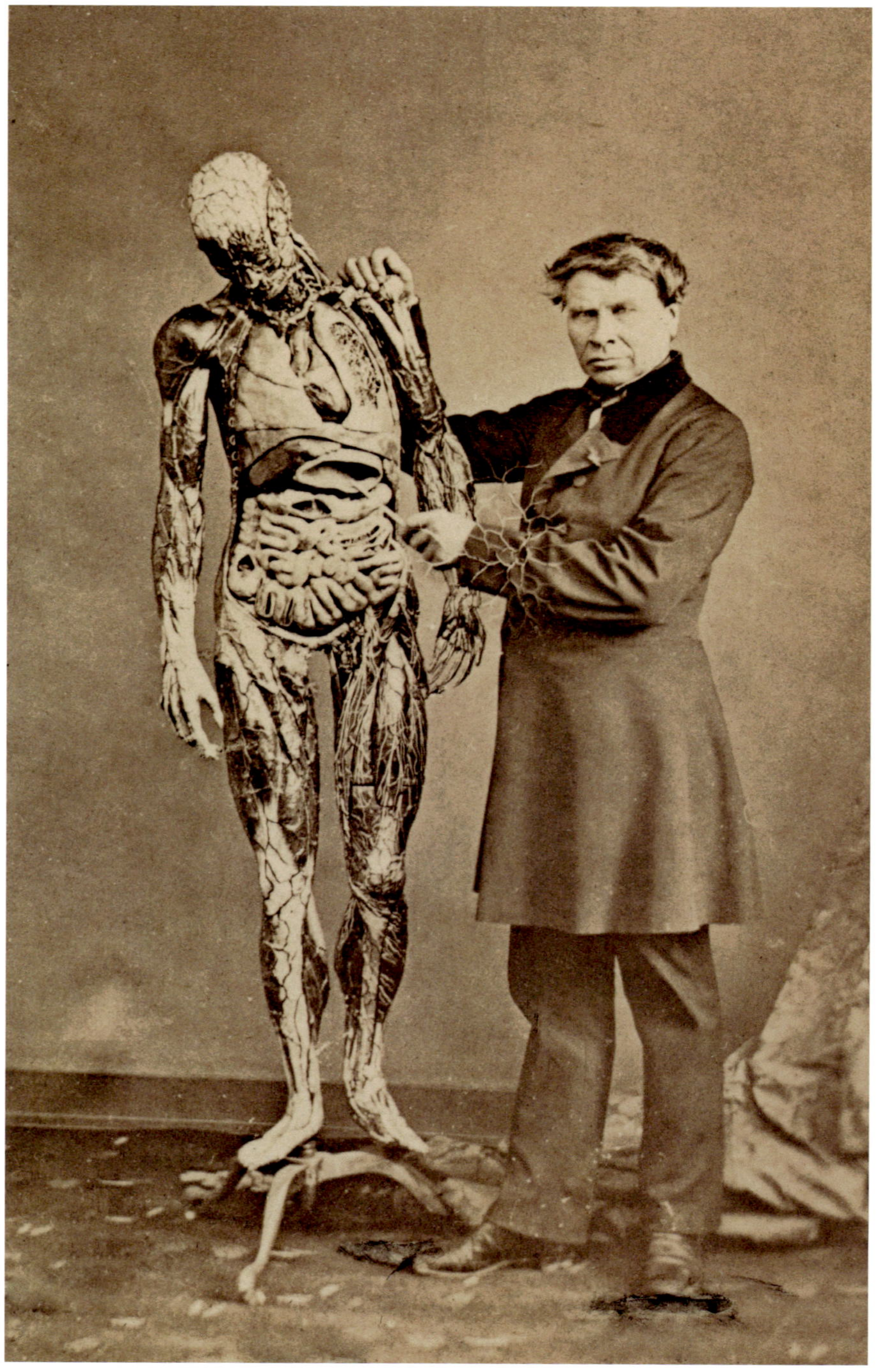

Plate 6.9

Louis Thomas Jerôme Auzoux, MD, with Anatomical Model
Carte de Visite, 1858

The most famous maker of papier-mâché medical models.

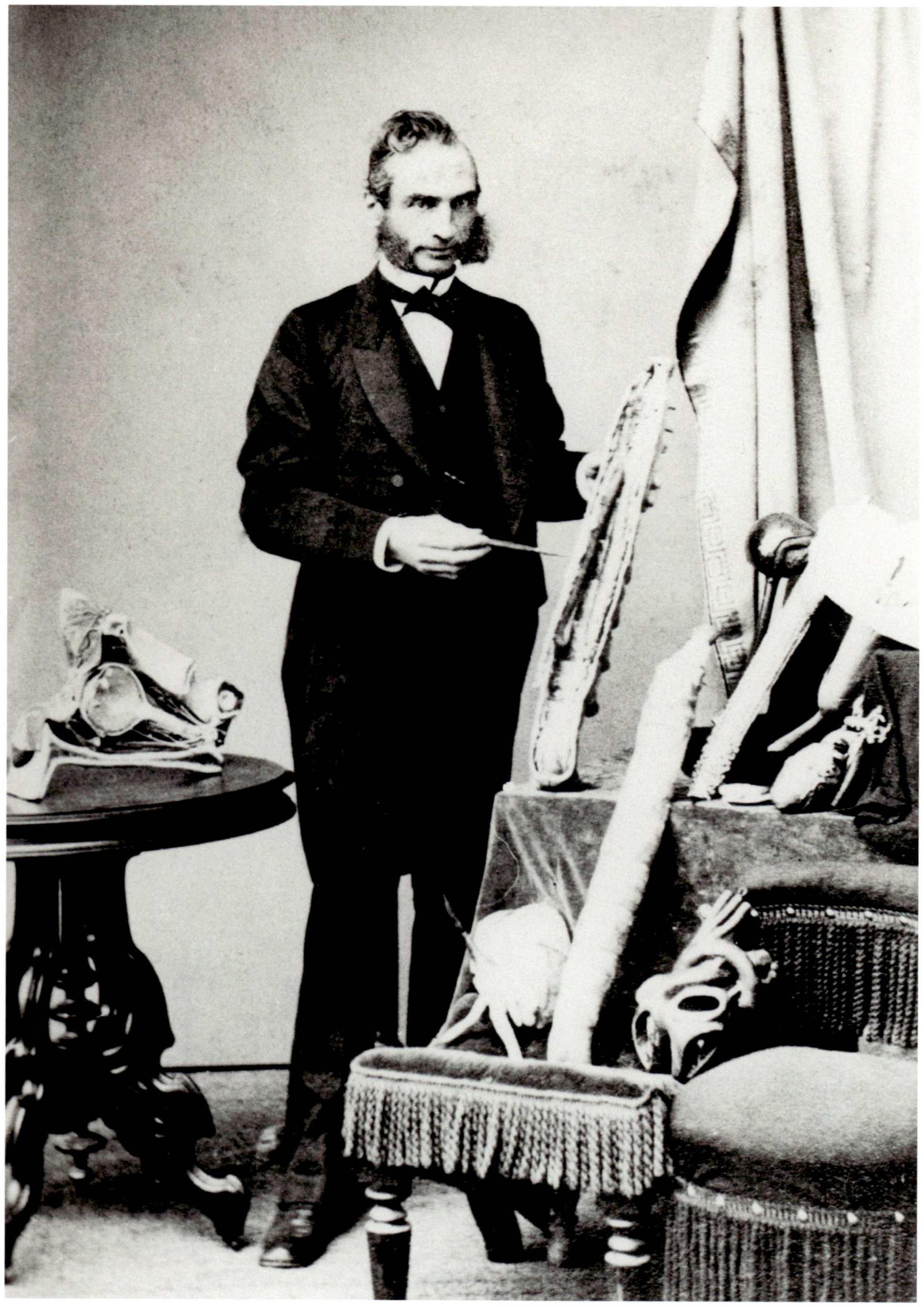

Plate 6.10

Parisian Physician with His Wax Model
Paris, 1865

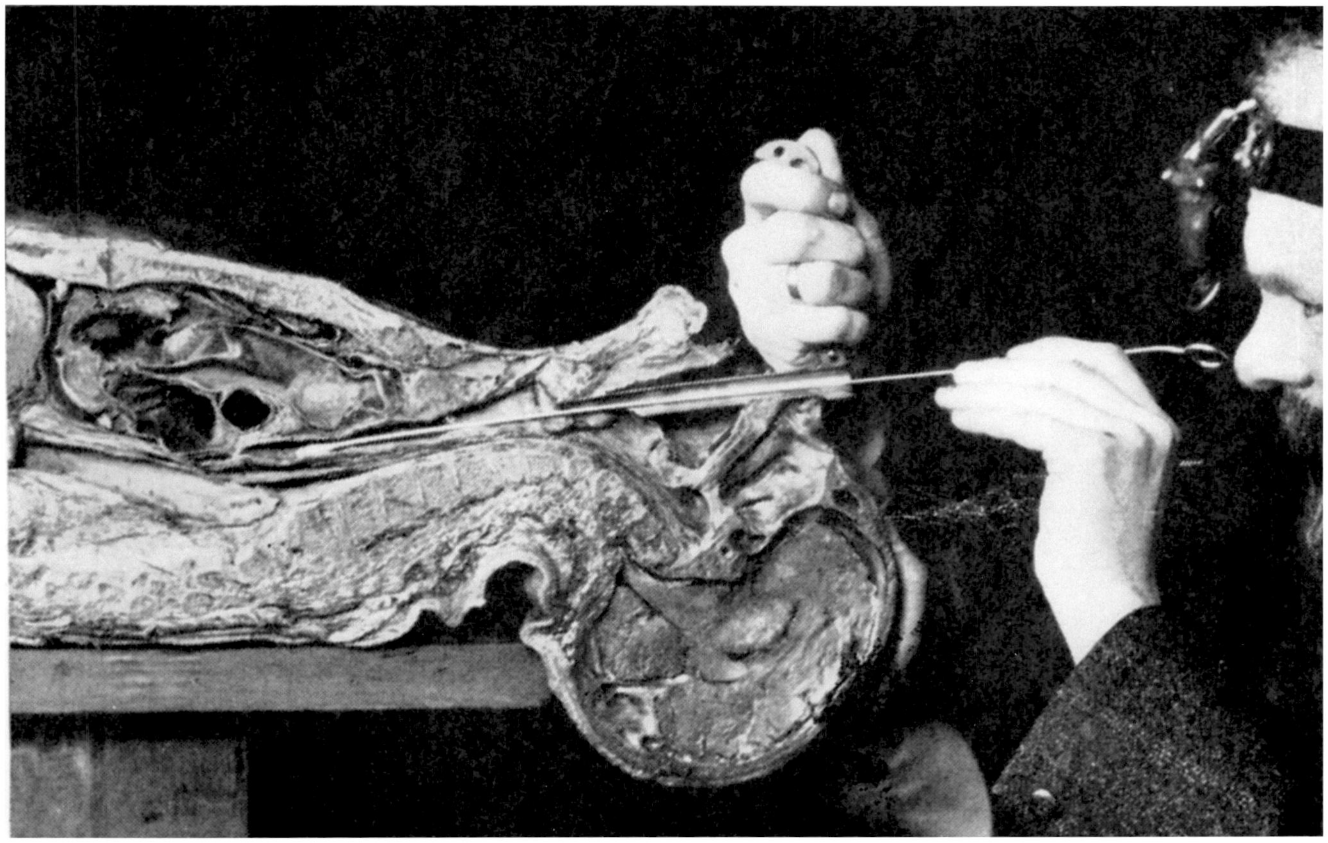

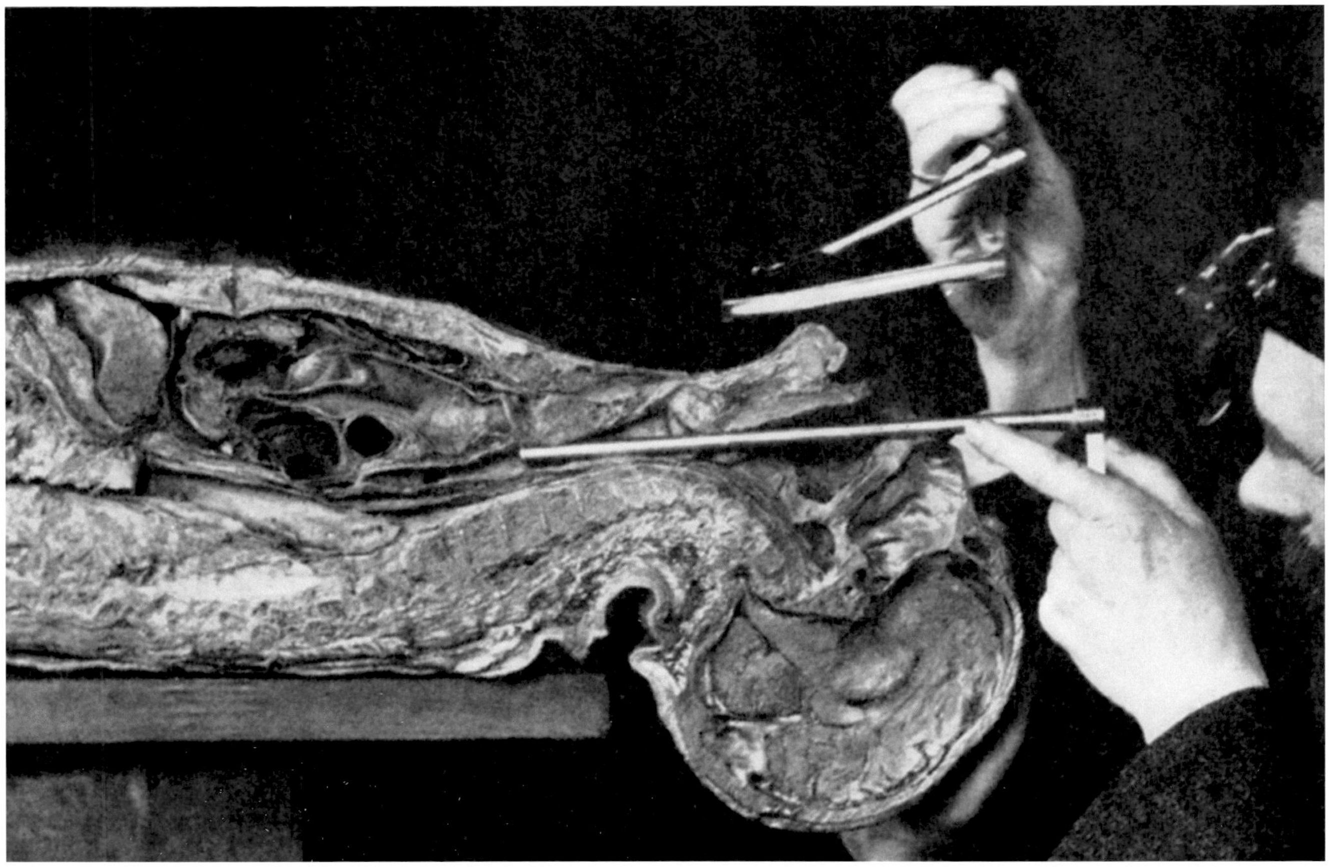

Plate 6.11

Professor Killian's Method of Tracheoscopy, Demonstration on a Cadaver
Published: *Medical Annual*, 1908

Introduction of the bivalve tube spatula for applying cocaine to the trachea.

Plate 6.12

Professor Killian's Tracheo-bronchoscopy, Demonstration on a Cadaver
Published: *Medical Annual*, 1908

The tracheobronchial tube is introduced and the tube spatula separated into two pieces is removed.

The figures indicate—

1. Superficial temporal artery.
2. Temporal branches of facial nerve.
3. Parotid gland.
4. Duct of parotid.
5. Zygomaticus major muscle.
6. Masseter muscle.
7. Orbicularis oris muscle, divided.
8. Lingual nerve.
9. Under surface of tongue.
10. Mucous membrane of floor of mouth, divided.
11. Genio-hyoid muscle.
12. Sublingual gland.
13. Anterior belly of digastric muscle.
14. Mylohyoid muscle.
15. Submaxillary gland.
16. Facial vessels
17. Buccinator muscle, divided.

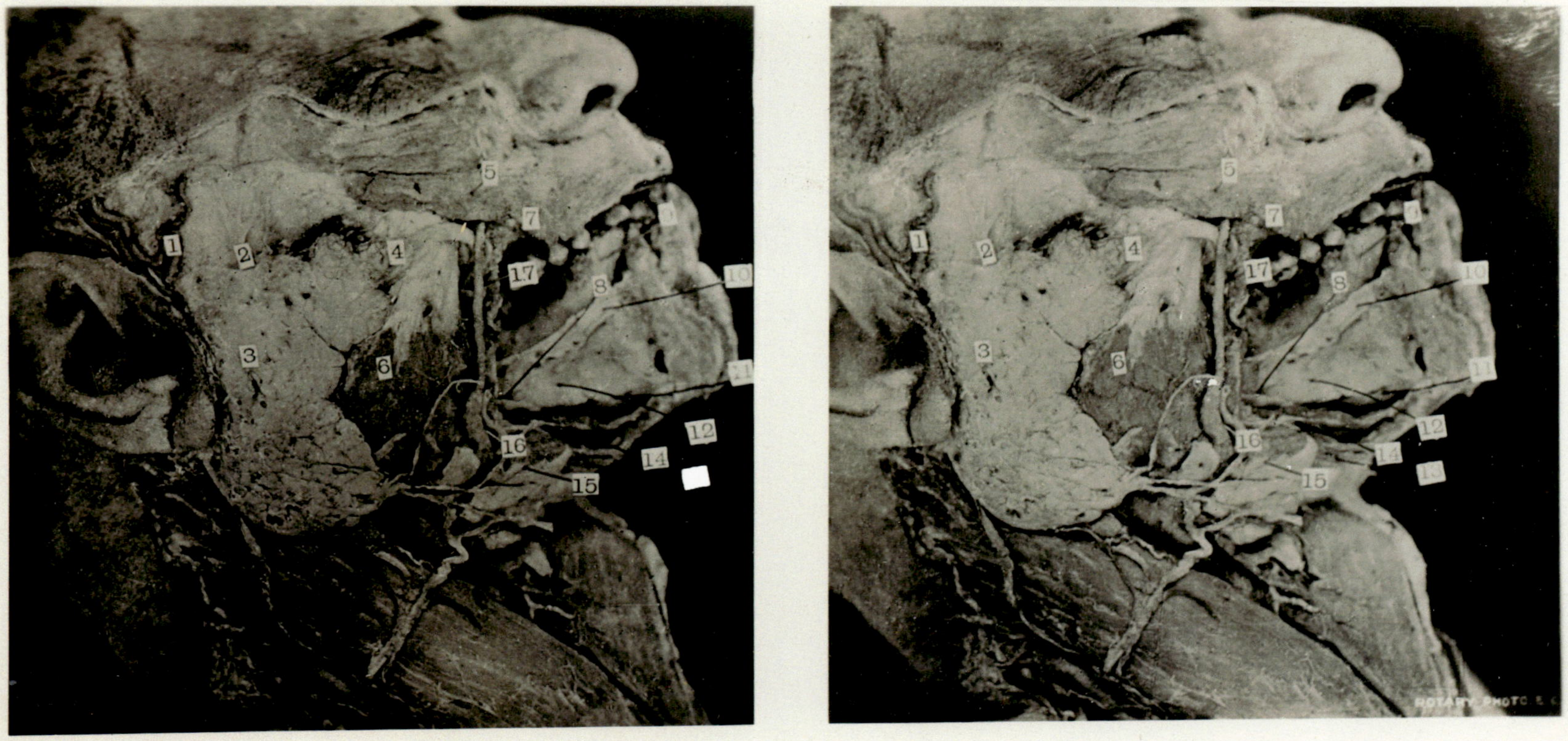

Plate 6.13

Head and Neck Dissection, Submaxillary Region
Edinburgh Stereoscopic Atlas of Anatomy, No.16, 1915

A portion of the horizontal ramus of the lower jaw has been removed.

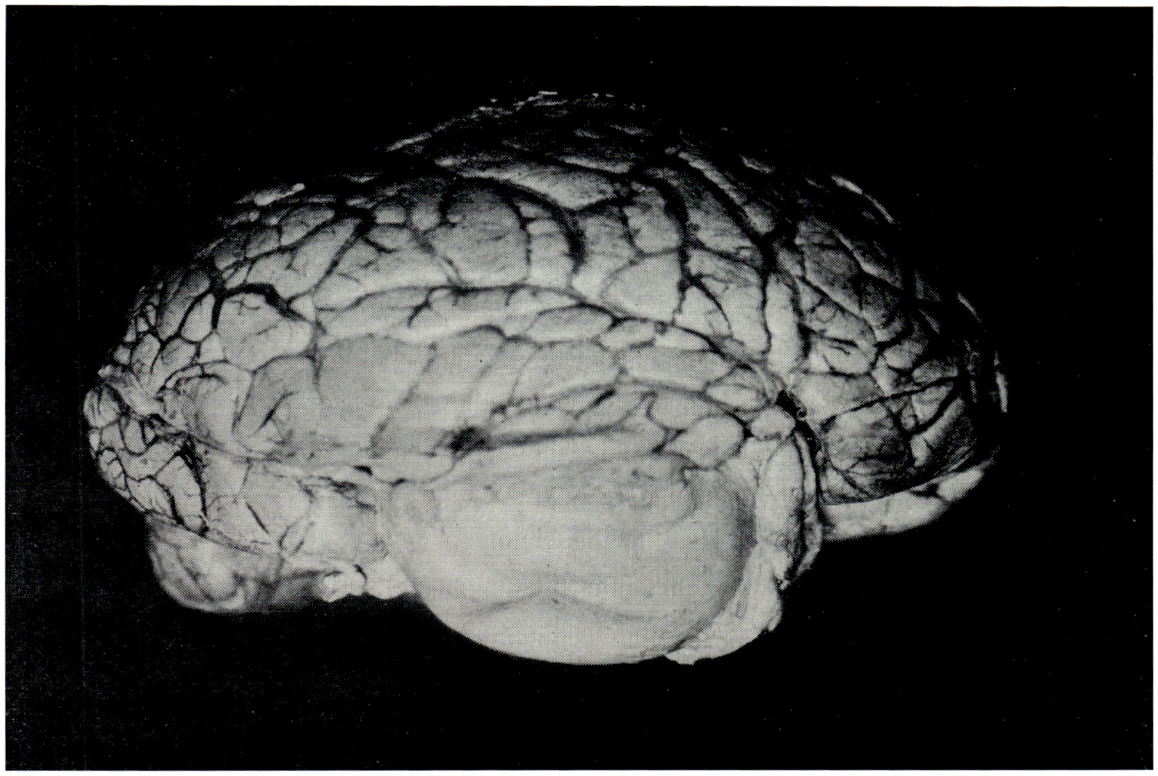

Plate 6.14

Otitic Brain Abscess
"Right Lateral Aspect of Brain Showing Abcess of Temporo-Spheroid Lobe"
Published: Dr. W.B. Waddington, *Quarterly Journal of Medicine,* 917

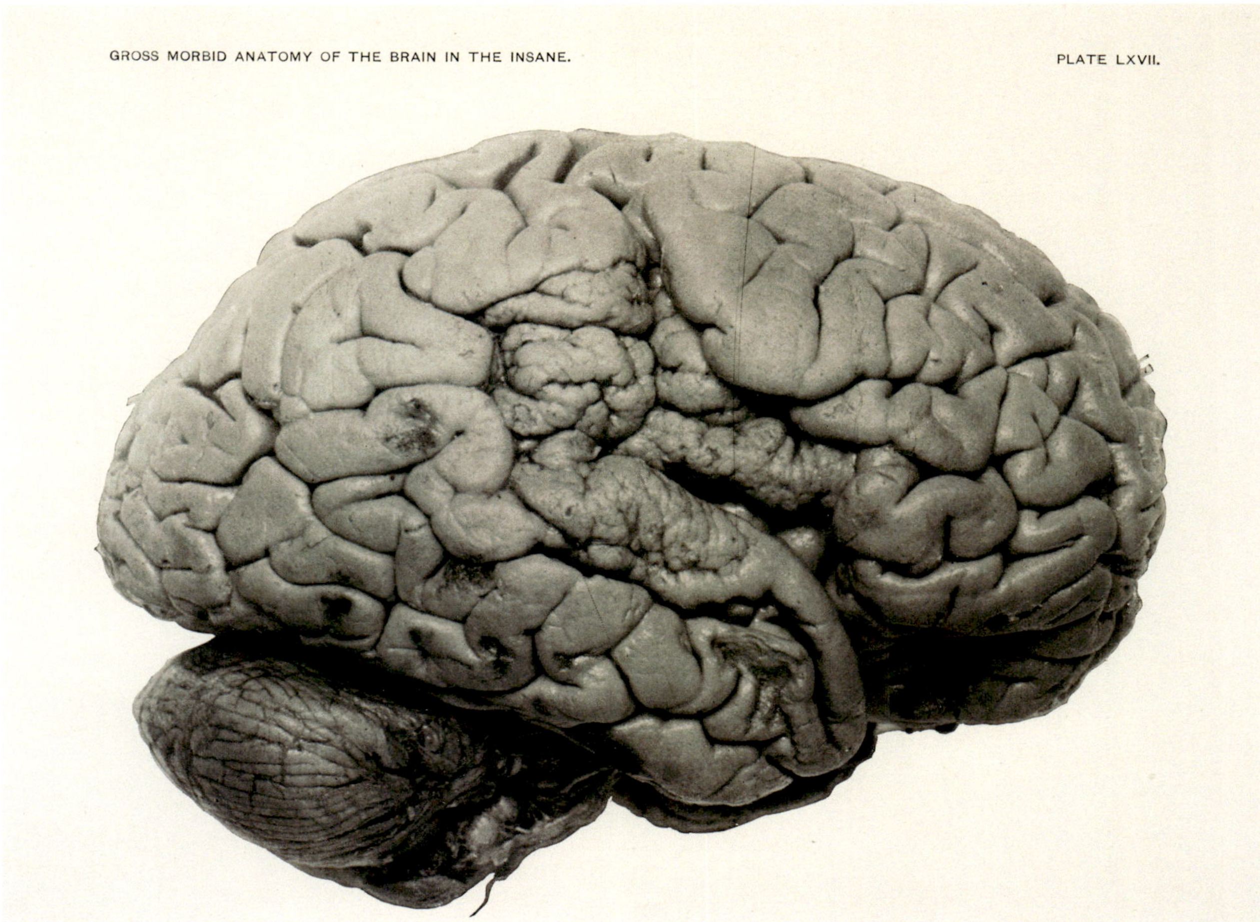

Plate 6.15

Sclerosis of a Portion of the Right Hemisphere in Epilepsy
Illustrations of the Gross Anatomy of the Brain in the Insane...
Plate 67, I.W. Blackburn, MD, 1908

In the center, an abnormally puckered, shrunken area is evident. 45-year-old patient.

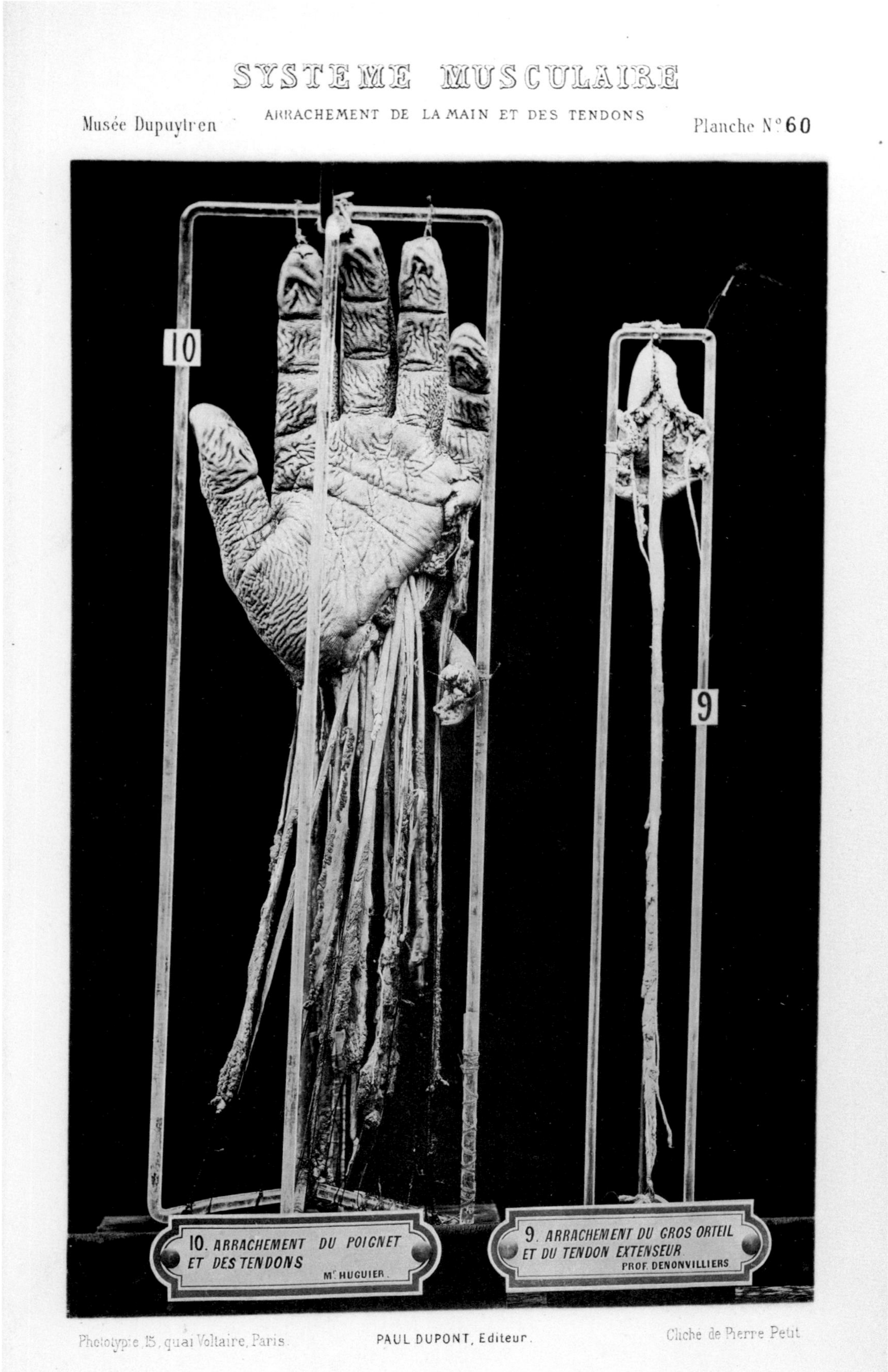

Plate 6.16

Attachments of the Tendons of the Hand
Catalogue des pièces du museé Dupuytren, Atlas du tome troisième
Plate 60, 1879

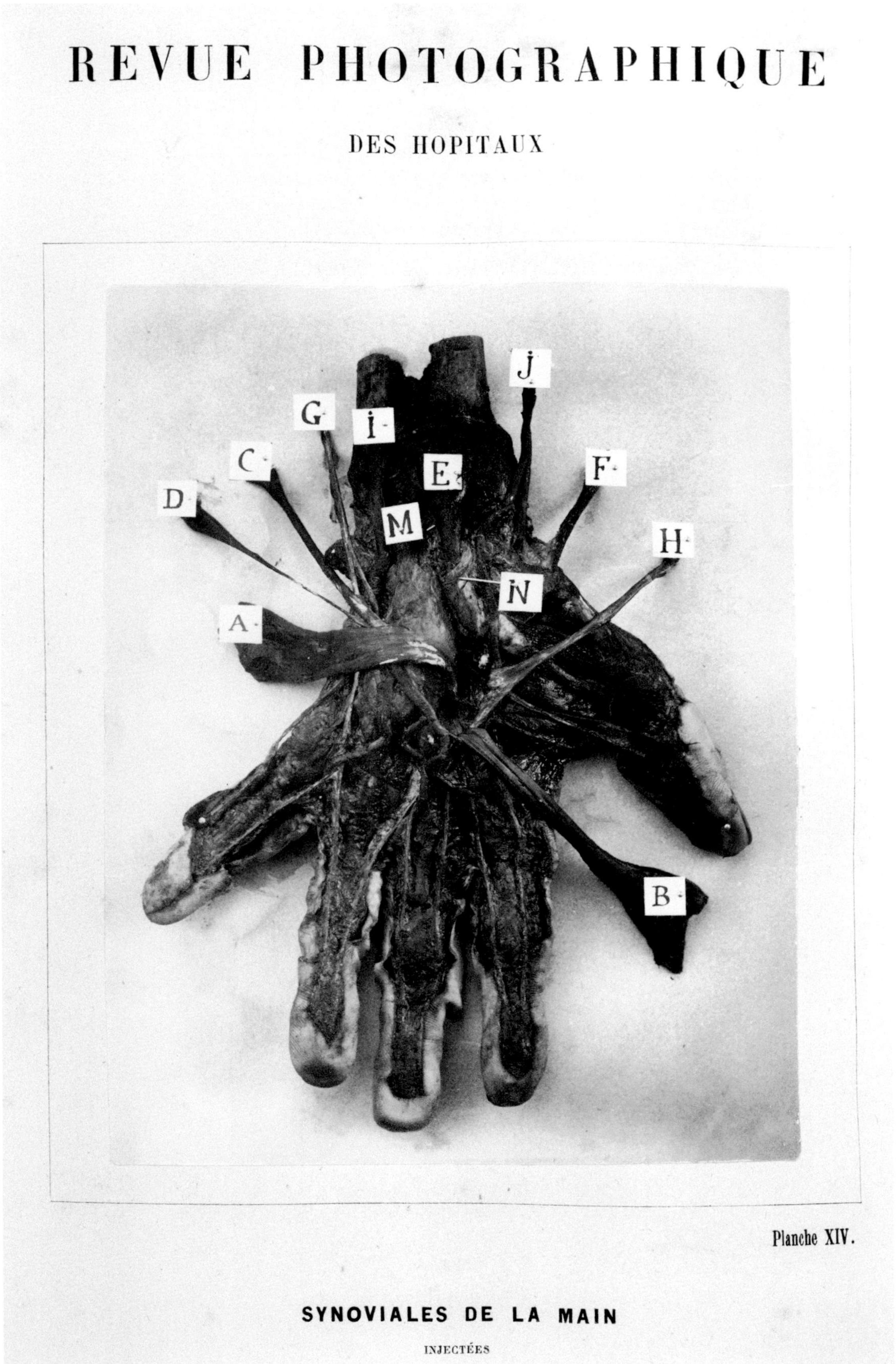

Plate 6.17

Dissection of the Hand to Show the Synovial System
Revue photographique des hôpitaux de Paris, Volume 2, Plate 14, 1870

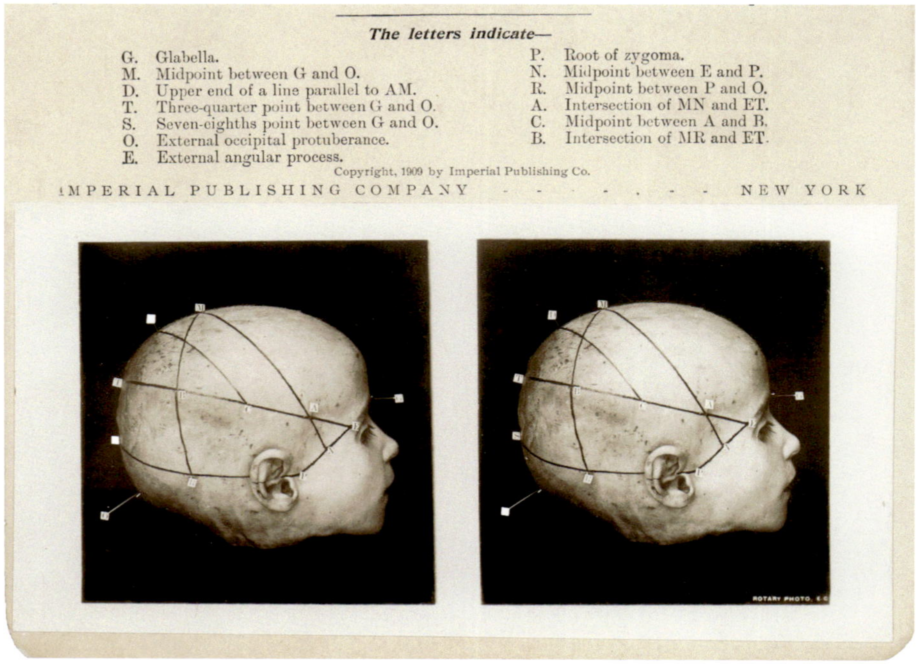

Plate 6.18

Cranio-Cerebral Topography, Number 1
Imperial Stereoscopic Anatomy of the Head and Neck, No. 1, circa 1929

Primary bony point landmarks with arbitrary lines drawn.

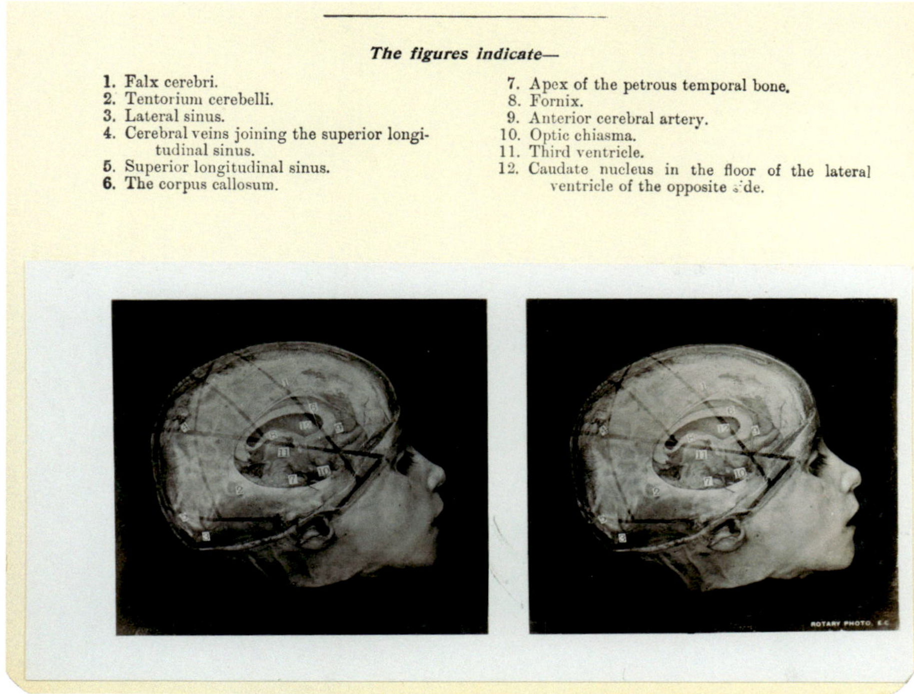

Plate 6.19

Cranio-Cerebral Topography, Number 8
Imperial Stereoscopic Anatomy of the Head and Neck, No. 8, circa 1929

Composite view to show relation of structures to the surface markings.

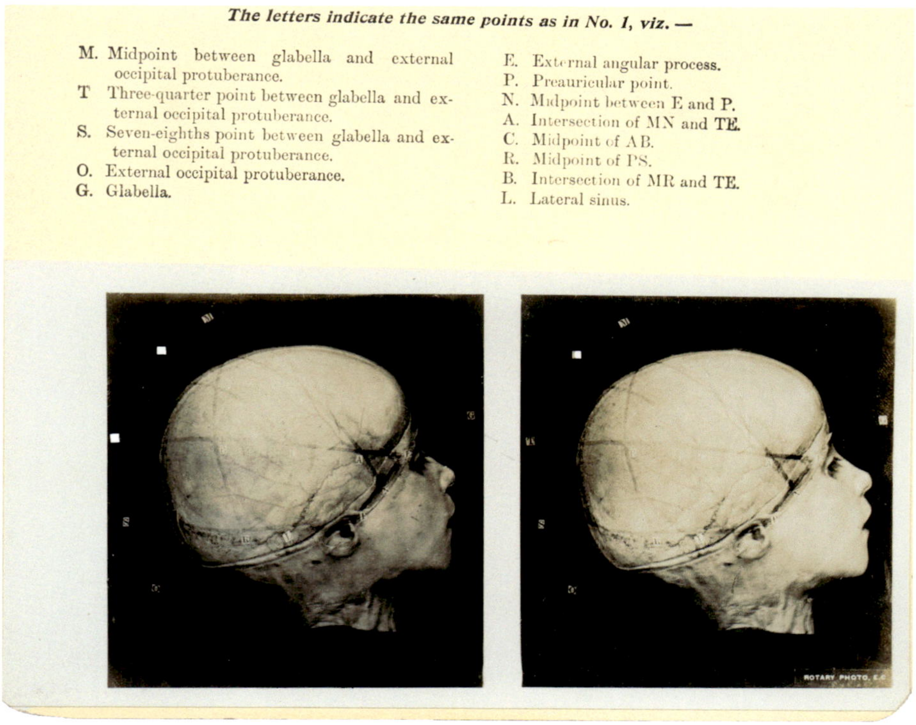

Plate 6.20

Cranio-Cerebral Topography, Number 4
Imperial Stereoscopic Anatomy of the Head and Neck, No. 4, circa 1929

View to show the relation of the surface markings to blood vessels.

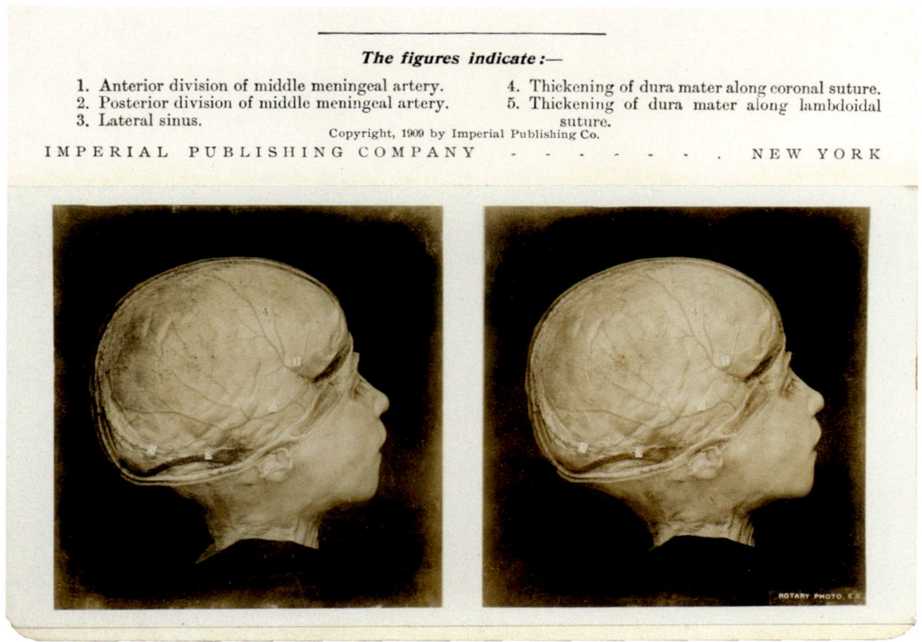

Plate 6.21

Cranio-Cerebral Topography, Number 3
Imperial Stereoscopic Anatomy of the Head and Neck, No. 3, circa 1929

Skull cap removed and outer surface of dura mater exposed.

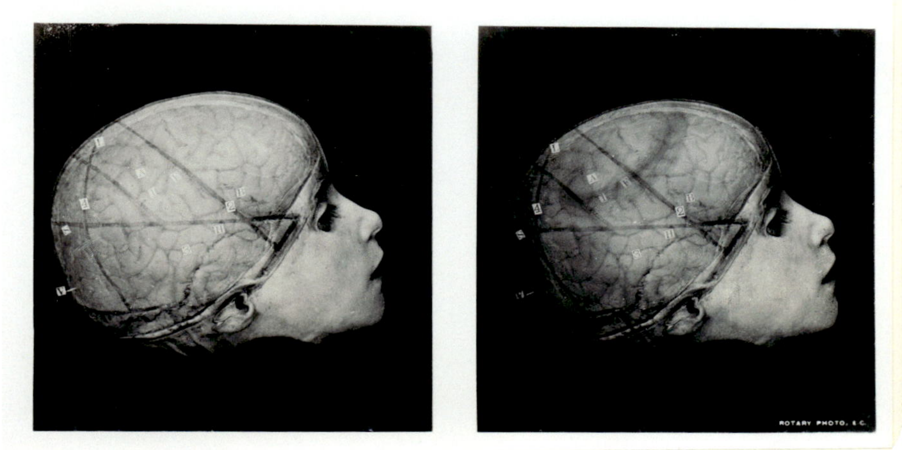

Plate 6.22

Cranio-Cerebral Topography, Number 6
Imperial Stereoscopic Anatomy of the Head and Neck, No. 6, circa 1929

Dura mater and membranes removed exposing the brain's sensory and motor areas.

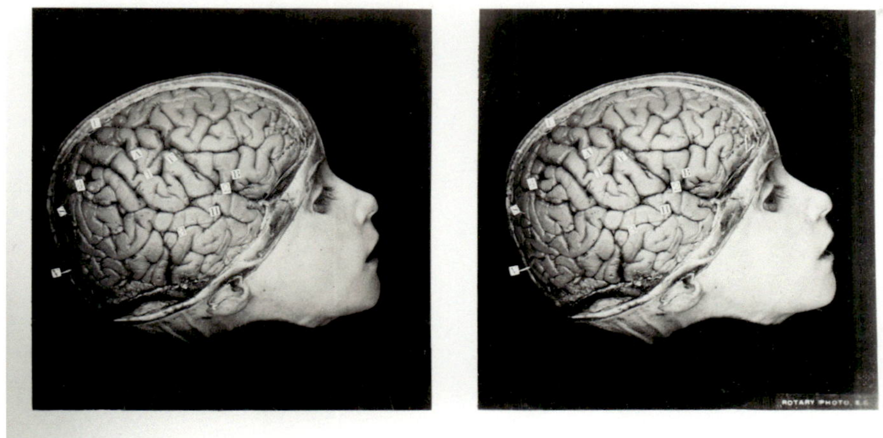

Plate 6.23

Cranio-Cerebral Topography, Number 5
Imperial Stereoscopic Anatomy of the Head and Neck, No. 5, circa 1929

Dura mater removed exposing the outer aspect of the hemisphere.

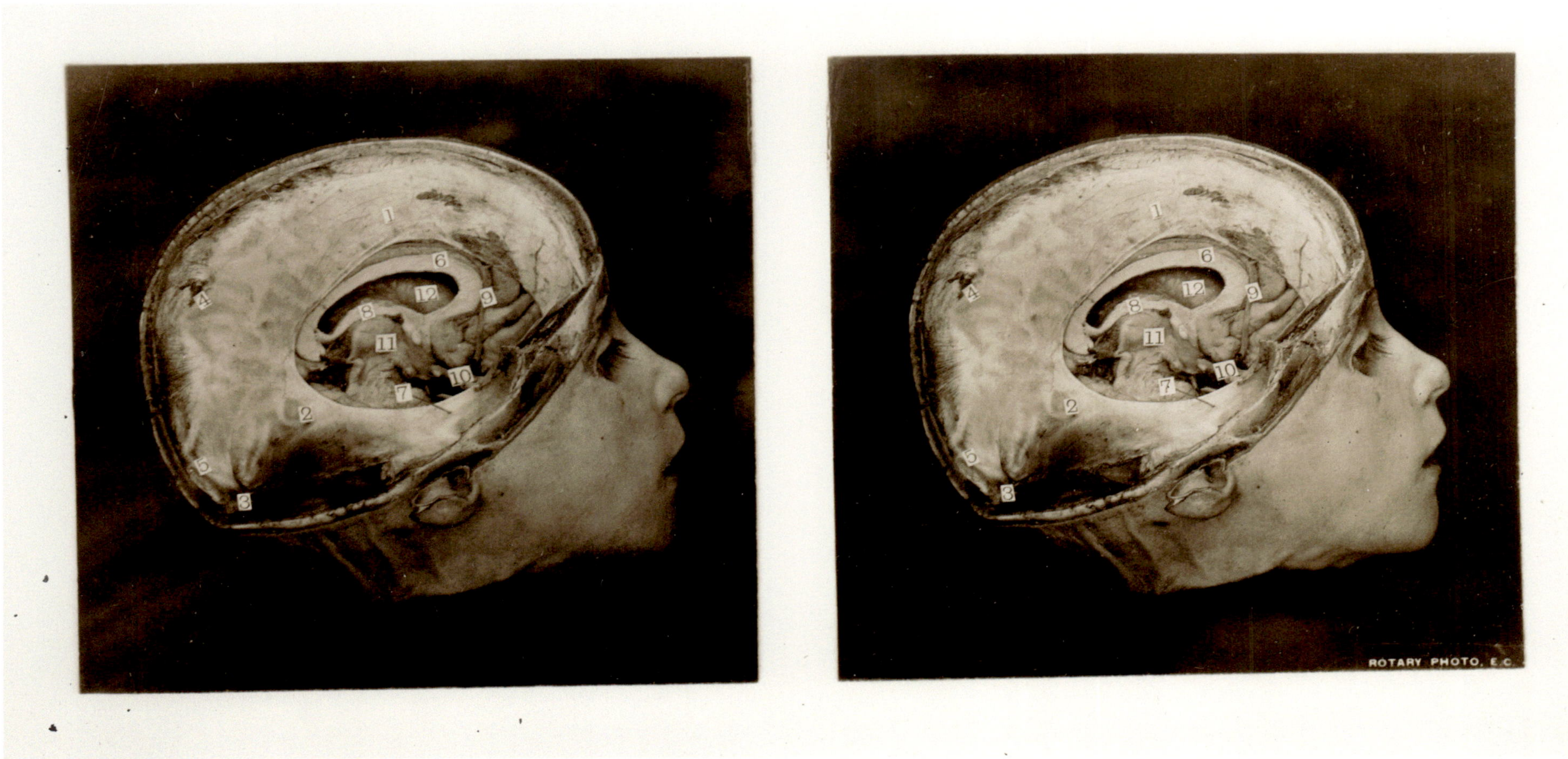

Plate 6.24

Cranio-Cerebral Topography, Number 7
Imperial Stereoscopic Anatomy of the Head and Neck, No. 7, circa 1929

Right cerebral hemisphere removed to show the membranes of the cranial cavity.

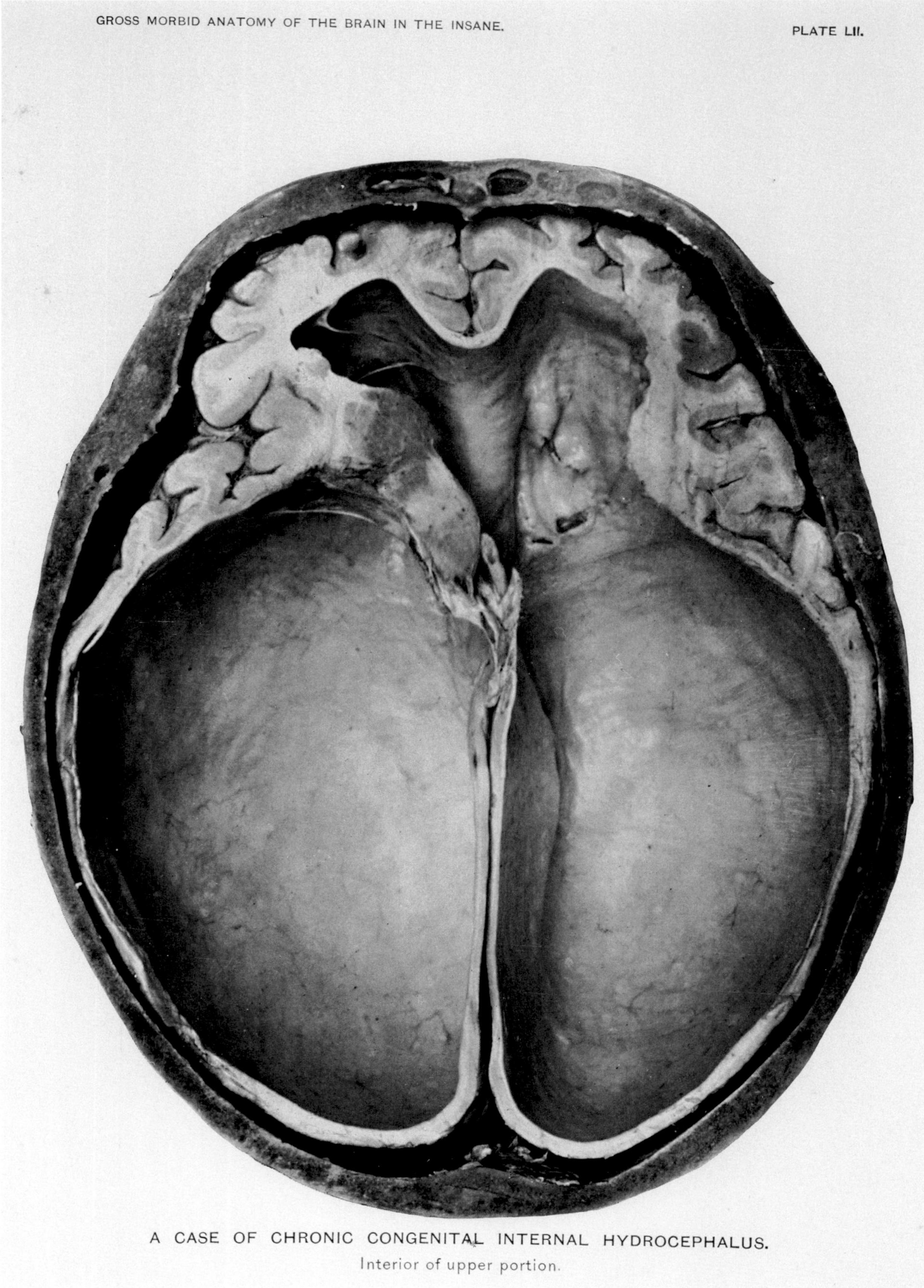

Plate 6.25

Chronic Congenital Internal Hydrocephalus: Interior of Upper Portion, 60-year-old man

Illustrations of the Gross Anatomy of the Brain in the Insane,
By I.W. Blackburn, MD, Plate 52, 1908

The preservation of the frontal lobes allowed the patient decades of intelligence.

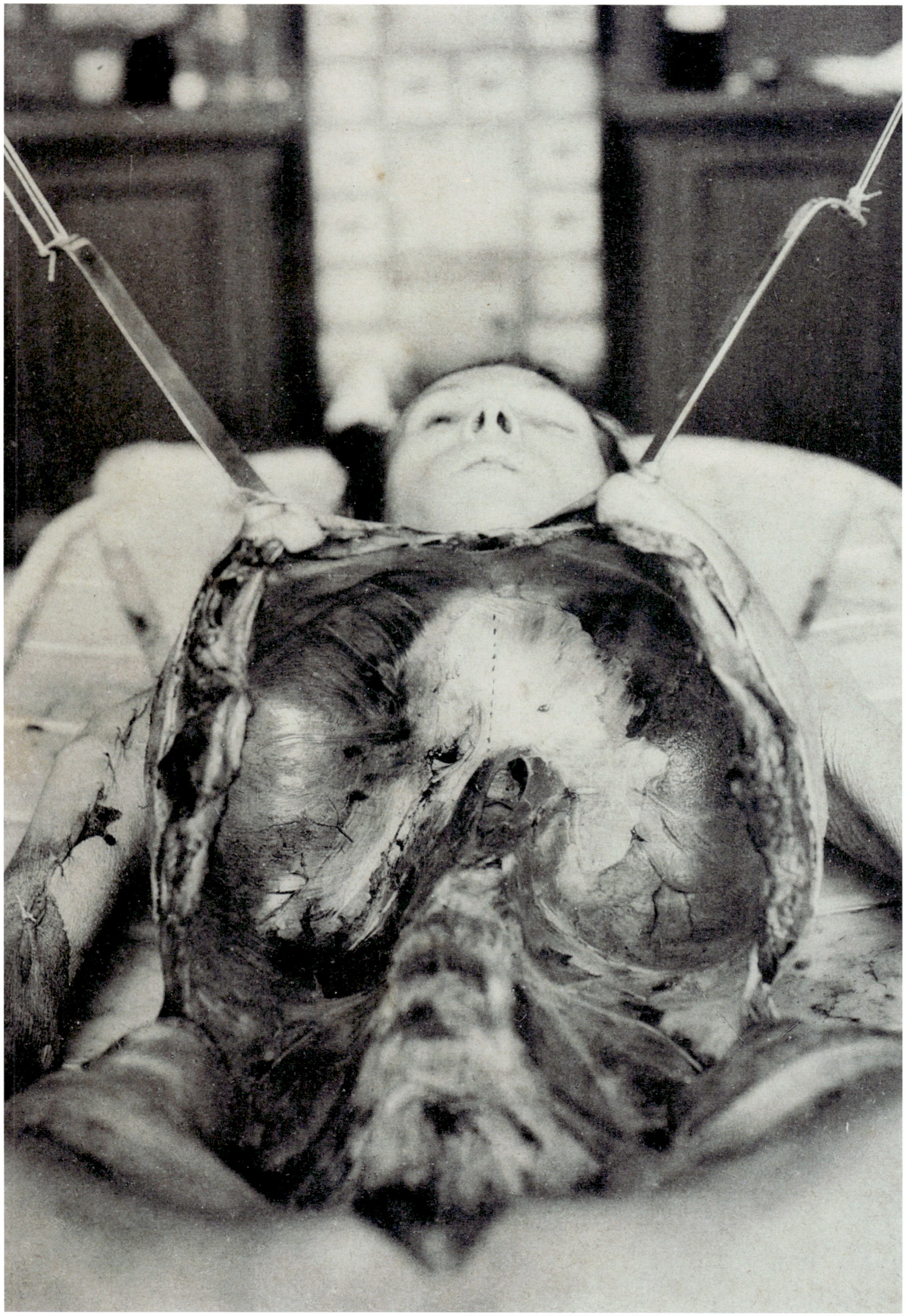

Plate 6.26

Dissection of the Normal Diaphragm
Haar Oorzaak, MD and Murk Jansen, MD, University of Leiden
De Physiologische Skoliose, 1912

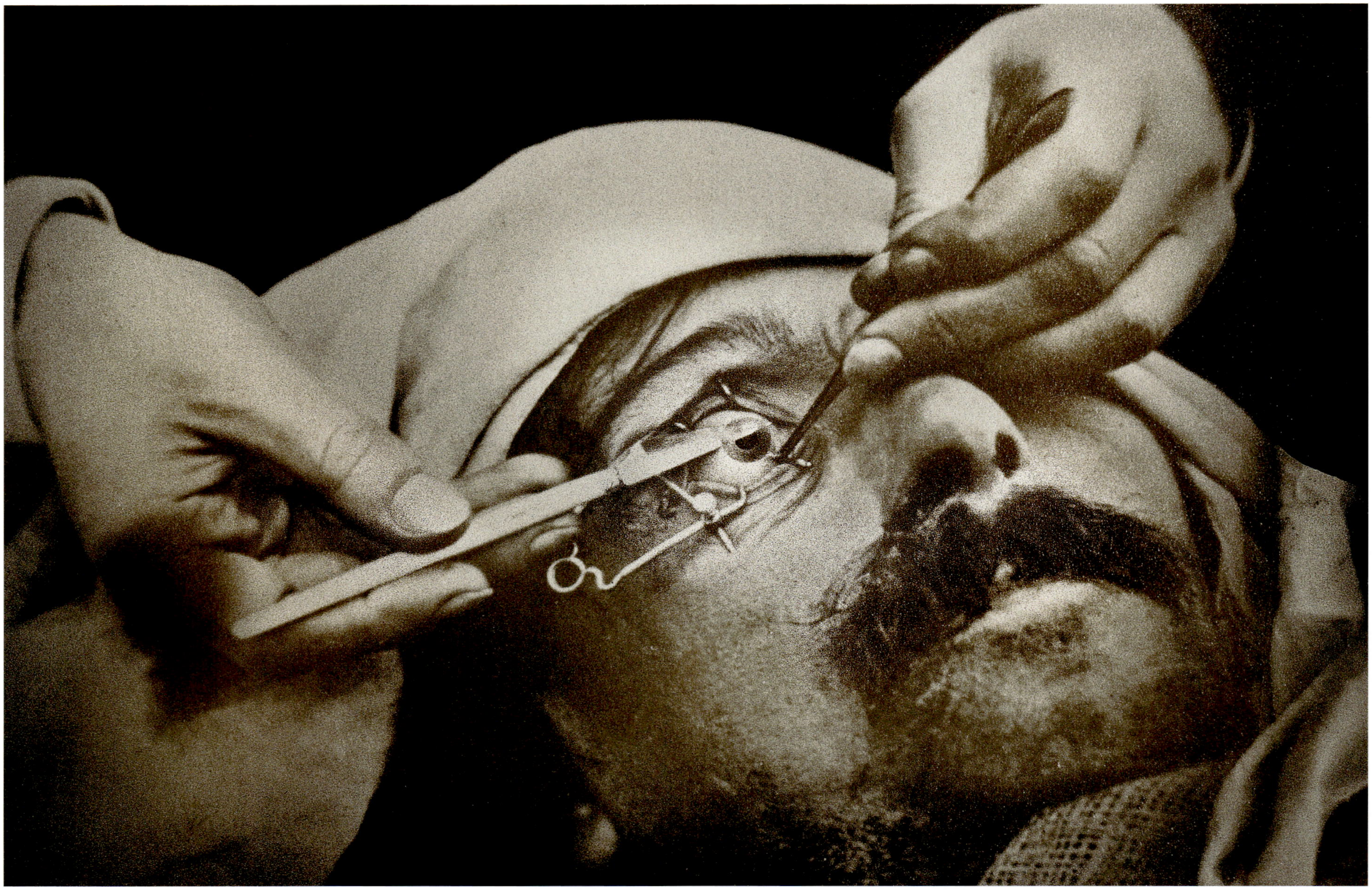

Plate 6.27

Superior Limbal Incision for Cataract Extraction Demonstrated on a Cadaver
Photographer, A. de Montméja, MD; Surgeon, Édouard Meyer, MD
Published as tipped-in albumen print in *Traité des opérations qui se pratiquent sur l'oeil, 1870*

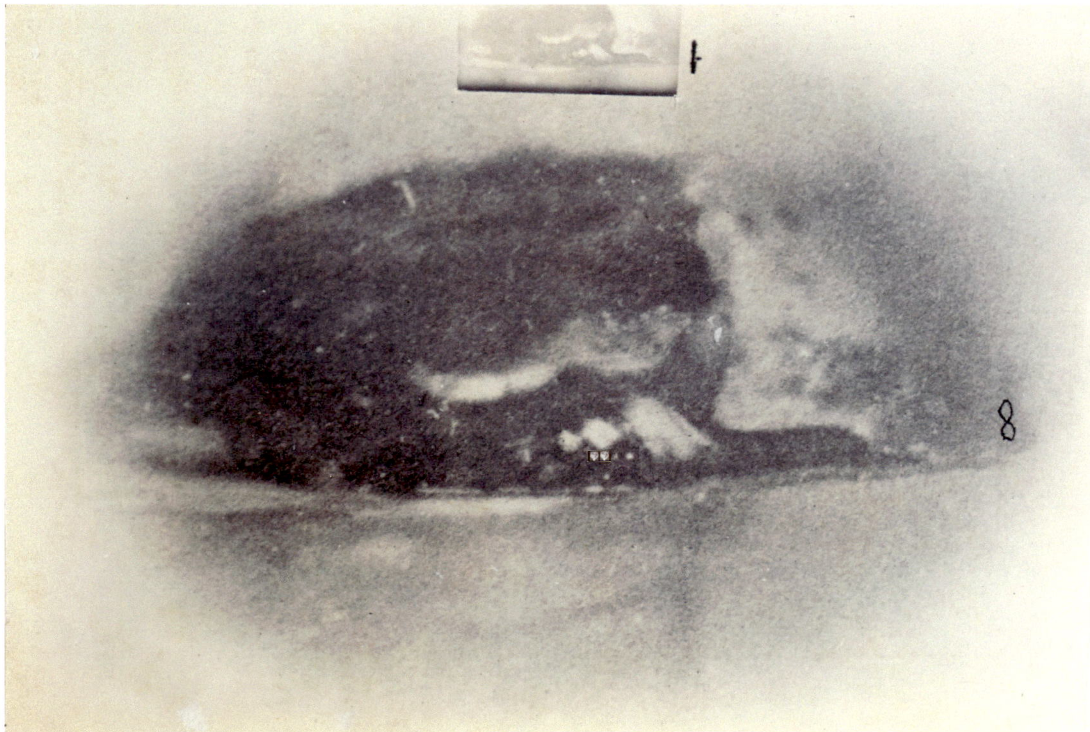

Plate 6.28

Retinal Photograph and "Optogram"
1868
Revue photographique des hôpitaux de *Paris*, Volume 2, Plate 10, published 1870

Image of a murderer depicted as the last image seen by the eye.

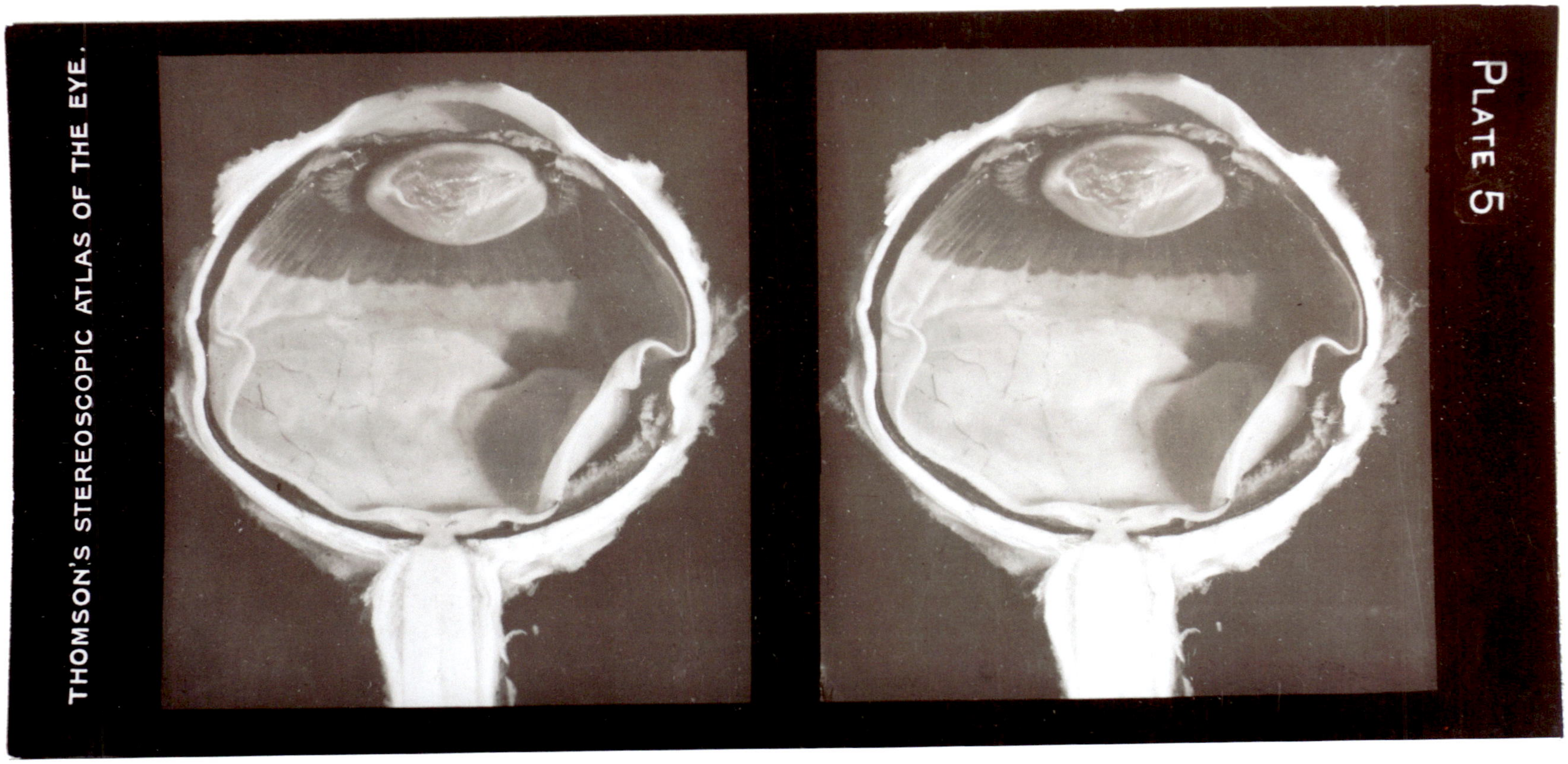

Plate 6.29

Retinal Separation
Stereoview, Arthur Thompson, MD
The Anatomy of the Human Eye, Plate 5, 1912

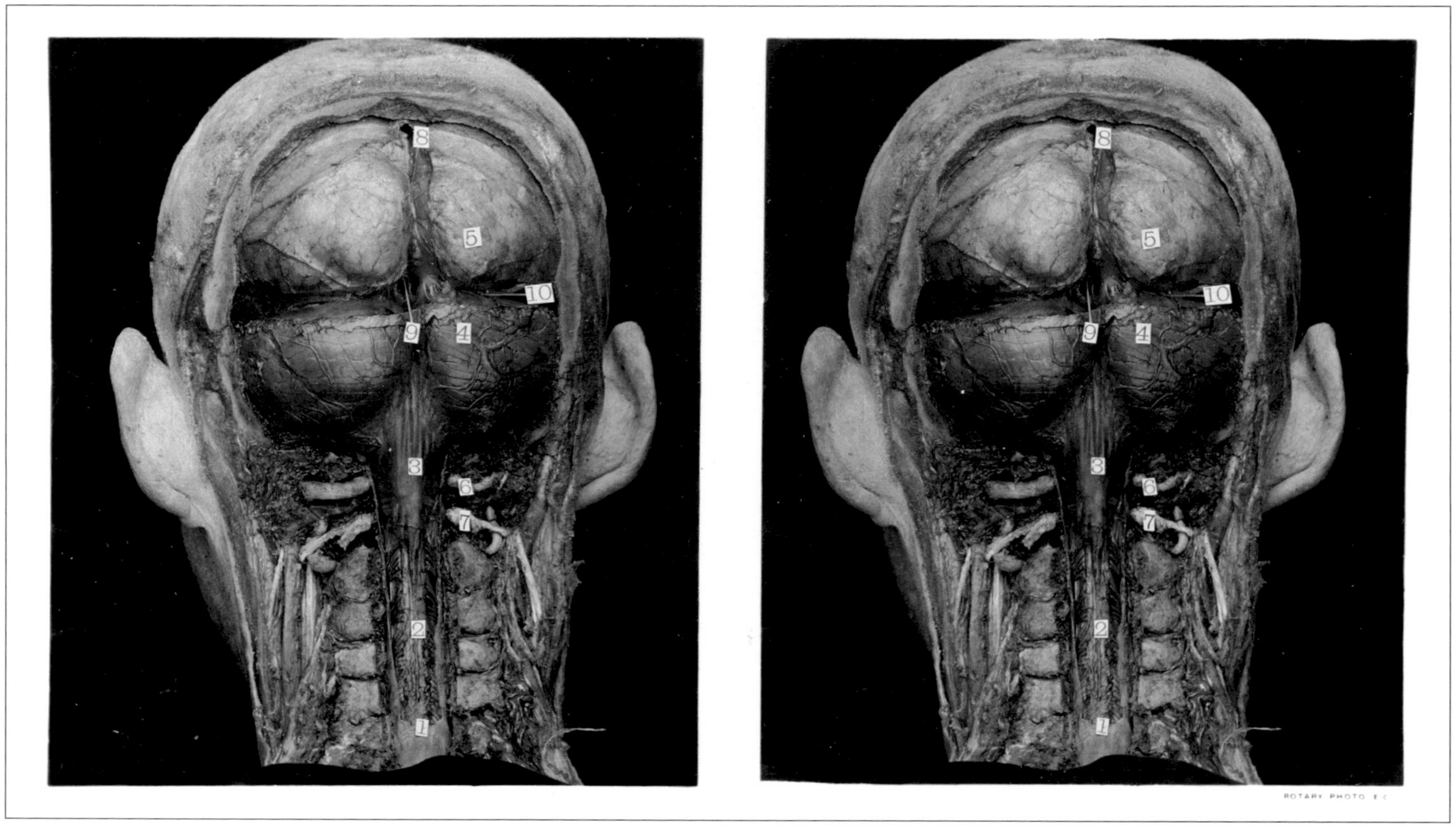

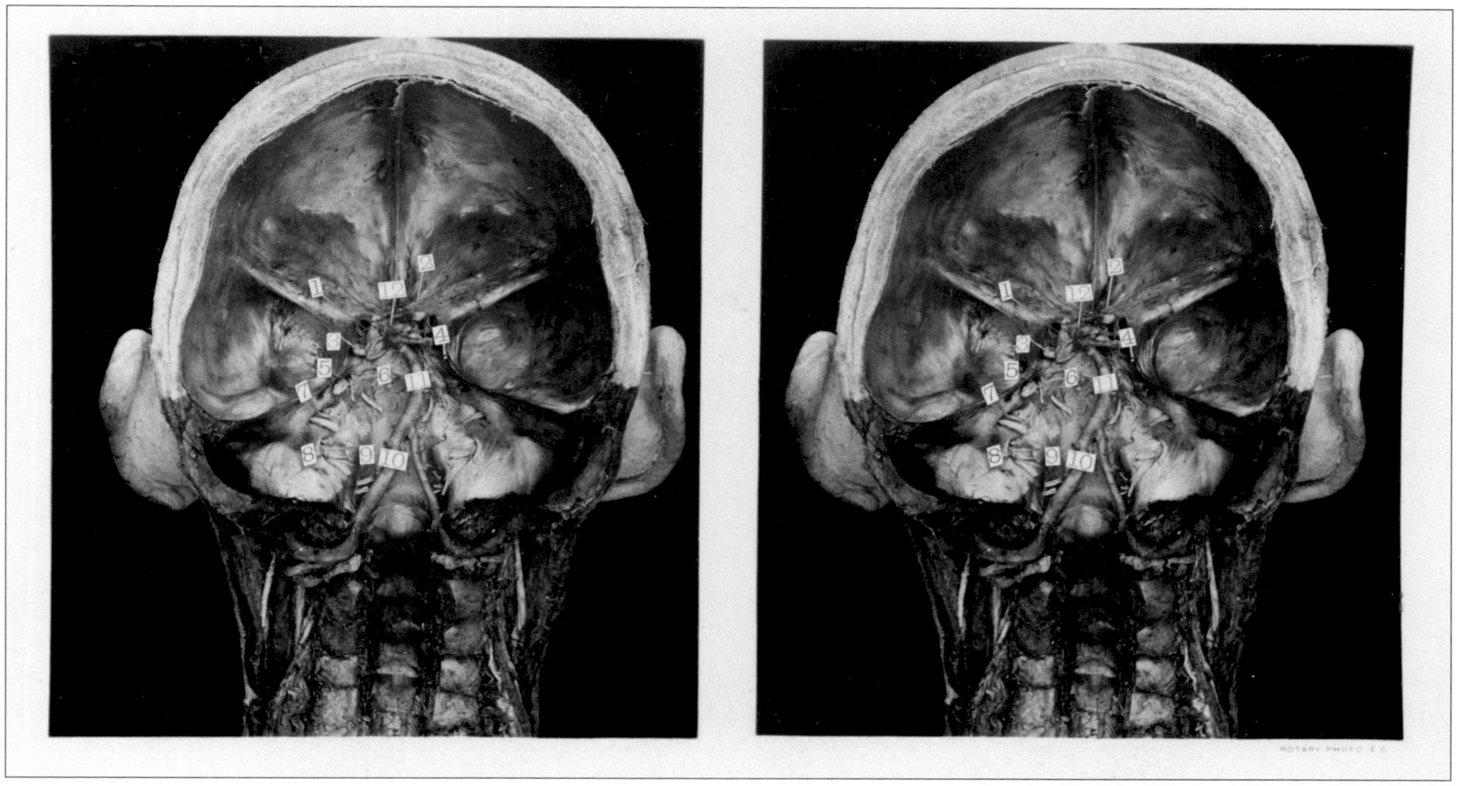

Plates 6.30

Head and Neck: Spinal Canal, Number 2
Edinburgh Stereoscopic Atlas of Anatomy,
1915

The posterior part of the skull has been removed as well as the dura mater over the cerebellum.

Plates 6.31

Head and Neck: Spinal Canal, Number 4
Edinburgh Stereoscopic Atlas of Anatomy,
1915

Showing the structures lying on the basis cranii.

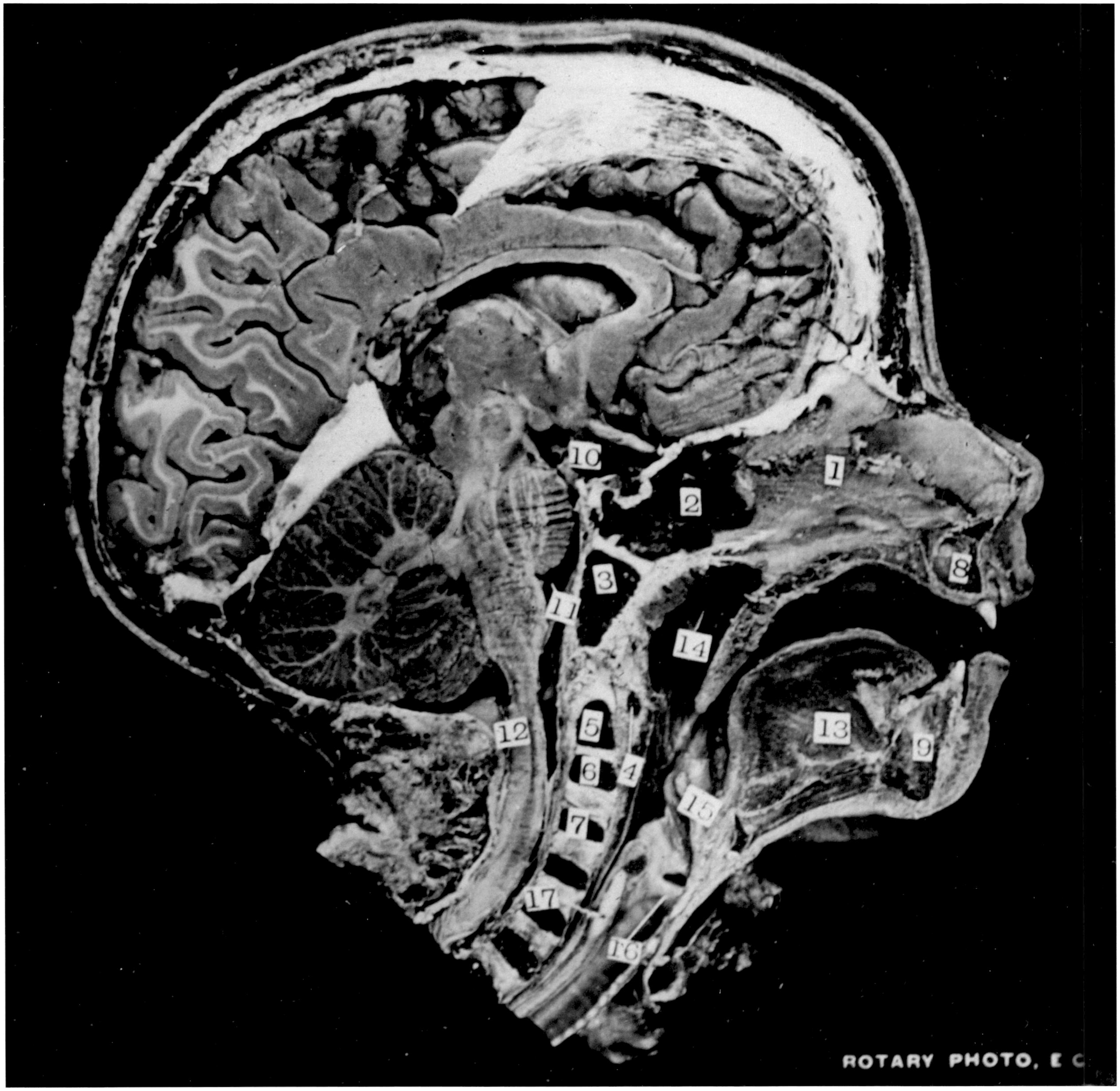

Plate 6.32

Head and Neck: Sagittal Section of the Head, Number 1

Edinburgh Stereoscopic Atlas of Anatomy, 1915

Section slightly to the right of the mesial plate in front.

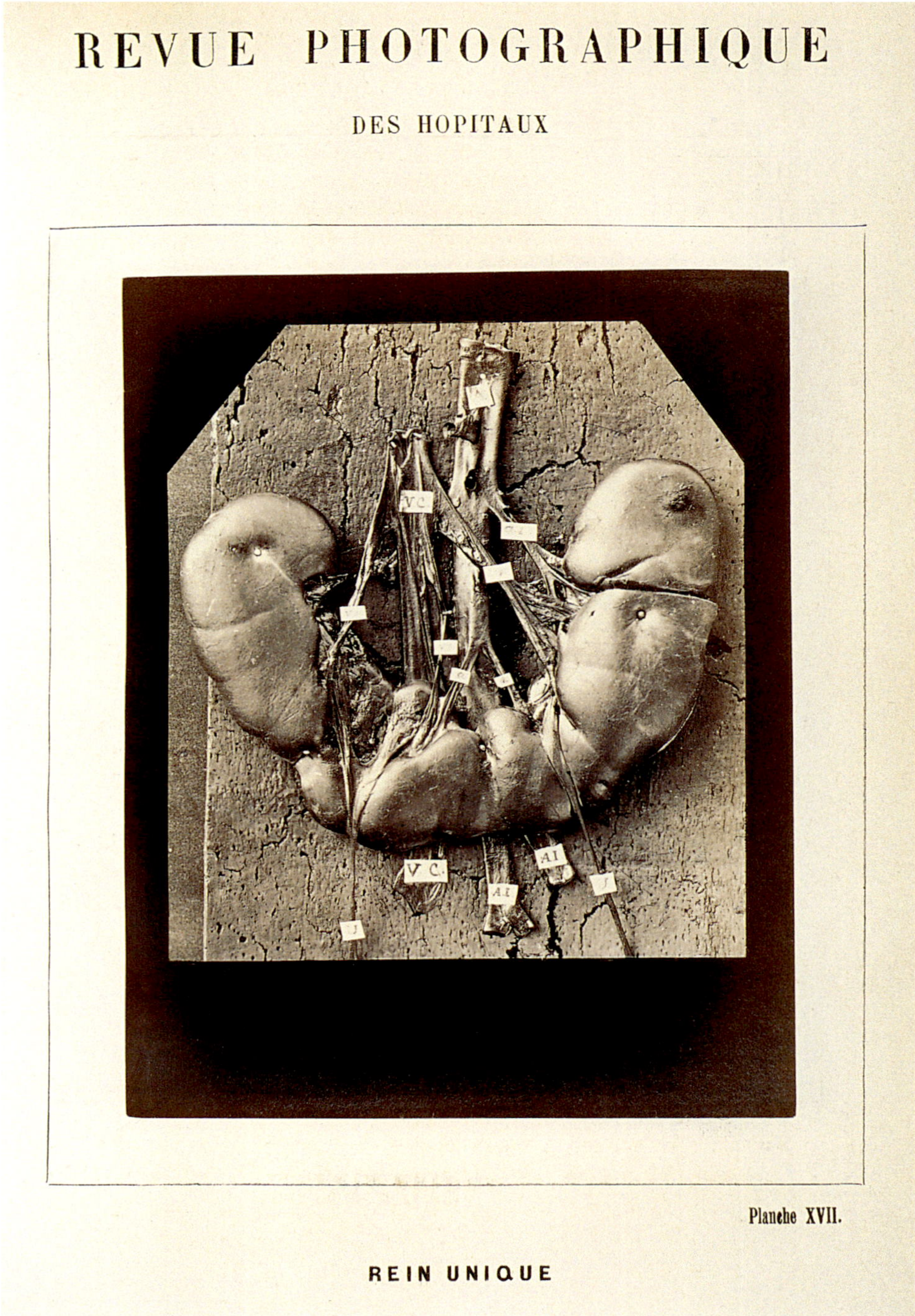

Plate 6.33

Horseshoe or Fused Kidney
Revue photographique des hôpitaux de Paris, Volume 4, Plate 17, 1872

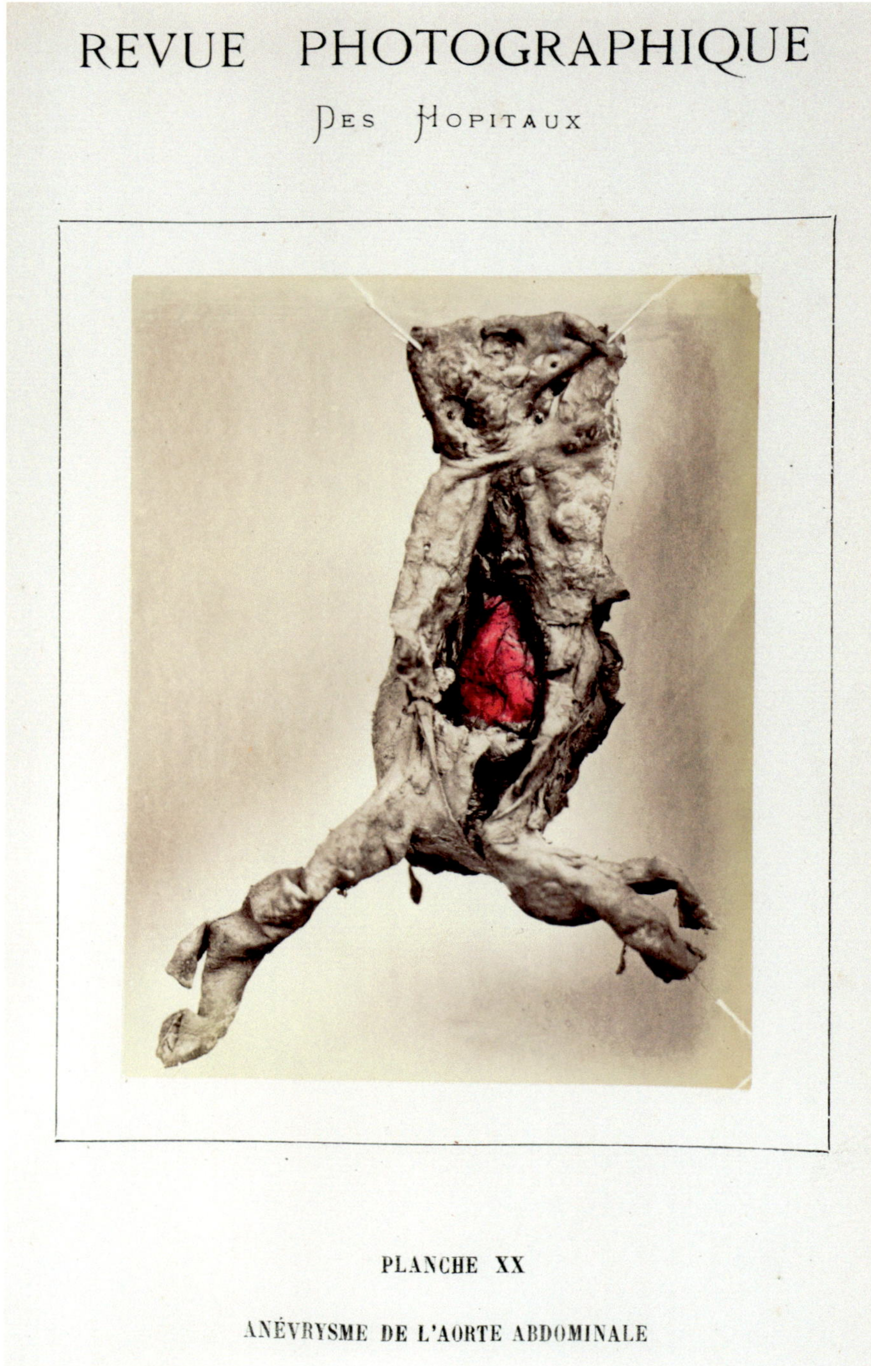

Plate 6.34

Aneurism of the Abdominal Aorta
Revue photographique des hôpitaux de Paris, Volume 1, Plate 20, 1869

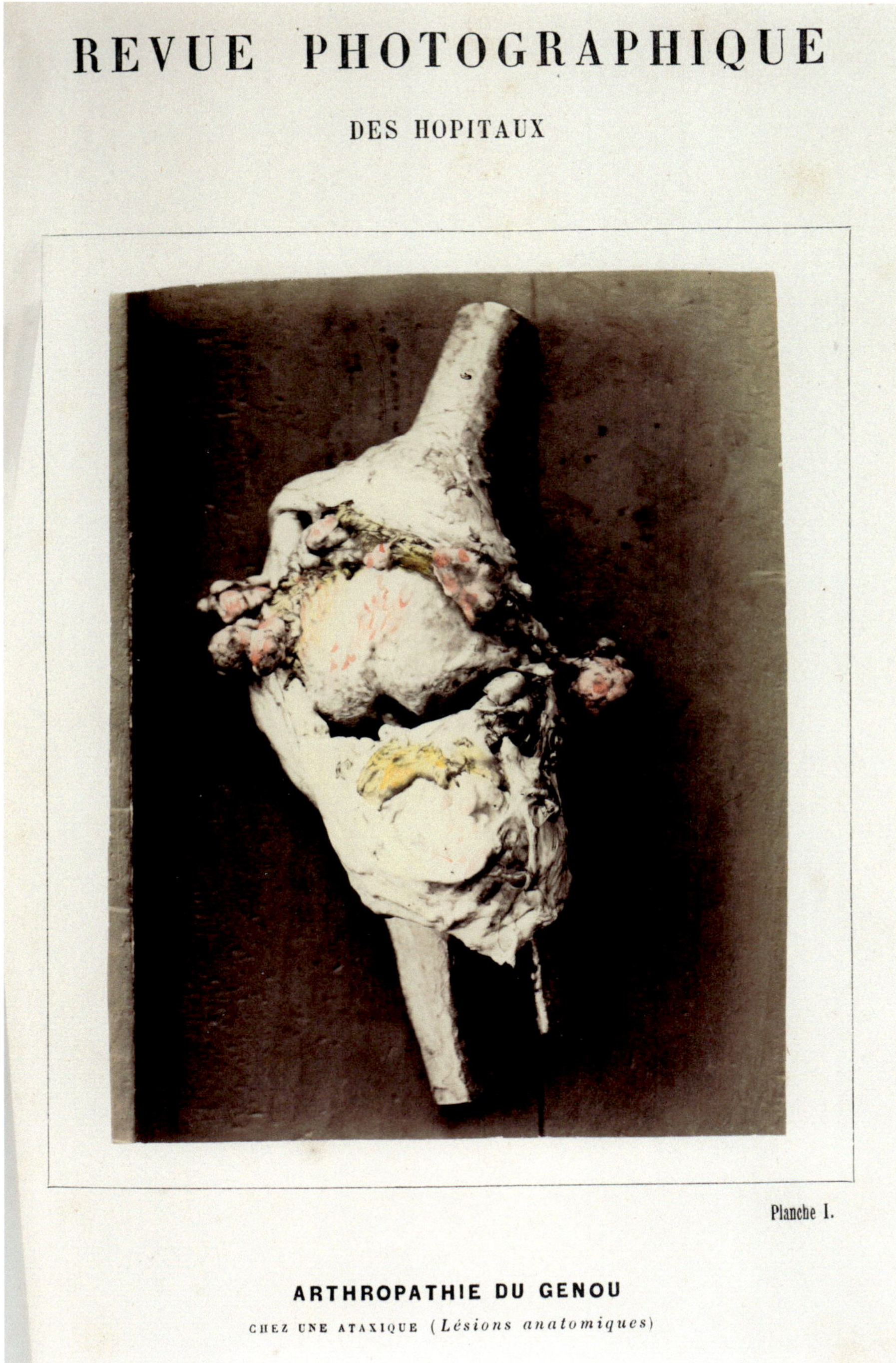

Plate 6.35

Dissection of an Arthritic Knee with Hand-Colored Synovial Joint Spaces
Revue photographique des hôpitaux de Paris, Volume 3, Plate 1, 1871

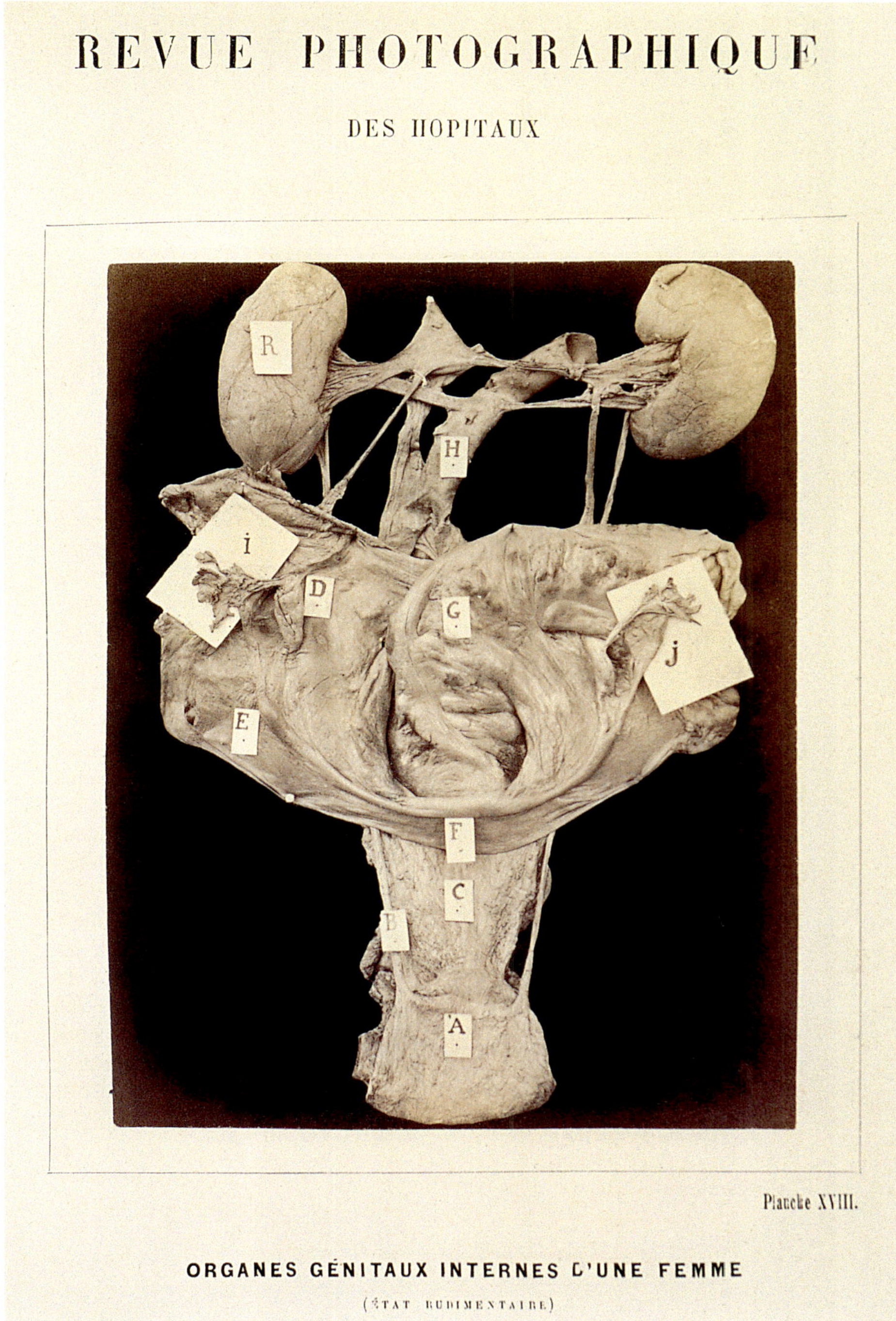

Plate 6.36

Dissection of the Female Urinary-Genital Organ System
Revue photographique des hôpitaux de Paris, Volume 2, Plate 18, 1870

Complete dissection of the system flayed out and exposed.

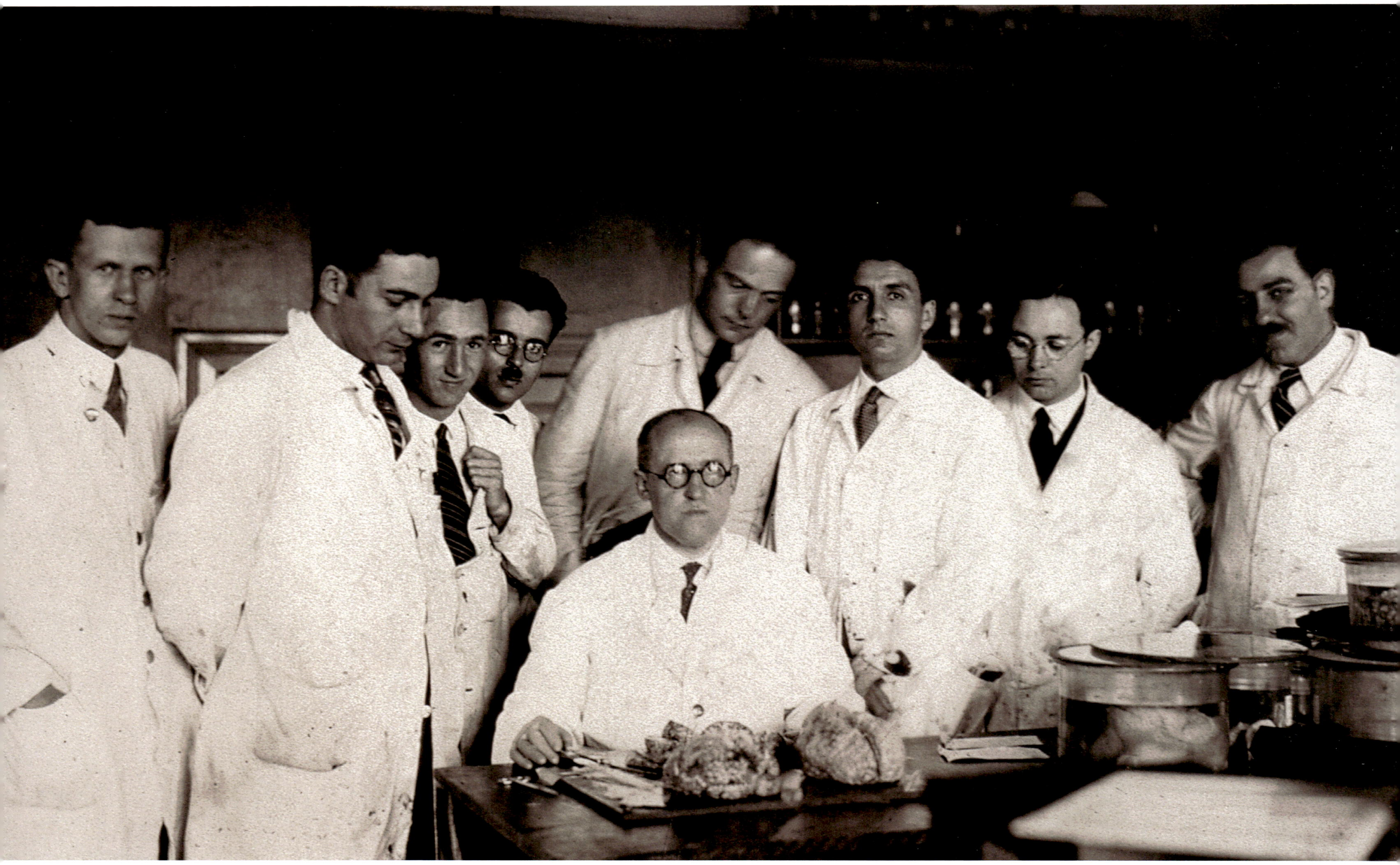

Plate 6.37

Brain Dissection by Neuropathologist
Circa 1945

A class surrounds the physician as he prepares to section a brain.

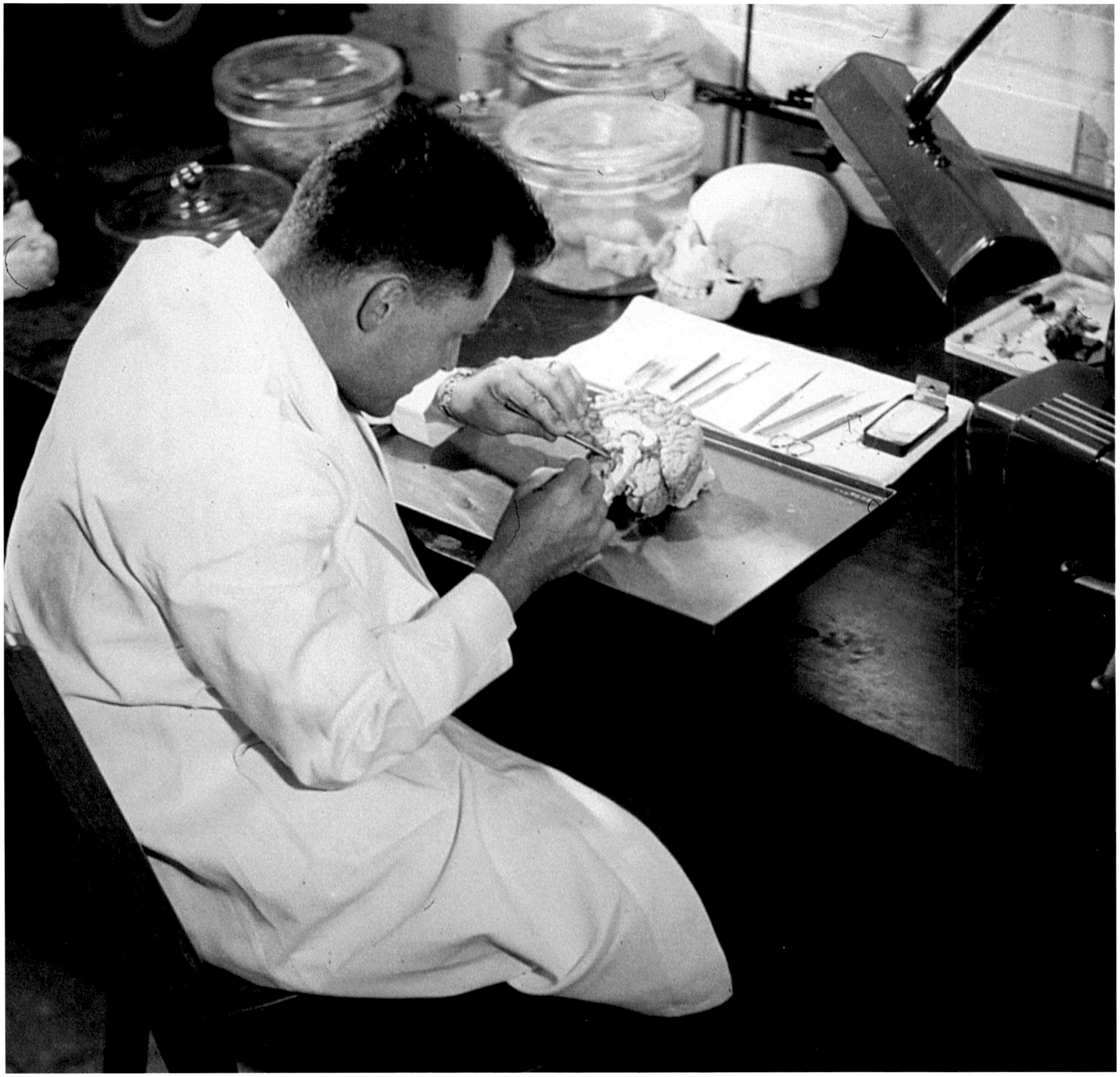

Plate 6.38

David L. Basset, MD, Preparing Neurological Specimens for Photographic Atlas
Using a new format of stereo photography, the seven photograph disc for the View-Master, 1948

His Stereoscopic Atlas of Human Anatomy *consisted of 221 View-Master Reels of every body region.*

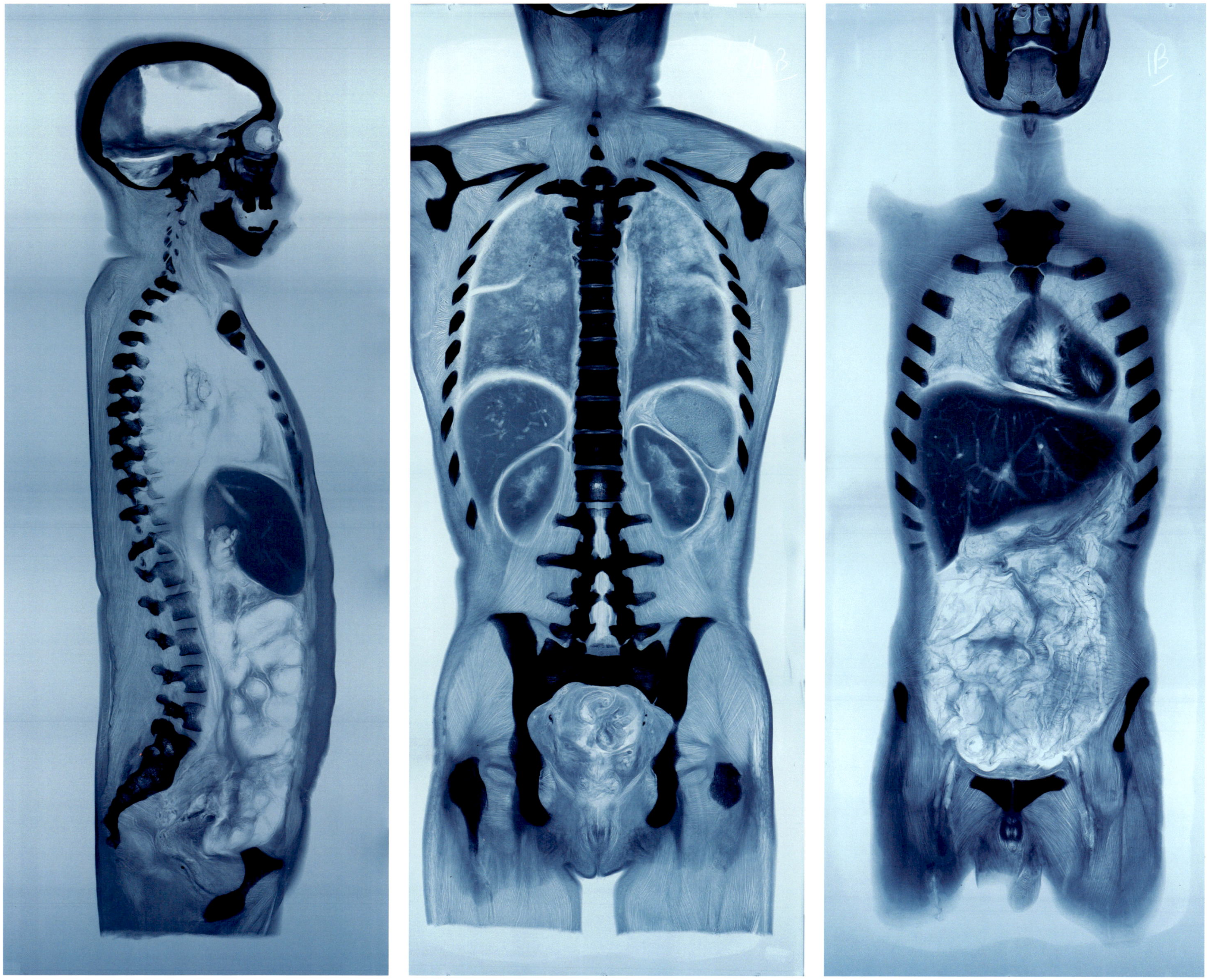

Plates 6.39

Photograms of Serial Upper Body Dissection
Cyanotypes, circa 1945

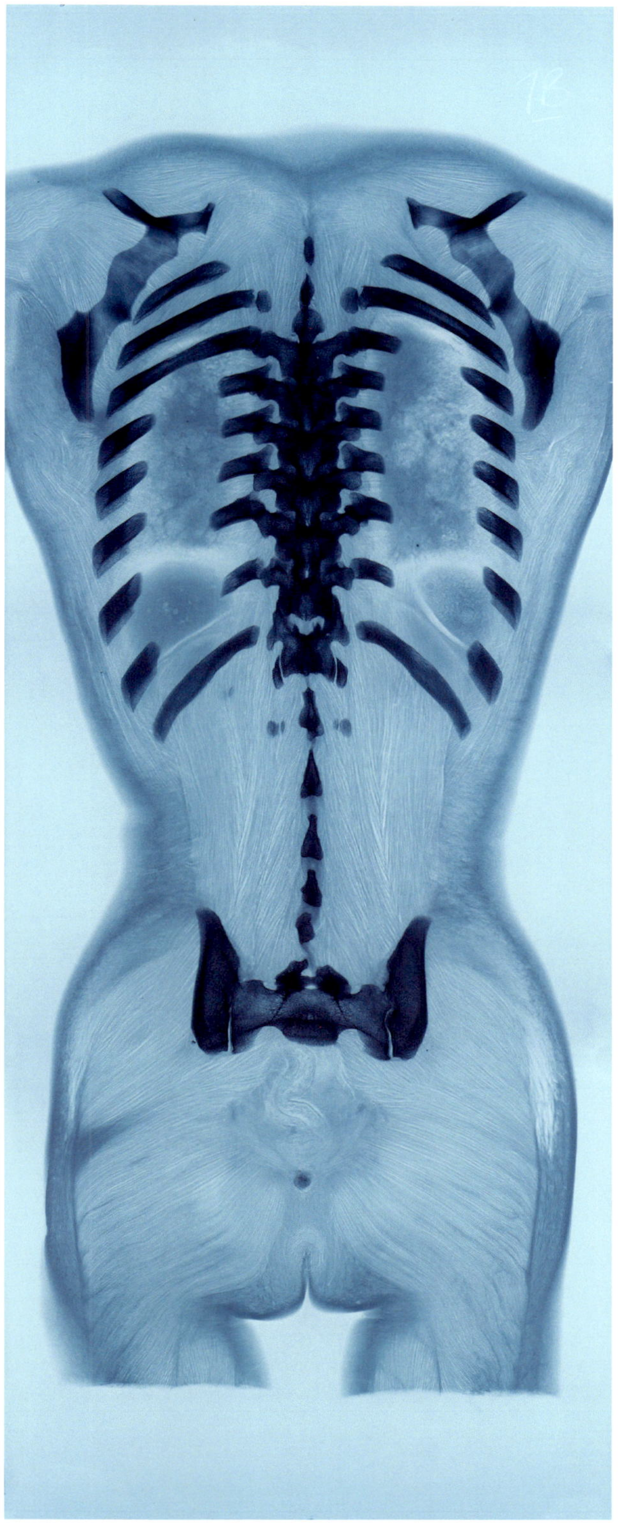

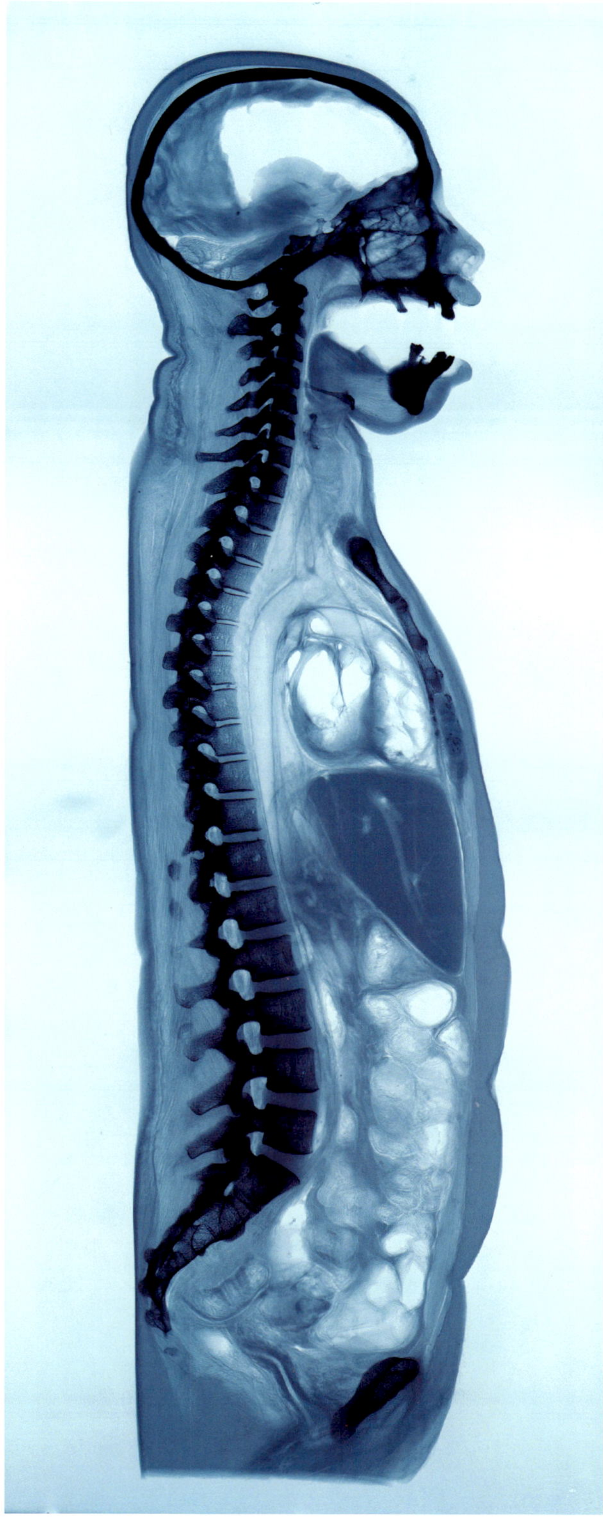

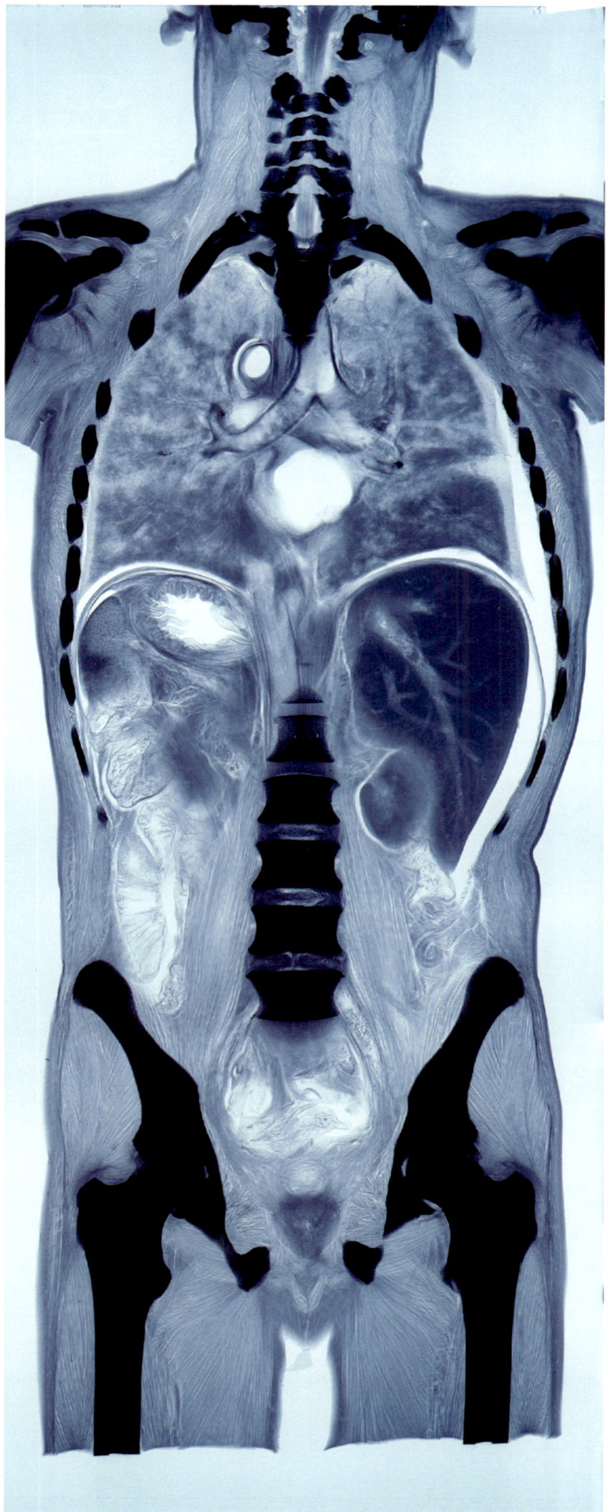

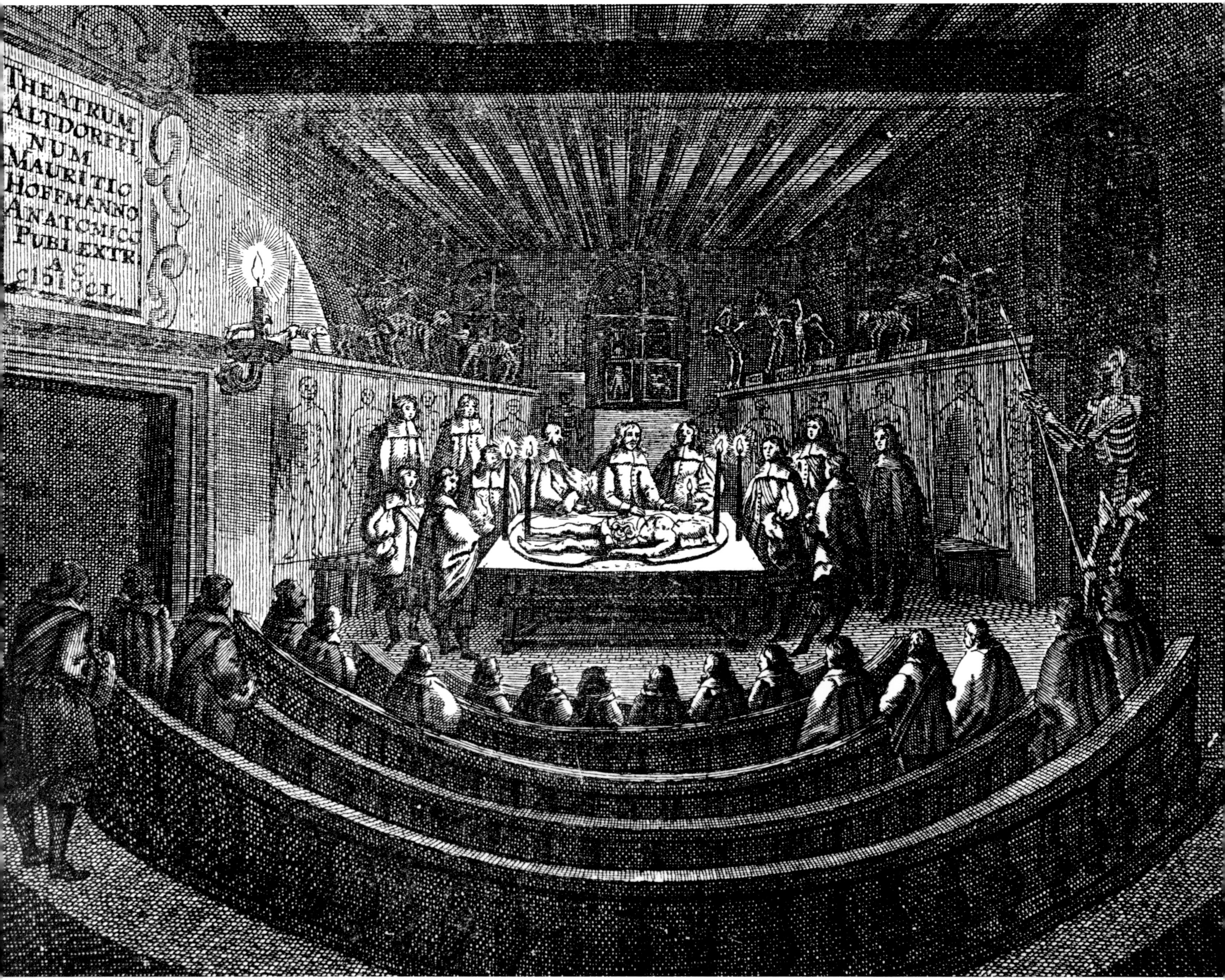
THEATRUM
ALTDORFFI
NUM
MAURITIO
HOFFMANNO
ANATOMICO
PUBL. EXTR.
A
CIƆIƆCL.

CHAPTER SEVEN

Autopsy & Forensic Photography

Determining the Cause of Death

Nineteenth-century autopsy photographs are very rare. An autopsy is a specialized surgical and medical examination by a pathologist to determine the cause and manner of death. While tens of thousands of dissection photographs survive from the nineteenth century, at best, there are one hundred autopsy images. No prior photographic historic text has shown the distinction between dissection and autopsy photographs. Many historians, and most collectors and curators, lump the images together as they seem to show the same subject matter—a group of physicians standing around a body in various stages of dissection.

An autopsy or postmortem examination is a highly specialized procedure that consists of a thorough examination of a corpse to establish the reason for death, and to evaluate any disease or injury that may be present. Autopsies are performed for either legal or medical purposes and are performed by a physician specialist or a pathologist trained in forensic analysis. The classic autopsy incision is Y-shaped, starting at both shoulders, extending to the pubic bone. The body is dissected, internal examination is conducted, and body fluids and tissue samples are collected. Specimens of most organ systems are taken for gross and microscopic evaluation. Depending on the circumstance, chemical and other specialized analytical methods are performed, looking for toxins or chemical abnormalities. Large organs such as the brain, heart, liver, etc. are weighed, and then compared to tables with established standard weights for sex, age, and other growth parameters. At the completion of the autopsy, the wound is sutured.

A forensic autopsy has legal implications and is carried out when the cause of death may be a criminal matter. Determination of death as a natural event, accident, suicide, or homicide is critical. According to the situation, minute puncture wounds are looked for, as are poisons, drugs, medication, and wound types. Clinical or academic autopsies are an essential aspect of medicine, as they ascertain the exact cause and mechanism of death.

Over the centuries, autopsies have helped physicians determine whether a particular medication, medical treatment, or surgical procedure has been effective. Autopsies also aid in identifying or characterizing the extent of disease states and determining if the medical diagnosis and treatment before death was appropriate. With the current trend of creating and inserting numerous medical contraptions, prosthesis, and electronic implants into the body, it is important to determine if they have resulted in any adverse health effects. There are side effects and unintended consequences of all medical procedures.

In the United States, the number of autopsies performed in hospitals has decreased every year since 1955. Critics, including pathologist and former JAMA (*Journal of the American Medical Association*) editor George Lundberg, MD, have charged that the reduction in autopsies is negatively affecting the care delivered in hospitals. They posit that when mistakes result in death, they are often not investigated and lessons therefore remain unlearned. Others contend that MRI, angiographies, and other specialized studies offer the modern physician a glimpse inside the body not previously achieved.

In this chapter is a rare photograph of the bodies on display at the Paris Morgue in the

1880s. It is the only surviving image of the unusual practice of displaying dead unidentified persons to the public, which was a custom common only to France. On Sunday, the morgue was open to the public. They came to view the bodies and to see who may have drowned in the Seine River that week, or simply been found dead or murdered. The clothes of the victim were displayed on a nearby stand. This chapter begins with autopsy scenes and images of pathologists in their laboratories, and ends with five twentieth-century autopsy images.

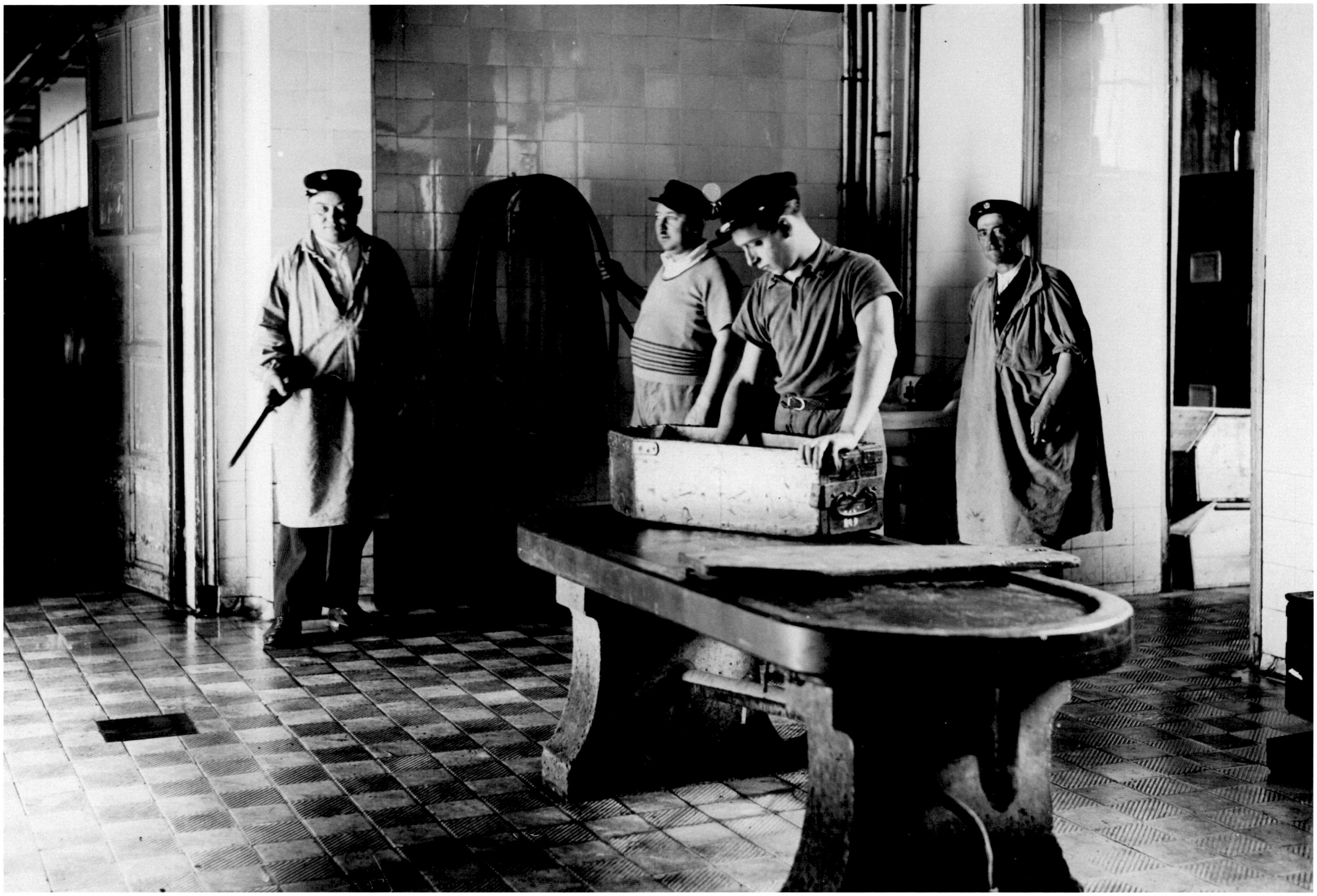

Plate 7.0

Workers in Paris Morgue with Child's Coffin
Circa 1900

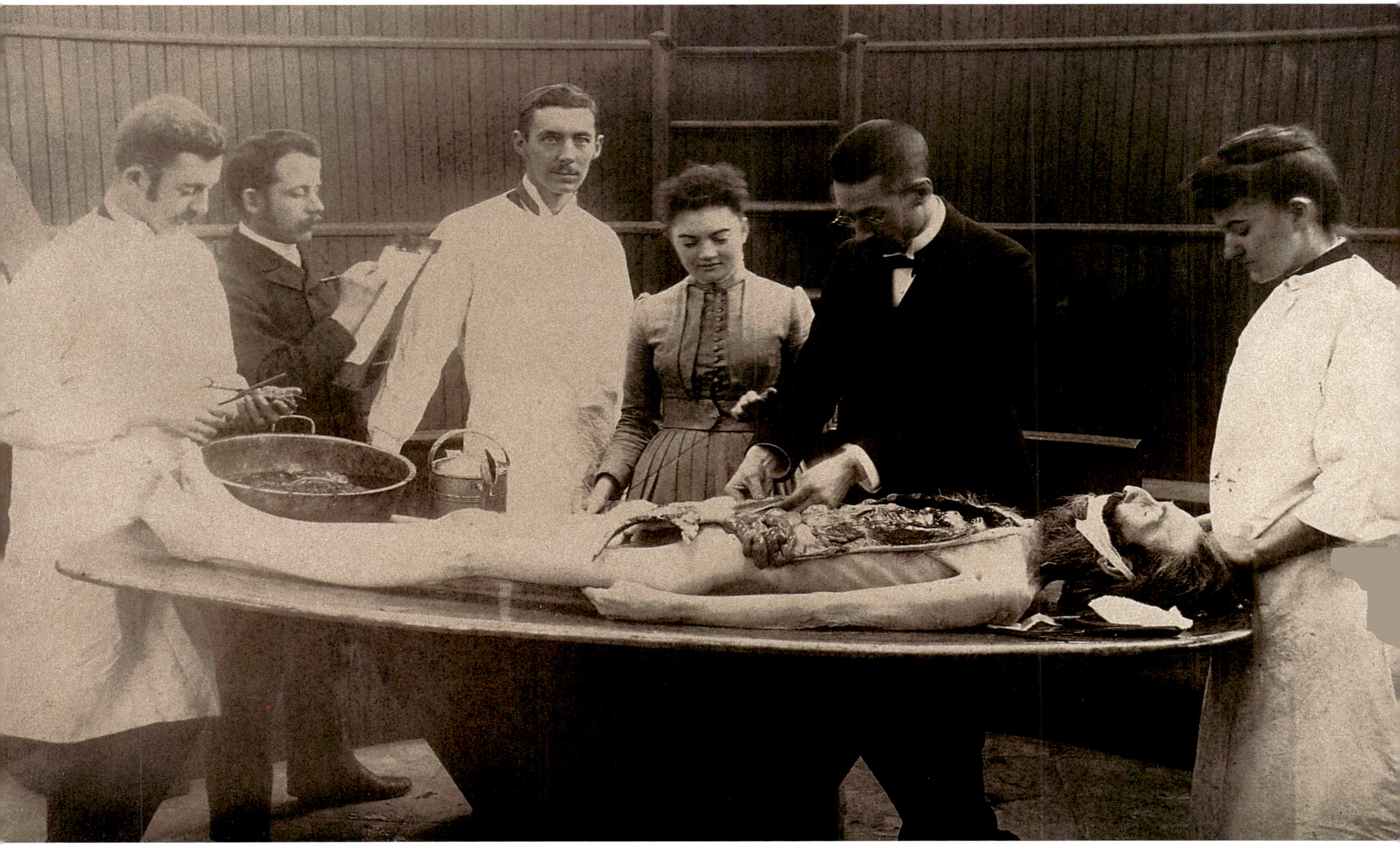

Plate 7.1

Autopsy, Classic Image of a Pathologist at Work
1880

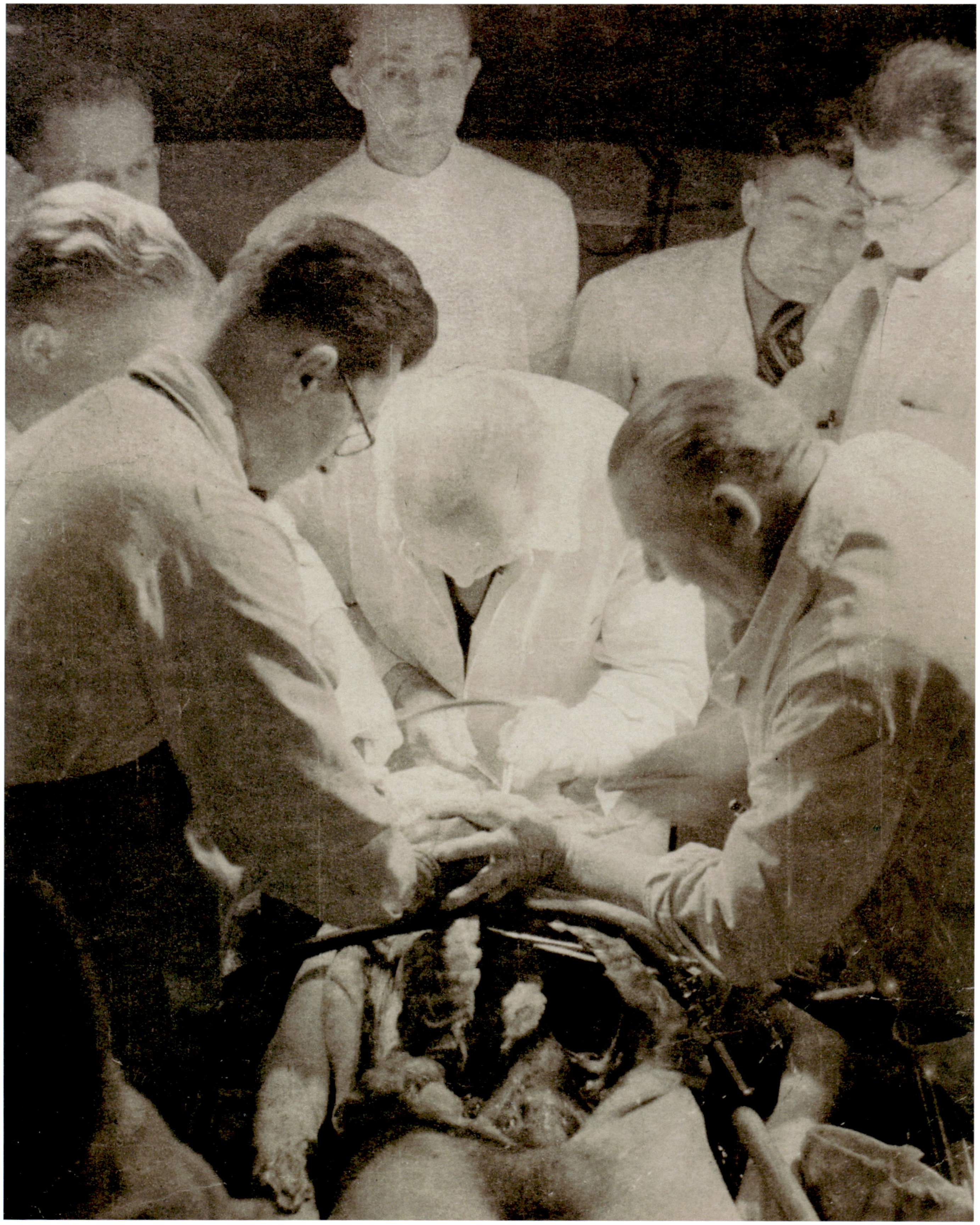

Plate 7.2

Autopsy, Group Around Table Watching Procedure
Circa 1925

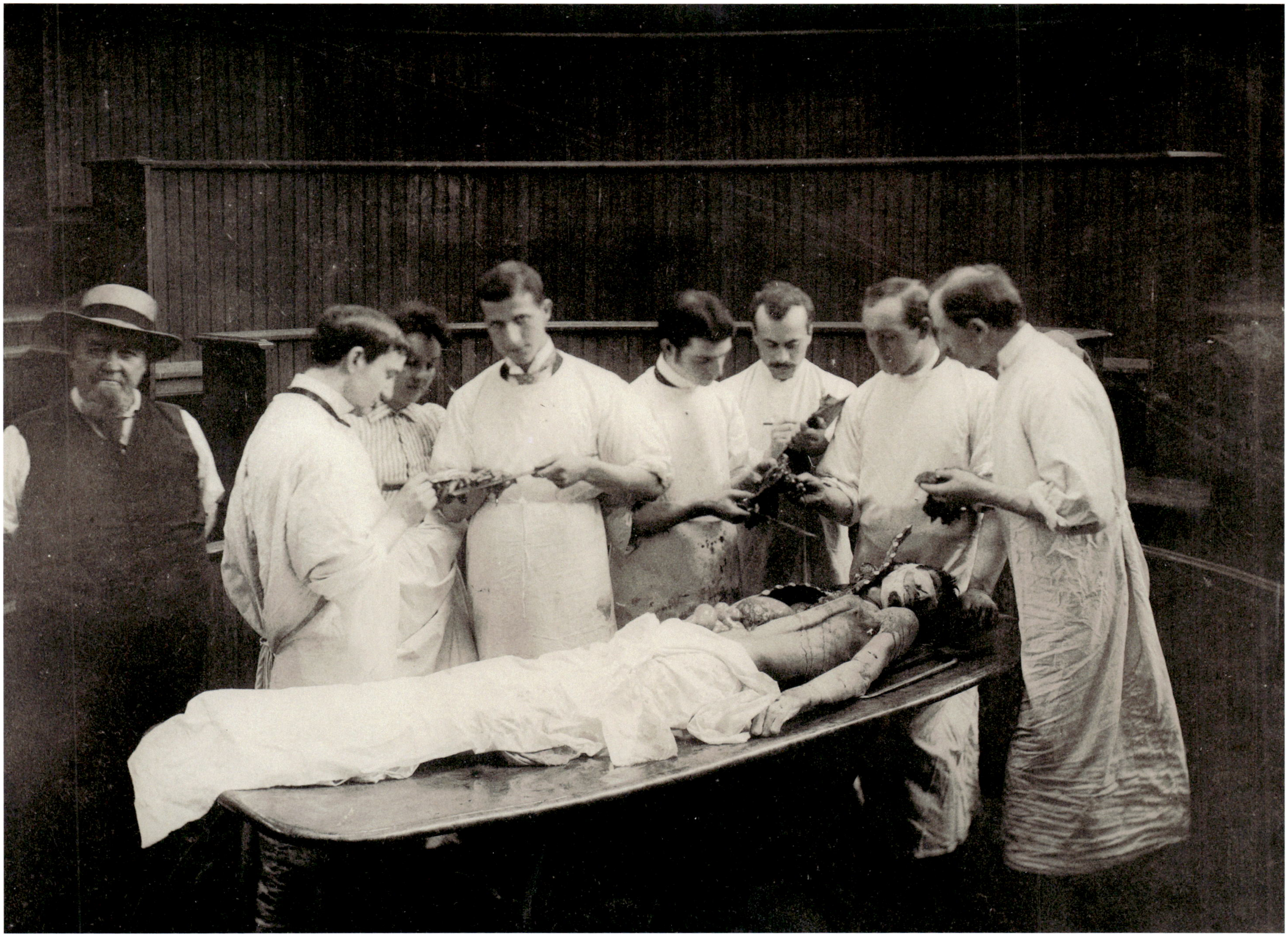

Plate 7.3

Autopsy, Cook County Hospital, Chicago
1890

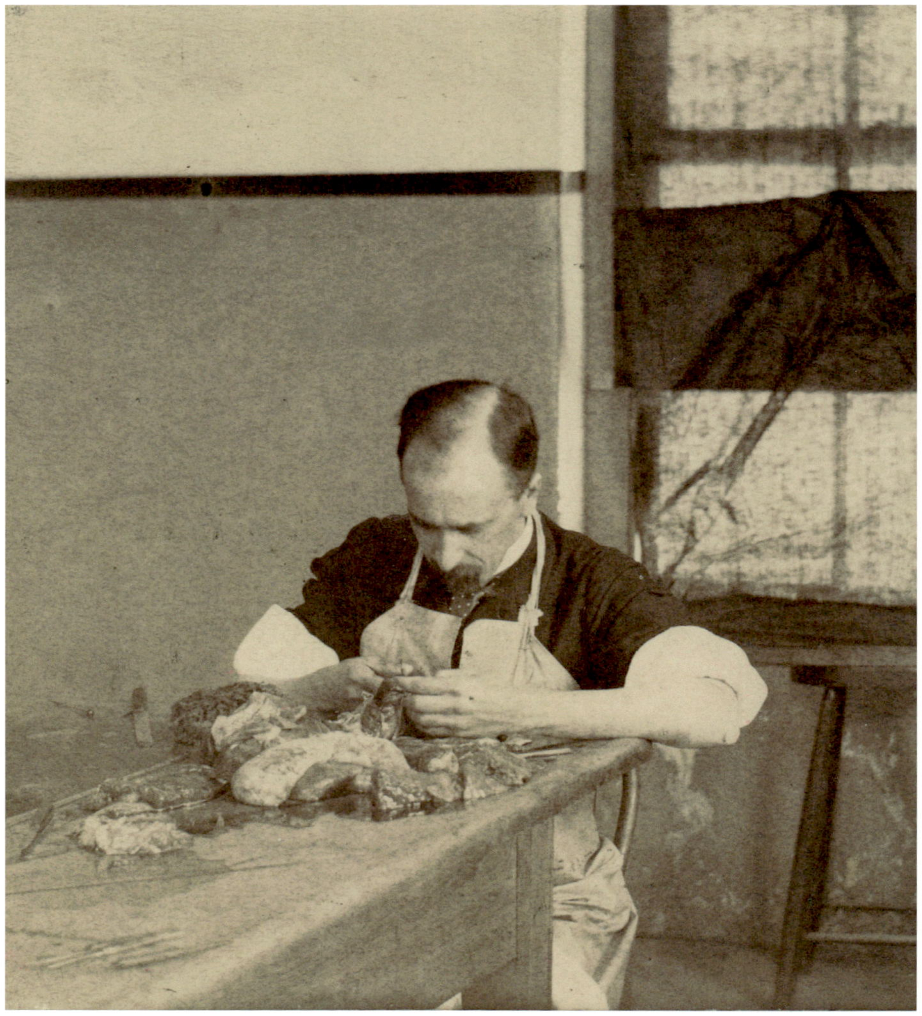

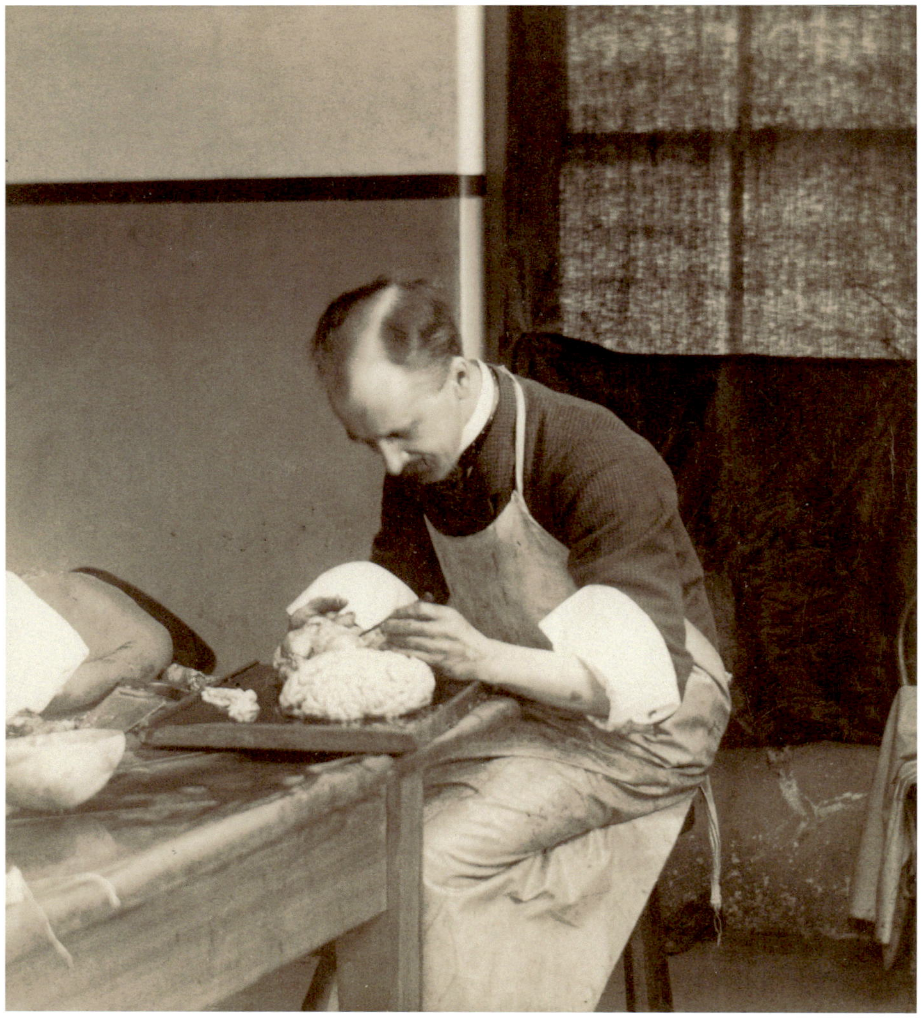

Plates 7.4 – 7.5

Sir William Osler, MD, at Work in the Blockley Mortuary
Philadelphia General Hospital, 1886 or 1889

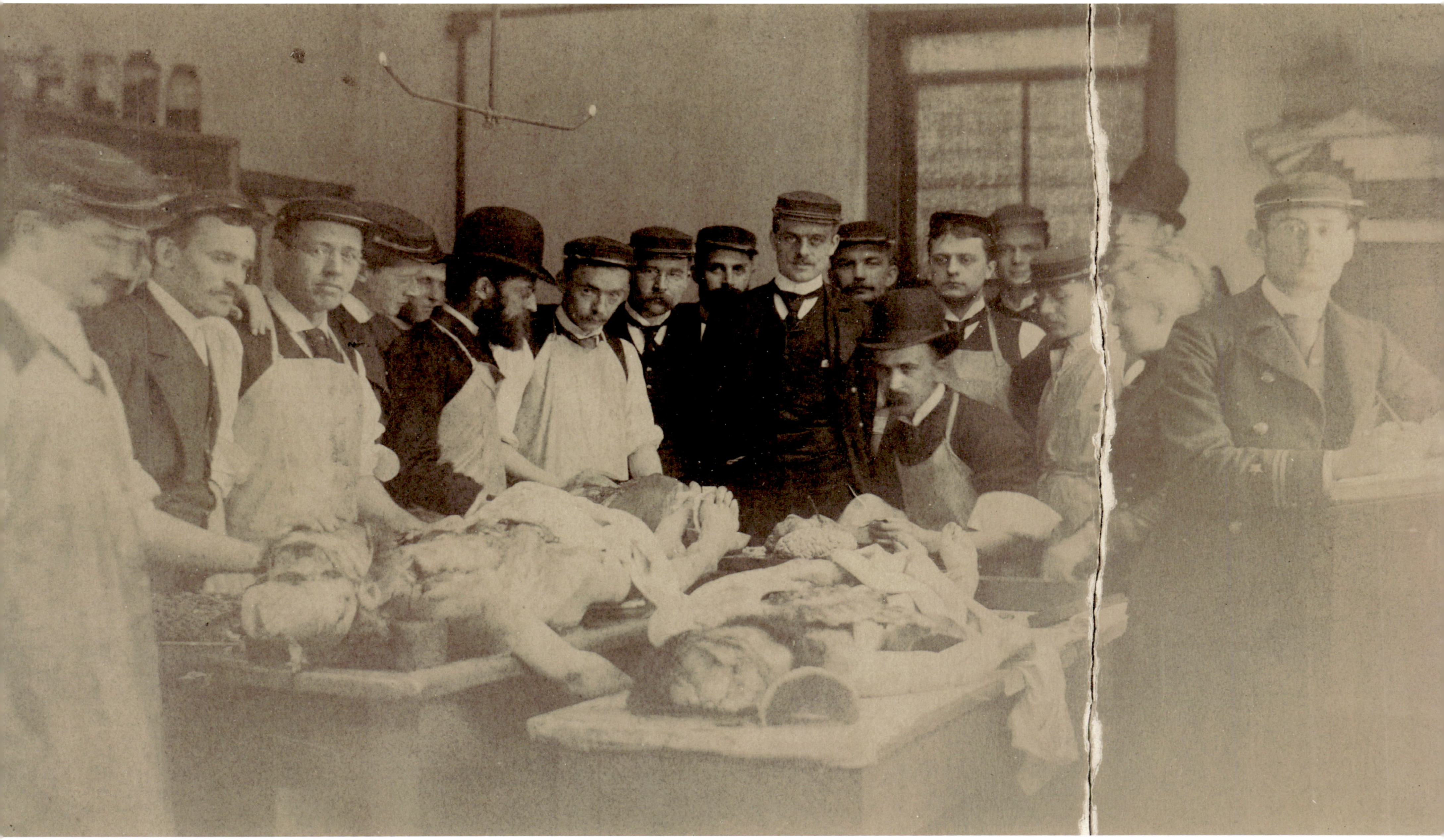

Plate 7.6

Sir William Osler, MD, Performing an Autopsy
Blockley Mortuary, Philadelphia, 1886 or 1889

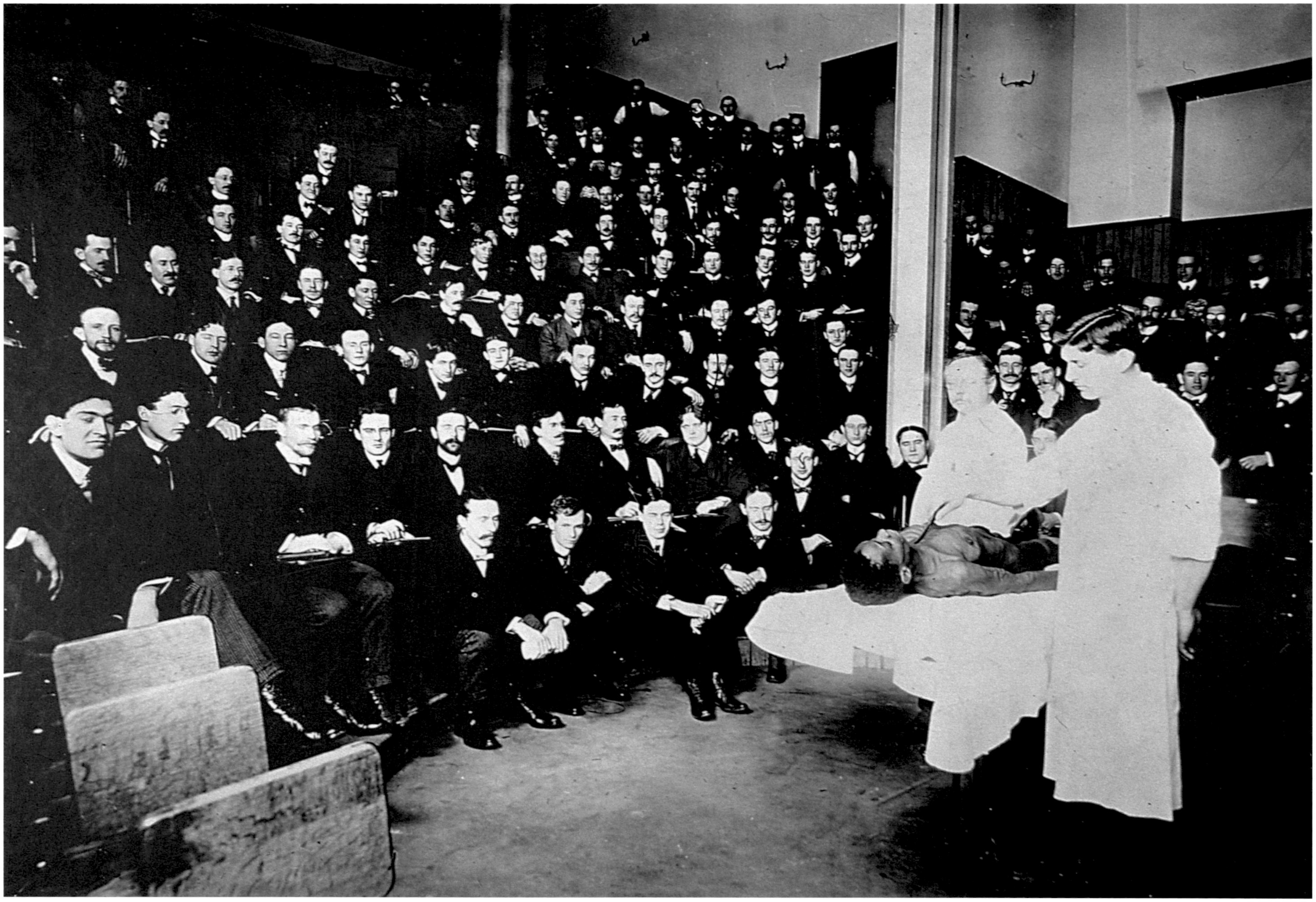

Plate 7.7

Autopsy Lesson
Unidentified School, circa 1910

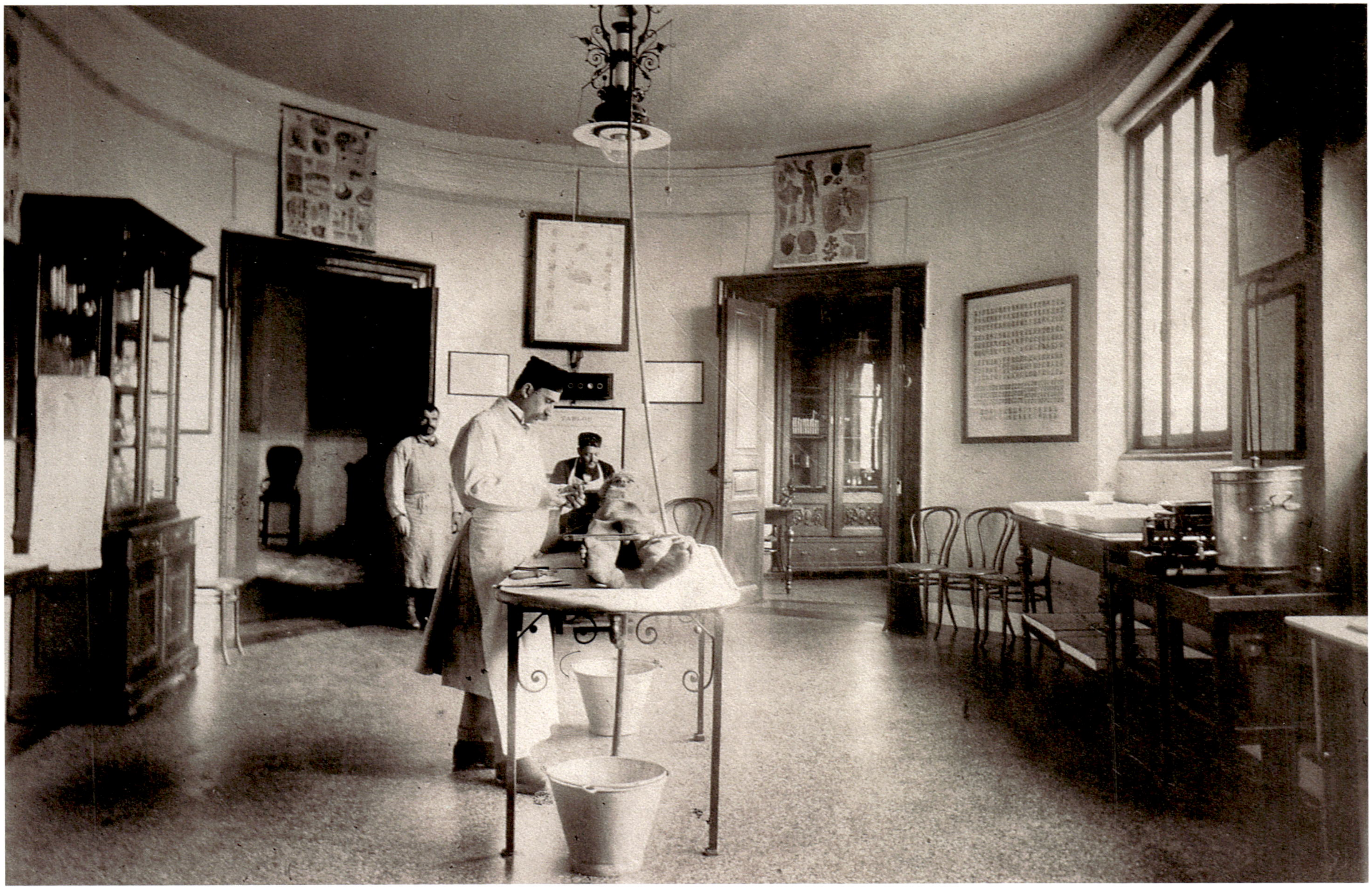

Plate 7.8

Autopsy at the Paris Morgue
Circa 1890

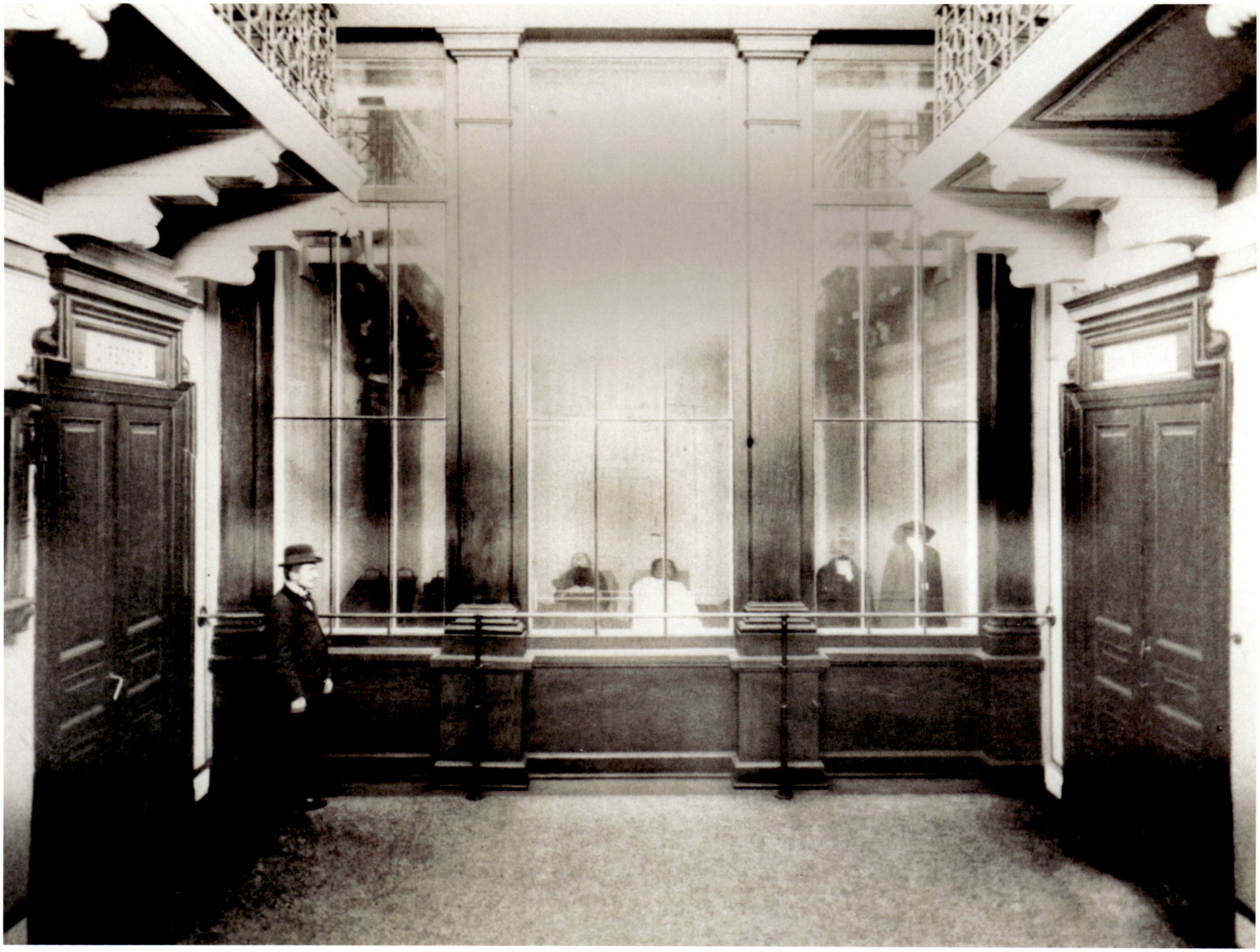

Plate 7.9

Viewing Room at the Paris Morgue
Circa 1890

The week's unidentified bodies on display.

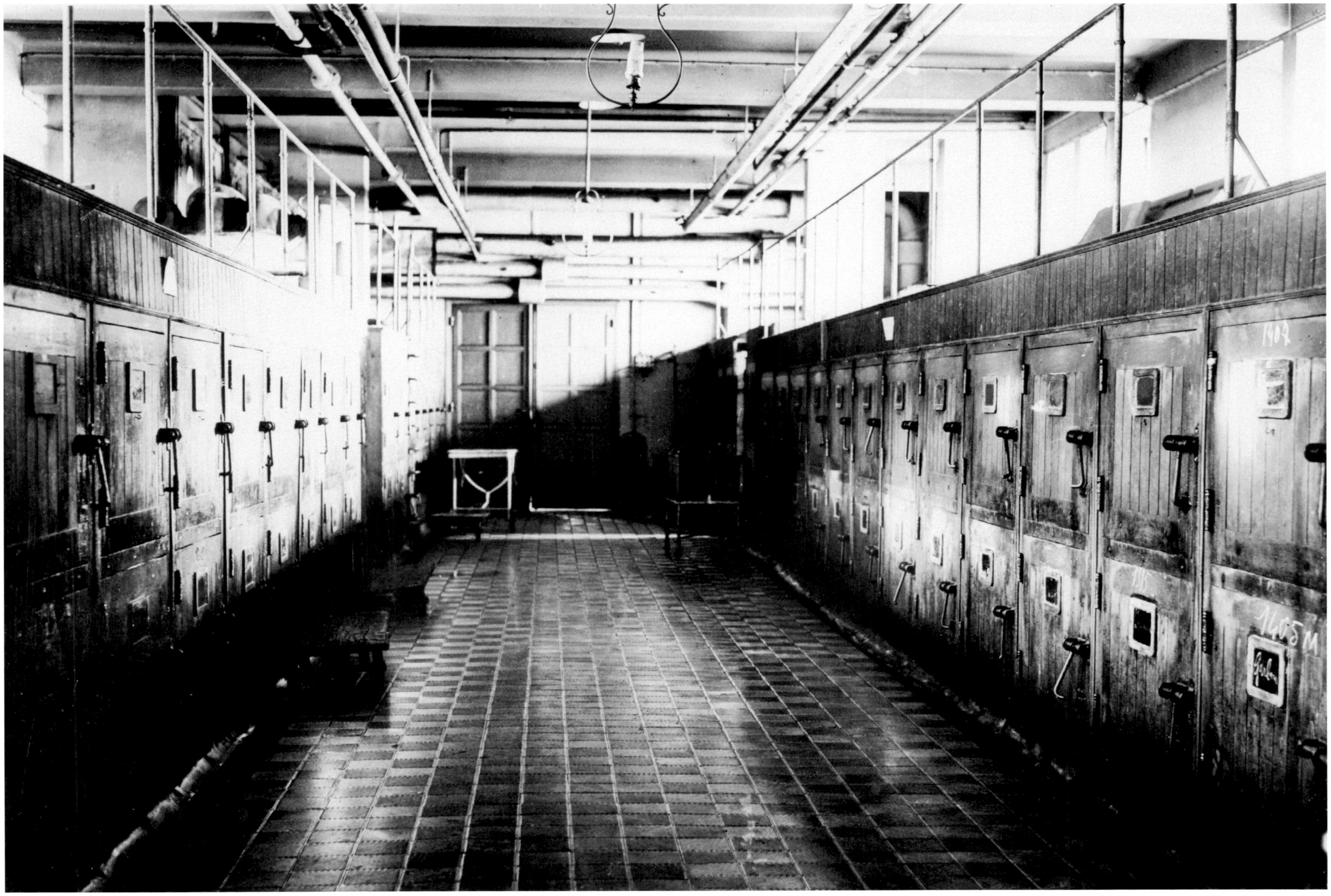

Plate 7.10

Body Storage Room at the Paris Morgue
Circa 1890

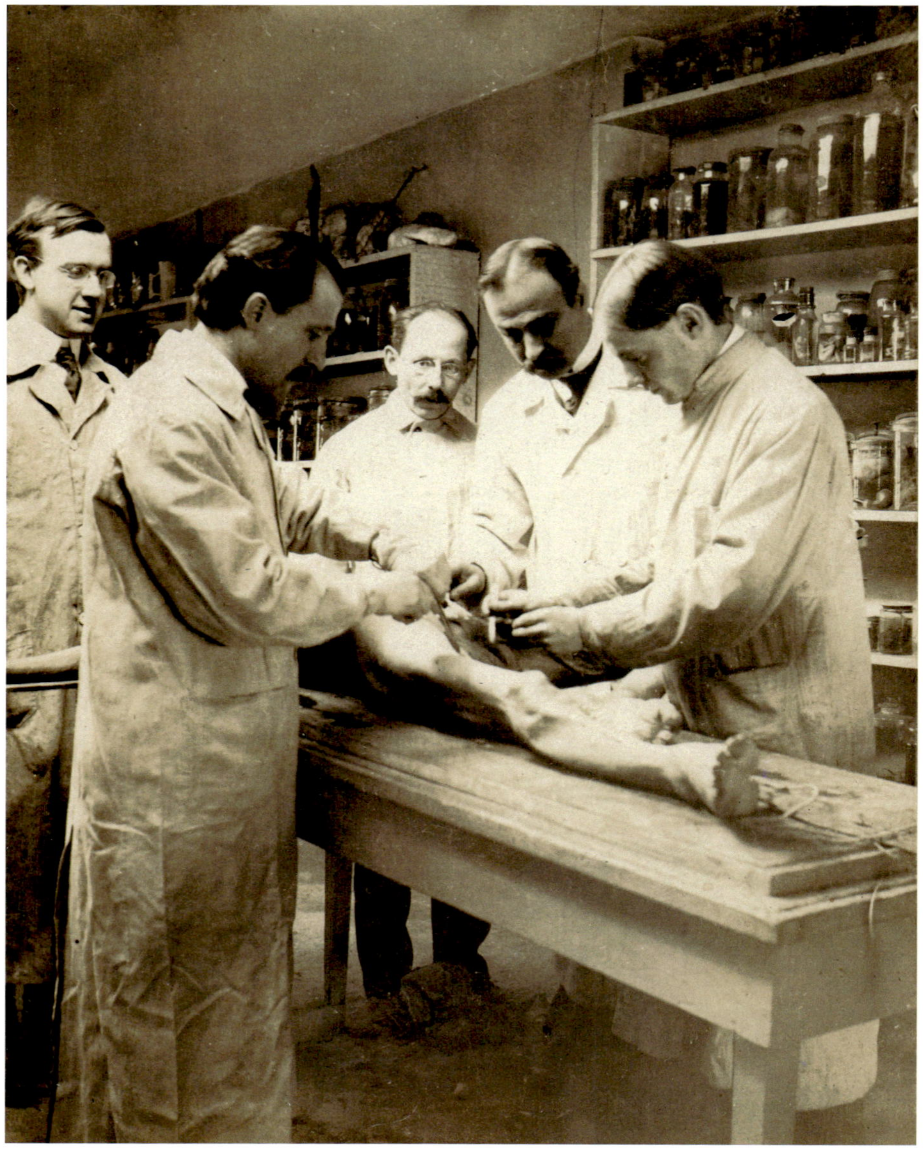

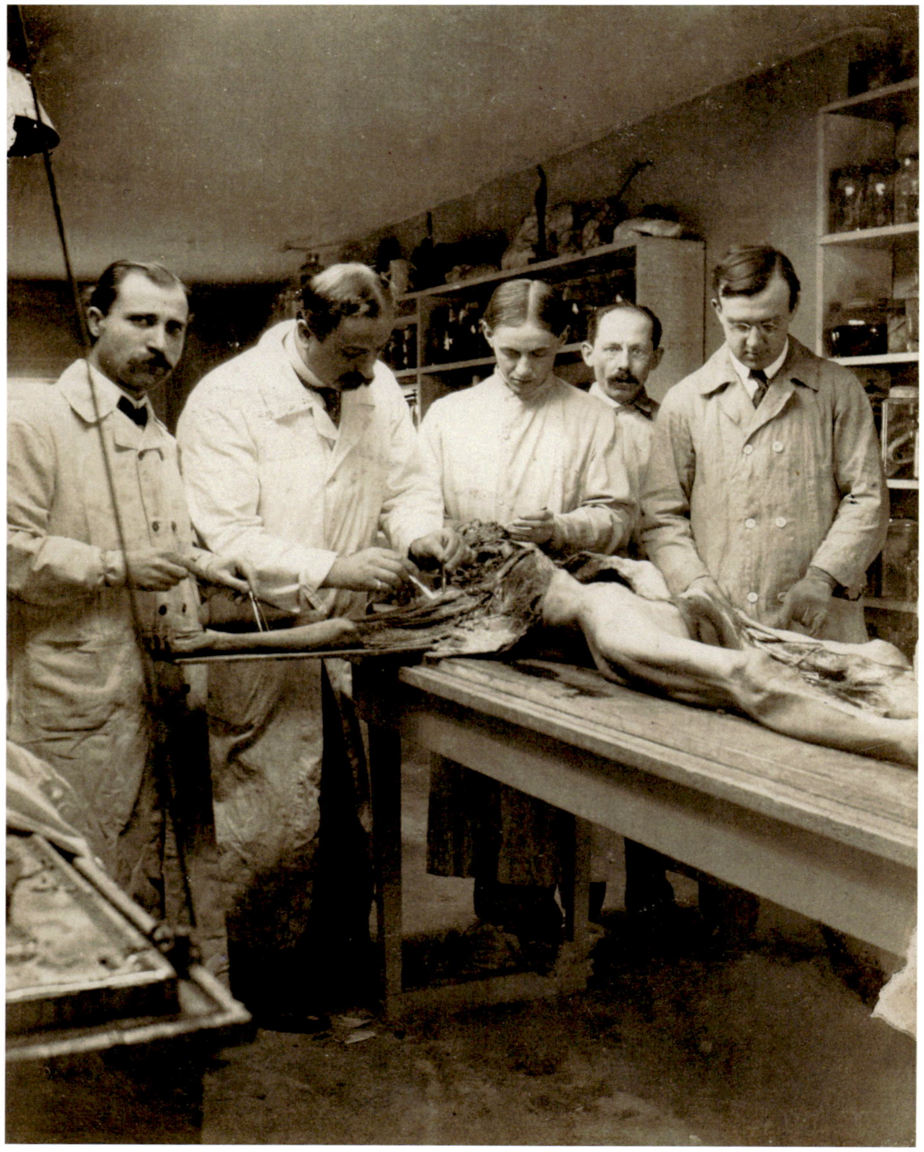

Plate 7.11 – 7.12

Autopsy, Shelves Along Wall Contain Pathological Specimens
1920

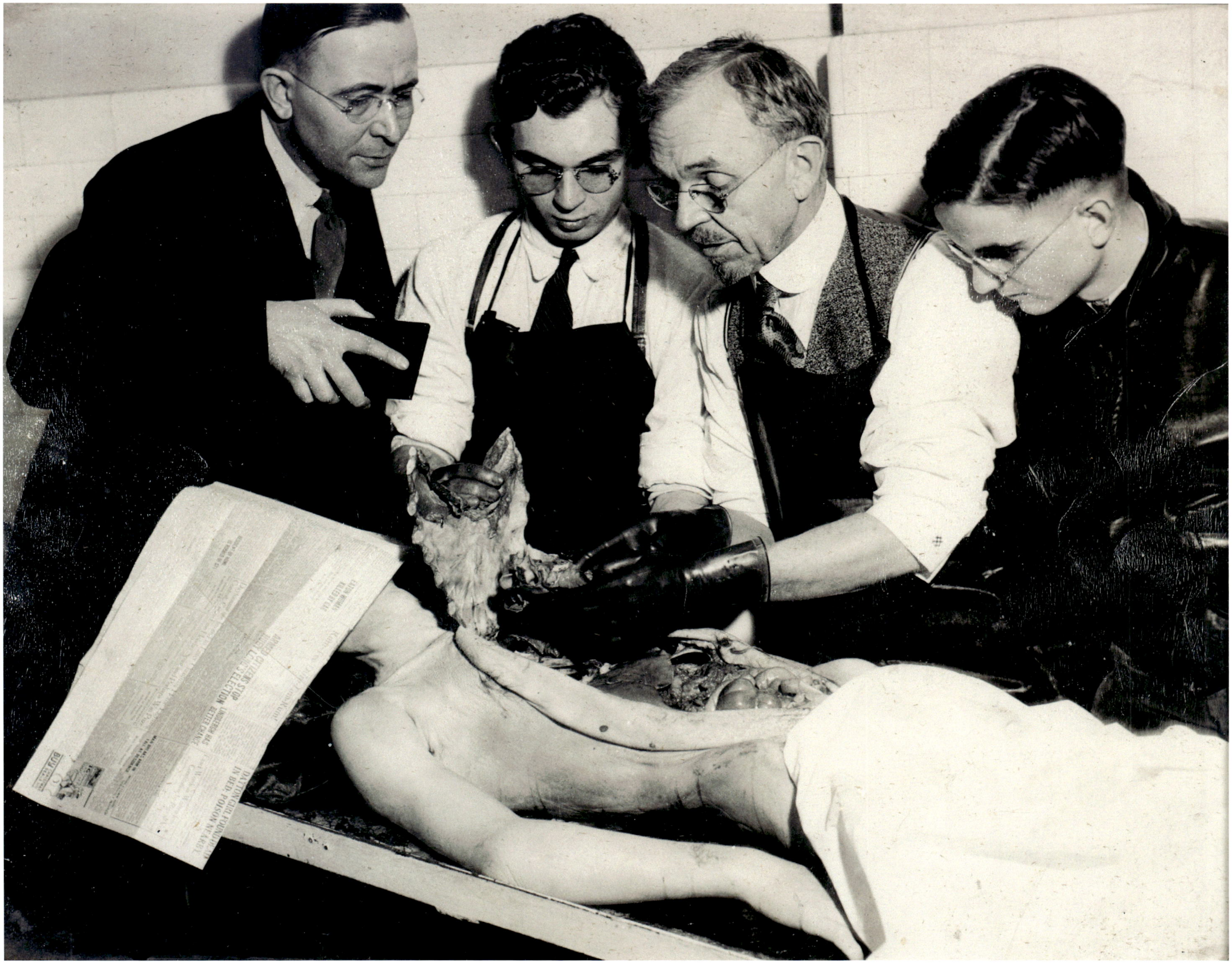

Plate 7.13

Howard Brundage, MD, Performing an Autopsy
1925

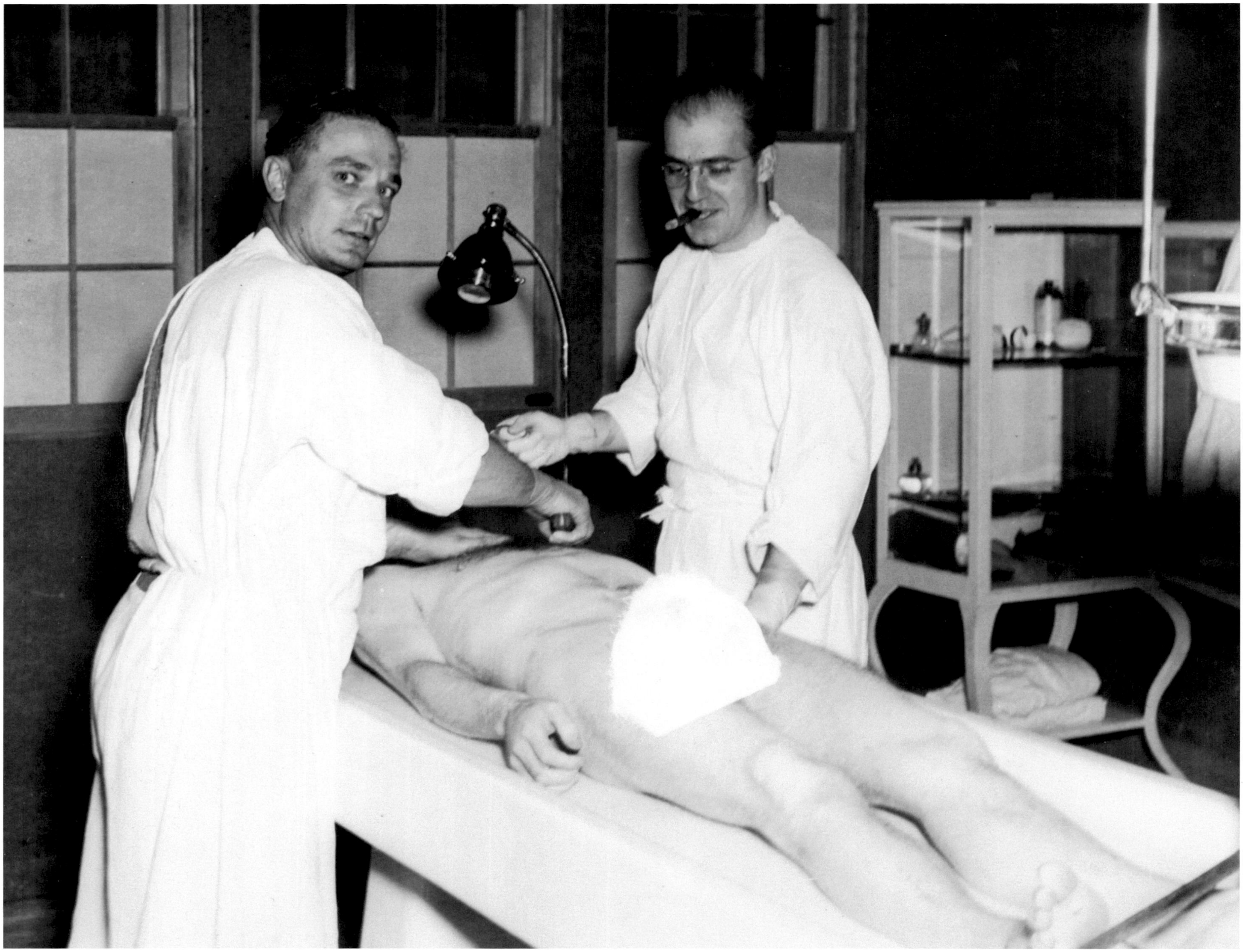

Plate 7.14

Military Pathologists Preparing to Autopsy a Deceased German Prisoner of War
Ft. Bragg, NC, 1944

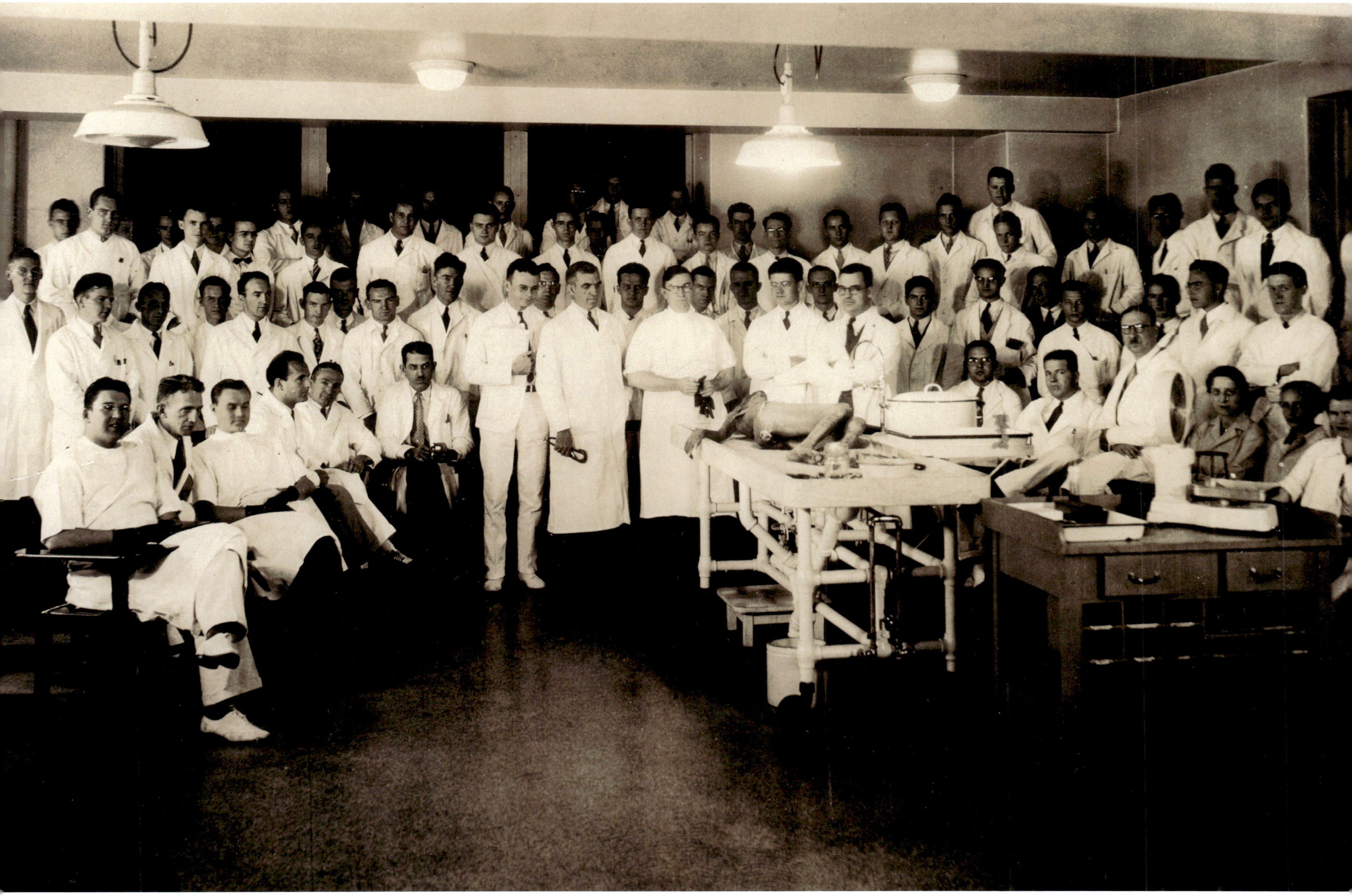

Plate 7.15

The One-Thousandth Autopsy At Duke Hospital
Durham, NC, 1934

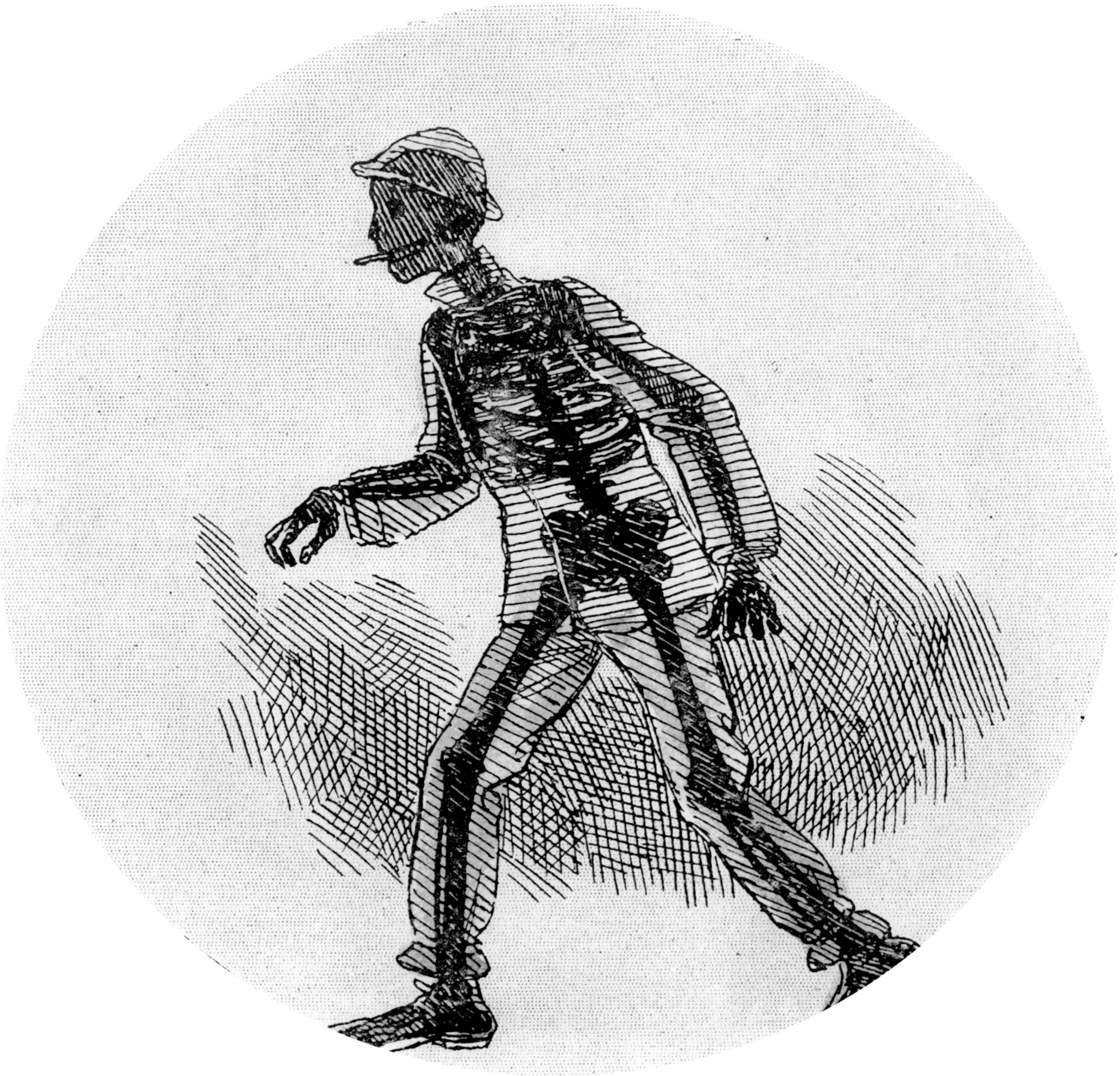

CHAPTER EIGHT

X-Rays – The Skeleton Becomes Visible in the Living

Removing the Mystery of Anatomy, Disease, and Wounds

Accepted as a miracle of science from the moment it was discovered, the X-ray rapidly spread worldwide and was used by professionals, as well as the public. The fact that people could see inside the living body revolutionized medical and popular thinking. Heralded as photography of the invisible, early X-ray photographs of skeletal body parts were produced and sold to the public and physicians. All were fascinated with the ability to see inside the body.

In November 1895, German physicist Wilhelm Conrad Röntgen discovered the X-ray. It was one of a series of amazing discoveries at the turn of the century and part of the greatest era for discovery and invention the world had ever seen. Dramatic innovations offered the middle and working classes new prospects for an improved life. The electric light, moving pictures, airplanes, automobiles, telephones, halftone prints, and phonographs were major accomplishments that set the stage for the twentieth century. Medicine also dramatically advanced and offered the public the promise of a long healthy life, even perhaps free of disease.

Medical innovations and discoveries of the time included the germ theory of disease, surgical asepsis, food and drug laws, antitoxin therapy, typhoid vaccination, knowledge of viral and vector parasitic diseases, along with the development of light therapy and a 'magic bullet' for syphilis. For a period of about thirty-five years, the last two decades of the nineteenth century and the first fifteen years of the twentieth century, more specific disease entities were described than at any other time in history. Medicine became an accurate science, offering specific diagnosis and, for the first time, specific therapy. Entirely new fields of disease entities were recognized such as endocrine and nutritional diseases. Psychiatry and subspecialties flourished as so many identified 'new' disease conditions and therapies required specialization.

The marvels of the X-ray were immediately understood and aside from the ability to see inside the body, its therapeutic benefits were recognized within the first year. Physicians no longer had to wait until a patient died, or a limb was amputated to evaluate a bony deformity, wound, or disease. By 1900, cancers and skin diseases were being treated by X-ray. The introduction of contrast media propelled radiography to new heights. Contrast media studies and stereoscopic X-ray photographs were innovations that offered physicians a peek into the body and its organ functions without opening it. Organs were correctly outlined and evaluated for function, defects, and disease states. The miracle of the X-ray also propelled surgery into the modern specialty, as accurate diagnosis and therapies were devised.

Medicine adapted with each new advance in photography. The X-ray was first used in research and clinical practice, then tailored to educate physicians. Print publications, motion pictures, television, video, computers, optical scanners, and other such innovations all became standards of medical education. Within a few months of Röntgen's discovery, X-rays were being reproduced as stereoroentgenograms. Various machines were also developed to be able to read large stereo X-rays.

The Southworth company of Troy, New York, taking a cue from the success of their stereophotographic publications, published a

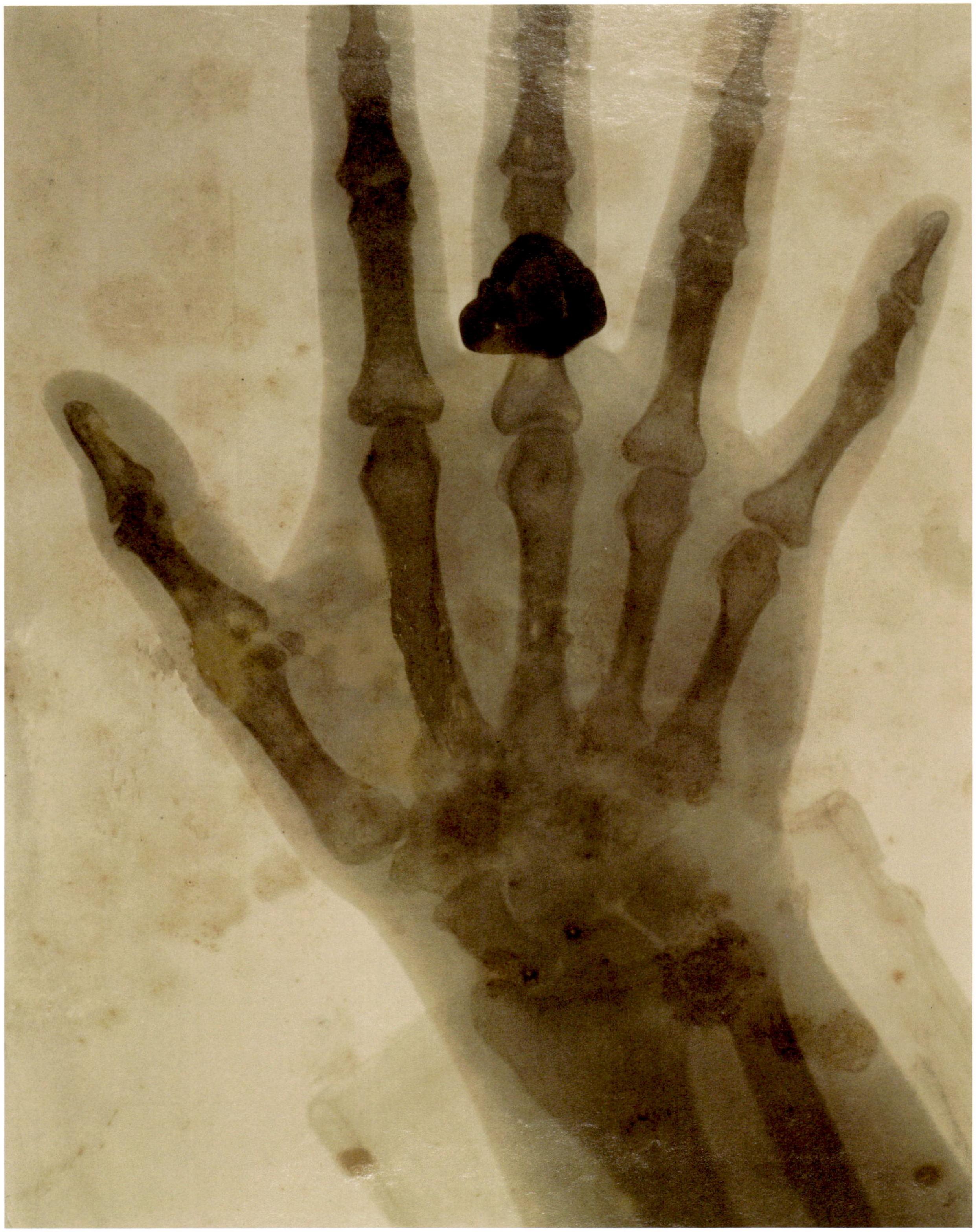

Plate 8.1

X-Ray of Hand, Emulating the First X-Ray
The original was of Conrad Röntgen's wife's hand, 1896

In X-ray's early years, it was popular to X-ray hands with rings.

new specialty journal in 1916, *The American Atlas of Stereoroentgenography: A Quarterly*, which offered original, tipped-in, three-dimensional X-ray stereo photographs, one of the most advanced methods of medical X-ray examination. The advantage of this diagnostic technique was that it allowed accurate determination of anatomic relationships, especially in difficult locales. The publisher's announcement noted, "The...Atlas... will be at once accorded a position among the foremost and most important clinical publications of the age." Until the end of the second decade of the twentieth century, another journal, *The American Quarterly of Roentgenology*, also published tipped-in original stereo-photographs in many issues. The use of original photographs was necessary, as halftone prints could not accurately reproduce many radiological conditions. The introduction of contrast mediums propelled radiography to new heights as organs were correctly outlined and evaluated for function, defects, and compression; also, disease states were readily identified. During the first decade of the twentieth century, Doctors Emil Beck, James MacKenzie, Charles Leonard, Kennon Dunham, and others perfected stereoroentgenography to study the head, lungs, heart, abdomen, and urinary tract, in addition to the ordinary studies of bones and joints that also used contrast media. In this chapter, X-ray photographs from several of the landmark stereographic publications will be presented. Also included are photographs from surgeons A. Wiesinger, Th. Deneke et al., as well as radiologist H. *Albers-Shoenberg, Rontgen-Atlas der Kriegsverletzungen (1914-1916)...Allgemain Krankerhauses St. Georg* in Hamburg. In this landmark work on surgical repair of wounds, the authors presented fifty-three full page and twenty-five stereoview cards of actual X-rays of wounds. Two cases are presented. The chapter ends with a series of X-rays of bullets lodged in various parts of the body.

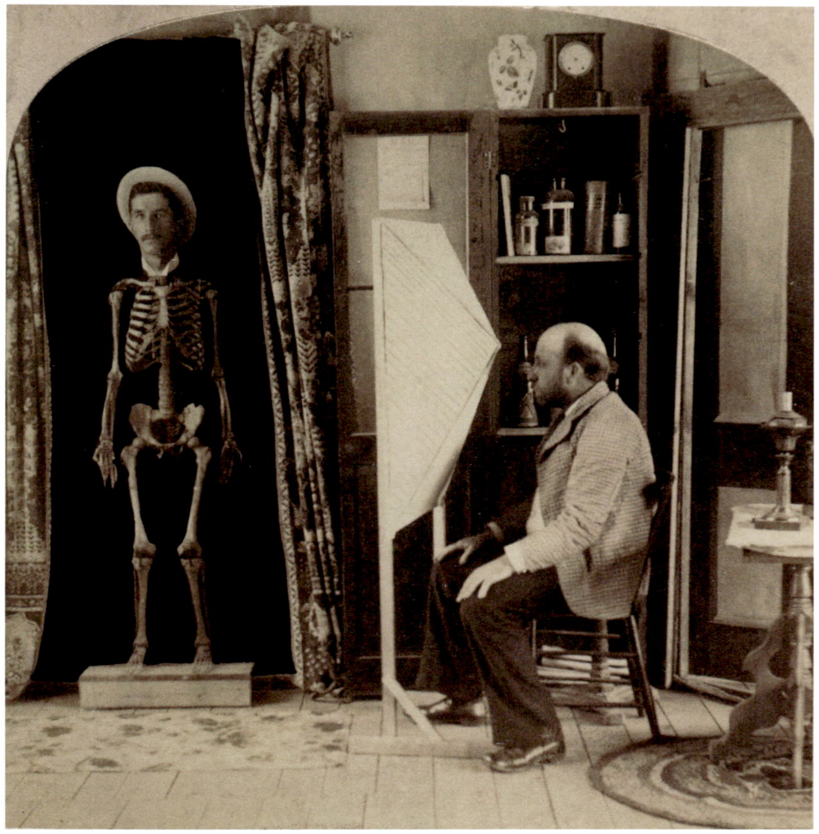

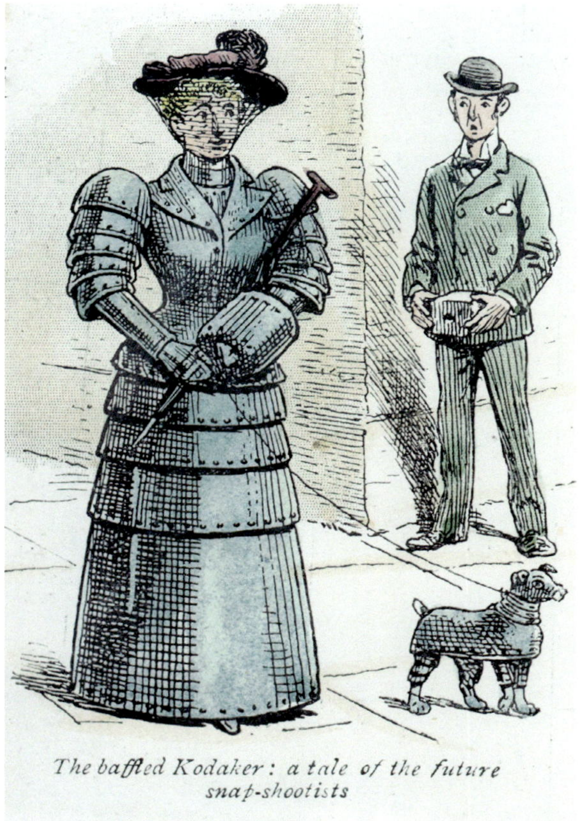

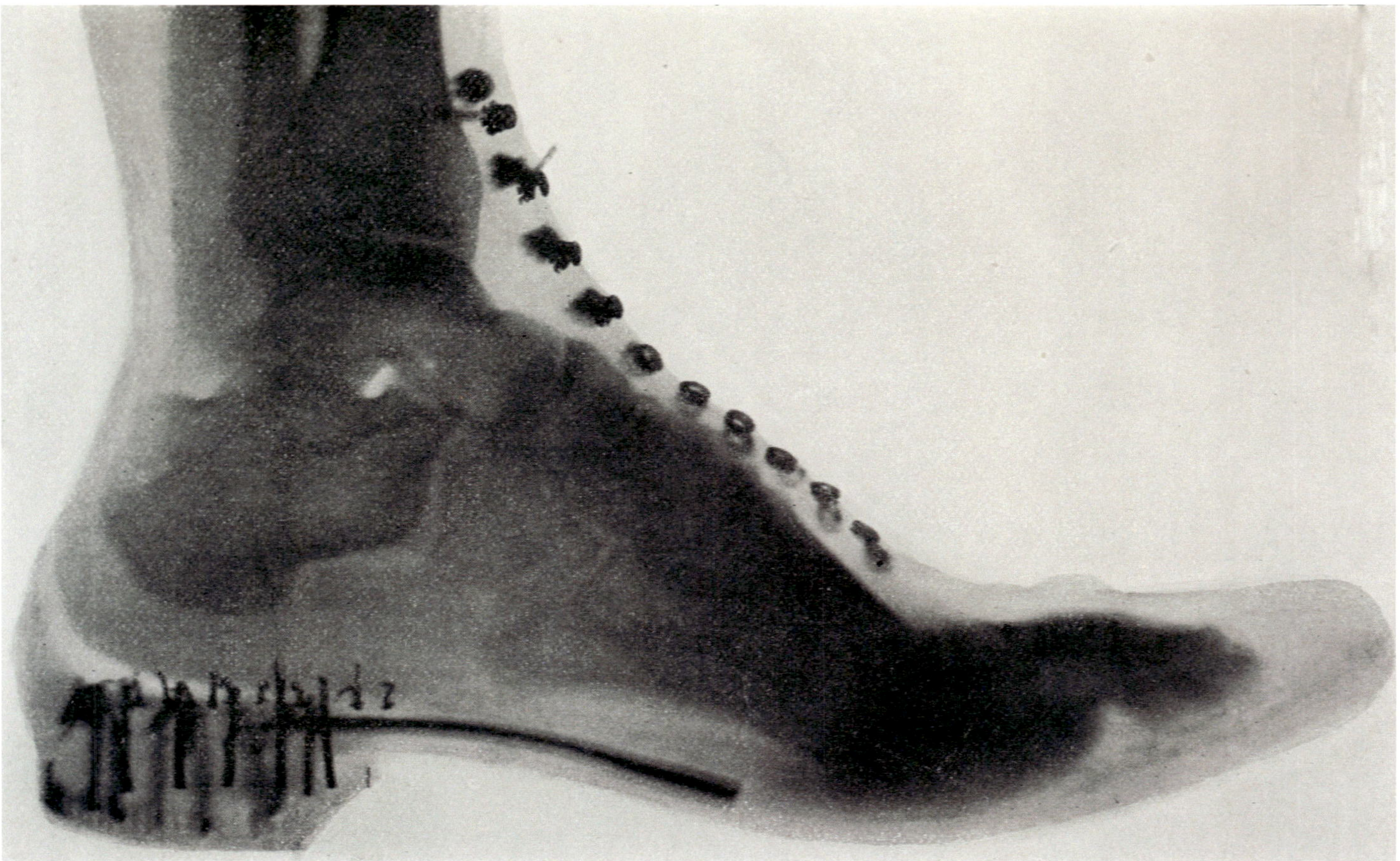

Plate 8.2

Seeing Inside a Man
1898

A popular stereoview that illustrated what the new device could do.

Plate 8.3

Wearing Metal Armor so Her Body Can't Be Photographed
1896

Numerous cartoons were published emphasizing the amazing phenomenon of the X-ray.

Plate 8.4

X-Ray of a Woman's Foot Wearing Shoe
1896

Photographs such as this were published in newspapers to show the new technology.

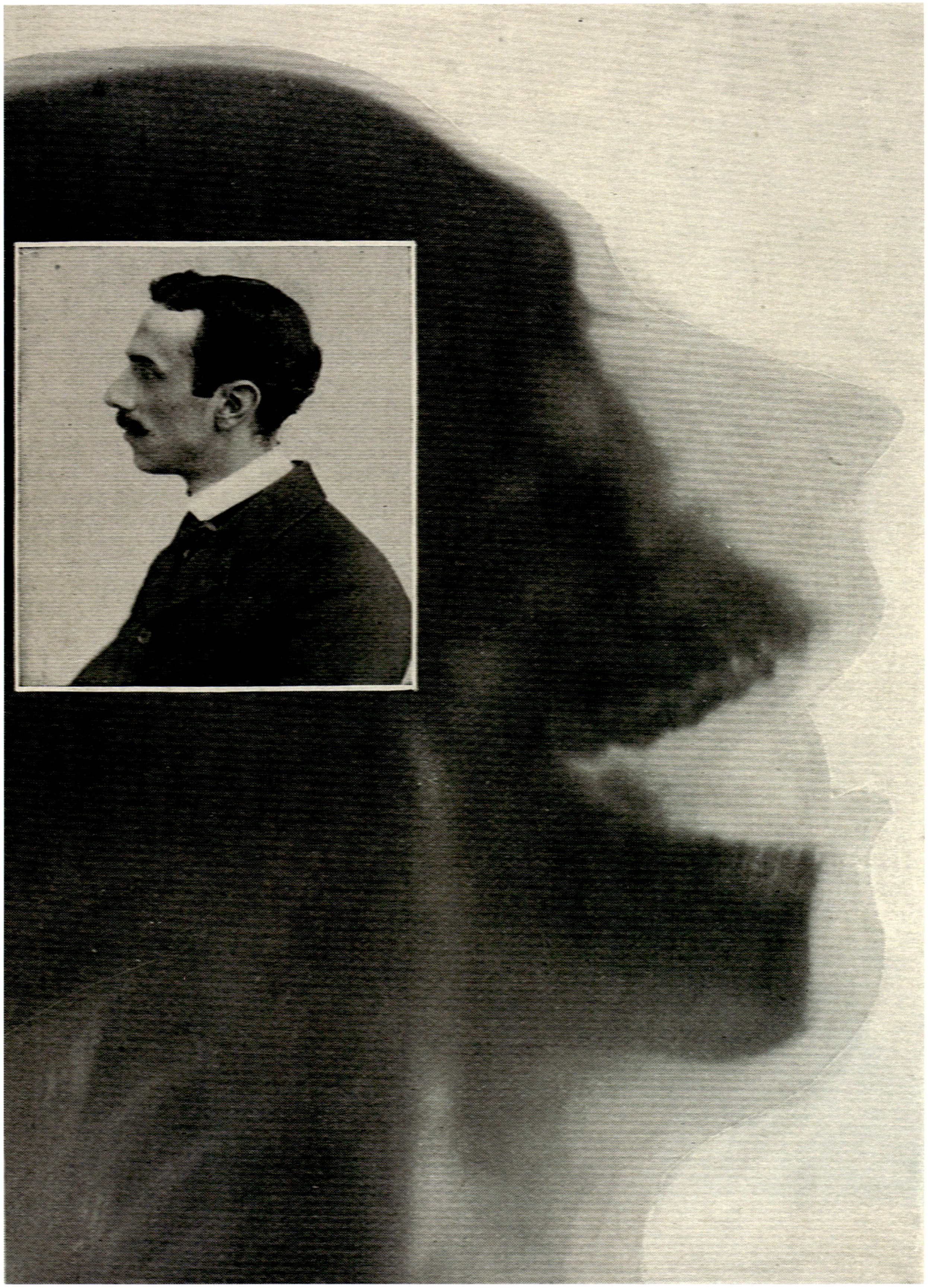

Plate 8.5

Early X-Ray of a Man's Head Showing the Comparison to the Subject
William J. Morton, MD, et. al., *The X-Ray or Photography of the Invisible,* Plate 66/67, March 1896

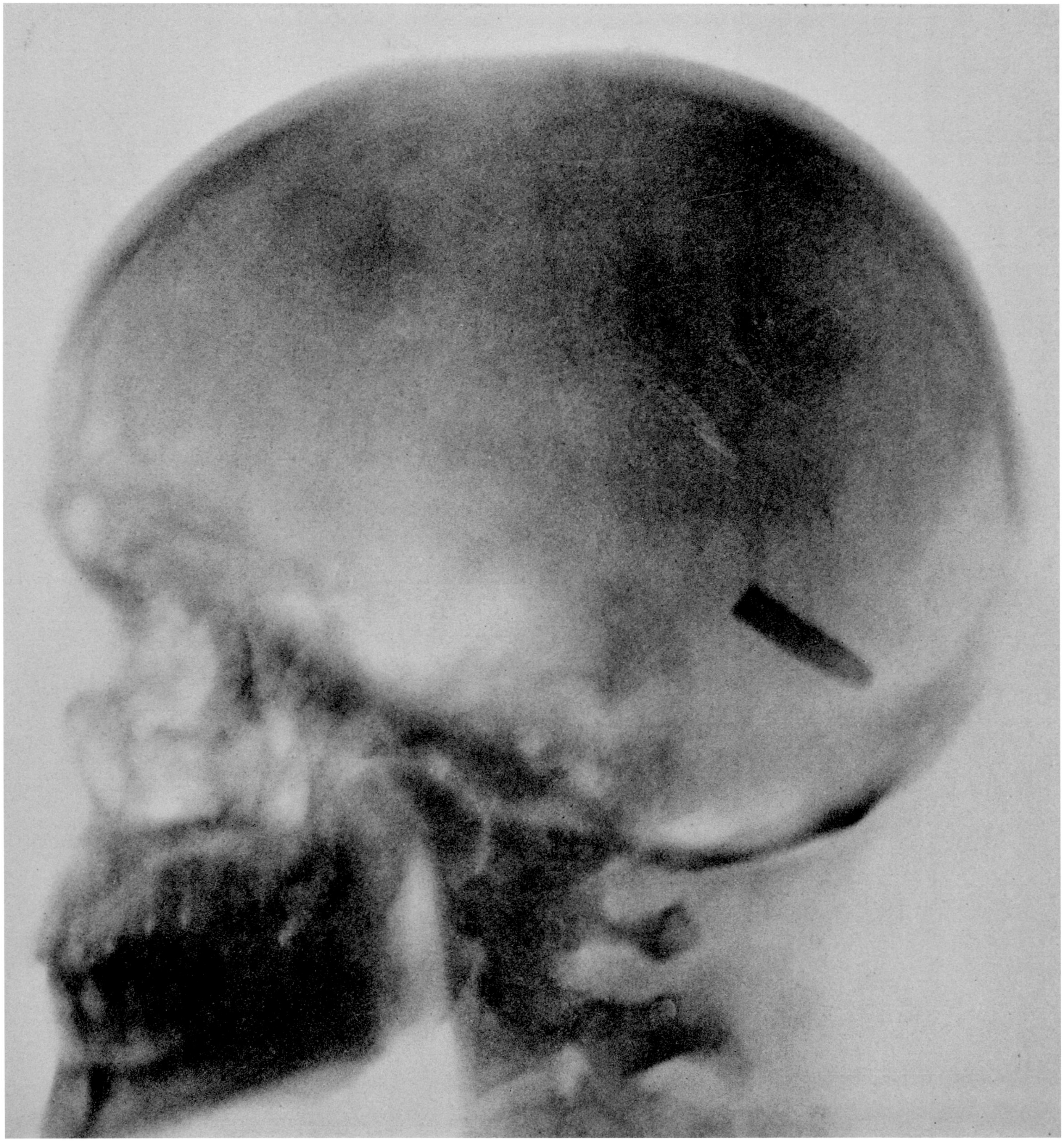

Plate 8.6

Spanish American War Soldier, Pvt. John Gretzer, Jr. with a Bullet in his Brain,
The Use of the Röntgen Ray by the Medical Dept. of the US Army in the War with Spain, Plate 11, 1898

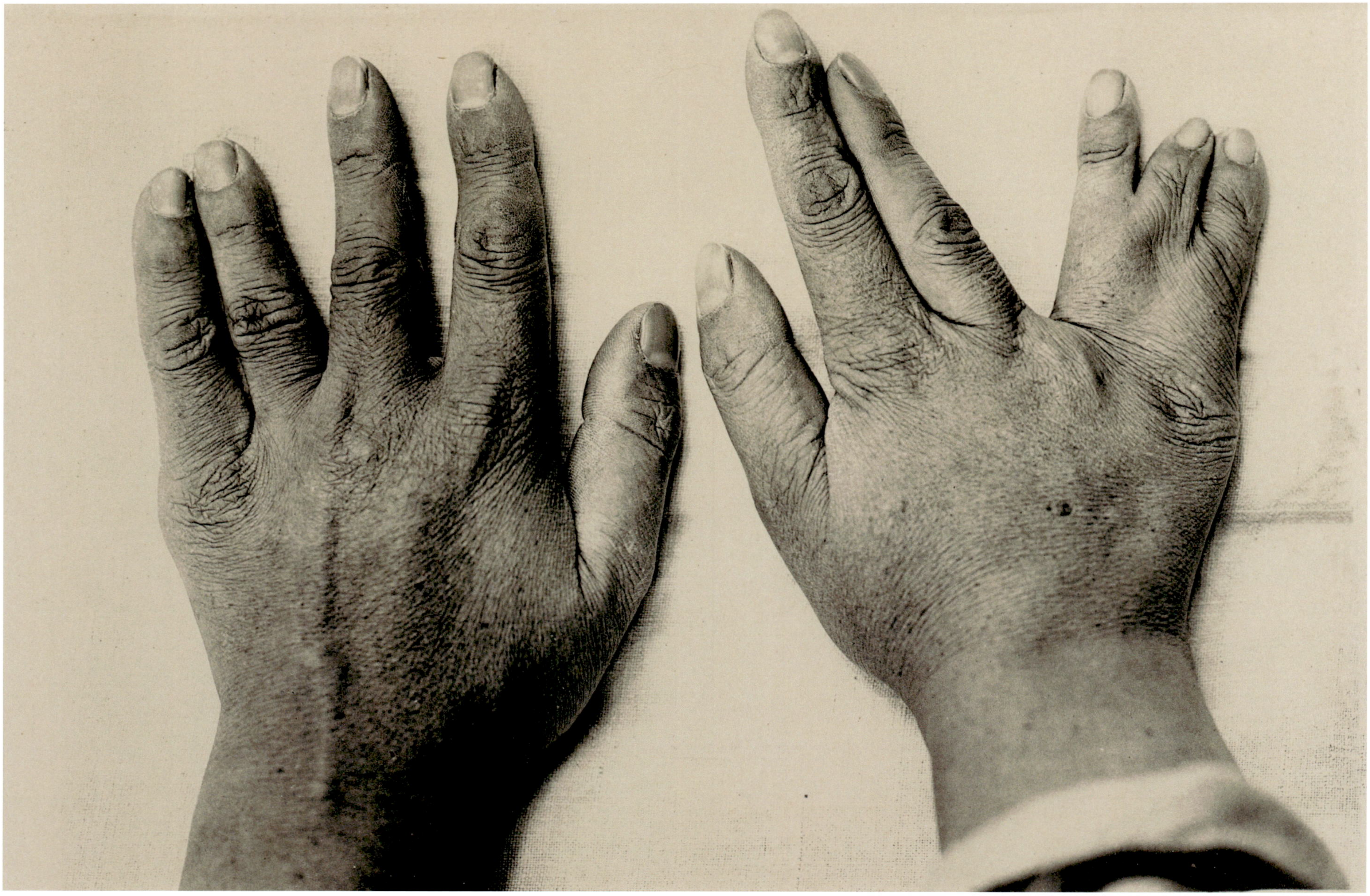

Plate 8.7

Marie M., 32 Years Old with Congenital Malformation of Both Hands
Archives des Laboratoires des hôpitaux d'Alger, Iconographie Médicale Algeriene, Volume 3, 1907

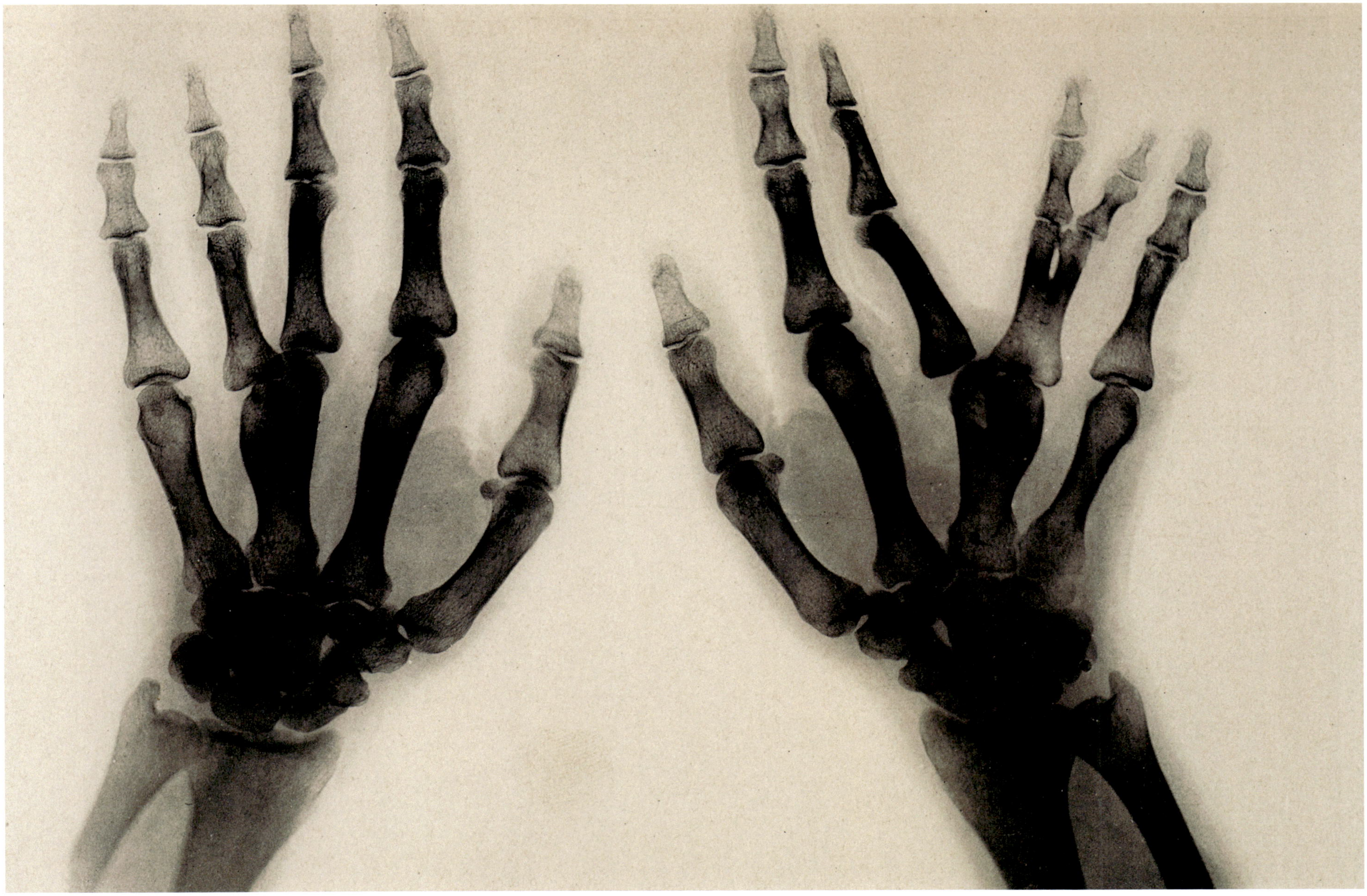

Plate 8.8

X-Ray of the Hands of Marie, M. With Fused Metacarpal Bones (She has four instead of five)
Archives des Laboratoires des hôpitaux d'Alger, Iconographie Médicale Algeriene, Volume 3, 1907

The X-ray allowed physicians to see exact skeletal deformities before the death of the patient.

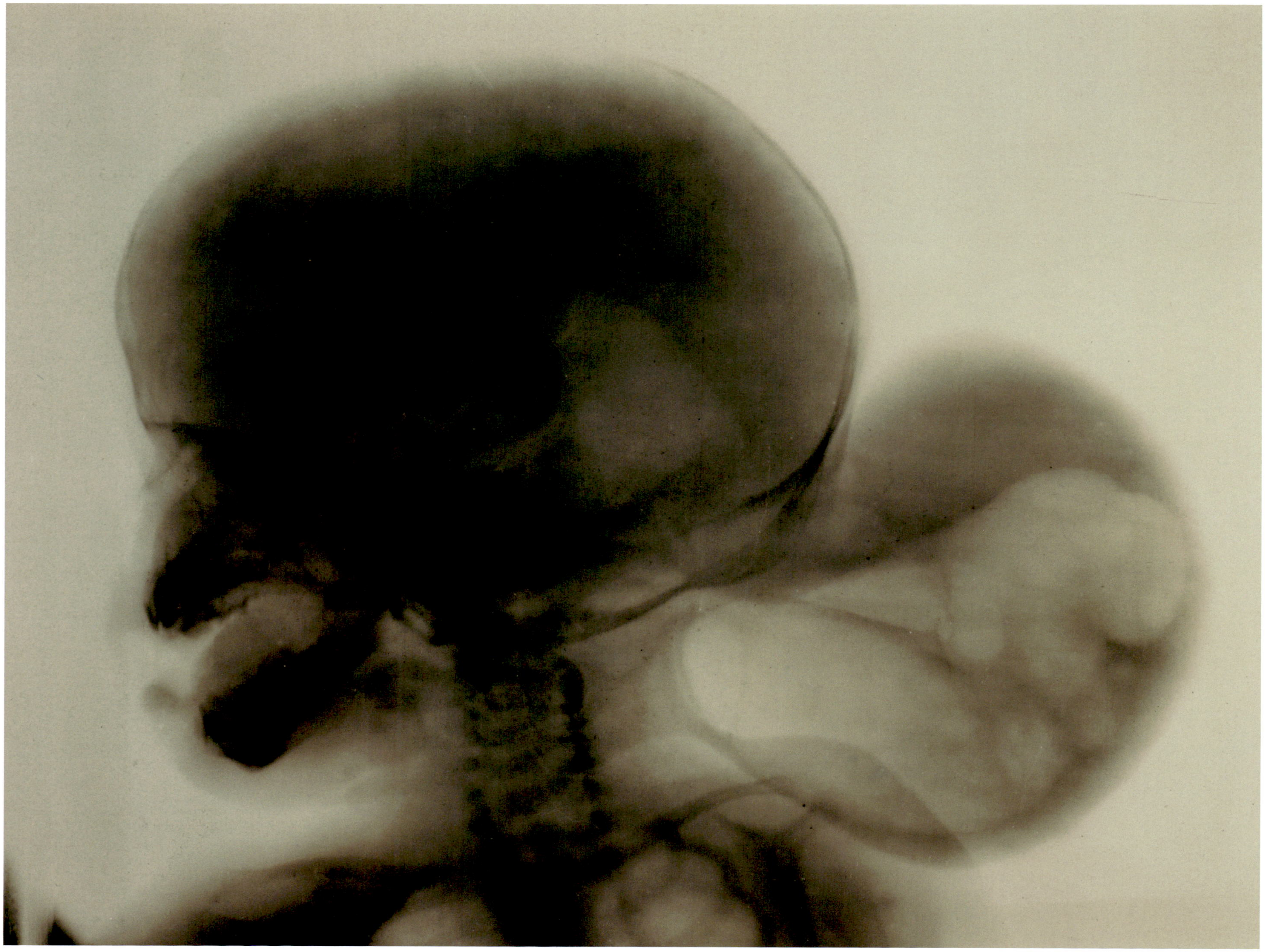

Plate 8.9

Spina Bifida in the Cervical Region
1928

From a series of X-rays taken at Johns Hopkins Hospital.

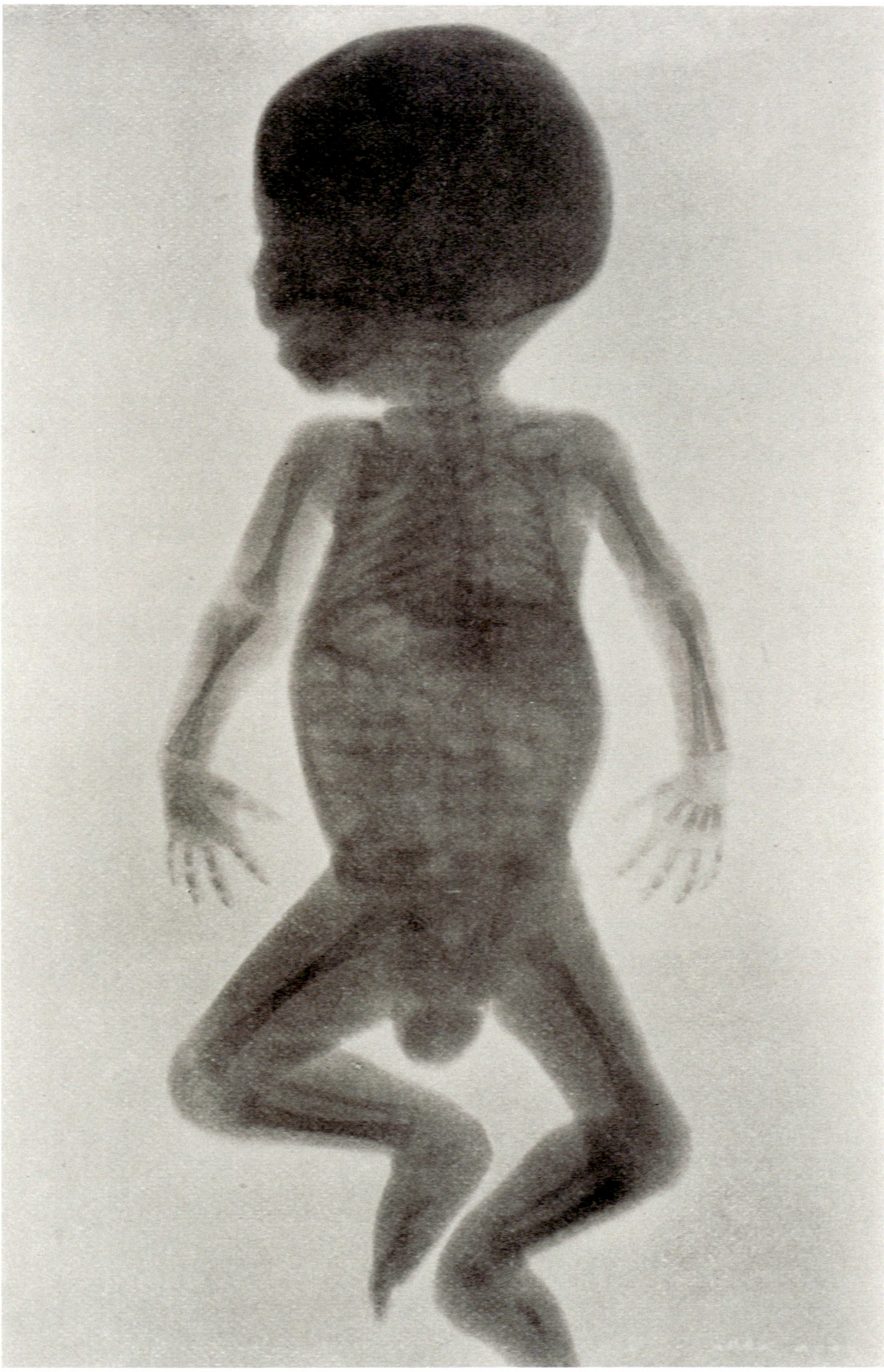

Plate 8.10

Life Size X-Ray of Nine-Week-Old Infant
1896

William Morton, MD, sold copies of this X-ray for $2.00, as reference works for other physicians.

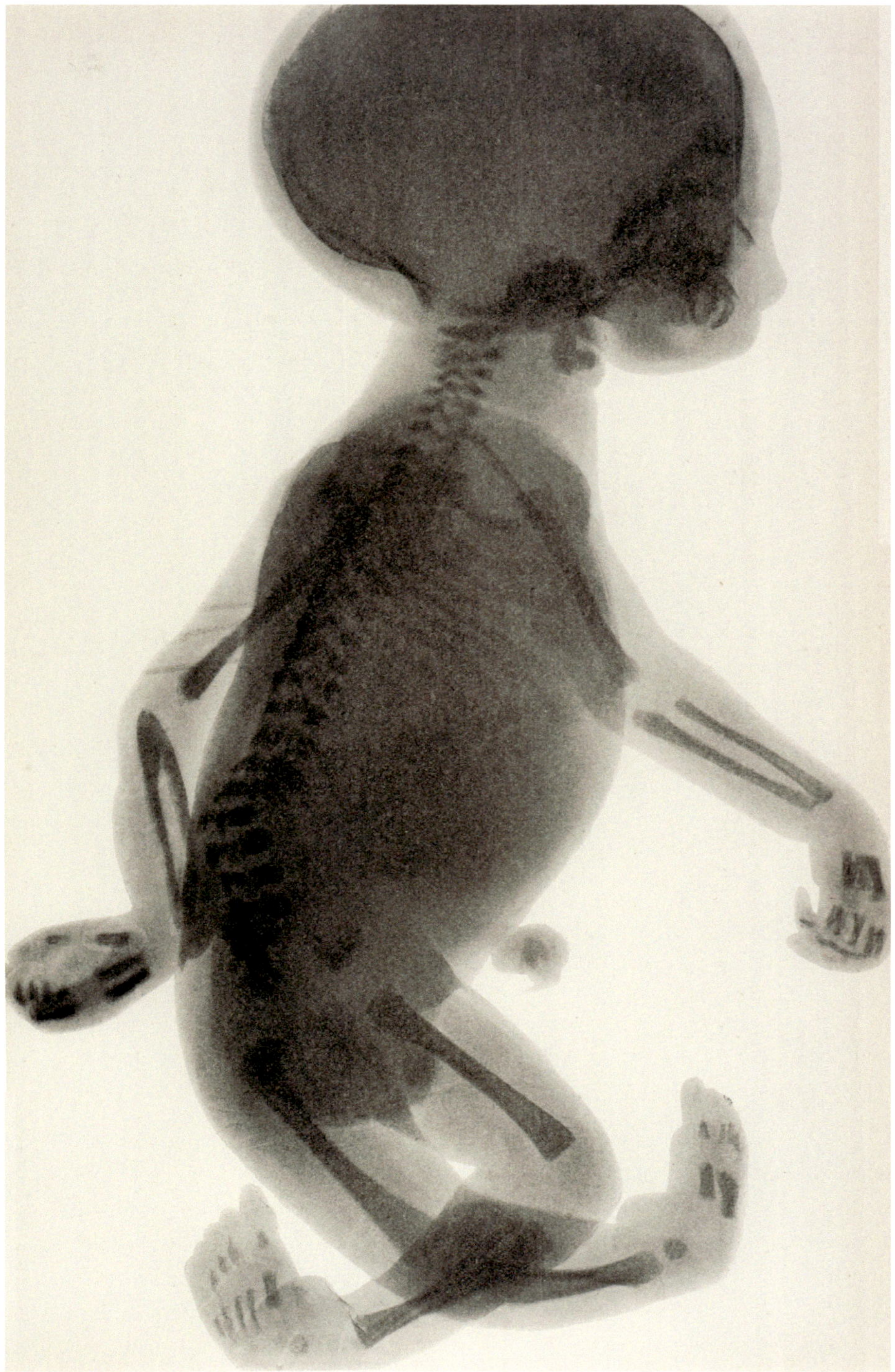

Plate 8.11

Case of Agnathia, Congenital Absence of the Jaw
1898

Patient of Bertram Rogers, MD, Bristol England, Published in *Journal of Pathology*, Volume V

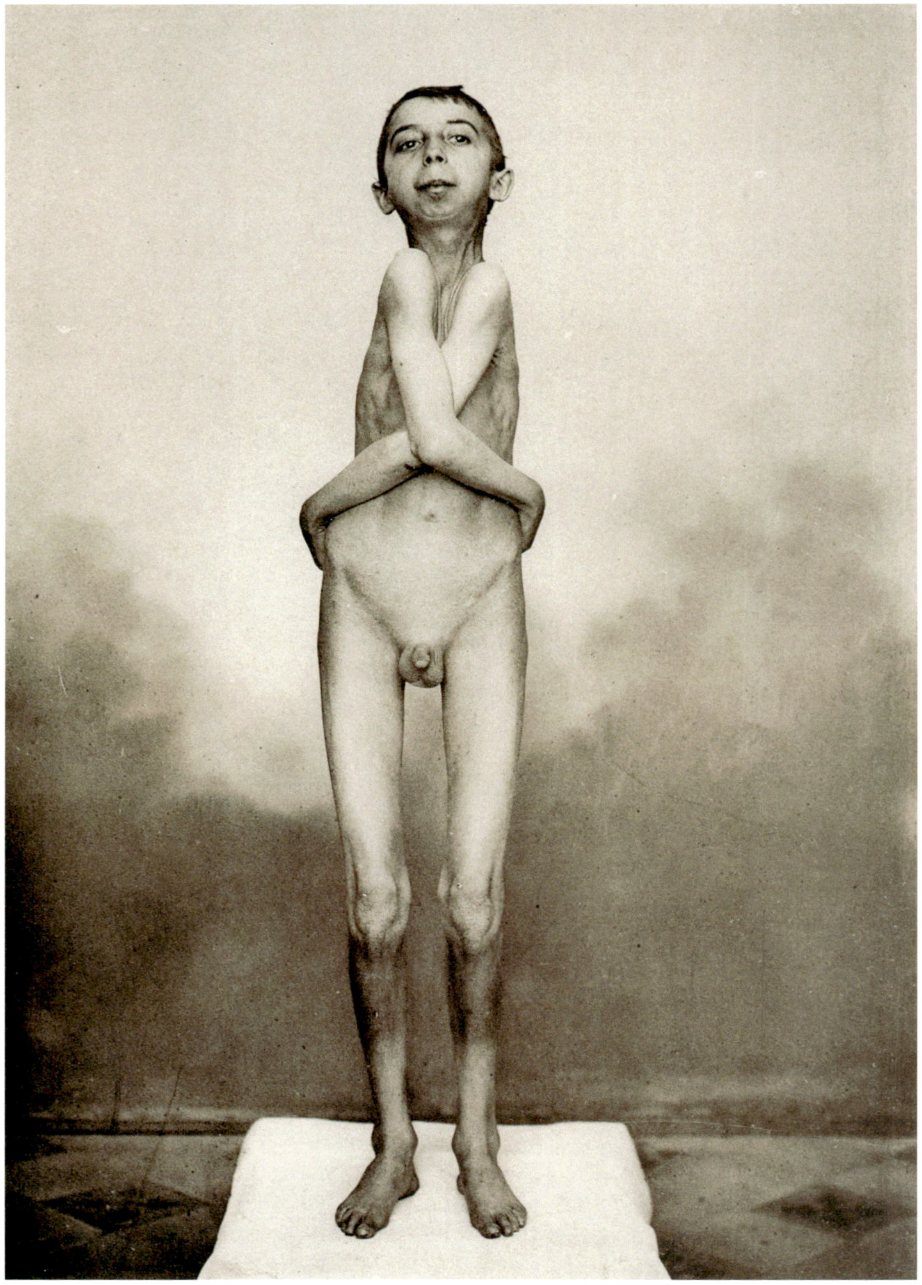

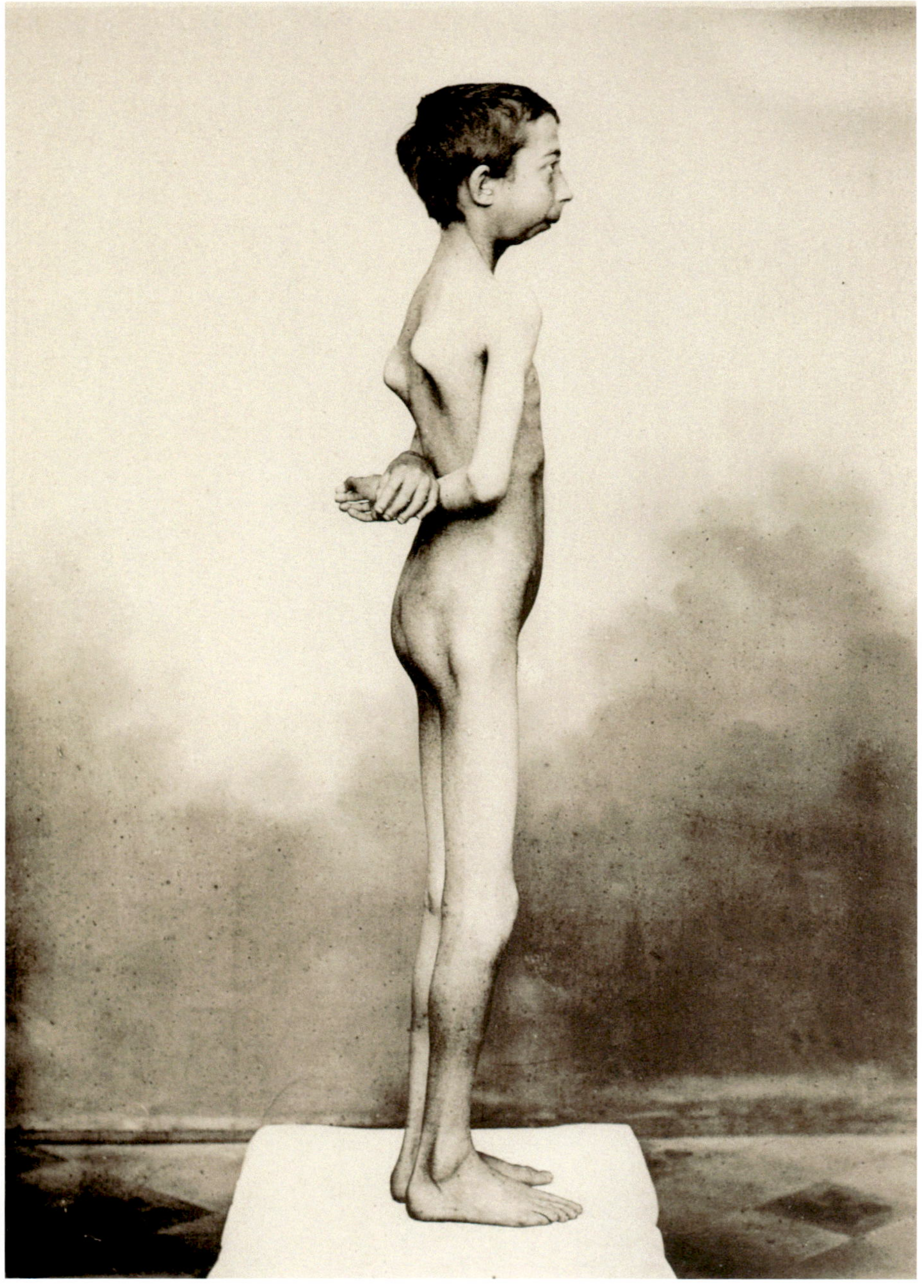

Plates 8.12 – 8.13

Fifteen Year-Old Boy with Cleido-Cranial Dystrophy Demonstrating Absence of Both Clavicles
Archives des Laboratoires des hôpitaux d'Alger, Iconographie Médicale Algeriene, Volume 3, 1907

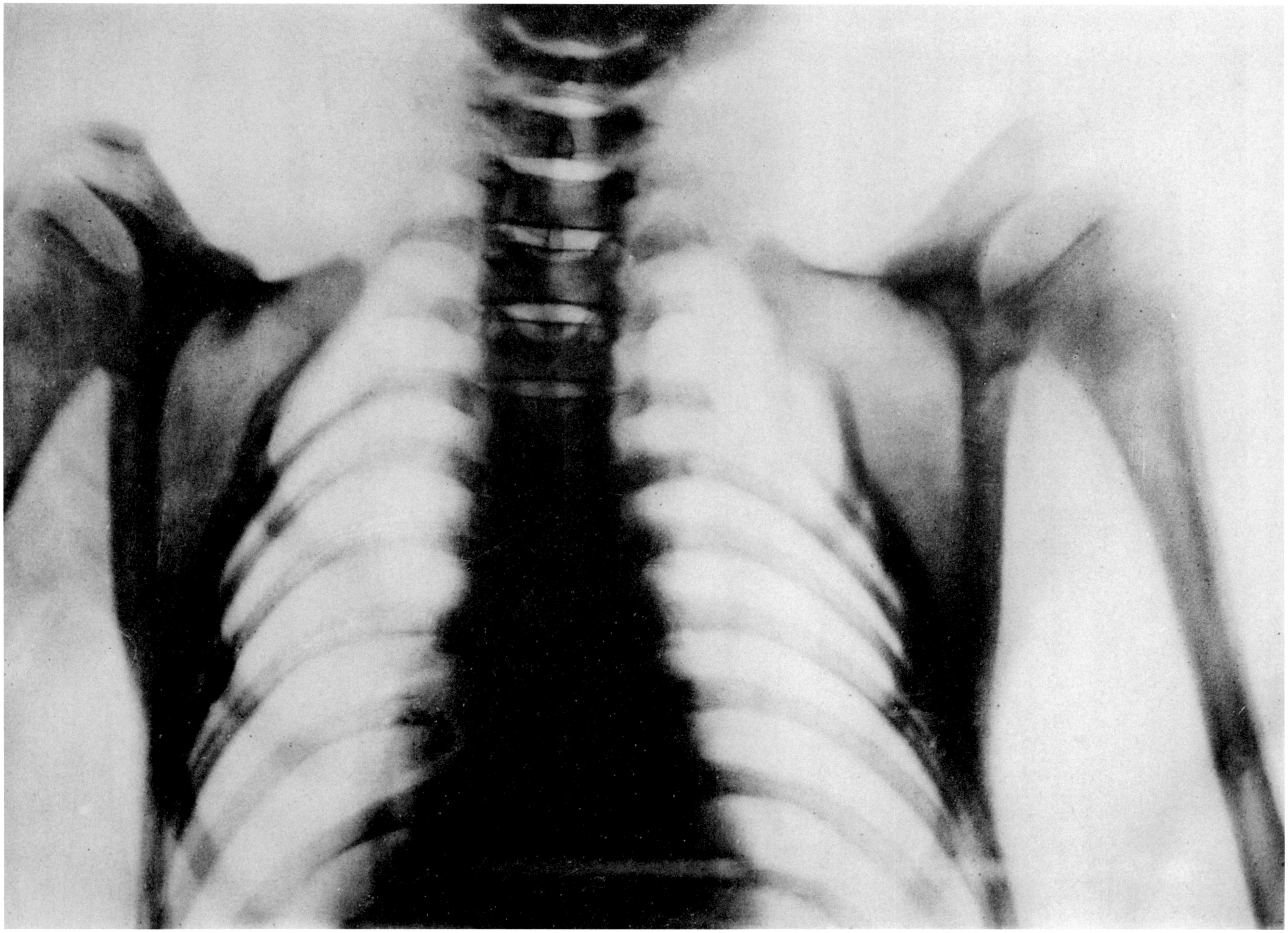

Plate 8.14

Chest X-Ray of 15-Year-Old Boy Showing Absence of Clavicles, Cleido-Cranial Dystrophy
Archives des Laboratoires des hôpitaux d'Alger, Iconographie Médicale Algeriene, Volume 3, 1907

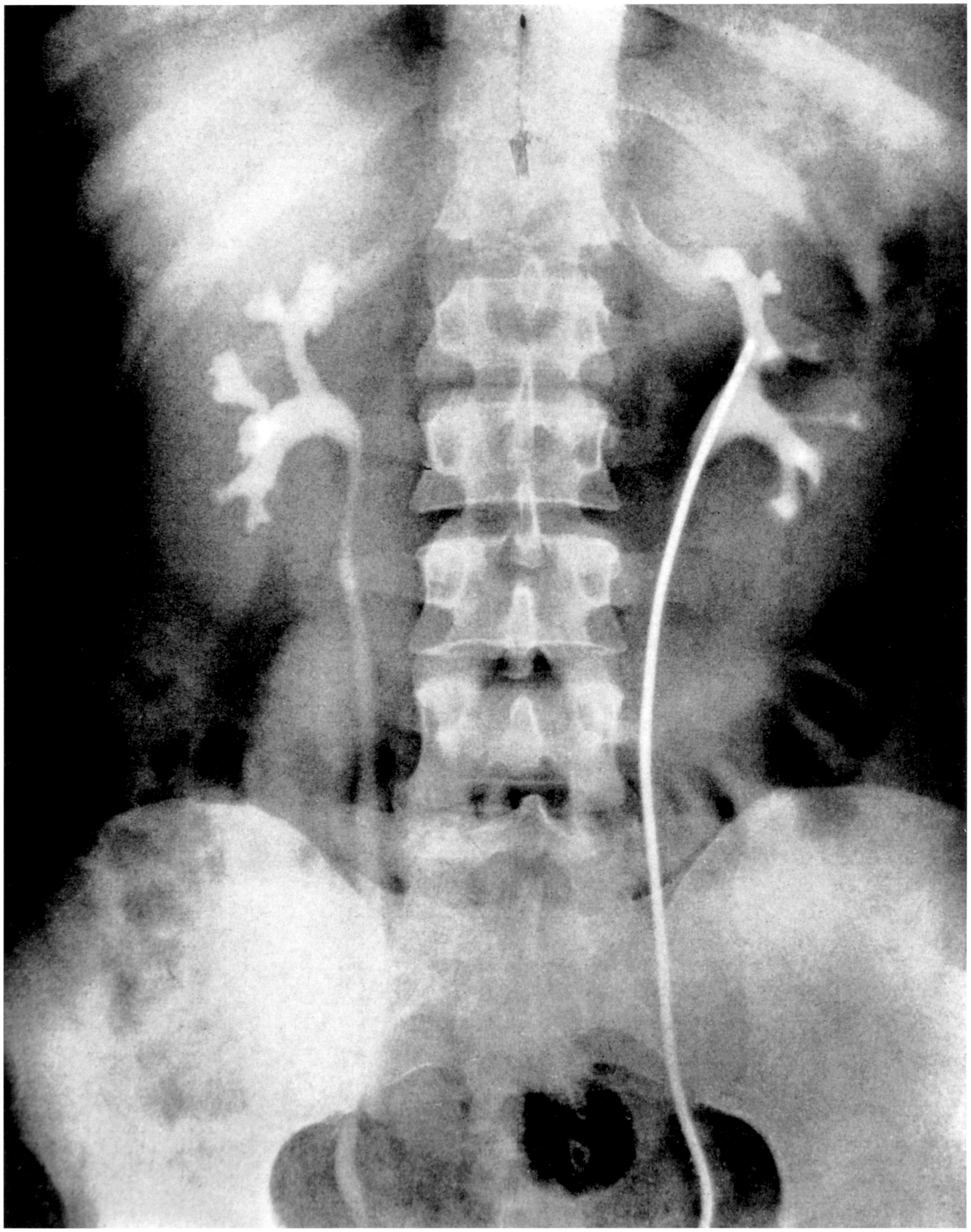

Plate 8.15

Normal Pyelogram, Catheter in Place, Outlining the Kidneys and Urinary Tract System
1920

The injection of contrast material in various organs offered specific diagnosis.

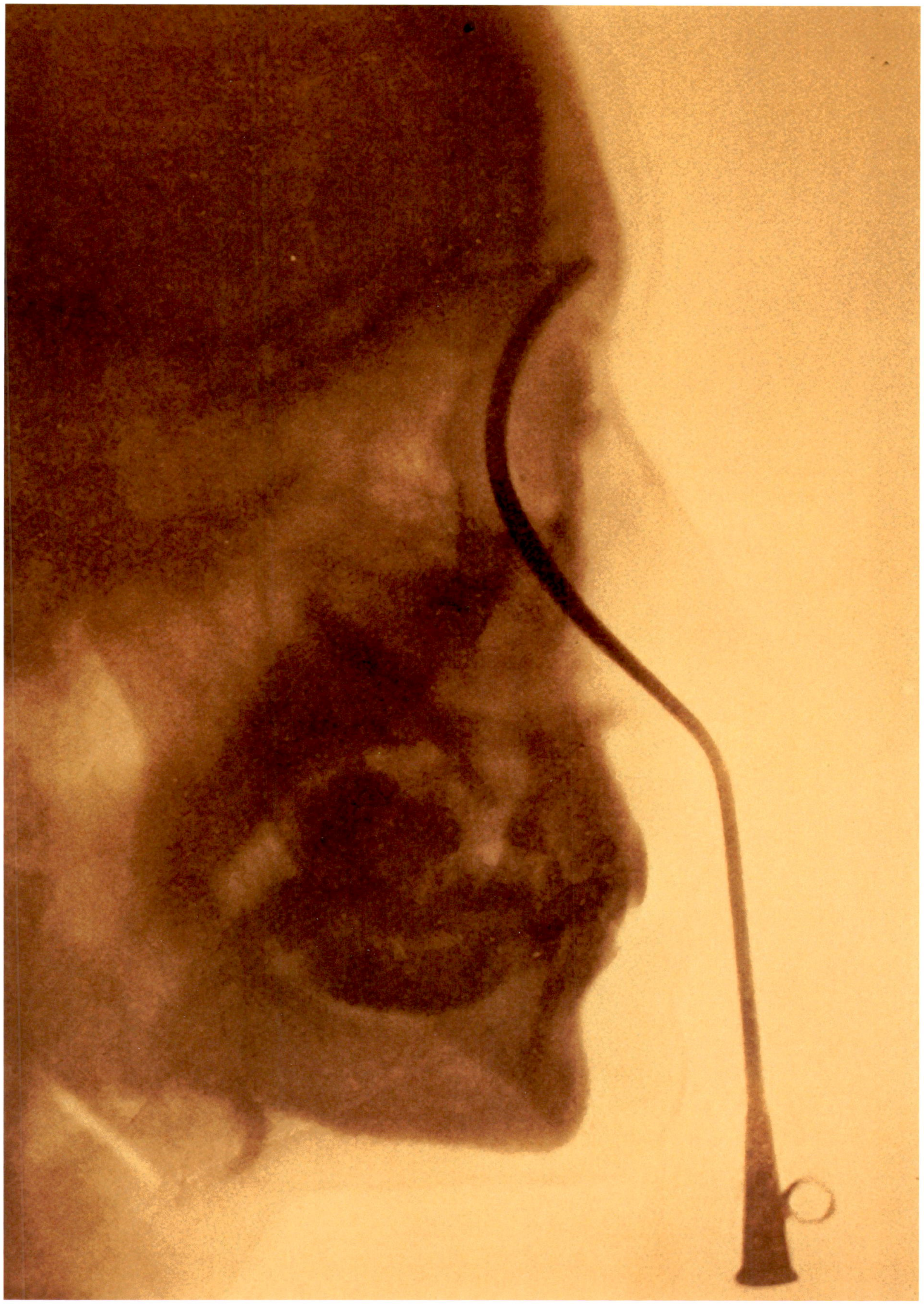

Plate 8.16

Metallic Probe in Sinus
1925

The use of catheters placed in various parts of the body allowed injection of contrast media.

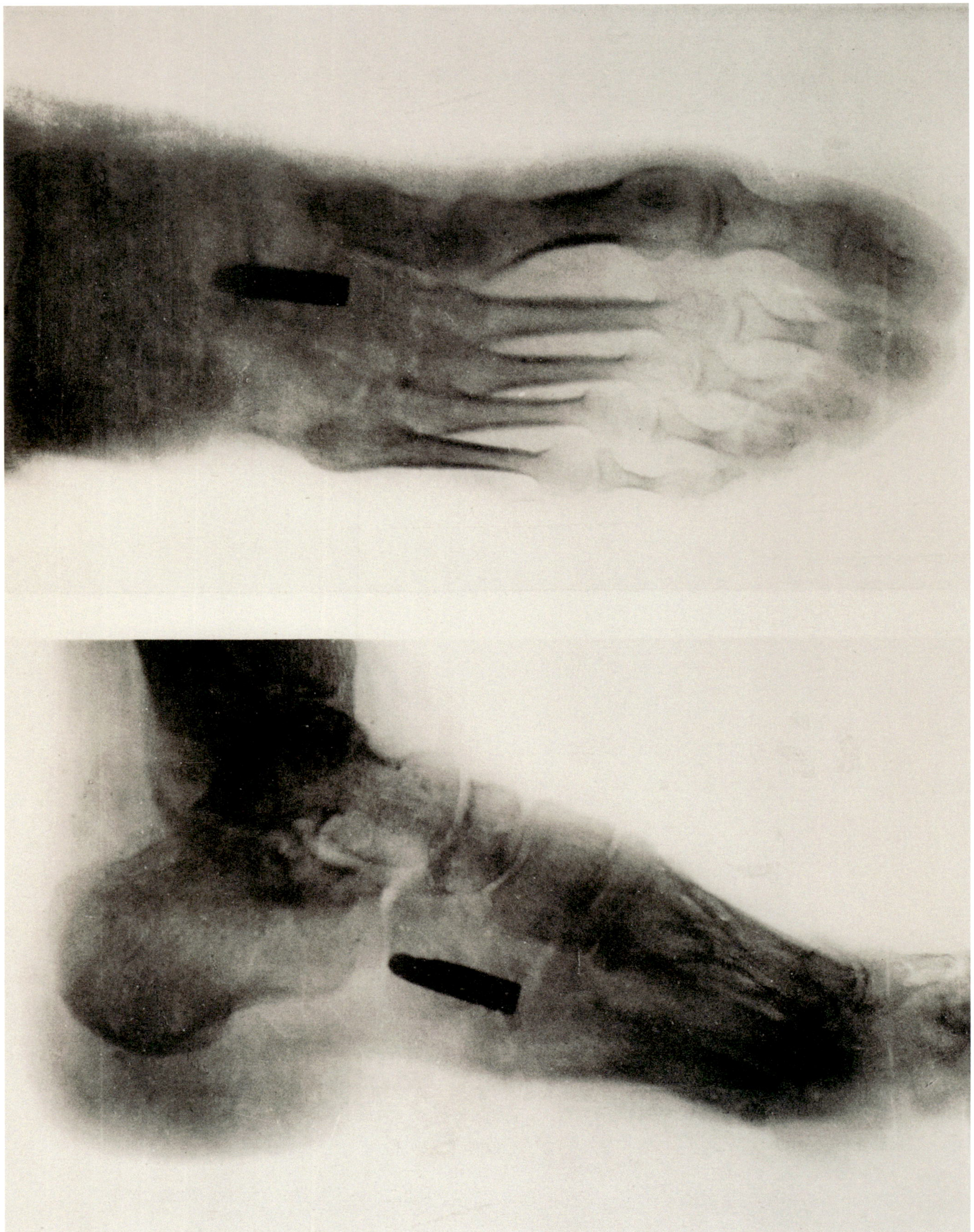

Plate 8.17

Bullet in the Foot, Pvt. John Taylor, Two Views Provide Exact Localization for Surgeons
1898

The Use of the Röntgen Ray by the Medical Dept. of the US Army in the War with Spain, Plate13

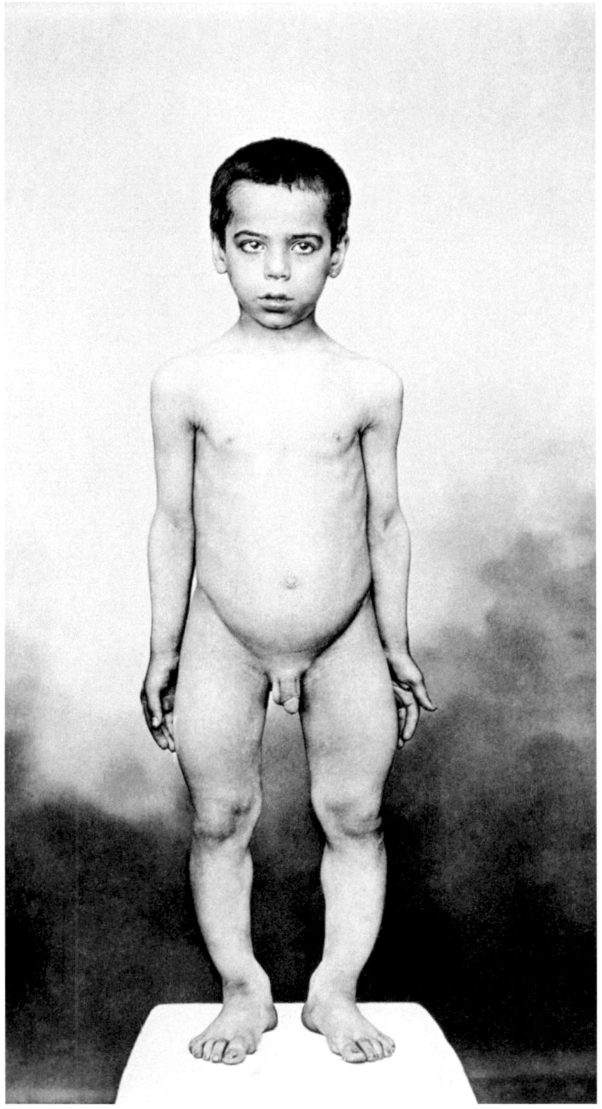

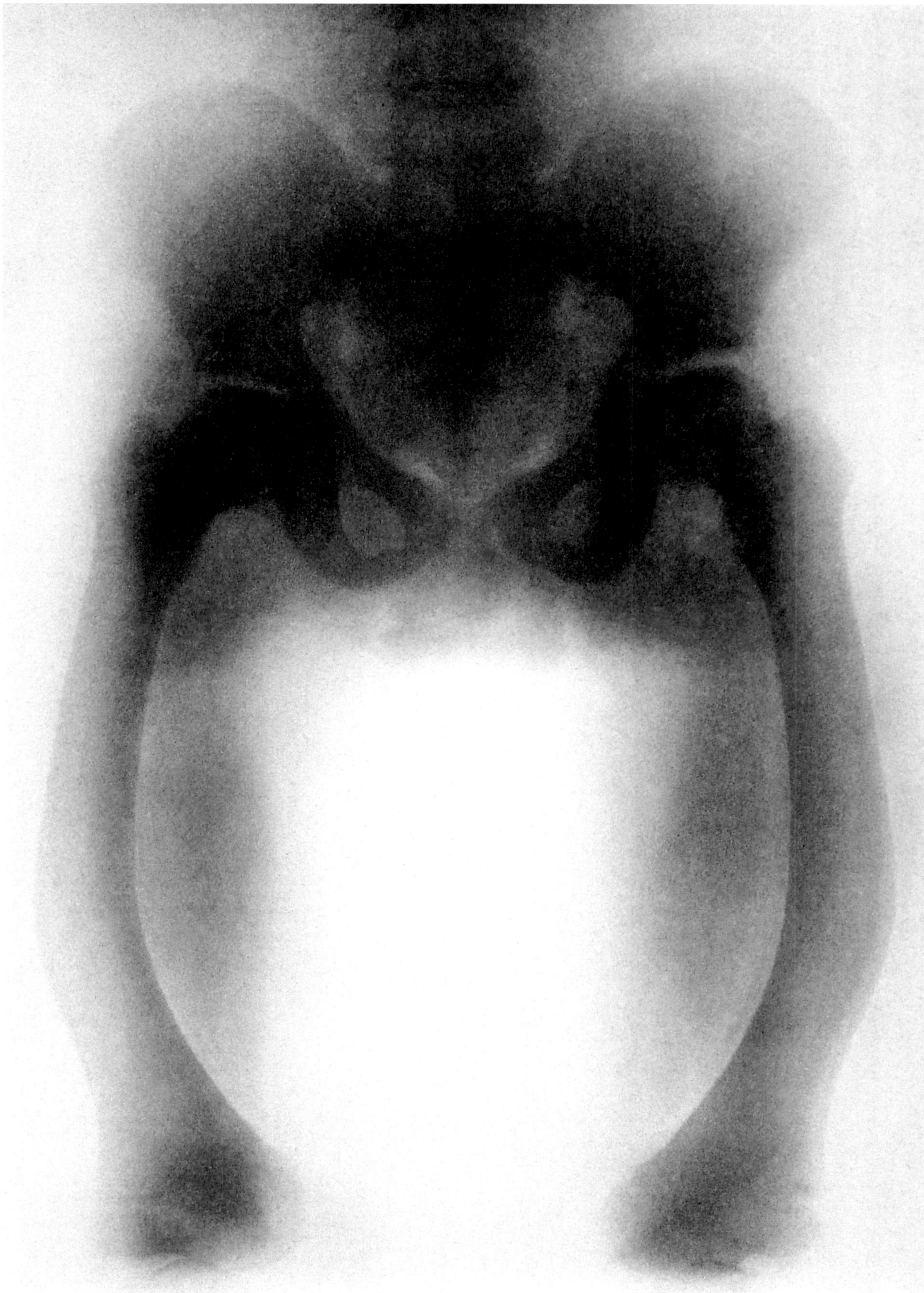

Plate 8.18

Young Boy With Rickets
Archives des Laboratoires des hôpitaux d'Alger, Iconographie Médicale Algeriene, Volume 2, 1906

Plate 8.19

X-Ray Showing the Classic Bowing of the Femur in Rickets
Archives des Laboratoires des hôpitaux d'Alger, Iconographie Médicale Algeriene, Volume 2, 1906

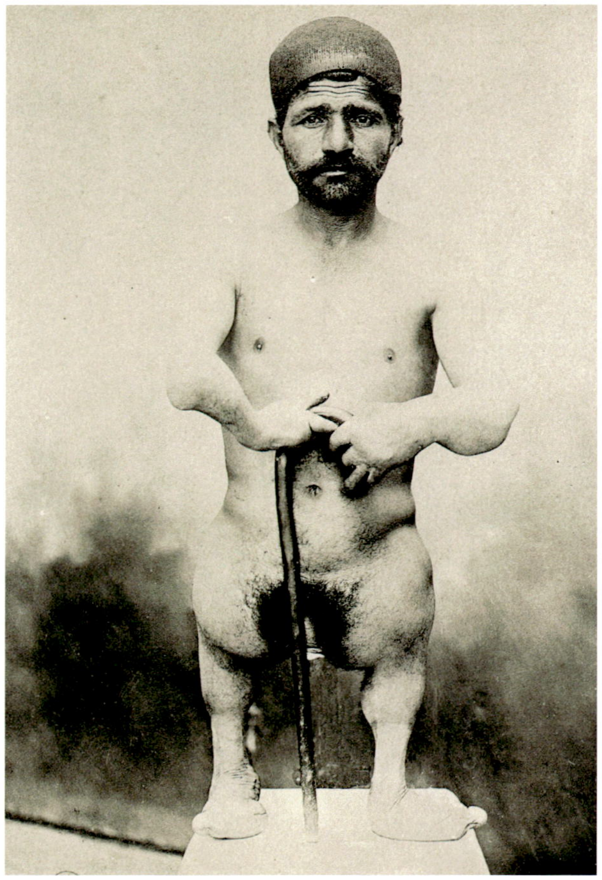

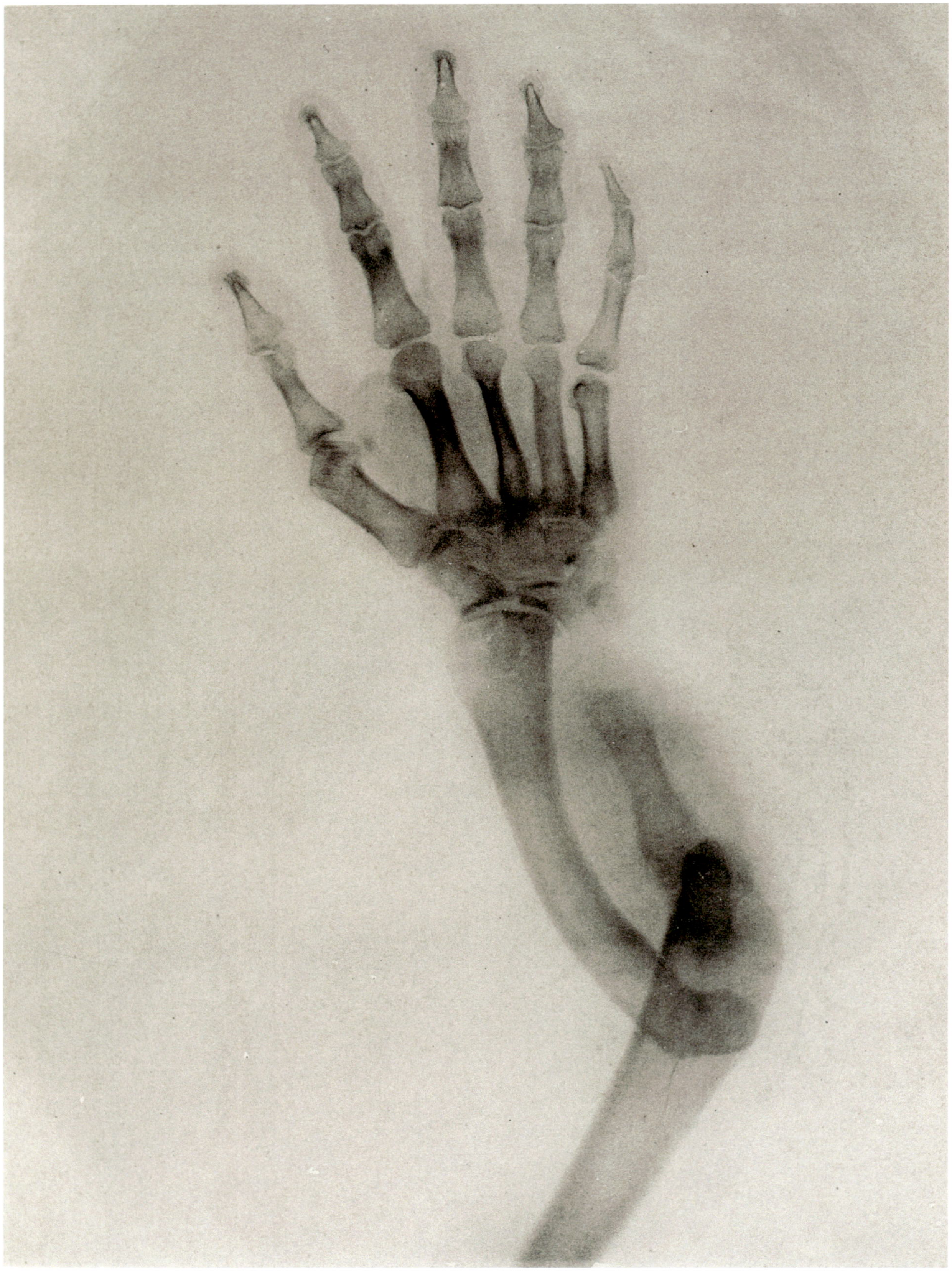

Plate 8.20

"Hemi Man," One of Four Members of a Family with Multiple Congenital Deformities
Archives des Laboratoires des hôpitaux d'Alger, Iconographie Médicale Algeriene, Volume 3, 1907

Plate 8.21

X-Ray of Patient's Left Arm with Major Malformations and Positions
Archives des Laboratoires des hôpitaux d'Alger, Iconographie Médicale Algeriene, Volume 3, 1907

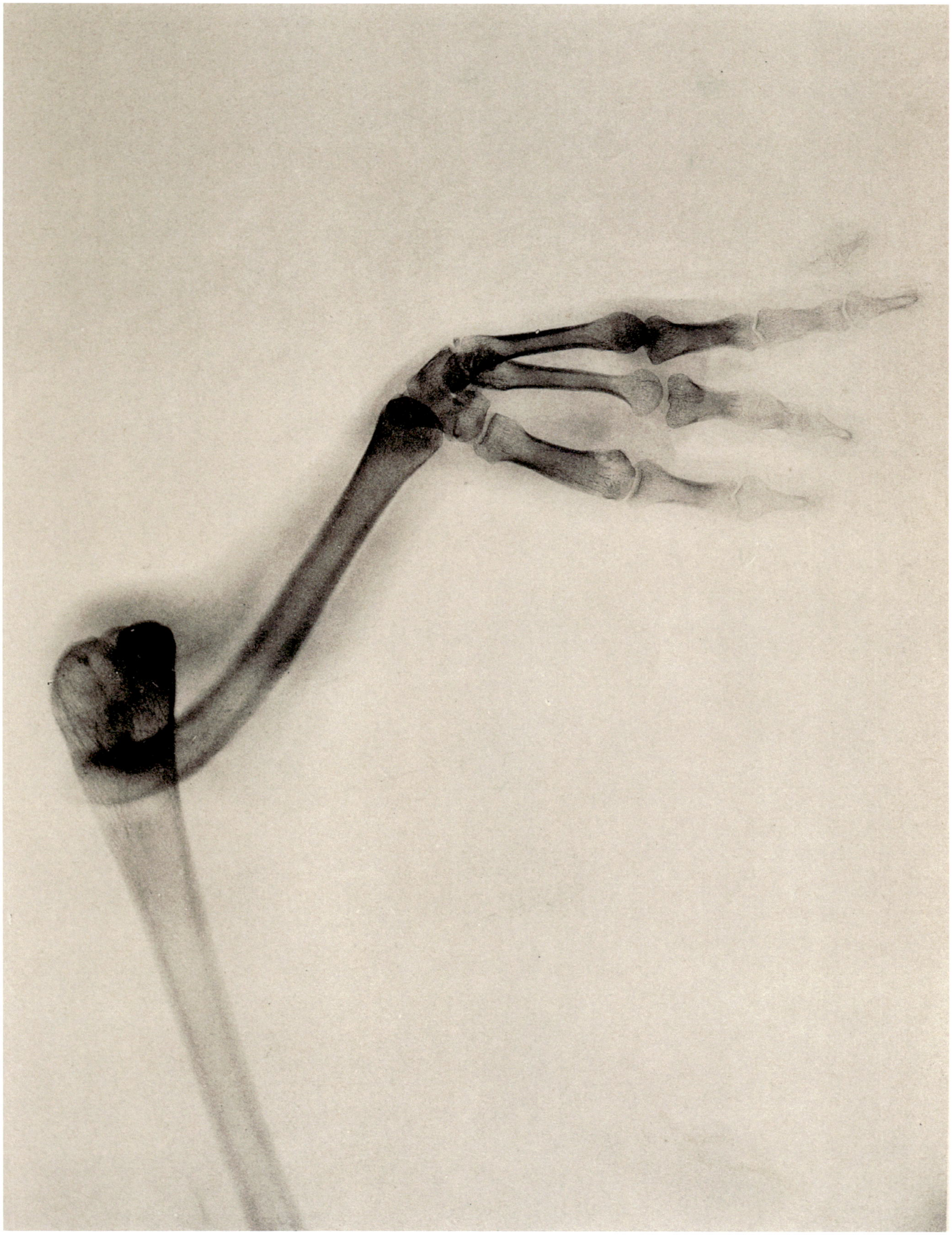

Plate 8.22

X-Ray of Patient's Right Arm with Malposition and Elbow Displacement

Archives des Laboratoires des hôpitaux d'Alger, Iconographie Médicale Algeriene, Volume 3, 1907

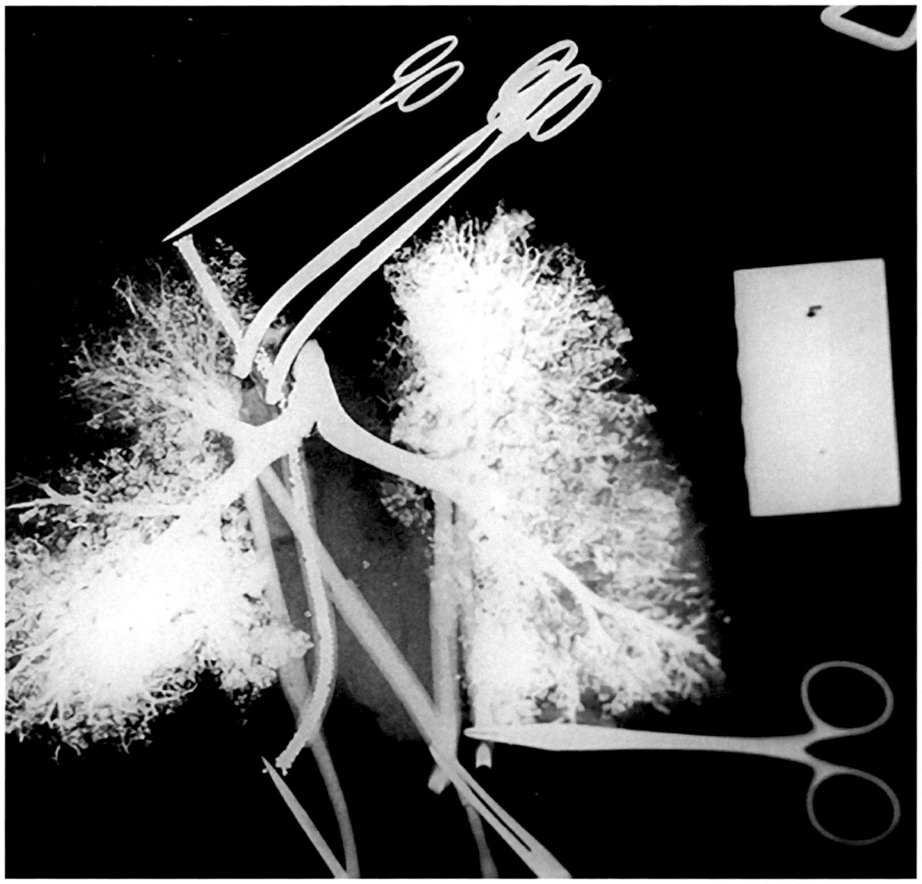

Plate 8.23

X-Ray of Cadaver Lung Injected with Bismuth

Kennon Dunham, MD, *Stereoroentgenography: Pulmonary Tuberculosis*, 1915

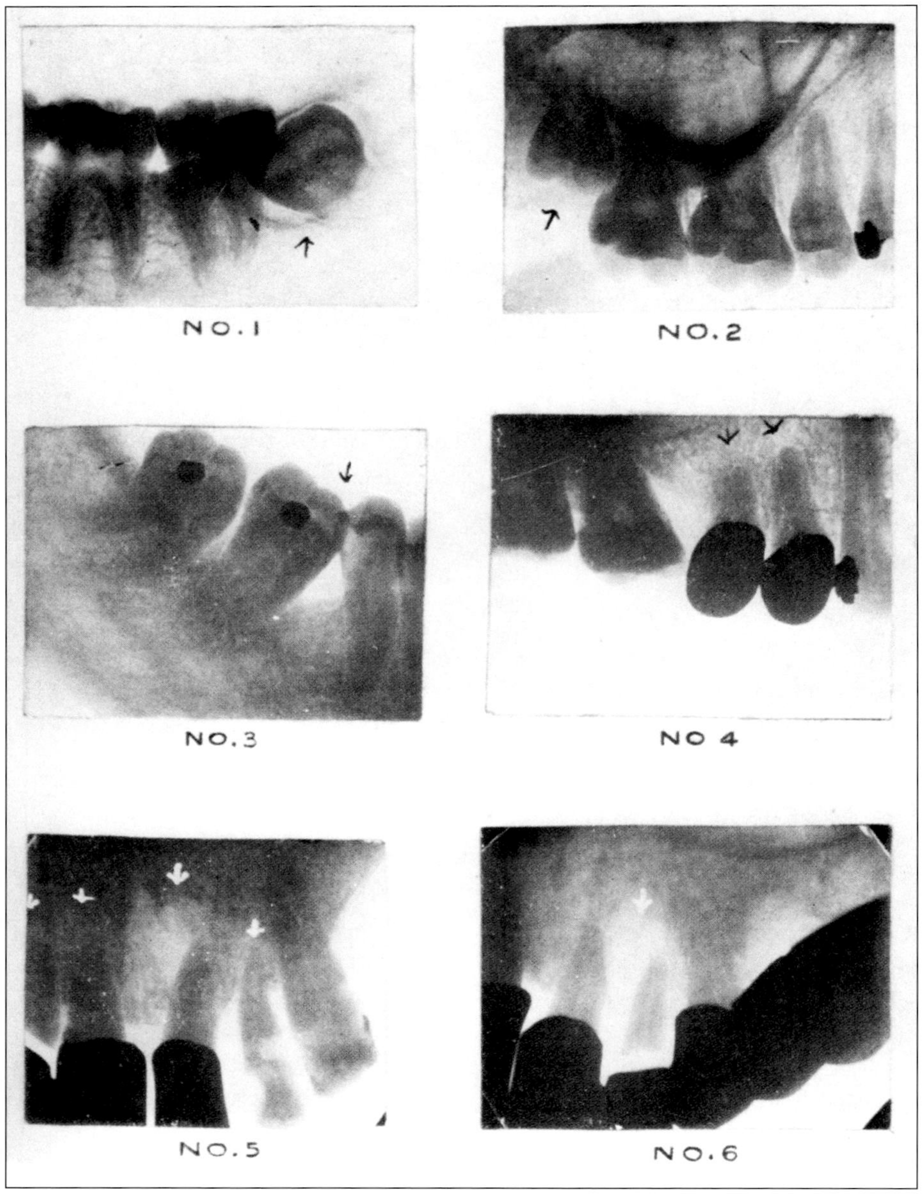

Plate 8.24

X-Ray of Teeth "Serious Mental Disturbances Caused by Painless Dental Lesions"

Dr. Henry Upson's concept of focal infection, *American Quarterly of Roentgenology*, December 1910

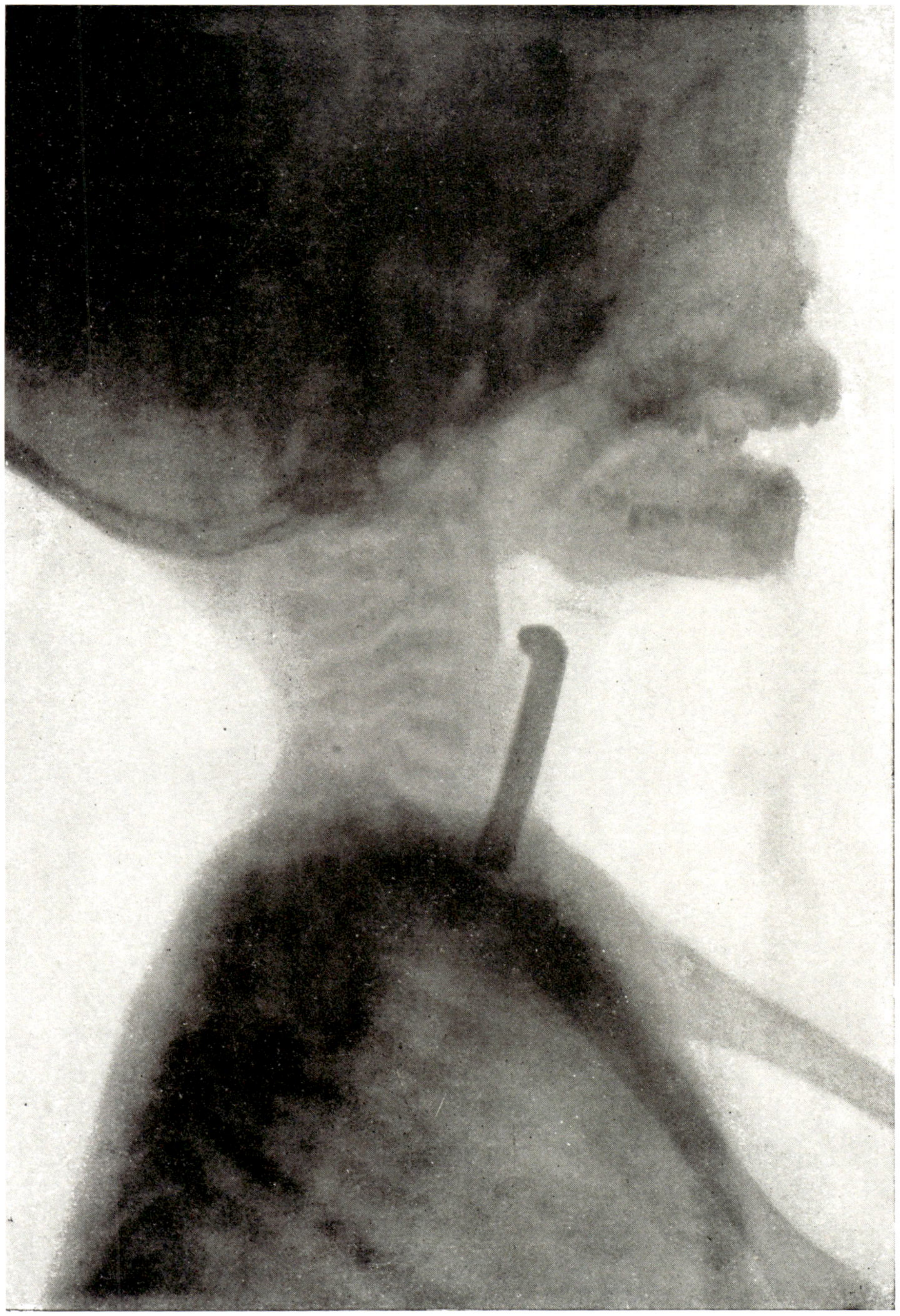

Plate 8.25

Intubation Tube in Larynx in 10-Month-Old Child with Diphtheria, Side View
International Medical Annual, 1898, Plate 1 (The use of antitoxin lessened the time of intubation.)

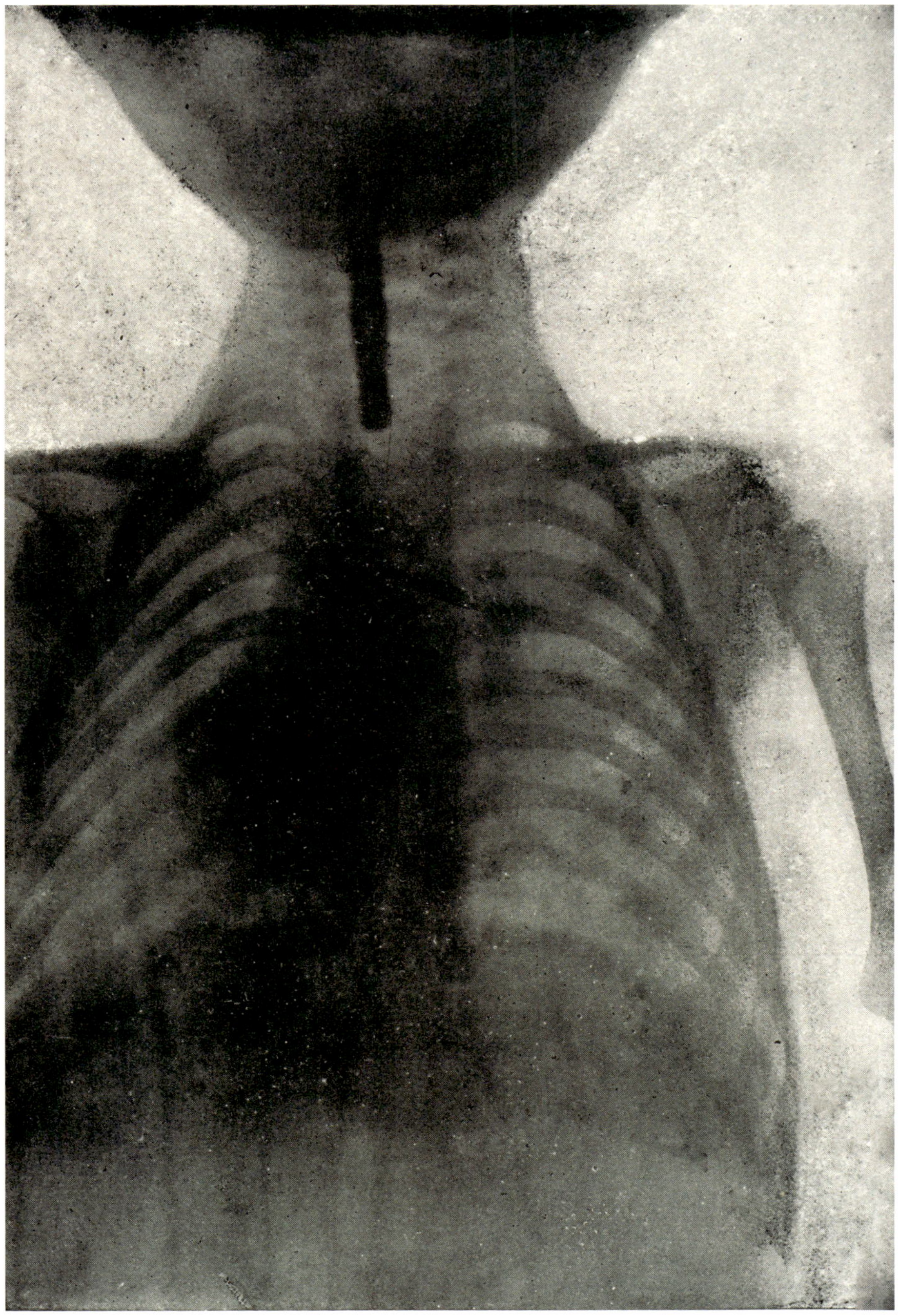

Plate 8.26

X-Ray of Intubation Tube in Place in a Case of Diphtheria, Anterior-Posterior View
International Medical Annual, 1898, Plate 2 (Antitoxin the first specific modern therapy developed 1890.)

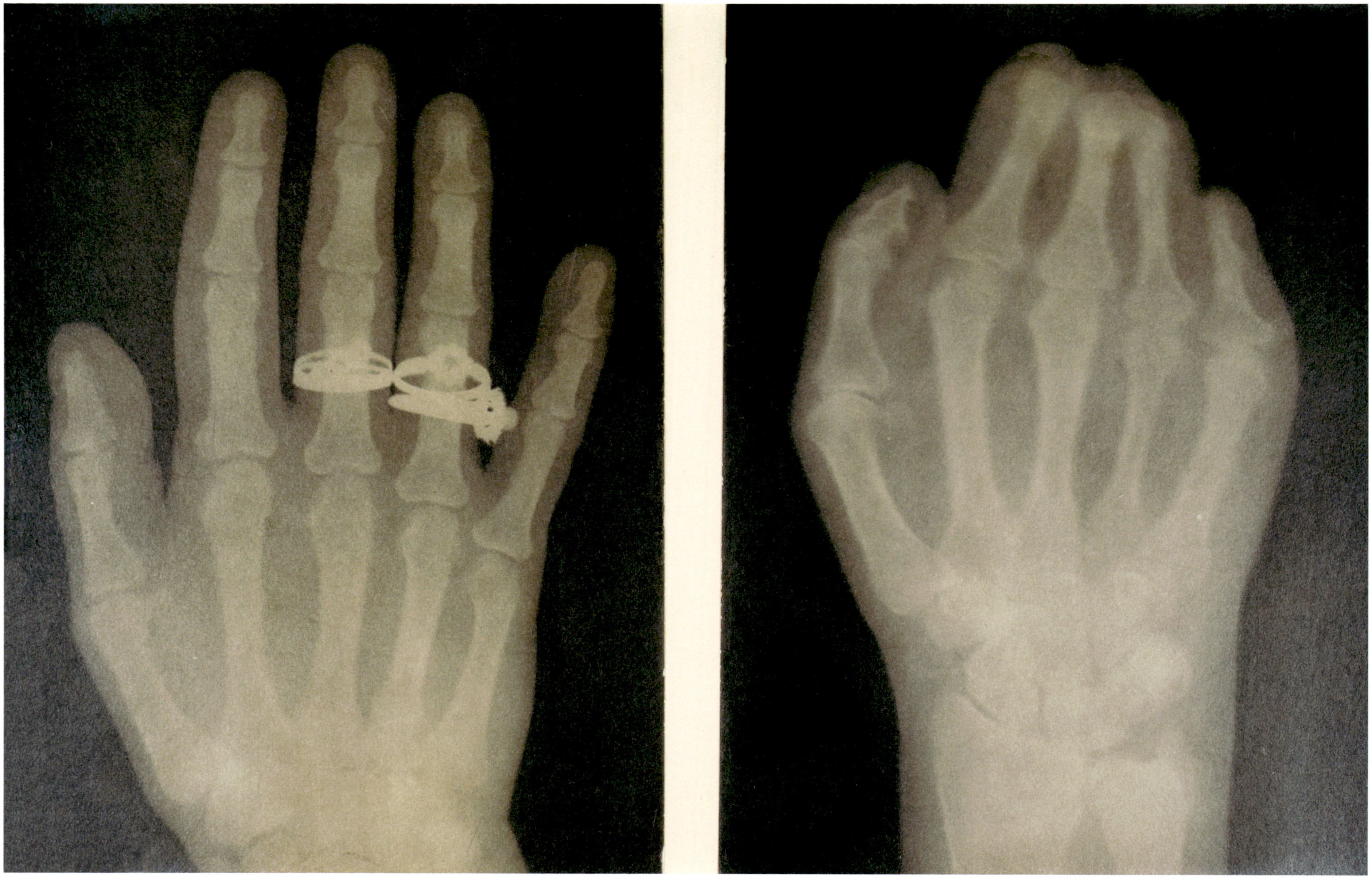

Plate 8.27

Early X-Ray Taken to Show a Normal and an Abnormal Hand with Destruction of Phalanges
Lyon, France, 1898

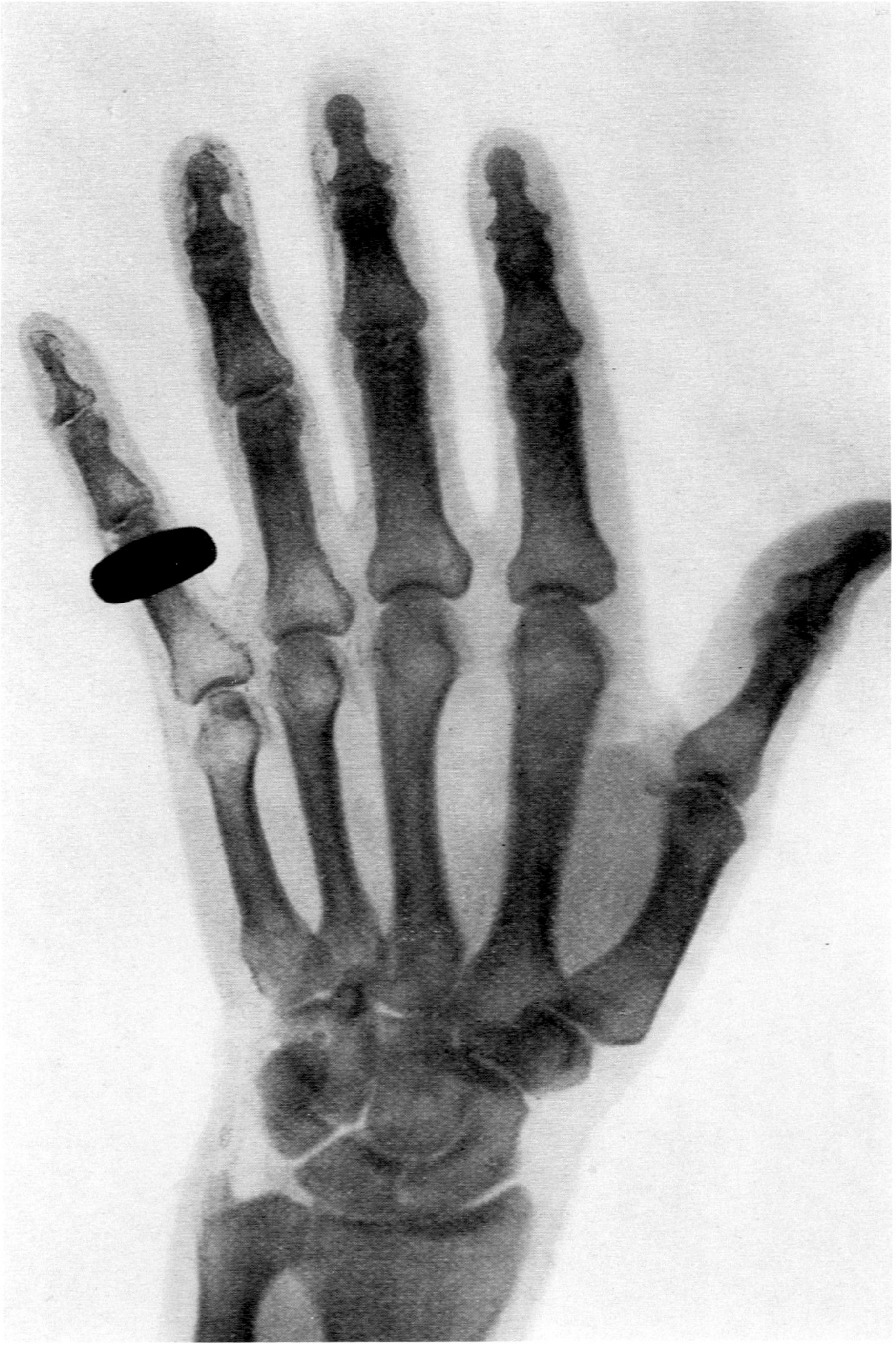

Plate 8.28

X-Ray of Hand With Ring in the Style of Röntgen
American, 1898

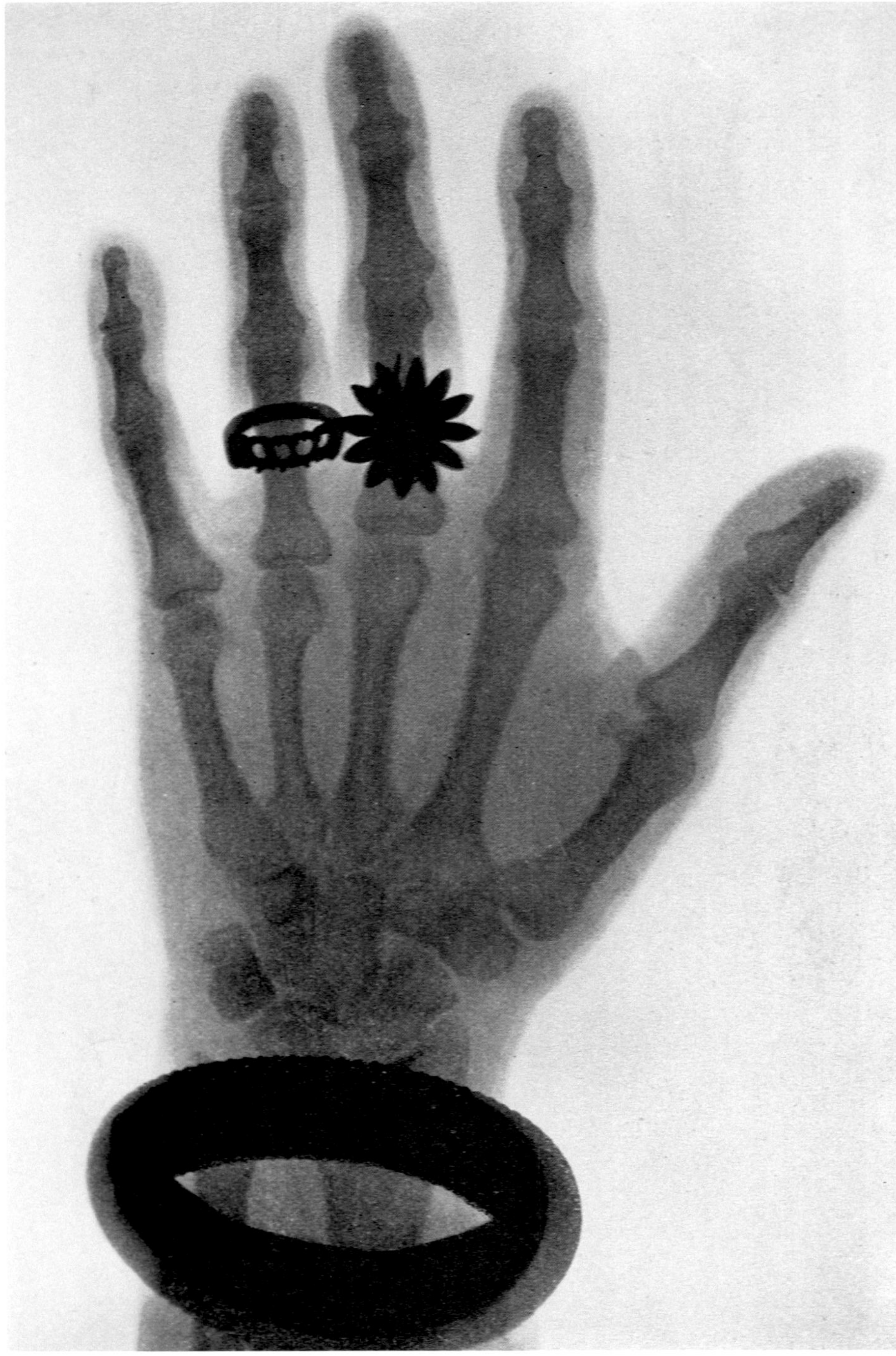

Plate 8.29

X-Ray of Hand With Ring and Bracelet in the Style of Röntgen
American, 1896

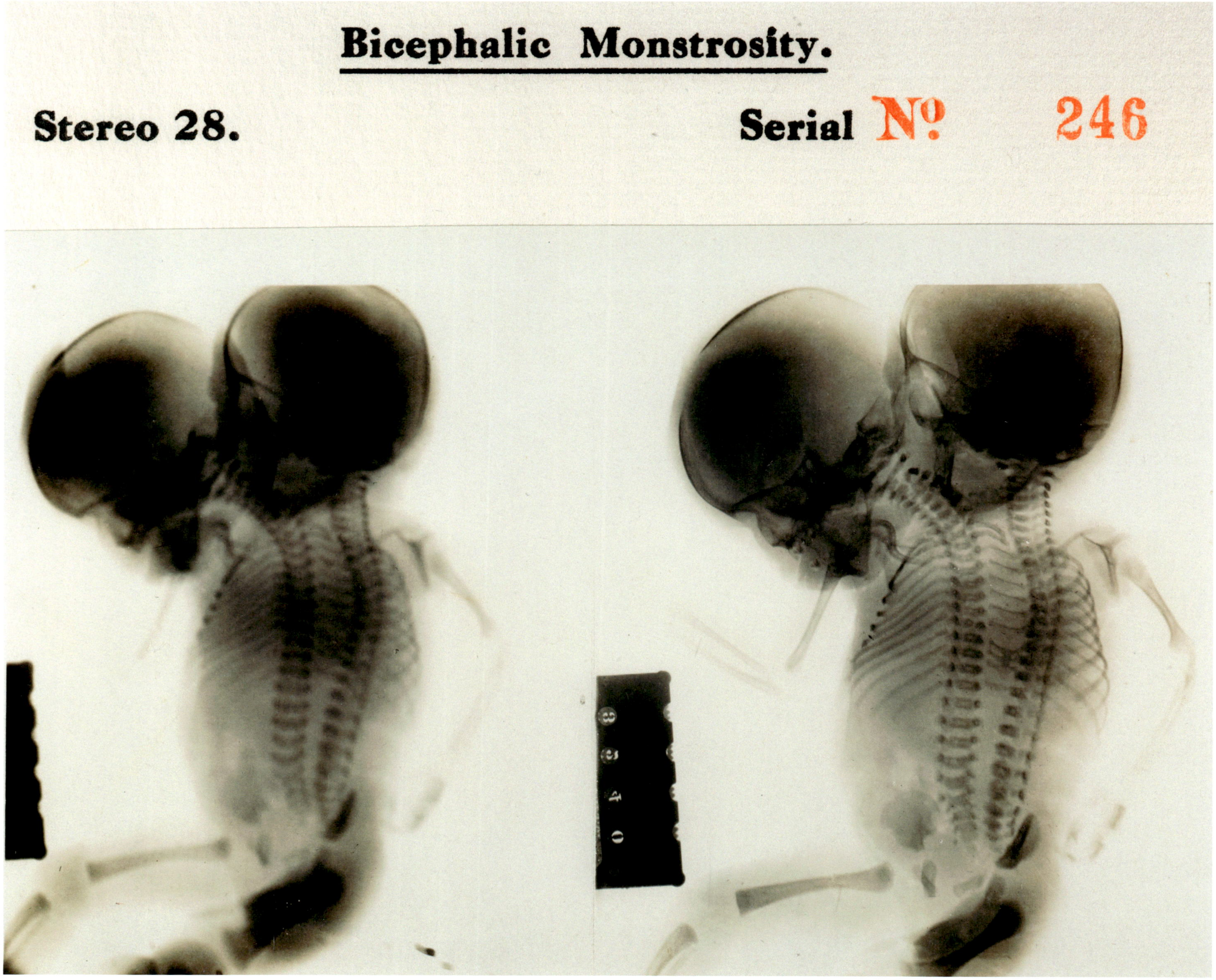

Plate 8.30

Bicephalic Monstrosity, Two Unconnected Spines with Identical Heads
A. Robert Taft, MD, *American Atlas of Stereoroentgenology*, Volume 1, December 1916

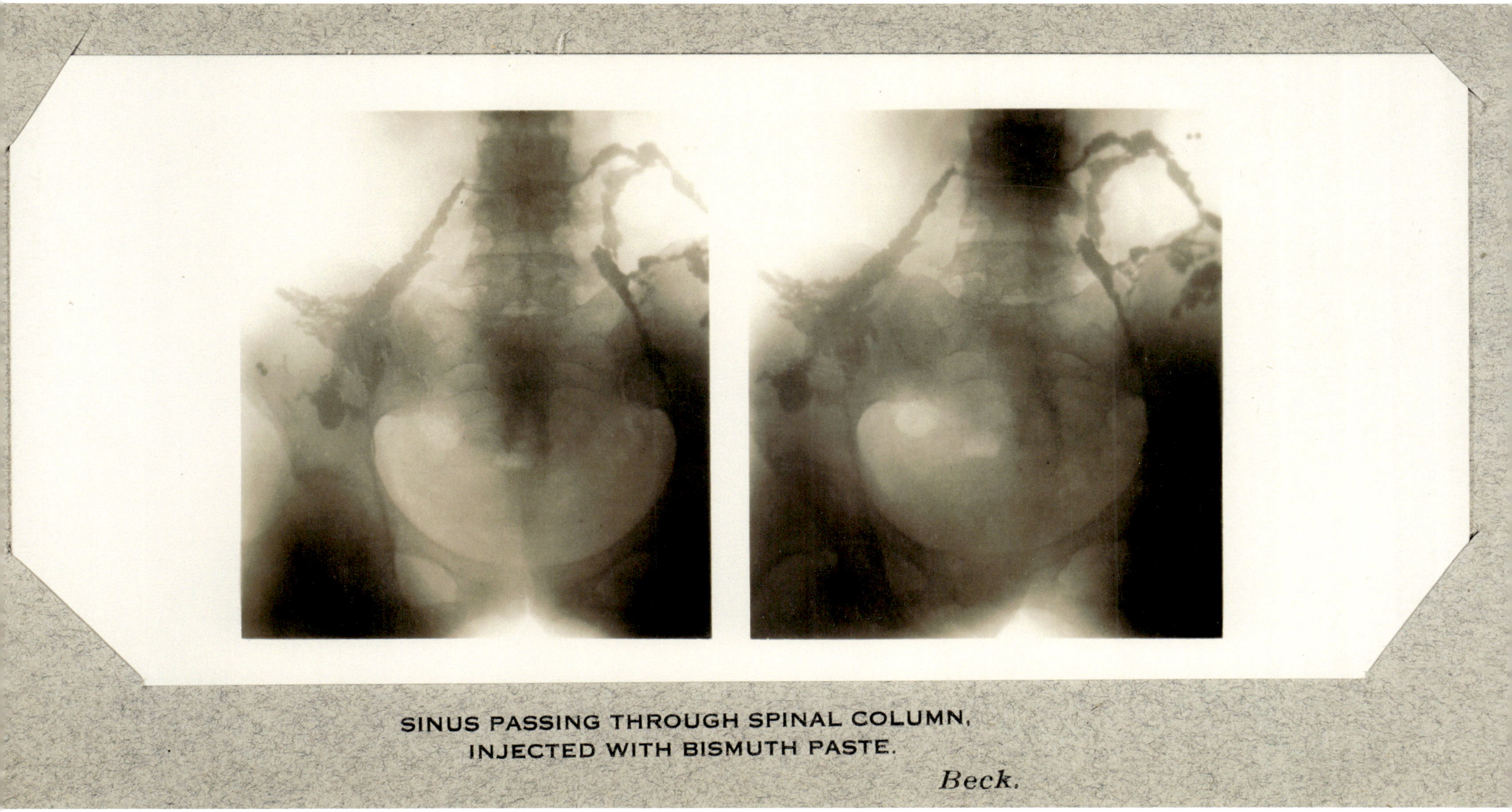

Plate 8.31

X-Ray of Pelvis, Tuberculous Sinuses Injected with Bismuth Paste Contrast Media to Outline Fistulas
Stereoview, Drs. Emil and Carl Beck, circa 1910

The Becks were the proponents of using bismuth in tuberculosis.

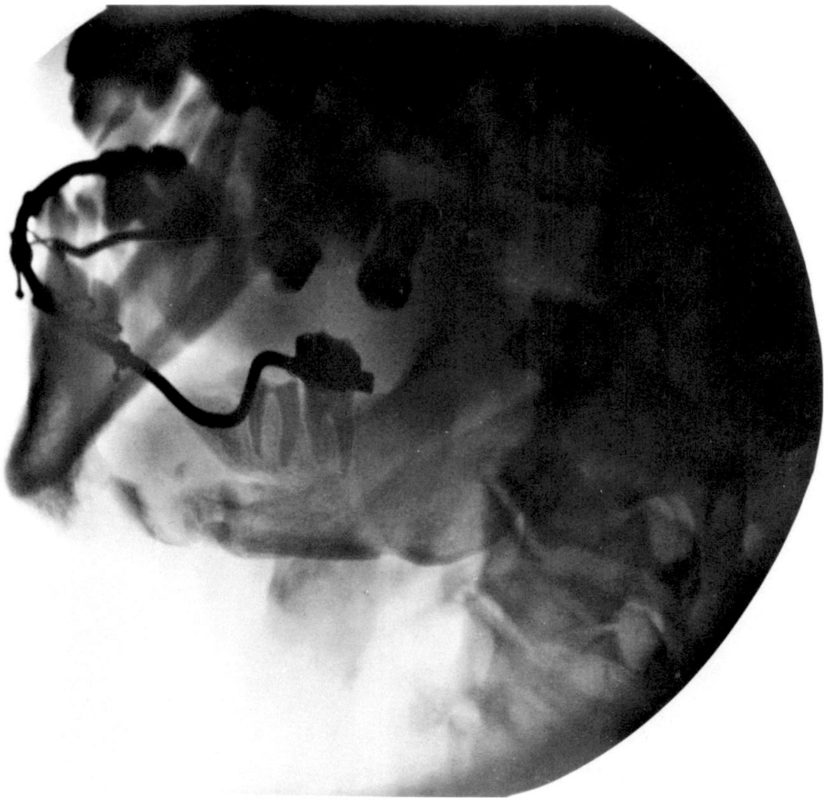

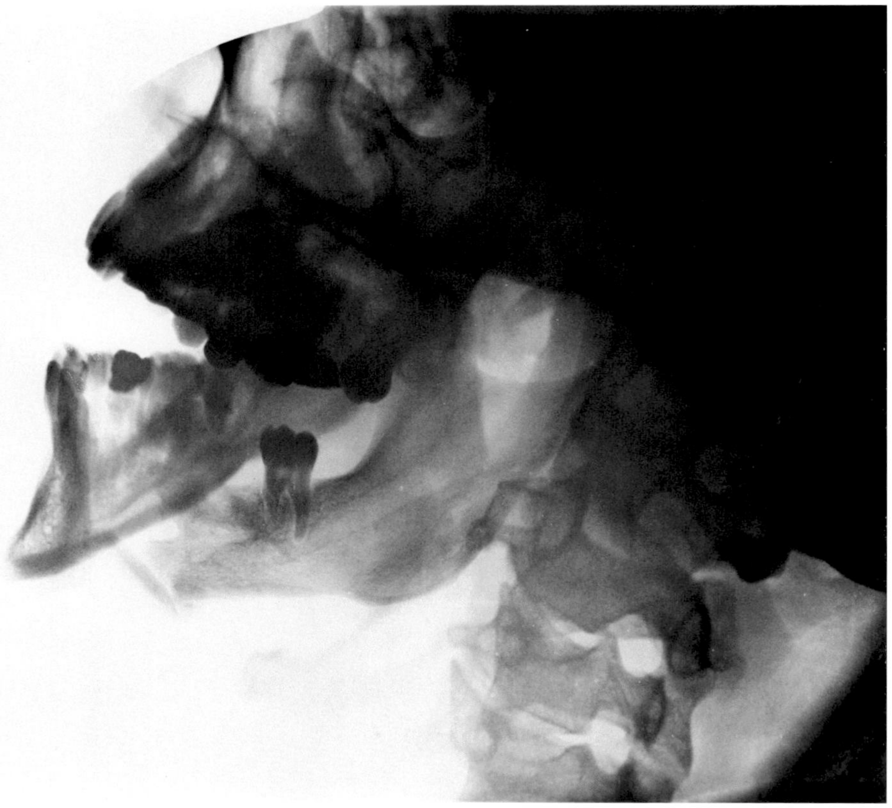

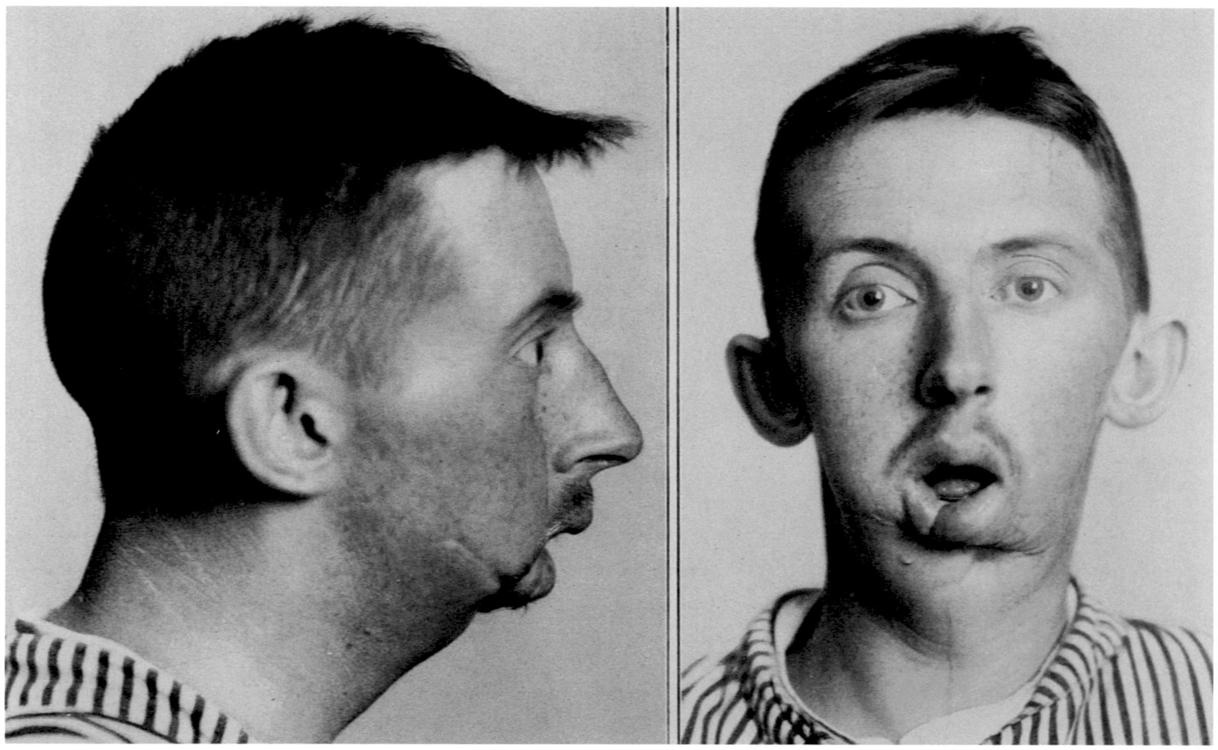

Plates 8.32 – 8.33

Wehrmann, R., X-Rays Showing Loss of Lower Jaw with Spring to Hold It Together
Röntgen-Atlas der Kriegsverletzungen, Plates 33 and 34
St. Georg, Hamburg, 1916

Plate 8.34

Wehrmann, R., Wounded in the Jaw by a Grenade Fragment in a Trench, January 26, 1915
Röntgen-Atlas of War Wounds, Photo Number 9
St. Georg, Hamburg, 1916

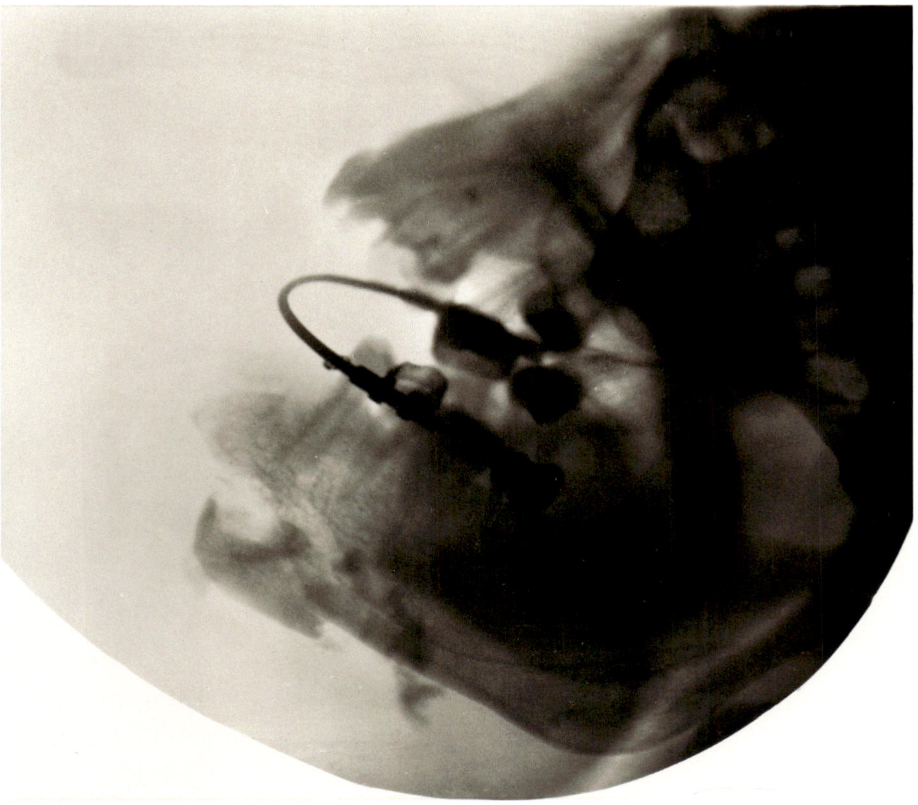

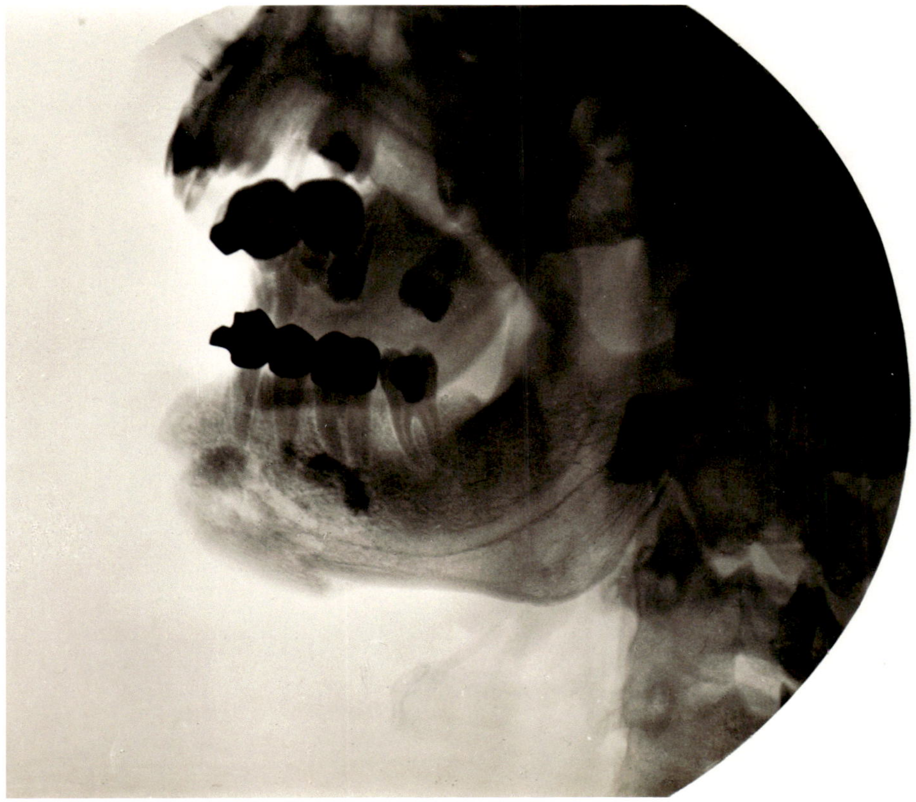

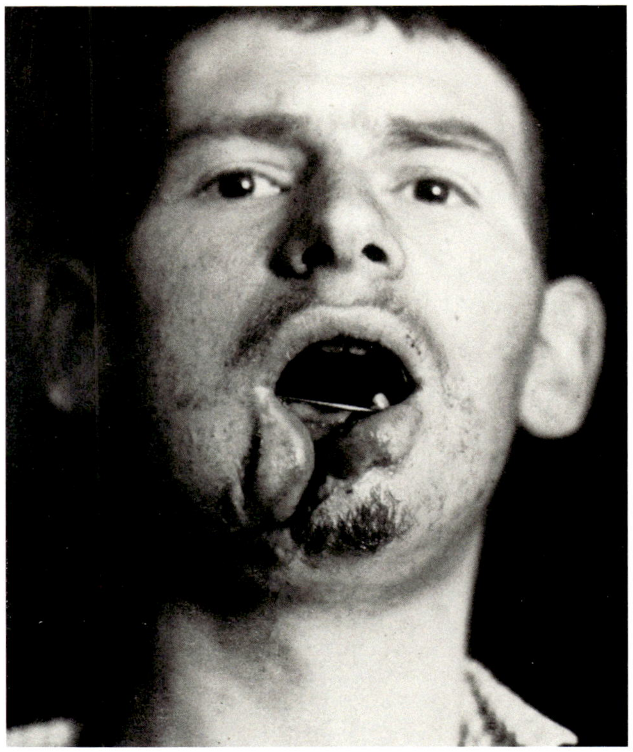

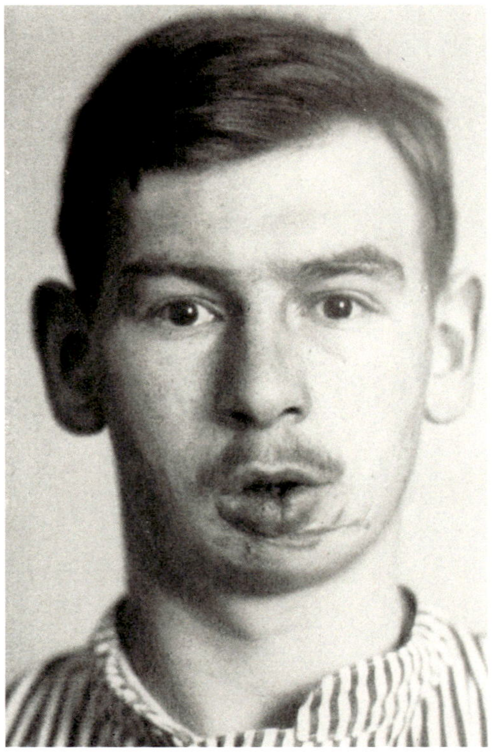

Plates 8.35 – 8.36

Infantry Soldier H, X-Rays Showing Loss of Lower Jaw, and at Healing with Teeth Capped
Röntgen-Atlas der Kriegsverletzungen, Plates 31 and 32
St. Georg, Hamburg, 1916

Plate 8.37

Infantry Soldier H, Wounded by a Bullet in the Jaw, November 5, 1914.
Röntgen-Atlas der Kriegsverletzungen, Photo Number 8
St. Georg, Hamburg, 1916

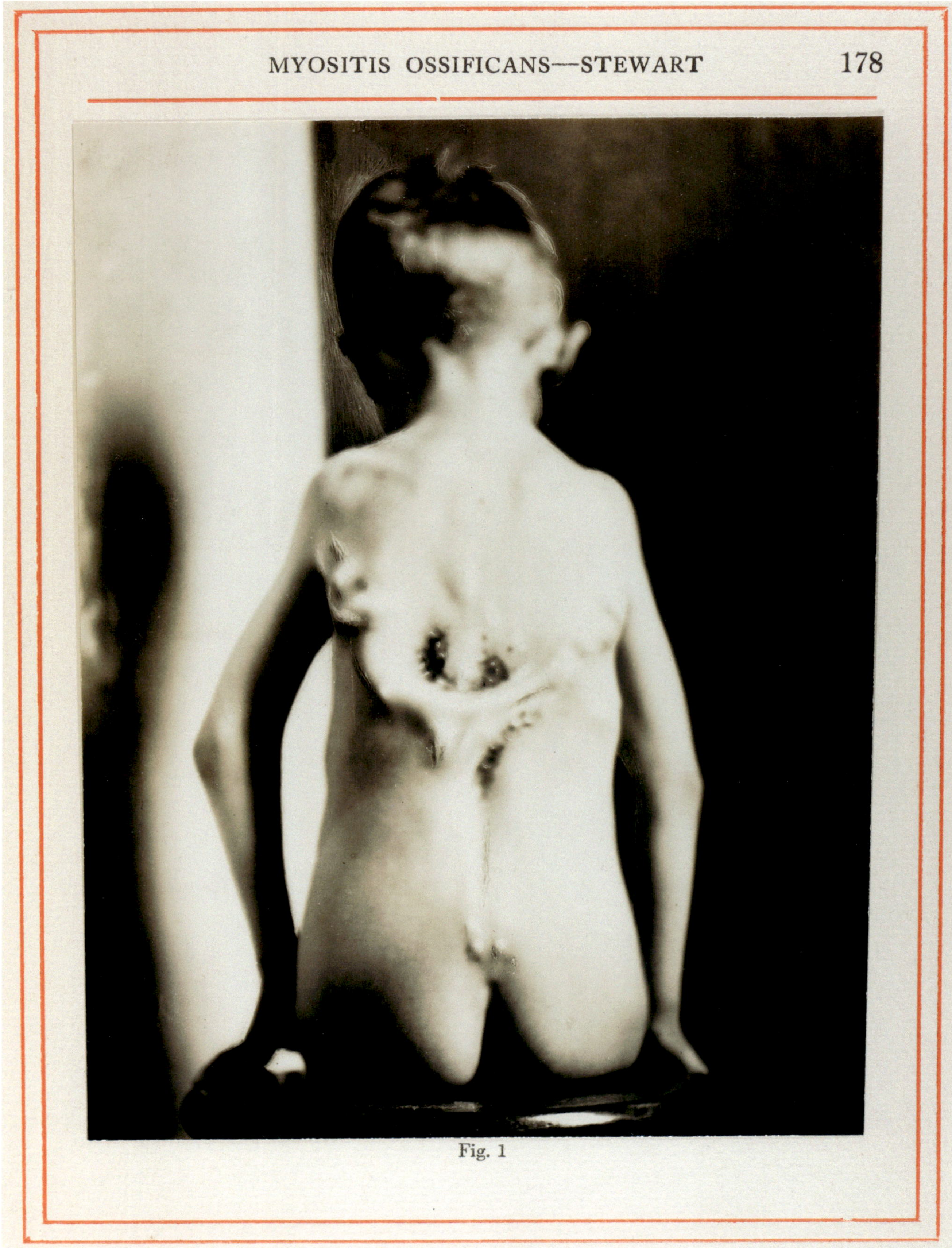

Plate 8.38

Eight Year-Old Boy, 'Frozen' due to Myositis Ossificans
Wm H. Stewart, MD, *American Atlas of Stereoroentgenology*, Volume 2, January 1917

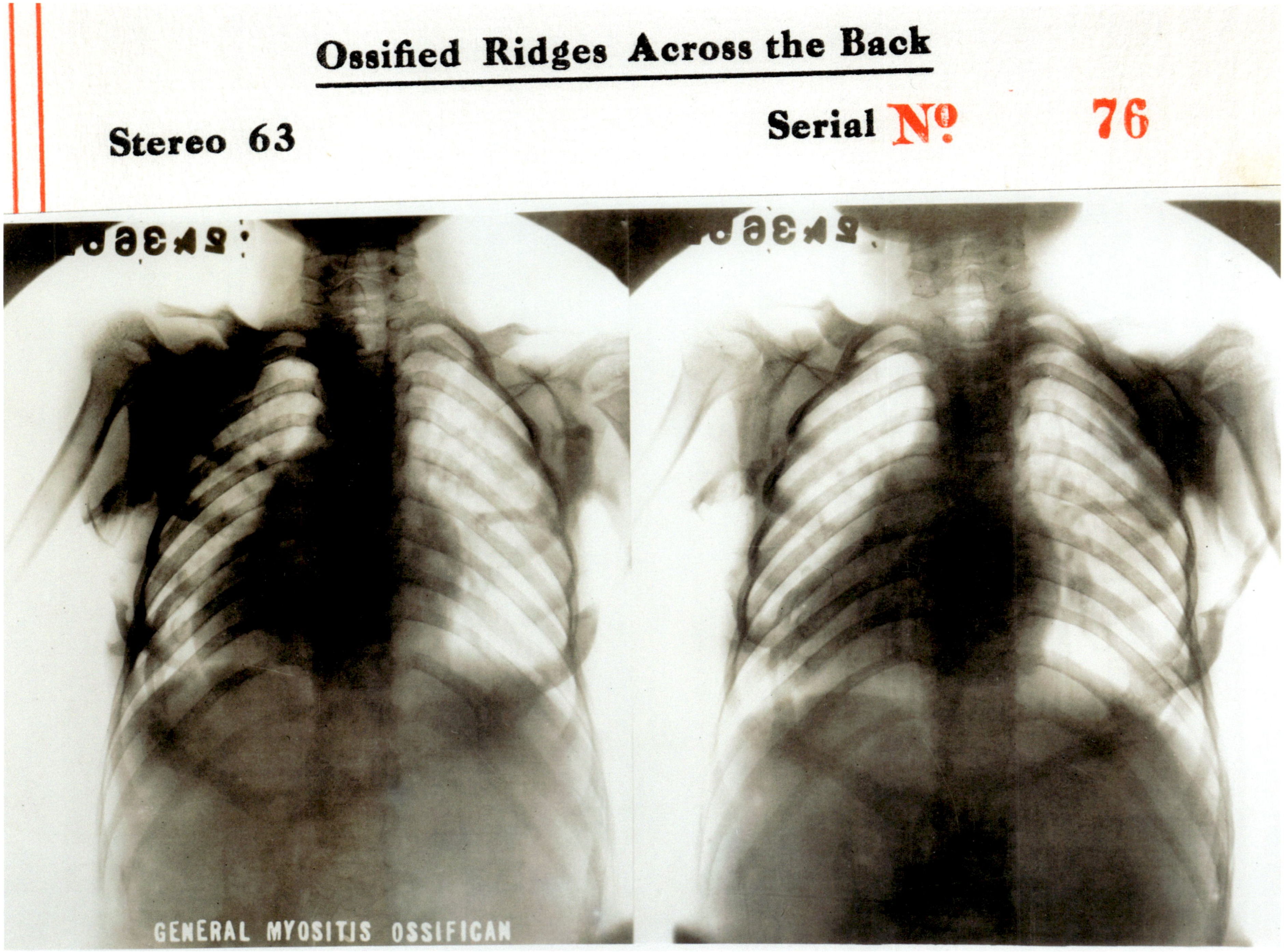

Plate 8.39

X-Ray Showing Ossified Ridges across the Back Involving All Muscles
Wm H. Stewart, MD, *American Atlas of Stereoroentgenology*, Volume 2, January 1917

This stereorontengram shows ossification of the back spreading forward freezing arms in position.

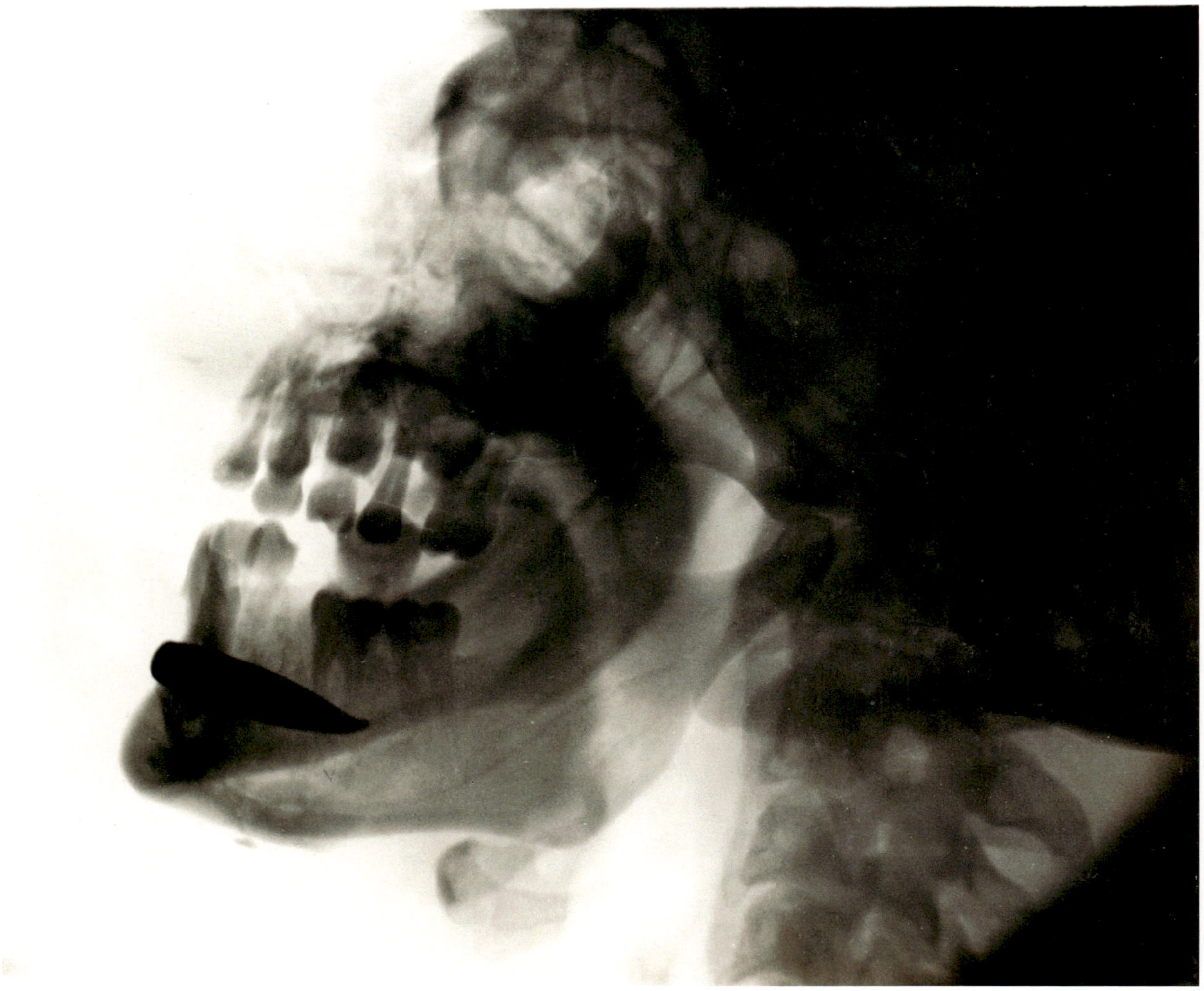

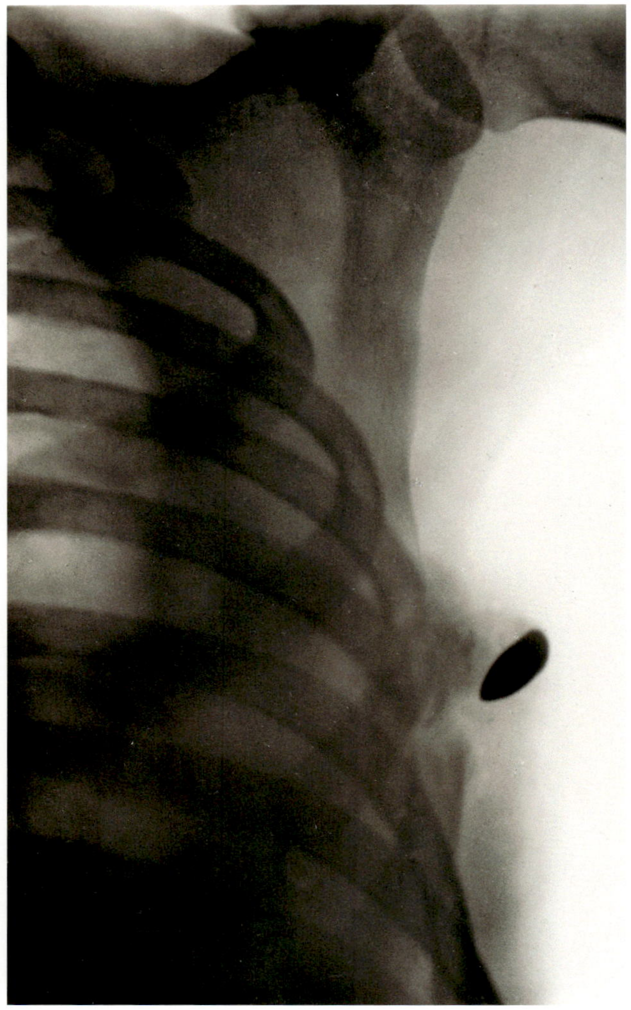

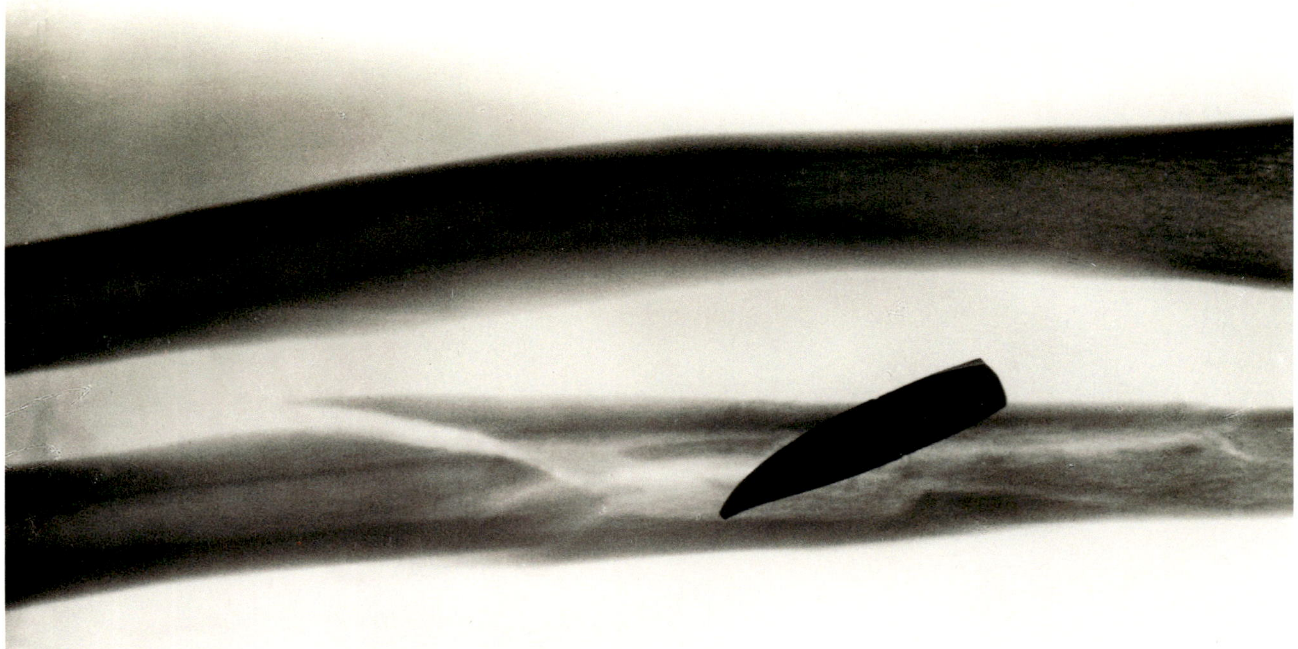

Plates 8.40 – 8.45

WWI X-Rays of Bullets in the Bodies of German Soldiers
Röntgen-Atlas der Kriegsverletzungen…
St. Georg, Hamburg, 1916

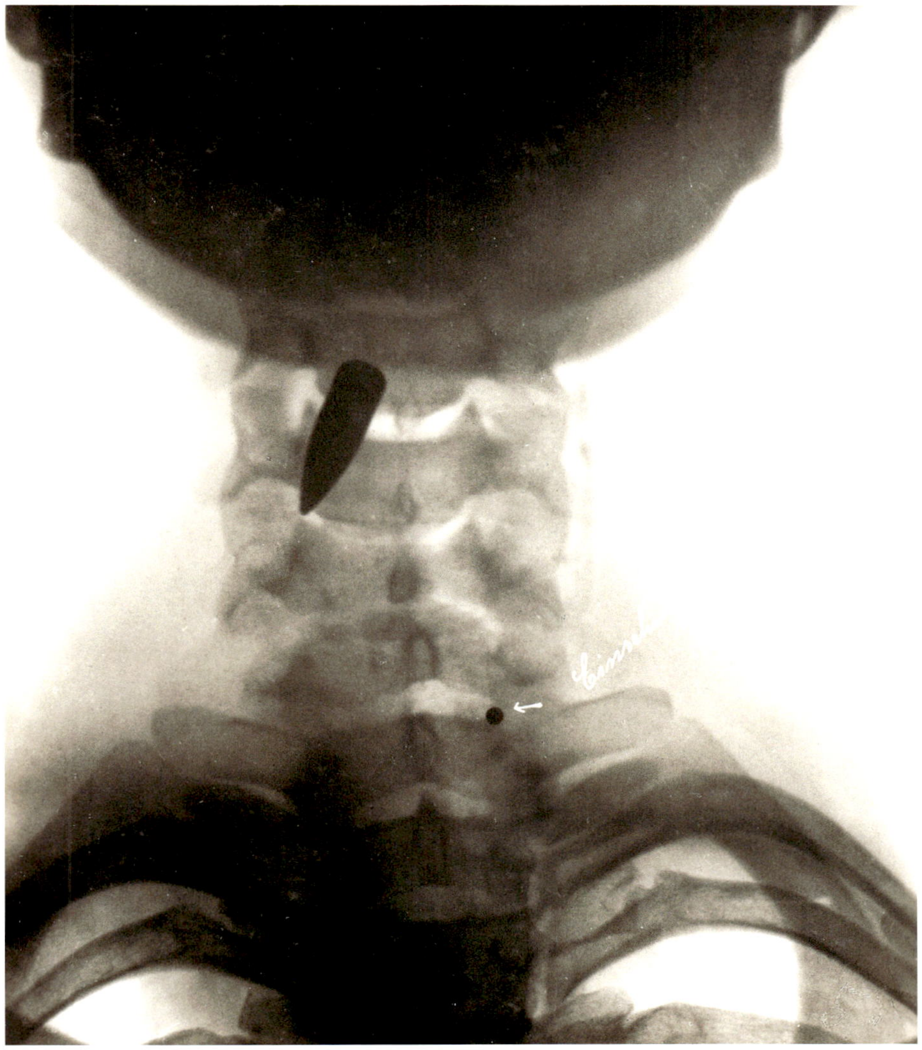

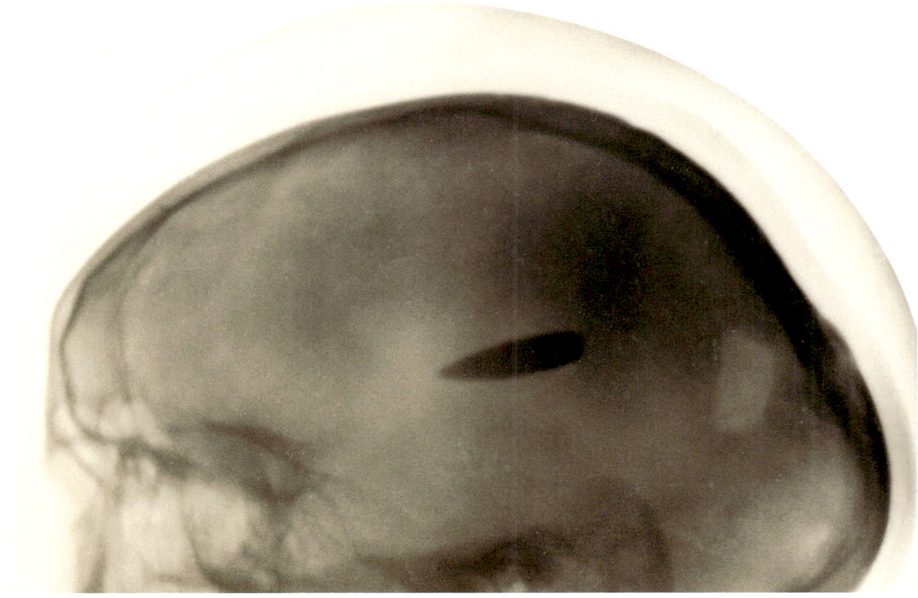

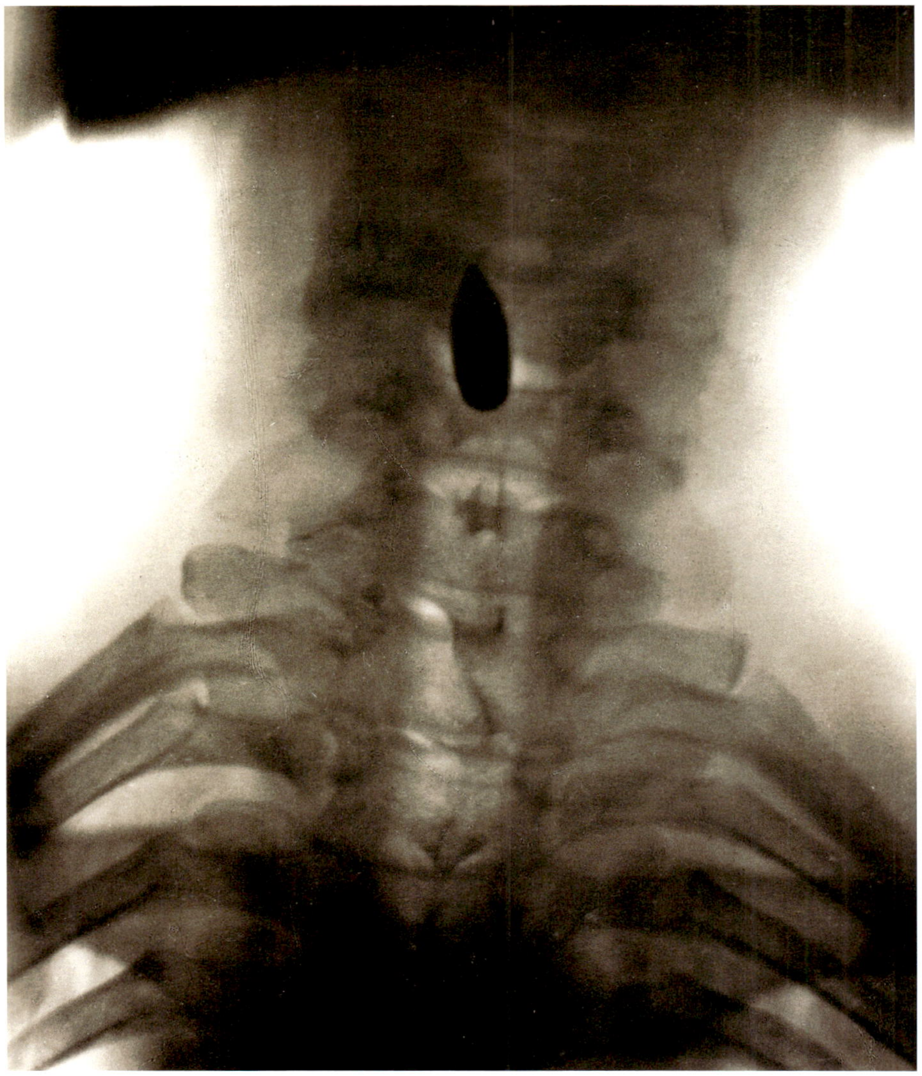

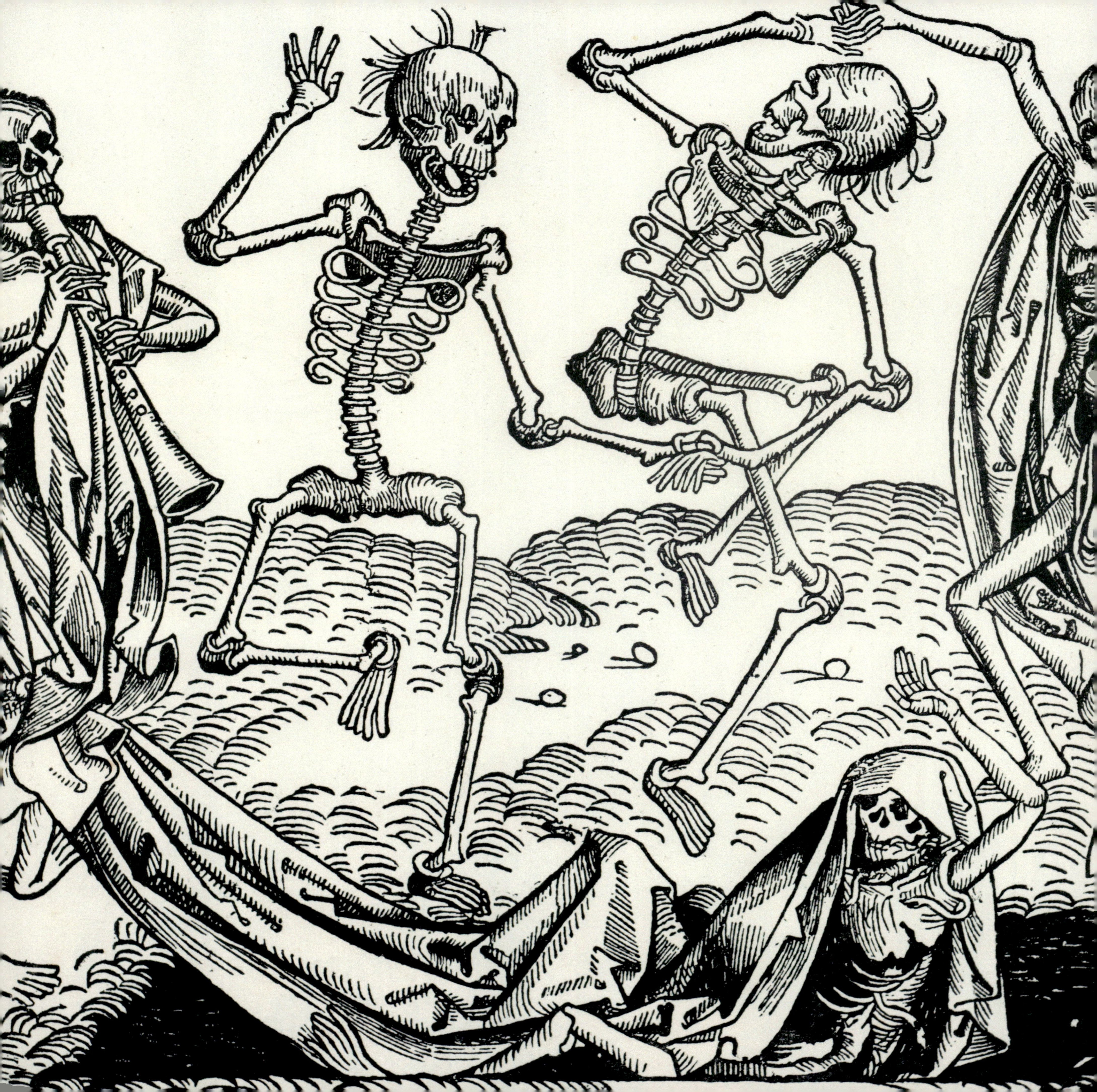

CHAPTER NINE

Gallows Humor & Group Identity Photographs

Doctors and Nurses at Play and Organizational Props

Medical gallows humor photographs are particularly American. They are photographs taken at a time when death was common and the value of life of the 'other' (those seen as inferior in race, intelligence and class) was often deemed a life of little value. Lynchings, executions, convict labor, forced sterilizations, immigrant exclusion acts, and laws that treated many as second-class citizens were the result of scientific and eugenic doctrines of the time.

It was only toward the end of the twentieth century that the value of an individual life was seen, while alive or dead, and that respect and honor for that individual was achieved. That respect is reflected in today's anatomy labs and dissection rooms. The images in this chapter show a wide range of the photographs that medical and nursing students at the turn of the century felt appropriate to record. Their familiarity with death is permanently documented.

Dealing with life and death situations stimulates some to take a lighter view by adding a humorous bent to the circumstance. Humor serves to relieve tension. A classic tale is of the executioner offering a cigarette to a man about to be shot by the firing squad. The man declines the cigarette because he is trying to give up smoking. In the early 1890s, amateur photographers proliferated, and photography lost the significance of being a once in a lifetime occurrence. Photography became a visual medium by which to express feelings, and medical students relieved the tension of facing the dead in their dissection classes by posing in macabre vignettes with their cadavers and/or a skeleton. The poses ranged from treating a skinned cadaver as a best friend or lover to dressing it up as if still living and interacting with it. In classic dissection table photographs, students often added a hat, cigarette, or pipe to the cadaver making it 'one of the boys.' Nonmedical fraternal organizations and arts associations also used bodies, skeletons, and skulls in their group photographs. This chapter highlights some of the classical posing conventions of the genre.

Sigmund Freud, in his 1927 essay "Humour" ("Der Humor"), puts forth the following theory of gallows humor: "The ego refuses to be distressed by the provocations of reality, to let itself be compelled to suffer. It insists that it cannot be affected by the traumas of the external world; it shows, in fact, that such traumas are no more than occasions for it to gain pleasure."

When I attended medical school in 1960, gallows humor was still evident. Fooling around and posing with body parts was still going on. One infamous story was that of a student handing his ticket and money to a New York State Thruway toll taker with a cadaver hand tucked into his sleeve. Genital parts were always popular for adventurous pranks. Examination of medical school yearbooks prior to 1950 may turn up photos that show, in general, what seems like disrespect, but is actually an aspect of gallows humor. The desensitization of the sanctity of life is perhaps most evident in these photographs.

It was only in the last decades of the twentieth century that modern concerns for the sanctity of life flourished. Reverence and respect for the dead, and their contribution to the education of physicians, elevated the cadaver to personhood. Ceremonies were established with proper burial or disposal of remains, along with readings honoring those who donated their bodies to help foster medical education.

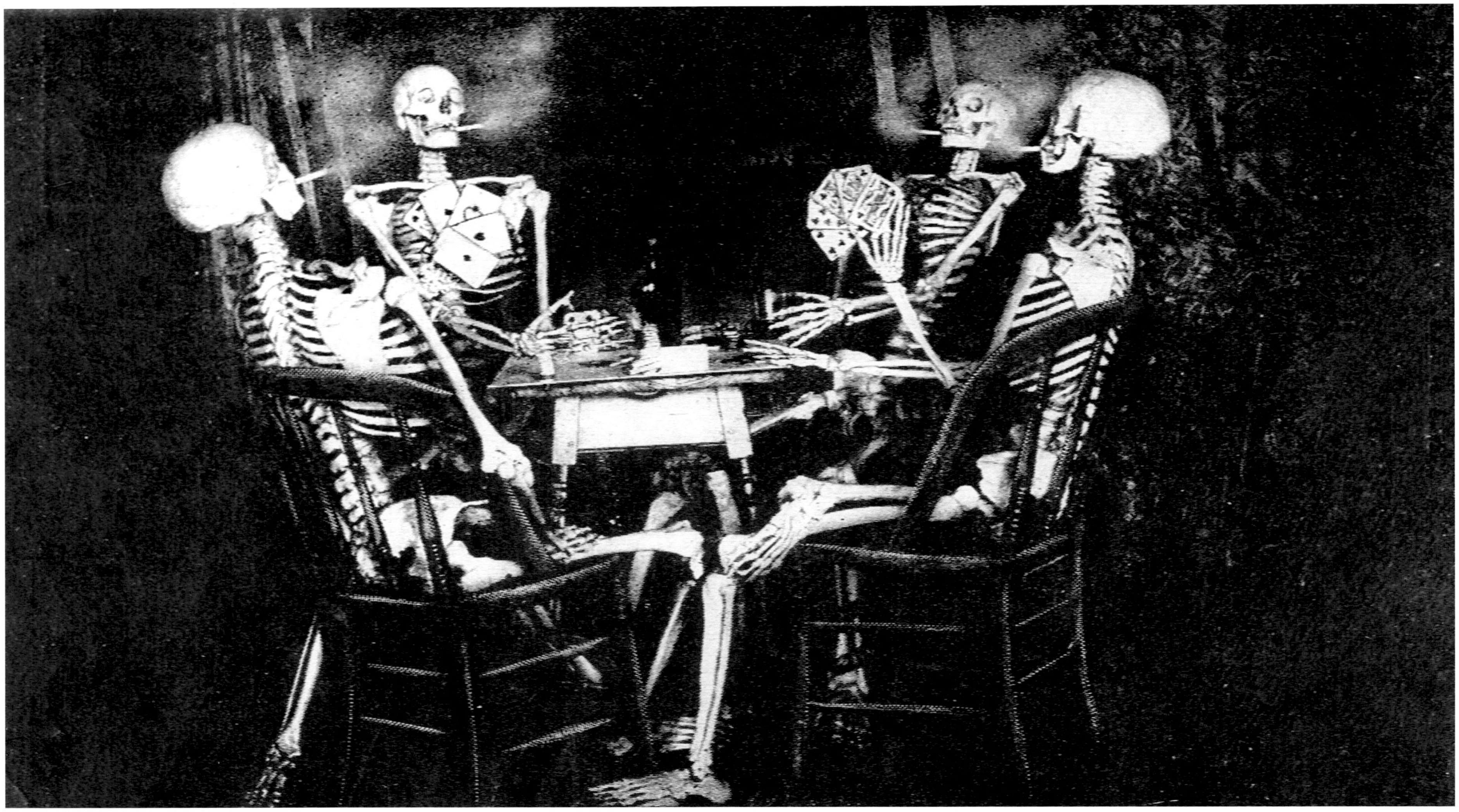

Plate 9.0

"Dead Game Sports"
Postcard, Photographer: Harry Murray, 1905

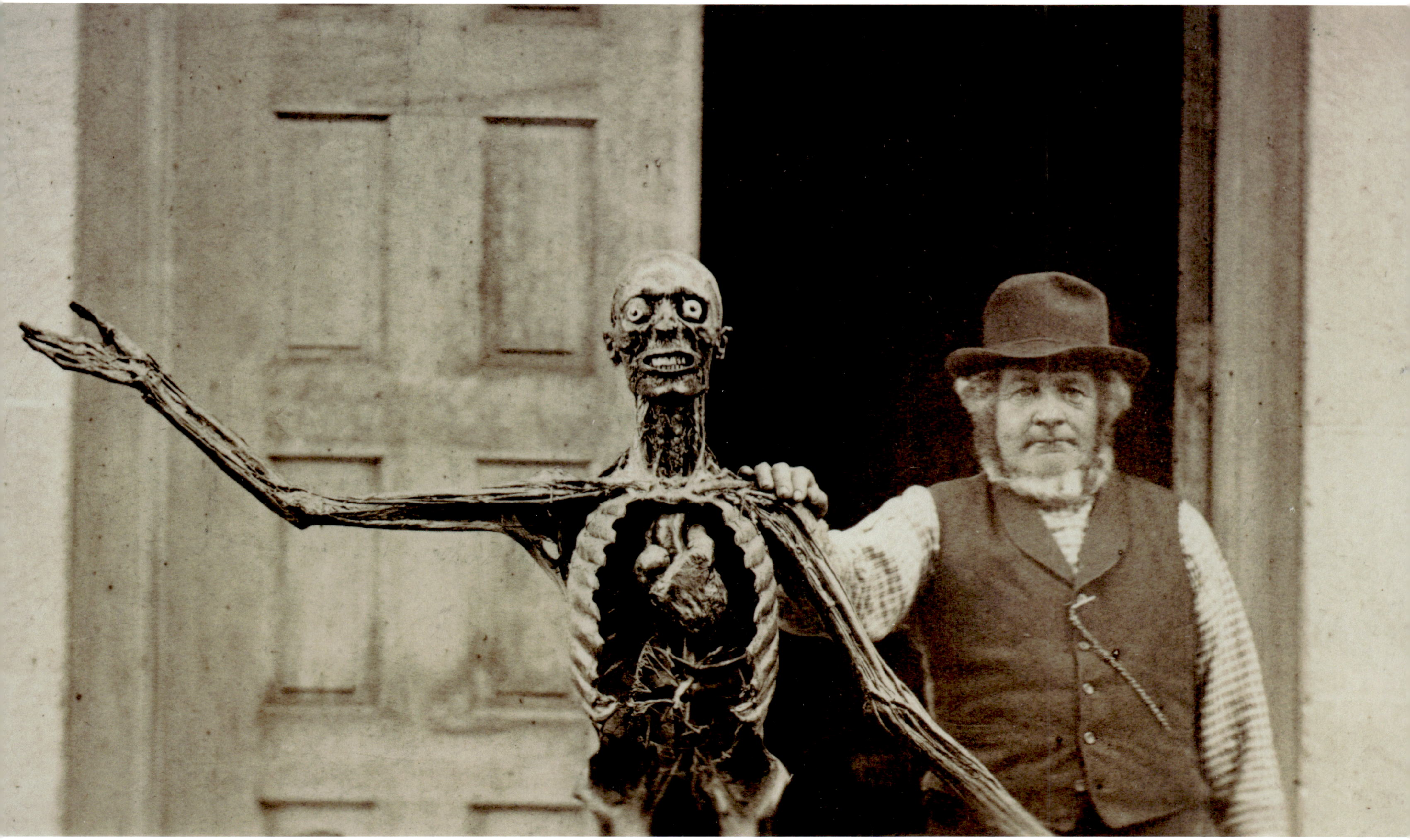

Plate 9.1

Welcome to Our Medical School
Cabinet Card, Photographer: Randall, Ann Arbor, Michigan, 1875

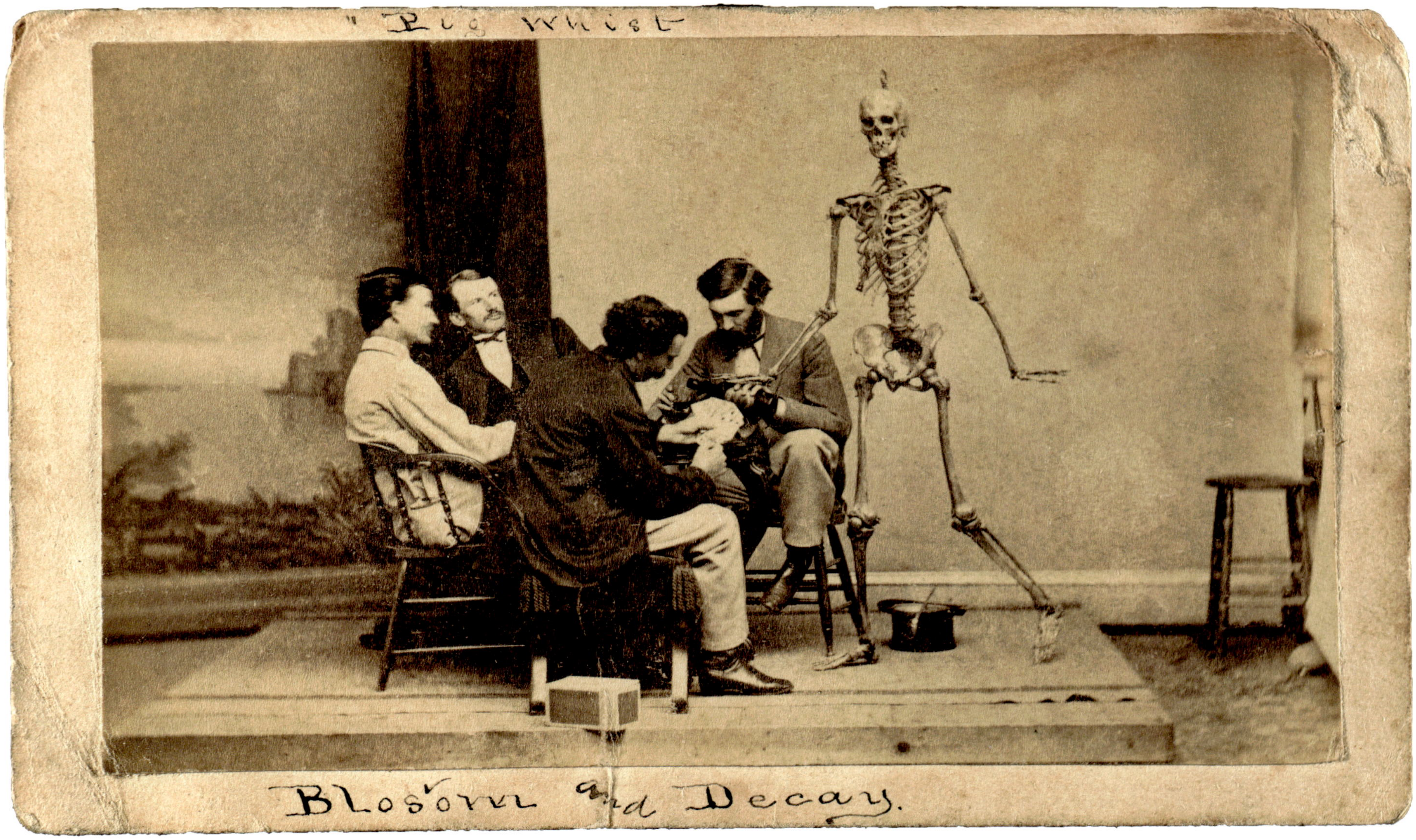

Plate 9.2

"Blossom and Decay"
Carte de Visite, Photographer: J. Haynes, Madison, Wisconsin, 1866

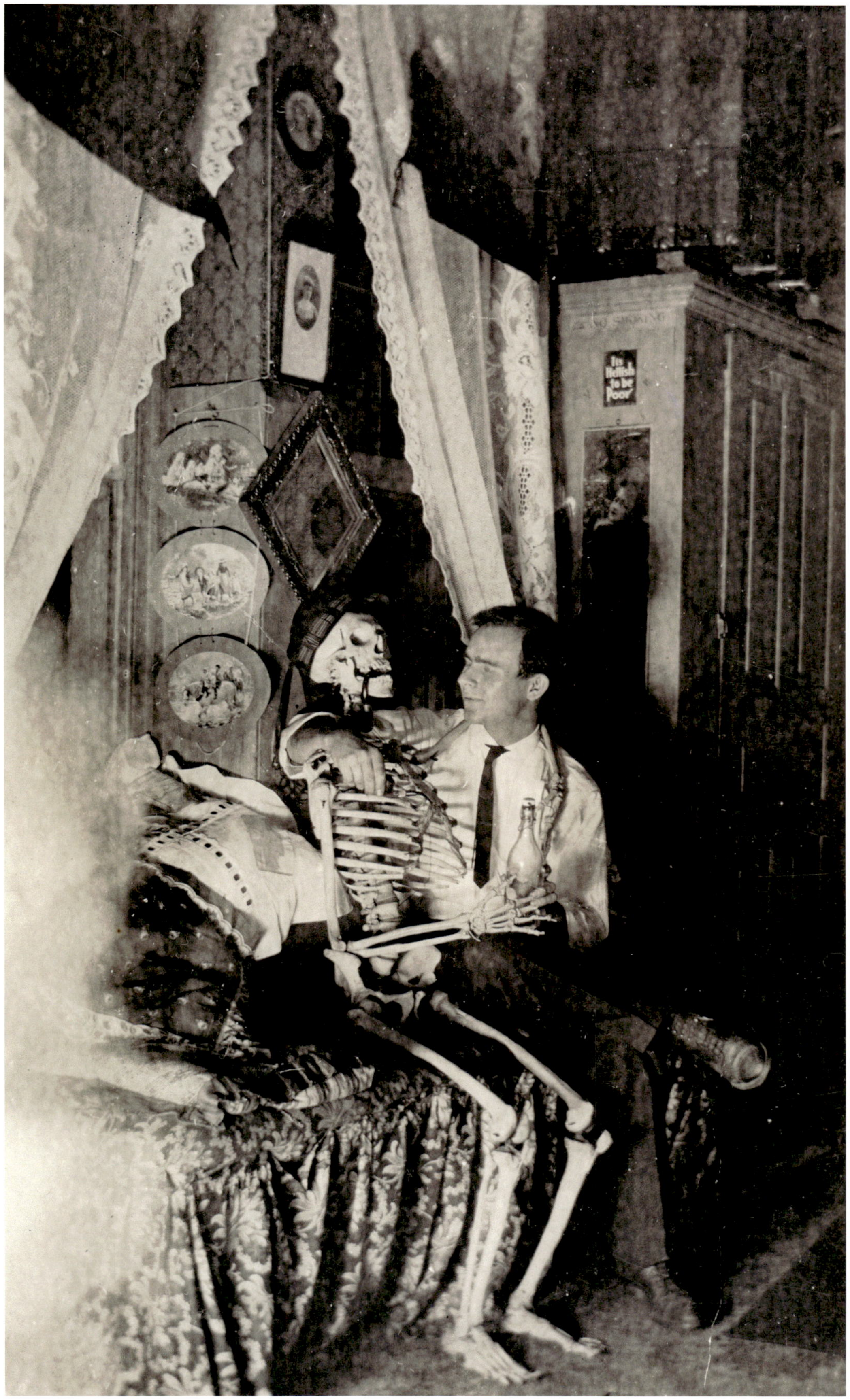

Plate 9.3

"Dear Mary, What do You Think of My Chum?"
Postcard mailed to Mary Miller, Mifflinburg, PA, 1908

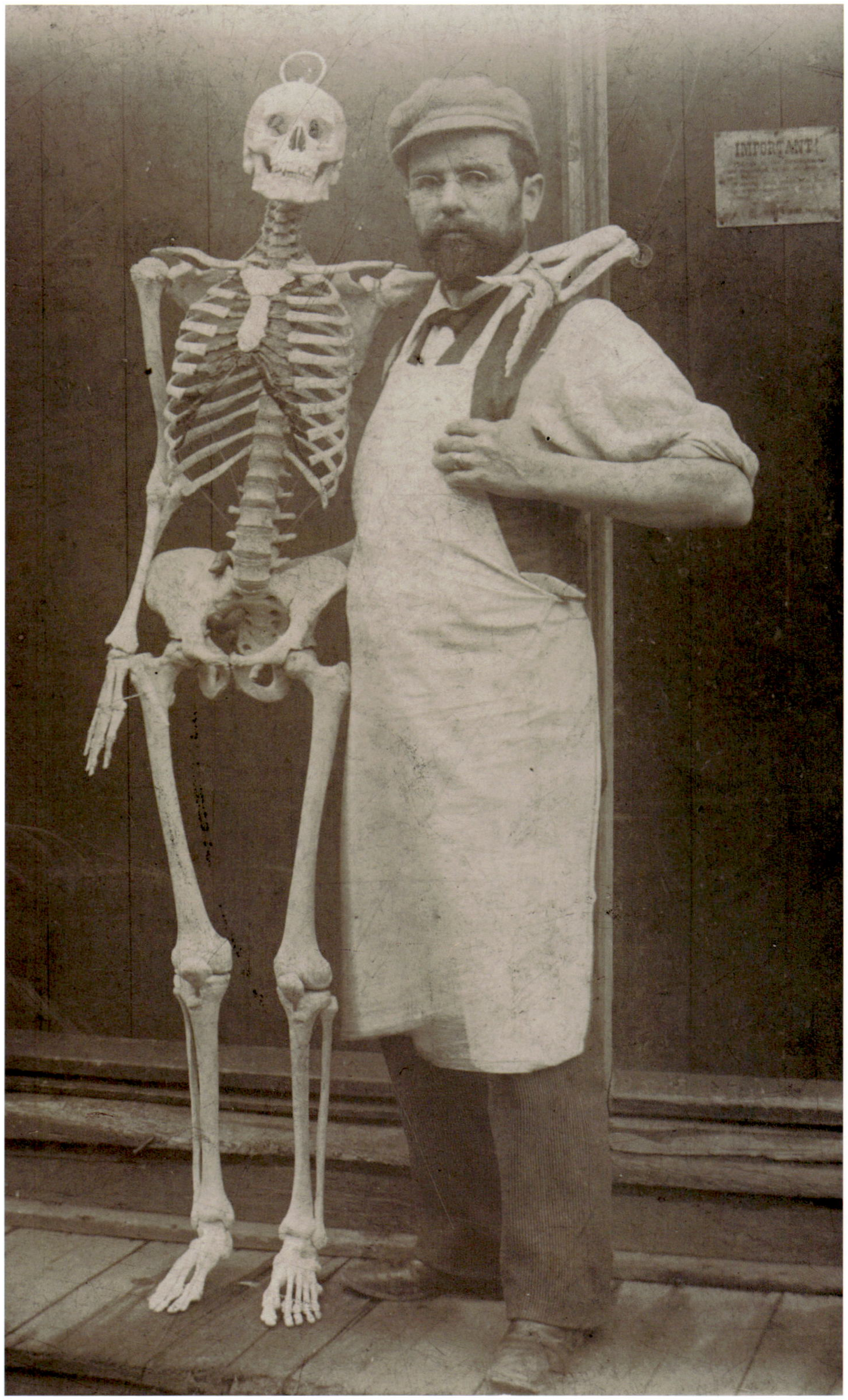

Plate 9.4

Helping My Pal
Circa 1920

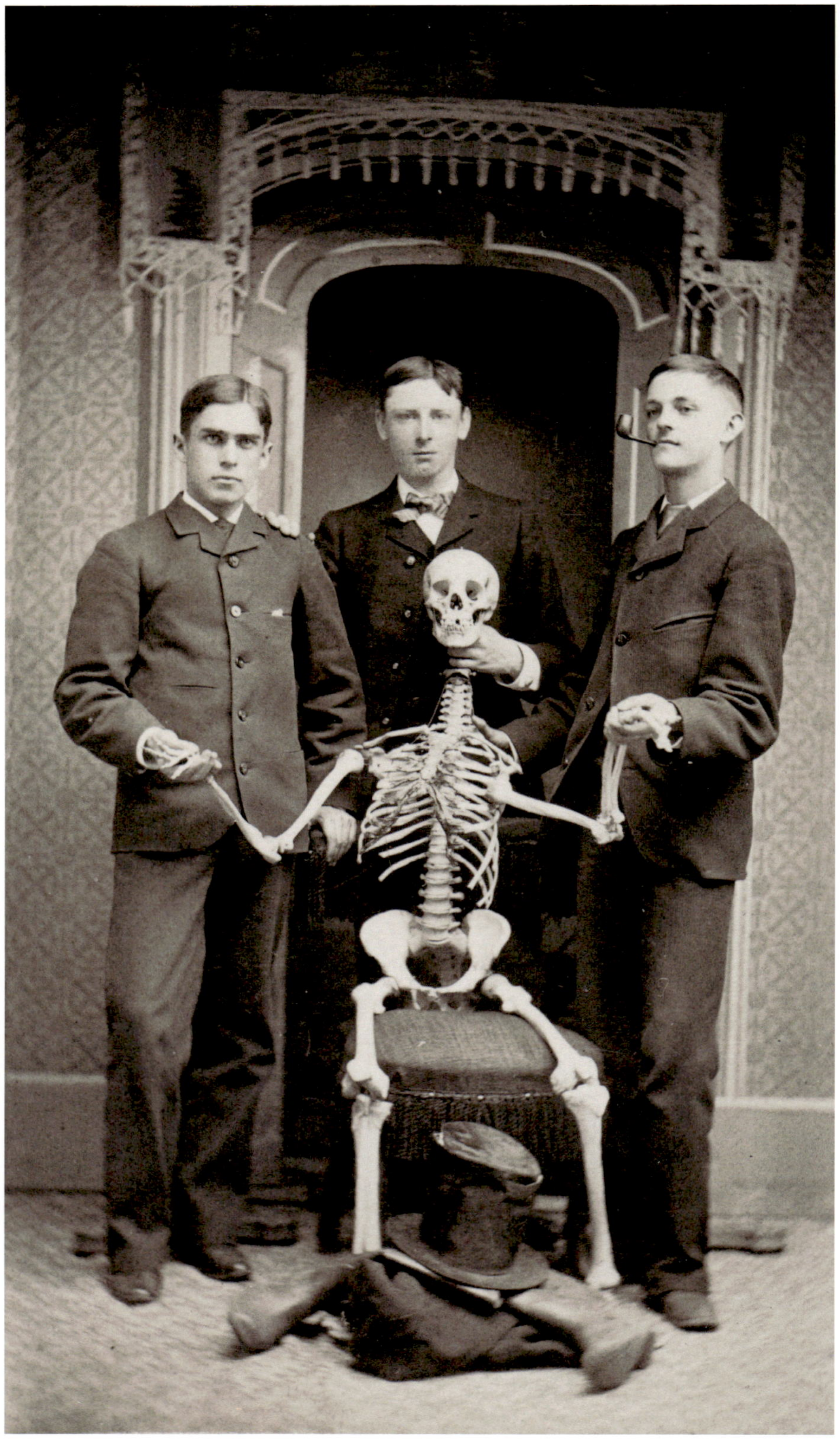

Plate 9.5

Four Friends
Circa 1900

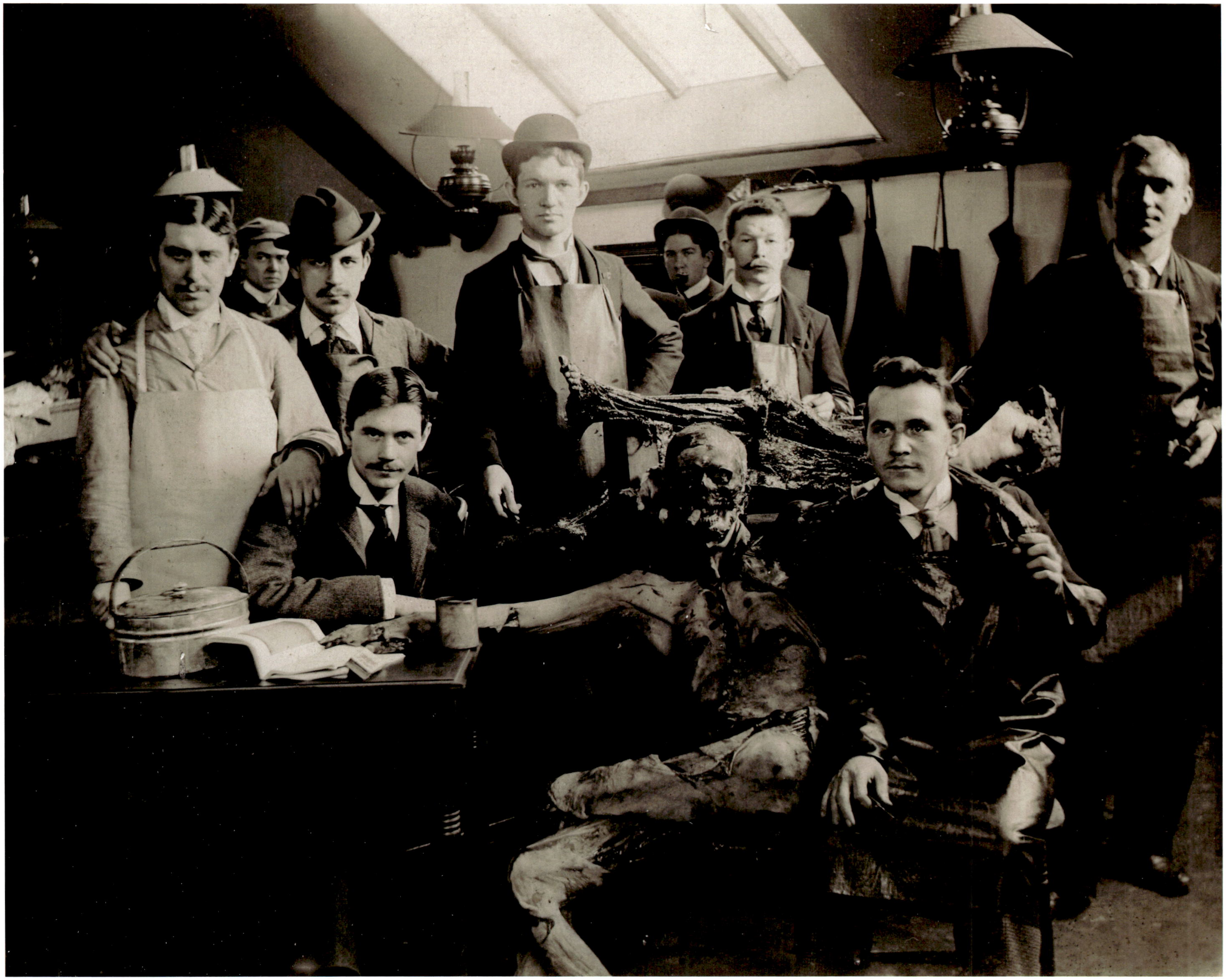

Plate 9.6

"Our Dissected Classmate"
Circa 1895

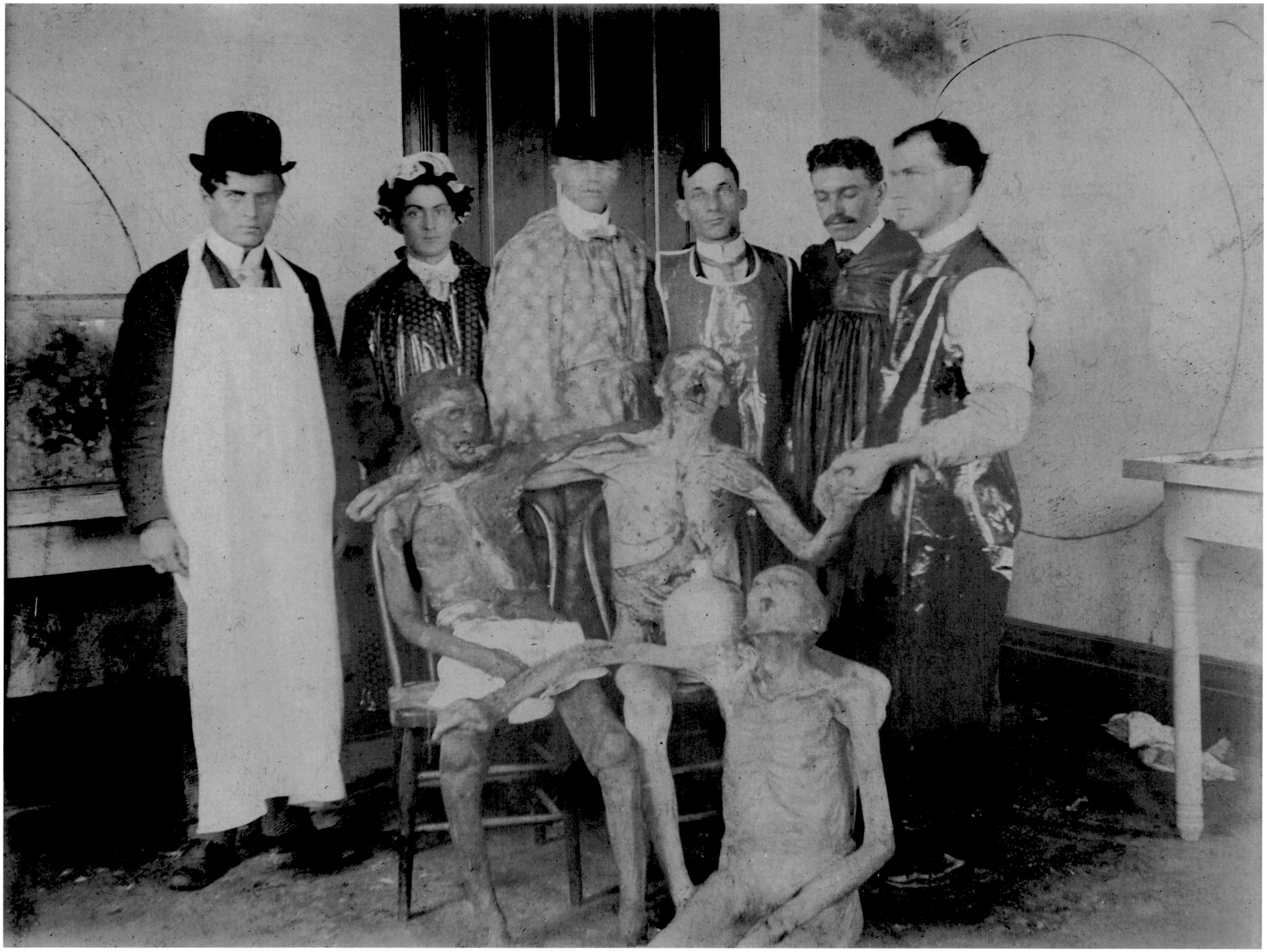

Plate 9.7

Group Portrait
Circa 1890

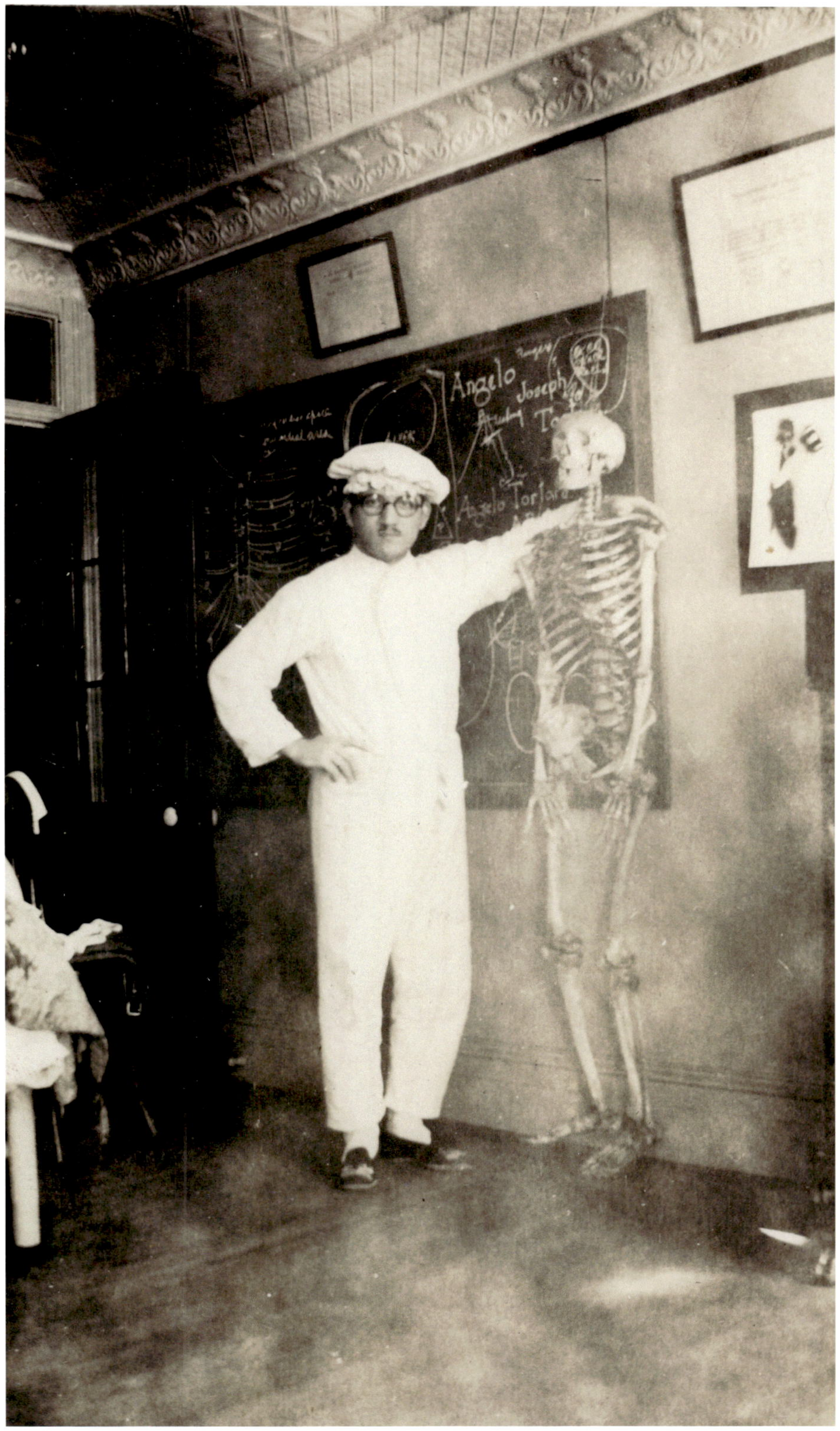

Plate 9.8

Angelo Tortora, MD, in Classroom
Snapshot, circa 1930

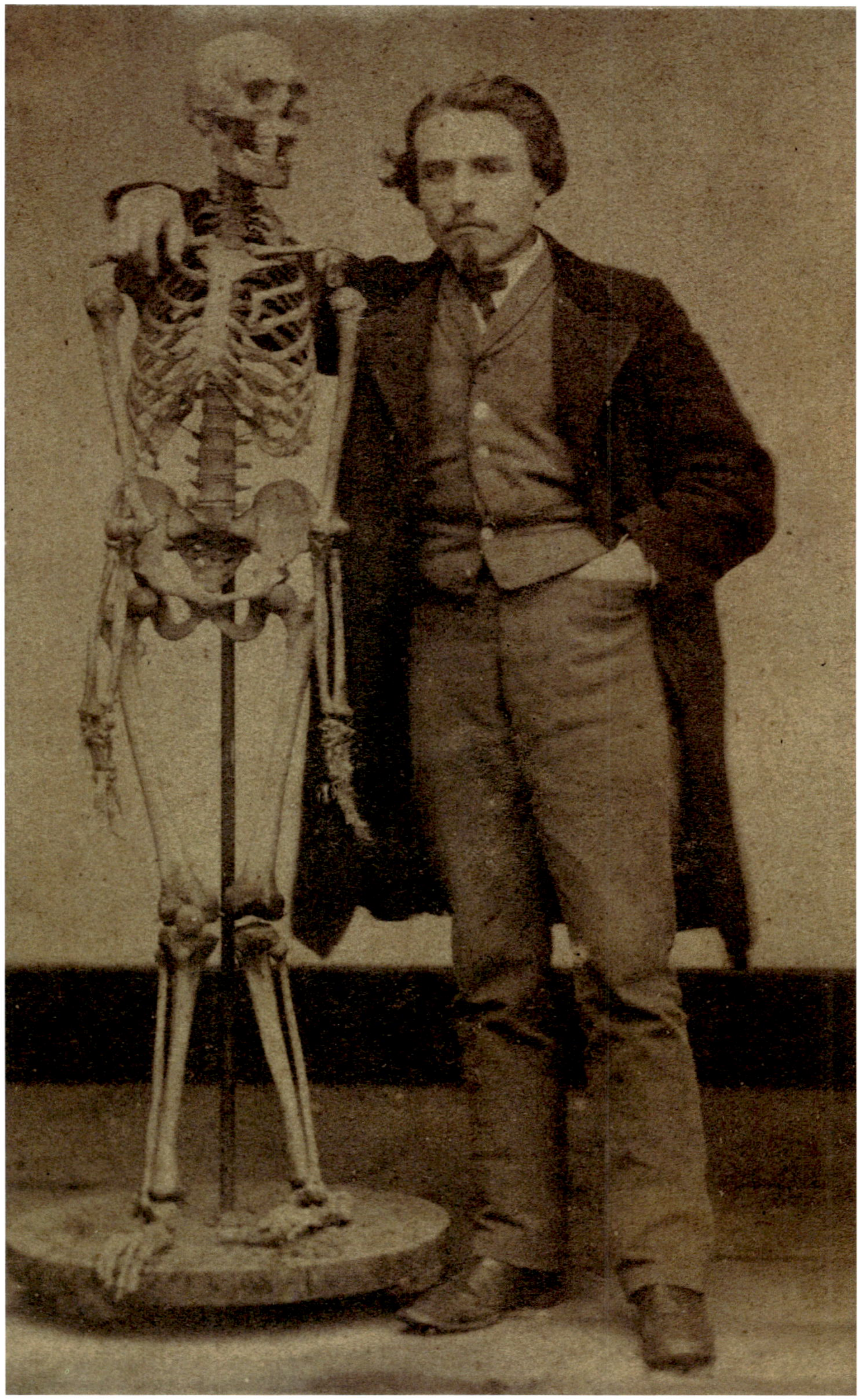

Plate 9.9

"My Best Friend"
Carte de Visite, October 1864

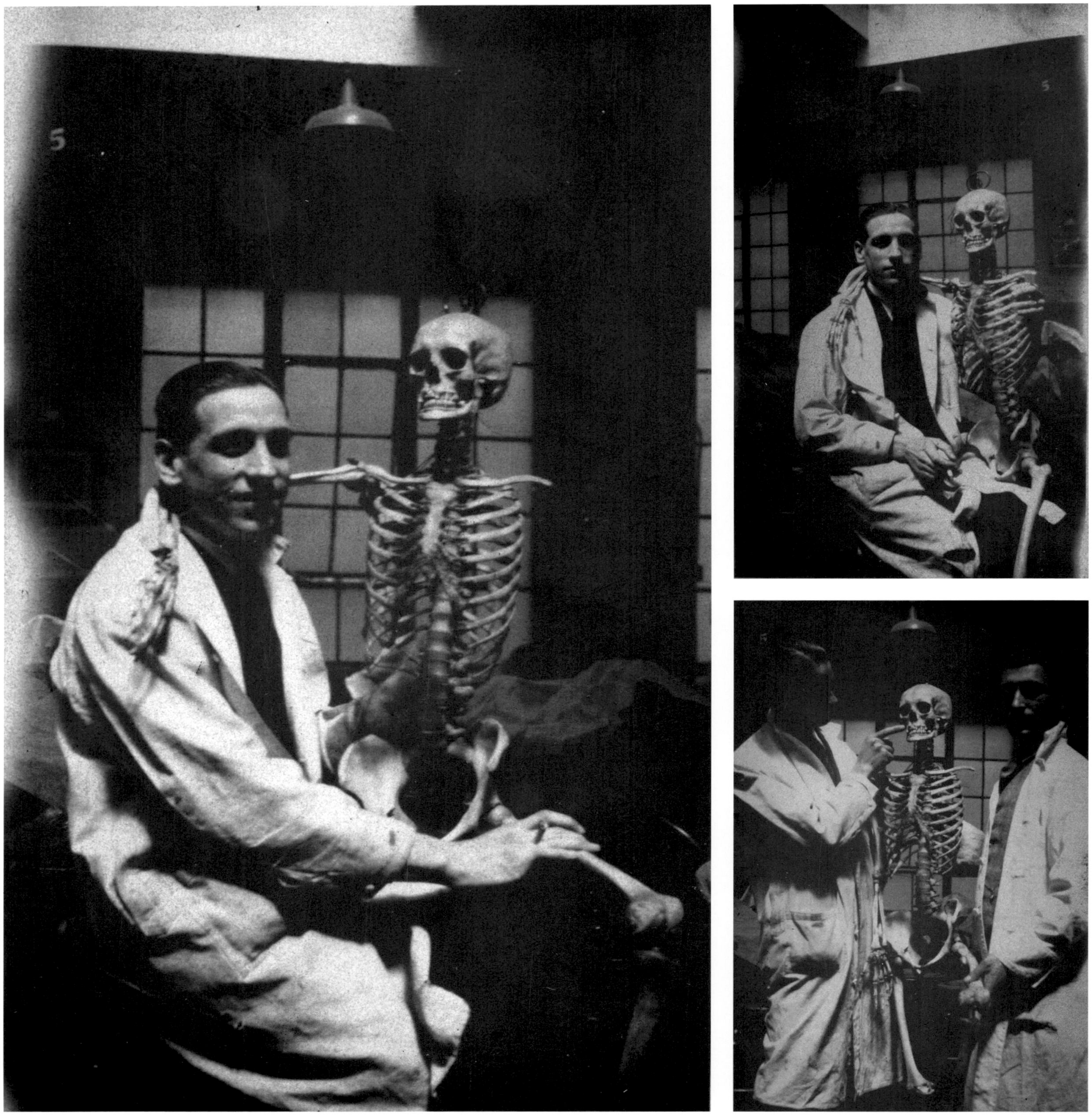

Plates 9.10 – 9.12

Student with his Skeleton
Snapshots, circa 1925

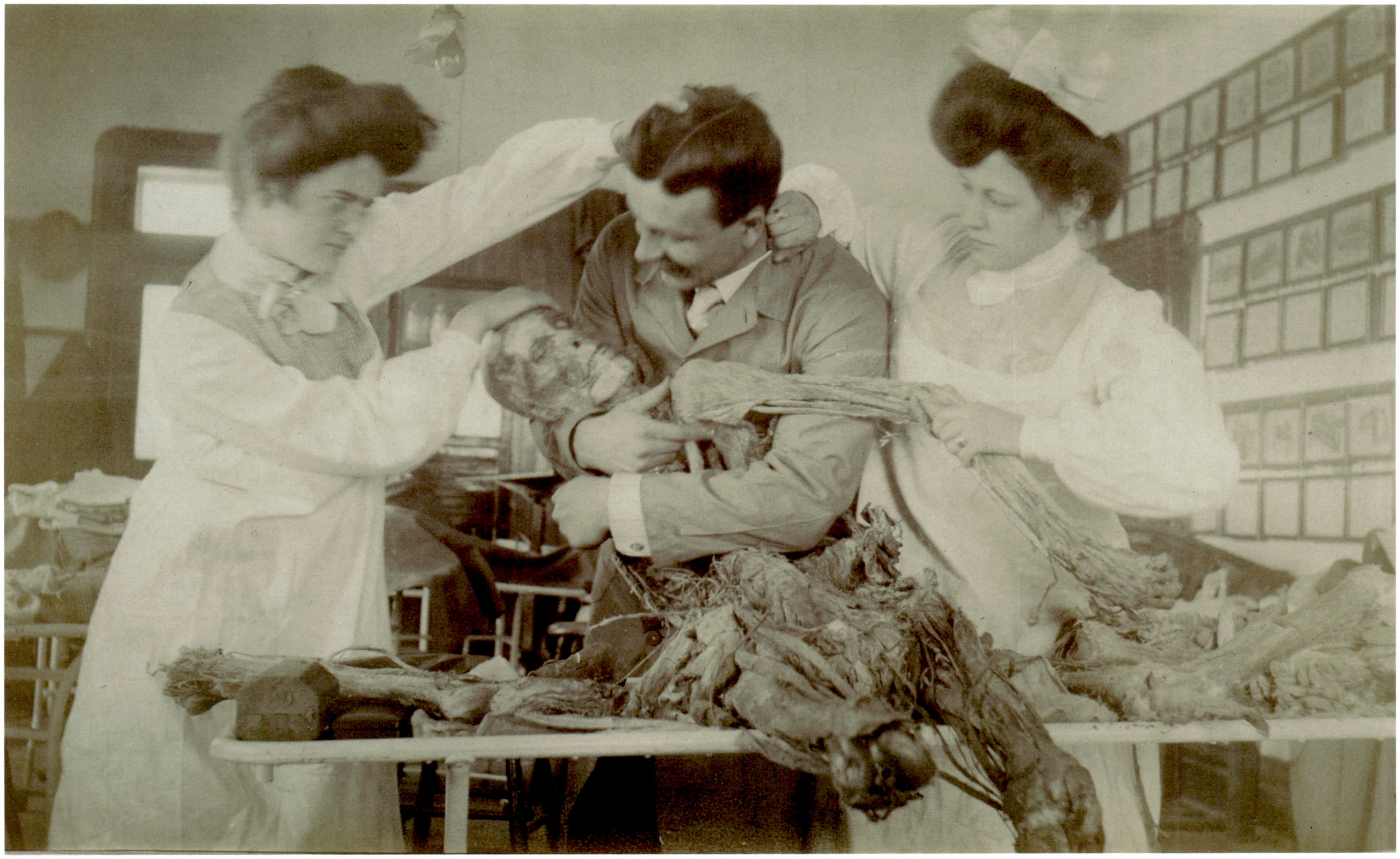

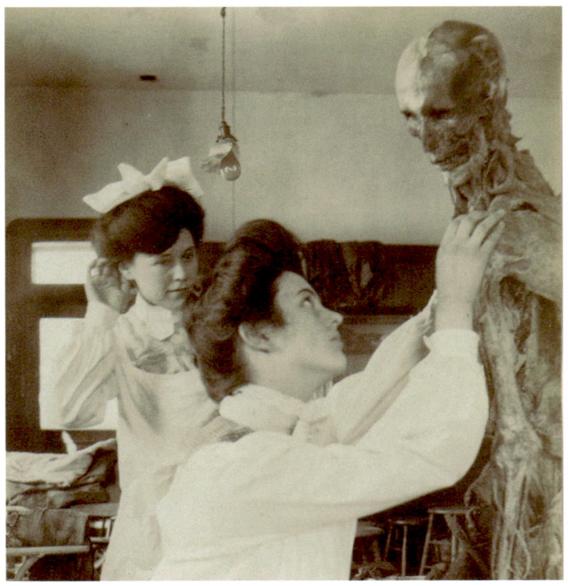

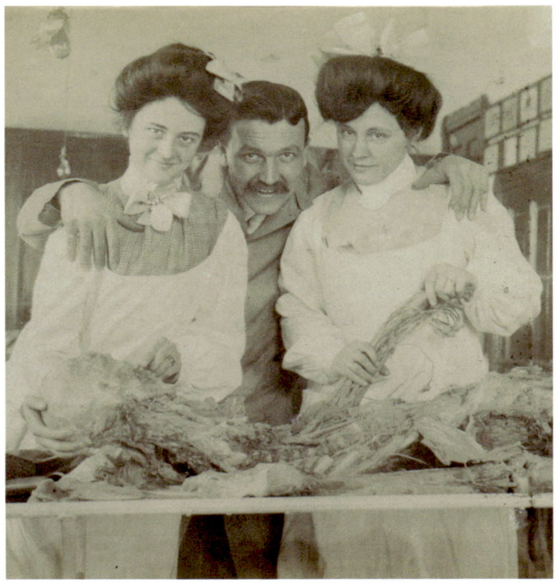

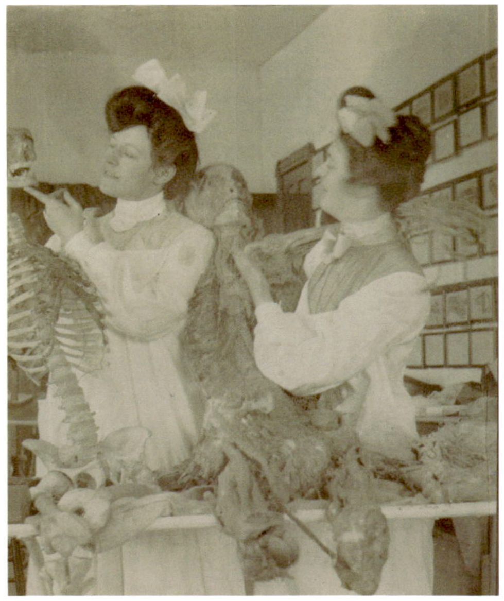

Plates 9.13 – 9.16

Nurses and Doctor at Play
Circa 1915

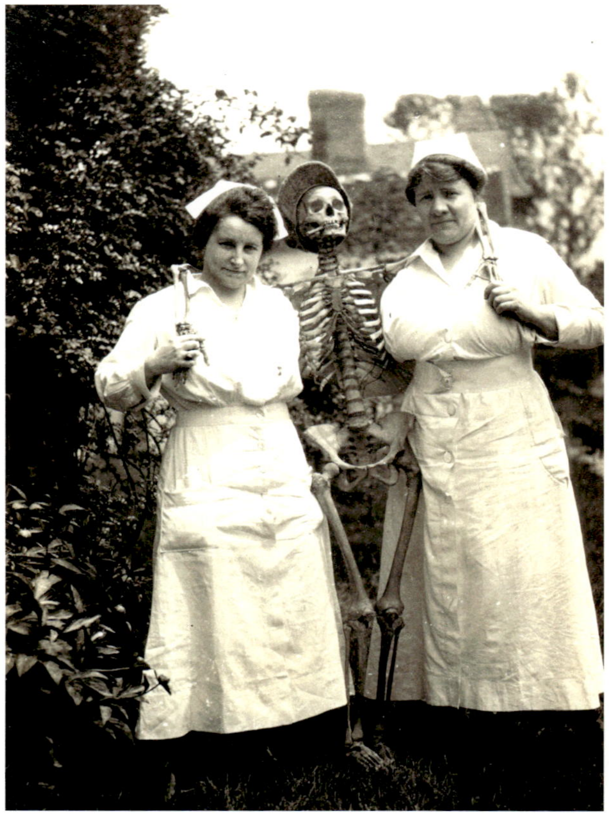

Plate 9.17

Nurses and Friend
Circa 1930

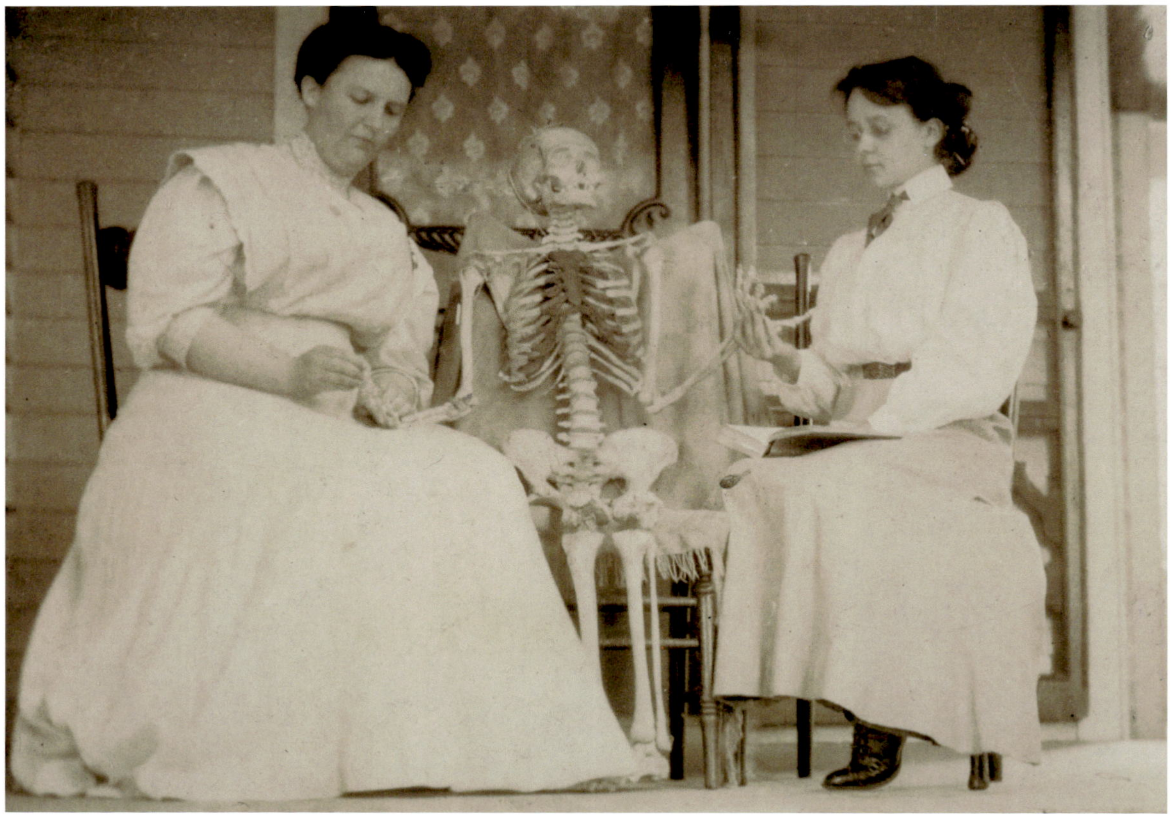

Plate 9.18

Nurses Examine Friend's Hands
Circa 1917

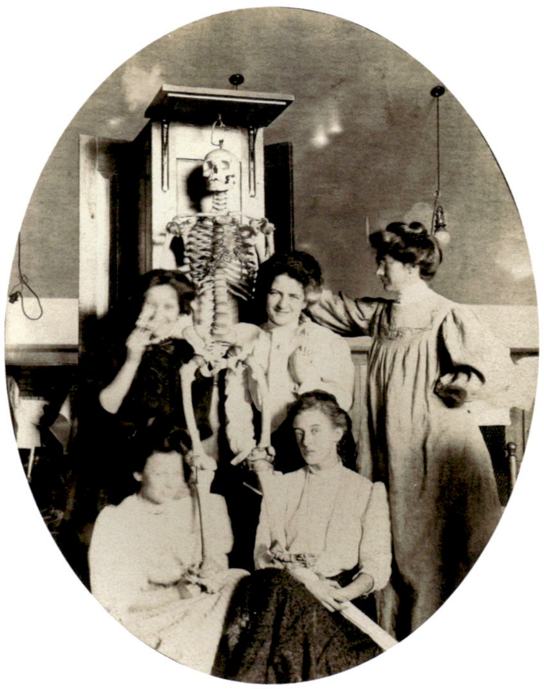

Plate 9.19

Nursing Students
Circa 1905

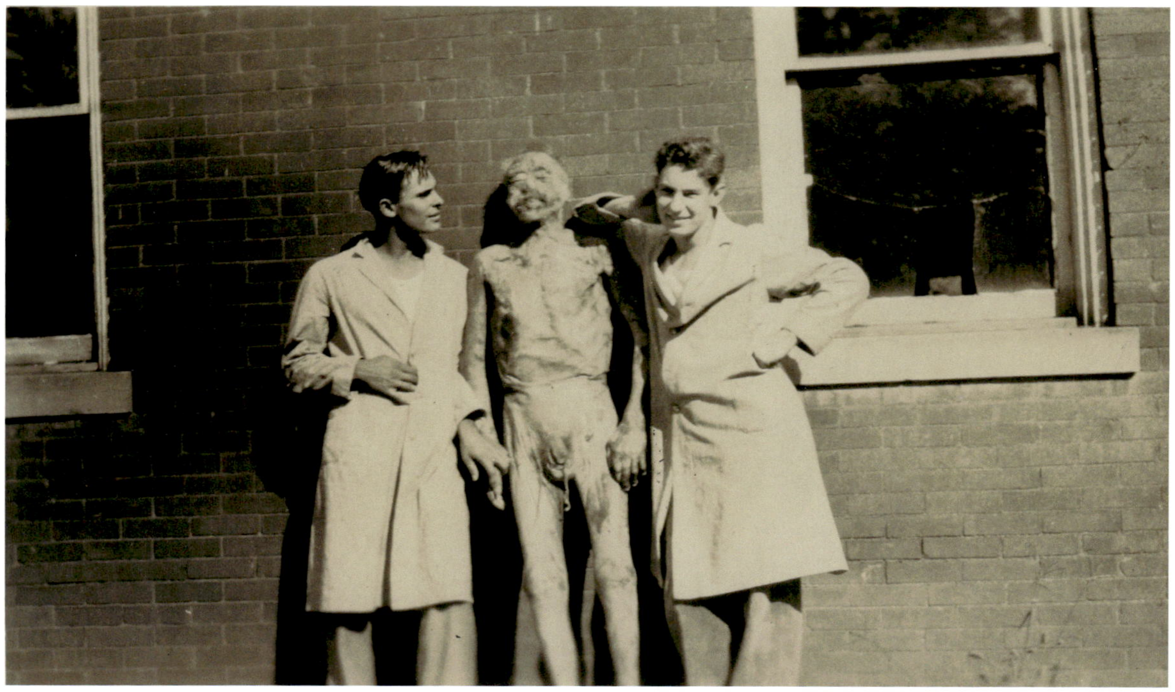

Plate 9.20

Our Pal "Louis Ward"
Circa 1920

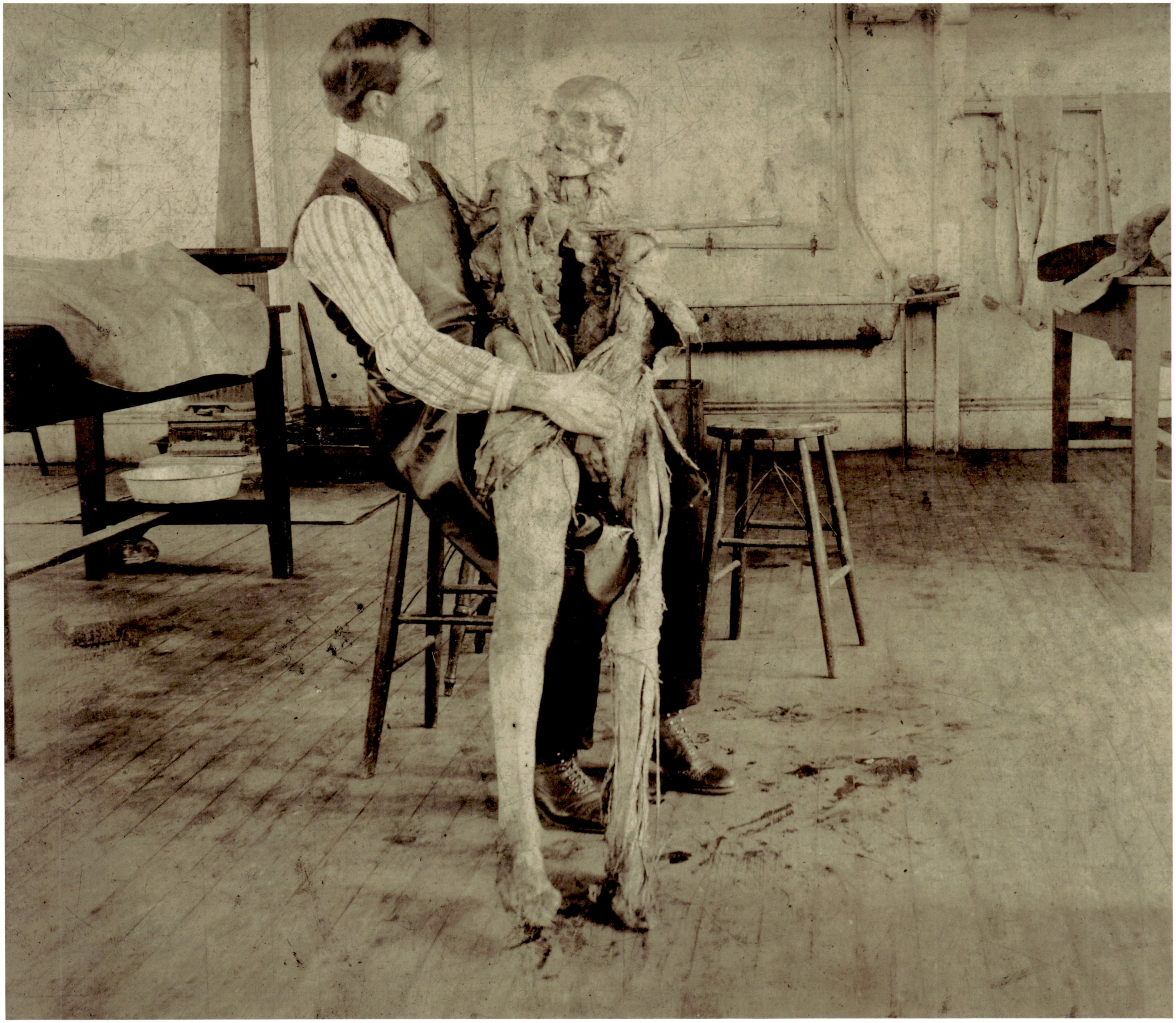

Plate 9.21

Dissection Mate

Circa 1898

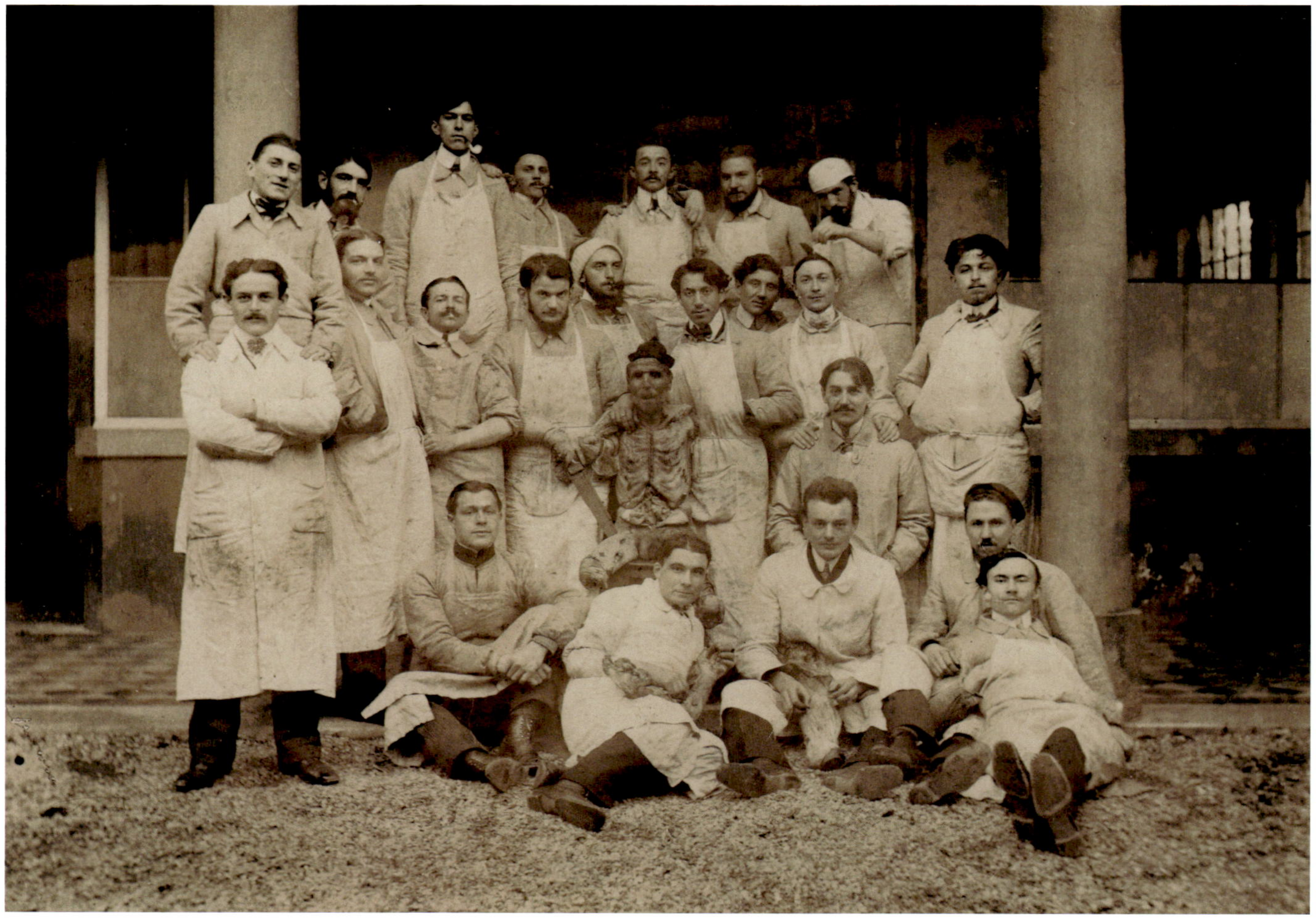

Plate 9.22

French Dissection Class
1909

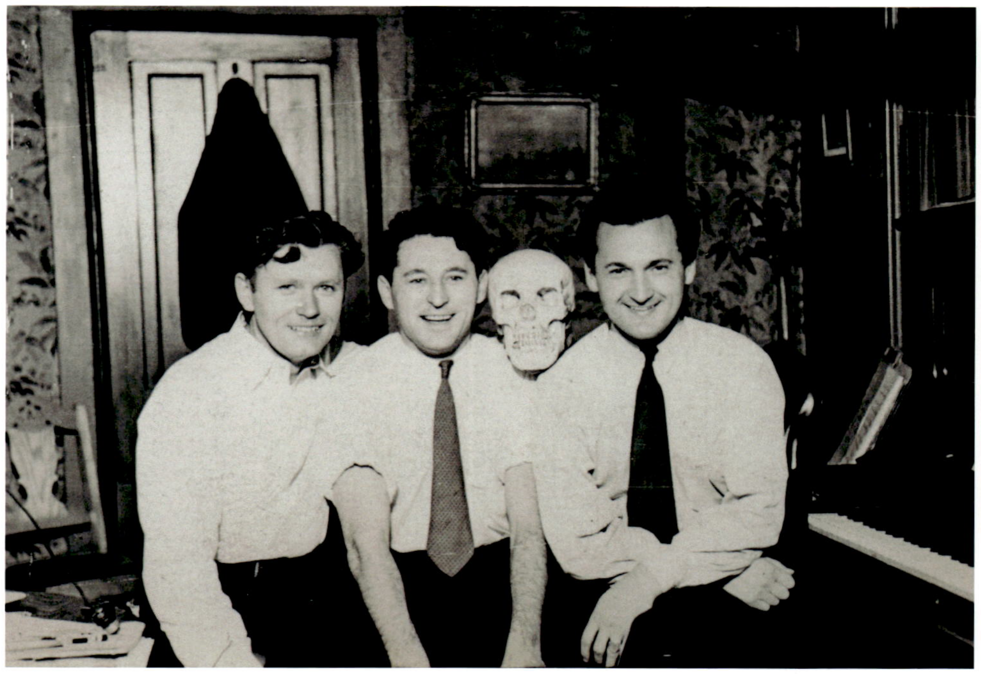

Plate 9.23

Four Friends
Circa 1935

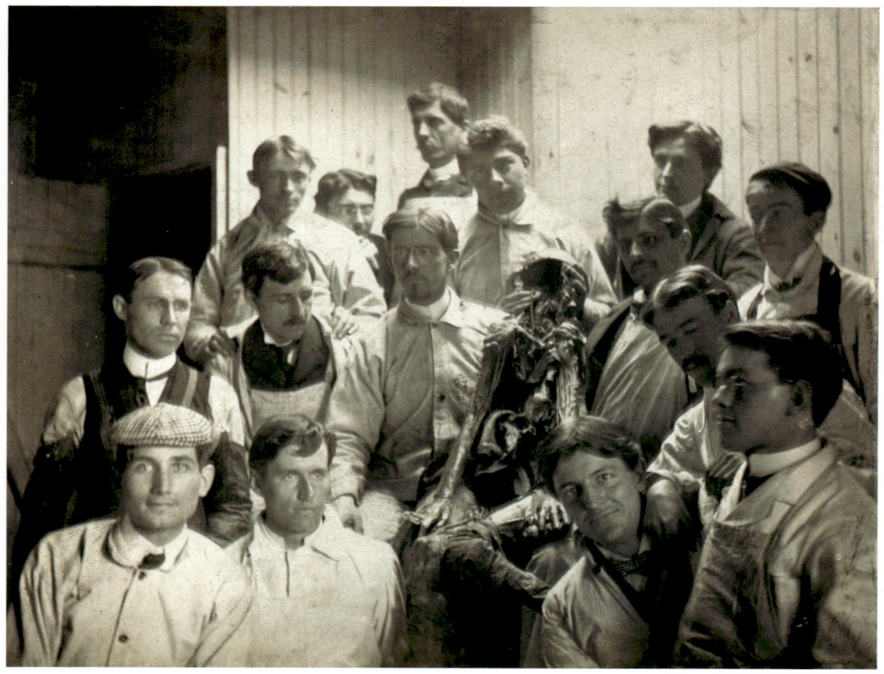

Plate 9.24

Dissection Class
September 1901

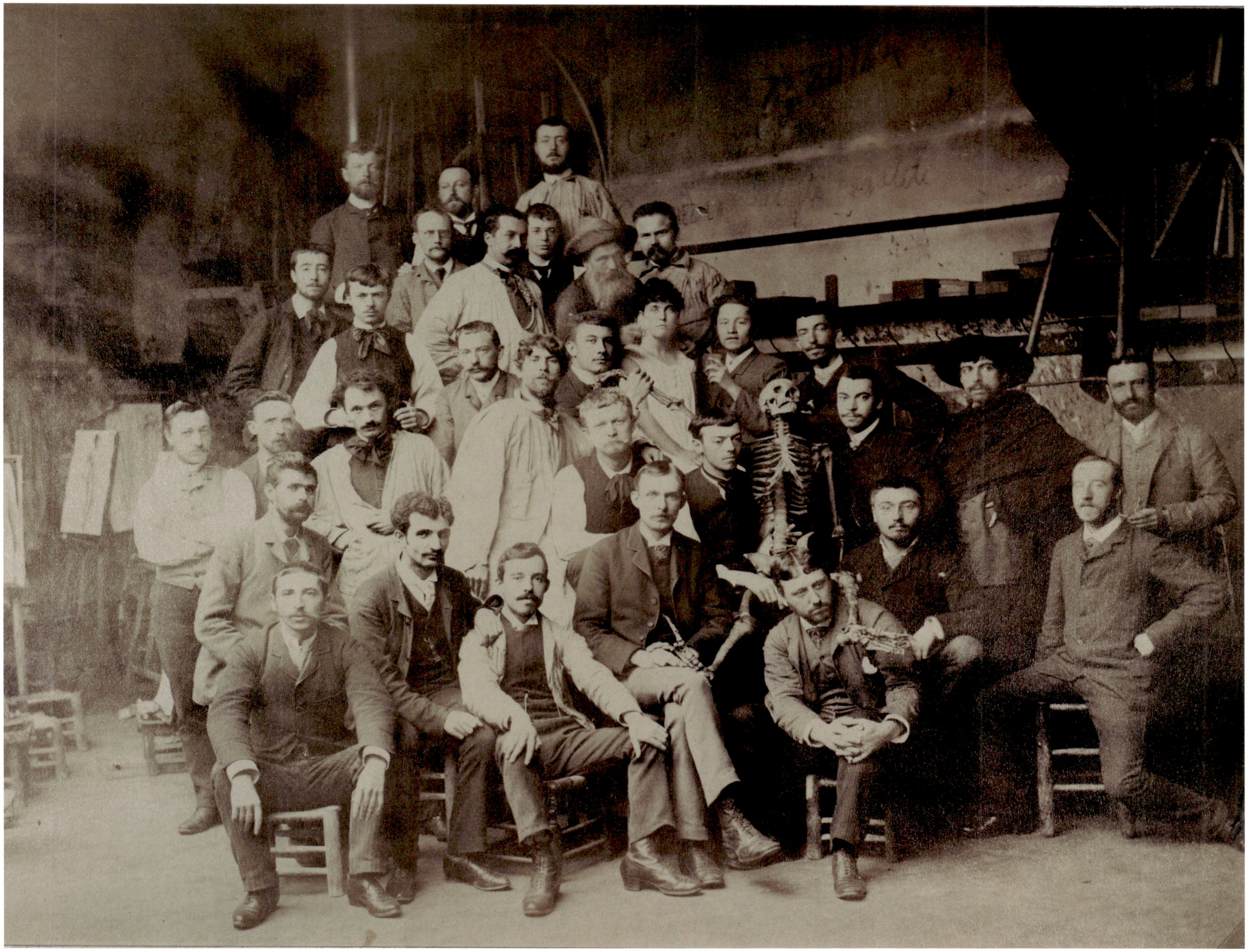

Plate 9.25

Painting Class at New York's Cooper Union
Photographer: Antonio Scilopti, noted artist in next to top row, 1899

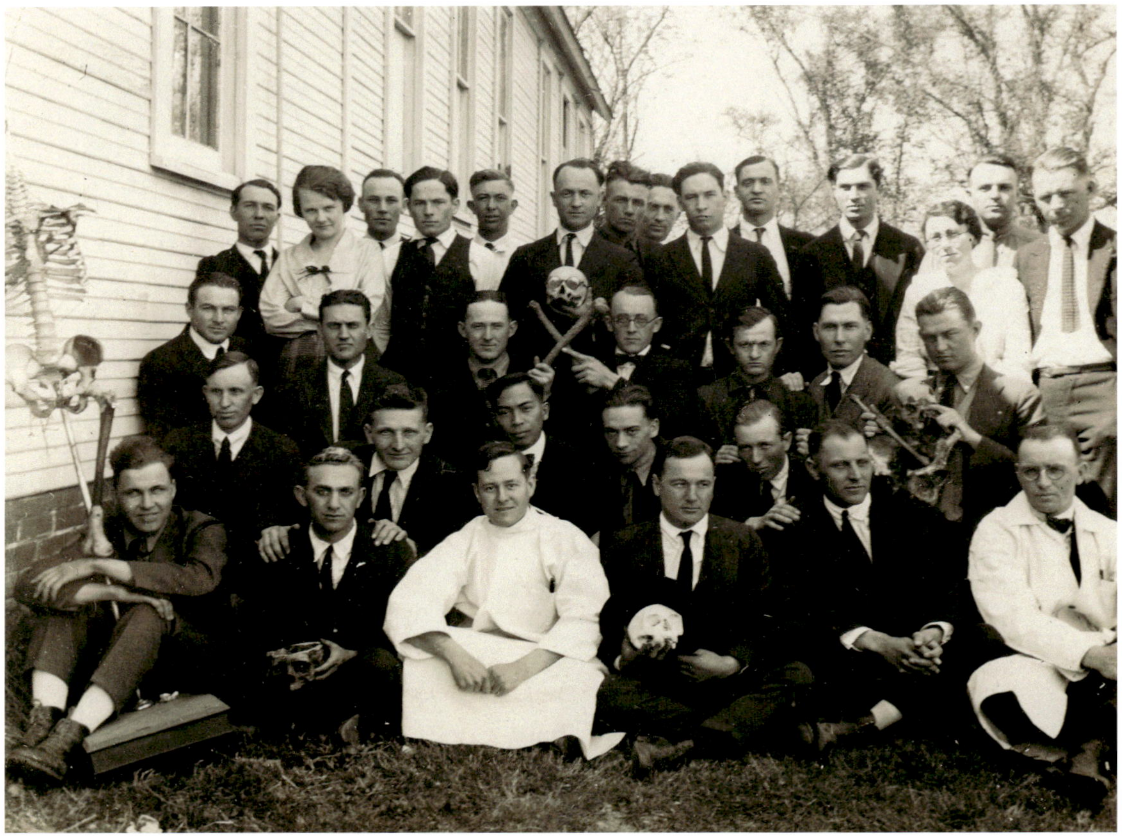

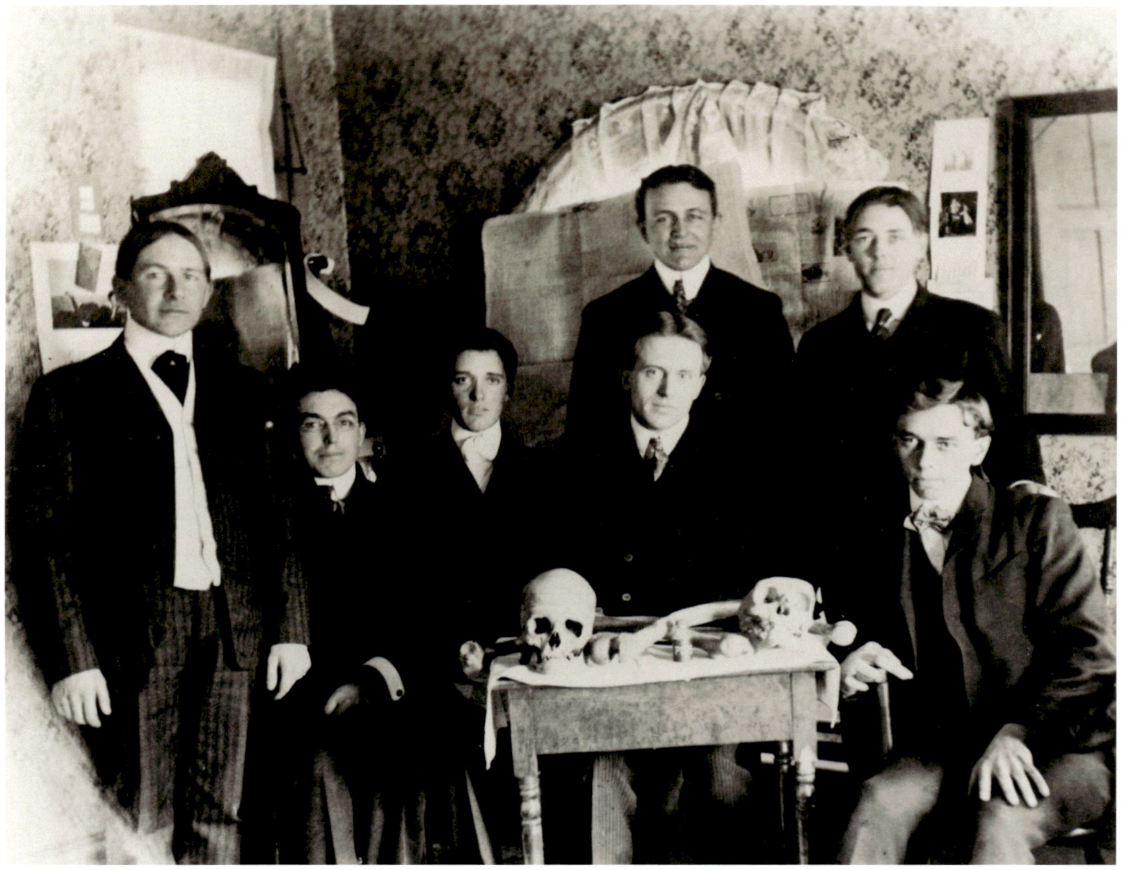

Plate 9.26

Anatomy Class, All Identified on Reverse
Circa 1910

Plate 9.27

St. Paul Minnesota Fraternity Group
Circa 1905

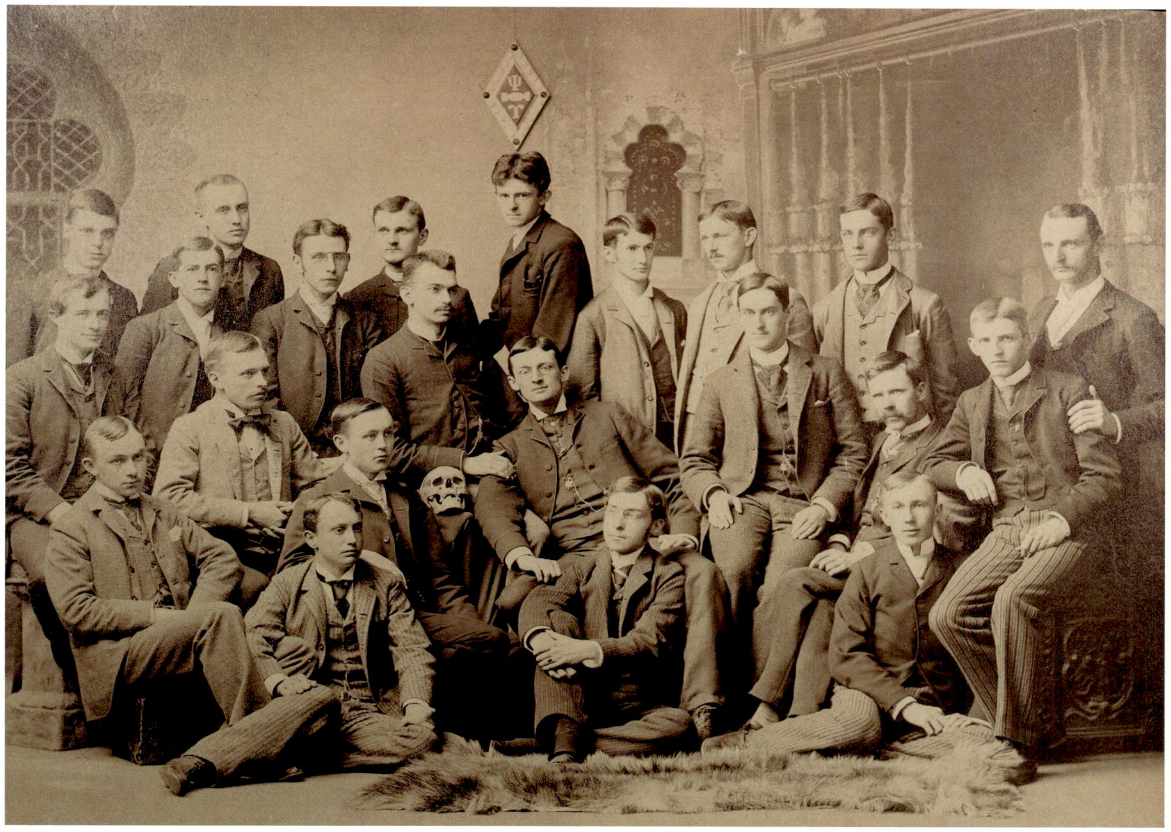

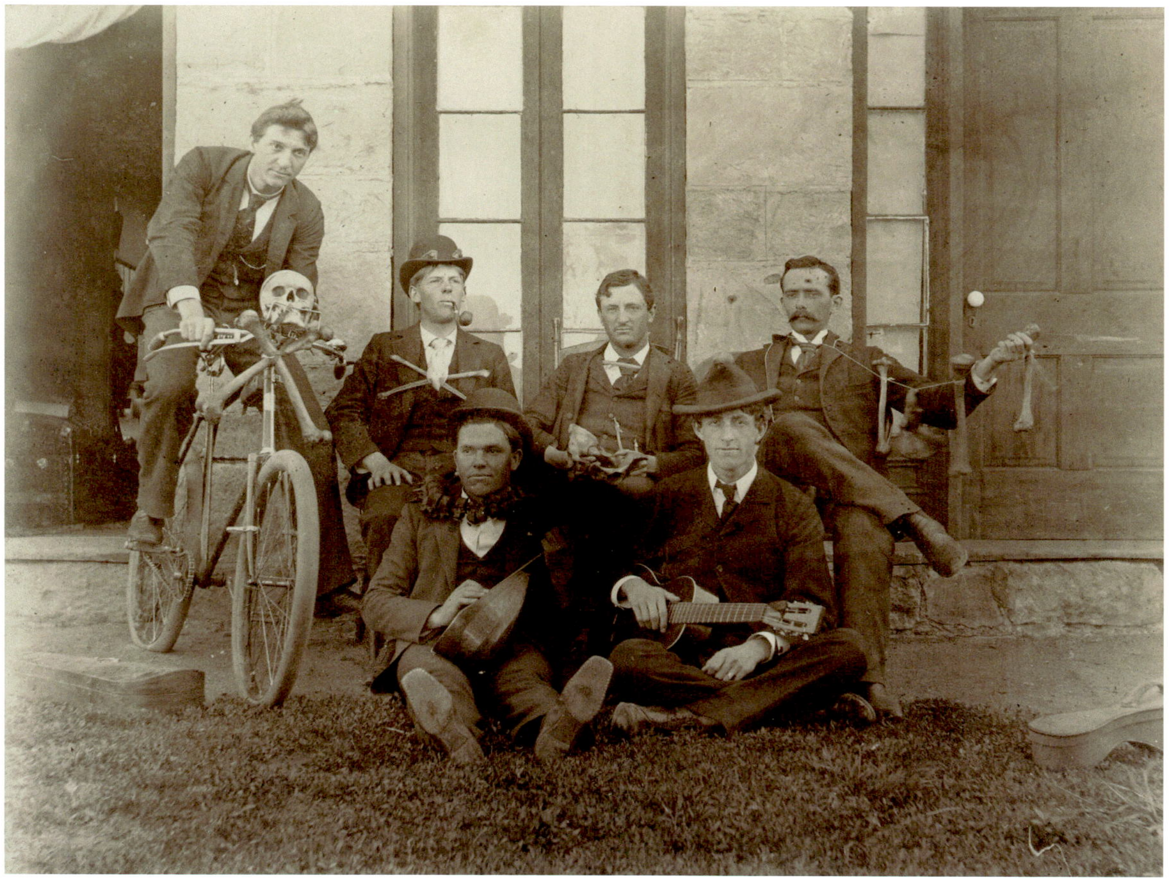

Plate 9.28

Psi Upsilon New York City Fraternity
Photographer: Pach Bros., 10" x 12" card photograph, 1989

Plate 9.29

Fraternity Group, with Bicycle
Circa 1900

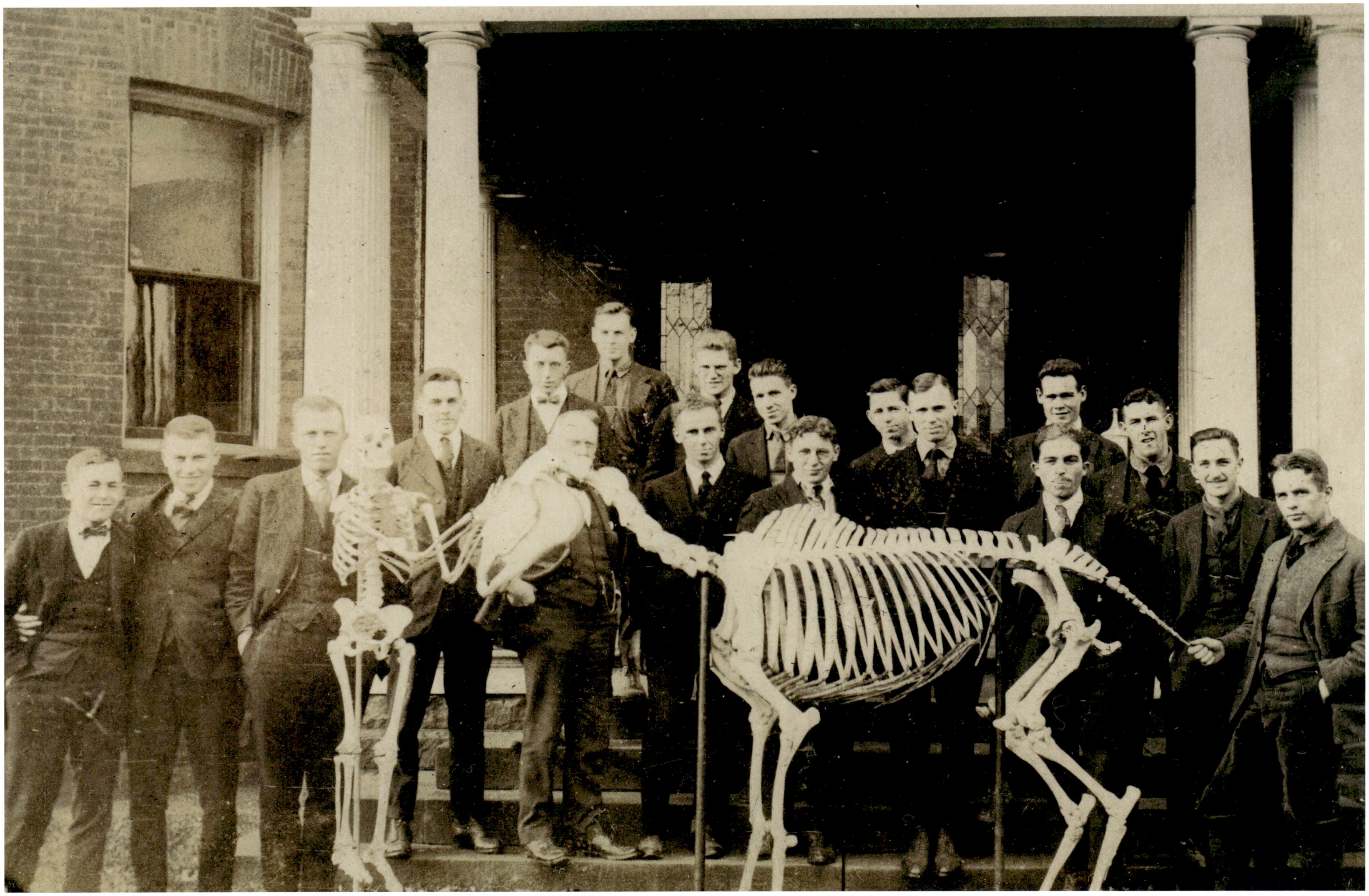

Plate 9.30

Veterinary Anatomy Class with Horse
1925

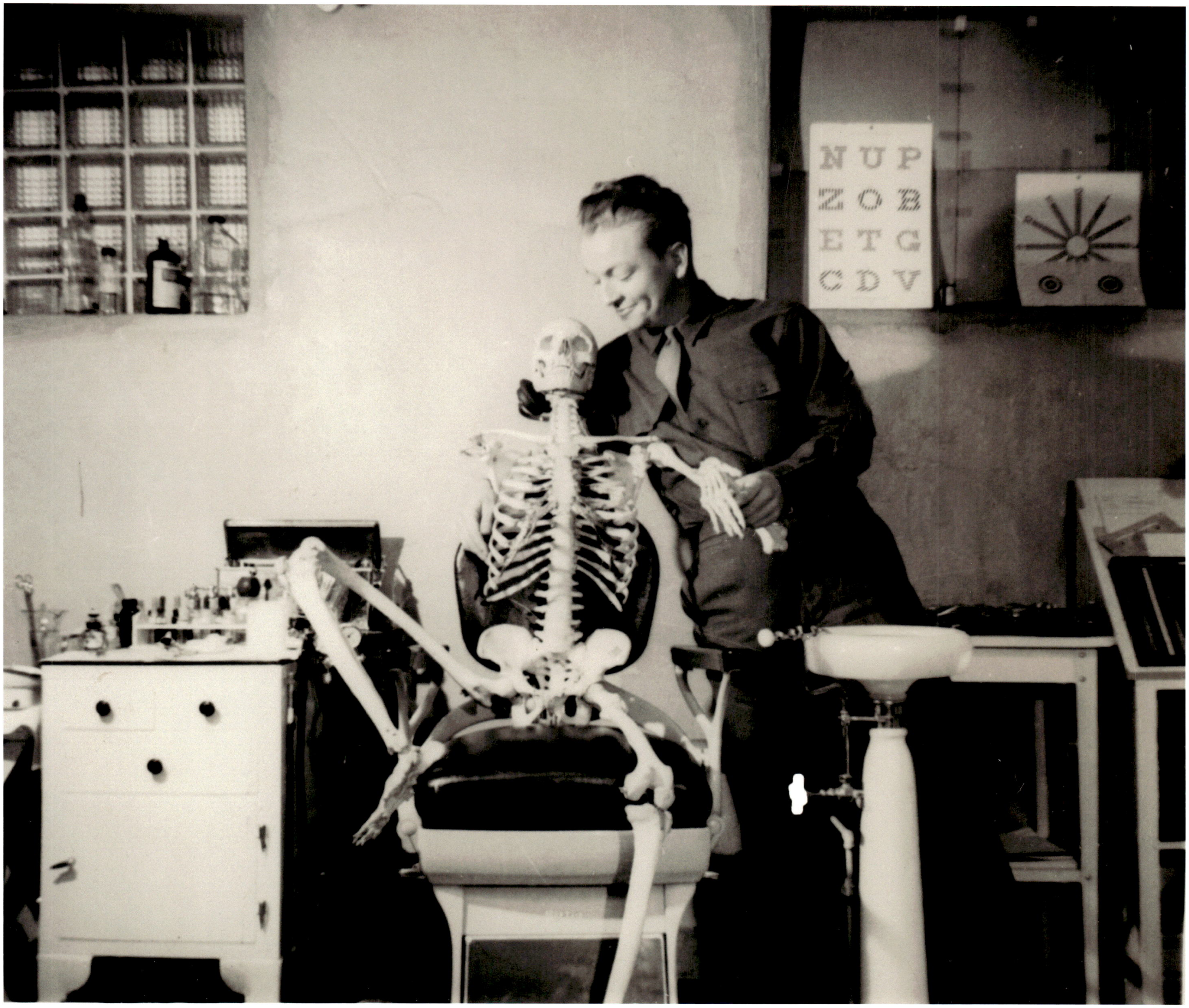

Plate 9.31

Army Physician with Patient
Circa 1935

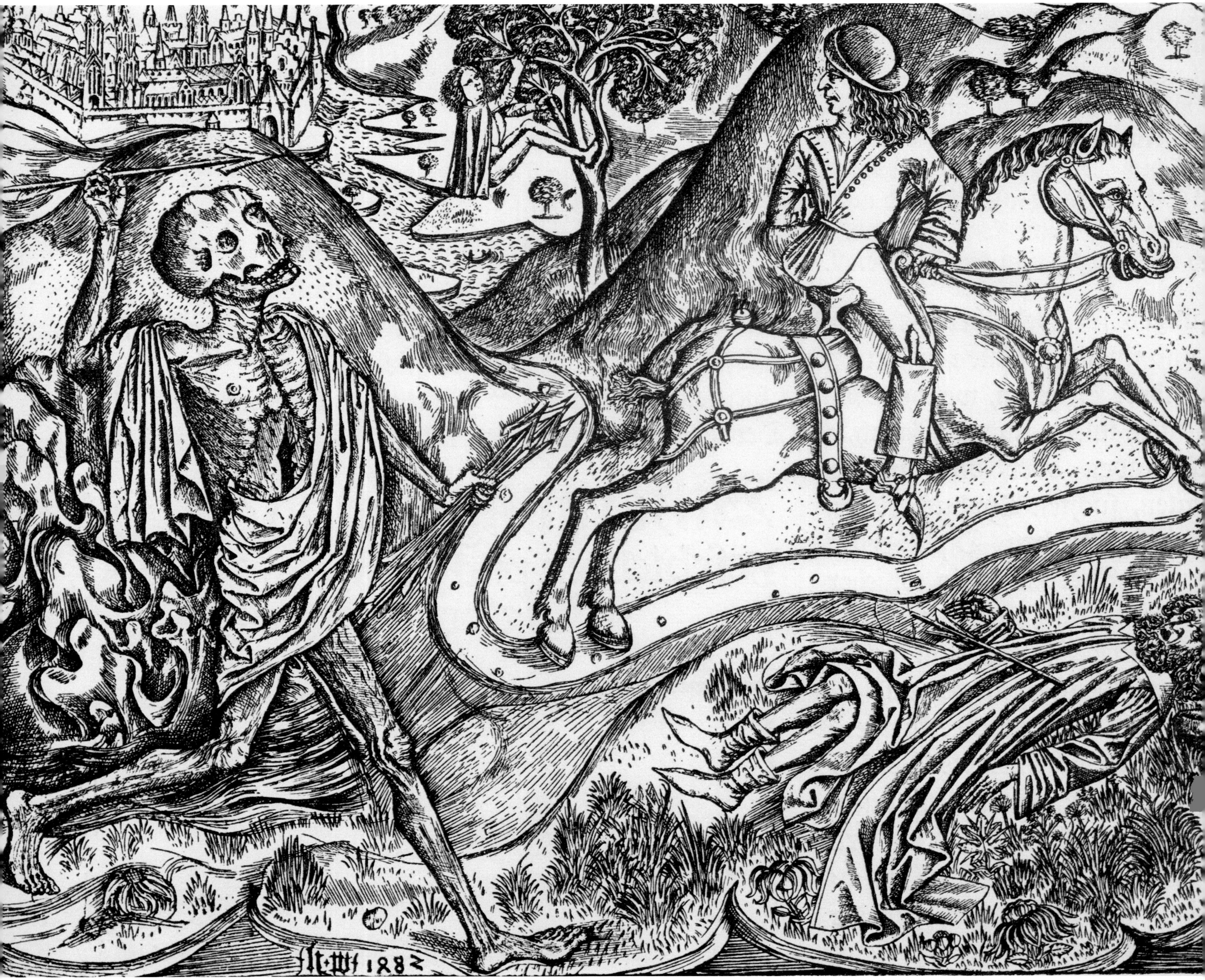

CHAPTER TEN

Popular Skeletal Views & Images for the Public

Indigenous Peoples – Ossuaries – War – Actors –Art

Skeletons, skulls, and bones have been the symbols of death since ancient times. The iconography can be seen in all aspects of life. It is found in religious, political, theatrical, artistic, and medical photographs. Fear of death has been one of mankind's greatest ancient and primal anxieties and features into the fascination with skulls and bones. Artists have used the imagery in many works that have influenced generations.

The use of skulls, skeletons, and body parts in nonmedical scenarios and cultural traditions, as well as use by autochthonous peoples is documented in this chapter. It begins with images of indigenous peoples using bones as part of their rites and rituals. Native Americans adorned themselves with body parts of their conquered enemies. A popular image sold to the public was of East Asian headhunters with stacks of trophy skulls. Common photographic postcards in the 1920–1950 time period featured the skeletons of indigenous peoples and exhumed graves. In the Victorian era, unwrapping mummies was in vogue. Europeans would bring back or buy mummies from Egypt and have unwrapping parties. Many mummies were destroyed in this mainly British upper class cultural phenomenon. Photographs of mummies were sold as souvenirs to tourists for their travel albums.

Boneyards, which became necessary, as well as noted catacombs and ossuaries are presented. The vast amount of land in the United States permitted everyone an underground burial. However, in Europe and in Asia only the rich could afford permanent burial plots due to the value of land. Ordinary people simply rented plots for a five or seven year period dependent on the area. When a family could not pay or would no longer pay, the bones were dug up. They were then thrown into pits called boneyards, or placed in catacombs, special large underground vaults often associated with a church. In some localities, such as Paris, due to the need for land, church cemeteries were dug up and the bones placed beneath the city in catacombs. Several French, German, and Italian cities had catacombs or other ossuary areas in church cellars and vaults. These ossuaries became popular tourist attractions and remain so to this day. Fascination in viewing the remains of the dead, the bones and skulls of our predecessors, results in long lines at these attractions. Photographers since the 1860s have exploited this fascination by photographing catacombs and selling the images to tourists. Among the ossuaries presented are those in France, Italy, Germany, England, Czechoslovakia, and Mexico.

Artworks often portrayed medical personnel, soldiers, and the common man fighting death represented by a skeleton. These images were popular renditions during wars and epidemics. To show the horrors of war, photographs of skulls and skeletons on the battlefield or in the aftermath of war have been a photographic genre since the American Civil War. Anti-war activists have used skulls and bones in their battle against conflict. At the turn of the century, posters and postcards were made using figure-ground technique. When viewed up close, the image showed a common scene and when viewed from a distance, a skull. Posing with a skull or skeleton was popular among thespians, especially those aspiring to play in Shakespeare's *Hamlet* as the Prince of Denmark. Artists and writers used the objects as props in a variety of contemplative poses.

The chapter ends with a peculiar use of skull and skeleton imagery that was very popular in

France in the 1860s: Les Diableries, a series of stereoview tissue cards, depicting life in a strange parallel universe called *Enfer*, or Hell. These visionary dioramas depict an imaginary underworld populated by devils, satyrs, and skeletons, very much alive, and, for the most part, having fun. Much of the subject matter was satirical and frequently mirrored the corruption and excesses of Paris's upper classes during the Second Empire under Napoleon III. Due to the vast number of interpretations of the use of bone symbols in photographs, this chapter can only touch upon the myriad forms that these objects have taken.

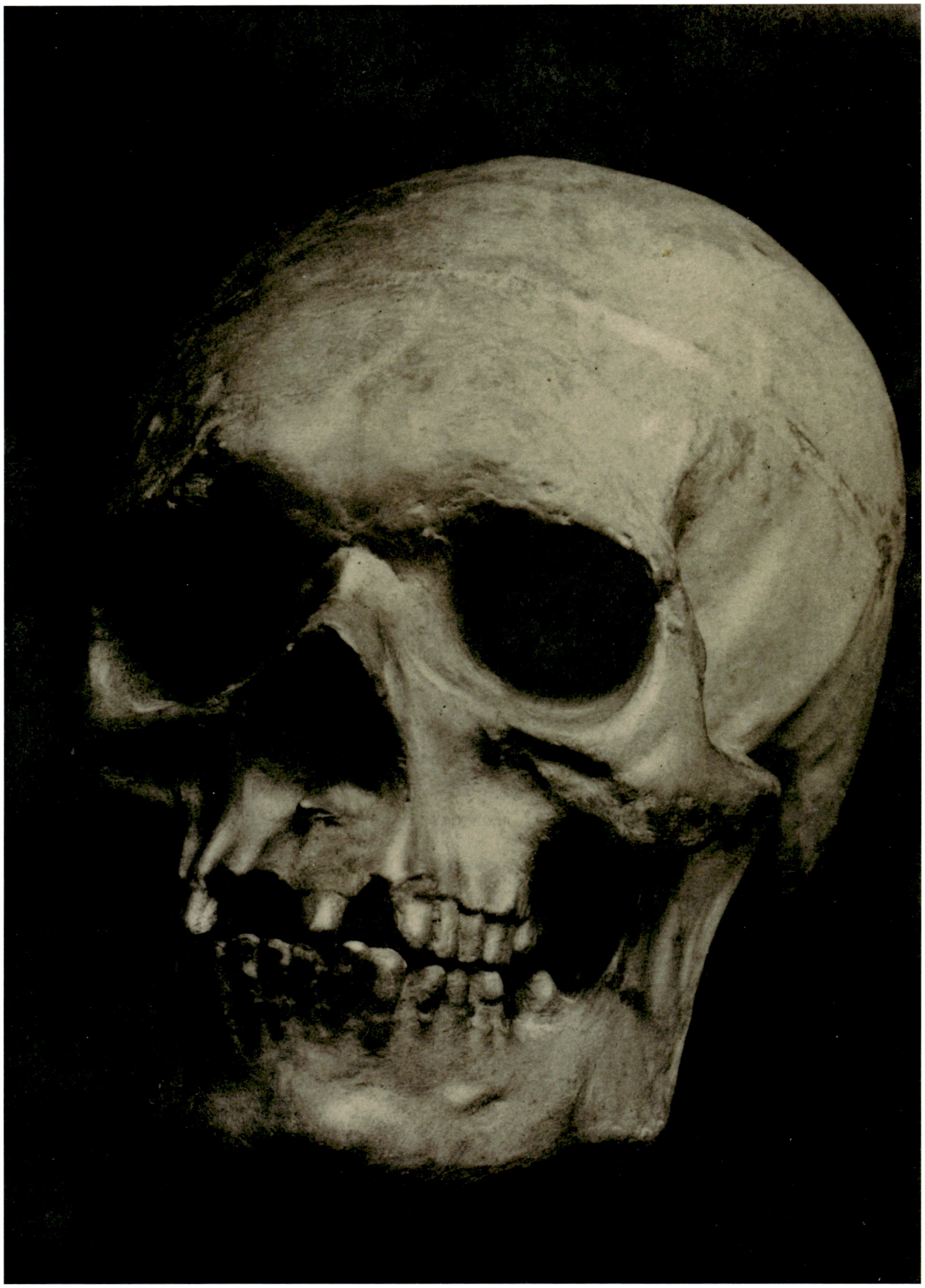

Plate 10.0

The Skull of Ludwig Van Beethoven *(1770-1827)*

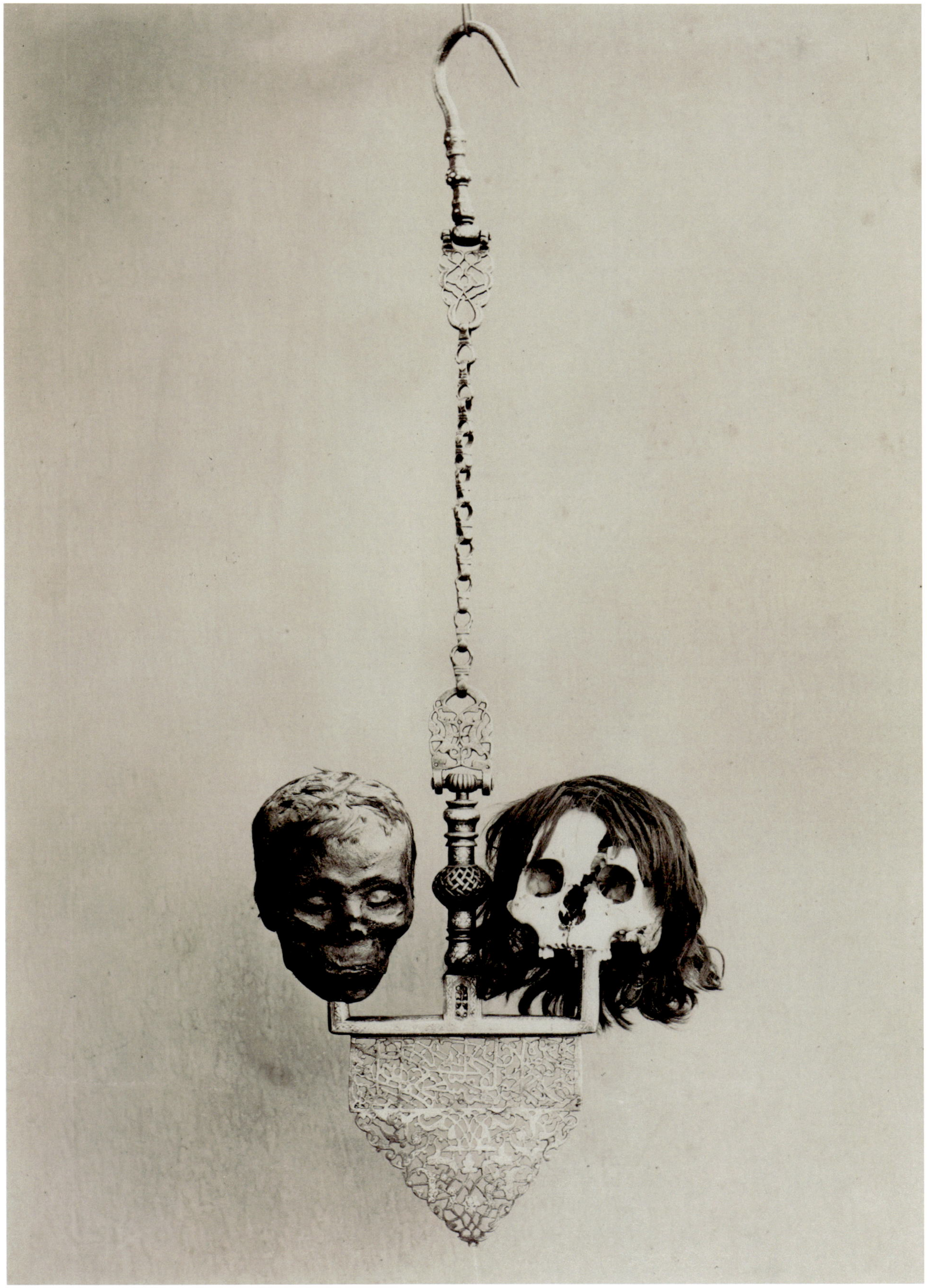

Plate 10.1

Mummified Head and Skull
South East Asian Religious Object, 1880

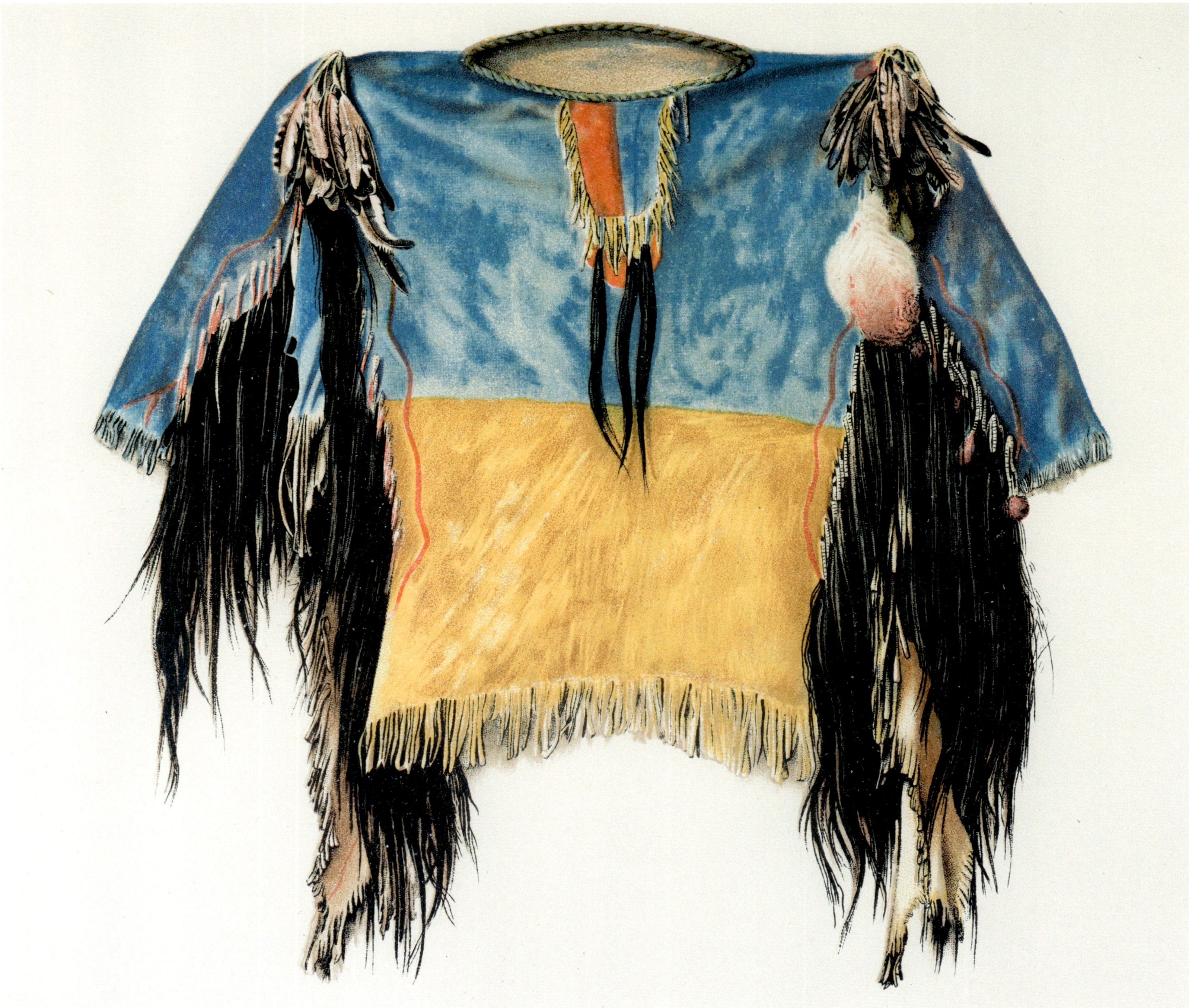

Plate 10.2

Sioux Scalp Shirt belonging to 'Little Big Man'
Circa 1875

Buckskin trimmed with human scalps, once worn by Crazy Horse.

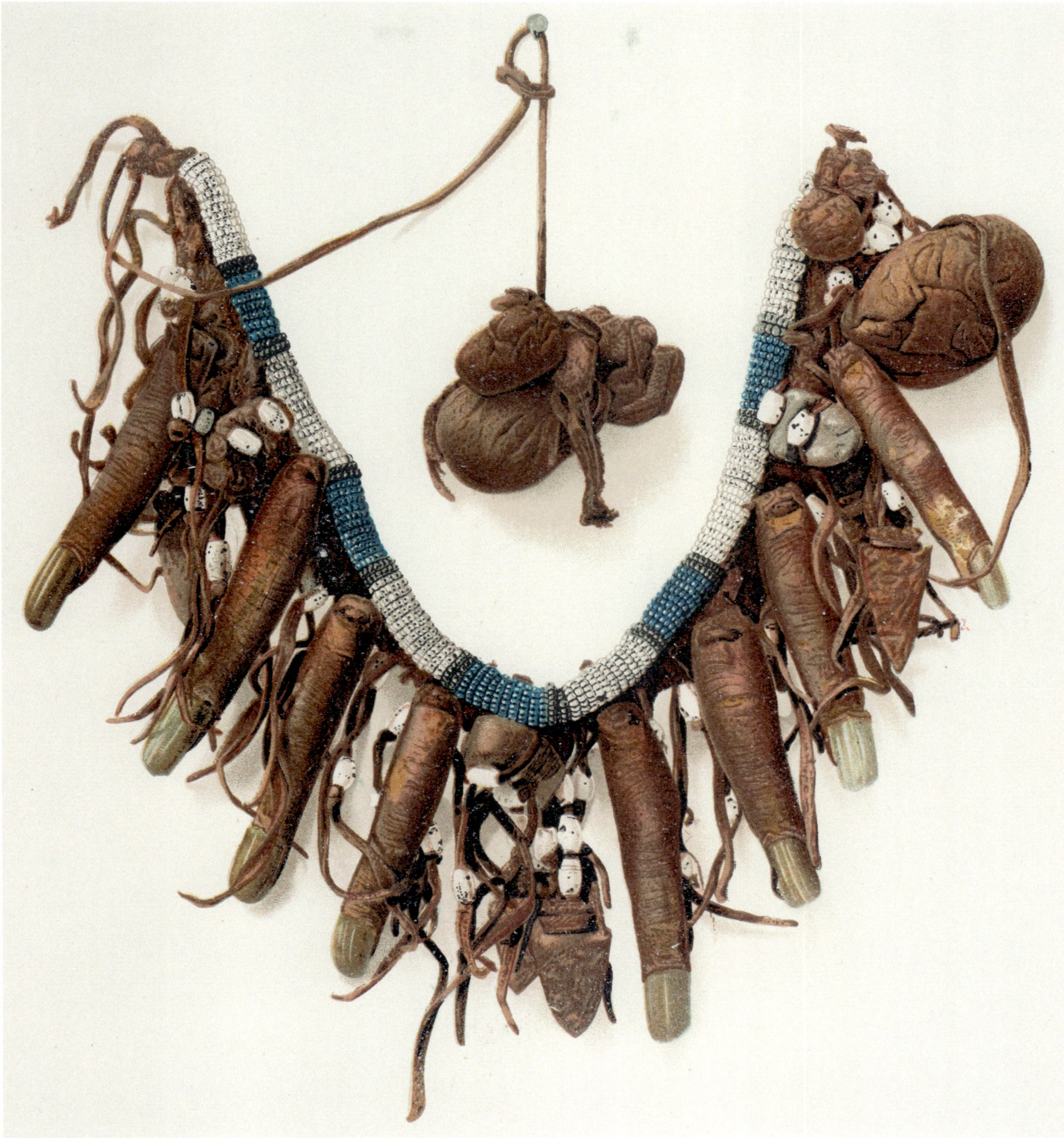

Plate 10.3

Cheyenne Necklace of Human Fingers Owned by High Wolf
Circa 1876

All the fingers are the left middle finger of enemies killed.

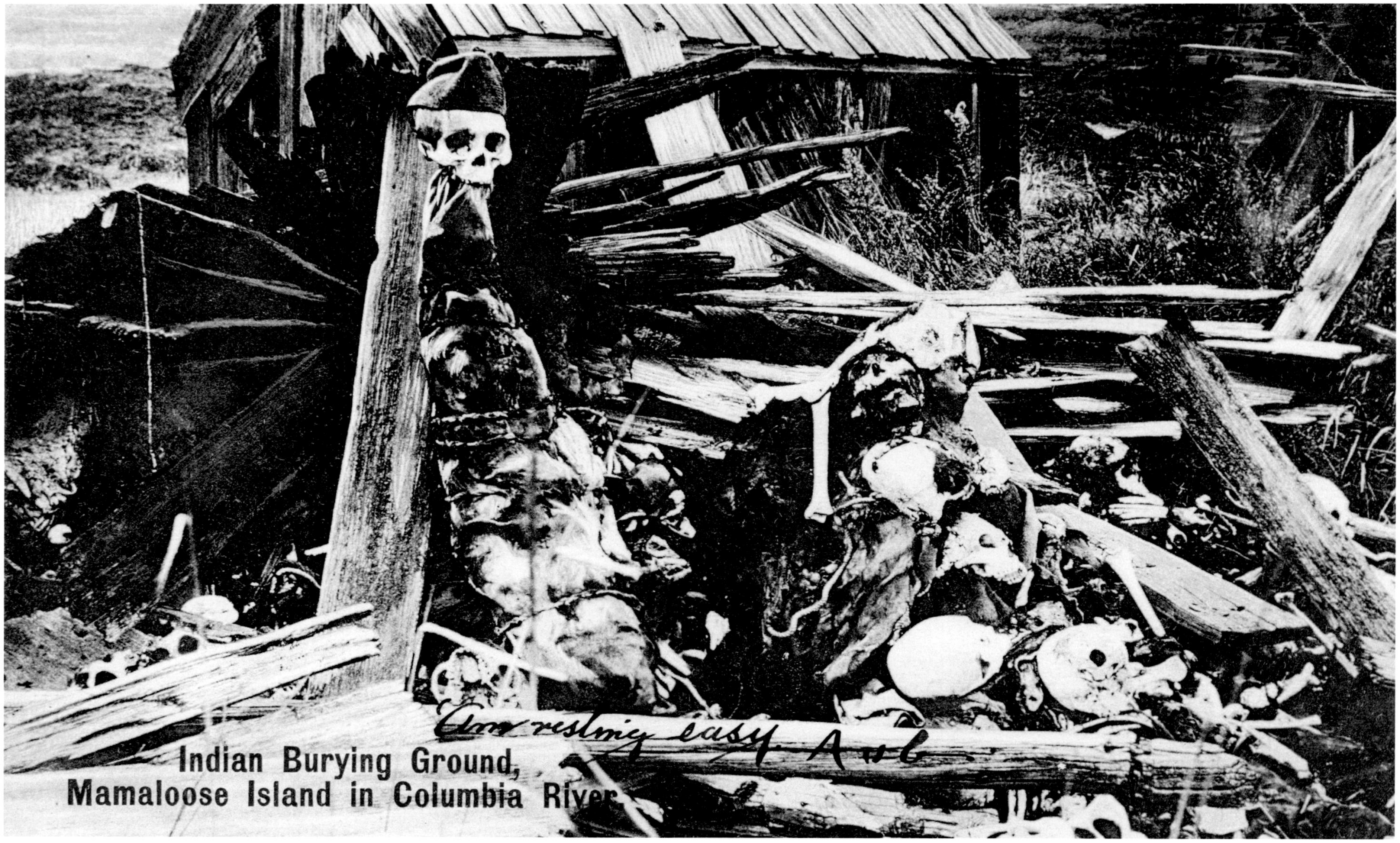

Plate 10.4

Indian Burying Ground, Mamaloose Island in Columbia River

Postcard with written note: "Portland Oregon the home of the busted banks," 1908

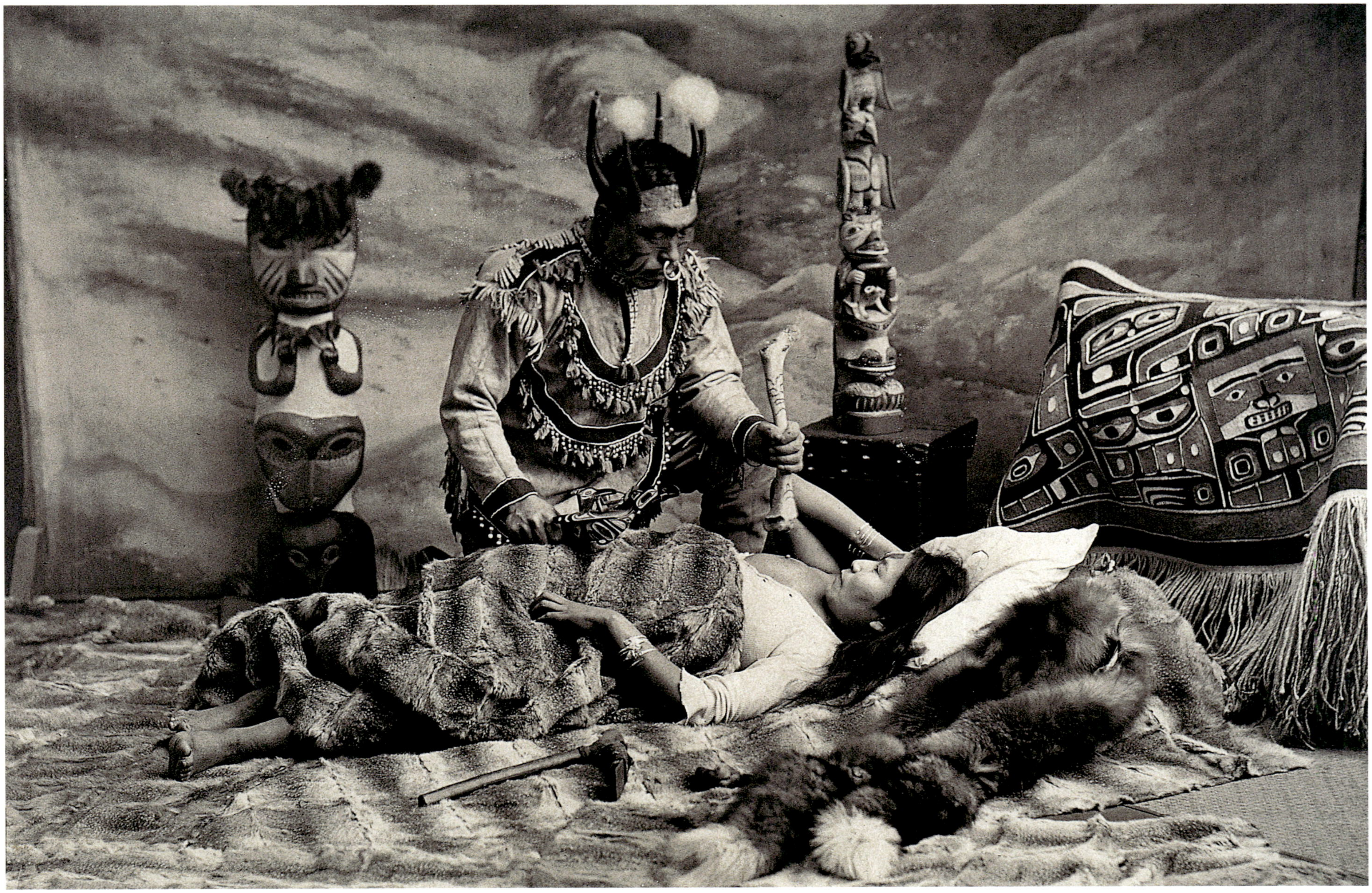

Plate 10.5

Skundoo, a Chilkat Shaman, Treating a Woman for Tuberculosis
Platinum Print, Photographer: Case and Draper, Juneau, Alaska, 1908

Plate 10.6

Menmaatre Seti I, Pharaoh of the New Kingdom 19th Dynasty of Egypt
1885

Son of Ramesses I and Queen Sitre, and the father of Ramesses II.

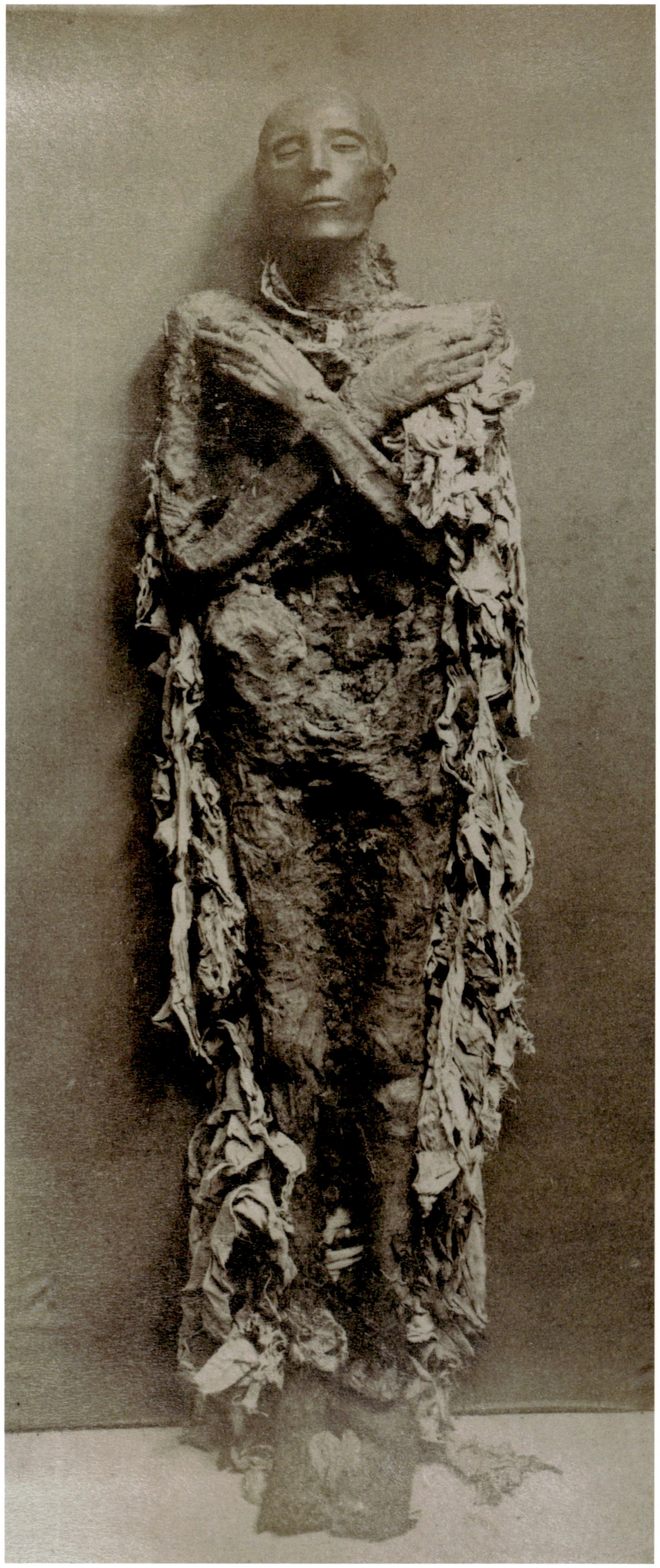

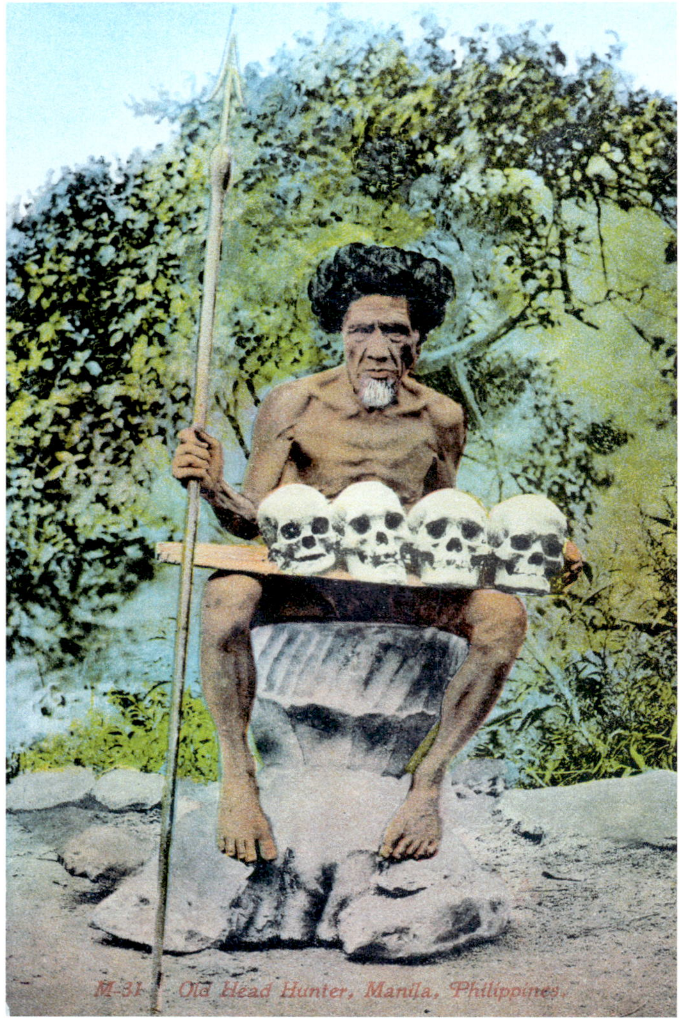

Plate 10.7

Head Hunter with Skulls, Manila, Philippine Islands
Postcard, 1900

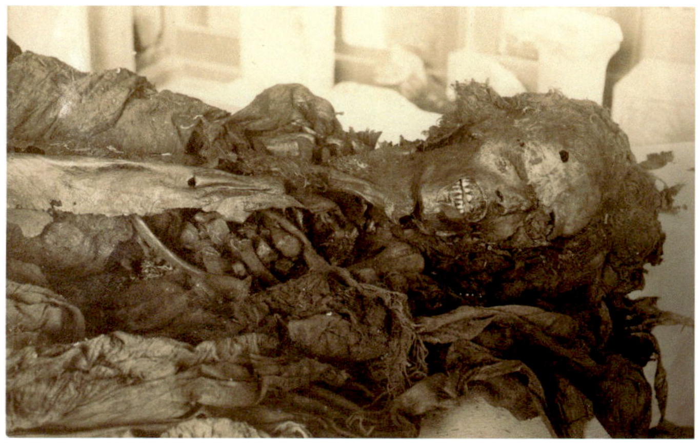

Plate 10.8

Unwrapped Egyptian Mummy
1870

Excavating Egyptian tombs was a popular 'sport' of the wealthy.

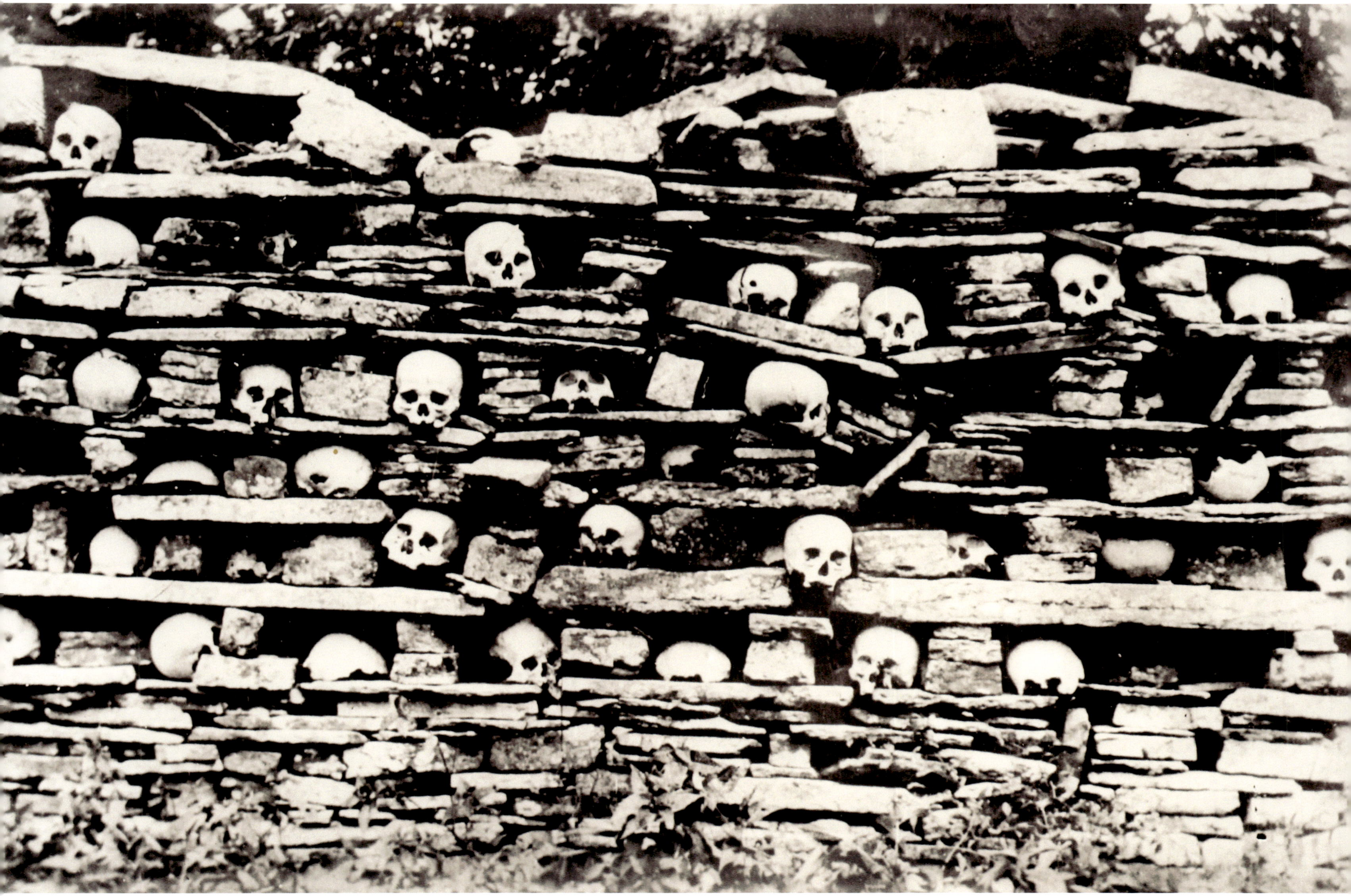

Plate 10.9

Head Hunters Trophy Wall, Formosa
Postcard, 1910

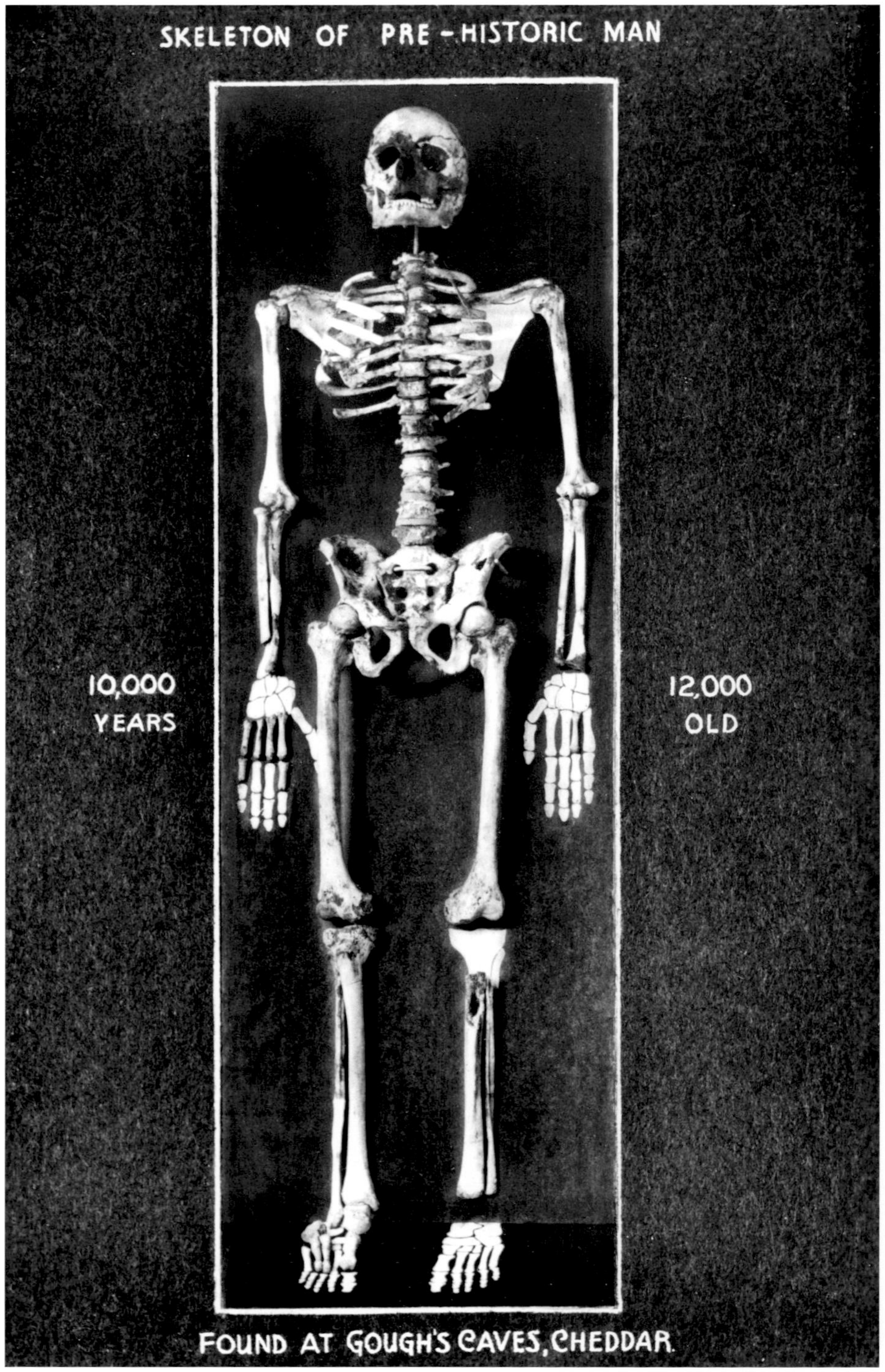

Plate 10.10

Skeleton of Prehistoric Man Found at Gough's Cave, Cheddar, England
Postcard, circa 1920

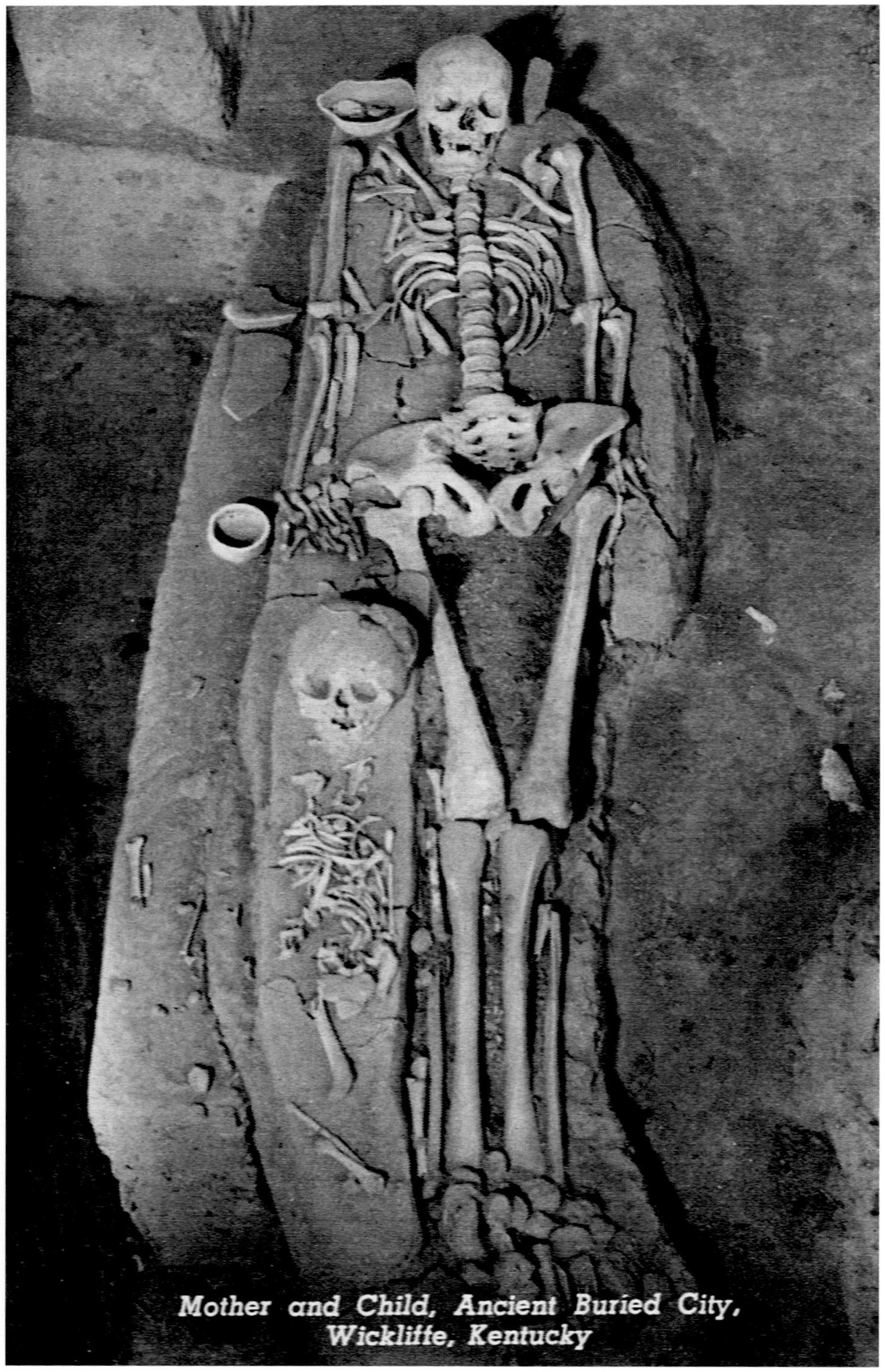

Plate 10.11

Mother and Child, Ancient Buried City, Wickliffe, Western Kentucky
Postcard, circa 1940

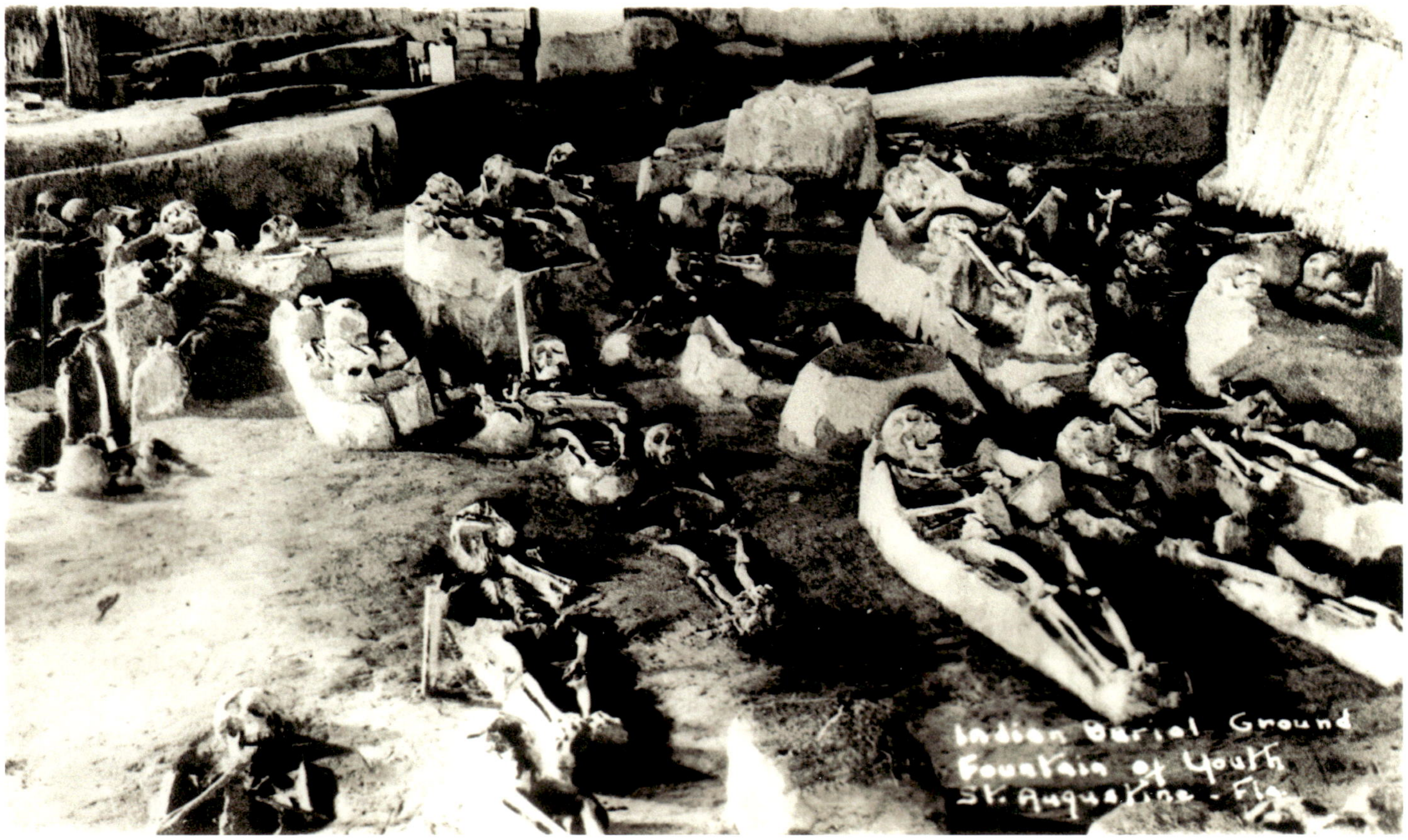

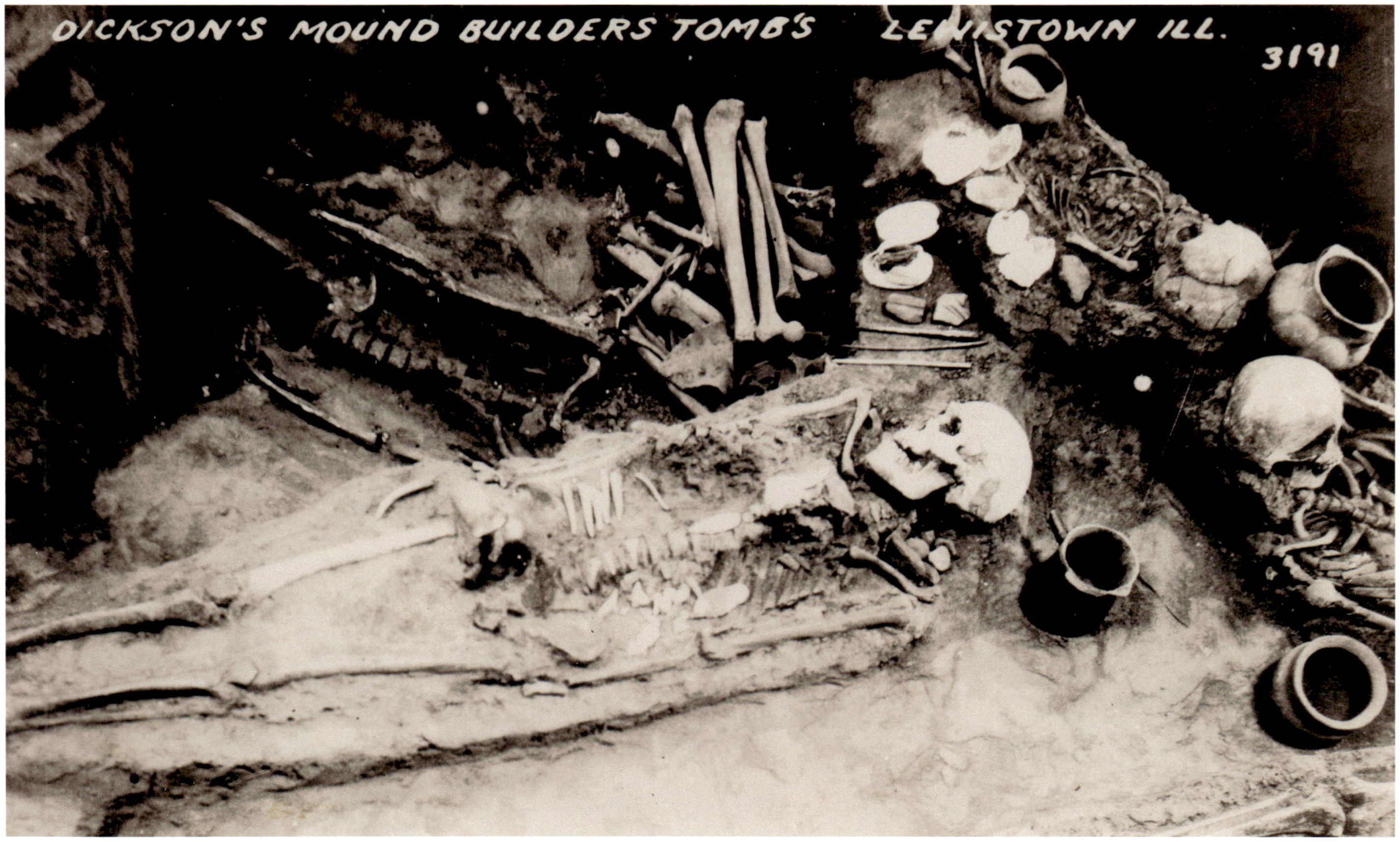

Plate 10.12

Indian Burial Ground, Fountain of Youth
St. Augustine, FL
Postcard, circa 1935

Plate 10.13

Dickson's Mound Builders Tombs
Lewistown, IL
Postcard, circa 1945

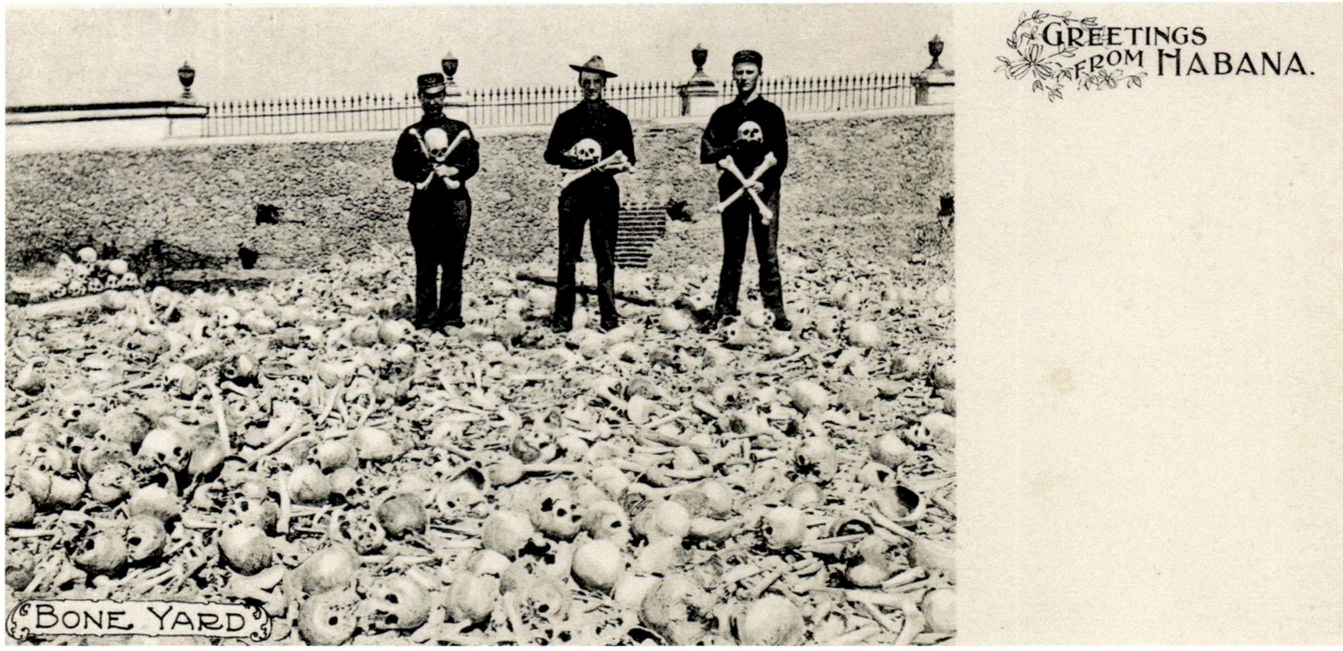

Plate 10.14

Spanish American War Soldiers in the Bone Yard, Havana, Cuba
Postcard, "Greetings from Habana." 1898

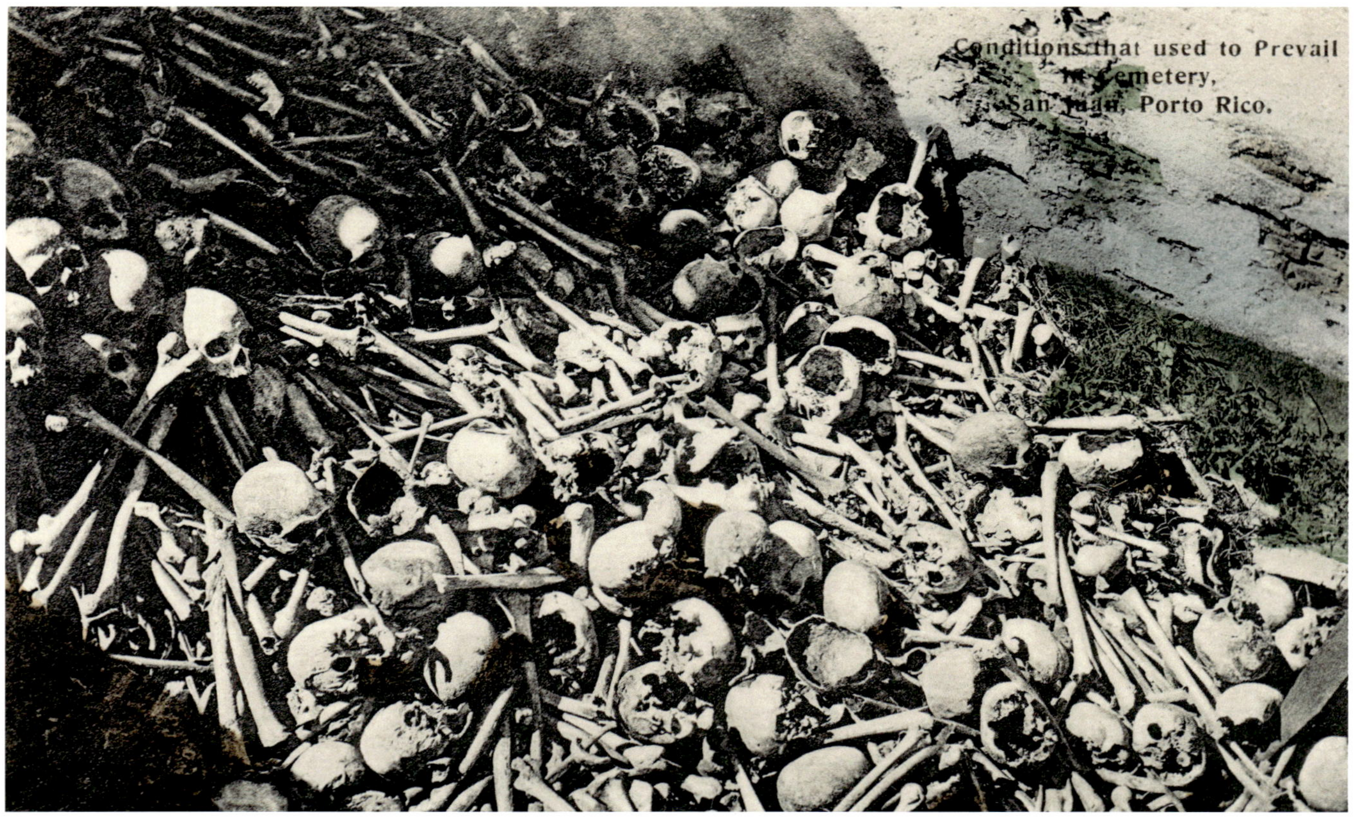

Plate 10.15

The Bone Yard in San Juan, Puerto Rico
Postcard, 1900

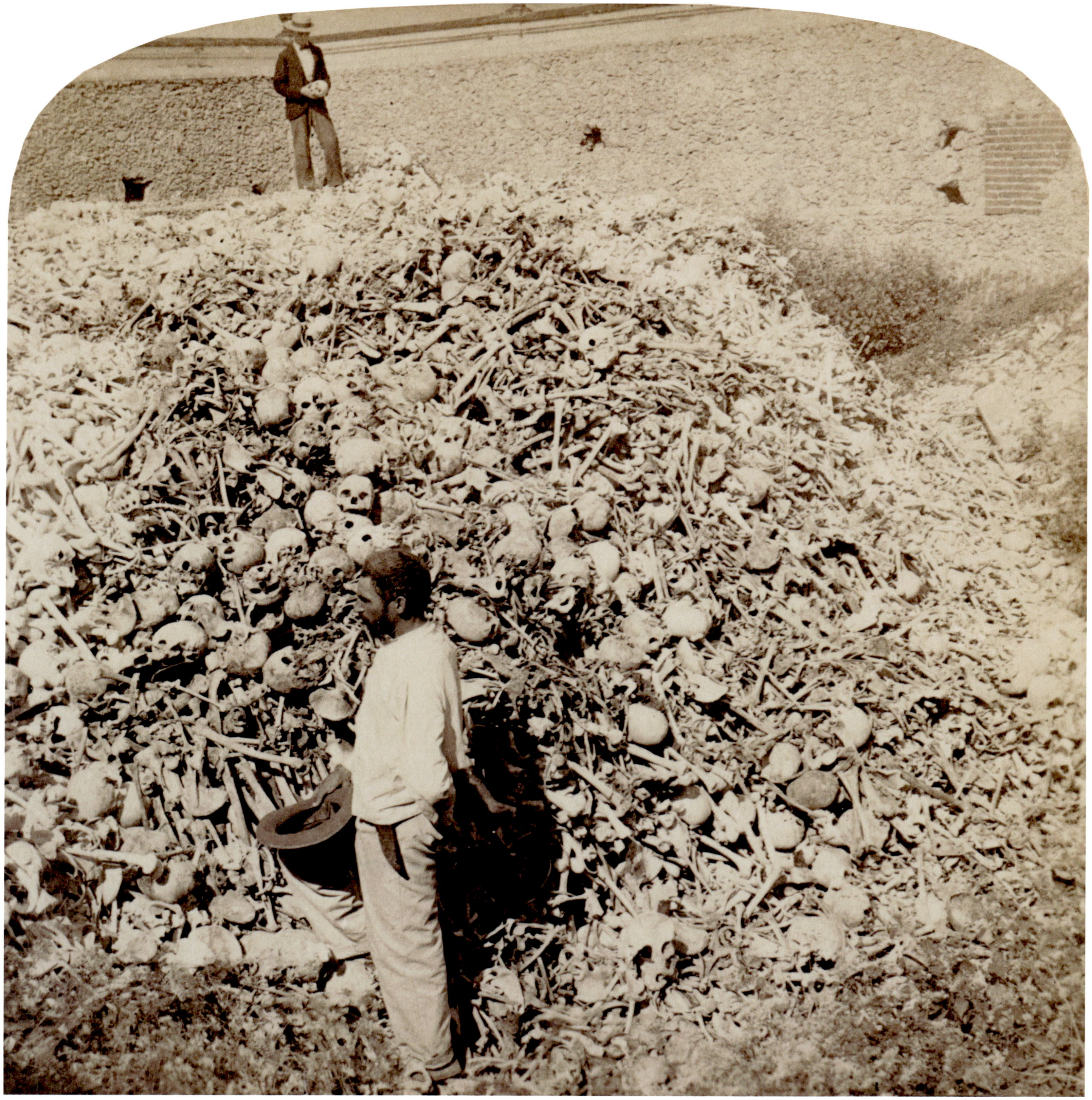

Plate 10.16

Two Hundred Thousand Skeletons in Havana's Colon Cemetery
Popular stereoview of the Havana Bone Yard, 1899

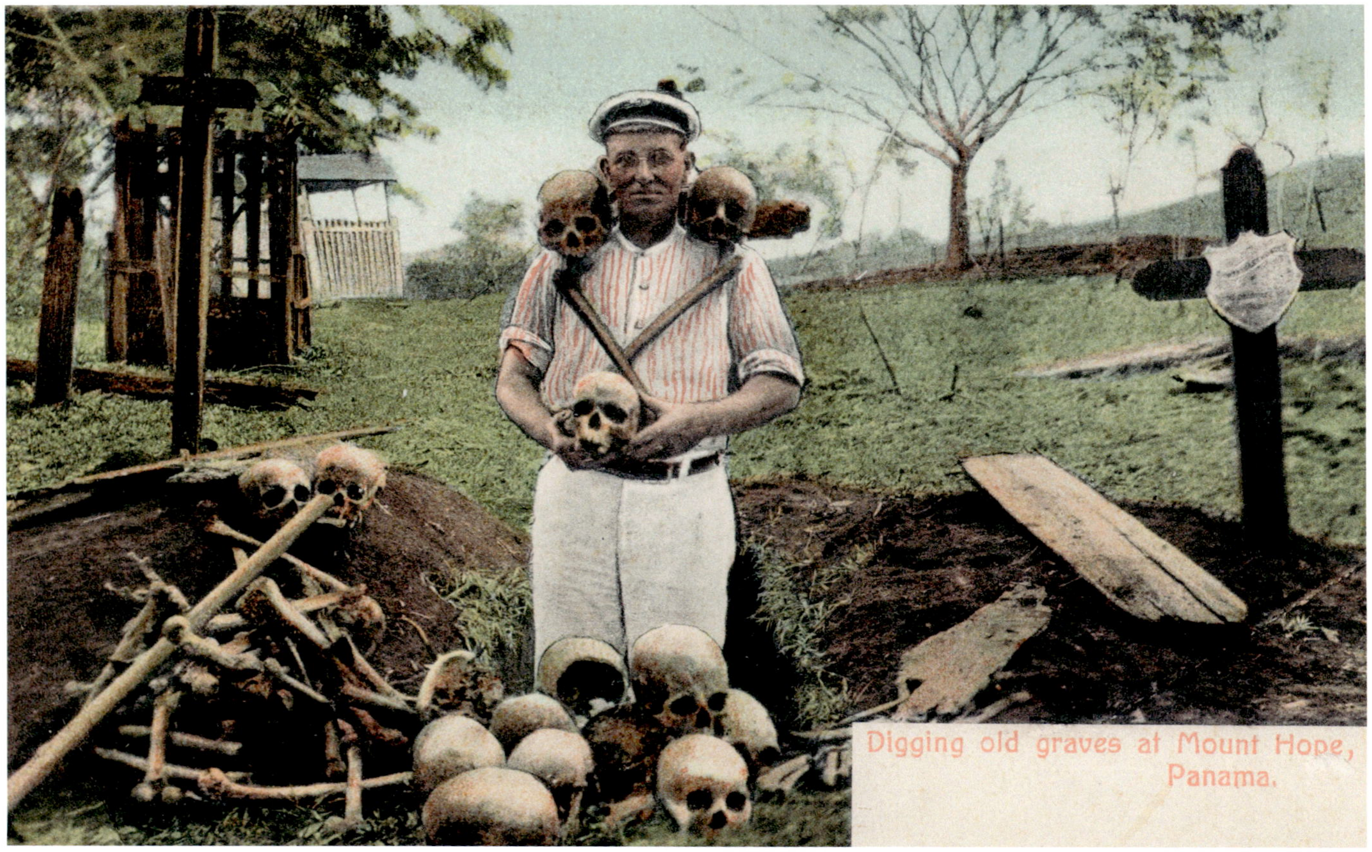

Plate 10.17

Digging Old Graves at Mount Hope, Panama
Postcard, 1907

An American hobby since the seventeenth century was hunting for skulls in old graves.

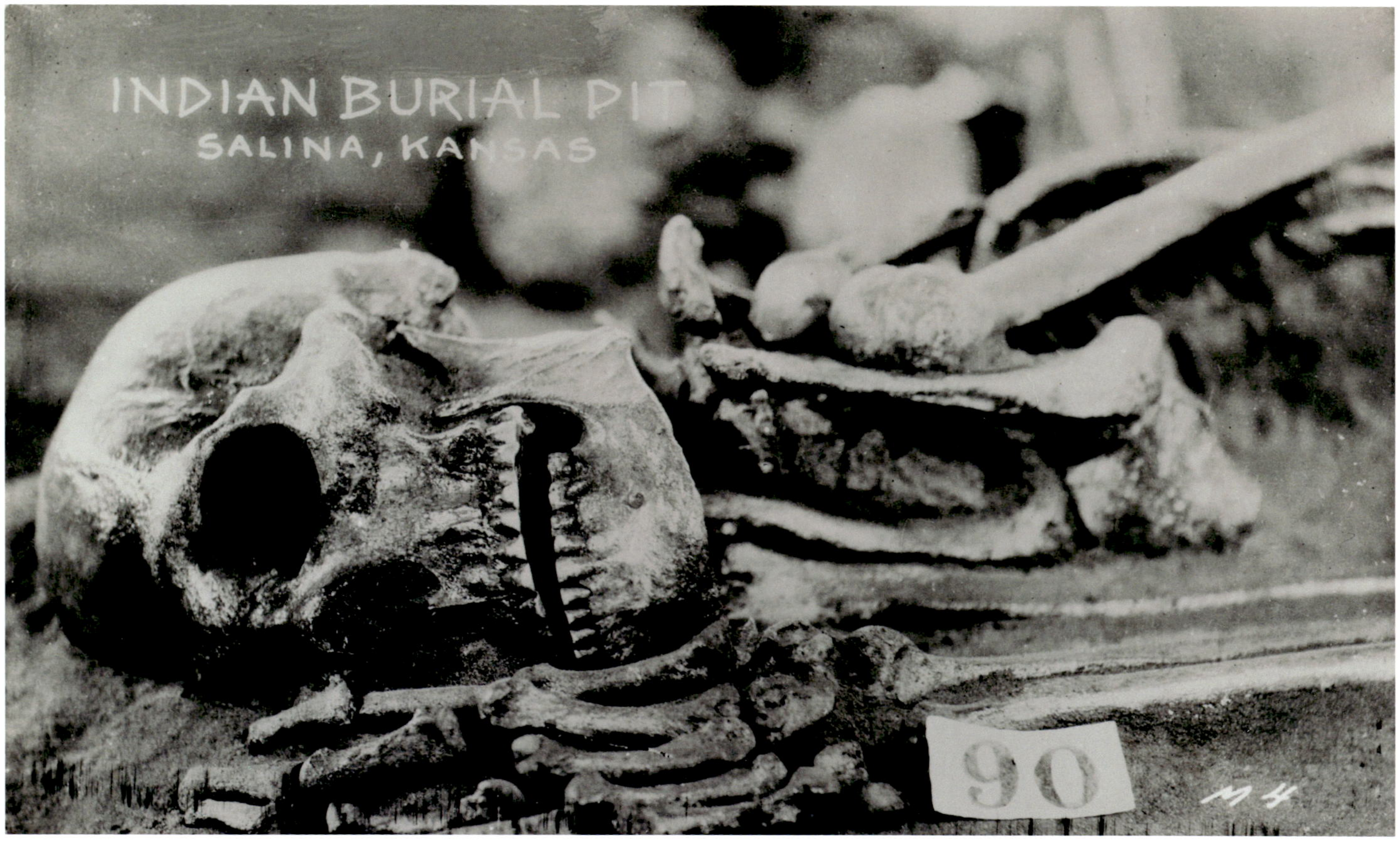

Plate 10.18

Indian Burial Pit, Salina, KS
Postcard, 1910

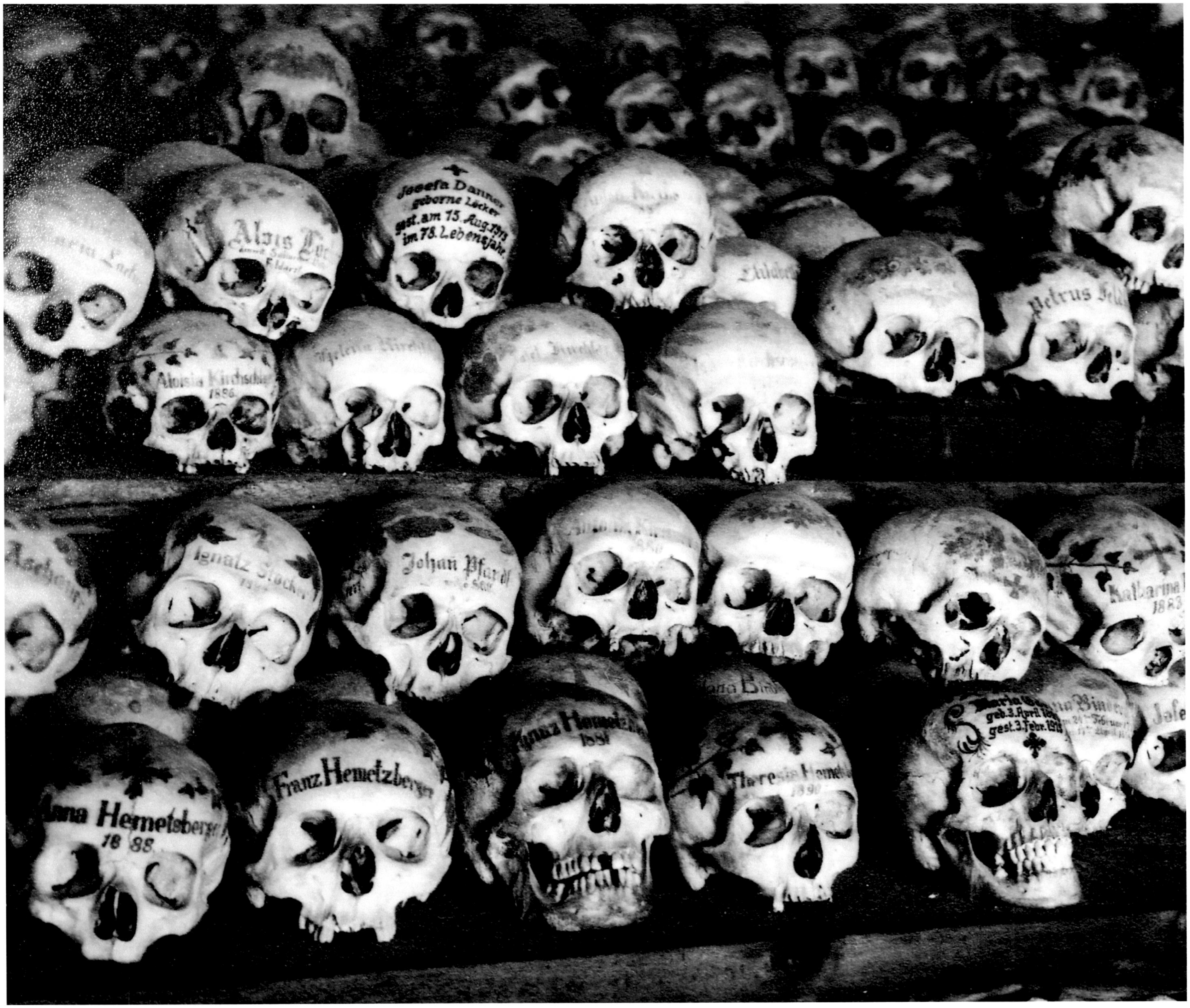

Plate 10.19

Ossuary at Hallstatt Beinhaus, Austria, 1910

Very unusual ossuary, as each skull is decorated, dated, and labeled with the name of the deceased.

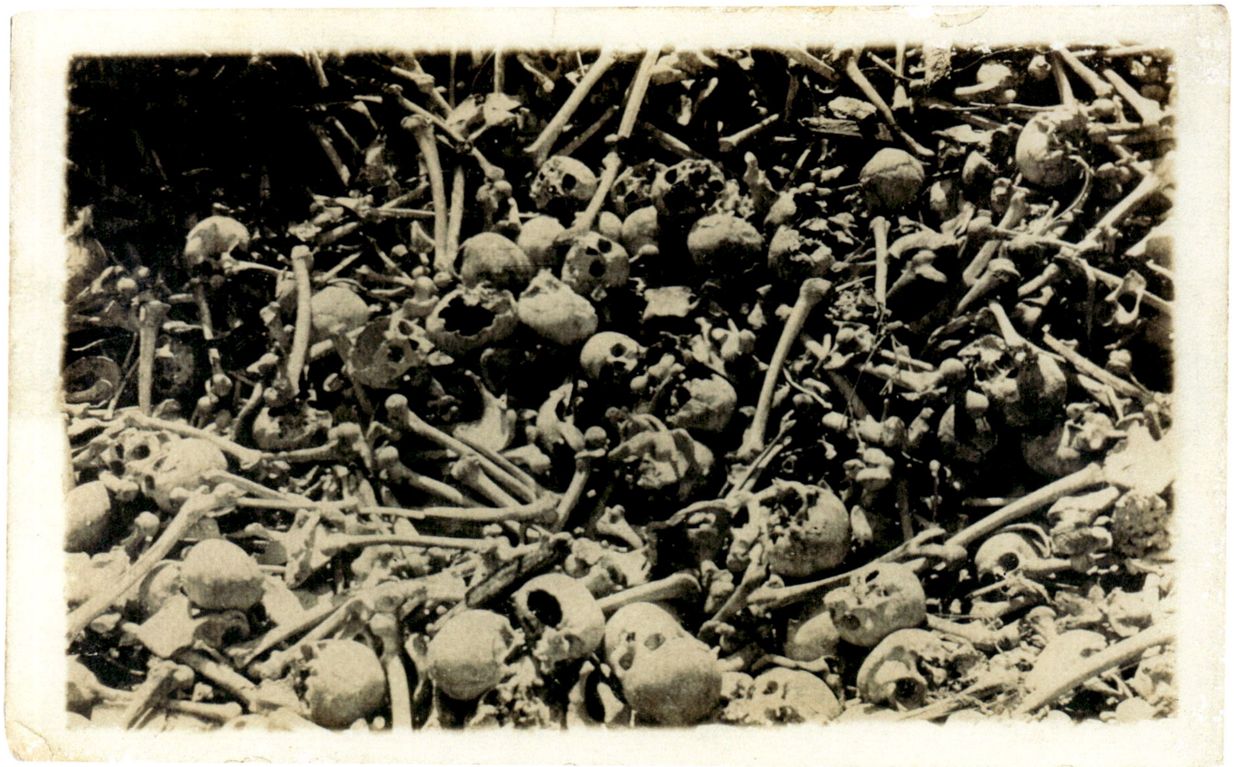

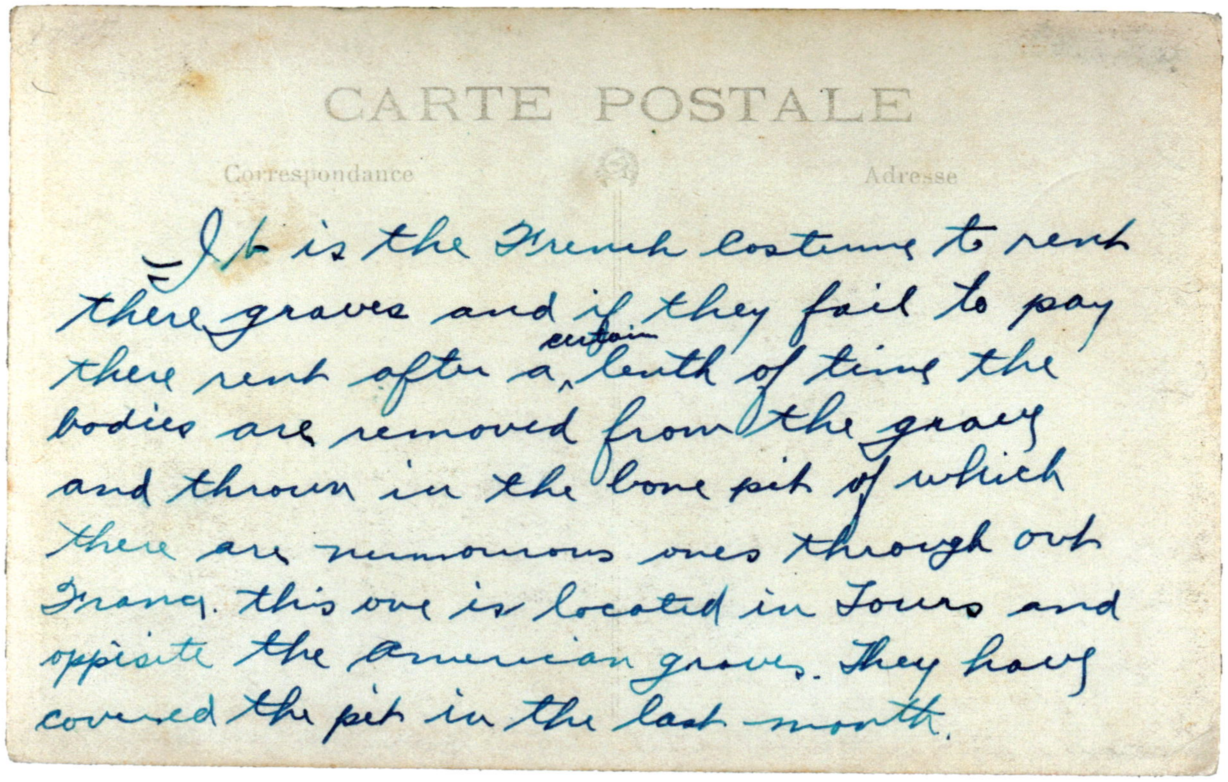

CARTE POSTALE

Correspondance

Adresse

It is the French costume to rent there graves and if they fail to pay there rent after a certain lenth of time the bodies are removed from the graves and thrown in the bone pit of which there are numorous ones through out France. this one is located in Tours and oppisite the American graves. They have covered the pit in the last month.

Plates 10.20

Bone Yard in Tours, France

Postcard with detailed explanation, circa 1920

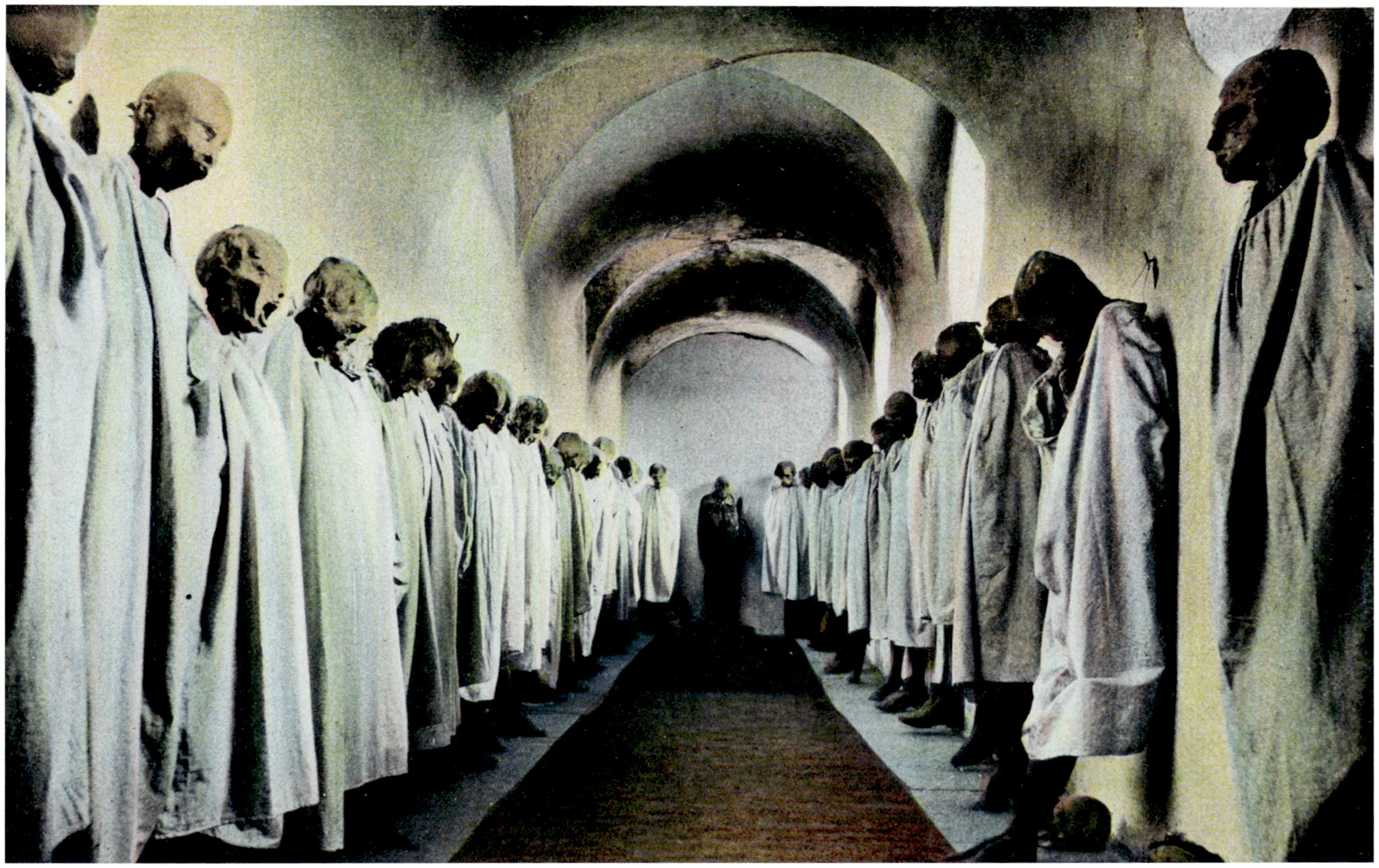

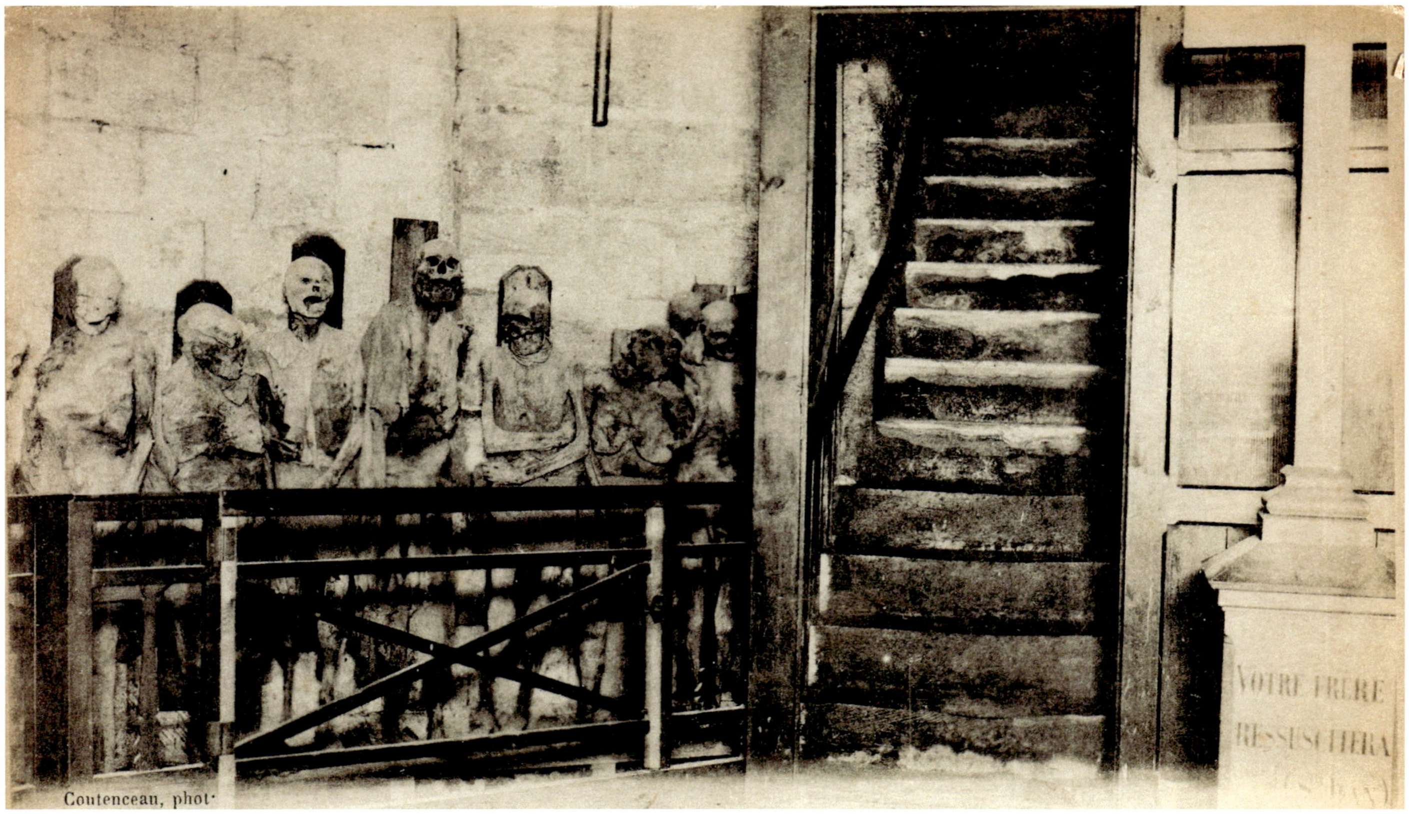

Plate 10.21

Mummified Corpses in Robes
Guanajuato Catacombs, Mexico
Hand-Colored Postcard, 1900

Plate 10.22

Mummies at Cave of St. Michel
Bordeaux, France
Postcard, 1912

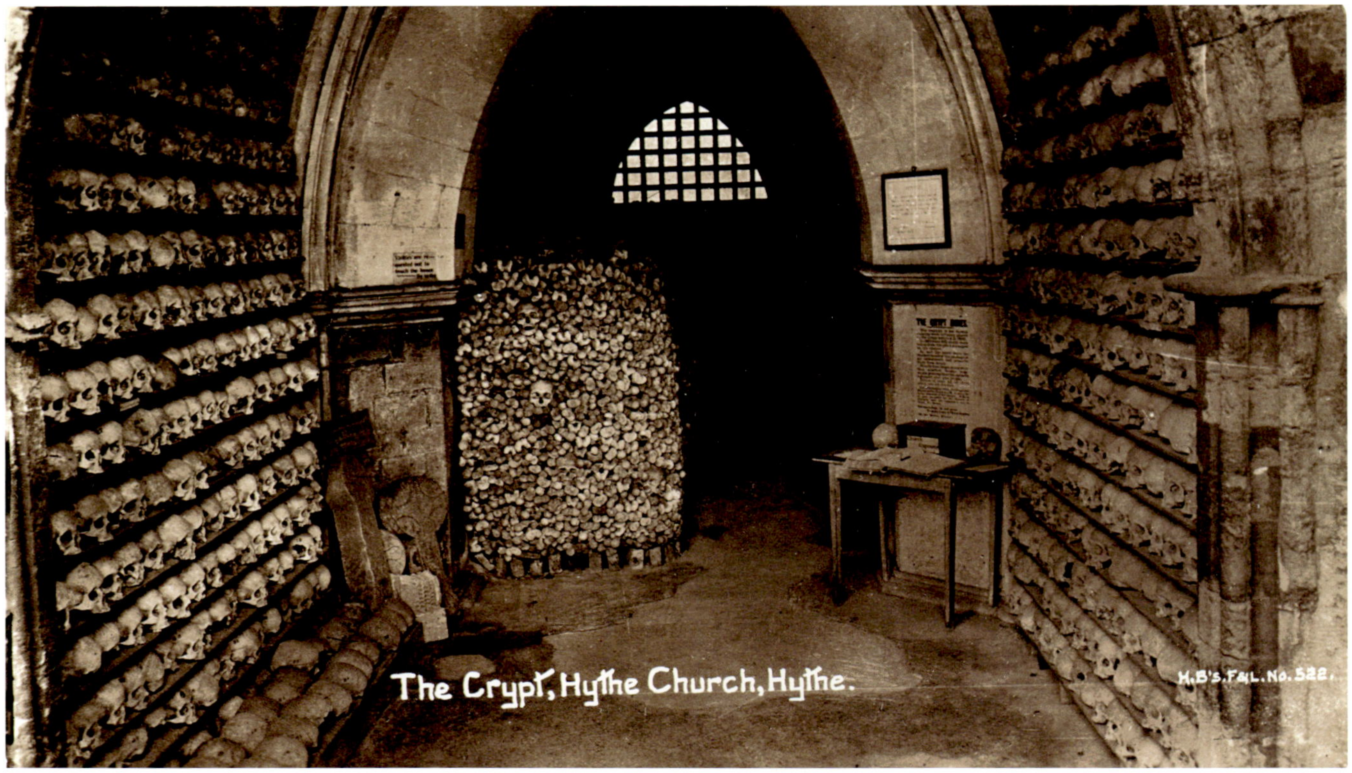

Plate 10.23

The Bone Crypt in the Parish Church, Hythe, England
1899

One of several popular postcards of the ossuary at Hythe.

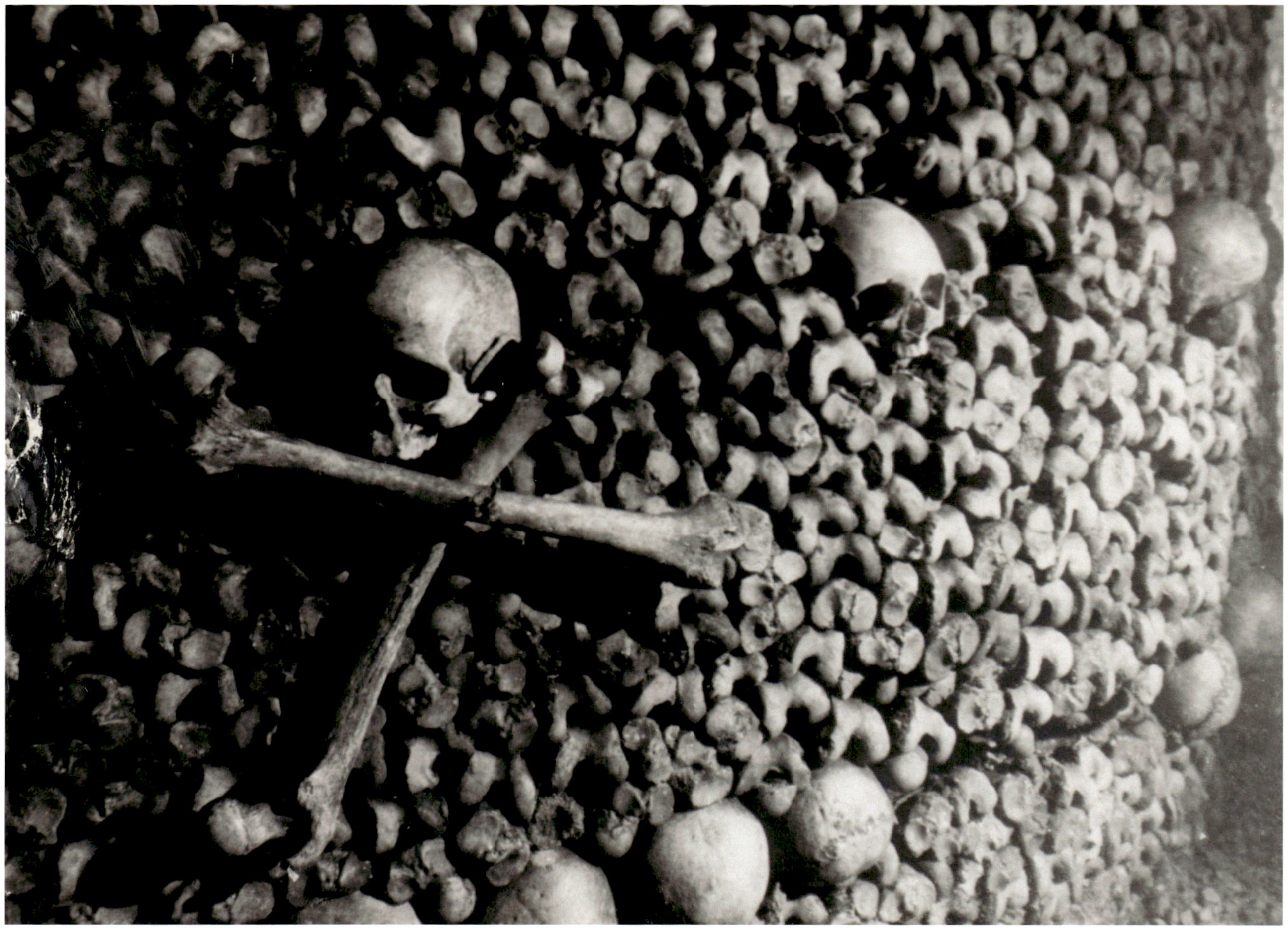

Plate 10.24

Close-up of Bones with Skulls, Paris Catacombs
Circa 1890

This design is typical of these catacombs, which is segregated by each churchyard that was dug up to create more land in the city.

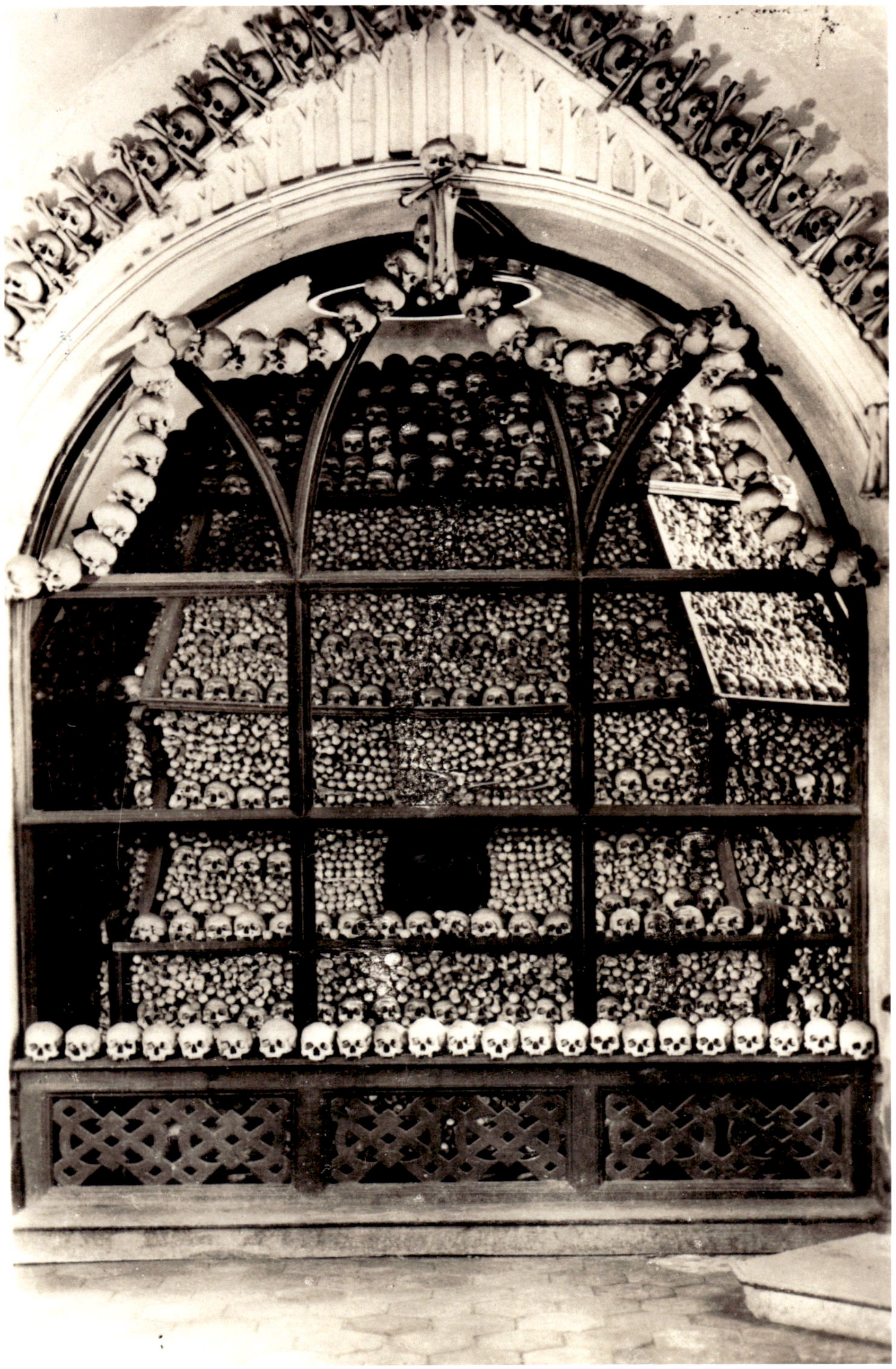

Plate 10.25

The Sedlec Ossuary in The Cemetery Church, Czech Republic
Postcard, 1920

This ossuary contains an estimated 40,000 skeletons and is very complex.

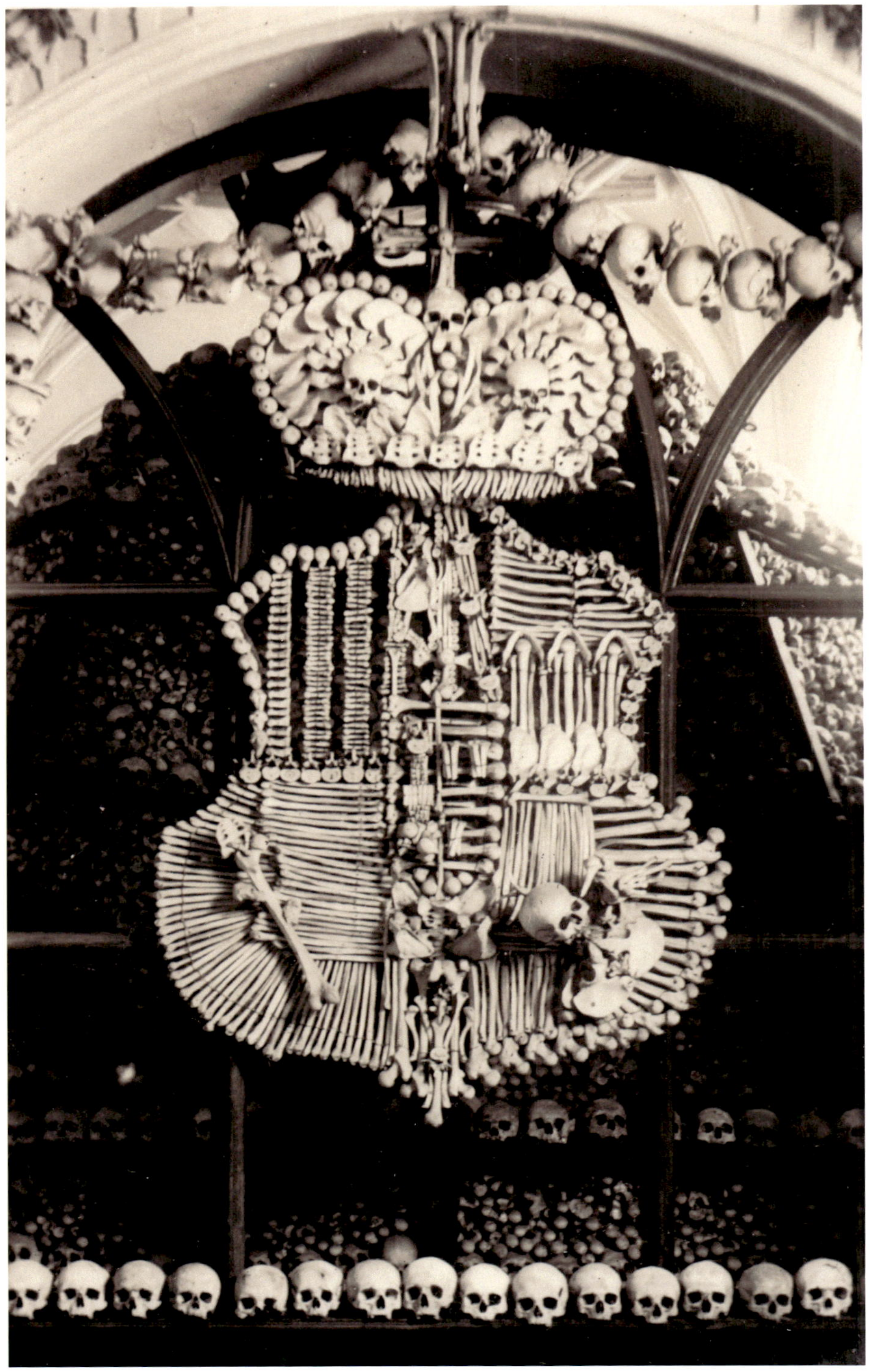

Plate 10.26

Elaborate Coat of Arms of the Schwarzenberg Family Made of Bones, Sedlec Ossuary
Postcard, 1920

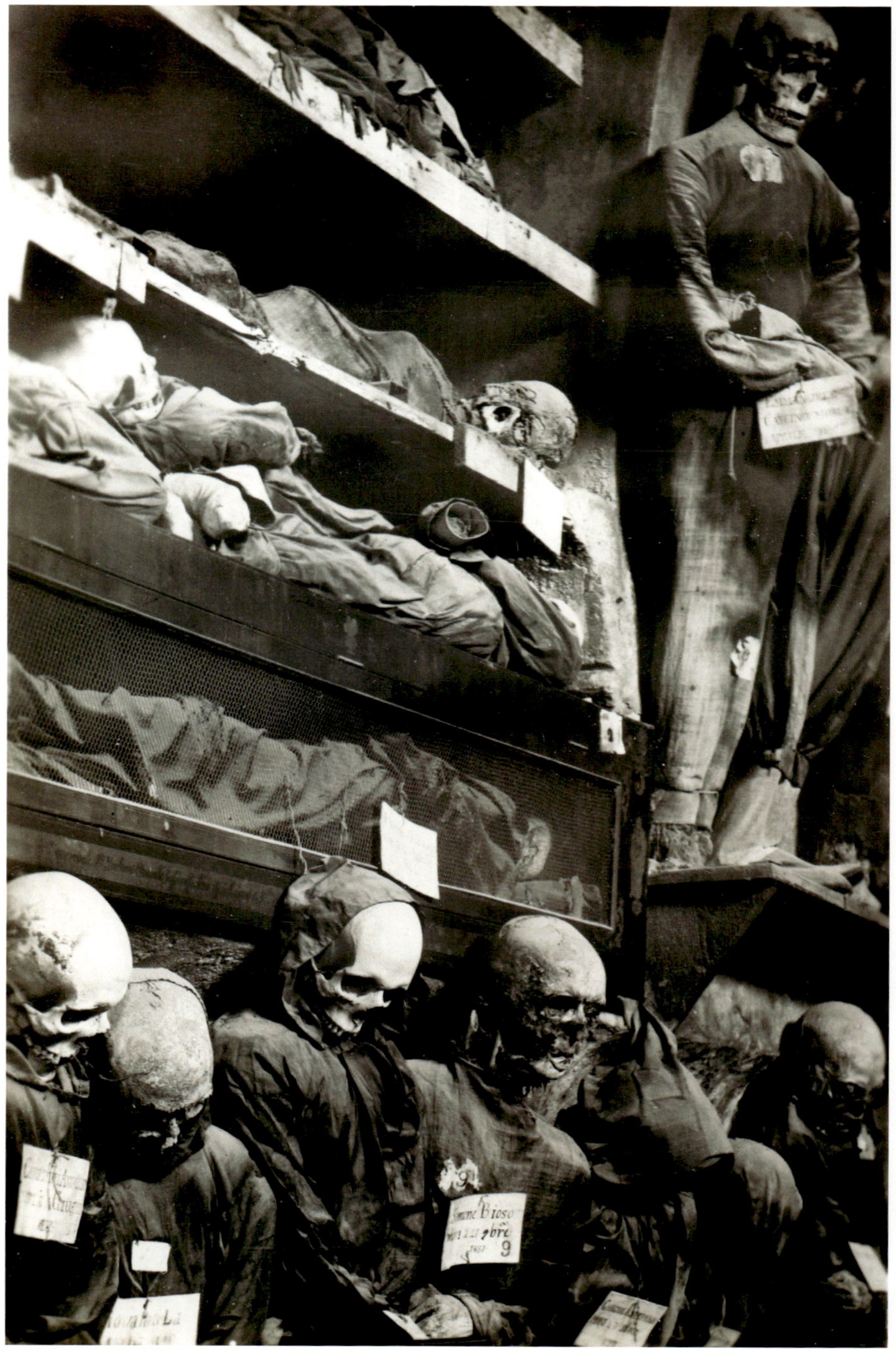

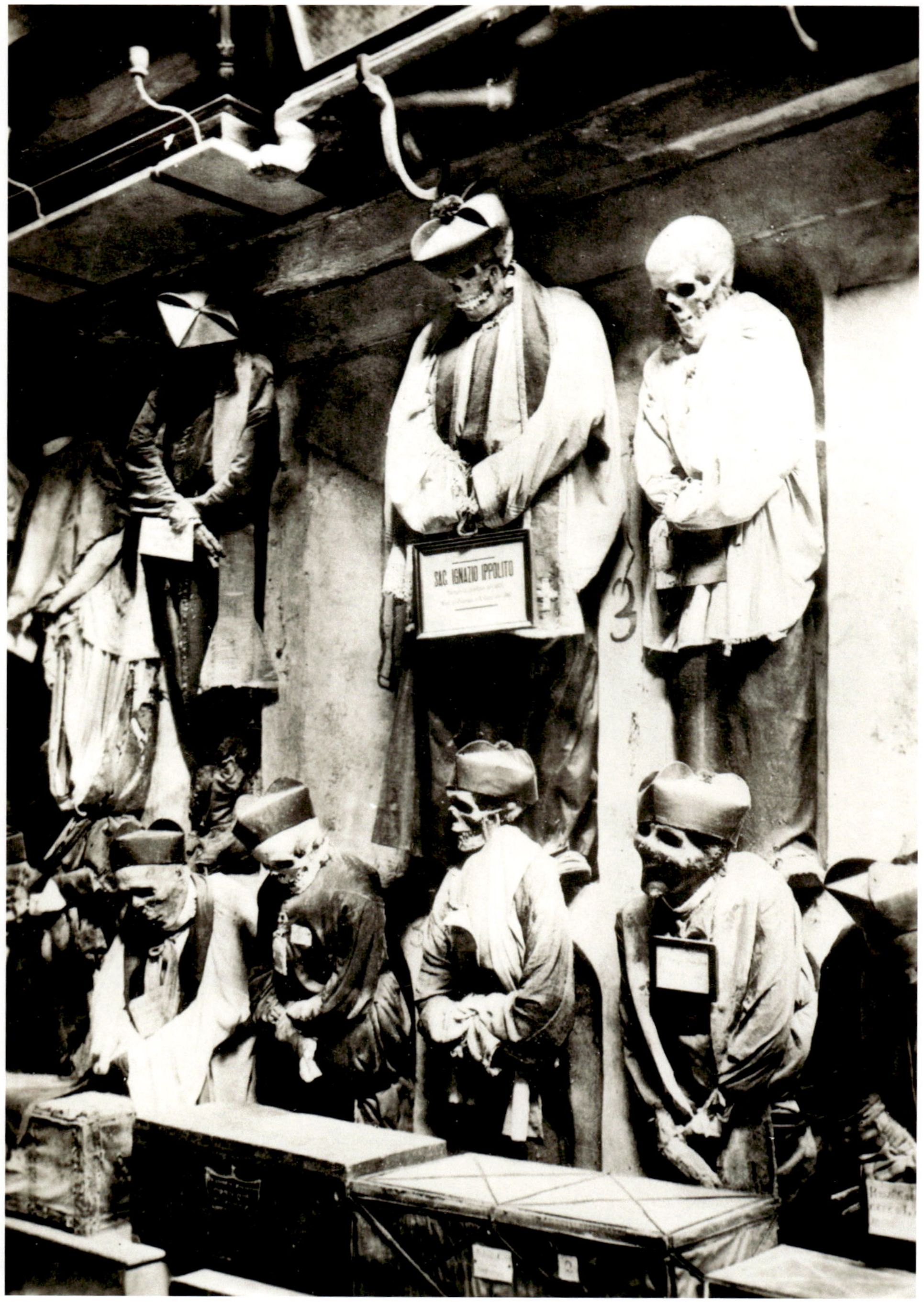

Plate 10.27

Catacombe dei Cappuccini, Polermo, Sicily
1900
A popular tourist attraction for centuries.

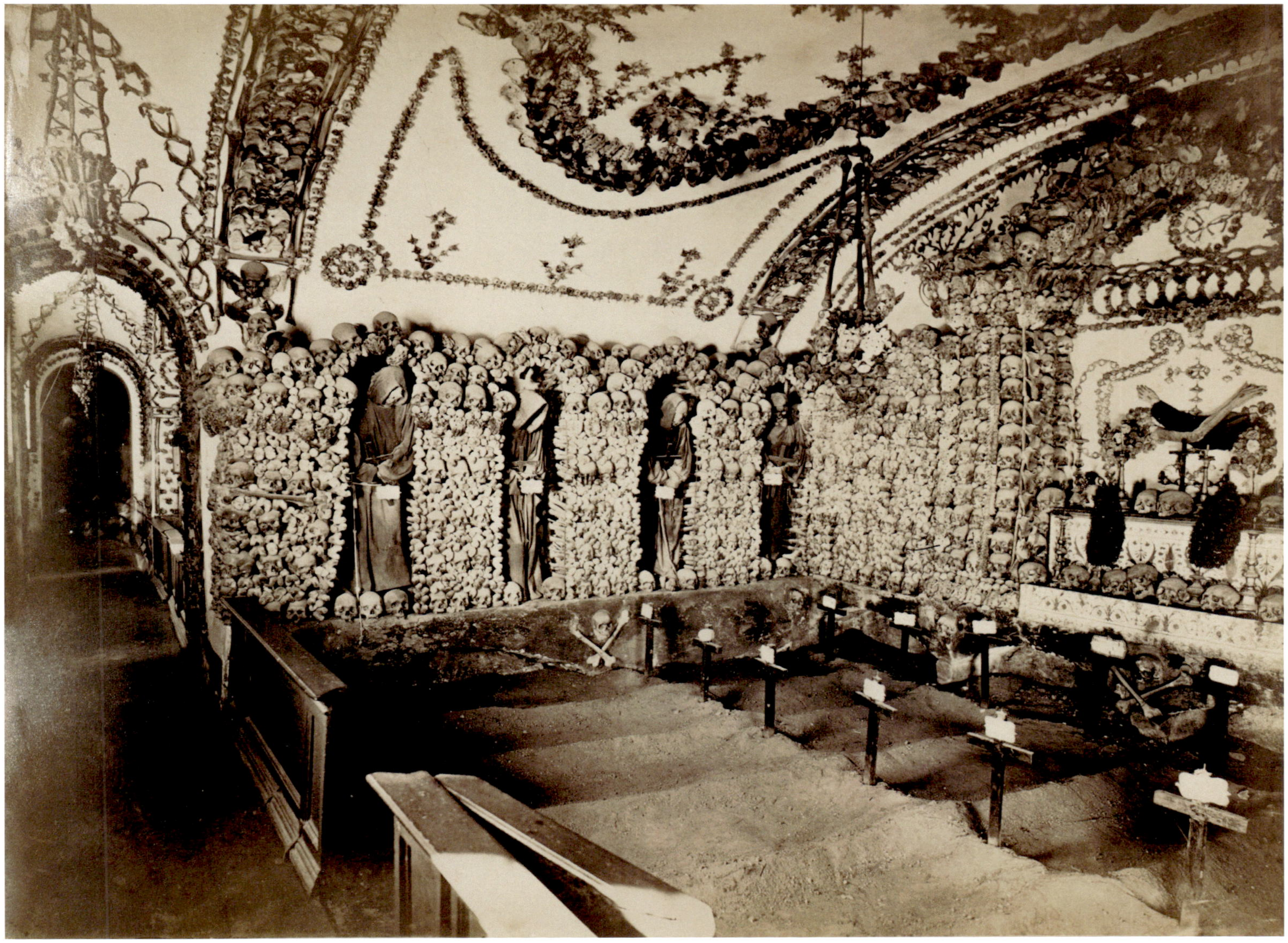

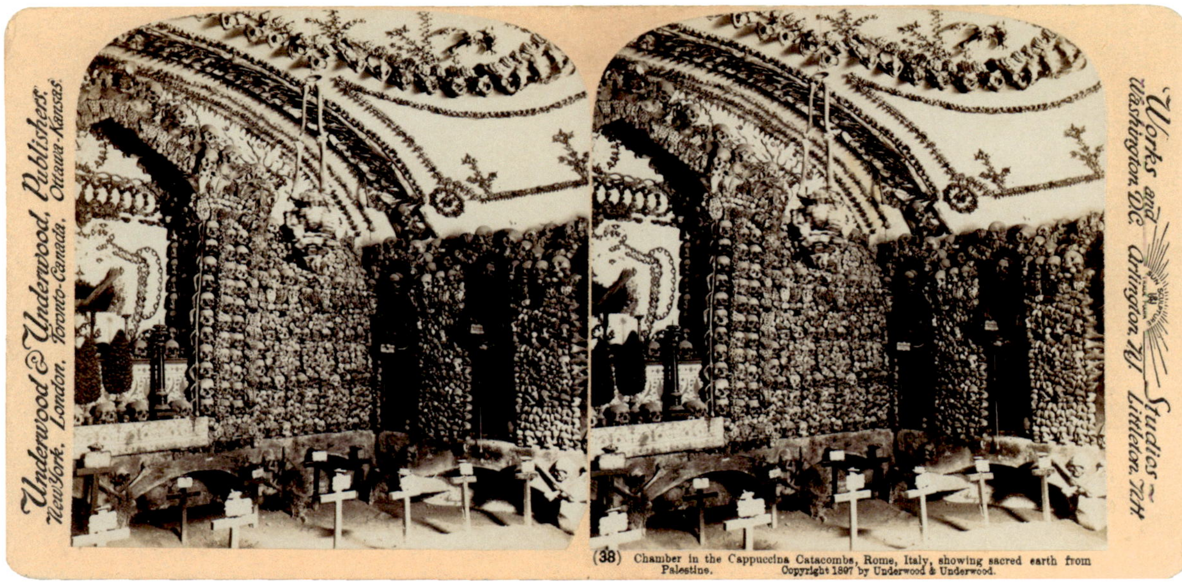

Plates 10.28

a. Chamber in Catacombe dei Cappuccina, Rome, 1885
b. Catacombe dei Cappuccina, Stereoview, Rome, 1897

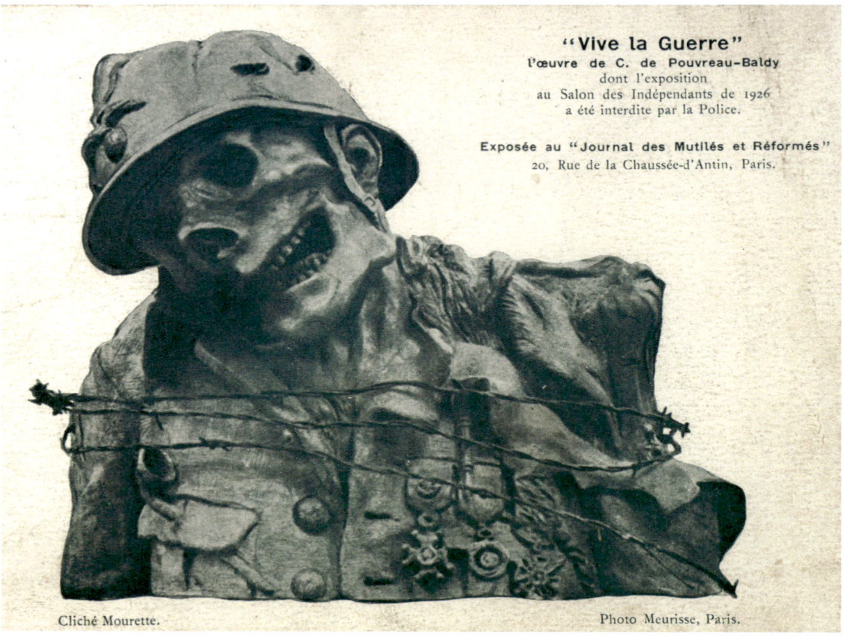

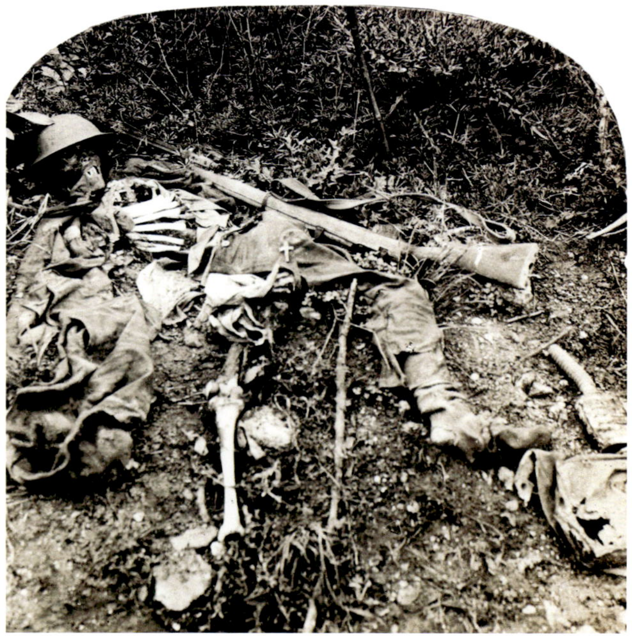

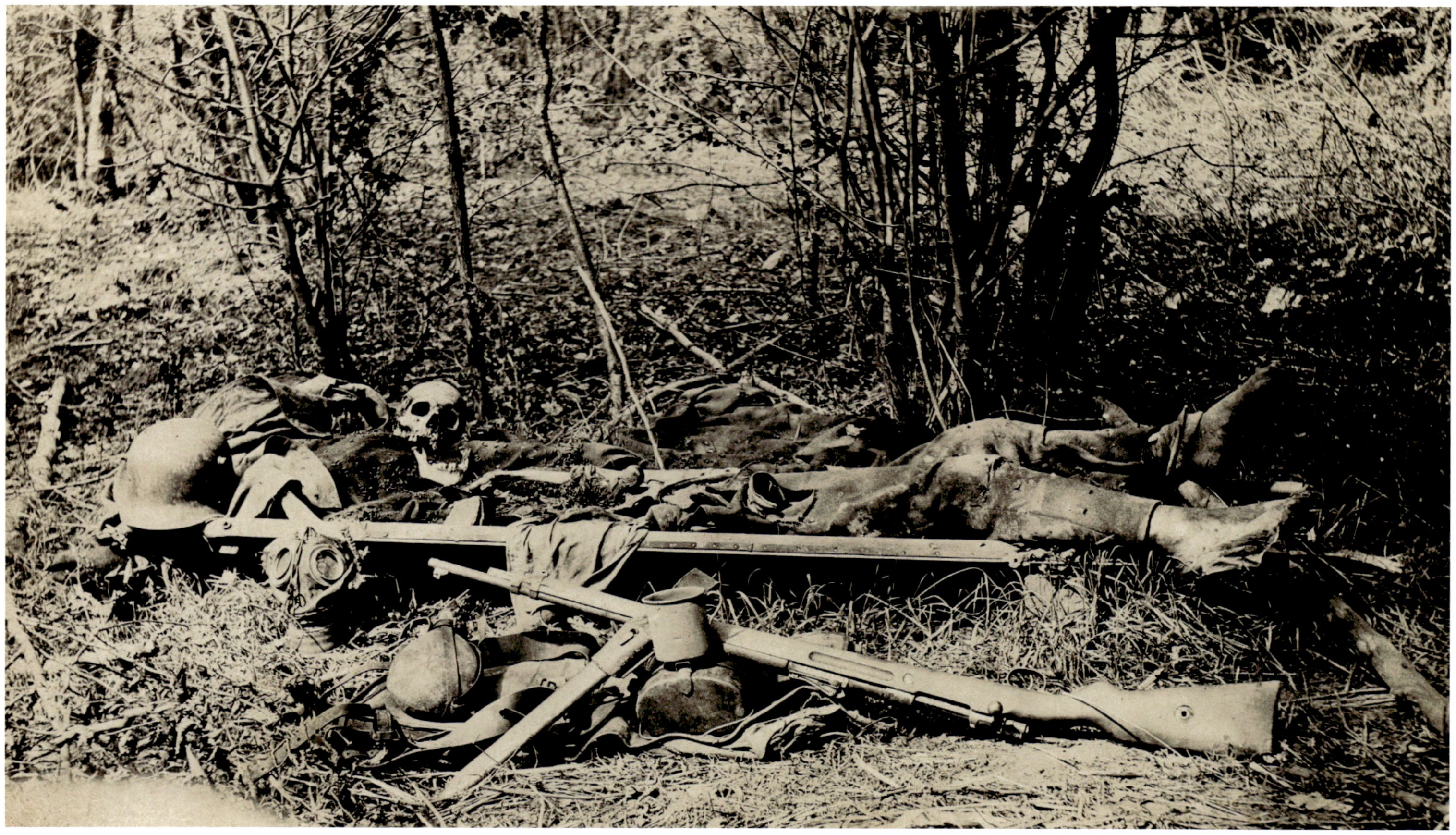

Plate 10.29

"Vive la Guerre"
World War I Exposition Postcard, 1920

Plate 10.30

Human Wreckage in No Man's Land
Stereoview of Remains of World War I Soldier
Postcard, 1918

Plate 10.31

Remains of a World War I Soldier
Postcard, 1918

The devastation of the war was depicted on several postcards.

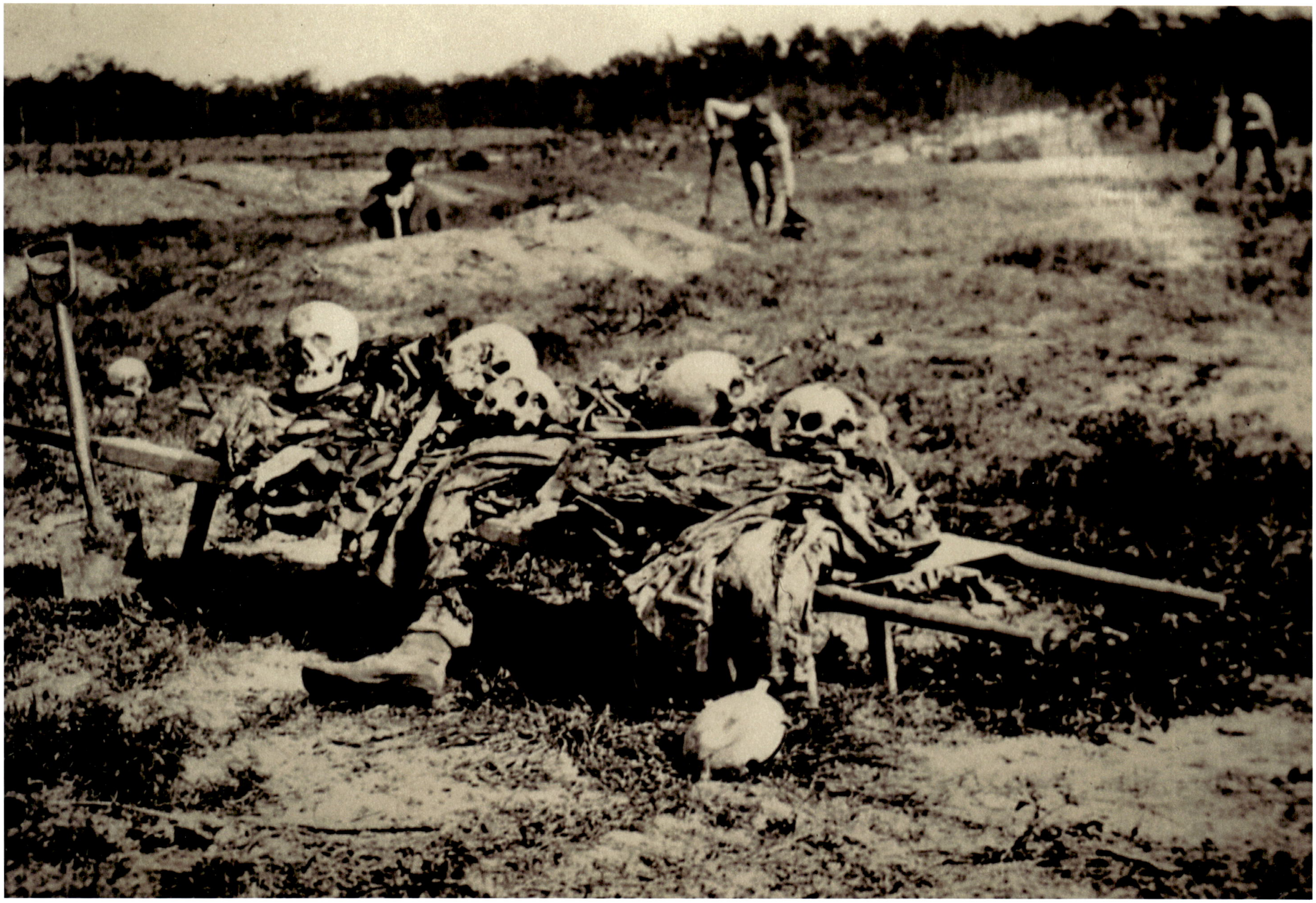

Plate 10.32

Picking Up the Remains of the Dead, Cold Harbor Virginia Battlefield, Civil War
1865
The Battle was fought from May 31 to June 12, 1864

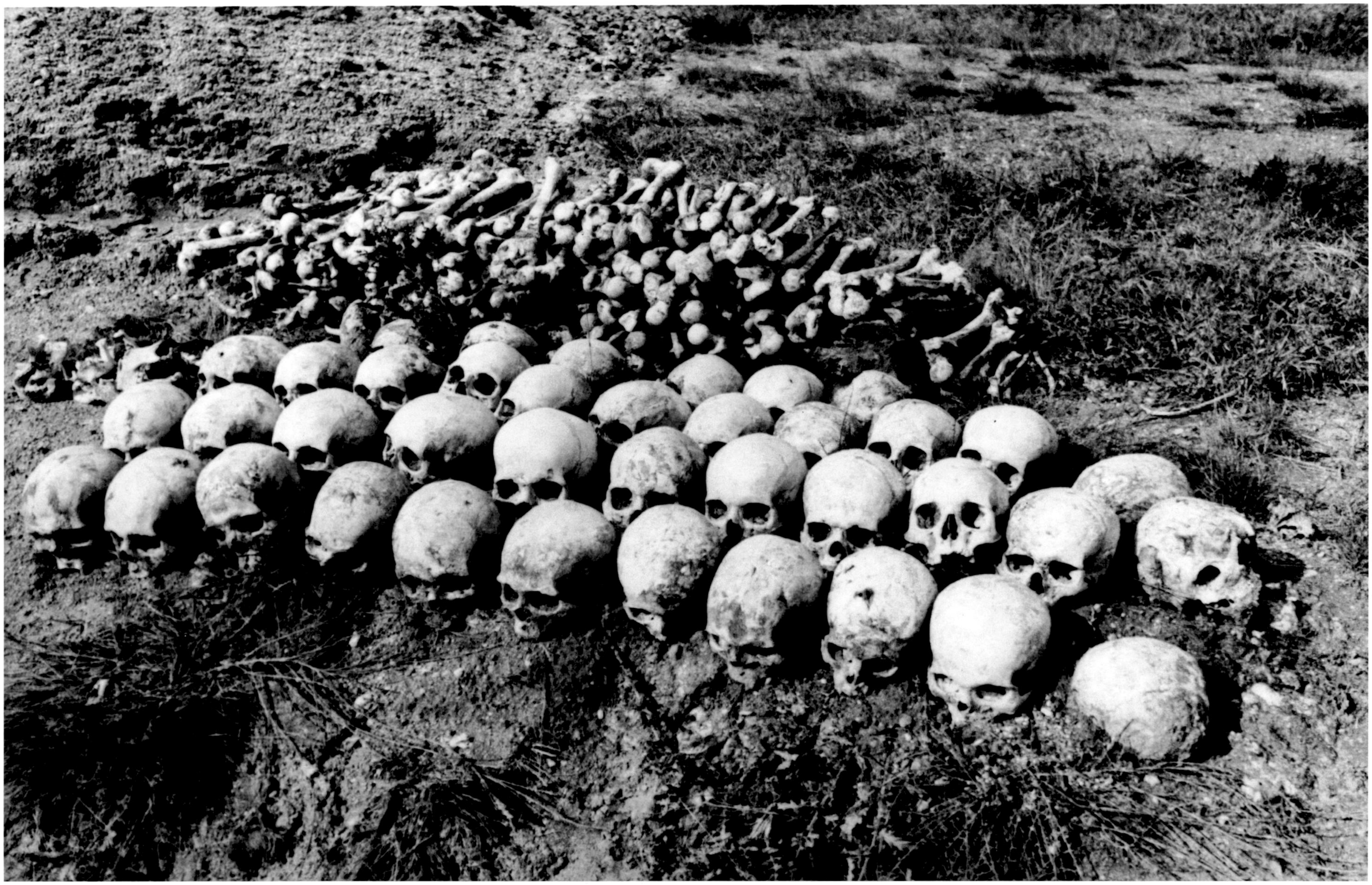

Plate 10.33

Exhumed Skulls and Femurs at Buchenwald Nazi Concentration Camp
1945

The camp was built in Germany near Weimer, in July 1937. It was one of the largest camps.

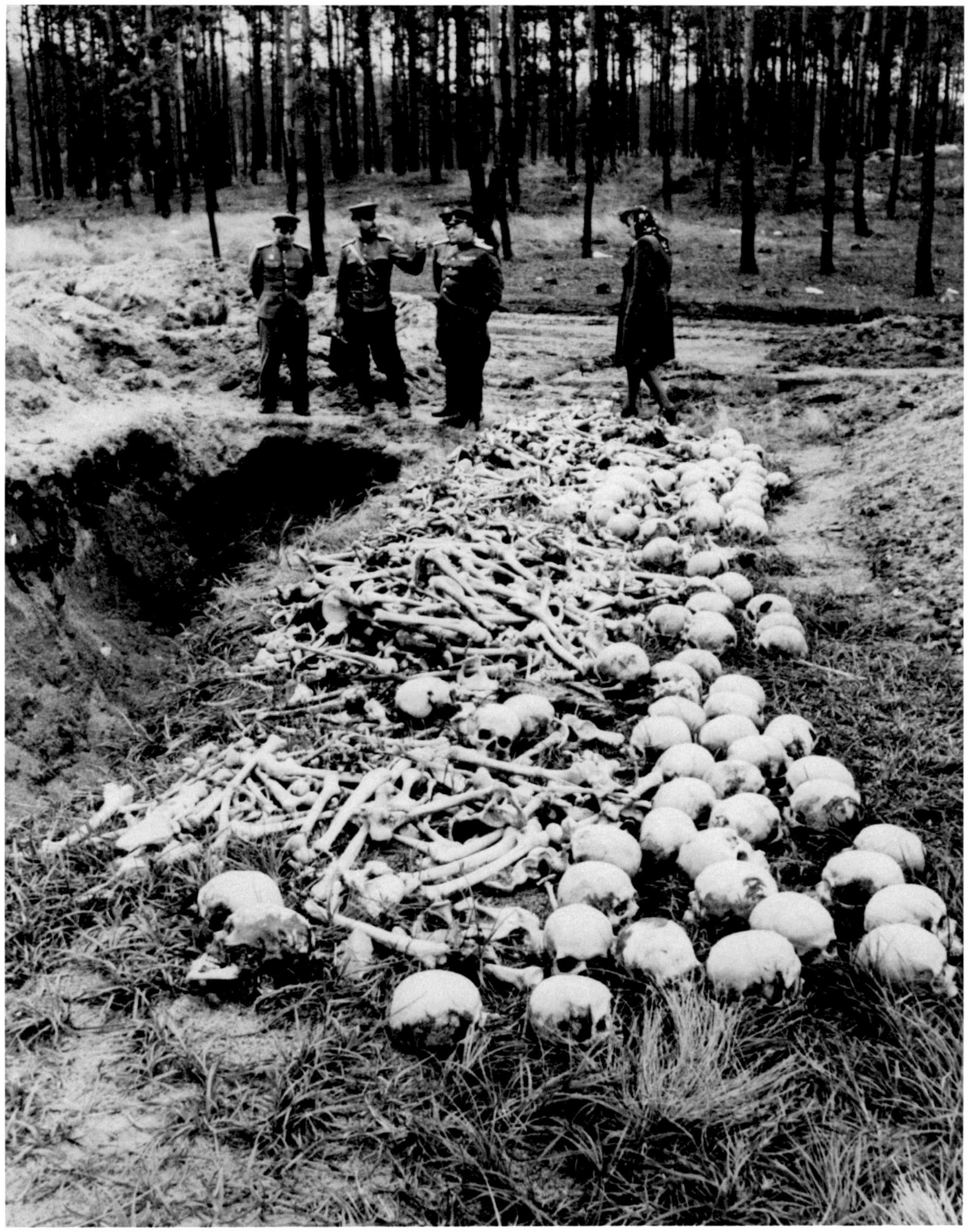

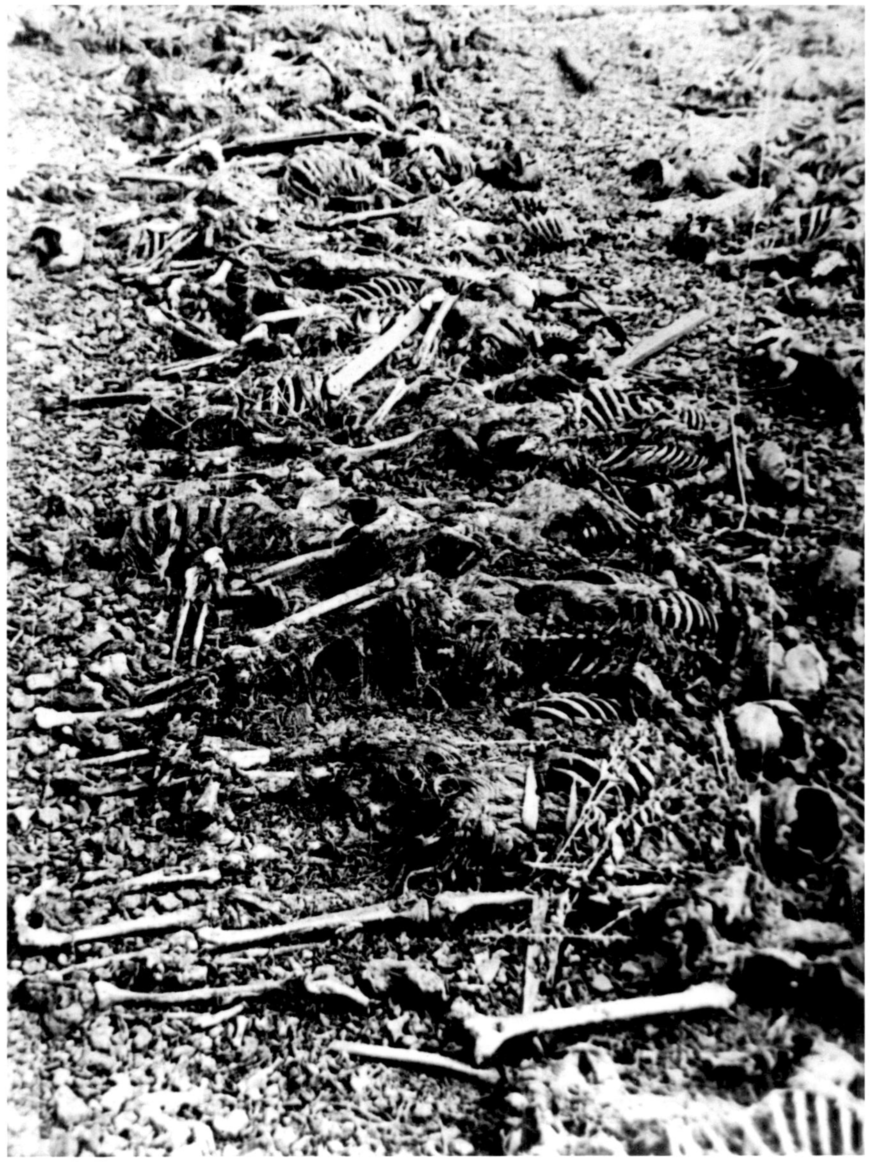

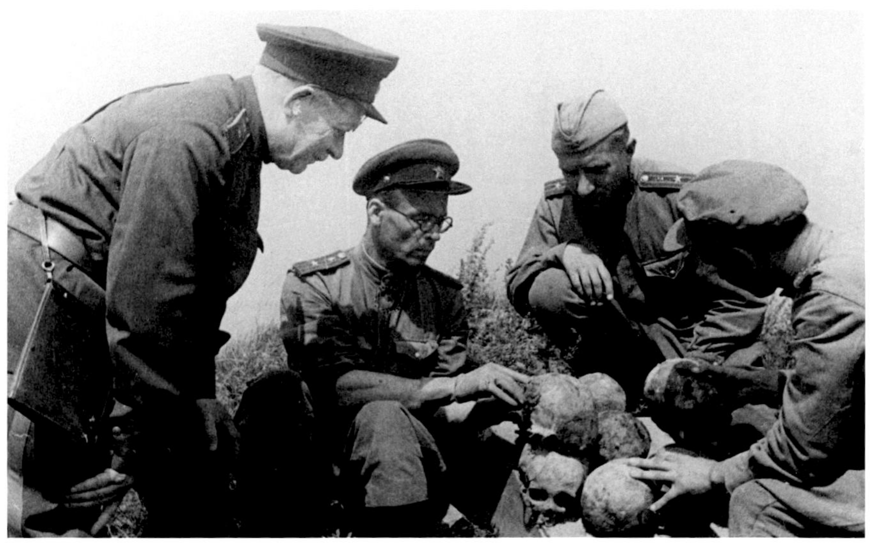

Plate 10.34

Russian Soldiers Stand at the Foot of a Pit of Exhumed Skulls and Bones
Buchenwald Nazi Concentration Camp, 1945

Plate 10.35

Skeletal Remains at German Extermination Camp Majdanek near Lublin Poland
August 1944

Majdanek is the least known of the Nazi's six extermination facilities built for murder not confinement.

Plate 10.36

Lt. General Budenko and Surgeons Examine Skulls from the Orel Execution Site
September 1943

Russian medical staff investigation at Nazi atrocity site after Battle of Kursk.

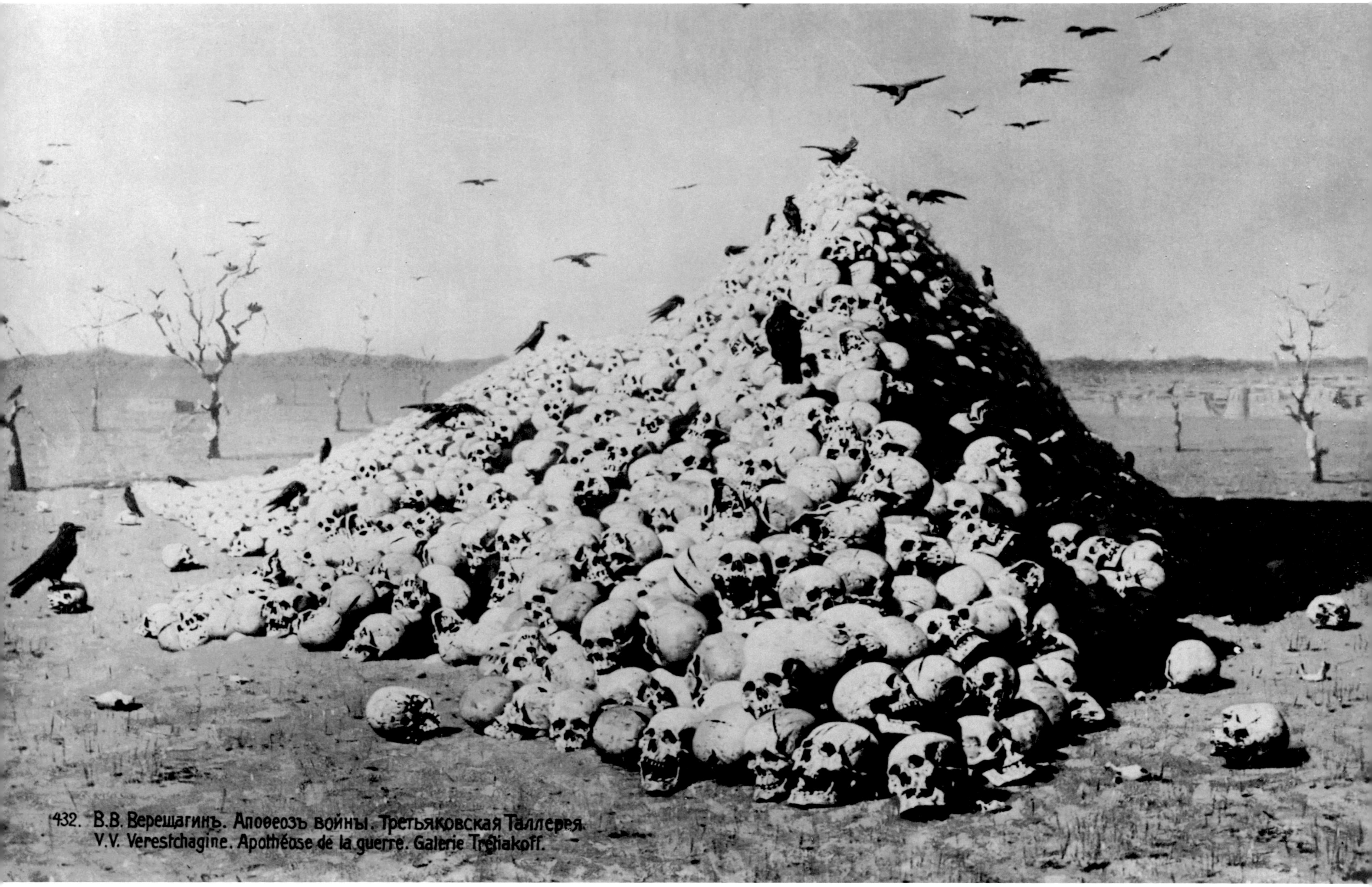

Plate 10.37

"The Results of War"
Postcard, V.V. Verestchagine, Gallery Tréhakoff, circa 1925

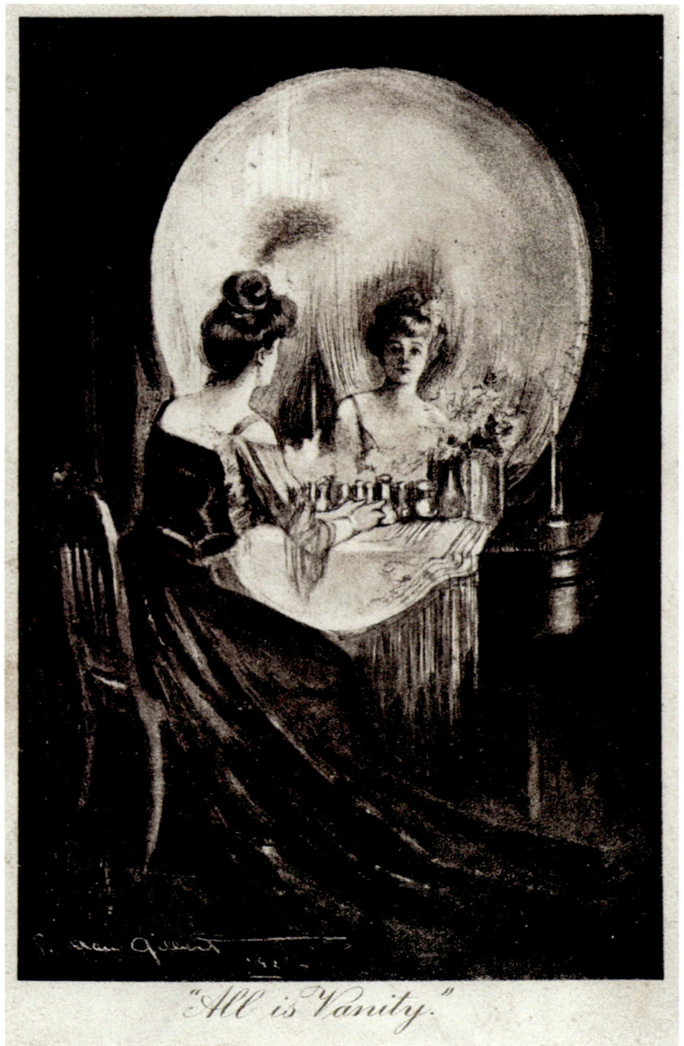

Plate 10.38

"All is Vanity"
Postcard, 1908

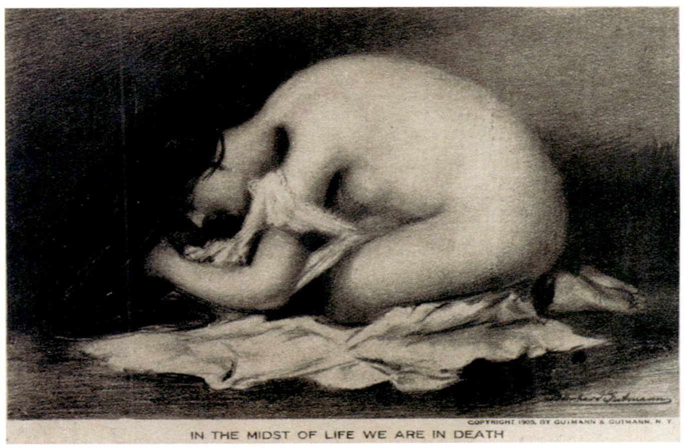

Plate 10.39

"In the Midst of Life, We are in Death"
1905

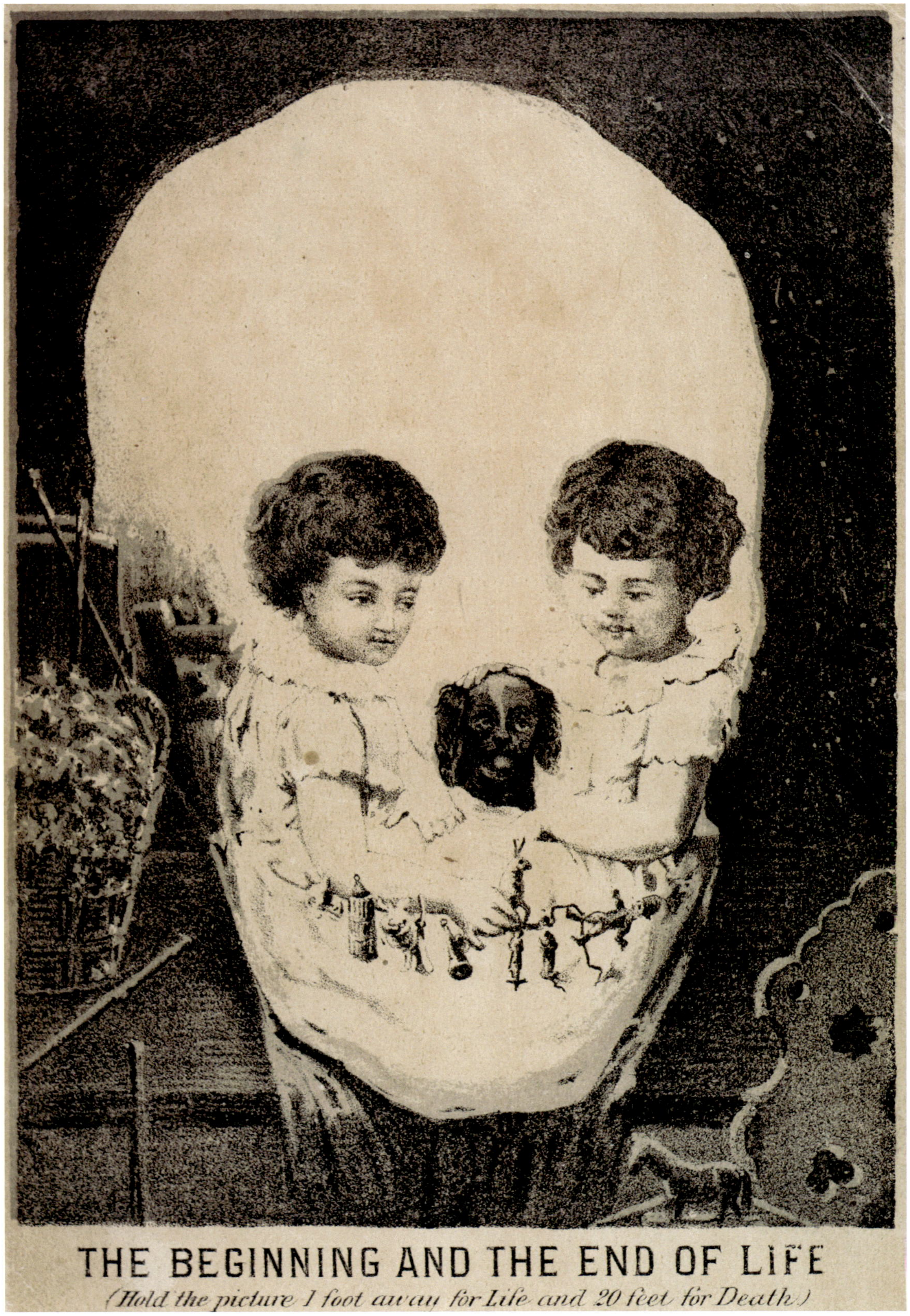

Plate 10.40

"The Beginning and The End of Life"
1905

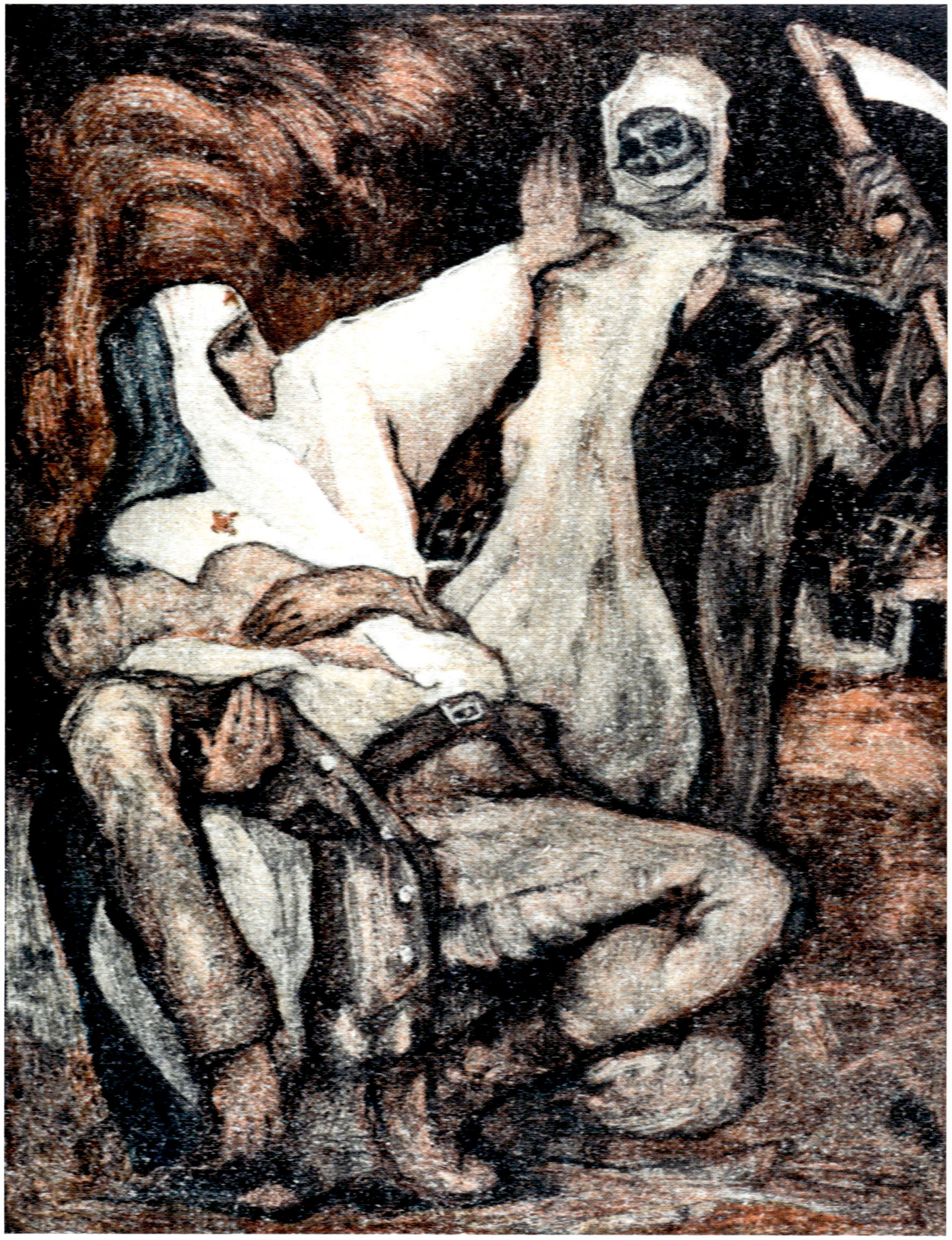

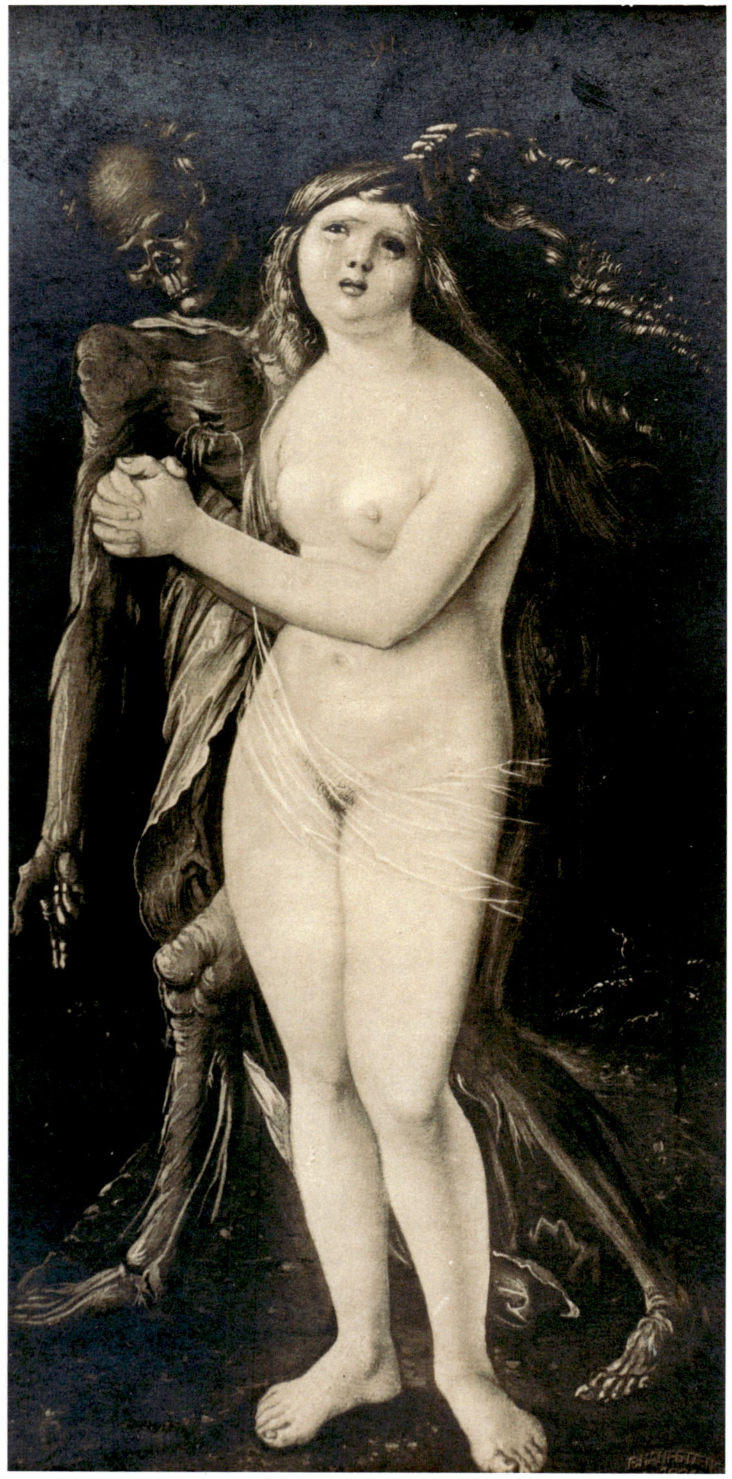

Plates 10.41

Artworks Depicting Skeletons Interacting With the Living
Postcards of Paintings, circa 1925

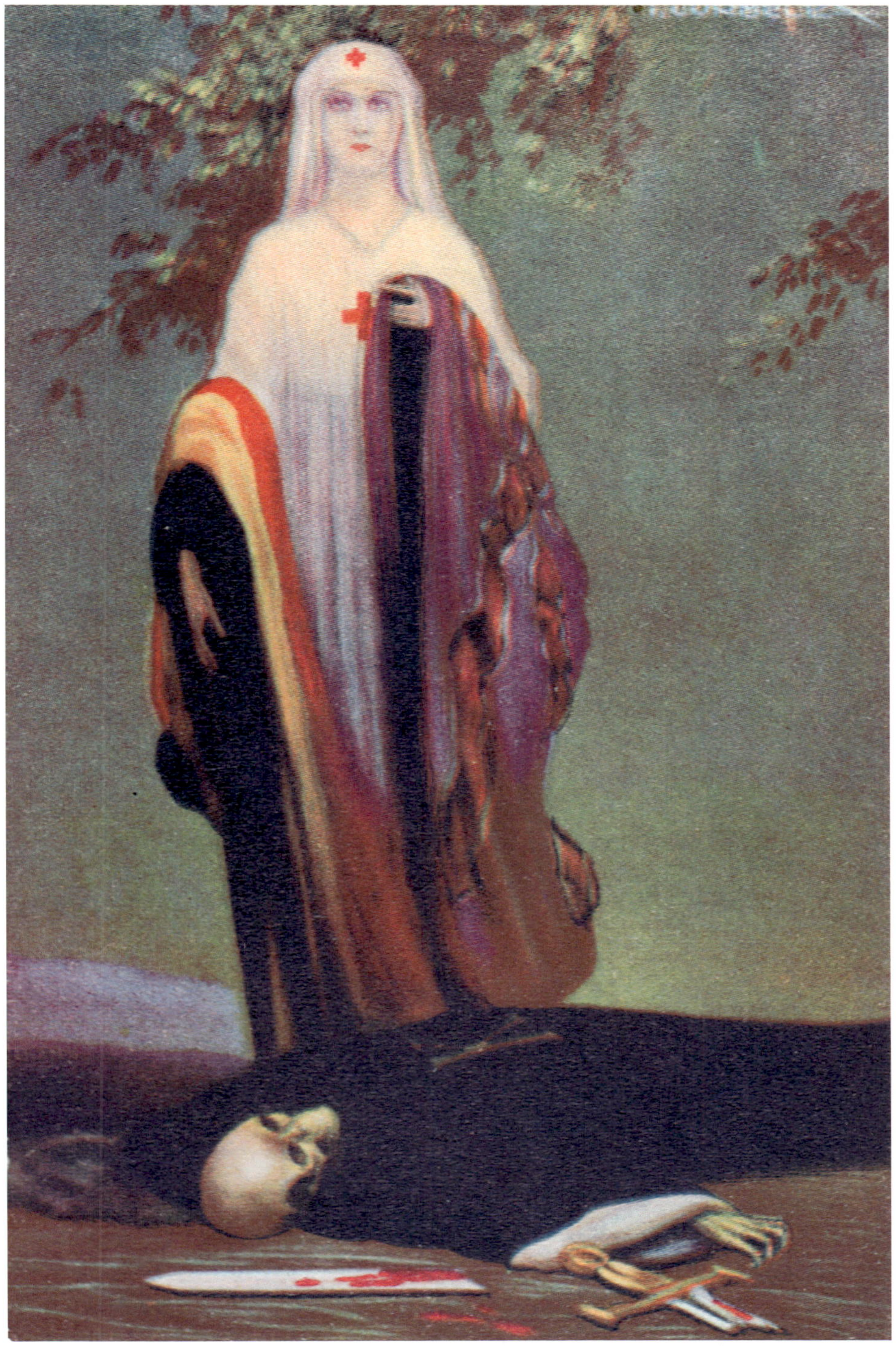

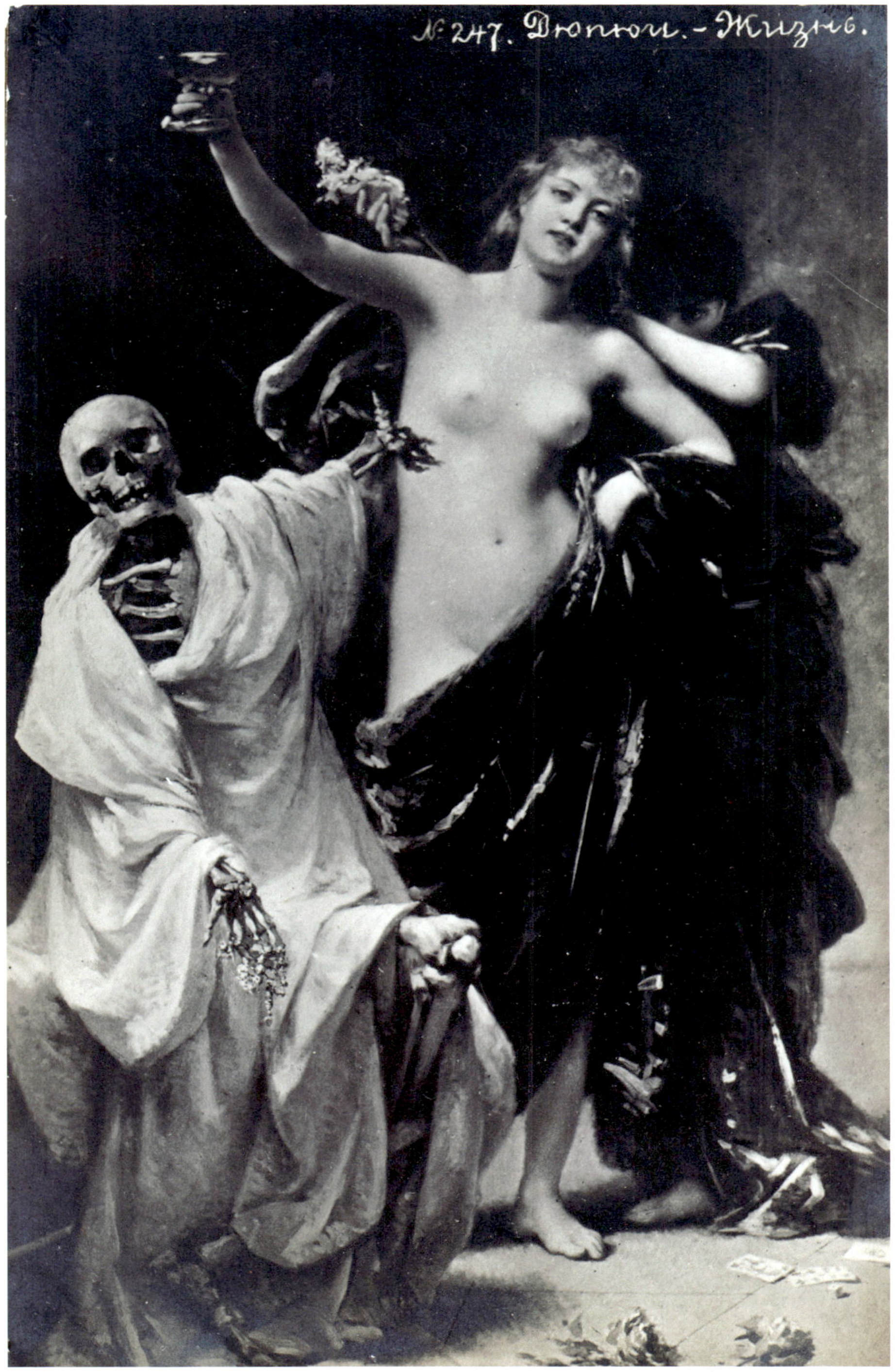
№ 247. Дюнтонъ. - Жизнь.

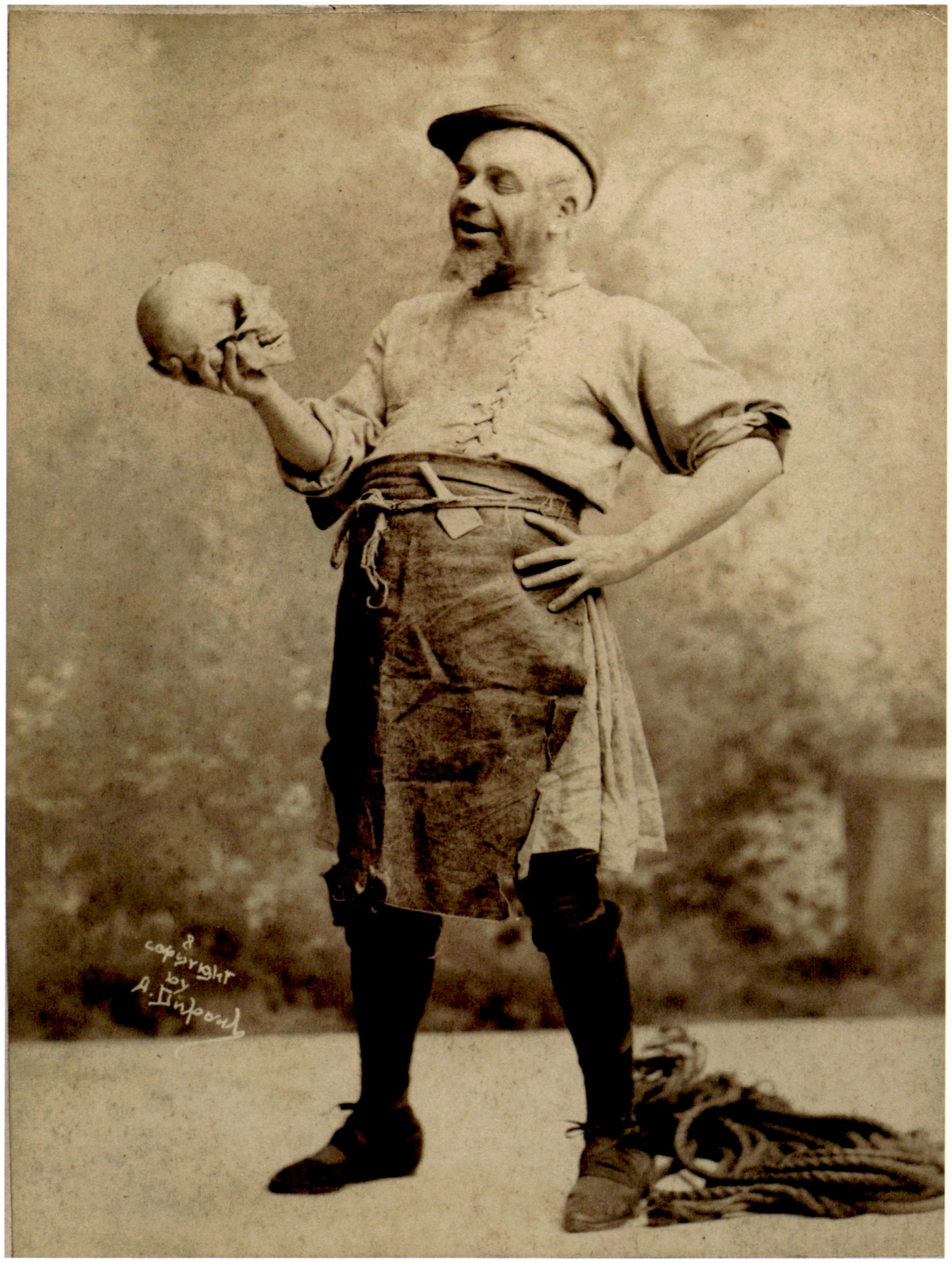

Plate 10.42

Actor Contemplating a Skull
Cabinet Card, Photographer: A Dupont, 574 Fifth Ave., NY, 1890

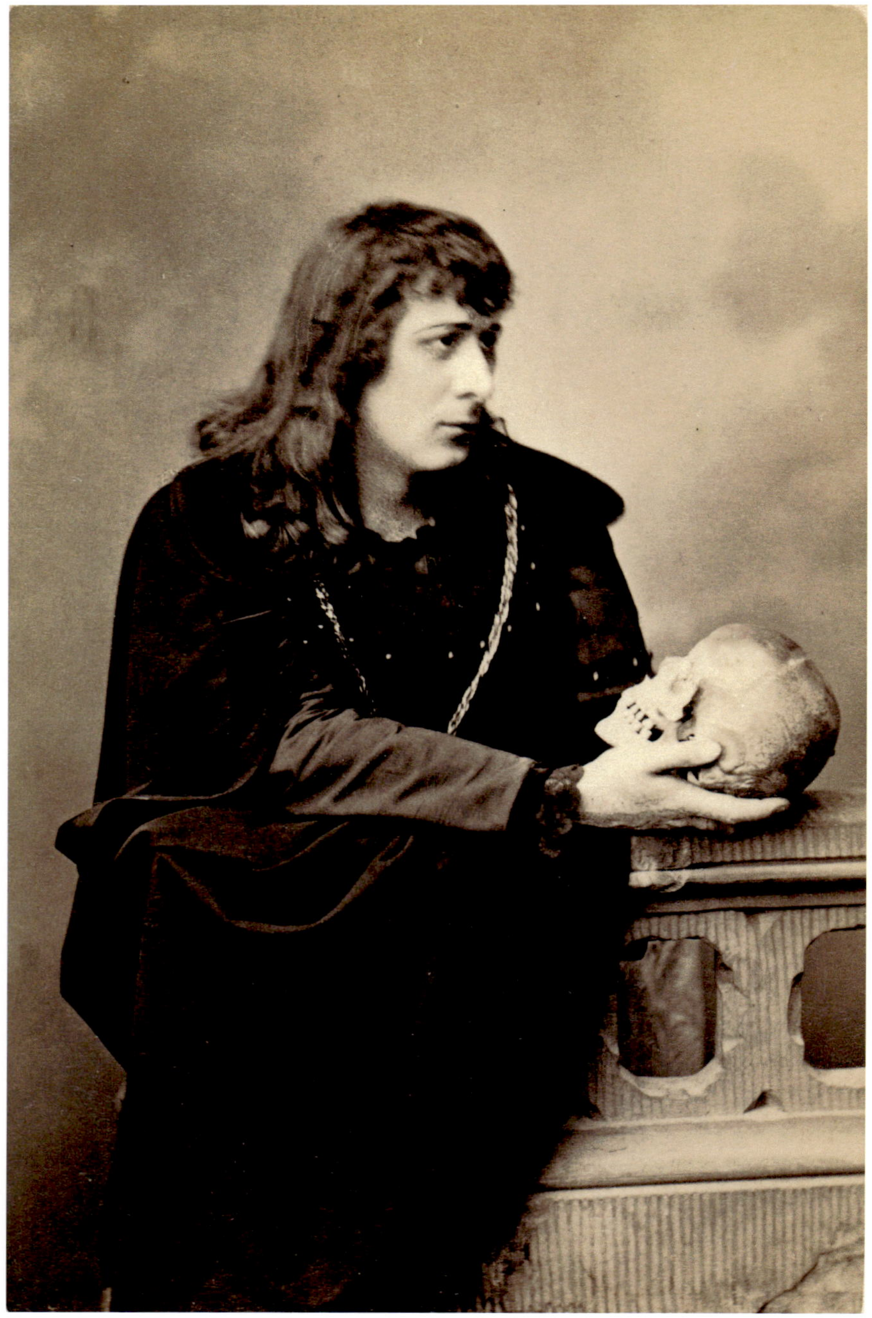

Plate 10.43

Actor in the Role of Hamlet
Cabinet Card, 1885

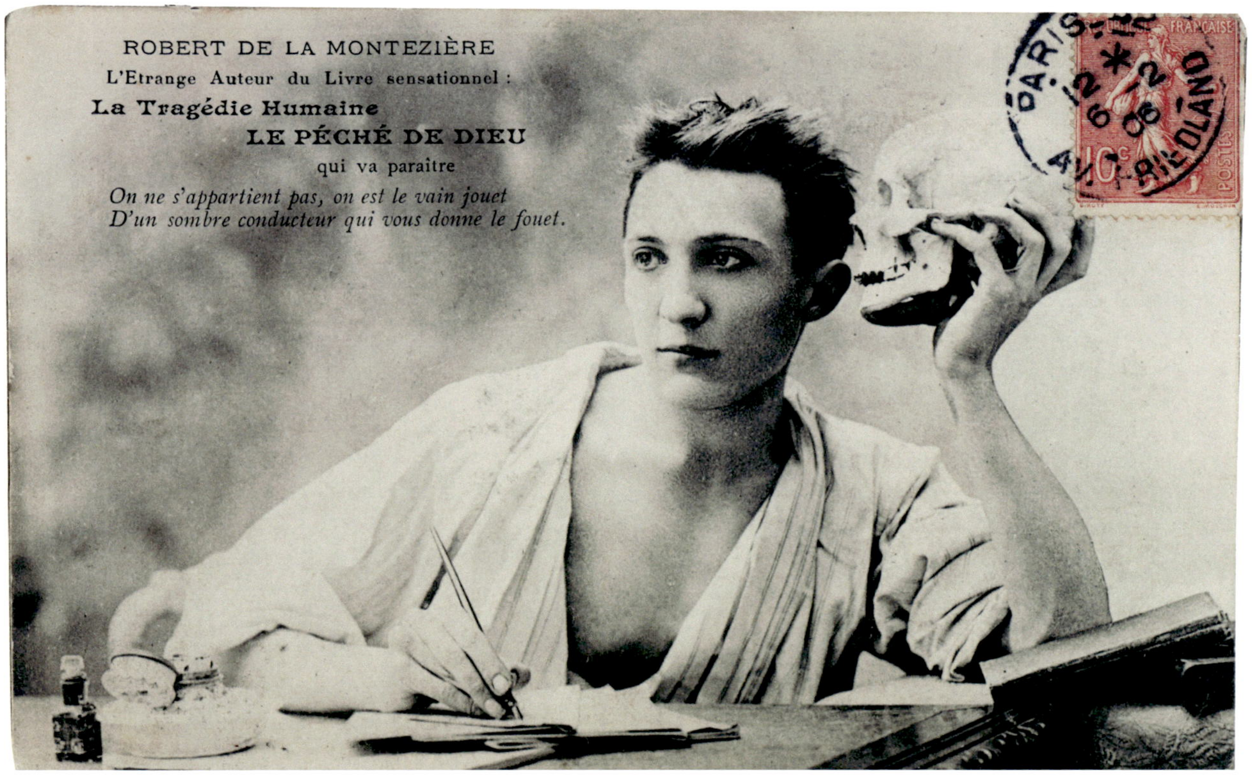

Plate 10.44

Robert de la Montezière, Writer *Le Péché de Dieu (The Sin of God)*
Postcard, France, 1908

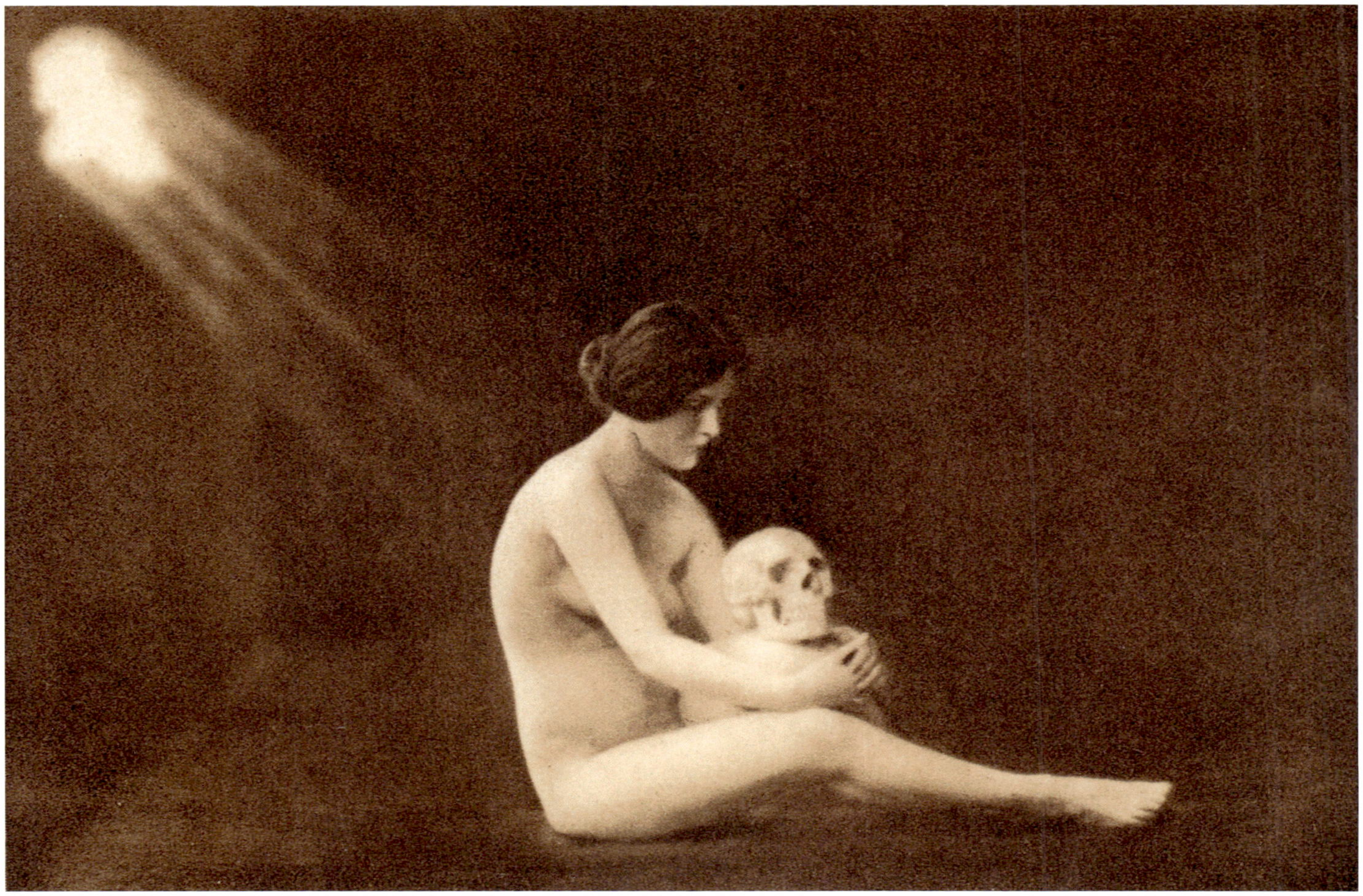

Plate 10.45

'Reflection,' Woman with a Skull
Postcard, German Modernism, 1930

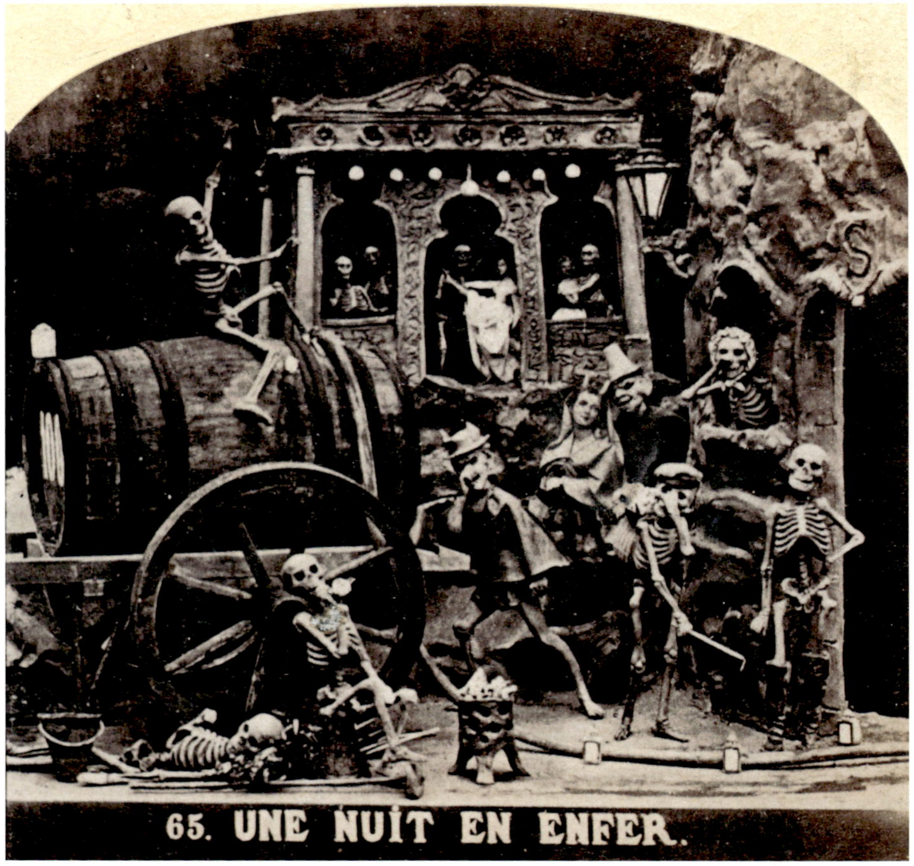
65. UNE NUIT EN ENFER.

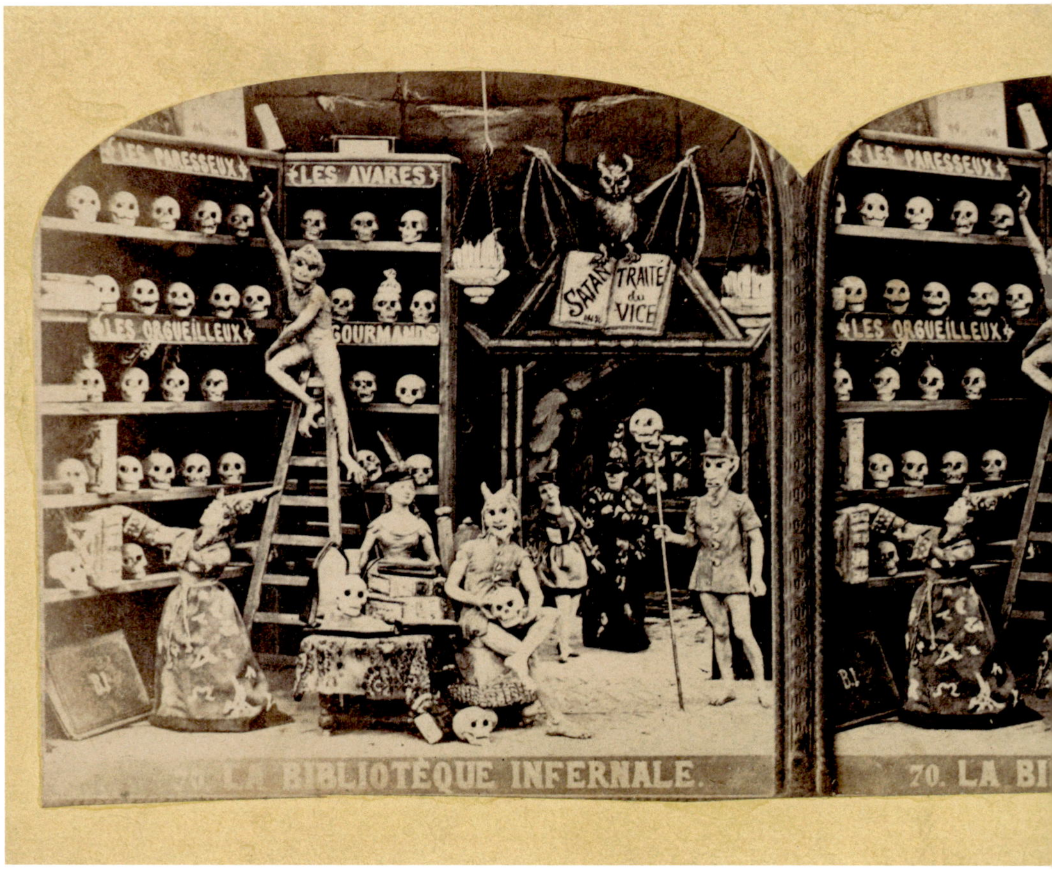
LES PARESSEUX
LES AVARES
SATAN
TRAITE du VICE
LES ORGUEILLEUX
GOURMANDS
LA BIBLIOTEQUE INFERNALE.
70. LA BI

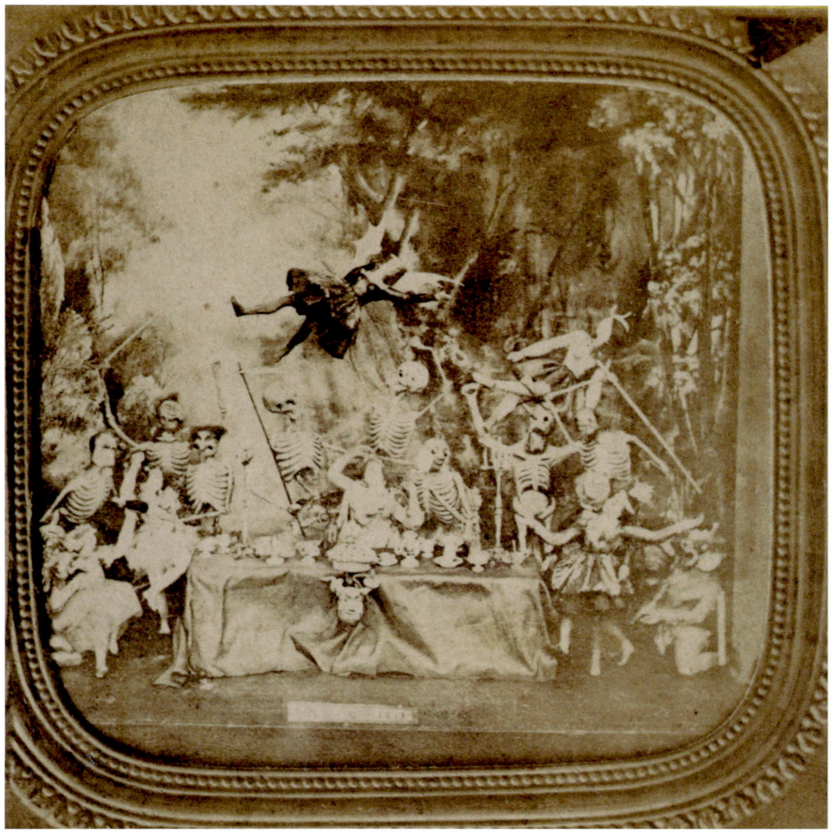

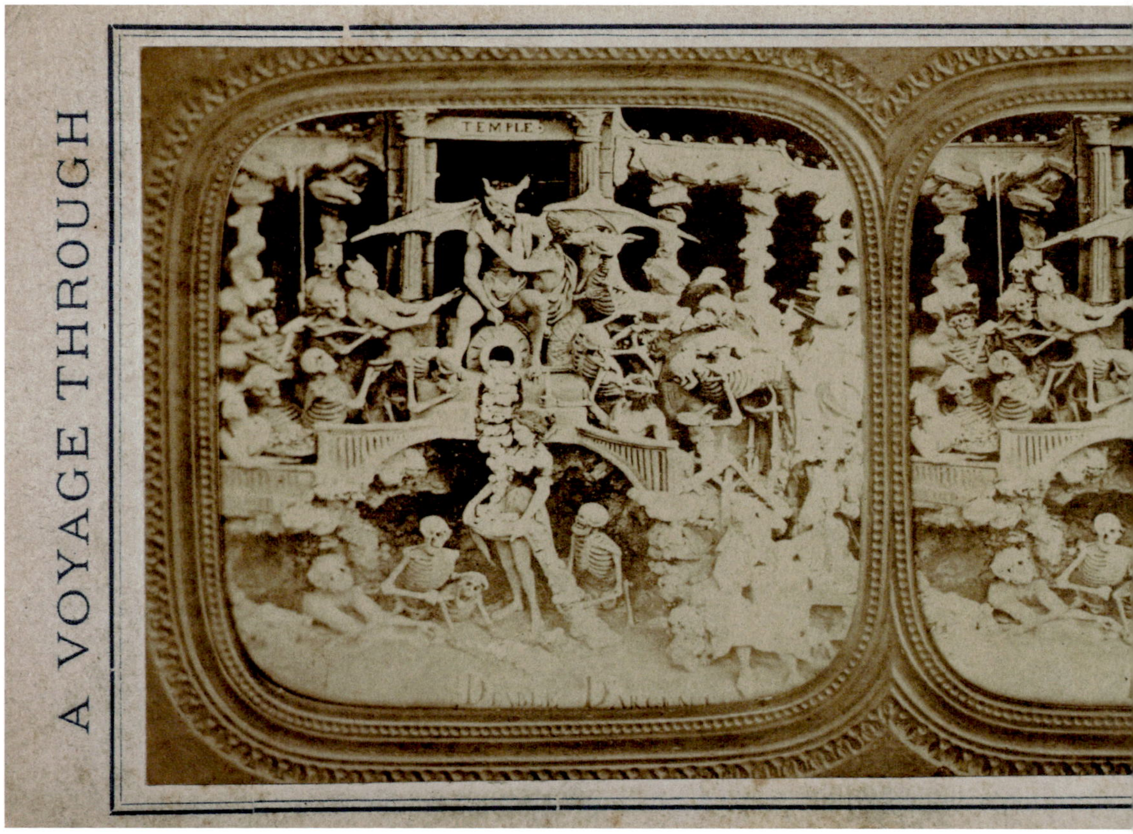
A VOYAGE THROUGH
TEMPLE

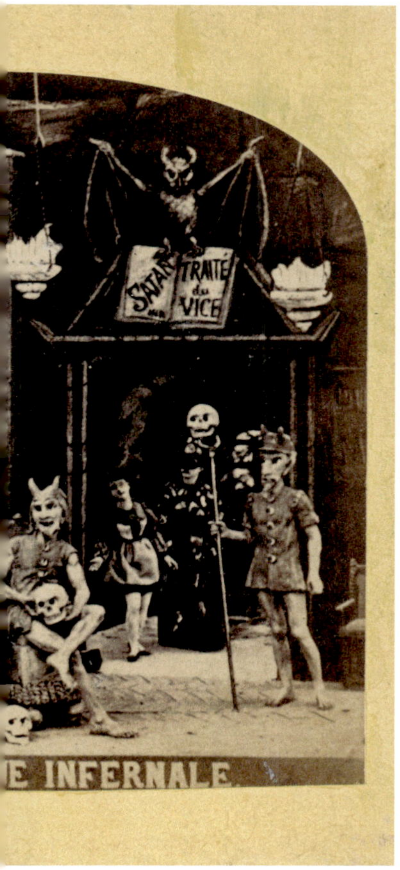

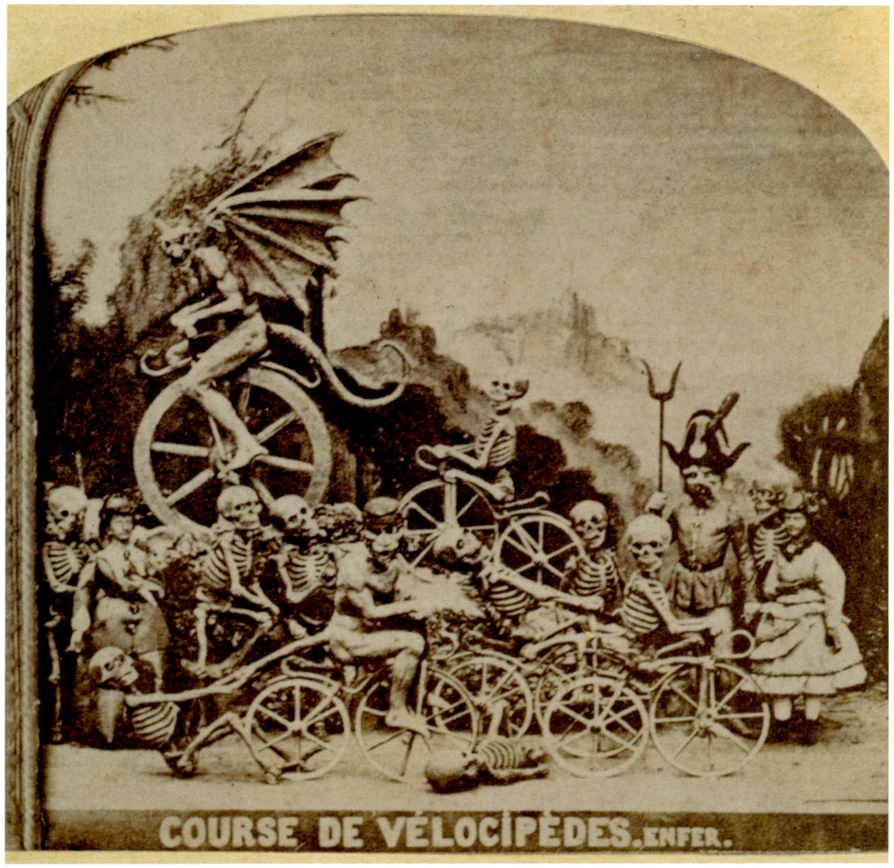

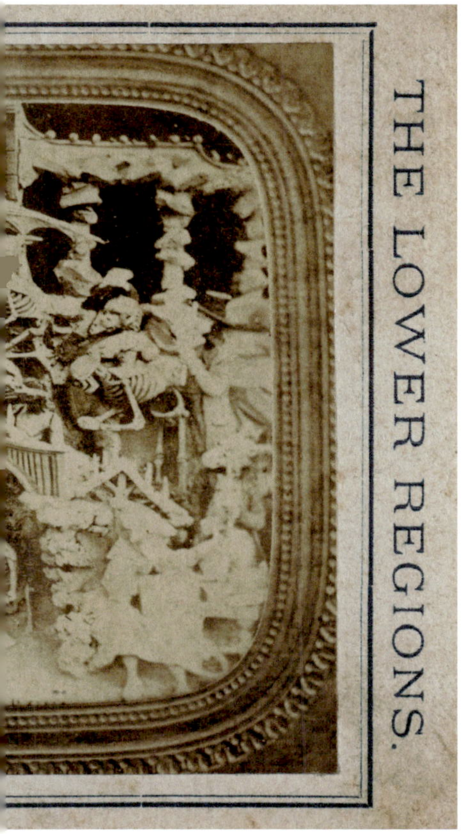

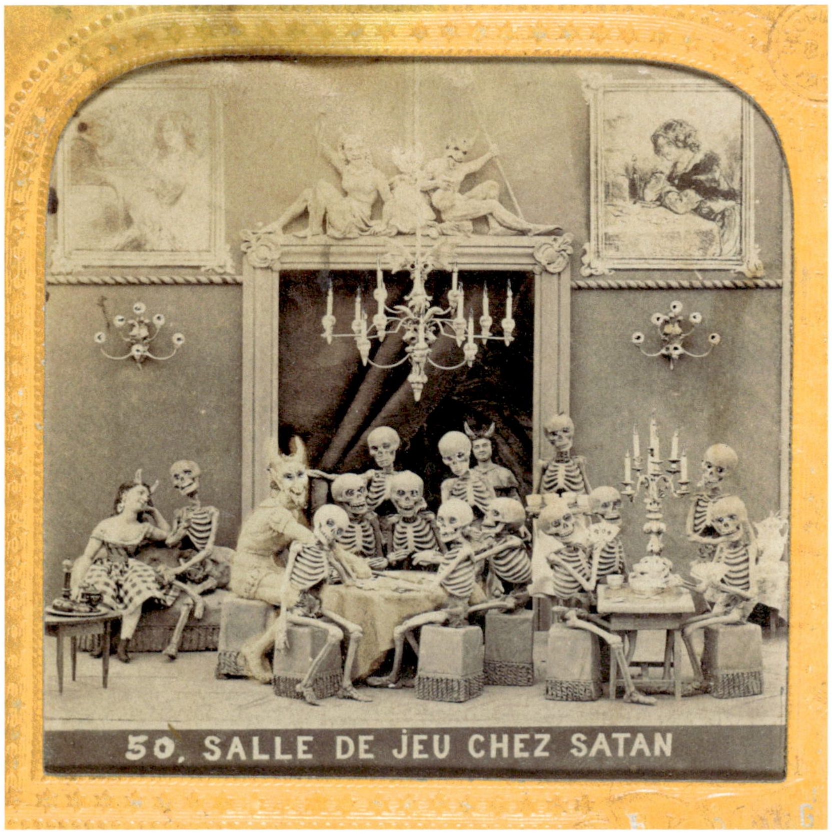

Plates 10.46-10.51

Les Diableries
Stereoviews Depicting Visionary Dioramas of "Hell", circa 1865

Clay figures made by artists presented in various tableaus and photographed in stereo.

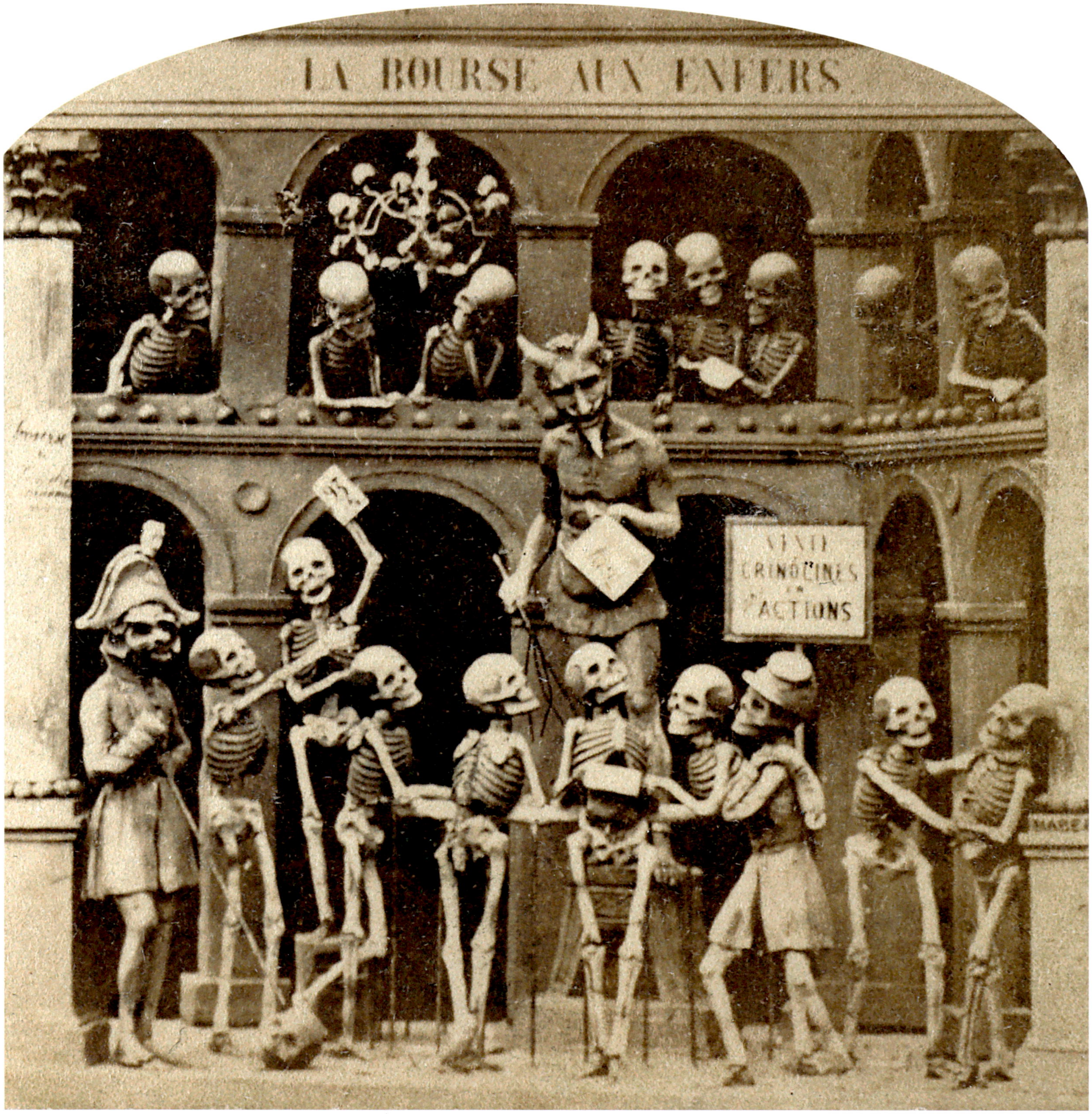

Plate 0.15

Stock Market of Hell
Les Diableries: Stereoview, circa 1865

Enhanced Captions

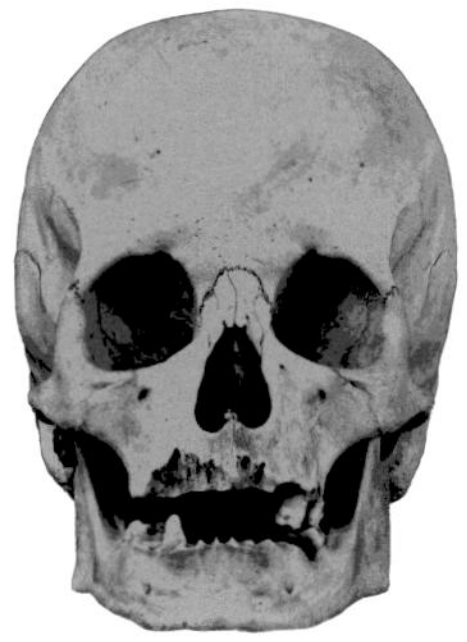

Plate 0.02 (page 6)
The first anatomical theater was built at the University of Padua in 1594. Other early theaters included the Anatomicum of Leiden (Belgium) built in 1596 and this theater at Bologna built in 1637. It was at Bologna, that after 1,500 years of neglect the study of human dissection was rekindled. Mondino De'Luzzi, "Mindinus," (1275–1326) the chair of anatomy at Bologna, performed necropsies and wrote the first book devoted to anatomy, *Anthomia*. It was initially written in 1316 and was published in 1487. The discovery of movable type and development of the printing press around 1450 resulted in publication of anatomical texts with images.

In 1491, Johannes de Ketham published the first anatomical illustrations in his *Fasciculus medicinae*. In 1543, a professor at Padua, Andreas Vesalius (1514–1564) published *De humani corporis fabrica libri septum*. With this magnificently illustrated epoch work, Vesalius earned the title of "The Father of Modern Anatomy." The book has been considered the greatest event in medical history since the work of Rome's Galen (A.D. 130–200) as it stimulated the emergence of medicine. Prior to Vesalius, anatomy was generally taught by reading Galen and dissecting animals.

In 1563, the new main building of the University of Bologna, a palace, The Archiginnasio of Bologna, was completed. An anatomical theater was built in 1595, and in 1636 was rebuilt in its current location in the palace. It was built out of spruce wood. The room was reconstructed during the following century and completed in its final form between 1733 and 1736. Elaborate decoration was added by some of Italy's noted artists. Sculptor Silvestro Gianno carved the full-size statues of Hippocrates, Galen, Tagliacozzi, and Mindinus. Plaques carved with figures on the higher level are also by Gianno. Also added, in the center on each side of the professor's chair, are two statues of *spellati* (skinned men) showing musculature. The *spellati* are the work of Ercole Lelli (1702–1766).

In 1742 Lelli began helping to prepare the noted wax anatomical displays for which the Bologna medical school was noted. The wax models made by Bologna's many masters survive today as the Museum of Human Anatomy of the University of Bologna. It is a must-do destination for medical and social historians. During World War II, on January 29, 1944, Allied bombing destroyed much of the Bologna theater. It was rebuilt after the war with the original pieces recovered from the rubble. The Leiden and Padua anatomical theaters also survive. In the United States, in 1825, a year before his death, Thomas Jefferson designed the anatomy theater at the University of Virginia. It was built in 1827 and demolished in 1938 to make room for a new library.

Chapter 1: Posing with Skulls & Bones

Plate 1.4 (page 29)
Anatomical medical school study group in rare identified image. Left to right: Dr. G. H. Chadwick, Portland, ME, Dr. J.J. Meyres, Utica, NY, Dr. S. Graves, Luzern, NY, Pointing to the top of the skull, Dr. C. S. Spencer, New Bedford, MA. An extraordinary image from photography's ambrotype era. The ambrotype was popular in America from 1854 to about 1865. The tintype invented at the same time was less fragile, cheaper, and was often mounted in similar cases.

Plate 1.19 (page 38)
This photograph of pioneer phrenologist Lorenzo Niles Fowler (1811–1896) in his study in London, England, provides a rare glimpse of a phrenologist's 'laboratory.' On a shelf over Fowler's shoulder is a porcelain phrenological head created by him as an aid to understanding and diagnosis. It became a ubiquitous phrenological icon of the era and has recently been reproduced in England and elsewhere as a souvenir for tourists.

In the early nineteenth century, for a brief time, phrenology was the major American psychiatric ideology. It remained an alternate diagnostic psychological evaluation methodology used by professionals and others until the early decades of the twentieth century. Phrenology continues to be promoted and studied by some psychologists and educational professionals, including the Belgian pedagogic professor Paul Bouts (1900–1999), who established Psychognomy, which combines phrenology with other fields, and the London psychologist Peter Cooper, who founded the London Phrenology Company in 1983.

Phrenology was founded in the 1790s by Viennese physician Franz Joseph Gall (1758–1828). Dr. Gall, a brain physiologist, postulated that mental faculties were innate brain functions, which could be localized, and that character traits had specific areas within the brain. He then hypothesized that the skull would have prominences over areas of the brain that were overdeveloped and concluded that a person's character could thus be read by palpating the skull. Gall's most famous disciple was Casper Spurzheim, MD, (1776–1832) who popularized his ideas and most importantly added the concept that proper moral influences could change personality. In 1828, phrenologist Richard Chenevix exclaimed, "Before it [phrenology], all we knew about the brain was how to slice it." In the 1830s, Spurzheim and another disciple, English physician George Combe (1788–1858), the most prolific English phrenologist of the 19th century, carried the phrenological doctrine to the United States. Spurzheim died in Boston only three months after arriving in America. His collection of 550 skulls and brain casts was bought by Harvard's leading physician, John Collins Warren (1778–1856), who added it to his anatomical collection. The Warren Anatomical Museum remains to this day at Harvard's Countway Library, Center for the History of Medicine.

In the 1830s, Lorenzo and Orson Fowler became the main proponents and promoters of phrenology in America. The Fowler brothers established a countrywide and Canadian lecture tour and began publishing numerous books and journals. In 1838, they established the noted *American Phrenological Journal*; it ceased publication in 1911. In the 1850s the Fowlers established a museum, The Phrenological Cabinet, at 131 Nassau Street in lower Manhattan, near P.T. Barnum's noted emporium. There the curious could examine craniums, casts, diagrams, and engravings of a wide range of heads of notables with their characteristics and personalities explained by their head shape. The Fowlers' beliefs and influence became renowned when America's best known and greatest clerical orator, Brooklyn's Henry Ward Beecher (1813–1878), who at first denounced phrenology, became a vocal advocate. Phrenology also became part of some religious groups' practice, others used it as a practical tool for self-improvement and scientists used it to identify "social misfits." Lorenzo Fowler's belief in abstinence to alcohol, as a critical ideal personality trait, also fit in with major religious doctrines and the temperance movement.

In the late 1850s, the Fowlers began a partnership with S.R. Wells. The Fowler and Wells name became synonymous with phrenological achievements and publishing in America. By 1860, however, the fever for phrenology had abated in scientific circles. Lorenzo Fowler and Wells traveled to England, where Fowler revived the interest in phrenology. In 1863, he established the Fowler Institute in London. Lorenzo remained in England the rest of his life using the country as a base to lecture in Europe. He examined the heads of notables, including Charles Dickens, Edgar Allen Poe, Walt Whitman, and Baron Rothschild. At his lectures, he would astonish his audiences by delineation of the personalities who would come forth from the audience. To the end of his life, Fowler continued his publishing endeavors; his *Mental Science as Explained by Phrenology* was published in the United States and England into the first decade of the twentieth century. In 1896, Fowler wrote that he had sold over 250,000 copies of his *Self-Instructor in Phrenology, Physiology and Physiognomy–A Handbook for the People*. Fowler's texts were the best selling of self-help texts of the nineteenth century. Phrenological analysis theoretically could show which areas of the brain were over- or under-developed and "repeated exercise of a mental faculty...could alter the size or strength of the corresponding area thus improving character."

His texts, with his added physiognomic analysis methodology, also appealed to those who wanted to identify, by visual cues, the personality and character of those with whom they interacted.

Plates 1.23, 1.24 (page 41)
The Promise Of The Profession
These, extraordinarily rare, half plate (four and a half by five inch) tintypes document the promise of medicine in the 1880s. F. R. Reynolds is depicted studying anatomy with his unidentified roommate. A sign at the foot of the table reads, "Before Taking Degree." In another photograph he is dressed in fine clothes, sporting a top hat, again posed with his skeletal parts but now the sign reads, "After Taking Degree." A newspaper article accompanying the photographs states Reynolds, a graduate of Chicago's Rush Medical College, 'practiced ophthalmology.' It is unusual to have documented photographs as very few people inscribed them, then or now, with who, what, where, and when. While the camera is ubiquitous today, historians in the next century may have fewer photographic documents, as most images are digital and not printed.

The accompanying article provides a contemporary picture of a successful, well-rounded, community-based ophthalmologist:

THE SENTINEL: SUNDAY MORNING, APRIL 18, 1897, WORKED FOR THE MEDICAL BILL.

Dr. F.R. Reynolds, a Well-Known Specialist in Northern Wisconsin

One of the most effective workers for the Medical bill before the legislature has been Dr. F.R. Reynolds of Eau Claire. He is a well-known, experienced specialist in diseases and surgery of the eye, ear, nose and throat. He was born in La Crosse, and is the son of the late Hon. B.W. Reynolds, one of the first settlers there, prominent in politics in this state, whose brothers were all extensive slave holders in the South, while he was a pronounced abolitionist.

The elder Reynolds was also a delegate to the famous Free Soil convention in Kansas, and held the office of receiver of public moneys in the St. Croix district under President Lincoln. He frequently contributed able articles of public interest to the press under the cognomens "Thunder," "Syntax," and "Rex."

His son, Dr. F.R. Reynolds, began his career as an associate of the late Dr. Bury for many years a well-known and successful specialist in Northwestern Wisconsin... , and after his death Dr. Reynolds became his successor. In 1891 he removed from there to Eau Claire as a more central locality and wider field of practice.

Besides being a graduate of Rush Medical College, he has taken special courses at the best institutions in Chicago and New York City and has studied in the best schools of London and Vienna. He is a member of the American Medical Association, Wisconsin State Medical Society, Inter-County Medical Society, Anglo-American Medical Association, Vienna; Eighth International Ophthalmological Congress, Edinburgh, etc.

He has invented a number of novel and useful instruments, increasing the armamentarium of specialists in his line of practice. He is noted for his kindness and consideration in treating the afflicted poor and needy for slight compensation and the needy without charge, and is said to be a constant worker and student in his profession.
He is a frequent contributor to a number of medical journals and the author of some able papers of professional and literary merit and reputed as an able translator of German medical literature into English.

He is credited with having performed many of the most delicate operations in the line of his profession with eminent skill and success. His high standard of medical ethics is said to be his ruling motive in the active efforts he has made for the medical profession of the state on behalf of the Medical Bill.

Plate 1.27 (page 43)
Sir William Osler (1849–1919), arguably the leading physician in the recent history of medicine, poses with doctors, nurses, and patients at the specialty hospital, which opened a new building in April 1887. Osler secured an appointment there thanks to America's leading neurologist, Silas Weir Mitchell

(1829–1914). One of the four founding professors of Johns Hopkins Hospital and Medical School, Osler created the first residency program for specialty training of physicians, and he was the first to bring medical students out of the lecture hall for bedside clinical training. He has frequently been described as the "Father of Modern Medicine." Born in Canada, Osler's legendary career includes his attendance at the proprietary Toronto School of Medicine. He transferred to Montreal's McGill University Faculty of Medicine where he received his MD degree in 1872. In 1874 he was appointed professor. In 1884 he left Montreal to become Chair of Clinical Medicine at The University of Pennsylvania. In 1889, he accepted the position as the first Physician-in-Chief of the new Johns Hopkins Hospital. In 1905, he was appointed to the Regius Chair of Medicine at Oxford, which he held until his death. Osler was created a baronet in the Coronation Honours List of 1911 for his many contributions to the field of medicine

Left to right: J.M. [sic] Cattell, Guy Hinsdale, F.P. Willitts, J.K. Mitchell,MD, George .E. de Schweinitz, MD, Miss C. Dalziel, Mrs. Green, Joseph Otto, Miss J. Dalziel. University of Pennsylvania at the orthopedic hospital and infirmary for nervous diseases, Summer 1887. (Transcribed from back of photo.)

Plate 1.28 (page 43)
Andrew Taylor Still, MD, DO, (1828–1917) a frontier physician and surgeon, was the founder of osteopathy and osteopathic medicine. He was also an author, inventor, and Kansas territorial and state legislator. Still believed that bones (osteon) were the starting point from which he was able to determine the cause of pathological conditions. In 1892, he founded the world's first osteopathic medical school, American School of Osteopathy (now A.T. Still University) in Kirksville, Missouri. Osteopathy was originally considered a medical sect, but in the mid-twentieth century, the DO degree was deemed equal to the MD degree as osteopathic physicians embraced standard medical and surgical education and practices. However, osteopaths received special training in musculoskeletal conditions, and are taught osteopathic manipulative medicine (OOM).

Plate 1.30 (page 45)
This can not be the current public institution, as The University of Washington School of Medicine was founded in 1946. These students are most likely attending a proprietary school in the state or are a group from the University of Washington (which was founded in 1861), attending a school anywhere in the country.

Plate 1.34 (page 46)
In the first decade of the twentieth century, photographs 14" x 17" and larger were produced with close-up and detail of two or three physicians who stand on the left bending forward and eagerly watching the dissection in Rembrandt's *The Anatomy Lesson of Dr. Nicolaes Tulp*. The painting, unquestionably the most renowned dissection painting of all time, was reproduced as the full scene and in parts. The photographs were framed and sold to medical establishments. The close-up photograph of the physicians leaning forward can be seen hanging on the far wall in this lounge. I went to medical school in Syracuse and Good Shepard Hospital was one of the institutions I studied in.

Chapter 2: Teaching Anatomy & Osteology

Plate 2.1 (page 50)
Joseph Mellick Leidy (1823–1891) spent two years in apprenticeships with physicians, then three years at the University of Pennsylvania Medical College, graduating in 1844. With the typical scientific curiosity of nineteenth-century geniuses, Leidy had varied interests and accomplishments in several fields and has been profiled as "the last man who knew everything." He was the foremost American human anatomist of his time and served as Professor of Anatomy at Penn from 1853 until his death in 1891. His 1861 anatomy textbook became the standard anatomical text for medical students for decades. He became the first president of the American Association of Anatomists in 1888 and 1889 and was a member of more than fifty societies. Leidy was the first in America to use the microscope as a tool in forensic medicine as he proved blood on a shirt was human and not that of an animal in a legal case. Leidy established the concept of parasitism in America. He is considered the father of American protozoology and parasitology, and described for the first time Trichina larvae in the pig, the source of human trichinosis. He advised cooking the meat fully. In 1885, he founded the department of Biology at Penn and served as its head. He also founded American vertebrate paleontology and put together dinosaur skeletons. He also realized the principles of Darwin's work prior to Darwin's publication. Leidy published more than 800 scientific articles, many of them groundbreaking.

Plate 2.3 (page 54)
In 1904, the Academy issued an album of assembled photographs on a card, 15" x 11". This image is one of two pages devoted to the anatomy department and anatomical subjects. The leading professors are highlighted with individual photographs. Various dissection scenes are shown.

Founded in 1798 as the St. Petersburg Medical and Surgical Academy, the school was ranked as one of the best educational institutions in the Russian Empire. In 1808, the name was changed to the Imperial Medical and Surgical Academy. A member of the Medical and Surgical Academy (Army Medical Academy) had the rights, liabilities, and benefits of a member of the Academy of Sciences. The academy received its present-day name in 1881. The institution played a major role in the development of Russian natural science and medicine.

Plate 2.6 (page 56)
This photograph is unusual, as it is identified in all pertinent aspects except the photographer. Eleven physicians pose in a variety of garb including rubber gowns and various smocks. All these physicians are older men who are seeking advanced training in general or a surgical subspecialty. The New York Post-Graduate Medical School and Hospital was established in 1882 and was one of the leading advanced training medical facility in the city. In 1948 the New York University Post-Graduate Medical School was formed, representing a union of the Postgraduate Division of the College of Medicine of NYU and the New York Post-Graduate Medical School.

Plate 2.9 (page 58)
In the audience are students from around the world. Viennese physicians were the founders and leaders in several specialties since the late nineteenth century and attracted so many American students there was a official group called "American Physicians Studying in Vienna," which served as a sort of ex-pat organization providing American students with their own society as they studied. This photograph is from the collection of a Jewish physician who was attending classes in Vienna. A year later, she was to leave amid the 'uprising.' Her album has photographs of the pro-Nazi population involved in the riots.

Plate 2.17 (page 62)
Identified on chalkboard: A.C. Walker, Rockdale, TX; B.C. Anderson, Beardstown, IL; Dwight Manson, Marion, PA; C. Sheley, Independence, MO; James Lyle, Pataskala, OH; G. Johnson, Seattle, WA; ??srm, Darborn, NY.

This class of older physicians comes from all over the United States indicating the institution they are attending has national standing. Most likely it is the New York Post-Graduate Medical School, The College of Physicians and Surgeons (Columbia) or The University and Bellevue Medical College (Current NYU). The prosector, dressed in a black cap and gown, poses with scalpel in hand, his students in white gowns. This was before the era when surgeon's donned white garb. Both gloves and white gowns were used by anatomists and pathologists more than a decade before they were standard in surgery as they wore them to protect themselves from the moist cadaver and the possible septic diseases in unembalmed cadavers and deceased patients.

Chapter 3: Dissection Photography

Plate 3.1 (page 73)
This stereoview was sold to the public for decades. The card perhaps elevated the fears of the public in their

concern about grave robbing, as the building the doctors are dissecting in front of looks like a mausoleum with its iron gate. This is not a classic 'rite of passage' dissection photograph of a medical school class as it clearly depicts physicians busy at work, not those interested in having the camera document their arrival in the profession. The photograph was presented as a secretive intrusion in the life of physicians.

Plate 3.2 (page 74)
This photograph had no identification, however dissection photographs can be identified as to location by comparison to identified images. Women were frequently photographed outdoors at the Kirksville school dissecting on this metal table in front of this door. Some of the Kirksville dissection photographs have osteopathy founder A.T. Still in the image; unfortunately, this one does not. The man is not identified.

Plate 3.3 (page 75)
Eight men, seven in hats, all with writing on their gowns showing the states they are from. Some have inscribed medical fraternity Greek letters. The students come from a wide range of Eastern states. Right to left: New York, Maryland, Pennsylvania, Virginia, Massachusetts, West Virginia, and South Carolina, and the student on the left distinguishes himself by inscribing, NY City. This photograph is unidentified, but its location was identified by comparison to other dissection images. The large anatomical posters, special table, window and its framing identified the image. In addition other photographs showed an identical style of posing with state letters written across the chest and an epigraph. The number on the left leg of the table denotes the year, 1915.

The statement "He/She lived for Others and Died for Us" was perhaps the most common inscription found on dissection tables. The cadaver is meant to be symbolic and perhaps mindful/reflective of the Christian symbolism of the body and philosophy of Christ's sacrifice. His believers note that "Christ died for humanity," and so the cadaver represents the body of Christ. During the 1890s through the 1910s, epigraphs were most popular and adorned dissection room tables, blackboards, and charts. The student's anxiety, and thoughts about mortality can often be seen inscribed about the dissection room. The statements related to gallows humor, allowing students to relieve situational tension and academic pressure.

Plate 3.4 (page 76)
The classic American way to affirm medical identity as well as the teamwork involved in dissection was to pose with your classmates around the cadaver. A photograph taken by a professional photographer of an individual was rare in the United States prior to the 1930s, however it was frequently done in Europe. The concept of the image was not so much a 'rite of passage' image, but that of a proud occupational portrait. By the 1920s, Kodak and other roll film cameras allowed the common person to be a photographer, and individuals were photographed with their cadaver by their classmates; however, the images are clearly snapshot in quality and posing.

Plate 3.6 (page 77)
Each medical school had a particular format for their dissections and the photographs. The University of North Carolina students dissected outdoors, as all their dissection scenes I have found from the nineteenth century are outdoor photographs. Dissecting outdoors helped dissipate the odor of the cadavers and lighting was good. Examination of North Carolina students show a range of outdoor posed images.

Plate 3.12 (page 80)
With rolled-up sleeves, a medical student joins his classmates in dissection. They wear derby hats. The message on a postcard, until 1908, had to be written only on the front of the card. The message: "Many are the joys and pleasures of a medical student."

Plate 3.13 (page 80)
In the classic style of Baltimore dissection photographs these dental students have their home state written on their aprons and include the year, '09. School identification is on the table: B. C. D. S. In 1908 the United States passed a law allowing postcard backs to be divided with space for an address and a short note. Prior to that year, postcards as seen here could only have text on the front. This fledgling dentist inscribes a typical satiric message: "Dear Bob, This is a cadaver. I will send a slice for the B.H. Think of this when you eat."

The Baltimore College of Dental Surgery was the first dental college in the world. Founded in 1840 by Horace Hayden, MD, and Chapin Harris, MD, it became the leading dental facility and awarded the DDS degree. In the nineteenth century, American physicians routinely traveled to Europe for graduate education; however, European dentists traveled to America for advanced training. American dentists were recruited by European notables for their expertise.

Plate 3.16 (page 82)
In large medical schools cadavers were placed on hooks, as shown, and a chain link system delivered the bodies into the main dissection room.

Plate 3.17 (page 83)
The photograph is unidentified except for the photographer's signature. The poster of the dissected man with his hand raised and the slanted windows identify the location.

Plate 3.22 (page 85)
Dr. William S. Huber (July 1865–May 1909) of Lebanon, PA, is on the right. Dr. Huber attended the University of Pennsylvania from 1884 to 1886. The three posters in the background, along with other items in the room and its construction, will help identification of unidentified dissection scenes.

Plate 3.23 (page 86)
An unusual image, as students wear skull and crossed bones on their gowns. The sign contains a variety of statements. This is Class No. 2 and lists two classes of students, "Lowers" and "Uppers" as well as the name of the Dem. (demonstrator) who stands on the right without a gown. The slogan about man is placed at the bottom of the blackboard instead of a prominent place on top. The most noted school in Kirksville was the school of osteopathy, and this may be a variant.

Plate 3.24 (page 86)
These students wear bakers-style hats popular in the 1910–1930 time period. The skeleton in the background adds a clever touch. That this was perhaps not the highest-level medical school is indicated by the sign on the wall advising that a $5.00 fine was the penalty for spitting.

Plate 3.26 (page 88)
Two cadavers are noted. The men on the right hold a black child seated on the other cadaver's legs. One of the students is from NB (New Brunswick), Canada.

Plate 3.40 (page 94)
Medical education was typically piecemeal until the turn of the century. It often consisted of an apprenticeship with a physician and formal classes elsewhere, either at a noted medical school or a small proprietary establishment. Dissection was a course offered in some localities only on a part-time basis as the supply of cadavers dictated. This casual group without gowns appear to be one of these schools. The 1910 Flexner report on the status of American Medical Schools and the influence of university medical establishments changed medical education by 1920. Formal university attendance became the only road to an MD degree.

Plate 3.46 (page 96)
Unusual in this early era-1906 image is that all the students are dressed in white. In Europe at this time, white had become the standard dress of physicians and also medical students.

Chapter 4: Special Bone Studies

Plate 4.1 (page 102)
Bizarre medical cases fascinate the public as well as physicians, especially when someone survives a devastating wound in a freak accident that should have resulted in immediate death. Despite authenticated documentation

and physical evidence, the incident still defies belief. Near Cavendish, Vermont, on September 3, 1848, twenty-five-year-old Phineas Gage (1823–1860), the foreman of a Rutland and Burlington railroad construction crew, was placing an explosive charge into boulders he and his coworkers were clearing. While tamping the explosive in place with a metal tamping rod, Gage inadvertently sparked an explosion. The resulting blast turned the tapered 3 ½ foot and 1 ¼ inch diameter tamping rod into a 13 pound projectile that exploded from the stone hole. Fortunately, it was the pointed end that first hit Gage's head; the long rod flew through his head and landed several yards away. Gage supposedly lost consciousness. While his coworkers thought he was certainly dead, he amazingly regained consciousness, got up and began walking. Placed in an ox cart, he was carried three quarters of a mile to his hotel, where he was examined later that day by a physician. Dr. John Martin Harlow noted his patient was "quite conscious and collected in his mind…. The scalp was everted, the bones extensively fractured and upraised, and the brain protruded." Harlow replaced some brain and bone fragments. In his examination, Dr. Harlow passed one index finger down its whole length through the wound and into the brain; another finger was passed from below upward from until they touched. It was standard practice in these pre-aseptic surgery days to explore a wound with one's fingers, looking for foreign matter. It was not until 1881, when doctors killed President James Garfield by repeatedly putting their fingers into his abdomen looking for a bullet and infecting him with septicemia, that evidence showed that cleanliness in treating wounds was paramount. Gage, despite his wound and his treatment, was very fortunate. Harlow dressed Gage's large upper head wound with adhesive straps and covered the opening with wet compresses.

A few days later, massive swelling and infection of the upper brain tissue occurred and Gage slipped into a semi-comatose state. About two weeks after the accident, Dr. Harlow open his localized brain abscess, releasing about eight ounces of pus. If it had extended into deeper brain tissue, Gage would have died. Gage miraculously recovered. His hemorrhaging ceased within two days, but infection resulted in a purulent discharge from the patient's head and mouth. Remarkably, Mr. Gage survived and grew stronger until the ninth day, when erysipelas set in. He became comatose and exhibited a large "fungous growth from the upper wound " and " inner canthus of the (left) eye." By the twelfth day, the growth was cut away and the brain abscess removed. Eight ounces of pus were drained. Although subsequently described as "demented," Mr. Gage was up and feeling generally better on the 21st day. Within one month he was walking about freely. By January 1st, 1849 he was leading a normal life. Physical examination at the time revealed that the wound had closed, Mr. Gage's left eye was blinded and exhibiting lid ptosis, and partial paralysis had frozen his face. The top of his head had a quadrangular prominence with a deep depression behind it. The physical wound had healed and the patient had recovered, but Mr. Gage was a changed man. Once an honest, amiable, and responsible worker, he became ill-mannered, profane, and unreliable. According to those who knew him, he was "no longer Gage." His intelligence, memory, speech, and physical function, however, were unaffected.

Perhaps the only circumstance more startling than this rapid recovery was Gage's return to a rugged frontier life. Deemed unfit to resume his duties as a foreman due to his radical personality change, Mr. Gage profited from his extraordinary story by exhibiting himself in backwoods carnival sideshows, accompanied by the fateful tamping rod. Within the last five years, two daguerreotypes have been discovered that show Gage posing with the bar that went through his head. His injuries are clearly seen.

By 1851, he had become a hostler at a stable and, by the following year, he traveled to South America where in Chile he was employed as a stage coach driver. Gage returned home from South America in 1860, when his health began to fail. He suffered an epileptic seizure the next year. More seizures followed, and Mr. Gage died on May 20, 1861, almost twelve years after his injury.

In 1866, Mr. Gage's mother gave medical practitioners permission to examine her son's skull. Examination revealed an opening measuring 1" x 2" at the skull's base, with most of the sphenoid bone missing. A somewhat larger piece of the frontal bone appeared to have been uplifted and replaced, although a 4" defect remained. Both skull and tamper were ultimately donated to Harvard University's Warren Anatomical Museum, where, despite the ravages of time, they remain to this day. Nineteenth-century pioneer neuroanatomists and neurologists spent a great deal of time trying to determine Gage's exact brain injury that allowed him to survive such a drastic wound. This photograph was published as a tipped-in albumen print in the *Descriptive Catalogue of the Warren Anatomical Museum*, (1870). Numerous articles have been written about Gage and the effect of his wound. In 1994, Hanna Damasio and her colleagues at the University of Iowa used neuro imaging to reconstruct Gage's skull. They concluded he had an injury to his left and right perfrontal cortex. But according to other scientists (Liebert et al,) using computer-generated three-dimensional reconstructions of a thin slice computed tomography scan of the Gage skull, the damage to Gage's brain was limited to the left hemisphere. In any event Gage's skull will continue to be studied and fascinate physicians.

Plate 4.2 (page 102)

Specimen 1108: A portion of cranium, with conoidal ball embedded in the ethmoid and frontal bones. The patient survived two months. The photograph was issued as a separate plate and was also published in *Photographs of Surgical Cases and Specimens taken at the Army Medical Museum*. This photographic series was published by the Surgeon General's Office in eight volumes, during the years 1869 to 1881. The first five volumes were issued to medical officers and medical schools and were produced as a set in 1871. Each volume contained 50 photographs, most 4" x 6", oval albumen prints matted on 9.5" x 11.5" board with a caption on the reverse. The final three volumes were kept at the Army Medical Museum and not issued.

Cpl. G.W.S., 12th Massachusetts Volunteers, aged 29 years, was wounded at the battle of Fredericksburg, December 13th, 1862, and admitted into the Camden Street USA General Hospital, Baltimore, Maryland, December 19, 1862. The globe of the right eye was destroyed. There was a slight slit in the lower lid. The patient did not complain of much pain. The functions of the other eye were not disturbed, although the organ was observed to be unnaturally prominent. Slight headache was the only brain symptom present. Neither the patient nor medical attendants suspected that the ball had entered the orbit. The wound healed kindly, and after three weeks the man went about the city habitually, with a pass. He continued well, with the exception of an occasional pain over the left eye, until February 6, 1863, when he had a chill, followed by febrile reaction. There was however, no marked cerebral symptoms, until February 10, 1863, when delirium came on. On the following day, the patient became comatose, and at midnight of February 11, 1863, he died.

The autopsy revealed a conoidal bullet, entering at the inner angle of the right orbit, through the orbital plates of the frontal and ethmoid bones, and resting between the sphenoid, ethmoid, and frontal bones, pressing back the left orbital plate of the latter. The ball was in contact with the dura mater, which had ulcerated at this point. At the base of the anterior lobe of the left cerebral hemisphere was an abscess containing two drachms of pus. A formation of callous encrusted the ball and diminished the opening originally made by it. There was no necrosis.

Plate 4.3 (page 103)

This atlas contains the most comprehensive collection of nineteenth-century studies of anatomical anomalies, many of them beautifully presented and artistically photographed. *Atlas dumuseé Dupuytren* was accompanied by a four-part text published by M. Houel, conservateur des collections de la faculty de Paris. The work contains 85 'Pantotypie' or 'Lithotypie' prints, a French style of mechanical permanent print similar to the heliotype. Many of the anatomical specimens were from the noted collection of Parisian physician Baron Guillaume Dupuytren (1777–1835). Dupuytren was France's leading surgeon and perhaps the richest physician of the time. (He is noted especially for his description and operation of finger contracture deformity.)

At his death, he left 200,000 francs to the Medical Faculty of Paris to establish a chair of pathological anatomy. However, the French government had already established that position so the faculty decided to found a pathological-anatomical museum instead. Parisian physician Mathieu J. B. Orfila (1787–1853), the pioneer toxicologist, was chosen to create the Dupuytren Museum, which was established in 1835. The museum continued to acquire collections of anatomical pieces from other physicians illustrating diseases and malformations until 1935. This extraordinary collection contains specimens dating back to 1820, including many related to skeletal deformities of the skull. The museum still exists and can be found at the School of Medicine in Paris at 15-21 rue de l'Ecole de Médecine.

By the end of the nineteenth century, the discovery of the X-ray in 1895 revolutionized the study of the skeleton, and the necessity for a museum of bone pathology was diminished. Boston physician/writer Oliver Wendell Holmes reminds us: "Every now and then a man's mind is stretched by a new idea or sensation, and never shrinks back to its former dimensions." Radiographs gave physicians the ability to look inside a living body and changed the study of medicine forever.

Plate 4.4 (page 104)

Deformities of bony structures were a phenomenon that fascinated physicians, and skeletal remains of patients with the extremes of bone diseases were preserved in various anatomical museums. This photograph is plate 41 from the 1879 *Atlas du museé Dupuytren.* The atlas was included with a four-part text published by M. Houel, conservateur des collections de la faculty de Paris. The atlas contains 85 'Pantotypie' or 'Lithotypie' prints, a French style of mechanical permanent print similar to the heliotype. This photograph shows two cases of extensive exostosis of bone that has grown into the eye orbit. The image on the left shows an ivory exostosis that has grown into the orbit from the ethmoid. The central image documents an extreme case of hyperostosis with exostosis. The image on the right shows a portion of skull from the vault (top) of the skull with extensive hyperostosis.

Philadelphia ophthalmologist George De Schweinitz, in his 1899 *Diseases of the Eye,* discussed exostosis of the orbit and notes: "Exostosis are very hard tumors having an ivory-like shell and a nucleus of spongy bone, their anatomical structure in general being like those of osteomas preceding from adjacent cavities. All osteomas grow slowly—the external exostosis more slowly than the boney tumors which originate from the frontal and ethmoid sinuses. They spring from the periosteum, and are generally found at the upper border of the orbit… they are recognized by their dense hardness, and evident connection with the bone. They may arise from injury; sometimes they are congenital and often their origin is obscure."

Hyperostosis—an excessive growth of bone—may lead to formation of a mass projecting from a normal bone: an exostosis. Almost any mass lesion in the orbit can result in exophthalmos, so a variety of extremely rare musculoskeletal diseases were brought to the attention of ophthalmologists. Exostosis in the orbit can also mimic symptoms of meningiomas of the optic nerve sheath. On the other hand, the growth patterns of neoplasms such as neuroblastomas and meningiomas can cause hyperostosis. Numerous syndromes and pathologic conditions are associated with hyperostosis. Hyperostosis of the skull occurs in Paget's disease, hyperostosis frontalis interna, frontometaphyseal dysplasia, fibrous dysplasia, craniodiaphyseal dysplasia and skeletal metastasis, periosteal chondroma, chondroblastoma, fibrous nonossifying fibroma, and fibrosarcoma of bone. Vitamin A intoxication has also been associated with skeletal hyperostosis. Recurrent episodes of inflammation was a cause of orbital bone hyperostosis in the pre-antibiotic era, as sinus infections were common. In most cases, the exostosis is a benign condition; however, as they can grow near nerves or tendons and press on them, they often need to be removed to prevent damage to those structures.

A unique subset of patients suffers from hereditary multiple exostoses (HME), an autosomal-dominant disorder manifested by multiple osteochondromas and frequently associated with characteristic progressive skeletal deformities.

John Hunter (1728–1793), considered one of the greatest surgeons of all time, founder of experimental and surgical pathology, was perhaps the first to comment on the condition now known as hereditary multiple exostoses. In 1786, he described a patient with multiple exostoses in his *Lectures on the Principles of Surgery*. Hunter gave his medical museum to London's Royal College of Surgeons. (Unfortunately, a highly explosive bomb destroyed most of the collection during a German air raid in May 1941.) In 1814, Boyer published the first description of a family with HME, which was followed by Guy's description of a second family in 1825. The name "multiple exostoses" was given to the condition by pathologist Rudolf Virchow in 1876. A number of synonyms have been used for this disorder, including osteochondromatosis. Most of the clinical aspects of the disease had been described by 1900. In 1943, Jaffe made a significant contribution by further elucidating the pathology of HME, differentiated the disorder from Ollier's disease—multiple enchondroma, whose lesions are primarily located within the metaphyseal regions of tubular bones. (About 25% of those with Ollier's develop chondrosachoma by age 40). The World Health Organization prefers the name multiple osteochondromatosis for HME. Chondrosarcoma is one of the very rare complications of the disease, occurring in less than one percent of patients as a benign exostosis develops into a malignant tumor.

The deformed skull in the center of the photograph represents an extremely rare condition, leontiasis ossea, also known as leontiasis or lion face, now called craniodiaphyseal dysplasia. It is an autosomal-recessive bone disorder that causes calcium to build up in the skull, disfiguring the facial features and reducing life expectancy. The calcium deposits decrease the size of cranial foramina, and can also decrease the hole in the cervical spinal canal, severely affecting cranial nerves and the spinal cord. The few cases recorded reveal most of the sufferers died in childhood. Among the clinical signs are dacryocystitis, seizures, and paralysis, each of which is a complication resulting from the narrowed nerve foramin. The condition is always brought to the attention of ophthalmologists as exophthalmos gradually develops, progressing to a complete loss of sight due to compression of the optic nerve by the overgrowth of bone.

Plate 4.7 (page 106)

Right Humerus and Elbow—Necrosis of the Entire Humerus following Gunshot Fracture of the Epiphysis

Private Barney White, Co.I, 37th New York Volunteers, aged twenty-two years, was wounded at Williamsburg, Virginia, May 5, 1862, in the right elbow, by a conoidal musket ball. A simple dressing was applied, and he was removed to Yorktown, and thence to Philadelphia, where he was admitted to Race Street Hospital. No report can be obtained on the progress of the case until March 13, 1863, when the patient was transferred to the McClellan General Hospital, at Nicetown, with extensive necrosis of the shaft of the humerus and caries of the elbow-joint. About the joint and arm, there were numerous sinuses, through which a profuse purulent discharge escaped, and, occasionally, small fragments of detached bone. Under a supporting treatment, the general condition improved. About the middle of July, it was first noticed that the patient's legs were edematous. General anasarca ensued, and an examination of the urine showed that it was highly albuminous. The patient became rapidly worse, and, on August 12, an uncontrollable diarrhea set in, accompanied by vomiting. The case terminated fatally, August 22, 1863. The characteristic lesions of Bright's disease were revealed at the autopsy.

Plate 4.8 (page 107)

During the nineteenth century, the Army Medical Museum (AMM) evolved into the world renowned Armed Forces Institute of Pathology (AFIP). The AFIP served as a consultant to military and civilian physicians, dentists, and veterinarians. The unique character of the AFIP rested in the expertise of its civilian and military staff of diagnostic pathologists whose daily work consisted of the study of cases that are difficult to diagnose owing to their rarity or their variation from the ordinary. The AFIP was a diagnostic court of last resort and even consulted on difficult cases from around the world. It played a critical role in the standardization of pathologic diagnosis of tumors. In 2011 the AFIP closed and was reformed and incorporated into the Joint Pathology Center. This case is one nineteenth-century example of a difficult diagnosis sent to the AMM.

"A Dry Preparation of the Left Hand and Lower Portion of the Fore-arm, showing Incipient Necrosis of most of the Metacarpals and Phalanges."

"The disease" for which this member was amputated, it would seem, was involved in considerable obscurity, several distinguished surgeons having differed widely in their diagnoses. One considered it a disease of the blood vessels, another elephantiasis, whilst a third, from a microscopic examination, called it chronic cellulitis. The subject of the operation was Mary McFarland, a housekeeper, aged twenty-seven, who had an inflammation in the first phalanx of the little finger about January, 1868. Some six months later the disease invaded the metacarpal bone, and gradually extended during the year 1869. Becoming pregnant in September, she gave birth to a healthy child, which was carried the full-term. Her appetite was insatiable. In March, 1870, she was examined by Doctors Parker and Hamilton and a number of others. In April, the hand becoming enlarged and painful, several punctures were made through the skin, one of which discharged daily from one to three pints of bloody serum. This discharge continued till the limb was amputated. When the punctures closed the hand would again become swollen and painful. In May a seton was introduced, but soon ulcerated, and the opening never healed. At no time was there any marked tenderness in working the joints. On August 10th the little finger was amputated. On the 15th it became necessary to remove the hand and a portion of the forearm. The pathological specimen was contributed to the Army Medical Museum by Dr. J.C.Snively, of Brooklyn, New York. A microscopical examination of several sections of the soft parts of the specimen was made at the Army Medical Museum on August 25, 1870, and nothing new observed except the ordinary exudation products of inflammation. It was then directed that the specimen should be dissected and prepared in the dry way. A cheesy abscess was discovered under the palmer fascia, and periostitis at the metacarpals and phalanges with incipient necrosis. On January 10, 1871, Dr. Snively reported the amputation to have healed promptly, but that within the last month a redevelopment of the disease has occurred, implicating the several glands extending to the axilla."

Plate 4.9 (page 108)
"Jesse Smith, a colored lad of sixteen, employed by a cattle drover, was sent to sleep, on the night of August 11, 1866, in a hay-loft over a stable. He rolled off the hay-mow during the night, and fell to the floor of the stable, and was found in the morning cold and unconscious. He was taken to the Freedman's Hospital in Alexandria, and was examined by Acting Assistant Surgeon R.N. Atwood, who found a contused crucial wound of the scalp and a depressed fracture of the right parietal eminence. The boy had recovered his consciousness; the symptoms of compression of the brain had passed away, and there was no evidence of compression. An expectant treatment was adopted, and no unsatisfactory symptoms appeared for twelve days. On August 22, there was severe headache, and, at night, a compulsion. On August 23, the pain had increased, and there was stupor, and it was decided to elevate the depressed bone. Assisted by Acting Assistant Surgeon A.W.K. Andrews, Dr. Atwood enlarged the wound and trephined. On removing a button of bone, there was a free escape of pus. A depressed fragment of the inner table extending three-fourths of an inch beyond the fracture in the outer table was, with much difficulty, elevated and removed. The patient was rendered insensible during the operation by sulfuric ether. On recovering from the anaesthetic, he had intolerable pain, and one grain of sulphate of morphia was given him. For the next ten days he took from two to three grains of the sulphate of morphia daily, and no other medicine. He was fed on bread and milk acidulated with vinegar, to suit his taste, and he had vinegar and water to drink. His bowels acted regularly. Three days after the operation, the brain began to protrude; and on the tenth day the cerebral hernia was half the size of a hen's egg. Gentle compression of the fungous mass was essayed, but immediately provoked violent convulsions. After the tenth day, the tumor began to diminish in size, and by the end of September it had disappeared, and the wound had cicatrized. At this period, the boy was brought to the Army Medical Museum and photographed. His health was perfectly good and his faculties unimpaired. The specimen of the disk and depressed fragment of the parietal (bone) was presented to the Museum by Dr. Atwood."

Plate 4.11 (page 109)
Reed Brockway Bontecou, MD, (1824–1907) Surgeon-in-Charge of Harewood US Army General Hospital in Washington, D.C., was the war's leading medical photographer. His images of wounded soldiers were taken during the war as medical teaching photographs and are not to be confused with the photographs of wounded soldiers taken after the war as pension verification images. Bontecou was the largest contributor of photographs to the Army Medical Museum and the six volume *Medical and Surgical History of the War of the Rebellion*.

Many of Dr. Bontecou's compelling artistic interpretations of war wounds have been recognized to be among the highest levels of photographic art. The images have universal appeal not only as photographs but as anti war documents. In this photograph, Bontecou has arranged photographs of bones, some from his collection, and some from the Army Medical Museums collection, on a page. The artistic result makes it look like a page of images of ancient pottery and artifacts.

Bontecou's images were included in a major exhibit on *Civil War Photography* (April–September 2013). curated by Jeff Rosenheim, Chief Curator of Photography at the Metropolitan Museum of Art. The images were presented in their own room at the museum. In 2011, to accompany exhibits of Bontecou's photographs, I published *Shooting Soldiers: Civil War Medical Photography by R. B. Bontecou*. The book details his personal life and Civil War exploits and presents wounded soldiers from over 100 regiments.

Plates 4.12, 4.13, 4.14 (page 111)
These photographs come from the collection of Irish surgeon Mr. Edward Stamer O'Grady (1838–1897). He used them as illustrations in the 1870s in the *Dublin Journal of Medical Science*. O' Grady was one of the great surgeons of his time and specialized in tumor and complicated shoulder operations. O'Grady began practicing at Dublin's Mercer Hospital in 1866. During the 1870s it had become a medical posing convention to have the patient shown with the bones that had been removed. Such an image was a great advertisement for the surgeon, as it showed a difficult successful case with a healthy live patient. This patient was operated by O'Grady on twice. After the first operation, a procedure to remove the head of the humerus, the patient proudly holds his humerus (upper arm bone) on the end of a wooden stick. The second operation performed March 3, 1874, shows the patient a few years older, a front and back view clearly showing the extensiveness of the procedure. Additional surgical scars are seen on the patient's shoulder in the front view and major skin incisions are seen on his back.

These photographs come from a large collection of cabinet cards of Irish surgeons cases from the 1870s that were used as illustrations in the Dublin journal. They are currently in the Burns Collection.

Plate 4.15 (page 112)
From *Cliniques Médical Iconographiques* by Haushalter, P, Spillman, L, et al., 1901 At the Faculty de Médicine de Nancy.

This elephantine-sized volume contains 62 plates and 398 photographs produced as phototypies, the French form of mechanical print. All common diseases of the turn of the century are illustrated. The section on hydrocephalus shows the cerebral and bony skull abnormalities as well as the failure in normal somatic growth. Plate 20 shows the differences in a hydrocephalic skull and a normal adult. The patients globular-shaped head and frontal and parietal bossing (prominence) are evident. The dilatation of the cerebral ventricles show the brain tissue destruction.

Plates 4.16, 4.17 (page 112)
Plate 19 of Spillman's text shows a ten-year old child with hydrocephalus and the hydrocephalic skull of a ten-year-old child. A normal ten year old is shown, and the lack of growth of the patient is evident. The ten year old's hydrocephalic skull is much larger than the normal adult shown. The cross section of the brain at the bottom of the photograph illustrates the significant loss of brain tissue in the condition.

Plate 4.18 (page 113)
Open sutures and fontanels allow for expansion of the brain. At birth the brain is only about twenty-five percent of its adult volume, at age one it is fifty percent of adult size and continues to grow until about age 20. Fontanels are membranous structures in contrast to sutures, which are the joints of the bones that make up the skull and are fibrous. Cranial sutures remain open and capable of growth until early adulthood. Premature closure of one or more of these openings results in conditions such as craniosynostosis. In contrast to normal skull growth, in which the skull expands uniformly to accommodate the growth of the brain, a premature closure of a single suture restricts the growth in that part of the skull and promotes growth in other parts of the skull where sutures remain open in order to accommodate brain growth Skull shape abnormalities occur depending on which suture or combination of sutures close.

Humans have six primary fontanels—two along the midline of the top of the skull and two on each side of the lateral vault and right and left mastoid or posterolateral fontanels. The two midline fontanels both participate in anteroposterior and medioateral brain growth. The fontanels on the lateral vault permit superoinferior brain growth. Two additional fontanels (metopic fontanels and sagittal or third fontanels) can also be present in humans. In humans, the sequence of fontanels closure is as follows: 1) posterior fontanels generally closes two to three months after birth, 2) sphenoidal fontanels close around six months after birth, 3) mastoid fontanels close from six to eighteen months after birth, and 4) the anterior fontanels are generally the last to close between one and three years of age (in one recent human sample, the anterior fontanelle was closed in most individuals by 31 months).

Plate 4.19 (page 114)
Over the centuries, the owners of manufacturing businesses attempted to increase production with the selection and creation of tools, machines, and work processes. But there was no interest in changes in the workplace environment for worker safety. In the twentieth century, that changed as physicians documented occupational diseases and organizations were formed not only to protect workers but also, in the process, to increase productivity. It was a win-win situation. This photograph of a display by a German occupational group shows the correct posture to sit at an office or school desk. The signs explain the correct divisions of body weight. The chair keeps the spine straight. During the twentieth century, employers and the public became more aware of the necessity of correct postures and actions at work and play.

Ergonomics is the scientific study of how one's body interacts with the environment when performing a task or activity. It helps fit workplace conditions and job demands to the capabilities of the working population. It is popularly described as Human Engineering or Human Factors Engineering. To supervise adherence to new laws regarding work conditions, the government established the Occupational Safety and Health Administration (OSHA). Worker safety is achieved through the evaluation and design of workplaces, environments, job tasks, equipment, and processes in relationship to human capabilities.

Ergonomics has become increasingly important in the computer age as many workers are using devices repetitively. Repetitive stress injuries have been addressed by OSHA. Statistics show that today the idea of ergonomics in the workplace is mostly commonly applied to posture and sitting position in front of the computer. Musculoskeletal injuries resulting from poor workplace ergonomics account for about a third of all lost workday injuries and illnesses. Carpal tunnel syndrome accounts for about 15 percent of all workplace injuries. Other conditions are tennis elbow, tendonitis, bursitis, strains, and sprains. As new occupations develop, new medical conditions will arise and appropriate scientific studies will find ways to ameliorate them.

Plate 4.21 (page 116)
In 1908, Alphonse Bedin, MD, a member of the Faculty of Medicine in Nancy, France, published *La photographie au laboratoire de medicine légale de la Université de Nancy*, a manuscript containing 77 heliotype prints. Dr. Bedin provided perhaps the widest variety of photographs of patients with mental disorders and self-inflicted wounds, ranging from gunshot injuries to masochistic mutilations to suicide attempts. The skull was obtained at autopsy. The more than fifty excoriations of the skull were the result of blows by an axe. "The patient's successful suicide was not the result of a cerebral hemorrhage, found incidentally at postmortem, but from a severe generalized infection that occurred two days after an attack of 'insanity.' "

Plate 4.24 (page 116)
This patient's head was frozen in position in life as indicated by the ossification of the occipital condyles to the first cervical vertebra. The base of the occipital bone has two small kidney shaped bony areas that attach to the first cervical vertebra. They are called the occipital condyles, and the first cervical vertebra (C1), the atlas. The atlas was displaced and firmly attached to the base of the skull.

Plate 4.26 (page 118)
"Photograph number 12—Lower third of Right Femur, the Outer Condyle split off by a Round Musket Ball. Amputation. Death.

"Pvt. Richard Williams, 28th Pennsylvania, was wounded September 16, 1862, at the Battle of South Mountain by a round musket ball, which entered near the outer hamstring and lodged between the condyles of the right femur. He was conveyed to Washington and admitted into Mount Pleasant Hospital, September 22, 1862. On September 30, his right thigh was amputated by flap incisions. At this date the limit was extensively swollen; the discharge of pus was profuse; abscesses had burrowed in the soft parts and irritated fever existed to an alarming degree. The case terminated fatally on October 2, 1862. The ball, which remained embedded in the cancellated structure of the femur, between the condyles, had caused a very oblique fracture into the knee joint, separating the external condyle from the shaft and breaking off a fragment from the anterior surface of the inner condyle."

Plate 4.27 (page 118)
At the battle of Chancellorsville, on May 3, 1863, Brigadier General Edmund K----, U.S.V.First Lieutenant, 1st U.S. Artillery, was wounded by two leaden balls from a spherical case shot, which entered the lower part of the left side. One of the missiles fractured femur and was impacted in its medullary canal, the other lodged in the vastis externus. The wounded man was conveyed to Washington on May 5th, and, on May 10th, Surgeon Basil Norris, USA, amputated the thigh by the circular method. Ptarmigan supervened, and death on May 28, 1863. This is one of the rare cases of a Civil War bullet lodged in the medullary cavity of a long bone in the Army Medical Museum."

Plates 4.28, 4.29 (page 119)
Photograph Nos. 136 and 137 Successful Secondary Amputation at the Left Hip-Joint

"Pvt. George W Lemon, Co. C, 6th Maryland Volunteers, age 30 years, was wounded at the battle of the Wilderness, May 5, 1864, by a musket ball shattered the left femur at the junction of the upper and middle thirds. He was left on the field until removed by the enemy to a neighboring temporary hospital. On May 13th, this hospital fell into the hands of the Union army, and Lemon was transported to Alexandria, Virginia. He was in a pitiable condition, suffering from diarrhea and bedsores as well as from the constitutional disturbance produced by the fracture of the femur. Apparently no attempt has been made to keep the limb in position. After the patient's admission to the Third Division USA General Hospital at Alexandria, and the employment of suitable dressings and medicines, his general condition was somewhat improved, but the suppuration from the wound in his thigh was copious, and the fractured extremities of the femur were extensively necrosed. Notwithstanding a careful supporting regimen his strength gradually failed, until in the autumn of 1865 seventeen months after the reception of the wound, operative interference of some sort seemed imperative. Accordingly, Surgeon Edwin Bentley, U.S.V. in charge of the Alexandria Hospital, decided to amputate at the hip-joint. The operation was performed by Dr. Bentley by the anterio-posterior flap method, on October 12, 1865. On November 15th, Dr. Bentley reported that the ligatures had come away and that the wound was granulating kindly. At the end of January, 1866, cicatrization was complete, and a colored drawing of the patient was made by Mr. Baumgras, one of the artists of the Army Medical Museum. On January 31, 1866, the patient was transferred to Harewood Hospital, Washington, DC, where a plaster cast of the perfectly sound stump was prepared. A photograph of the patient was also taken. On February 3, 1866 Lemon was discharged (from) the service of the United States at his own request. The exarticulated femur is preserved at the Army Medical Museum."

Plates 4.31 (page 121)
James A. Garfield was shot at Washington, July 2, 1881, by a ball from a pistol in the hands of an assassin. He died September 19, 1881. The autopsy revealed the course of the missile. It had entered over the tenth intercostal space, three and a half inches to the right of the vertebral spines, fractured the right eleventh rib, passed to the left through the body of the first lumbar vertebra, and lodged in the adipose connective tissue, immediately below the lower border of the pancreas, about two inches and a half to left of the spinal column, and behind the peritoneum. It had become completely encysted.

Garfield's death was not due to the two bullet wounds he received. He was killed by the physicians who tended to him. Disregarding the principles of antiseptic and asepsis technique that were in vogue in Europe, they dug their unwashed fingers into his abdomen trying to find the bullet in the then-mistaken belief by physicians that bullets had to be removed. The physicians, by their constant probing, opened his wound up to a five-inch gash and lacerated his liver creating a pool of pus that led to septicemia. The photographs show the bullets had passed through his abdomen, not causing serious injury and lodged safely in an area near his spine.

For over two months, doctors continued to probe Garfield's wound with dirty, unsterilized fingers and instruments. In their desperation, they asked noted inventor Alexander Graham Bell for his help. He devised a special metal detector invented specifically for the purpose of finding the bullet lodged inside Garfield. Bell failed, as Garfield was on one of the first metal mattresses and also a metal bed frame, which interfered with the device. Garfield died a slow painful death as his doctors infected him. He died of septicemia and pneumonia. President Garfield and Kennedy share a special bond aside from being assassinated; they are the only presidents to die before they reached the age of 50. After Garfield died, American physicians embraced asepsis as a surgical and wound treatment modality. The difference between antiseptic treatment and asepsis is that in asepsis all aspects of treatment involve sterilization of hands and instruments as well as patient prep. In antiseptic treatment, the treatment is topical; the idea is to kill the bacteria present by using germ killing solutions.

Plate 4.33 (page 122)
An unknown Mexican was killed by an arrow wound in an Indian fight, which occurred twenty-five miles northwest of Fort Concho, Texas, February 22, 1868. The arrow perforated the frontal bone. When the skull was opened, an incision was found to extend across the opposite hemisphere, touching the dura mater just above the tentorium. The dura mater was stained, but no mark could be found on the skull. At the post-mortem, the arrowhead was found in the brain. When the Mexican was hit, he seized the arrow shaft with both hands and pulled it out, then dropped, and, of course, remained unconscious until he died, about six hours subsequently. The arrowhead has been removed from the cavity of the skull and fastened at the point of entrance. Apart from the lesion, the skull is a highly interesting specimen.

Plate 4.34 (page 123)
Pvt. Martin W -----, Troop E, 4th Cavalry, was killed by Indians, on September 30, 1870, twenty miles from Fort Concho, Texas, while on duty as one of the mail-stage guards from Fort Chanbourne. The escort being attacked by a band of Comanches, this soldier was wounded by an iron-headed arrow, which entered the squamous portion of the temporal bone and penetrated the left cerebral hemisphere to a depth of an inch or more, causing intracranial bleeding that was speedily fatal. The pathological specimen, with the history, was forwarded to the Army Medical Museum by Assistant Surgeon W.M.Notson,U.S.A.. A puncture of the thin calvaria, without fissuring, is well indicated. Internally there is no splintering. The vitrous table is as cleanly divided as the outer table.

Plate 4.36 (page 125)
This cranium was picked up on the first Bull Run battlefield by Dr. F. Schafhirt and presented to the Army Medical Museum. It is numbered 3251 of the Surgical Section. It displays a fracture caused by a musket ball that entered at the right fronto-parietal suture and temporal ridge, and fractured the os frontis in a long fissure, which runs in front one inch above the orbits, and downwards through the greater wing of sphenoid and squamous portion of the left temporal into the mastoid process. One fissure branches off above the left orbit and downward through the maxillary sinus. Another fissure passes posteriorly from the wound of entrance and upwards through the right to the left parietal protuberance. Another fissure downwards through the right auditory meatus has divided the petrous bone. Yet another fissure passes backwards through the upper portion of the right temporal into the occiput. The ball passed out at the upper part of the occipital near the inter-parietal suture.

Plate 4.37 (page 126)
Two unfortunate patients of two of Paris's noted physicians, Guerin and Verneuil. Young girls and women working in the match-making industry using white phosphorus suffered from a chronic progressive painful melting away of the lower jaw. The treatment was surgical excision of the jaw because of the pain and infections. The resulting agony and disfigurement due to bony necrosis was recognized by medical authorities and was considered the most devastating of occupational diseases. Multiple spontaneous fractures of long bones as well as serious infection of the necrotic parts complicated the disease.

The first friction match was invented in 1826 in England and called the Lucifer match. In 1832, in Germany and Austria, a readily ignitable match using white phosphorous was developed. It was named the Congreve match but throughout most of the century, the matches were called lucifers, and their makers were indeed working with the devil. The disease, phossy jaw, is slow to develop and it was not until 1844, twelve years after the matches were first produced, that cases were recognized first in Vienna and then other countries. The first symptom is a toothache, and then the appearance of a dull red spot on the buccal mucosa. The red spot is pathognomonic; it only occurs in this disease. Soon a hole and sinus develop and an foul infection. Eventually teeth are lost, pieces of bone fall out, and the infection can spread to the brain. Because of the slow onset and seemingly innocuous start in an era where tuberculous skin and bone infections were common, it was at least two decades before the disease was fully appreciated. In 1855, a safety match using harmless red phosphorous replaced white phosphorous in matches in some factories. White phosphorous is still used today in military projectiles.

The story of phossy jaw is a good example in the difficulty in achieving safe public health measures worldwide. In the 1870s, countries began outlawing the use of white phosphorus in match- making. Finland was the first to start the process in 1872. In 1906, a Berne Convention was signed by European nations, and by 1910, all Europe countries had outlawed the substance. In 1919, India and Japan followed suit, and by 1925 the League of Nations convinced China to agree to the ban. The United States curiously did not follow the ban believing America's clean factories were better than those in Europe. In 1931 the U.S. government finally placed a prohibitive tax on white phosphorous matches and forbid their export and import.

Plate 4.38 (page 126)
The left-hand figure, specimen 4740, section 1, Army Medical Museum, is from the case of Private O. Payne, Co. D, 20th colored troops, aged 19, wounded near Petersburg, June 7, 1864. Accidental gunshot wound through tendo-achillis, passing through the ankle joint, and emerging almost

at a right angle with the line of entrance, literally grinding the astragalus and part of the oscalis. On June 27th, Acting Assistant Surgeon A. Transue amputated the leg by the circular method, dividing the bone one and a half inches above the malleoli. Extensive necrosis of the shafts of the tibia and fibula followed the amputation and the pain and discharge from the stump became so great as to necessitate a reamputation, which is performed by the circular method, three inches below the knee, on March 18, 1865. The patient recovered and was a pensioner in 1880.

Figure on the right, specimen 4172, Section I, Army Medical Museum, is from the case of Private W.H. Miller, I, 72d Pennsylvania, aged 21, who received a shot fracture of the tarsal and metatarsal bones of the left foot, at Antietam, September 17, 1862. Circular amputation at lower third of leg was performed on September 28th, by Acting Assistant Surgeon P. Middleton. In March, 1864, Miller was admitted into Christian Street Hospital, Philadelphia. His general health was feeble, both bones of the stump were necrosed in nearly their whole length, and the soft parts were in a very unhealthy condition, necessitating a reamputation at the lower third of the thigh, which was performed by the antero-posterior flap method by Acting Assistant Surgeon R. J. Lewis, on April 28, 1864. The stump healed readily, and the patient was discharged from service on September 8, 1864. He was a pensioner in March, 1879.

Plate 4.39 (page 127)
During the Civil War, medical directors asked for surgeons to send photographs of interesting cases to the museum for documentation, teaching purposes, and eventual publication. These photographs, almost all cartes de visite, were assembled and at war's end over 1,000 were in the museum's collection. They were numbered after the war most often with the designation XXX Cont. Phot. If another number appeared on the card, that referred to the same patient depicted on another card. In the beginning of the twentieth century about 900 of the card photographs were dispersed. The museum continued to add to the contributed photograph collection until the early twentieth century.

Plate 4.44 (page 130)
Charles McBurney, MD, (1845–1913) is best known for his work on appendicitis. However, his innovative jaw spring mechanical device became a standard therapy during World War I. Trench warfare head wounds were common, and facial injuries included jaw destruction that necessitated repairs with McBurney style springs (see X-ray). In 1888, McBurney was appointed surgeon-in-chief of New York's Roosevelt Hospital. It was there that he did his most famous work on appendicitis. He presented his report on operative management to the New York Surgical Society in 1889 in which he described the point of greatest tenderness in appendicitis, which is now known as McBurney's point. In 1891, he described the indications for appendiceal surgery, and in 1894 he made another significant contribution to abdominal surgery when he described the incisional area that must be used to easily find the appendix. While the incision had been previously described by Louis L. MacArthur, it was McBurney who popularized it. To this day McBurney's point and McBurney's incision are known by every surgeon.

Plate 4.46 (page 132)
Leontiasis ossea, also known as lion face disease, results from overgrowth of bone. It is not a disease itself but a symptom of other conditions, among them Paget's disease, fibrous dysplasia, hyperparathyroidism, and renal osteodystrophy. Aside from general thickening of bones, these conditions cause also create exostosis, local areas of new bony growth that interfere with function. When the skull thickens, it compromises the cranial cavity with resultant increased pressure on the brain.

Plate 4.49 (page 133)
This cranium was presented to the Army medical museum by Surgeon Jerome B. Green, U.S.V., and is numbered 830 of the Surgical Section. A musket ball entered at the center of the left branch of the coronal suture, and passed out at the posterior inferior angle of the right parietal bone, the opening of entrance being three fourths of an inch, and that of the exit one and one-fourth inches in diameter. There is a fracture of the right orbital plate of the frontal, of the squamous portion of the right temporal, and of the body of the right superior maxilla, probably by contre-coup. A fracture of the occipital bone extends from the opening of the exit to the right jugular foramen. The frontal suture remains distinct, though the skull is that of a middle-aged man. The specimen is believed to have come from the 12th Army Corps hospital after the second battle of Bull Run.

Plate 4.50 (page 133)
Private Edward Volk, D, 55th Ohio, was wounded at the second battle of Bull Run, August 30, 1862, by a musket ball, which struck the forehead at a point half an inch above the right eyebrow, comminuting and carrying away both tables of the os frontis, to the extent of one and one-fourth by two and one-fourth inches. He remained on the battlefield six days, and was then conveyed to Washington, and admitted to the Emory Hospital on the sixth of September. About one-third of the missile was found lodged against the fractured edge of the frontal bone. One-half of the plates which compose the frontal sinus were found in broken fragments piercing the brain, and were carefully removed. About a teaspoonful of brain matter escaped. There was no cerebral disturbance, and the patient was perfectly conscious. On September 8, hernia cerebri appeared, the protrusion being about the size of a walnut. On the 19th, the edges of the wound were healthy; the hernia had receded, but diarrhea supervened, which greatly reduced the patient's strength. He died on the 25th of September. The postmortem examination revealed a more extended fracture than was supposed to exist, reaching through the parietal to within two inches of the occipital bone. The orbital plate of the right superior maxillary was fractured and depressed, and a fissure an inch long ran down the body of the bone. The specimen was contributed by Surgeon W. Clendenin, USV.

Chapter 5: Skull Studies

Plates 5.1, 5.2 (page 140)
In the early nineteenth century, phrenological principles were adopted by French psychiatrists, who incorporated the concepts into their diagnoses and theories. An aspect of phrenological science was a systematic effort to identify biological causes of crime. It was the first 'scientific' endeavor to understand crime. Phrenologists generally agreed that disorders in two mental faculties, combativeness and acquisitiveness, were the cause of criminal behavior. Physician phrenologists were major promoters of designating criminality as a form of mental illness. Under this medical model of criminality, the criminal was considered sick and therefore not responsible for his action. This concept ultimately became a part of the foundation for the legal defense of insanity. Predicting and analyzing criminal behavior along with developing criminal identification systems were major objectives of French medical and social scientists for most of the nineteenth century. Some French physicians established new related sciences in an effort to predict behavior and identify abhorrent behavior. Among these physicians were neurologist Paul Broca (1824–1880), of Paris, who developed criminal anthropology, and Alexandre Lacassagne (1843–1924), of Lyon, a medical legal specialist who studied tattooing and criminality and helped establish scientific criminology.

The first photographic psychiatric case study and analysis was published in 1863. It was in the field of criminal identification by phrenology and was an analysis of the well-known murderer, Martin Dumollard. The photograph was not an image from life, but of the plaster cast of his decapitated head. The image was accompanied by a detailed engraved outline of his head with a phrenological localization map. The discovery of this publication and photograph offers visual evidence of one of the early nineteenth-century psychiatric treatment ideologies. The Dumollard publication was issued only one year after the noted first medical photographic text produced by French pioneer neuropsychiatrist Duchenne de Boulogne (see Plates 7.3 & 7.7). Phrenologic photographic study dominated mid-century psychiatric publications. The first psychiatric book with engraved illustrations made from photographs was Eliza Woodson Farnham's 1846 *Rationale of Crime, and its Appropriate Treatment: Being a Treatise on Criminal Jurisprudence, Considered in Relation to Cerebral Organization* (see volume 3 number 2). An experienced prison matron, Farnham believed that "crime was the result of overdeveloped faculties of destructiveness and undeveloped faculties of reason or spirituality, —all subject to correction." In her opinion, crime was a disease subject to phrenological treatment. Her conclusions, shared by many professionals, were not popular in many segments of the legal and medical community because phrenologically diagnosed defects destroyed accountability. However, phrenological analysis of the heads of criminals, as represented here, was an important part of serious science in the nineteenth century and, combined with physiognomy, led to racial, criminal, and eugenic principles.

Martin Dumollard has a distinct place in the history of criminology and psychiatry. He is recognized as France's first serial killer. The term 'serial killer' was coined in the 1980s by FBI agent Robert Ressler, to describe a killer who killed repeatedly and obsessively on separate occasions. Dumollard, 'The Monster of Montluel,' was arrested after a servant girl escaped his attack. The clothes of several young women were found in his cottage and he killed between six to twelve women over a ten-year period. Dumollard told police that he picked up the women in Lyon and lured them into the woods promising them a lucrative job. He killed them by stabbing, crushing their skulls, or choking them, at times using a noose. After confessing, Dumollard led police to the bones of several of his victims. His wife, Marie, helped him with some of the crimes and was given a life sentence. Dumollard was guillotined in 1862. Curiously, he was said to have drunk the blood of several of his victims, and his exploits have been reported by those interested in the history of vampires. In 1886, Professor of Neurology and Psychiatry at the University of Vienna, Richard von Krafft-Ebing, MD (1840–1902), published *Psychopathia sexualis,* which described 238 case histories and put some vampire myths to rest. Many of his case studies concerned a violent eroticism triggered by blood. It has been hypothesized that one inspiration for the psychopathic or psychotic killer's mind is 'physical dominance mixed with blood.' The supposed 'blood lust' of Dumollard has served as a basis for numerous theories of the vampire, serial killer, and sexual criminal. This recently unearthed photograph and phrenological study is a significant historic record in the pioneer analysis of the psychopathology of the criminal mind.

Plate 5.3 (page 141)

During the 1870s to 1890s, scientists around the world were working at a fever pitch, attempting to gain insights into human nature and behavior by analyzing body types, shapes, sizes, and configurations. Their hypothesis was that clues to intelligence, human ability, criminality, and other traits could be found in anatomic topography. Scientists of the era believed human intelligence and other characteristics were determined by skull capacity (see Plates 5.9, 5.10 & 5.11). This capacity was thought to vary from race to race and according to a host of other socio-cultural determinants. Caucasians were thought (by a profession dominated by Caucasians) to have the largest cranial capacity and therefore to be the most highly civilized. During the half-century between the 1840s and 1890s, skulls were aggressively collected from around the world and examined by renowned medical scientists, anthropologists, neuropsychiatrists, and various educators. Collectively, they established craniology as a precise science, using exacting measurements that were reproducible, accurate, and scientifically based. In 1885, neuropsychiatrist Georg Konrad Rieger of Wurzburg, Germany, published *Eine Exacte Method der Craniographie,* which included a series of artistic heliotype photographs of skulls, demonstrating how to measure them. This image is number 10 in Rieger's book and illustrates the correct parameters of measurement.

Though craniology is an extremely precise process, the original theory that motivated much of the early research has since been disproved. With the exception of major congenital deformities, skull capacity has nothing to do with intelligence. Skull measurements persisted in the United States as part of the eugenics movement until about 1930. Under the Nazi regime, German scientists continued to use cranial measurements for eugenic purposes, attempting to validate their ideology of a Germanic 'master race.' Nazi researchers were sent around the world to examine isolated indigenous peoples as well as all European ethnic groups.

Rieger was extremely critical of Freud and psychology. As a proponent of the first biological psychiatric movement with emphasis on anatomy, Rieger and his colleagues thought mental illness meant psychosis and hospitalization. In 1896, he expounded on his disgust with Freud's methodology and ideology with its emphasis on sexuality and an extreme doctor-patient relationship. It was designed for "people who were unhappy and not mad." In the first decades of the twentieth century, it was biological psychiatry that fell into disrepute as psychoanalysis and its proponents gained a foothold on the profession that was to last for six decades. Office-based private practice psychiatry became established, and treating psychosis was largely abandoned for treating neurotics.

Psychiatric and criminal science often used the same concepts and methodology throughout much of the nineteenth century, with the basic idea of both professions being identification and incarceration of appropriate personalities. At one extreme were the criminally insane individuals who were especially important to identify. Roughly paralleling the rise of craniology was the science of anthropometry, which used cranial measurements for an entirely different purpose. In the early 1880s, Alphonse Bertillon, Chief of the Identification Bureau of the Paris Police, developed an identification system based on a series of selected body measurements, especially the skull. His methodology (Bertillonage) became the standard for positively identifying criminals, and lasted until the acceptance of fingerprinting in the 1920s.

Plate 5.8 (page 145)

This 1926 photograph of Dr. Arthur MacDonald (1856–1936) using his brain-measuring device on a United States Congressman is vivid evidence of the interest of scientists and the government in determining policies toward the mentally defective. A fascinating aspect of mental health care is the range of professionals who practice or discover value in outmoded and discredited therapies or ideologies. Some of these individuals have influential friends or government positions and can influence lawmakers.

What seems today an outlandish episode in psychological governmental studies is some of the work and experiments of MacDonald, a Washington, D.C., experimental psychologist. He describes himself in a book given to Johns Hopkins Hospital chief psychiatrist Adolf Meyer, MD, as a "Specialist in anthropological psychiatry in the Government Hospital for the Insane." (The Government Hospital for the Insane, St. Elizabeth's Hospital, was established in 1855.) He was the typical government scientist of his era, relying on government funding authorized by special congressional legislation. In 1926, Senator Royal Copeland introduced a bill authorizing MacDonald to measure the brains of U.S. Congressmen and Senators. (Copeland, a New York physician, had been Professor of Ophthalmology and Otology at the University of Michigan Medical School, Dean of New York Homeopathic Medical College and Director of Flower Hospital, and then President of the New York Board of Health. He made his public reputation calming New Yorkers during the flu epidemic of 1918, and, in 1922, he was elected to the U.S. Senate.) Senator Copeland supported MacDonald's work, stating that it was necessary to have "anthropological, psycho-physical, and statistical" examinations of members of both houses. Through measurement of the heads of America's solons, the weight of the brain would be determined, and under the bill would be made public, but without naming the legislator. Unofficially, MacDonald had already examined seventy-five U.S. Representatives and twenty-five Senators.

While MacDonald may appear to some as a wild, demented scientist, he did indeed have a substantial background in experimental psychology. In 1893, he was among the second group of members joining The American Psychological Association, established in 1892 by Clark University President Granville Stanley Hall (1844–1924). The majority of the twenty-six founding members were leading scholars interested in psychology and associates of Hall at Clark or Johns Hopkins University. Not all the original members were experimental psychologists; several were philosophers, two were psychiatrists, and two were teachers. By the end of 1893, there were 42 members.

MacDonald, as a federal employee, had most of his publications printed by the government printing office, and they reflect not only his intellectual interest, but also his need to acquire funding for his projects. Among these publications are: *Abnormal Man, Being Essays on Education and Crime and Related Subjects, with Digests of Literature and a Bibliography. Bureau of Education Circular of Information No. 4* (1893), with sections on alcoholism, genius and insanity, criminal sociology, charitological literature, criminology, education and crime, and including a 228 page bibliography; *A Plan for The Study of Man, Bills to establish a laboratory for the study of the criminal, pauper and defective classes* (1902); and *Statistics of Crime, Suicide, Insanity, and Other Forms of Abnormality*, and *Criminological Studies, with a Bibliography, in connections with bills to establish a laboratory for the study of the criminal, Pauper, and defective classes* (1903). The U.S. government's active restriction of undesirables of all sorts culminated in the Immigration Act of 1924, the notorious Johnson-Reed Act, which included the National Origins Act and the Asian Exclusion Act. In essence it stopped the active immigration policies set up in the late nineteenth century. The annual number of immigrants who could be admitted from any country was limited to two percent of the number of people from that country who were already living in the United States in 1890.

Plate 5.9 (page 146)
Ascertaining capacity of cranial cavity by means of water: Filling the skull, first part. The skull is covered with putty and placed, with the frontal portion depressed, in the hands of an assistant while the operator nearly fills the skull as rapidly as possible from a half-gallon measure with a spout held in the left hand.

Plate 5.10 (page 147)
Ascertaining capacity of cranial cavity by means of water. Filling the skull, second part: The skull, placed with the base in a horizontal position, is held in the hands of an assistant while the operator fills accurately to the edge of the foramen magnum from a reservoir through a small stop-cock. Before this is done, the skull is brought close to the pan which is to receive the water so that no time may be lost in emptying.

Plate 5.16 (page 152)
In 1885, John Shaw Billings, MD, and Washington Matthews, MD. published "On a New Craniophore for Use in Making Composite Photographs of Skulls," including in the article a series of photographs showing various aspects of the device including a series of composite photographs. This photograph No. 4 shows everything ready to make the exposure. The anterior and middle frames are lowered, and the velvet screen has been dropped.

Plate 5.18 (page 152)
The exposure. Two of the frames are lowered. A black velvet background is hung on the posterior frame, a piece of white card board is hung on one of the frames to illuminate dark side of skull. The attachment for securing the skull at the auditory meatuses is here employed. The skull is in position for taking picture of norma basialis. The small velvet screens for concealing the brass work are seen beside the craniophore.

Plate 5.19 (page 152)
Apparatus for taking composite photographs of skulls. No 5. The adjustment. The frames on object stand bearing the adjusting wires are raised. The attachment for securing skull at foramen magnum is here employed. The skull is in position for a picture of norma facialis.

Plate 5.30 (page 159)
At the time, when presented in 1926, the "Peking Man" claimed to be the most important discovery of its kind and hundreds of thousands of years old. It is shown here partially freed from the rock in which it was found embedded in a limestone cave at Choukoutien, forty miles from Peking. This photograph shows the dome of the skull, the development of which is said to demonstrate that the "Peking Man" was definitely above the "Java Ape Man" in brain capacity but below the "Neanderthal Man." The skull was ultimately found to be a hoax after many years of study.

Plate 5.31 (page 159)
Scottish-born South African physician, paleontologist, and anthropologist Dr. Robert Broom (1866–1951), examining an early hominid skull. Dr. Broom found numerous hominid fossils in the caves of South Africa, which he assigned to the subfamily Australopithecinae. Raymond Dart had first described this group in 1925. His claim that it represented the earliest ancestor of man was ridiculed until Broom's discoveries confirmed that the group was bipedal, erect, had manlike dentition, and relatively small brains. He found his specimens at Sterkfontein Caves, South Africa. Broom published, in 1946, *The South Africa Fossil Ape-Men, The Australopithecinae*, in 1949, *Sterkfontein ape-man Plesianthropus*, and in 1950, *Finding the Missing Link.*

Plate 5.32 (page 159)
Ancient relics at University of Pennsylvania Museum. Louis S Shotridge, Tlingit Indian, and assistant at U. of Penn. Museum exhibiting a skull, whalebone awl, and ivory beater excavated near Point Barrow, Alaska, by an archaeological expedition from the museum. The skull is part of the collection of skeleton material which was found by the expedition and which is typical of an early population known to archaeologists as the Thule Culture, the prehistoric Eskimo (Inuit) culture that developed along the artic coast.

Plate 5.33 (page 159)
Dr. Jesse Walter Fewkes (1850–1930), an archeologist and anthropologist, was chief of the Bureau of Ethnology, The Smithsonian Institution, Washington, D.C. He is examining some skulls recently unearthed near Washington. The skulls are supposedly those of Indians who once roamed this part of the country. The skull shown is not one of American's oldest men, but each skeleton that comes to Dr. Fewkes must be examined with infinite care so that the scientific information can be determined. In 1889, Fewkes became leader of the Hemenway Southwestern Archaeological Expedition and became an expert on Hopi and Zuni cultures. He explored the cliff dwellings at Mesa Verde in Colorado and lamented the poaching and destruction of the sites. Fewkes was one of the first voices for government preservation of ancient sites in the American Southwest. By the mid-1890s, vandalism of these sites was widespread. In the *American Anthropologist* for August 1896, Fewkes described a large cliff dwelling called Palatki, or "Red House," situated in the Red Rock country southwest of Flagstaff, Arizona, and appealed for protective legislation:

"If this destruction of the cliff-houses of New Mexico, Colorado, and Arizona goes on at the same rate in the next fifty years that it has in the past, these unique dwellings will be practically destroyed, and unless laws are enacted, either by states or by the general government, for their protection, at the close of the twentieth century many of the most interesting monuments of the prehistoric peoples of our Southwest will be little more than mounds of debris at the bases of the cliffs. A commercial spirit is leading to careless excavations for objects to sell, and walls are ruthlessly overthrown, buildings torn down in hope of a few dollars' gain. The proper designation of the way our antiquities are treated is vandalism. Students who follow us, when these cliff-houses have all disappeared and their instructive objects scattered by greed of traders, will wonder at our indifference and designate our negligence by its proper name. It would be wise legislation to prevent this vandalism as much as possible and good science to put all excavation of ruins in trained hands."

Chapter 6: Special Dissections – Anatomical Studies

Plate 6.2 (page 163)
In the first decade of the twentieth century, brain pathologists in various countries published atlases and texts with photographs hoping to bring to a wider audience the evidence of disordered brain anatomy, as evidence of/or etiology of insanity and criminal behavior. 'Criminal behavior' is among the most serious social deviations that have always affected society. Attempts to explain this type of behavior, in terms of defective biology, are an ancient quest and its etiology remains a critical issue throughout the world. During

the last third of the nineteenth century, after discovering that specific disease states were caused by specific pathological changes, some practitioners concentrated on dissection of the brain as a way to identify these changes in mental disease and behavior. Several schools of criminal behavior were established. The idea of the degenerative theorists was that individuals can 'devolve' over time. Self-abuse and excesses of life were believed to eventually result in anatomical cerebral degenerative changes leading to criminal behavior. Anatomists working on this theory looked for the supposed areas of cerebral degeneration. Another group, the criminal anthropologists, taught that some criminals were not only unchangeable, but were born criminals and identifiable as such. They were described as having long arms, low brows, and an apelike appearance. These individuals were atavistic primitive people who could not conform to the laws of society and may also have some identifiable cerebral anatomic changes. If such malformations were found, it would validate the criminal anthropologist's theories. In practicality, these findings were evident in the obviously visibly mentally retarded.

By the end of the first decade of the twentieth century, 'feeblemindedness' as a cause of criminality was established. By the end of the nineteenth century, many neurologists and neuroanatomists had attempted, or were attempting, to find the 'holy grail' of brain changes that would explain personality and behavior. They were, perhaps, 'scientific' kin to the phrenologists, who at the beginning of the century, palpated the skull for supposed cerebral features and associated behavior. The neuroscientists, in a more sophisticated manner, were attempting to identify internal features and behavior. Abnormal anatomy was the main theoretical theme of the first biological psychiatric movement.

In South America around the turn of the century, Argentina had a special social problem. A massive influx of Italian immigrants, some of whom were criminals, flooded the country. (It has been stated that about half of Argentineans living today have Italian heritage.) In an effort to positively identify the Italians, Argentineans needed a quick and accurate system. They were the first major country to officially use the fingerprint for identification purposes instead of Bertillonage, the eleven-step measurement system of Parisian Police expert Alphonse Bertillon used by most countries. As criminal identification was of particular concern, Argentinean pathologists and psychiatrists dissected and studied the brains of criminals looking for areas of degeneration or anatomical defects to support the various behavior theorists In Buenos Aires, in 1904, pathologist F. Perez, MD, published a series of photographs of serial sections of the heads and brains of recently executed criminals. Self-abuse and excesses of life were believed to eventually result in anatomical cerebral degenerative changes leading to criminal behavior. Anatomists working on this theory looked for the supposed areas of cerebral degeneration. However, no major anatomic differences were found between the brains of normal people and criminals. While brain changes in criminals were found to be a scientific dead end in the early twentieth century, pathologists had better luck in demonstrating cerebral changes in the mentally ill. Neuroscience and neurobiology have been integrated into modern psychiatric concerns. For more than two decades, psychiatric disease research has been expanded to include advanced brain imaging techniques to find areas of excess functioning—firing of neurons or areas that are accumulating or being depleted of various substances. Development of specific drugs to effect these areas have been of prime concern to researchers.

Plate 6.3 (page 164)
Perhaps the most astounding dissection in the history of medicine is that of the entire nervous system by Rufus Benjamin Weaver, AM, MD, in 1888. Generations of medical professionals and neurologists used Dr. Weaver's dissection to learn neuro-anatomy. The dissection, a recognized national medical treasure, has been visited and admired by generations of physicians. It is still on display at the entrance to the library at Philadelphia's Hahnemann Medical College.

A knowledge of anatomy is the basis of medical practice and while many of the mysteries of the body had been unraveled over the centuries, Dr. Weaver was the only physician to literally unravel the entire human nervous system. Dr. Weaver was Demonstrator of Anatomy and Lecturer on Surgical Anatomy at Hahnemann Medical College. He was born in Gettysburg on January 10, 1841, and was educated at the Pennsylvania College in Gettysburg, receiving a BA in 1862. He earned his MD in 1865 at the Pennsylvania Medical University in Philadelphia. He studied anatomy under Dr. D. Hayes Agnew, also in 1865, and in 1867–68 under anatomist Dr. Joseph Leidy (see Plate 2.1).

At medical school, Weaver became a close friend of Dr. Amos Russell Thomas. When Dr. Thomas became associated with Hahnemann Medical College in 1869 (he was ultimately to become its dean), Dr. Weaver took a position in the department of anatomy. He remained at Hahnemann preparing anatomic preparations and teaching for almost seventy years. By 1921, Hahnemann's staff boasted that the "Weaver Anatomic Museum at the college had the most complete collection of Anatomical, Surgical and Pathological specimens for teaching purposes in the world." The neurologic dissection fondly known as "Harriet" was its pride and joy.

Harriet Cole was an African American cleaning lady who worked at Hahnemann, cleaning Dr. Weaver's laboratories. She willed her body to the institution before she died of tuberculosis at age 35. Dr. Weaver, seeking a proper disposition of her remains, consulted other anatomists and studied the world s dissections. He realized that no dissection of the entire nervous system had ever been attempted. Most dissections were of regions of the body but never of an entire body system. Dissected human preparations had been the backbone of anatomical study until the wax preparations of Azoux and others. Weaver wanted to make the world's most difficult dissection, and he did. None had been recorded before or after his feat. After proper preservation, on April 9, 1888, Weaver began to filet and debone Harriet. The bones of the skull and vertebral column where carefully chipped away. It took two weeks just to dissect the base of the skull. All the nerves were preserved by moistening, then by rolling gauze around them, then cotton with alcohol and then rubber cloth. The dura was left intact and open, the brain removed, and the cavity filled with curled hair. By late June, working eight to ten hours a day, he had completed the dissection. All the nerves were dissected to their terminal filaments. Now came the difficult process of mounting. It looks in the photograph that the nerves are resting directly on the surface; they are not. 1,800 pins and fine ligatures support the dissection. In the end, the intercostal nerves, although perfectly dissected, were sacrificed because of the impossibility of spreading them out properly with the proper curvature. The final display is 82" tall x 35" x 8 1½" deep.

The photograph was taken by Weaver in 1888 and reproduced as a mechanical print, a collotype, by Wells & Hope, Philadelphia. The resultant image was 8 ¾" x 19" on a 14" x 24" paper. It was intended to hang on physicians' office walls. In the February 1889 issue of *Hahnemannian Monthly,* Dr. A. R. Thomas, Dean of the College, published a report of Weaver's accomplishment along with a photograph of "Harriet."

The seeming impossibility and exquisiteness of the preparation was appreciated. Harriet was displayed at the Chicago World's Fair in 1893 and won a Gold Medal. She was then put on display in Dr. Weaver's anatomy museum at Hahnemann and attracted visiting physicians around the world.

Harriet wasn't Weaver's only claim to fame. Weaver, a Gettysburg resident, lived there through the Civil War. His father witnessed, from his rooftop, the famous battle July 1 to 4, 1863, including Pickett's charge and its repulsion. Weaver's father, because of his local knowledge and civic interest, worked with state officials in reburying the Union dead. He, along with his son, kept careful records of the burial places of both sides. Many Confederates had been buried helter-skelter in isolated places. After the elder Weaver died, the records and responsibility fell on Dr. Weaver to deal with the dead of Gettysburg. Dr. Weaver was the only person who knew the location of the graves and was knowledgeable in handling the dead. Also, he knew the farmers who owned the land where the graves were. When Confederate General Robert E. Lee requested that Southern State societies bring their boys home so they could have, "a permanent resting place among their kindred," Dr. Weaver complied with the Southern appeal. He worked 18-plus hour days for three summers 1871,'72,'73, and when necessary personally paid farmers to dig on their land. Dr.. Weaver exhumed and identified about 3,320 Confederate bodies. He sent them home for proper burial in Southern soil: 74 to Charleston, 137 to Raleigh, 101 to Savannah, and 2,935 to Hollywood Cemetery in Richmond. He charged $3.25 "to disinter, box, and ship each body to Virginia." When it was all done, despite his time and great expense, Weaver was paid a token $380.00 of his $6,500 bill. In 1985, Weaver's story was retold in a book on Hollywood Cemetery.

Dr. Weaver's legacy was perhaps greater than Harriet or his battlefield work. Students regularly recognized him as the "greatest man on the faculty of Hahnemann." In 1921, it was noted that since 1869, he had taught all of the 3000+ graduates of the medical school. However, Weaver and Harriet were inseparable. When Dr. Weaver died in 1936, *Time* magazine carried his obituary under the title 'Harriet' and told their story.

(Hahnemann Medical College was founded in 1848 and merged with the Medical College of Pennsylvania in the 1860s. After several associations, it functions today as part of a greater Philadelphia medical system now known as MCPHU.)

Plate 6.14 (page 173)
In reviewing medical history, we see that each time period has diseases that attract major attention then fade from prominence as medical science progresses. Prior to the discovery of antibiotics, brain infections and abscesses were common complications of numerous conditions especially, ear infections. Even if operated upon by the best surgeons in the pre-antisepsis era, brain abscesses most frequently resulted in death. It was noted, however, that very rarely a "brain abscess may disappear by natural drainage or undergo calcareous transformation."

Modern neurosurgery developed in the four decades from 1880 to 1920. In the nineteenth century, Europeans were the leaders. Among the pioneers of the field were Ernst von Bergmann (1836–1907) of Berlin, Victor Horsley (1857–1916) of England, and William Macewen (1848–1924) of Glasgow. Macewen was the first to use neurological signs and symptoms to direct brain operations. In 1893, he published a landmark treatise on surgery of brain abscess: Pyogenic Infective Diseases of the Brain and Spinal Cord. While the foundations of neurology and neurosurgery were laid in Europe, by the end of the second decade of the twentieth century, American neurosurgeons had made the leading advances, led by Harvey Cushing (1869–1939), and Walter Dandy (1886–1946).

Space-occupying cerebral lesions of the era were frequently found to be abscesses. Diagnosis from tumor was difficult if there was the presence of long-standing common chronic infections, such as tuberculosis, mastoiditis, sinusitis, labyrinthitis, etc. In the January 1918 issue of the *Quarterly Journal of Medicine*, edited by the distinguished William Osler, MD, et al., W. B. Warrington, MD, published a review of brain abscess. He notes that, in 1855, English otologist Joseph Toynbee, MD, (1815–1866) and, in 1867, German internist Anton Biermer (1827–1892) recognized the various causes of brain abscesses. Ear infections were recognized as a leading cause of a wide variety of cerebral disease, and otologists helped define otitic cerebral syndromes. (Curiously, Toynbee died as a result of self-experimentation in his attempt to find a cure for tinnitus—he inhaled a combination of prussic acid and chloroform.)

Regarding the etiology of brain abscess, Waddington notes that infecting organisms obtain access to the cranial cavity in several ways:
1) Most commonly from "pus foci in the neighborhood of the brain." Occurring through injury of the skull, through septic thrombosis, through sinus infection directly or from nasal infections, from carbuncles of the scalp or face, from orbital infections. The most common cause of brain abscess is of otitic origin, Mastoid infections resulting frequently from infection through the Eustachian tubes being very common. About one half or more of all brain abscesses are due to ear infection. Syphilis and tubercular infections of the bones of the skull are another common cause of abscess as the disease spreads into the underlying cerebral tissue.
2) The second important etiological factor is… spread "via the blood stream from a distant site of suppuration." The most common source is septic disease of the lungs, notable from "bronchiectasis."
3) "Pyemic abscesses may follow a general infection of the blood…"
4) "In rare cases abscess of the brain is found unassociated with any other site of pus formation. Thus the organisms of influenza, typhoid fever pneumonia…have been found with… no other lesion discovered. For practical purposes otitic abscess occurs either in the temporosphenoid lobe or the cerebellum."

This photograph of the right lateral aspect of the brain showing an abscess of the temporo-sphenoidal lobe is typical of otitic brain abscess. (Infection of the labyrinth often leads to cerebellar area abscess.) Other fatal complications associated with middle ear infection, frequently seen in this time period, are lateral sinus thrombosis, extra-dural abscess, and purulent leptomeningitis. Waddington notes that, "simple purulent otitis, may, especially in children, give rise to grave cerebral symptoms, such as fever, drowsiness, rigidity of the neck, and alteration in the pupils; all these symptoms may disappear on simple puncture of the tympanic membrane." By the end of the 1960s, antibiotics made advanced otitic infections and otitic abscesses uncommon in developed countries.

Plate 6.15 (page 173)
In the first decade of the twentieth century, brain pathologists in various countries published atlases and texts with photographs hoping to bring to a wider audience the evidence of disordered brain anatomy, as evidence of/ or etiology of insanity and criminal behavior. In Buenos Aires, in 1904, pathologist F. Perez, MD, published a series of photographs of serial sections of the heads and brains of recently executed criminals. Self-abuse and excesses of life were believed to eventually result in anatomical cerebral degenerative changes leading to criminal behavior. Anatomists working on this theory looked for the supposed areas of cerebral degeneration. However, no major anatomic differences were found between the brains of normal people and criminals. While brain changes in criminals were found to be a scientific dead end in the early twentieth century, pathologists had better luck in demonstrating cerebral changes in the mentally ill.

Several brain atlases were produced in the 1885–1900 time period. The earliest photographic atlas was published by John Call Dalton, MD, (1825–1889), in 1885. Dalton's precise photographs of vertical and horizontal sections equal modern works. Dalton was the first American to devote his time exclusively to study physiology and was America's first neurophysiologist. He studied the long-term effects of cerebellar lesions and the consequences of raised intracranial pressure during his career was professor at universities in Buffalo, Vermont, and New York City. In Europe, cerebral atlases were produced by Sir William MacEwan ,MD, (1893), Gustaf Retzius, MD, (1896), and Carl Wernicke, MD, (1897).

Photographs of brain pathology were a popular subject in neuropsychiatric journals and texts. In 1908, I. W. Blackburn, MD, pathologist to the Government Hospital for the Insane, Washington, D.C., (St. Elizabeth's) and professor of morbid anatomy in the medical department of the University of Georgetown, produced a landmark atlas on brain anatomy in mental illness: *Illustrations of the Gross Anatomy of the Brain in the Insane: A Selection of Seventy-Five Plates Showing the Pathological Condition Found in Post-Mortem Examinations of the Brain in Mental Diseases.* Blackburn analyzed hundreds of photographs of pathological subjects taken at the hospital to produce this remarkable text.

"This photograph is Plate 67, Autopsy No. 2283, Sclerosis of a Portion of the Right Hemisphere in Epilepsy. Case # 10008 F.B.; aged 45; male; colored; nativity, Maryland. Mental disease, chronic epileptic insanity; duration of disease, since admission, ten years. The patient had frequent and severe

convulsions, followed by excitement, during which he was combative and threatening. The patient's mother stated that he had no convulsions before he was 23 years old. Death due to injury to the head during an epileptic seizure.
Brain: Weight 1,355 grams. Some subplial hemorrhage with laceration of the brain cortex over the right temporal lobe and a small softening, with loss of substance, on the outer surface of the temporal lobe. In the region of the right Sylvian fissure was a large depression, over which the cortex presented a nodular puckered appearance… The lesion found is not infrequently met with in epileptics. It belongs to the class known as the scleroses and may affect almost any part of the brain. It would seem that these lesions are in some instances developmental defects. The microscope shows almost complete absence of nerve cells in the depressed regions and an irregular arrangement of these elements in the projecting nodules or granulations."

The Government Hospital for the Insane was established in 1855. It was the first large-scale, federally run psychiatric hospital in the United States. It was founded by Congress in 1852, largely as the result of the efforts of Dorothea Dix. In 1916, its name was officially changed to St. Elizabeth's (the Colonial-era name for the tract of land on which the hospital was built). The hospital had been casually known by this name since the time of the Civil War, when soldiers in their letters home to loved ones and patients of U.S. Army hospitals temporarily located on the grounds during and after the war were reluctant to refer to the institution by its full title. The hospital has treated over 125,000 psychiatric patients. In 1987, control of the institution passed from the federal government to the District of Columbia.

Plate 6.25 (page 178)
This case of late onset hydrocephalus—excess cerebral spinal fluid in the brain cavities resulting in increased cerebral spinal fluid pressure causing destruction of brain tissue—may be due to one of the infective processes that was common at the time. Other causes includes cerebral trauma with hemorrhage that results in blockage of the outflow passages. A recently recognized condition "Normal Pressure Hydrocephalus" occurs in patients over 60 years and results in cerebral ventricle enlargement and destruction of brain tissue without a rise in pressure. This patient was unusual as he had a normal childhood and development and then developed hydrocephalus.

Dr. Blackburn notes: "The patient had at one time considerable intelligence. He had some musical ability, some education, and could converse with ordinary facility on current topics. Later in life he became deeply demented, physically helpless, blind, and partially deaf. He had some spasticity of the lower limbs and beginning contractures of knee and hip. He never had convulsive attacks, and, aside from the brain disease, his health had always been fair… The chief interest in this case was the preservation of a degree of intellect, doubtless correlated with the better preservation of the frontal lobes, the thickness of the skull, and the preservation of an almost normal shape of the face."

Plate 6.27 (page 180)
French ophthalmology, which had flourished in the eighteenth century with the innovation of cataract extraction surgery (1746), languished in the first fifty years of the nineteenth century due to the more than 600 physicians killed during the French Revolution and Napoleonic Wars. France's lack of interest and personnel in ophthalmologic practice was solved by an influx of two waves of foreign ophthalmologists. The first, in the 1830s, included Sichel, Rognetta, Furnari and Carron du Villards. As there were no university eye clinics, the foreigners provided the principal ophthalmic services in Paris. The second wave of immigration, in the 1860s, awoke French nationalism when they realized that foreigners dominated Parisian ophthalmology. The foreign practitioners who moved to Paris included de Wecker, Liebrech, Galezowski, Landolt, and von Graefe's favorite student of 1862–3, Édouard Meyer. Meyer had spent three years in Berlin under von Graefe (1859–62) before migrating to Paris in 1863. Von Graefe's letter of recommendation lauded and praised Meyer's potential and skill far above those he had given to other students he had recommended to practice in Paris, in 1863, Louis de Wecker and Richard Liebreich. His belief in Meyer's abilities would be continually reaffirmed.

Édouard Meyer (1838–1902) had come from Dessau in Saxony (Germany) to Paris and had been required by French law to pass a second doctoral examination. He received his second medical degree after writing a thesis, "On Strabismus and the Success of Tenotomy." He bought the practice of Julius Sichel (1802–1868), who for the longest time had the busiest practice in Paris. Both came from Jewish backgrounds. Meyer married into the French branch of the Rothschild family, which endowed him with an elevated social status and an elite patient base. In 1870, Meyer collaborated with Chief of the Ophthalmology Clinic at Hôpital Saint Louis, A. de Montméja, publishing *Traité des opérations qui se pratiquent sur l'oeil* (1870–2). The book is a true masterpiece and an icon of medical and photographic history. It was the first photographic surgical teaching text in any field. The next similar, close-up photographic surgical teaching text was not published for about another thirty-five years, in the first decade of the twentieth century. Meyer's book was a 'true how to do it,' containing 271 pages of text with 190 woodcuts and twenty-two full-page tipped-in albumen photographs. The first image (front cover) shows the correct position for cataract surgery; the other twenty-one images are detailed close-ups of operations. They clearly and carefully show the surgeon how to place sutures and make incisions for a variety of operations. Meyer included twelve photographs related to cataract surgery and incisions, four for iris surgery, three for strabismus surgery, one for canthoplasty, and one for lacrymal sac catheterization. Ophthalmologist A. de Montméja, one of the first medical photographers, took the photographs. His landmark publications are documented in Volumes 1 and 2.
In 1869, Dr. de Montméja founded the first journal devoted to medical photography, *Revue photographique des hôpitaux de Paris*. In 1868, de Montméja had collaborated with his colleague at the Hôpital Saint-Louis, dermatologist Alfred Hardy, on the first photographic atlas of skin diseases published in France. When Meyer's photographs were taken in 1870, close-up photography of live surgery was impossible. The slow speed of the cameras required Montméja and Meyer to use cadavers and for de Montméja to retouch the negatives to emphasize critical points of the operations. Their 1870 photographic text is one of the most valuable of all nineteenth-century medical texts. Incredibly, ophthalmic historians have not recognized de Montméja's seminal role in medical photography.
Meyer's next book, his 1873, *Traité practique des maladies des yeux* became a classic and was reprinted in six languages for over 30 years (English, German, Italian, Spanish, Greek, Russian, and Polish). The book went through four editions in France, the last published in 1895. Meyer's book was so popular and successful because it relayed the seminal accomplishments of Albrecht von Graefe to a wider audience. As a result of the influence and presence of immigrant ophthalmologists, the French finally established a Chair of Ophthalmology at the University of Paris in 1873.

Meyer's next contribution to the profession was in the 1880s, when he founded the journal *Revue Genérale d'Ophthalmologie* with Swiss ophthalmologist Henri Dor (1835-1912). Meyer followed the principles of his teacher von Graefe throughout his text. He described the three types of procedures for cataracts in extreme detail and showed exquisitely specific photographs of the critical parts of some of the procedures. This photograph shows a superior limbal incision:
1) Extraction by limbal or linear incision either with or without an iridectomy. In this operation, after the corneal incision, the lens is manipulated and removed by spoon or pressure (propulsion or expulsion), each of which Meyer photographed for his book. Linear extraction created a smaller corneal wound than limbal incision and was used for soft or small cataracts.
2) Discission (or needle operation) was done with or without iridectomy. In this procedure, the lens capsule is opened to admit aqueous humor into the lens matter to promote eventual absorption of the lens. It is used in the treatment of soft 'juvenile' cataracts and is rarely done after the teenage years. Discission can be done with one, or more easily with two needles, and does not need an incision into the eye, just one or two stab wounds. Ether anesthesia was used as

the procedure was performed on the young. If large pieces of lens matter escaped into the anterior chamber instead of simply being absorbed in place then another operation was necessary by linear extraction or suction, if the material appeared soft enough, through a puncture wound. Lens material could swell to large size and cause glaucoma. Pain was the deciding factor in whether to reoperate.
3) Depression (reclination or couching). In this operation, a special needle is introduced into the eye that dislodges the cataractous lens into the posterior chamber.

Plate 6.28 (page 181)
The most fascinating phenomena associated with the retina is the supposition that the last image seen is permanently preserved. The concept of the retina as film had caught the imagination of the public, scientific community, criminologists, and especially writers. This 'last image of life' has been labeled an optogram. The idea that one's eye preserves the very last moment of life gained a very powerful hold on the Victorian imagination, whose concern with death and dying rituals were an important aspect of their culture. In particular, it was suggested that optograms might be obtained from murder victims to help identify their assailant. Because of the propagation of this theory, murderers would sometimes destroy the eyes of their victims for fear that their image would be recorded. A recent example is the notorious case of the Russian serial killer who carved out the eyes of his victims so he couldn't be identified.

The first published photograph of a retina was taken in 1868 and printed in the second volume of *Revue Photographique des Hôpitaux de Paris*. In 1870, Dr. Vernois presented an unusual medical-legal case of Dr. Bourion, "Étude photographique sur la rétine des sujets assassins," with which he hoped to end the concept of sustained retinal imagery. While his studies were admirable and well intended, a fascinating idea, even if proven wrong, lingers in the mind of those who would rather believe. Vernois reported Bourion's case:
For a number of years in American journals there has been discourse on "a new application of medical-legal photography" that if one could daguerreotype the retina of a murdered person, they could recover an image of the person who committed the crime. In January 1869, Bourion addressed the Secretary General and members of the Medical Legal Society and presented a photograph taken by Gallard of the retina of a woman who was killed on June 14, 1868, representing the moment of her death. Her infant child was also killed. Her eyes were removed on June 16 at 6:00 AM and photographed the same day at 6:00 PM... The eyes were thus removed fifty or fifty-two hours after death. On examination no definite object was identified.

Bourion, in order to find out if the concept of retinal imagining was true, did sixteen experiments on animals, including dogs and rabbits. The animals were held in a dimly lit room and a specific well-lit object was held in front of their eyes. They were immediately killed by strangulation or by administration of prussic acid. Photographs were then taken of the eye and retina and examined. The photographs did not show any object, nothing precise or identifiable, just anatomical structures. It should be noted that at a conference in Heidelberg in 1869 and in *Revue des cours scientific*, Helmholtz stated that stereoscopic photography may give better resolution to the images and solve the problem. Others suggested amplification by microscopy. Bourion concludes that science does not support the idea that the retina functions as film and records events at death. There is no medical legal use of retinal images after death.

In the early 1860s, British photographer William Warner believed murder victims' eyes could be photographed to identify their murderer. In April 1863, when Emma Jackson was killed in London, Warner convinced Scotland Yard Detective James Thomson to try photographing the retina; unfortunately, she had already been buried. *The Lancet* editors reported that they were glad the detective did not "attempt such an absurd impossibility." Over the next decade, numerous researchers and photographers reported on the impossibility of the phenomenon in the *Scientific American, Harper's Weekly, The Photographic News, The British Journal of Photography*, and other magazines and newspapers from Moscow to San Francisco. Despite a mounting body of evidence to the contrary, each generation of scientists considered the possibility that new technology would allow visualization of the supposed optogram. The stimulus to restudy possible imagery on the retina was invigorated in 1876. Franz Christian Boll (1849–1879), working with frog retinas that had been detached from the choroid, noted that the reddish-purplish color of a retina when held to light changed to yellow. He recognized that there must be a fundamental chemical-physiological change in the retina, probably due to a chemical reaction to light, much like silver solutions react to light in film. He named the retinal substance, Sehpurpur or 'visual purple.' It was ultimately renamed 'rhodopsin.' Boll next showed that 'visual purple' was present in the area of the retina where there were rods. Most importantly, he recognized the visual purple reaction was a two-way function; it bleached, then regenerated to be bleached again and again.

Investigations at the Physiological Institute of University of Heidelberg by physiologist William Kühne (1837–1900) and Carl Anton Ewald (1845–1915) proved rhodopsin existed in the human retina. In 1881, Kühne published his results in *Beobachtungen zur Anatomie und Physiologie der Retina*. In order to observe and fix the retinal photo-chemical reaction, the eye had to be extracted very quickly after death. His experiments grew out of an accidental observation of the shape of a gas flame from his laboratory on the retina of a frog. His experiments were mainly performed on rabbits, which he had placed in front of a bright window, but included one infamous human subject. An optogram appears as a white image on a red or purple background, though unlike a photographic negative, the light and dark areas are not reversed. They were not very stable and soon faded. Kühne described one of his experiments:

"An albino rabbit, after being kept 15 min. in the dark, was decapitated; one eye was removed from the head under sodium light...and fastened onto the edge of a cork by means of needles... (The eye was placed in a) dark chamber with the cornea pressing softly against the diaphragm. The image was visible on the sclerotic, on one side of the optic nerve... that I was sure that it fell on the more deeply coloured division of the retina and could readily mark its place in the appropriate quadrant. Thereupon the yellow curtain was removed from the pane and the eye after five minutes' exposure was taken away, divided along the equator and examined in feeble gaslight... There was evident on the retina a most distinct brighter diffused spot, the small dimension of which corresponded to those of the image previously seen by me, and the position of which made me already sure that it was the optogram."

On November 16, 1880, at Germany's Bruschal Prison, Erhard Gustav Reif, a murderer who had drowned his children, was decapitated by guillotine and, by previous arrangement, his eyes were quickly removed for study. (The Germans used the guillotine through the end of WWII, killing more by this method than the French had from the 1790s Reign of Terror until the death penalty was abolished in France in 1981). The prisoner, Reif, had been especially prepared before his execution for the examination of his eyes: he was kept in a dimly lit room screened with yellow and red glass to keep rhodopsin in the eyes from bleaching and his eyes were bandaged seven minutes prior to execution. Ten minutes after execution, the eyes were removed and the retina prepared for examination. Kühne drew what he thought was a retinal optogram.

The discovery by Max Schultz of the cone's special function, in 1866, and the silver nitrate method of retinal fixation and staining by Camillio Golgi in 1873 played an important role in the study of retinal anatomy and function. In the 1870s, Schultz and others drew diagrams of the foveal area and its cell layers. In the 1880s photographs replaced the diagrams. It was not until the last half of the twentieth century that cone pigments and physiology were isolated and identified.

News of the German experiments reached London and it was suggested that the detectives investigating the Jack the Ripper murders in 1888 try the technique. It was never used. Dozens of reports continued to appear in various publications during the twentieth century. In 1975, police in Heidelberg, Germany, invited the physiologist Evangelos Alexandridis at the university's Department of Clinical and Experimental Ophthalmology to re-evaluate Kühne's experiments to see if they might play a useful role in forensic investigations. The experiments were not successful.

Plate 6.38 (page 187)
Since the presentation of photography in 1839, each advancement in photographic processes results in its adaptation by physicians for education, research, or documentation. During every decade since 1839, there seems to be miraculous advance. Today, digitized video and computerized photographs are the latest fad, but in the 1940s View-Master stereoscopic photographic reels visualized in a special holder were the rage. They allowed the viewer to see realistic three-dimensional images. Photographer William B. Gruber, of Portland, Oregon invented the View-Master in the late 1930s. Gruber wanted to create a modern stereoscope by using the newly developed Kodachrome color film. It was the most realistic imagery of the era. In 1938, he formed a partnership with postcard publisher Harold Graves. The idea was to offer an alternative to the popular scenic postcard. View-Master was presented at the 1939 World's Fair in New York. Within a short time, three-dimensional views of tourist spots such as the Grand Canyon were particularly popular. View-Master reel sales soon eclipsed Graves's postcard business. During World War II, over 100,000 viewers and 6 million reels were sold to the U.S. government for personnel training. Ultimately, a special stereo camera was offered to the public to create their own reels. By 1950, the Kodachrome 35mm stereo photograph in various forms— reels or individual stereo slides—become the 'modern' educational tool in numerous professions and a new modality for documenting family celebrations.

Dr. David L. Bassett, associate professor of anatomy at Stanford University, dissects and prepares a brain for engineer-photographer William B. Gruber. Harold Graves's Sawyer Publishing Company published the Bassett's book. Gruber used a three-dimensional color camera and photographed the specimen for the latest modern photographic educational tool, his View-Master three-dimensional reel book. Each reel contained about seven stereo photographs, which were placed in a viewer. A separate text described each image.

Bassett's *A Stereoscopic Atlas of Human Anatomy* was a monumental project that would take about fifteen years and about 2,000 photographs to complete. Begun about 1948, it was completed in 1962. The final atlas consisted of eight sections containing 221 View-Master reels and 1,551 color stereo photographs of the human body. The first volumes issued were of neurological significance. The first section, *The Central Nervous System*, was published in 1952. It cost $27.50 and contained 238 stereoscopic color views, mounted in thirty-four reels accompanied by 400 pages in four volumes of descriptive text and diagrams. In 1954, *The Head and Neck* was issued. It contained 350 stereoviews on fifty View-Master reels. Five volumes of more than 700 pages accompanied the reels; it sold for $38.50.

The brochure advertising the texts heralded that "One can review a single dissection, a series of steps in a carefully planned sequence of dissections or an entire area in full perspective and in natural colors comparable to those in real life. A fully labeled sketch based on a tracing of the original photograph accompanies each plate and furnishes a topographic guide of each view... This is a new and magnificent way to present anatomy to both student and physician alike. The only objection to this work would be its cost., but considering the excellence of the material...one can hardly object to that at the present time."

Bassett's neurological stereo-photographic atlas joined the list of other pioneer photographic neurological texts, among them:
Philadelphian Dr. John Dean's *Smithsonian Contributions to Knowledge, No 173., Photographs: The Grey Substance of the Medulla Oblongata and Trapesium* (1864), in which he presented serial photographic sections on cards with an accompanying text;
Philadelphia neurophysiologist John Call Dalton's (1825–1889) *Topographical Anatomy of the Brain* (1885), in which Dalton presented heliographic photographic plates of gross anatomy of the brain; and Parisian Jules Bernard Luys's (1828–1927) *Iconographie photographique des centres nerveux* (1873). Luys's various photographic neurological works include histological brain and spinal cord sections, as well as photographs of hypnotic subjects. Parisian Jean Martin Charcot (1825–1893), who initiated his career as a pathologist before turning to neurology, published numerous photographic papers as well as atlases depicting gross neurological anatomical and histological specimens on a variety of subjects.

Hundreds of thousands of subjects have been produced on View-Master reels. Disneyland, TV, Movie and other popular subject reels are collected. View-Master reels are still produced today at times as an adjunct to a medical text. Bassett's and Gruber's neurological text, however, is a pioneer work in the medical field. A modern shortened version of the atlas is still produced.

Chapter 8: X-Rays: The Skeleton Becomes Visible in the Living

Plates 8.12, 8.13, 8.14 (page 216–217)
At the end of the nineteenth century, the leading neurologic institutes were French, and French neurologists led in describing new syndromes. Parisian neurologist Pierre Marie (1853–1940), a pupil and ultimately successor to Charcot at the Salpêtrière, isolated and named several common neurologic disease states, such as progressive muscular atrophy (with Charcot), and acromegaly, both in 1886. He also described several rare cor ditions such as the one pictured here, cleideo-cranial dysostosis, a congenital syndrome characterized by absence of both clavicles and delayed ossification of the skull.
As in any disease state there are wide variations in the manifestations of the condition. The main neurologic component relates to the bony condition of the skull. At birth, the fontanels are excessively large and the sutures are widely open. As a result of the open fontanels, large bosses develop and the skull assumes a globular shape and at times with a "hot cross bun" type of deformity. Occasionally, the fontanels remain open into adulthood. Facial bones are underdeveloped and sinus malformations are common. Deformity of the bones of the vertebra column, fingers, and toes also occur. Fortunately, for most patients, mental and general health is unaffected.

At the end of the nineteenth century, the French led the world not only in neurology but also in medical photography and in the publication of neurological photographs. Important French medical iconographic journals were published from 1869 through 1918. The journals were published not only in France but also in the French colonies. This photograph was one of four of the patient published in *Archives des Laboratories des Hôpitaux d'Alger: Iconographie Medicale Algierienne (Photographique et Radiographique)*, Volume 3, 1907, by Dr. C. Aboulker, Chirurgien des Hôpitaux.

This 15-year-old boy is the sixth of seven children. His brothers have the same condition. His parents were related (first cousins). His particular condition includes: 1) complete absence of the clavicles as verified by X-ray and by unusual acrobatic ability; 2) aplasia of the thyroid gland with hypothyroid symptoms; 3) abnormal cranio-facial appearance with overdevelopment of the superior cranio facial area and underdevelopment of the mandibular and maxillary areas; 4) all fingers are deformed—shortened and stubby with a sclerosing skin condition and 5) hypospadias.

Today the search for the specific cause of a disease has evolved into identification and isolation of a specific gene and its protein components. We are now experiencing another era of "disease discovery." Disease states are further subclassified and renamed for the specific gene or protein anomaly.

In the twentieth century, physicians' hope for a long healthy life was significantly achieved, as life expectancy rose from forty-seven to seventy-six years. Neurology has more to gain from gene therapy than most other specialties, as many neurologic disease states are hereditary and, whether from accident or disease, neuronal damage is currently irreversible. Gene therapy can offer growth of nerve tissue and regeneration of neurologic pathways. At start of the twenty-first century, we now hope gene therapy will one day offer the promise of conquest of disease and that death could be postponed indefinitely.

Plates 8.18, 8.19 (page 221

The X-ray allowed physicians to evaluate aspects of disease states that had previously only been available for study after a patient died. The use of the X-ray for locating renal calculi was a well-known significant diagnostic accomplishment. Another important early use of radiographs was evaluation of the skeleton. Chronic renal disease could result in bony anomalies, and various states of osteomalacia in children (rickets) and adults were popular subjects for physicians to document by photography. With the discovery of the X-ray, a radiograph of the bony defect became an essential part of the medical record of those with rickets. By 1905, the X-ray had advanced to the point that skeletal photographs were accurate and detailed.

These photographs were published in 1906 in *Archives des Laboratoires des Hôpitaux d'Alger: Iconographie Médicale Algérienne, (Photographique et Radiographique)*, by Dr. Binet of Algiers' Clinique des maladies des enfants. The journal, established in Algeria in 1905, consisted of three parts. The first, Articales Originaux, contained the text of an article. The second, Revue Analytique, featured descriptions of laboratory studies of interesting cases and reviews of books. The third section, Partie Iconographie, contained heliotype prints of the cases presented in the original articles, or simply of interesting medical cases. The editors of the journal were Dr. E. Murat, chief of medicine and of the laboratory services, Dr. E. Bordet, chief of the radiological service, and J. Bringau, preparator of the laboratory. France, the birthplace of photography, produced more medical photographic publications than any other country in the nineteenth century. French physicians continued their interest in medical photography in their colonies as well, and produced this journal in Algeria. In the early twentieth century, only France and its Algerian colony were publishing medical photographic journals. The noted neuropsychiatric periodical *Nouvelle Iconographie de la Salpêtrière* was produced for longer than any other early photographic journal. It was established in 1888 and published for thirty years, until 1918. In addition to neurological and psychiatric subjects, it contained numerous photographs of children with rickets and patients with unusual syndromes.

Plate 8.23 (page 224)

In 1915, Kennon Dunham, MD, published *Stereoroentgenography: Pulmonary Tuberculosis*. In this work Dunham produced a series of forty-one stereoviews of X-rays of the lung in normal and diseased states as well as X-rays from cadavers. Dr. Dunham states, "I know that the X-ray has its limitations. What I bring you is my own creative work towards the diagnosis of tuberculosis. What I have done is adopt a practical technique (stereoroentgenography) to replace individual judgments with scientific accuracy." Dr. Dunham notes physical examination of patients by physicians with instruments such as the stethoscope are variable. The X-ray examination creates a permanent record that can be reviewed and interpreted by various physicians.

Stereoroentgenographs create a three-dimensional view. Dr. Dunham's cadaver lung preparations were injected with bismuth to demonstrate various aspects of lung and chest anatomy. This technique also generated an image that was more easily read. This is a view (one half of the stereoview) of a lung removed from the body with "bismuth injected into the trachea, a (rubber) tube filled with 'shot' passing through the descending vena cava has been brought out of the heart, and the aorta can be seen passing over the left bronchus. There has been an unsuccessful effort made to inject one of the arteries of the upper right and one of the arteries of the upper left. This stereo has been produced to show the upper, middle and lower divisions on the right, with their respective trunks, and the upper and lower divisions of the left side, and to again call attention to where the bronchi are given off, because this is most important when trying to make out the location of the various lobes on these stereoscopic plates." Various forms of Dr. Dunham's stereoroentgenographs remain a part of modern radiological diagnosis.

Plate 8.24 (page 224)

Various physical abnormalities were evaluated as causes of mental illness. The discovery of the X-ray in 1895 offered an innovative parameter to further examine the body and its dysfunctions. In the December 1910 *American Quarterly of Roentgenology*, Henry S. Upson, professor of neurology in Cleveland's Western Reserve Medical School, penned "Serious Mental Disturbances Caused by Painless Dental Lesions." Upson's paper provides an interesting look at medical and psychiatric concepts of the era. Upson's basic premise was that irritation or toxins created by various tooth maladies caused psychiatric disease and that the X-ray offered a new tool for psychiatrists to evaluate patients. The idea that focal infection and toxin production was the etiology of a long list of conditions influenced medical practice until the 1940s. Tonsillectomies and adenoidectomies were routine procedures for removing possible sources of toxins. Tooth and related sinus infections were readily identifiable conditions, which were the source of severe maladies from endocarditis to meningitis. Upson believed that treating dental lesions could be an effective therapy for psychiatric disease. Pain was an common indicator of tooth disease, but a painless serious condition could also be present and X-rays were necessary tools to identify problem areas.

Upson presented this series of X-ray images to show the tooth abnormalities in his patients. Figure No. 1 and 2 are from a 19-year-old woman with a mental state known as psychasthesia, as described by Pierre Jane in *Les Obsessions et la Psychastheie* (1903). Extraction of the impacted teeth resulted in dramatic improvement of her condition. The second patient, a 27-year-old woman, suffered from 'nervousness' and had one episode of a mental breakdown. Figure No. 3 shows her impacted lower second molar and No. 4 her two abscessed bicuspid teeth. Extraction resulted in physical and mental improvement. The third case is a severely psychotic 55-year-old man; Figures No. 5 and No. 6 show extensive alveolar abscesses. No headache or pain, other than a moderate toothache, was noted. Extraction of all his teeth was performed with immediate improvement in his sleep pattern and his delusions.

Upson notes: "Ever since the study of insanity began psychoses have been known to develop as a consequence of diseases of all kinds.... Poisons are among the commonest causes of mental aberration of all kinds. In view of the fact that morphin deadens pain, that diphtheria causes neuritis, that the toxins of malaria and typhoid produce delirium, poisons are thought to have usual or habitual effects... The same diversity obtains among cases of insanity caused by psychic shock, fatigue, worry, and every kind of mental stress as among those caused by poisons.

"Considering the cases due to physical causes, any organ of the body possessed of sensory nerve-endings is capable, under the influence of irritant disease, of setting up a mental aberration, varying in severity according to the circumstances of the disease and the equation of reaction of the patient... No doubt irritants of some kinds and in particular locations are more often efficient and enduring causes of aberration than others...the fact that pelvic, digestive, and other disorders are causative in cases of manic-depressive insanity creates a strong probability that other irritants besides the dental ones will be found."

The discussion of Upson's paper by nine physicians, several of whom were prominent, such as Philadelphia's Henry Pancoast, supported his view. Several physicians commented that impacted teeth, especially molars, were a cause of a variety of mental and neurological conditions from headaches to neuralgias to mental distress. They appreciated Upson's "awakening the profession in this condition of the teeth as a possible cause of nervous affections."

Upson's approach of tooth focal infection as a cause of mental disease was 'scientific' for the era. Another psychiatrist would give tooth removal therapy one of the darkest blots in the history of psychiatry. John Hopkins Medical School's first professor of psychiatry Adolf Meyer, MD, (1866–1950), "The Dean of American Psychiatry," supported and referred patients to one of his prior associates, Dr. Henry Cotton (1877–1933), Superintendent of the New Jersey State Hospital at Trenton, 1907–1930. He posited he could cure up to eighty-five percent of patients with his therapy. His treatment, however, was the most mutilating and invasive of all therapies in the history of psychiatry, and one of the most fatal. Modern medicine had provided the germ theory of disease, as well as the ability to safely perform a wide range of surgery. Cotton apparently felt these advances gave him carte blanche. He started his treatment by removing all of a patient's teeth—this also served to make feeding easier, especially if the patient was violent. If the mental symptoms did not abate, next came surgery on the tonsils, adenoids, and sinuses followed by spleen, stomach, and, in women, the cervix and uterus. His piece de resistance was removal of the colon so "autointoxication" could not occur. Although Cotton initially had a team of surgeons to perform his operations, he eventually learned and carried out the procedures himself. In publications, he admitted to a thirty percent mortality rate for his abdominal cases. Meyer was Cotton's patron, so when a Hopkins associate, Phyllis Greenacre, MD, presented Meyer with a study documenting the true nature of the damage Cotton was inflicting (an actual morality rate of about forty-five percent), he suppressed the report. Meyer allowed Cotton and his followers in other institutions to continue killing hundreds and mutilating thousands. Cotton so firmly believed in his theory that he removed the teeth of his own sons prior to their starting Princeton University. Some time after graduation they each committed suicide. Psychiatric practice has been noted for drastic treatments throughout the ages, some of which, although seemingly helpful at the time, were ultimately proven harmful.

Plate 8.31 (page 229)
X-ray of Pelvis, Tuberculous Sinuses Injected with Bismuth Paste Contrast Media to Outline Fistulas
Stereoview, Emil and Carl Beck, MD. The Becks were proponents of using bismuth in TBc.
This X-ray shows a tuberculos fistula entering the spine. Tuberculosis was the leading cause of death in the United States until about 1940. Most today think of tuberculosis as a lung disease, but at the turn of the century it affected all organ systems. It eroded the skin of the face so severely the condition was called lupus vulgaris. It caused bony and cartilaginous destruction. And it entered the body, literally drilling holes into every organ. Sinus and fistulas destroyed the spinal skeleton causing severe kyphosis. Hips, shoulders, and even skulls were destroyed. Patients would lose pieces of their skull as small pieces of bones were eroded through sinuses. In the first decade of the twentieth century, surgeons Carl Beck, MD, (1864–1952) and his brother Emil Beck, MD, (1866–1932) became known for their 1906 bismuth paste injections for roentgenological diagnostics for tuberculous sinus and fistula diagnosis and especially therapy. In 1908, Emil published *Further Studies in the Treatment of Tuberculous Sinuses, Fistulous Tracts and Abscess Cavities.* After further studies, he published *Bismuth Paste in Chronic Suppurations: Its Diagnostic Importance and Therapeutic Value*. The book became a classic in medicine and roentgenology.

Plates 8.32, 8.33, 8.34 (page 230)
Wehrmann's X-rays show the metal splint that was applied at a front-line hospital before the patient was transferred to St. Georg Hospital in Hamburg. When the patient arrived on May 7, the fragments were in good position. However, the jaw presented with pseudoarthrosis and the patient wound was abnormally scarred. On August 12, an operation was performed and the scar was excised. The jaw was laid open and bone exposed. A 2 centimeter piece of tibia was removed and then the fragment sculpted to fit the patient's defect and it was bolted to the remaining jaw bone with a suture. On December 1, the metal splint was removed and the patient was discharged.

Plastic surgical repairs of the face were advanced during World War I as a result of the necessity to treat horrendous facial wounds received as a result of trench warfare, which left the head vulnerable to artillery as well as to small arms fire. Dental and maxillofacial surgeons were particularly innovative in their procedures. As America did not enter the war until 1917, the last year of the fighting, American maxillary surgeons made their major contributions after the war ended in November 1918. In 1916, the personnel of the St. Georg Hospital in Hamburg published an extensive survey of war wounds and their treatment, *Röntgen-Atlas der Kriegsverletzungen herusgegeben von den lietenden Arzten der Lazarettabeilungen des Allgemein Krankenhauses St. Georg in Hamburg*, by Deneke, Th., MD, H. Albers-Schönberg, MD, et al. This large atlas contains patient histories, original X-ray photographs, as well as original stereoroentgengraphs. The purpose was to show unusual wounds as well as wounds frequently encountered. The cases were selected from a large group treated at the Hamburg Hospital. In the Burns Collection is a war-time album assembled by German dentists at a military hospital in Lublin, Poland, that documents all of the aspects of facial rehabilitation. The pictures illustrated in all these volumes show severe, but in the larger picture, relatively minor wounds of the face. Many wounds were so horrific they required a face mask to be able to go out in public.

Plates 8.35 8.36 8.37 (page 231)
Infantry soldier H. treated by metal springs, cheek flap, jaw and teeth prosthesis. He was discharged March 24,1915.

The patient was hit by a bullet in the jaw on November 5, 1914. After initial treatment at a field hospital, the patient was admitted to Hamburg's St. Georg Hospital on November 20. Both jaws were fragmented and the teeth were pushed into the gums. The central part of the lower jaw was missing—the front teeth soft parts as well as part of the jaw bone. The fractured ends of the jawbones were coming out of the mouth. The patient had been fitted at a front-line hospital with a metal spring. On January 15, 1915, a major repair was attempted. A flap was made from both sides of the check. The lip was resected and the cheek flaps and lip were put into normal position. The surgeons tried to use a bone transplant, but it was rejected so a lower jaw prosthesis was put in place and the defect repaired. At another procedure, teeth were replaced and then gold caps put on the teeth. The fragments had been permanently fixed. On March 24, 1915, the patient was discharged.

Chapter 10: Popular Skeletal Views & Images for the Public

Plate 10.3 (page 261)
Cheyenne Necklace of Human Fingers Owned by High Wolf, 1876
The fingers are those of the left hand middle fingers of enemies killed.
Full text in Appendix.

Plate 10.5 (page 263)
This photograph is perhaps the only surviving visual legacy that symbolizes the killing of Native Americans by pulmonary disease and the treatment provided by their healers. The Chilkat, a branch of the Tlingit tribe, lived on the panhandle of the South Alaskan coast. Most of the Northwest coast tribes, Tlingit, Haida, Kwakiutl, Tsimshian, Nootka, and Haisla, existed in relative solitude and immunologic isolation until the last third of the nineteenth century. The Yukon and Alaska gold rushes combined with the fishing and timber industries brought the white man and his diseases in unprecedented numbers. Many villages were forced to be abandoned as the native population succumbed in horrifying numbers to the common childhood diseases, as well as smallpox and cholera. It is estimated that over eighty percent of these Native Americans perished, a population drop from 500,000 to 100,000. Tuberculosis, the number-one killer in the nineteenth century, proved to be the final straw, completely decimating the surviving population. This rare photograph depicts Skundoo, a well-known shaman, treating a woman for tuberculosis whose exposed chest testifies to her disease state. Skundoo has his symbolic paraphernalia consisting of a crown headdress, robes, and bone rattle. In the background is a Tlingit pillow and carved poles. These are not totem poles but represent guardian figures associated with the power of the shaman. Tuberculosis triumphed as hundreds of thousands died. Shamans and shamanism eventually disappeared as an important cultural force. It is only with the recent attempt to restore the cultural integrity of the Native American that shamanism has resurfaced. All known historic shaman

healing scenes (less than a dozen known) are posed, as these healing sessions usually occurred at night in the patient's home. The posed images, however, allow a glimpse into the nature of this healing practice. Skundoo holds a thighbone in his hand.

Plate 10.14 (page 268)
The Colon Cemetery in Havana, was the site of the celebrated 'bone yard.' A single grave in the cemetery cost about $10 to rent for five years. At the end of the five years, if the remains were not claimed, the bones were thrown into the bone yard by the cemetery authorities. After the United Sates conquered Spain in the short four-month Spanish American War (April 25, 1898– August 12, 1898) American troops visited the Cuban grave yards, especially the huge one in Colon Cemetery and took photographs among the bones. The soldiers often removed skulls and bones and drove through the streets of Havana displaying them. Their commander, General Brooke, ordered the practice to stop and gave instructions for the pit to be covered over, ending a popular tourist destination. In the card shown, American soldiers stand on the thirty-foot-deep pile holding up bones in the shape of the skull and crossbones. Photographs were taken and sold commercially as souvenir postcards to send home to their loved ones.

Plate 10.19 (page 271)
The Hallstatt ossuary is small single room building in a corner of the cemetery beside the Catholic church in this Upper Austrian village. Housed in the vault are 1,200 skulls; 610 are painted with decorations, dated, and labeled with the deceased names. The skulls are presented on a flat surface beneath which are the femur and other large bones of the deceased. Hallstatt is a small picturesque village with a population of about 900 and is a World Heritage Site for Cultural Heritage. It is a popular tourist attraction owing to its small-town appeal and can be toured on foot in about ten minutes. The beinhaus is a very current part of oberösterreichische (upper Austrian) life. It is not a mere relic of times past, but supposedly a tradition still taking place to this day. The town is so picturesque that the Chinese mining company China Minmetals Corporation built a full-scale replica of the entire town in Huizhou, Guangdong province.

Plates 10.25, 10.26 (page 275)
The Sedlec Ossuary (Czech: Kostnice v Sedlci) is a small Roman Catholic chapel, located beneath the Cemetery Church of All Saints in Sedlec, a suburb of Kutná Hora in the Czech Republic. It looks just like an average old medieval Gothic church. However it is one of the most amazing and unique churches in the world as it contains an unusual ossuary artistically decorated with more than 40,000 human skeletons. It is also known as the Church of Bones or as the Bone Church. It has several spectacular features made of bones, includinga A huge chandelier in the center of the ossuary with at least one of every human bone and an impressive artwork fashioned into the coat of arms of the Schwarzenberg family. Garlands of skulls drape the vault.

The importance of the cemetery as a burial ground was established in the thirteenth century. In 1278, Henry, the abbot of the Cistercian monastery in Sedlec, was sent to the Holy Land by King Otakar II of Bohemia. He returned with a small amount of earth he had removed from Golgotha (the hill where Jesus was crucified and whose biblical meaning is 'the place of the skull') and sprinkled it over the abbey cemetery. Word of this pious act soon spread, and the cemetery in Sedlec became a desirable burial site throughout Central Europe. In the mid-1300s, during the Black Death, and after the Hussite Wars (1419–1434), many thousands were buried in the cemetery, and it had to be greatly enlarged. A Gothic church was planned in the center of the cemetery. Because of the mass graves unearthed during construction, a lower chapel was to be used as an ossuary for skeletal remains found in the graves. The ossuary has been a tourist attraction for at least two centuries and continues to be among the most visited sites of the Czech Republic, attracting over 200,000 visitors yearly.

Plates 10.27, 10.28 (page 276–277)
One of the most popular ossuaries and the one most frequently photographed in the nineteenth century is Rome's Cappuchina Catacombs. The crypt comprises several tiny chapels located beneath the church of Santa Maria della Concezione dei Cappuccini on the Via Veneto near Piazza Barberini. It contains the skeletal remains of 3,700 bodies believed to be Capuchin friars buried by their order. As is the case with most of the noted church ossuaries in Europe, skulls and bones decorate the walls. They are nailed to the walls in intricate patterns; many are piled high among countless others, while others hang from the ceiling as light fixtures. Five of the six rooms in the crypt feature unique displays of human bones believed to have been taken from the bodies of friars who had died between 1528 and 1870. These rooms were photographed and made into 8" x 10" prints and sold to tourists, who placed them in the albums they assembled during their 'European Grand Tour.' A series of stereoview cards were also sold. The key to interpreting the crypt's displays of funereal art lies in the Christian belief in the resurrection of the body and everlasting life.

The six rooms of the crypt are 1: The Crypt of the Resurrection, featuring a picture of Jesus raising Lazarus from the dead, framed by various parts of the human skeleton. 2: The Mass Chapel, an area used to celebrate Mass, which does not contain bones. 3: The Crypt of the Skulls. 4: The Crypt of the Pelvises. 5: The Crypt of the Leg Bones and Thigh Bones. 6: The Crypt of the Three Skeletons. The center skeleton is enclosed in an oval, the symbol of life coming to birth. In its right hand it holds a scythe, a symbol of death that cuts down everyone, like grass in a field. In its left hand it holds the scales, symbolizing the good and evil deeds weighed by God when he judges the human soul.

While the ossuary in Rome contains skeletal remains, perhaps the most fascinating ossuary in Italy is in Palermo, Sicily. The Catacombe dei Cappuccini (Catacombs of the Capuchins) contain about 8,000 mummies that line the walls. About 4,000 are friars; the rest are local citizens. The halls are divided into categories: Men, Women, Virgins, Children, Priests, Monks, and Professionals. Some bodies are better preserved than others. Some of the Palermo mummies simply stand (an illusion, they're pegged to the walls) in all their finery, while others take part in elaborate mosaics. Swirling loops of skulls provide the backdrop to monks arranged in teaching poses designed to motivate visitors to reflect on their own mortality. Many are set in poses; for example, two children are sitting together in a rocking chair. The coffins were accessible to the families of the deceased so that on certain days the family could hold their hands and they could "join" their family in prayer.

The process began in 1599 when they mummified a recently dead brother, Silvestro of Gubbio, and placed him into the catacombs. The bodies were dehydrated on the racks of ceramic pipes in the catacombs and sometimes later washed with vinegar. Some of the bodies were embalmed and others enclosed in sealed glass cabinets. Monks were preserved with their everyday clothing and sometimes with ropes they had worn as a penance. Originally the catacombs were intended only for the dead friars. However, in the following centuries, it became a status symbol to be entombed into the Capuchin catacombs. In their wills, local luminaries would ask to be preserved in certain clothes, or even to have their clothes changed at regular intervals. Priests wore their clerical vestments, others were clothed according to the contemporary fashion. Relatives would visit to pray for the deceased and also to maintain the body in presentable condition. The catacombs were maintained through the donations of the relatives of the deceased. Each new body was placed in a temporary niche and later placed into a more permanent place. As long as the contributions continued, the body remained in its proper place, but when the relatives did not send money any more, the body was put aside on a shelf until they resumed payment.

The last mummy was added in 1920. It was that of a young girl, Rosalia Lombardo, who was just two years old when she was mummified. She was mummified by professor Alfredo Salfia. In March 2009, Piombino-Mascali D, et. al., published the rediscovered method of mummification of professor Salafia in the *Virchows Arch*. 454 (3): 355–7. The embalming procedure, performed by him, consisted of formalin to kill bacteria, alcohol to dry the body, glycerin to keep her from over-drying, salicylic acid to kill fungi, and the most important ingredient, zinc salts to give the body rigidity. (The formula is one part glycerin, one part formalin saturated with both zinc sulfate and chloride, and one part of an alcohol solution saturated with salicylic acid.) Rosalia was remarkably preserved with this formulation.

Plate 10.36 (page 281)
Orel, (Oryol in Russian), 200 miles southwest of Moscow in Western Russia, was occupied during World War II by the German Wehrmacht on October 3, 1941. It was freed by Russian forces on August 5, 1943, after the famous tank Battle of Kursk. The prison in Oryol was used by both the Russians and then the Germans for criminal as well as political prisoners. This photograph's period caption states these Russian senior medical officers were examining victims of the Nazis. Decades after the war it was acknowledged that on September 11. 1941, about a month before the Nazis arrived, Joseph Stalin ordered the execution of 160 political prisoners in the Medvedev forest outside Oryol. This episode is analogous to the massacre of 22,000 Polish officers and intellectuals in the Katyn Forest in April 1940. When the graves were discovered by the Polish and then announced to the world by the Germans in 1943, they blamed the Russians on it, although the Nazis had killed tens of thousands in the surrounding areas. By the end of 1943, it became a hot political topic as the Russians denied their guilt, arguing that the Nazis were trying to drive a wedge in Polish, Allied, and Russian relations. For decades, the Russians maintained and argued it was a Nazi atrocity. After the fall of the Soviet Union, in 1990, Soviet declassified documents verified that 21,857 Polish internees and prisoners were executed by the People's Commissariat for Internal Affairs (NKVD) and the Soviet secret police in April and May 1940. It usually takes about 100 years for the truth of events to be exposed; all participants and their children have to be dead before governments and families tell the truth about assassinations and other notable events. The fall of the Soviet Union occurred only fifty years after these World War Two events.

Plates 10.46-51 (page 289)
The Diableries, which translates roughly as "Devilments," is the name of a series of stereoview tissue cards made in Paris starting in the 1860s and then reproduced in other countries as ordinary photographic print stereoviews. A French traditional national fascination with all things diabolical resulted in the new sensation. This series of visionary dioramas depicted life in a strange parallel universe called ENFER – Hell. The scenes were made in clay by three sculptors, Louis Alfred Habert, Pierre Adolph Hennetier, and Louis Edmond Cougny, and then photographed with a stereo camera. The series was originally published by Francois Benjamin Lamiche but was later taken over and expanded by the publisher Adolph Block. A total of seventy-two scenes were published by Block. A group of lesser quality views were subsequently published by a competitor of Adolph Block's named Jules Marinier.

The resulting stereo pair of prints was made on thin albumen paper, and water-colors were reverse painted (on the back of the prints.) The eyes of each skeleton were then pricked out with a sharp instrument, and small pieces of red gel, or blobs of reddened varnish, were applied to the pricked holes. Behind this pair of prints was added a layer of tissue paper. When held to the light, the colors were seen and the eyes of the skulls and skeletons were red. Much of the subject matter was satirical and mirrored the corruption and excess of Paris during the Second Empire. This series consists mainly of American views copied from the French originals and sold to the public in the 1870–1890 time period. Collecting these has become very popular.

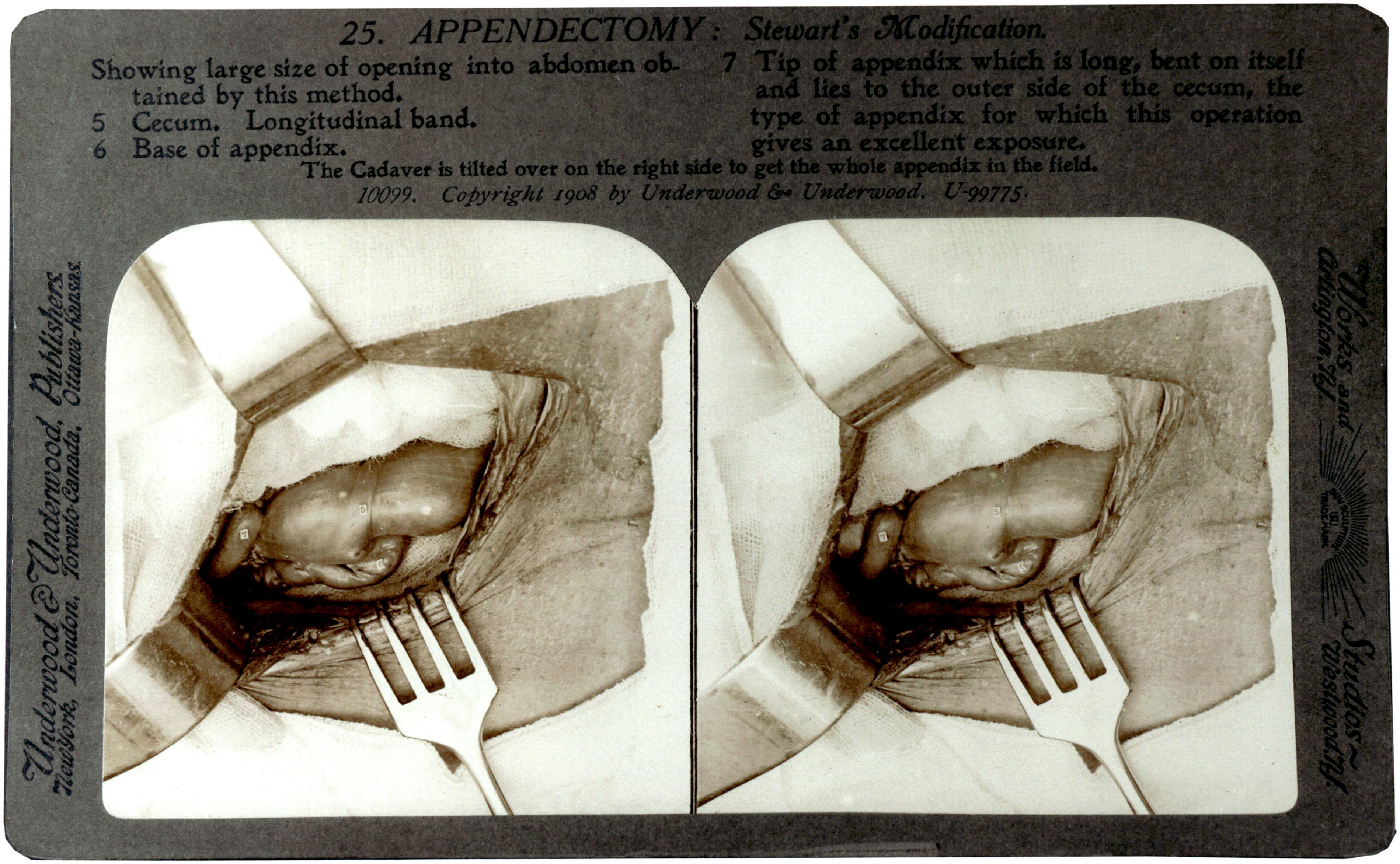

Plate 0.14

Teaching Appendix Surgery with Photographs of a Prepared Cadaver
Locating the appendix using Stewart's Modifications
Stereoview, 1906

Operative Surgery Through the Stereoscope – Appendectomy by Frank Hartley, MD – Surgeon to New York Hospital and Professor of Surgery and Chief Clinical Instructor in Operative Surgery at the College of Physicians and Surgeons and Alfred S. Taylor, MD, Instructor in Operative Surgery at the College of Physicians and Surgeons. Published by Underwood and Underwood, New York, 1908. This teaching set on operative technique using a cadaver allowed careful preparation of each aspect of the operative scene. The authors in their series of thirty-five images show three different appendectomy operative techniques, that of Drs., McBurney, Weir, and Stewart.

This Appendectomy series is perhaps the earliest work using stereo photographs to teach surgery. It was published about the same time as the first of the famous Stereo-Clinic series produced by Johns Hopkins Hospital surgeon Howard Atwood Kelly. The Stereo-Clinics however show real time surgical procedures, on live patients, by various leading surgeons. It should be noted as cadavers became regularly available in the early twentieth century they were regularly used to practice surgical techniques. Hartley and Taylor note: The stereographs are put forth in the belief that they simplify the teaching of operative surgery by removing it largely from the realm of words to that of pictorial representation, and that the well known shortcomings of photographs are almost entirely eliminated by the introduction of the stereo effect… The operations must be done upon the cadaver to obtain the quiet and blood-free field requisite for an exposure (time) of proper length and to bring out the details in the depth of cavities. Much time is consumed in making the necessarily very clean dissection, composing the picture, draping it, placing the numbers for identifying the structures, etc., so that few stereographs can be taken at one sitting. In any series, some of the photographs are apt to be unsatisfactory and must be replaced by those taken from another subject. For these reasons an operation is likely to be shown in stereographs taken from two or even three subjects. (See Special Dissections Plate 6.27 for an 1870 photograph of an eye operation on a cadaver.)

Appendix

The Appendix includes an overview of photographic formats, original articles, and relevant tables. The articles presented in the Appendix supplement the text with original documents. These first-hand accounts of events are fascinating. Paraphrasing or summarizing would have provided only a taste, without the benefit of the full meal. These reference works are presented as they offer additional insight into the mindset of the era.

PHOTOGRAPHIC FORMATS

A brief description of the main photographic processes discussed in this book (see page 314).

ORIGINAL ARTICLES

The Dying of Death, 1899 by Joseph Jacobs (see page 314).

This turn-of-the-century essay is an intellectual pillar clearly illustrating the changing attitude surrounding death and dying practices. It hastened the decline of long-established social rituals, including public and prolonged grieving, contemplating death, and spending time in death-related endeavors, including post-mortem/memorial photography.

This article by Joseph Jacobs first appeared in the *Fortnightly,* then was reviewed and republished in Volume 20 of *Review of Reviews* in 1899. These comments by the reviewer seem so accurate in today's world that they are worth considering.

The Barred Road to Anatomy, 1908 by Howard Kelly, MD (see page 315).

Howard Kelly, MD, was one of the greatest innovators, surgeons, and publishers of his time. He was one of the "Big Four" founding fathers of John's Hopkins University Hospital and Medical School.

This article, published in the July 1908 *Johns Hopkins Hospital Bulletin*, volume XIX, number 208, provides a vivid description of what anatomizing was like in the United States and also includes several detailed episodes of resurrectionist activities in the United States and Great Britain that resulted in the establishment of anatomy/dissection laws.

Necklaces of Human Fingers, 1892 by Captain John G. Bourke, Third Cavalry, U.S. Army (see page 318).

Disrespect for Native American culture was the core American attitude until the last third of the twentieth century. Desecration of the burial grounds of Native Americans, as well as other indigenous people around the world, to acquire skulls for study was a goal of American nineteenth- century scientists in their effort to define the races of humankind. Indigenous people used human body parts as part of religious ceremonies, decorations, or symbols of triumph and victories or even as bereavement objects.

This article appeared in 1892 as part of *Medicine-Men of the Apache* Smithsonian Papers, Ethnology. This detailed essay provides insight into the minds and attitudes of the American servicemen, as well as the government, in dealing with Native American peoples and their sacred objects in the late 1870s. Captain Bourke surveys of the use of fingers, nails, hands, arms, skulls, and other body parts in various indigenous peoples around the world, as well as their use in European cultures. He puts into cross-cultural context the Cheyenne finger necklace. Captain Bourke represents the change in attitude of anthropological thinking at the end of the nineteenth century.

Why The Stereoscope, 1916, by E.W. Caldwell, M.D. (see page 321).

Medical periodicals have always been a forum for presenting leading ideas and research projects. Part of the introduction to the first volume of *The American Atlas of Stereoroentgenology*, January, 1916, provides an explanation of one of the great advances of photography and medicine—the use of a three-dimensional image to provide realistic experience of the human body. Caldwell describes the problems the untrained have in viewing X-ray "shadows" as ordinary photographs.

In the first decade of the twentieth century, John Hopkins Hospital's surgeon Howard Kelly, MD, began publishing *The Stereo Clinic*. Each volume showed in detailed close-ups how to perform a specific operation. It revolutionized surgical training and operative competence worldwide. In 1916, the publication of stereographic radiological photographs that could be easily seen in the common stereoviewer, expanded the ability to understand radiological cases and procedures. During the later twentieth century, other stereoscopic formats such as the View-Master, which shows seven stereo-photographs on a small round disc, have continued the tradition of medical stereographic photographic publishing. The reels are simply inserted into a flap in the book cover.

Photographic Formats

HARD IMAGES

Daguerreotype

This process, the first practical form of photography, was presented to the world in 1839 by Frenchman Louis J.M. Daguerre (1787–1851). Popular between 1839 and 1860, it consisted of an image developed on a polished, silver-coated, copper plate. This fragile surface made it necessary to protect the plate behind a glass cover and within a special case. Daguerreotypes and ambrotypes were both produced in standard sizes that had a direct relation to the plate size used in the camera.

Ambrotype

This form of photography was popular between 1854 and 1865, and the process used a silver sensitive collodion-based emulsion placed on glass. Like the daguerreotype, the surface was fragile and the image was cased.

Tintype

Although most popular between 1858 and 1890, tintypes, patented in 1856, were produced as a novelty until about 1940. In this process, the silver collodion emulsion was affixed to a thin sheet of iron. This surface was not fragile and tintypes could even be sent through the mail. The term 'tintype' comes from the use of tin snips to cut the large sheets of iron.

PAPER PRINTS

Calotype

The calotype, the first type of paper print made from a paper negative, was patented by England's Henry Fox Talbot (1800–1877) in 1841. The process was not used in America. In 1851, the collodion wet plate process was developed using glass negatives. The photographic glass negative had to be wet when inserted into the camera and immediately developed while still wet. The paper albumen print was its most popular form. In 1871, Englishman Richard Maddox, MD, invented the gelatin-based silver print, known as the dry plate process. By the late 1870s, instantaneous stop action photographs were possible. Photographs were usually taken by professionals or serious amateurs until 1888 when Eastman Kodak introduced roll film with a special camera that allowed anyone to take a photograph. The age of amateur photography began, and physicians began to personally document their lives and patients.

Carte de Visite

Most popular between 1860 and 1880, the CdV was a 2 ¼" x 3 ½" paper print pasted onto a card 2 ½" x 4". It was this style photograph that necessitated the creation of the photograph album, so the public could easily compile cards of relatives or celebrities.

Cabinet Card

Popular between 1875 and1895, the cabinet card is a 4"x 5" print pasted on a 4 ¼" x 6 ½" card. Cabinet cards were also assembled in albums.

Stereoview

Popular from 1850 to 1930, stereoviews were produced in all the different photographic processes. It consisted of two images, each about 3" x 3 1/2", taken with a special twin-lens camera. The prints were mounted side by side on a cardboard about 3 1/2" x 7". The card was inserted into a stereoscope viewer, which created a 3-D effect and gave the viewer the illusion of actually viewing the scene. This three-dimensional effect was useful in teaching medical subjects.

Albumen Print

Popular 1850s to 1880, the albumen print was a sharp, detailed paper print produced by the wet plate process. An egg white based emulsion was the key to this process, hence the name "albumen" print.

Gelatin-Silver Print & Other Silver Print Processes

Popular in various forms from 1880 to today.

Original Articles

The Dying of Death, 1899
Joseph Jacobs

[Mr. Jacobs does not mean that the certainty of death is gone. He talks about the spirit of the times:] "Death as a motive is moribund ...perhaps the most distinctive note of the modern spirit is the practical disappearance of the thought of death as an influence directly bearing upon practical life... [D]eath has lost its terrors and is often regarded as the last and best friend. Death was king throughout medieval Europe." The Church was strong in the dread [of death]. But now "the Church in all its sections is devoting its attention more and more to this life than any other. Death is regarded no longer as a king of terrors, but rather as a kindly nurse who puts us to bed when our day's work is done. The fear of death is being replaced by the joy of life. The flames of hell are sinking low, and even heaven has but poor attraction for the modern man. Full life here and now is the demand; what may come after is left to take care of itself."

[Jacobs suggests several causes for the change.] "Life in [the past] was shorter and more precarious. Modern progress has increased the security and extended the duration of life. Nowadays death comes later, with more warnings of his approach, and takes us less by surprise. We are now more willing to go, less eager to stay... [T]he hurry-scurry of modern life leaves no one to meditate among the tombs. The increased number of interests lowers the intensity of any single one and prevents us from being able to concentrate our attention on the subject... We have so much to think about we cannot think much about anything... A robust and active man at forty expects to live till fifty and believes in his continued existence after forty-nine. But he thinks as little of fifty as he thinks of death....

Mr. Jacobs finds in town life a greater readiness to forget the disappearance of friends. And town life is on the increase... it is generally the old who pass; and their passing is so often lengthened by modern medicine as to make the end seem merciful. The stress and strain of modern life regard the cessation of life with much more equanimity than of old." Nirvana is one of the attractions of Buddhism to modern minds. Town life begets sameness of life. And "...with this dying out of individuality, the belief in personal immortality tends to fade simultaneously. We are getting more humble; we are realizing the possibility that the universe can manage to get along without us. The world forgets us while we live; we are getting to fear or think that God may forget us when we die... People in short do not think about the grave perhaps at all... Jacobs finds signs in the increasing popularity of cremation..."

[Jacobs notes the changing attitude of the church—where the clergy in the past worked to prepare us for another life, now acts as a social regenerator.]

"At the root of half the socialism of the day is the thought that this life is the only with which men have practically to do." The keen thirst for pleasure, the hot chase after wealth, the increasing popularity of suicide are cited as further proof of "the dying of death..."

[T]he loss of belief in personal immortality results often either in a feverous quest after present enjoyment, or with the higher and better mind, there is "increased social activity and a striving to make things better all around." ...One of the striking things about the Old Testament is the complete absence of death as a motive from its pages... Death was dying two thousand years ago but he revived to rule the world almost to this present day.

The Barred Road to Anatomy, 1908
Howard Kelly, MD

Two skeletons, that of Burke and one exhumed by Liston, gaze hollow-eyed from museum cases; and a new verb, "to burke," is created. A man of this generation has to explore anatomical annals for an explanation, and his wildest imagination cannot conceive the deadly feud and storm, the backwash of invective, the fear, the hate and the tumult, which that man when living gave rise to, nor believe that the verb was born—strange contradiction —of murder and avarice, of lofty ambition, and disinterested care for suffering humanity.

Let a seeker's eyes grow accustomed to the shades of wilful obstruction and groundless prejudice which obscured the period in which those men lived, and the ghostly forms and protesting voices of Knox and Liston, Cheselden, Monro and Syme will be seen and heard. Let him tread the Anatomical Road, mid shadows still more dense, and discern heroes, such as Servetus, Etienne and Versalius, disheartened, hindered and eventually martyred in a just and eager desire to know more intimately the secrets of our human habitation.

To take a fairer view of the resurrectionary period, picture to yourselves this present theater crowded with students ready for an anatomical demonstration. No one is absent, for it is the greatest day of the year, that on which the dissection of the one corpse allowed yearly will be made. Next week the subject will have to be only a dog or a cat, or perhaps a monkey; for there is nothing else available from which to demonstrate the "human" anatomy.

Imagine a great surgeon having to take part in an affair such as this:

A rare osteological specimen has been buried without dissection, and high fees offered to resurrectionists have failed to secure the subject. Moreover, the relations, suspecting a theft, have kept careful watch every night by the grave.

One afternoon, two gentlemen drive up to an inn near the cemetery and ask that their horse be cared for. They also say that a man will bring a box and put it in the carriage. In a short time this man appears, the stable boy alone noticing a resemblance to the driver. In an hour's time the two gentlemen return, pay their bill and drive away.

Night comes on, the grave-watchers appear. They find no neatly turfed grave, but disordered earth and an empty coffin. The great anatomist has not been daunted by the failure of his men. He has sent for London's best resurrectionist, and visiting the little cemetery with him has in 30 minutes, and in broad daylight, accomplished the most daring piece of resurrecting work ever attempted.

And this anatomist was Liston himself; But few who gaze at a skeleton labeled No 3489 in the Royal College of Surgeons Museum can realize the feud waged over it, or the circumstances which determined its theft.

On this side of the Atlantic also, Science was chafing at the fedders she bore, and this present generation of students may not know that here also the resurrectionists flourished. Dr. John Collins Warren gives a little glimpse of things as they were in 1781.

There were, he says cheerfully, plenty of bodies before the Revolutionary War, but few after, and we had perhaps only two paupers a year. One night, hearing that a man without relatives had been buried, he and Dr. William Ingalls, of Boston, went with a party to exhume the body. As they were lifting it over the wall, a man came along. One of the party, pretending to be intoxicated, addressed him and picked a quarrel with the stranger, while another, coming up, took the stranger's part and led him away. Young Warren was left to finish tidying the grave, while the others drove off. When his father saw what a fine subject had been procured he was very much pleased, and "the body lasted the whole course of lectures."

More schools were formed in New England, and Boston had not subject [cadaver] to spare, so these were obtained from New York at great expense and fear of detection. Dr. Warren once sent two strong students to exhume a body, and the [cemetery] watch sprang upon them. One student escaped in the wagon, but the other had a series of escapades on his way to the watch-house. He escaped once by jumping over a wall into a marsh. Seized again, as he was walking with ostentatious slowness he again escaped into an empty house and dived beneath some shavings, where he lay till morning. But the excitement and hard running brought on hemorrhage of the lungs, and he soon after died a martyr to legislative prejudice. Two or three times Warren feared his house would be attacked, and most of the bodies had to come from New York at $25 each, the students combining to buy them at $5.00 each for their schools.

Dr. Joseph McDowell, founder of the McDowell Medical College in St. Louis and a firm believer in ghosts, tells an interesting story:

"A German girl died with a very unusual disease, and we determined to get her body for dissection. We dug it up and laid it in the college; but the Germans heard of it and made things lively, and I heard that they were coming that night to search for the body. I went down to hide it and threw it over my shoulder to carry it to the top loft to conceal it in the rafters. At the top of the first flight, out went my lamp. I put down the corpse and relighted it. Out again, and as I felt for another match I saw my dead mother standing a little way off and beckoning to me. I followed to the loft, hid the body and came down in the dark. At the window where my mother had appeared were two armed Germans, talking, and when I got to the dissecting room door I saw six more down in the hall. My only chance of hiding was in the dissecting room. Opening the doors the Germans came upstairs, I saw my mother standing by the table from which I had taken the corpse. I had no light, but light came from her. I lay down and covered my face with the sheet and the men came in. They uncovered four bodies and came to me. I thought of jumping up and scaring them, but heard a voice say, "be still, be still," they were looking for a girl, my feet were sticking out at the end of the table. "Here's a fellow who died in his boots; I guess he is a fresh one," said a German; and they did not even uncover my face, but left the house."

About 27 years ago, the janitor of the old Atlanta Medical College went "resurrecting" in the white burying ground with an aid who was intoxicated. On coming back the aid gave away the object of their mission to the ferry man on the Chattahoochee River. In a few days the sheriff arrived with a search warrant and went to the dissecting room, with the son of the deceased, who had come to identify his father. He stood by the very body he had come to seek and declared it was not his father, as he could not recognize him. So the officers departed, quite satisfied.

On another occasion a body was sold to the anatomist by an undertaker, who promised to put a bag of sand in the coffin and nail it down. In the evening a weeping widow came to claim the body, and the coffin was buried [the] next day. The supposed widow and some students came at night to dig up the corpse, and to their disgust found only a bag of sand.

The usual custom was to check the box or case right through and post the check on to the surgeon. One Porter said to a doctor, "Lor', that's the heaviest trunk I've ever lifted." "Books," answered the doctor, gravely, "students always carry so many." "He must have a lot of books this time," muttered the Porter as he shouldered the trunk. But "Pickles," not "Books" was the usual label for exhumed bodies.

Once, in Baltimore, two students exhumed a man. They boldly set him up between them on the wagon seat and drove off. It was a chilly night; and they stopped at an inn to drink, leaving their silent companion propped against the wagon side. Bye and bye came along a genial policeman to pass remarks on the weather; but the corpse was past remarks, and answered not a word. The policeman playfully prodded him in the stomach to make him speak, and was horrified by a heavy body falling on him and an icy face being pressed to his. He promptly fainted. People gathered around, but the students hastily put the body back and drove hurriedly away. Worse luck attended them. Their speeding wagon tried to climb a lamp-post and lost a wheel. Their hurry to be off had aroused suspicion, and they were overtaken and did the best thing possible—took to their heels.

The "Hubbardton Raid" which occurred in 1830, was one of those resurrectionary episodes, in which the sturdy sons and fathers and husbands of Hubbardton, armed with weapons from the farm, marched in a body to the old college to rescue from its dissecting room the body of the wife of one of their number, whose new-made grave was found empty.

The most fascinating incident of those irregular days, however, was the burking case of 1886.

In Baltimore, an old white woman was murdered, deliberately by three negroes, and her body was sold to the University of Maryland Hospital for $15.00, of which an equal division was made among the three conspirators.

It seems strange that the one who incited the crime and arranged its most fiendish acts should have been the one by whom it was betrayed, but such is the case. For some reason or other he became frightened after the crime had been fully accomplished and called upon the president of the university to tell him that a body lately received bore suspicious marks of violence! This was the beginning of the end. The president of the university inspected the body, agreed with his informant, and then called in the police. The informant was arrested on suspicion, broke down and confessed every detail of the crime. In the trial that followed, the chief murderer was sentenced to be hanged. The sentence was duly executed.

This and other occurrences that linger far back in the people's memory brought about the establishment of the State Board of Anatomy in a bill approved March 22, 1890.

This obstructive condition of affairs prevailed when masters of surgery like Knox and Liston, Cuvier, Bichat, Bell and our foremost men here, were doing their best work; yet Edinburgh, where most body-snatching had to be done, could boast an anatomical school which had existed nine years before the birth of Vesalius, and four before that of Ambrose Paré. Think of exacting an acquaintance with practical anatomy from every student when the price in gaining such knowledge frequently meant prison, or, at least, opprobrium. The law, at the instance of a dissatisfied patient, would fine the doctor in court for not displaying "proper skill" in a science to which the road was barred. Prison on the one hand, a monetary loss on the other. The condition, also, of the man who would write of surgery seems little advanced beyond that of Cortesius, the Italian anatomist in the early part of the 17th century who, "though he had a great desire to finish his work on practical anatomy, only twice in 24 years could procure an opportunity of dissecting a human body."

The urgent necessity for bodies to supply the ever increasing schools in London and Edinburgh, in order to keep students together and give the professor a fair chance of success, led to the resurrectionists being countenanced by the doctors; but the odious slander that men like Knox, Liston and Brookes could wink, even connive, at murder and assault, to gain bodies for dissection, is most ridiculous and improbable.

Frequently unauthorized persons would offer bodies or propose to sell their own or those of their relations after death.

One day Sir Astley Cooper received a letter which is still in the library of the Royal College of Surgeons. It ran: Sir. I have been informed you are in the habit of purchasing bodys and allowing the person a sum weekly; knowing a poor woman that is desirous of doing so, I have taken the liberty to call and know the truth. On the back Sir Astley has written: The truth is that you deserve to be hanged for such an unfeeling act.

...The great objection to relatives being dissected was the association of dissection and criminality; for at first only the bodies of murderers, and, later on, those of general criminals, were allowed to the schools. Also, there was the old superstition about preserving the body intact which dates from earliest times. ....But it need hardly be said that when once popular feeling was aroused it was fanned and fed by rumors of the wildest, crudest improbabilities—pretty much of the type which nowadays figure as facts, which the supposed "eminent surgeon" has taken the trouble to impart to the smallest recorder of the smallest local paper, concerning operations in a hospital.

In the latter half of the 17th century, apprentices and gravediggers were the only body-stealers; but eventually necessity forced the surgeons to deal with gangs of professional "resurrectionists," and even to obtain subjects from Ireland, as an advertisement offering to deliver such bodies "sweet and sound" in a medical paper of the period testifies.

Burke and his associate Hare were welcome visitors, for they were clever at their work—a work begun originally in all honesty. An old man in their lodging-house had died owing them much rent, and the idea of selling the body to recoup themselves occurred. A sack of tan filled the coffin, and Burke and Hare themselves followed the funeral with outward solemnity and sorrow—with $37.00 in their pockets for the corpse sold.

$37.00 to men whose spare cash often amounted to 37 cents! How to get more the same way became the question. Lodgers could not conveniently die every week; cemeteries, with their "mort-safe" iron railings, were difficult to rob. A devilish inspiration came. Those who would not die should die, but die—oh yes!—so carefully that no anatomist's suspicions should be aroused. Down the dirty court in old Edinburgh came a decent body full of years and whisky. She is enticed by simulated friendliness to come to the lodging-house. More whisky is given until she is drunkenly unconscious. Burke and Hare exchange glances; then the former throws himself on her prostrate body, and Hare holds her mouth and nose. There is a prolonged struggle, but the "burking" has been effectual, and $60 is secured from the anatomists.

Like two bloodthirsty spiders, Burke and Hare now lurked by the door of their infernal den. Two more tippling women, easily enticed by the idea of drink, are "burked"—a twin tragedy attending the death of one, for a daughter is seeking her mother through the town, but not yet has suspicion fallen on the lodging housekeepers nor their mistresses. The girl comes seeking, directed there by one who has seen her mother with Hare. She is met with sympathy and kindness, with assurance that she will meet her mother at Mid-Calder. Why not go there? But a glass of whisky before starting? Another? The eyes grow sleepy, the speech thick; soon the girl lies in drunken slumber in that fatal little room, with its "one window opening onto a dead wall," and she finds her mother through grim Murder's portals.

Victim after victim followed, but all chosen from the tramp class, who, having nothing about them to excite notice on their appearance in town, gave no cause for alarm on their disappearance from it. But avarice made the men reckless; a very beautiful young woman was enticed in by drink and "burked". As she lay on the dissecting table, a student exclaimed that he had seen her a day or two before. Rigid enquiries were made of Burke, whose powers of lying were equal to the emergency and satisfied the surgeons.

Soon after, Burke and Hare seemed to be rendered absolutely reckless by their easy earnings and unlimited whisky. An harmless idiot lad named Daft Jamie, whom everyone in Edinburgh knew, was "burked" and taken to the Surgeons' Square. Could the students, who had seen him daily, fail to recognize him?

History is silent, but suspicion began to fly in ever narrowing circles over the stronghold of murder held by Berkeley. The balladmonger, after the arrest of Burke and Hare, kept popular sympathy alive for the inoffensive, gentle idiot…

Some said that there existed in the city a gang of famished wretches, who, having once tasted human flesh, were keen on tasting more. People were afraid to go out at night, and when the body of another victim was traced to Surgeons' Square, the fury against the doctors ran so high that their lives were endangered. It must be admitted even by their warmest defenders that they had shown at any rate an unwise amount of confidence in the men of professedly low morals who had supplied the anatomical schools. Knox specially suffered, his house being attacked by the mob.

"Down the close and up the stair,
But and ben wi' Burke and Hare;
Burke's the butcher, Hare's the thief,
Knox the man that buys the beef,"

sang the rabble, but there was not a single circumstance in Knox's life to which blame could be attached.

Great ingenuity was used in the grave-robbery and equal ingenuity in protecting tombs, but the watch paid to protect was often in league with the thieves. Curiously, there were two indictable crimes, felony and misdemeanor, in these thefts, the first incurring the severer punishment. If the body were taken it was simply misdemeanor, but to steal the shroud and quaffing was felony. Why then take the trouble to dig the coffin up? Weeping relatives, as they stood by the apparently untouched grave congratulated themselves that no spoiling of their dead had been effected, but unwittingly shed tears over empty coffins. Just notice that recently disturbed square of turf some fifteen feet off. From it down to the coffin-head, a sloping tunnel has been made and coffin drawn up by iron hooks. The shroud has been carefully put back, the tunnel refilled, the square of turf replaced. The formidable iron grille over the tomb was useless. Spring guns were often set in the churchyards, but what unfeeling custodian could hinder a weeping widow from seeking the grave of her husband—and loosening the wires? Disconsolate relatives were useful even before death came. Was a pauper dying in workhouse, hospital or lodging house? Was there no one to close his eyes, to pay for his humble grave; no one to whisper a prayer into his dying ear? A respectable countryman comes seeking his prodigal nephew; a minister, not of godly mien, comes to read to the dying; a benevolent citizen who has heard of the case (He has indeed!) offers to pay for the funeral; a motherly old soul (smelling strongly of whisky) says she will nurse the dying man. But nepotism and piety, philanthropy and pity, are all simulated by these resurrectionists in disguise, and they in a few hours will have transferred the body to the anatomist, and eventually attend a pauper's funeral.

At first there were few professional resurrectionists; yet soon, both in London and in Edinburgh, gangs of low men had the anatomists at their mercy with regard to supplies. So much had to be paid down at the beginning of the session, and so much for "finishing" on delivery. Among Sir Astley Cooper's papers is an entry: May 10,1827 Paid Hollis, Vaughan and Llewellyn, finishing money £6.60 ($30).

If the preliminary payment were refused, things went badly with the surgeon. To arouse suspicion two dead bodies in an advanced stage of decomposition were once for this reason dropped at night near the door of Brookes, the London teacher. Two young ladies fell over them and raised such a commotion that, had it not been for the police, Brookes would have fared badly. Another time a body was brought him in his absence which he had neither ordered nor paid for. Fortunately it was discovered before night that the subject was very much alive—and a burglar.

Here is an ingenious trick. A man would steal a body which lay awaiting the inquest. He would sell it to an anatomist and then give information to the police where to find it. These naturally restored it to the coroner, who, after the inquest, restored it to the relations, and if the man was friendless these generally turned out to be confederates of the man who had stolen it, and thus the corpse was once more sold to the schools.

The severest encounters in graveyards were between gangs of rival resurrectionists, or "sack-'em-up" men. Often shots would be fired and men severely wounded.

In 1822 an undertaker Bridgman invented a patent wrought-iron coffin and obtained permission from the ecclesiastical and state courts to sell it on payment of a fee on each one, and another man who made patent coffins put some of Southey's ballad, The Surgeon's Warning, on his advertisement. The Surgeon fears he will be dissected as he himself has dissected others, so implores a metal coffin for himself:

"And my 'prentices will surely come
And carve me bone from bone,
And I, who have rifled the dead man's grave,
Shall never rest in my own.
"Bury me in lead when I am dead,
My brethren, I entreat,
And see the coffin weigh'd, I beg,
Lest the plumber should be a cheat.
"And let it be soldr'd closely down,
Strong as strong can be, I implore,
And put it in a patent coffin,
That I may rise no more.
"They burst the patent coffin first,
And then cut through the lead,
And they laugh'd aloud when they saw the shroud,
Because they had got at the dead.
And they allow'd the sexton the shroud,
And they put the coffin back,
And nose and knees they did then squeeze
The surgeon in a sack.
"So they carry the sack pick-a-back,
And they carved him bone from bone,
But what became of the surgeon's soul,
Was never to mortal known."

In England matters were rapidly growing too bad to endure much longer. Twenty two bodies were seized ready for shipment to Edinburgh. These were in casks and labelled "Bitter salts." There was no reason to suspect violence, as the bodies were uninjured; but the resurrectionists had realized the marketable value of a corpse otherwise then for anatomical purposes, and it was a well-known fact that camp-followers extracted the teeth of the dead in war time, and the bodysnatchers often made as much as $250 by extracting the teeth of subjects. But something more was done, savoring of fairy-tale ogres. Listen:

Police reports, 1732: "John Loftas, the Grave-Digger, committed to prison for robbing of dead! corpses, has confessed to the plunder of above 50, not only of the coffins and burial cloaths, but of their fat where bodies afforded any."

"I have made candles of infants' fat," Southey makes his repentant surgeon say. Truly it would be a grim chain of circumstances if, from a candle so composed, a light from the dead were given to reveal the secrets of the dead in the dissecting room.

"Come now," said some doctors to the public, "The government bars the road to necessary knowledge; mourning relatives erect still greater impediments; who will make a voluntary offering of his body for the sake of suffering humanity?" So, in Ireland and London, lectures were given on Anatomy to the public, and at once popular prejudice gave way to common sense and many promised their bodies. Jeremy Bentham left his for dissection, and so did Messenger Monsey, a physician, also ninety-nine gentlemen in Dublin.

But it was Murder's bloody hand which set the seal on the Act of 1832 which took the barriers down. The story of Burke and Hare was repeated in London; Avarice, glutted with gold, grew careless, and the two murderers drew the teeth of their 12-year-old victim before selling the body. The demonstrator noticed this and sent for the police. The resurrectionists were hanged and the Anatomy Bill rolled slowly through Parliament, its clauses at first so narrow, and favoring London so much, that The Lancet dubbed it "A Bill to Prevent Country Surgeons from Studying Surgery."

Necklaces of Human Fingers, 1892

Captain John G. Bourke, Third Cavalry, U.S. Army

This necklace of human fingers... belonged to the foremost of the medicine-men of a brave tribe—the Cheyenne of Montana and Wyoming. They were the backbone of hostility to the whites, and during the long and arduous campaign conducted against them by the late Maj. Gen. George Crook, which terminated so successfully in surrender of 4,500 of the allied Sioux and Cheyenne, at Red Cloud and Spotted Tail agencies, in early spring of 1877, it was a noted fact that wherever a band of the Cheyenne was to be found there the fighting was most desperate...

It may interest some readers to listen to a few words descriptive of the manner in which such a ghastly relic of savagery came into my possession. On the morning of the 25th of November, 1876, the cavalry and Indian scouts (Sioux, Shoshoni, Arapaho, Pawnee, and a few of the Cheyenne themselves), of Gen. Crook's command, under the leadership of the late Brig. Gen. Ranald S. Mackenzie, then colonel of the Fourth Cavalry, surprised and destroyed the main village of the Cheyenne, on the headquarters of the Powder River, in the Big Horn Mountains, Wyoming. The onslaught was irresistible, the destruction complete, and the discomfited savages were forced to flee from their beds, half naked and with nothing save their arms and ammunition. More than half of the great herd of ponies belonging to the savages were killed, captured, or so badly wounded as to be of no use to the owners. The cold became so intense that on the night after the fight eleven papooses froze to death in their mothers' arms, and the succeeding night, three others. This blow, the most grievous ever inflicted upon the plains tribes, resulted in the surrender, first of the Cheyenne, and later on of the principal chief of the Sioux, the renowned Crazy Horse; after which the Sioux troubles were minimized into the hunt for scattered bands. Undoubtedly, among the bitterest losses of valuable property suffered by the Cheyenne on this occasion were the two necklaces of human fingers which came into my possession, together with the small buckskin bag filled with the right hands of papooses belonging to the tribe of their deadly enemies, the Shoshoni. These were found in the village by one of our scouts—Baptiste Pourrier, who, with Mr. Frank Gruard, was holding an important and responsible position in connection with the care of the great body of Indian scouts already spoken of. From these two gentlemen I afterwards obtained all the information that is here to be found regarding the Cheyenne necklace.

The second necklace, consisting of four fingers, was buried, as Gen. Crook did not wish to have kept more than one specimen, and that only for scientific purposes. Accordingly, the necklace here depicted was sent first to the US Military Academy at West Point, New York, and later to the National Museum in Washington, where it was believed it could better fulfill its mission of educating students in a knowledge of the manners and customs of our aborigines.

The buckskin bag, with the papooses' hands, was claimed by the Shoshoni scouts, who danced and wailed all night, and then burned the fearful evidence of the loss sustained by their people.

The necklace is made of a round collar of buckskin, incrusted with the small blue and white beads purchased from the traders, these being arranged in alternate spaces of an inch or more in length. There are also attached numbers of the perforated wampum shell beads of native manufacture. Pendant from this collar are five medicine arrows, the exact nature of which, it was, of course, impossible to determine from the owner himself. Both Frank and Baptiste agreed that an arrow might become "medicine" either from having been shot into the person of the owner himself or into the body of an enemy, or even from having been picked up under peculiar circumstances. The owner, High Wolf or Tall Wolf, admitted as much after he had surrendered at the Red Cloud agency and had made every effort to obtain the return of his medicine, which was this necklace.

The four medicine bags to be seen in the picture are worthy of attention. They were carefully examined under a powerful glass by Dr. H. C. Yarrow, U.S. Army, in the city of Washington, and pronounced to be human scrota. The first of these contained a vegetable powder, somewhat decomposed, having a resemblance to hoddentin; the second was filled with killikinnick; the third with small garnet-colored seeds like the chia in use among the Apache, and the fourth with a yellow, clayey-white vegetable matter not identified. The fifth, also, remained unidentified.

Besides the above, there are artificial teeth, resembling those of the fossil animals abundant in the Bad Lands of South Dakota, but cut out of soft stone.

The fingers—eight altogether—are the left-hand middle fingers of Indians of hostile tribes, killed by High Wolf. I obtained the list and could insert it here were it worthwhile to do so. The fingers have not been left in the natural state, but have been subjected to very careful and elaborate antiseptic treatment in order thoroughly to desiccate them. They were split longitudinally on the inner side and after the bone has been extracted the surface of the skin, both inside and out, received a treatment with a wash or paint of ocherous earth, the same as is used for the face. I was told that the bones were not replaced but that sticks were inserted to maintain the fingers in proper shape.

Of the reason for making use of such a trophy or relic, there is not much to be said; even the savages know little and say less. From the best information that I have been able to gather, it would seem to be based partly upon a vainglorious desire to display the proofs of personal prowess, and partly upon the vague and ill-defined, but deeply rooted, belief in the talismanic or "medicinal" potency possessed by all parts of the human body, especially after death. It was such a belief which impelled the Mandan, Aztecs, and others of the American tribes to preserve the skulls of their dead as well as (among the Aztecs) those of the victims sacrificed in honor of their gods. As has been shown in another place, the Zuñi and others take care to offer food at stated periods to the scalps of their enemies.

The use of necklaces of human fingers or of human teeth is to be found in many parts of the world, and besides the fingers themselves, we find the whole arm, or in other cases only the nails. The Cheyenne did not always restrict themselves to fingers; they generally made use of the whole hand, or the arm of the slaughtered enemy. In a colored picture drawn and painted by one of themselves I have a representation of a scalp dance, in which the squaws may be seen dressed in their best, carrying the arms of enemies elevated on high poles and lances. There is no doubt in my mind that this custom of the Cheyenne of cutting off the arm or hand gave rise to their name in the sign language of the "Slashers," or "Wrist Cutters," much as the corresponding tribal peculiarity of the Dakota occasioned their name of the "Coupe Gorge" or "Throat Cutters."

The necklace of human fingers is found among other tribes. A necklace of four human fingers was seen by the members of the Lewis and Clark expedition along the Shoshoni at the headwaters of the Columbia, in the early years of the present century. Early in the spring of 1858 Henry Youle Hind refers to the allies of the Ojibwa on Red River as having "two fingers severed from the hands of the unfortunate Sioux." In Eastman's "Legends of the Sioux," we read of "Harpsthinah, one of the Sioux women, who wore as long as she could endure it, a necklace made of the hands and feet of Chippewah children." We read that in New Zealand, "Several rows of human teeth, drawn on a thread, hung on their breasts." Capt. Cook speaks of seeing fifteen human jaw bones attached to a semicircular board at the end of a long house on the island of Tahiti. "They appear to be fresh, and there was not one of them that wanted a single tooth;" and also, "the model of a canoe, about three feet long, to which was tied eight human jaw bones; we had already learnt that these were trophies of war." Capt. Byron, R.N., saw in the Society Islands, in 1765, a chief who "had a string of human teeth about his waist, which was probably a trophy of his military prowess."

"The wild Andamanese, who live only on the fruits of the forest and on fish, so far revere their progenitors that they adorn their women and children with necklaces and such like, formed out of the finger and toe-nails of their ancestors."

Bancroft says that the Californians did not generally scalp, but they did cut off and keep the arms and legs of the slain enemy or, rather, the hands and feet and head. They also have a habit of plucking out and preserving the eyes. Kohl assures us that he has been informed that the OJibwah will frequently cut fingers, arms, and limbs from their enemies and preserve these ghastly relics for use in their dances. Sometimes the warriors will become so excited that they will break off and swallow a finger.

Tanner says of the Ojibwah: "Sometimes they use sacks of human skin to contain their medicines, and they fancy that something is thus added to their efficacy. "…

The Mohawk "place their foe against a tree or stake and first tear off all the nails from his fingers and run them on a string, which they wear the same as we do gold chains. It is considered to the honor of any chief who has vanquished or overcome his enemies if he bite off or cut off some of their members, as whole fingers."

The Cenis (Asinai) of Texas, were seen by La Salle's expedition in 1687–1690, torturing a captive squaw. "Then they tore out her hair, and cut off her fingers."

In volume 2 of Kingsborough's Mexican Antiquities, in the plates of the Vatican manuscript, is to be seen a representation of an Aztec priest or other dignitary holding out in his hands two human arms. In plate 76 of the same is a priest offering up a human sacrifice, the virile member of the victim cut off.

Teoyamaqui, the wife of Huitzlipochtli, The Aztec god of war, was depicted with a necklace of human hands. Squirt also says that Darga or Kali, the Hindu goddess, who corresponds very closely to her, was represented with "a necklace of skulls" and "a girdle of dissevered human hands."

The Hindu goddess Kali was decorated with a necklace of human skulls. In the Propaganda collection, given in Kingsborough, are to be seen human arms and legs.

"On the death of any of the great officers of state, the finger bones and hair are also preserved; or if they have died shaven, as sometimes occurs, a bit of their mbugu dress will be preserved in place of the hair."

The principal war fetiches of Uganda consist of dead lizards, bits of wood, hide, nails of dead people, claws of animals, and beaks of birds." [Henry M.] Stanley saw them displayed before King Mtesa.

"Some of the women in Gippsland wear round the neck human hands, which, Mr. Hull says, were beautifully prepared. He moreover informs me that they sometimes wear the parts of which the 'Lingam' and 'Priapus' were the emblems." "The Gippsland people keep the relics of the departed. They will cut off the hands to keep as a remembrance, and these they will attach to the string that is tied round the neck."

Smyth also relates that the women of some of the Australian tribes preserve "the hands of some defunct member of the tribe—that of some friend of the woman's, or perhaps one belonging to a former husband. This she keeps as the only remembrance of one she once loved; and, though years may have passed, even now, when she has nothing else to do, she will sit and moan over this relic of humanity. Sometimes a mother will carry about with her the remains of a beloved child, whose death she mourns." The Australians also use the skulls of their "nearest and dearest relatives" for drinking vessels; thus, a daughter would use her mother's skull, etc.

"One of the most extraordinary of their laws is that, a widow, for every husband she marries after the first, is obliged to cut off a joint of a finger, which she presents to her husband on the wedding day, beginning at one of the little fingers."

In the Army and Navy Journal, New York, June 23, 1888, is mentioned a battle between the Crow of Montana and the Piegan, in which the former obtained some of the hands and feet of dead warriors of the first-named tribe and used them in their dances.

Catlin shows that the young Sioux warriors, after going through the ordeal of the sun dance, placed the little finger of the left hand on the skull of a sacred buffalo and had it chopped off. "The sacrifices [of American Indians] at the fasts of puberty sometimes consist of finger joints." In Dodge's Wild Indians is represented (Pl.vI,13) a Cheyenne necklace of the bones of the first joint of the human fingers, stripped of skin and flesh. I have never seen or heard of anything of the kind, although I have served with the Cheyenne a great deal and have spoken about their customs. My necklace is of human fingers mummified, not of bones.

Fanny Kelly says a Sioux chief: "He showed me a puzzle or game he had made from the finger bones of some of the victims that had fallen beneath his own tomahawk. The bones had been freed from the flesh by boiling, and, being placed upon a string, were used for playing some kind of Indian game."

The necklace of human fingers is not a particle more horrible than the ornaments of human bones to be seen in the cemetery of the Capuchins in Rome at the present day. I have personally known of two or three cases where American Indians cut their enemies limb from limb. The idea upon which the practice is based seems to be the analogue of the old English custom of sentencing a criminal to be "hanged, drawn, and quartered."

Brand gives a detailed description of the "hand of glory," the possession of which was believed by the peasantry of Great Britain and France to enable a man to enter a house invisible to the occupants. It was made of the hand of an executed (hanged) murderer, carefully desiccated and prepared with a great amount of superstitious mummery. With this holding a candle of "the fat of a hanged man" burglars felt perfectly secure while engaged in their predatory work. The belief was that a candle placed in a dead man's hand will not be seen by any but those by whom it is used. Such a candle introduced into a house kept those who were asleep from awakening.

The superstition in regard to the "hand of glory" was widely diffused throughout France, Germany, Spain, and Great Britain. As late as the year 1831 it was used by Irish burglars in the county Meath....

An Aztec warrior always tried to procure the middle finger of the left-hand of a woman who had died in childbirth. This he fastened to his shield as a talisman...

"Scrofula, imposthumes of the parotid glands, and throat diseases, they say, may be cured by the contact of the hand of a person who has been carried off by an early death; "but, he goes on to say, any dead hand will do, "provided it is of the same sex as the patient and that the part affected is touched with the back of the left hand. A footnote adds that the superstition still prevails in England in regard to the hand of a man who has been hanged.

The use of dead men's toes, fingers, spinal vertebrae, etc. in magical ceremonies, especially the fabrication of magical lamps and candles, is referred to by Fromann.

Grimm is authority for the statement that in both France and Germany the belief was prevalent that the fingers of an unborn babe were "available for magic." In England witches were believed to "open graves for the purpose of taking out the joints of the fingers and toes of dead bodies....in order to prepare a powder for their magical purposes."...

[Captain Bourke relates several more obscure practices by Europeans relating to the use of hands, fingers and teeth etc. of the dead. He sums his analysis up in a surprisingly modernist tone]: "The proceeding practice is strictly in line with the 'medicinal' and 'magical' values attached in Europe to human, teeth, skin, etc... it will be shown that the ideas of the people of Europe of that period were only in name superior to those of the savages of America, the islands of the South Seas, and of Central Africa."

Why the Stereoscope

By E. W. Caldwell, MD, New York

This question is asked so frequently that a brief answer seems proper as a preface to the stereoscopic plates which follow:

The stereoscope enables us to obtain from pictures of objects, a better perception of distances and space relations than is possible with single pictures. A stereograph consists of two pictures, which are slightly different and which correspond to the two different views seen by the two eyes when looking at the same object. The stereoscope enables us to view these two pictures as one, and we thus obtain artificially, the advantages that our two eyes give us when viewing objects directly.

If we close one eye and look at a collection of familiar objects, we can make a fair estimate of the relative distances from us, because of the difference in size of the images of the near and distant objects. If, however, we look at entirely unfamiliar objects with one eye closed, it is exceedingly difficult, without changing the point of view, to tell much about the size or space relations of the objects of viewed. If, however, we use both eyes, each eye sees the objects from a slightly different angle and by an unconscious process of triangulation, we are able to make a fair estimate of space relations, even of unfamiliar objects.

In Roentgenograms we have really shadow pictures. We do not find in these shadow pictures, differences in dimensions in the shadows which would enable us to make an estimate of the relative distance of the objects from us or from the source of X-Ray. The lights and shades in the Roentgenogram are determined more by the thickness and X-Ray opacity of the objects than by their space relations. For this reason the shading in a Roentgenogram may often be misleading to those who are accustomed to viewing photographs and who have not learned to appreciate the difference between a Roentgenogram and a photograph.

For this reason, in estimating the space relation of objects shown by X-Ray shadows, we are especially in need of some help such as may be obtained from the stereoscope. Elihu Thompson, in 1896, proposed that stereoscopic affects might be obtained from a pair of X-Ray shadows made with the source of X-Rays in two positions corresponding to the positions of the two eyes. Very shortly thereafter, Sir James Mackenzie Davidson of London, brought this into common practice. Davidson even went so far as to devise a method of obtaining stereoscopic effects in fluoroscopic examinations. Since Davidson's pioneer work, stereoscopic methods in Roentgenology have been practiced by most competent Roentgenologists the world over.

The history of the stereoscope begins with the experiments of Sir Charles Wheatstone on binocular vision about 1829... The special devices for enabling us to see two stereoscopic pictures as a single view full of perspective, have been developed by Wheatstone, Brewster and others. Wheatstone's reflecting stereoscope enables us to view pictures of practically any size. For this reason this instrument is used largely when viewing original X-Ray plates, the dimensions of which are too great for the more convenient Brewster stereoscope. Brewster's stereoscope, a description of which was published in 1843, is a very convenient instrument, which enables us to obtain stereoscopic vision with pictures not exceeding very much in width, the pupillary distance of our eyes, or about two and three-eights [sic] inches. Pictures of this size have the advantage of being very convenient, and the Brewster stereoscope has made the enormous development of stereoscopic photography possible.

The Roentgenograms in the accompanying pages have been reduced from the originals down to the size appropriate for use with the ordinary Brewster stereoscope. The ability to use the stereoscope varies considerably with different individuals and seems to depend upon some accommodation factors of the eyes. Some individuals are unable to obtain satisfactory stereoscopic vision at all, but to most people it translates a flat picture into something approaching an actual view, sometimes even with an exaggerated depth and perspective.

We cannot hope to obtain in the Stereo-Roentgenograms quite as much perspective as we see in stereo-photographs, because in the former we have not the many accessory aids to perception of distance which are found in photographs of familiar objects. In the Stereo-Roentgenogram we must depend entirely upon the help of two eye vision effect for our perception of distance and the stereoscope is, therefore, more of a necessity in Roentgenography than in ordinary photography. It goes a long way toward making Roentgenograms intelligible to the untrained, who have not learned to discard life long but unconscious habits and training in looking at pictures, and who, therefore, easily fall into error in viewing X-Ray shadows as pictures.

Primary Sources & Bibliography

Ariès, Philippe. *Images of Man and Death*. Cambridge, MA: Harvard University Press, 1985.

Ball, James Moore. *The Body Snatchers*. New York, NY: Dorset Press, 1989.

Ball, James Moore, *The Sack-em-up Men*. London: Oliver & Boyd, 1928.

Barkan, Elazar. *The Retreat of Scientific Racism: Changing concepts of race in Britain and the United States between the World Wars*. New York: Cambridge University Press, 1992.

Bondeson, Jan. *Buried Alive: The Terrifying History of Our Most Primal Fear.* New York: W.W. Norton & Co., 2001.

Borden, W. C. *The Use of the Rontgen Ray by the United States Medical Department of the United Sates Army in the War With Spain (1898)*, Washington: Government Printing Office, 1900.

Burns, Stanley B., *Sleeping Beauty: Memorial Photography in America*. Altadena CA: TweleveTrees Press, 1990.

Burns, Stanley B., MD, and Burns, Elizabeth A., *Sleeping Beauty II: Grief, Bereavement and the Family in Memorial Photography: American & European Traditions*. New York, NY: Burns Archive Press, 2002.

Burns, Stanley B., MD, *Sleeping Beauty III: Memorial Photography: The Children*. New York, NY: Burns Archive Press, 2002.

Clarke, Edward H., et al. *A Century of American Medicine*. Philadelphia: Lea, 1876.

Dain, Bruce R. *A Hideous Monster of the Mind: American Race Theory in the Early Republic*. Cambridge, MA: Harvard University Press, 2002.

Dickey, Colin. *Cranioklepty: Grave Robbing and the Search for Genius*. New York: Unbridled Books, 2009.

Di Gregorio, Mario. *T.H. Huxley's place in Natural Science*. New Haven: Yale University Press, 1984

Doolin, William. D*ublin's Surgeon-Anatomist and other Essays: A Centenary Tribute*. Dublin: Royal College of Surgeons in Ireland, 1987.

Edwards, Elizabeth, ed. *Anthropology & Photography: 1860–1920*. New Haven: Yale University Press, 1992.

Erony, Susan and Rafter, Nicole. S*earching the Criminal Body: Art/Science/Prejudice*. Albany: University Art Museum, 2000.

Fabian, Ann. *The Skull Collectors: Race, Science, and America's Unburied Dead*. Chicago: University Of Chicago Press, 2010.

Farrell, James J. *Inventing the American Way of Death, 1830–1920*. Philadelphia: Temple University Press, 1980.

Henschen, Folke. *The Human Skull: A Cultural History*. New York: F.A. Praeger, 1966.

Hunter, Ronald. *The Diseases of Occupations*. London: English Universities Press, Ltd., 1956.

Jones, D. G., and Whitaker, M.I. *Speaking for the Dead: The Human Body in Biology and Medicine, Medical Law and Ethics, 2d edition*. Farnham, Surrey, UK: Ashgate Publishers, 2009.

Lassek, A.M. *Human Dissection: Its Drama and Struggle*. Springfield IL: Charles C. Thomas Publisher, 1958.

Long, Esmond R. *A History of Pathology.* New York, NY: Dover Publications, Inc.,1965.

Mai, Francois Martin. *Diagnosing Genius: The Life and Death of Beethoven*. Montreal: McGill-Queens University Press, 2007.

May, Brian, et al. *Diableries: Stereoscopic Adventures in Hell*. New York: Shelter Harbor Press, 2013.

Maulitz, Russell C. *Morbid Appearance: The Anatomy of Pathology in the Early*

Nineteenth Century. Cambridge: Cambridge University Press,1987.
Mitford, J. *The American Way of Death*. New York: Simon and Schuster, 1963.
Morton, Leslie. *A Medical Bibliography (Garrison & Morton): An Annotated Check-List of Texts Illustrating The History Of Medicine*. London: A Grafton Book, Andre Deutsch Ltd., 1970.
Numbers, Ronald L., ed. *The Education of American Physicians: Historical Essays*. Berkeley: University of California Press, 1980.
Proctor, Robert. *Racial Hygiene: Medicine under the Nazis*. Cambridge, MA: Harvard University Press, 1988
Richardson, Ruth. *Death Dissection and the Destitute*. London/NewYork: Routledge & Kegan Paul, 1987.
Robbins, Alexandra. *Secrets of The Tomb: Skull and Bones and the Ivy League, and the Hidden Paths of Power*. New York: Little Brown and Company, 2002.
Rothstein, William G. *American Physicians in the Nineteenth Century: From Sects to Science*. Baltimore: The Johns Hopkins University Press, 1972.
Rufus, Anneli. *Magnificent Corpses*. New York: Marlowe & Company, 1999.
Sykora, Katharina. *Die Tode der Fotographie -1*. Munchen: Wilhelm Fink Verlag, 2009.
Spencer, Frank. "Some Notes on the Attempt to Apply Photography to Anthropometry During the Second Half of the Nineteenth Century," in Edwards, Elizabeth, *Anthropology and Photography*. New Haven: Yale University Press, 1992.
Tomlinson, Stephen. *Head Masters: Phrenology, Secular Education, and Nineteenth Century Social Thought*. Tuscaloosa: University of Alabama Press, 2005.
Warner, John Harley and James M. Edmonson. *Dissection: Photographs of a Rite of Passage in American Medicine 1880-1930*. New York, NY: Blast Books, 2009.
Wieczorek, Alfried and Wilfried Rosendahl, eds., *Mummies of the World: The Dream of Eternal Life*. New York: Prestel, 2010.

Journal Articles

Breuning, Gerhard von. "The Skulls of Beethoven and Schubert," *The Beethoven Journal*, Vol. 20, Numbers 1 & 2, Summer and Winter 2005.
Buklija, Tatjana. "Cultures of Death and Politics of Corpse Supply: Anatomy in Vienna 1848–1914," *Bulletin of the History of Medicine*, No. 82, Fall 2008.
Cockle, Richard. "The Nez Perce long for the return of Chief Old Joseph's stolen skull," *The Oregonian*, May 14, 2006.
Garment, A., et al. "Let the dead teach the living: The rise of body bequeathal in twentieth-century America." *Academic Medicine* 82:1000–1005, 2007
Hagner, Michael. "Skulls, Brains, and Memorial Culture: On Cerebral Biographies of Scientists in the Nineteenth Century," *Science in Context*, Vol. 16, Nos. 1 & 2, 2003.
Halperin, E. "The Poor, the Back, and the Marginalized as the Source of Cadavers in United States Anatomical Education," *Clinical Anatomy* 20:489–95. 2007
Hildebrandt, S. "Capital punishment and anatomy: History and ethics of an ongoing association," *Clinical Anatomy* 21:5–14. 2008.
Hulkower, Michael "From Sacrilege to Privilege: The Tale of Body Procurement for Anatomical Dissection in the United States," *The Einstein Journal of Biology and Medicine*, p23–26, 2011.
Korf, H. W., and H. Wicht. "The public cadaver," *Nature* 428:805. 2004.
Lewis, Jason E., D DeGusta, M. R. Meyer, J.M. Monge, A.E. Mann, et al. "The Mismeasure of Science: Stephen Jay Gould versus Samuel George Morton on Skulls and Bias." *PLOS Biology* 9(6). 2011.
Maraldi, Nadir, et al. "Anatomical Waxwork Modeling: The History of the Bologna Anatomy Museum," *The Anatomical Record*, 261:5–10, 2000.
Piombino-Mascali D, et. al, "The Salafia method rediscovered," *Virchows Archiv* 454 (3): 355–7. March 2009.
Renschler, Emily and Janet Monge. "The Samuel George Morton Cranial Collection: Historical and New Research," *Expedition*, Vol. 50, No. 3.
Rushton, J. Philippe and C. Davison Ankney. "Whole Brain Size and General Mental Ability: A Review," *International Journal of Neuroscience* 119 (5): 692–732. January 1, 2009
Tward, A., and H. Patterson. "From grave robbing to gifting: Cadaver supply in the United States," *JAMA* 287:1183. 2002.
Wade, Nicholas. "Scientists Measure the Accuracy of a Racism Claim." *The New York Times*. D4, June 14, 2011.
Zimmer, Carl. "A Twist on our Ancestry," *The New York Times*, DR, October 29, 2013,

Websites

Native American Netroots, www.nativeamericannetroots.net

PhysOrg.com, www.physorg.com/news/2011-06-samuel-morton-skulls-center-controversy.html

Wikipedia, en.wikipedia.org/wiki/Catacombe_dei_Cappuccini

Huffington Post, www.huffingtonpost.com/keith-thomson/for-sale-beethovens-skull

Other Burns Archive Publications

Mirror, Mirror: The Burns Collection Daguerreotypes

Shooting Soldiers: Civil War Medical Photography by R.B. Bontecou

Sleeping Beauty III: Memorial Photography: The Children

Ophthalmology: A Photographic History 1845-1945 (4 Volumes)

Deadly Intent: Crime & Punishment Photographs from the Burns Archive

News Art: Manipulated Photographs from the Burns Archive

Nephrology: A Photographic History 1840-1950 (4 Volumes)

Geisha: A Photographic History 1872-1912

Seeing Insanity: Photography & The Depiction of Mental Illness

Patients & Promise: A Photographic History of Mental & Mood Disorders (4 Volumes)

Photographic History of Early Podiatry: Selections from The Burns Archive

Photographic History of Nineteenth Century Dermatology: Selections from The Burns Archive (4 Volumes)

Oncology: Tumors & Treatment, A Photographic History 1845-1945 (4 Volumes)

Respiratory Disease: A Photographic History, 1845-1945 (4 Volumes)

Sleeping Beauty II: Grief, Bereavement and The Family In Memorial Photography, American & European Traditions

The American Dentist: A Pictorial History

American Surgery: An Illustrated History

A Morning's Work: Medical Photographs from The Burns Archive & Collection 1843-1939

Forgotten Marriage: The Painted Tintype & The Decorative Frame, 1860-1910, A Lost Chapter in American Portraiture

Harm's Way: Lust & Madness, Murder & Mayhem

Face of Mercy: A Photographic History of Medicine at War

Photographie Et Medecine: 1840-1880

Sleeping Beauty: Memorial Photography in America

Masterpieces of Medical Photography: Selections from The Burns Archive

Early Medical Photography in America: 1839-1883

The Bristol Gallery of Medical History

American Medical Publication With Photographs (Monograph) New York State Journal of Medicine

Civil War Medical Photography, (Monograph) New York State Journal of Medicine

Acknowledgments

We want to express our sincere appreciation to several unique individuals who helped with various aspects of this project. First and foremost is Alice Lease Dana, who provided invaluable online searches and editorial comments. She also helped edit each section and her insightful comments helped keep us on track in our busy life. Jason Lawrence "J" Burns's assistance with research, editing and proofing was exceptional. Amber Jolliffe Maykut brought fresh eyes and enthusiasm to the project. James M. Edmondson, PhD, Chief Curator of the Dittrick Medical History Center and Museum of Case Western Reserve University, provided insight into the history of dissection photography and reviewed the manuscript.

Several archivists, curators, and librarians helped with research. We particularly want to thank them for the time and effort put in and give special thanks to: Christopher Lyons, Head, Osler Library of the History of Medicine, McGill University, Montreal Canada; Eric Boyle, Chief Curator National Museum of Health and Medicine. Kristin Rodgers, MLIS, Curator Health Sciences Library Medical Heritage Center; Joan E. Hostetler Indiana Photographers Project. The photograph collecting community, especially several collectors, provided information regarding photographs in their collections and their knowledge of the use of skulls and skeletons. Among them are Howard McManus, George Whitely IV and his wife Susan, Steve and Mary Degenero, Matthew Stockton, Andrew Daneman, and Keith de Lellis. I want to thank several collectors who supplied images to this work: David Chow, Thomas Harris, David Winter, and especially Ben Ziegler. Ziegler as a dealer supplied me with many of my historic photographs over the years and also provided some fascinating images from his own collection for this work.

We acknowledge and are grateful to Pete Schiffer, Schiffer Publishing Ltd., for his support and appreciation of The Burns Archive and our work. He generously offered us an unfettered opportunity to produce this volume. We would also like to recognize and thank our editors, Douglas Congdon-Martin and Catherine Mallette, as well as the staff of Schiffer who made this project a pleasure to produce.

This has been a demanding undertaking and we have greatly benefited from the input and assistance of this fascinating group of individuals who helped to produce this book.

About the Authors

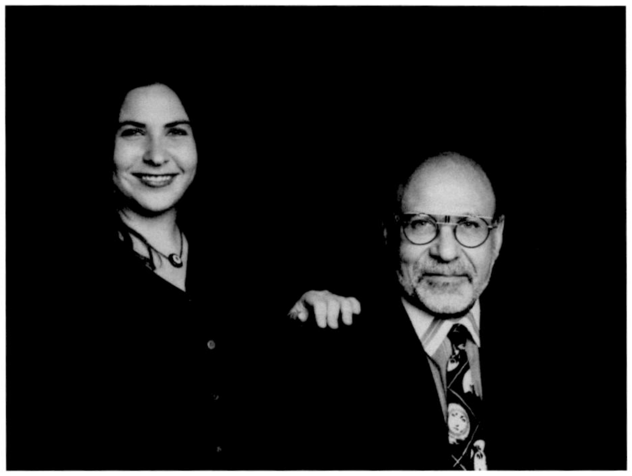

STANLEY B. BURNS, MD, FACS, a New York City ophthalmologist and clinical professor of medicine and psychiatry, New York University: Langone Medical Center, is an internationally distinguished author, curator, historian, collector, and archivist. In 1975, he began collecting historic photography with an emphasis on unique photographs not available anywhere else. Dr. Burns's collection of over one million vintage photographs (1840-1950) has been generally recognized as the most important private comprehensive collection of early photography. It includes world-class comprehensive compilations of ethnology, crime, medical, war, post-mortem, Judaica, African American, and early hand-colored photography. In 1977, he founded The Burns Archive to share his discoveries, and he began his writing career. The Archive is a stock photography and publishing entity and is a valuable photographic resource for both researchers and the media. Photographs from his collection have been featured in numerous publications, films, and major museum exhibitions. Dr. Burns has previously written 43 photo-historical texts and more than one thousand journal articles, and has curated more than one hundred exhibitions. He is also a consultant for Hollywood feature filmmakers, documentarians, and television. Some of his film credits include Adrian Lyne's *Jacob's Ladder*, Robert Altman's *Dorothy Parker and her Vicious Circle*, Tom Cruise's *The Others*, and Al Pacino's *Looking For Richard III*. Currently he is a consultant on *The Knick*, an HBO/Cinemax TV series directed by Steven Soderbergh, about a New York City hospital in the year 1900. He has produced and written an award-winning documentary, *Death in America* (1997), that is currently used in various educational systems. Dr. Burns is generally recognized as the leading promoter of a second history of photography, a history based on the use of photography by people and professions. The time-honored history of photography is that of the innovations, innovators, and fine art photographers. This book is another step in the appreciation of photography by subject rather than by artist. For the last four decades, he has concentrated on collecting photographs relating to unexplored areas of history and photography. Dr. Burns has collaborated with his daughter on several photo-historical projects, exhibitions, and two award winning books: in 2002, *Sleeping Beauty II: Grief, Bereavement and The Family in Memorial Photography, American & European Traditions*, published for their exhibit at Paris's Musée d'Orsay, and in 2006, *Geisha: A Photographic History, 1872-1913*.

ELIZABETH A. BURNS is the Creative and Operations director of The Burns Archive and Press. She has curated numerous exhibitions and created several books, including *Sleeping Beauty II* and *The Burns Archive Medical Specialty Series*. In 2002, under a grant by the United States and Israeli governments, she produced *A Dream Deferred, The African American Experience 1848-1928*, the inaugural exhibition at the Center for the Study of the United States in Haifa, Israel. Other exhibits she has prepared feature works on hand-painted Japanese photography, early flight, spirit photography, criminology, Lewis Hine and the playground, African American history, The SS Ben Hecht (1947 Palestine blockade runner), and a special exhibition on respiratory disease titled *BreathTaking Photographs*. In 2006, she co-wrote *Geisha: A Photographic History, 1872-1913* with Dr. Burns. She is currently working as a consultant medical photo-historian on the HBO/Cinemax series *The Knick*.

For more information about the authors and The Burns Archive, please visit:

www.burnsarchive.com

Library of Congress Control Number: 2014950508

Designed by John P. Cheek
Type set in Humanist 521 Lt BT/Sabon LT Std

ISBN: 978-0-7643-4746-7
Printed in China

Published by Schiffer Publishing, Ltd.

4880 Lower Valley Road
Atglen, PA 19310
Phone: (610) 593-1777; Fax: (610) 593-2002
E-mail: Info@schifferbooks.com

Photo Credits:
All photographs are the property of Stanley B. Burns, MD, except for the kind courtesy of the following individuals and institutions:
Ira F. Brilliant Center for Beethoven Studies, San Jose State University: Plate 10.0
The Metropolitan Museum of Art, Gilman Collection, Purchase, Harriette and Noel Levine Gift, 2005 2005.100.270/ copy print by Sotheby's, Inc.: Plate 0.3x
National Museum of Health and Medicine: Plates 4.8, 4.20, 4.23, 4.25, 4.26, 4.33, 4.34, 4.36, 5.16, 5.19
Osler Library of the History of Medicine, McGill University: Plates 1.26, 7.4, 7.5, 7.6
Elizabeth A. Burns: Plates 1.2, 1.11, 1.13, 1.16, 2.11, 2.17, 2.25, 2.31, 2.38, 5.29-33, 6.3, 7.10, 10.46-51,
Jason L. Burns: Plates 1.5, 1.8, 1.10, 1.28, 1.34, 6.26
Thomas Harris Collection: Plates 3.14, 3.28, 5.28, 9.5, 9.19, 10.42,
Winter Works on Paper, Inc.: Plates 6.40-6.45
Ben Ziegler: Plates 3.29, 5.23, 5.25, 9.6, 9.7, 9.13-16, 9.20, 9.21, 9.22, 9.30

Front Cover:
A Student's Dream, Photographer: A.A. Robinson, 1906
Robinson made a series of 8" x 10" photographs depicting students on the dissecting table surrounded by cadavers and/or skeletons. The photographs were popular images of their time and were sold to students throughout the United States. The images graphically represent one of photography's theoretical concepts—that of "magical substitution." Magical substitution is the phenomenon when the viewers place themselves in the depicted scene. It is one source of empathy, as one contemplates even fleetingly what it would mean to be in that situation.

Back Cover:
Rare Commemorative Daguerreotype of Advancement to Professorship, 1855
Professor of Chemistry Charles Linnaeus Allen, MD (1820-1890), at left, studying anatomy with Daniel Avery, MD. This photograph was taken to commemorate Allen's appointment as professor at Vermont's Castleton Medical College. Dr. Allen ultimately became professor of medicine and dean of the medical school in 1860. During the Civil War he was a brigade surgeon and head medical purveyor in Hilton Head, South Carolina. He was also in charge of Hospital No. 2. After the war, he practiced in Rutland, Vermont, until his death in 1890.